Livestock in a Changing Landscape

Volume 1

About Island Press

Since 1984, the nonprofit Island Press has been stimulating, shaping, and communicating the ideas that are essential for solving environmental problems worldwide. With more than 800 titles in print and some 40 new releases each year, we are the nation's leading publisher on environmental issues. We identify innovative thinkers and emerging trends in the environmental field. We work with world-renowned experts and authors to develop cross-disciplinary solutions to environmental challenges.

Island Press designs and implements coordinated book publication campaigns in order to communicate our critical messages in print, in person, and online using the latest technologies, programs, and the media. Our goal: to reach targeted audiences—scientists, policymakers, environmental advocates, the media, and concerned citizens—who can and will take action to protect the plants and animals that enrich our world, the ecosystems we need to survive, the water we drink, and the air we breathe.

Island Press gratefully acknowledges the support of its work by the Agua Fund, Inc., The Margaret A. Cargill Foundation, The Nathan Cummings Foundation, Betsy and Jesse Fink Foundation, The William and Flora Hewlett Foundation, The Kresge Foundation, The Forrest and Frances Lattner Foundation, The Andrew W. Mellon Foundation, The Curtis and Edith Munson Foundation, The Overbrook Foundation, The David and Lucile Packard Foundation, The Summit Foundation, The Summit Fund of Washington, Trust for Architectural Easements, The Winslow Foundation, and other generous donors.

The opinions expressed in this book are those of the author(s) and do not necessarily reflect the views of our donors.

About the French Agricultural Research Centre for International Development

Most of the research conducted by the French Agricultural Research Centre for International Development (CIRAD) is in partnership in developing countries. CIRAD has chosen sustainable development as the cornerstone of its operations worldwide. This means taking account of the long-term ecological, economic, and social consequences of change in developing communities and countries. CIRAD contributes to development through research and trials, training, dissemination of information, innovation, and appraisals. Its expertise spans the life sciences, human sciences, and engineering sciences and their application to agriculture and food, natural resource management, and society.

About the Food and Agriculture Organization of the United Nations

The mandate of the Food and Agriculture Organization of the United Nations (FAO) is to raise the levels of nutrition, improve agricultural productivity, better the lives of rural populations, and contribute to the growth of the world economy.

About the Swiss College of Agriculture

The Swiss College of Agriculture (SHL) in Zollikofen is the specialist Swiss institution for agriculture, forestry, and food business within Bern University of Applied Sciences. It offers three BSc programs and one MSc program in agriculture, forestry, and food science and management. In addition SHL conducts applied research and renders services in Switzerland and around the world. At SHL, sustainability is at the foundation of the degree programs and projects. With up to 500 students, the campus close to Bern is a manageable size. Thus the synergies between education, research, and services can be optimally leveraged.

About the International Livestock Research Institute

The Africa-based International Livestock Research Institute (ILRI) works at the crossroads of livestock and poverty, bringing high-quality science and capacity building to bear on poverty reduction and sustainable development. ILRI is one of 15 centers supported by the Consultative Group on International Agricultural Research (CGIAR). It has its headquarters in Nairobi, Kenya, and a principal campus in Addis Ababa, Ethiopia. It has other teams working out of offices in West Africa (Ibadan, Bamako), Southern Africa (Maputo), South Asia (Delhi, Guwahati, and Hyderabad in India), Southeast Asia (Bangkok, Jakarta, Hanoi, Vientiane), and East Asia (Beijing).

About the Livestock, Environment and Development Initiative

The Livestock, Environment and Development (LEAD) Initiative, established in 2000, is an interinstitutional consortium within the FAO. The work of the initiative targets the protection and enhancement of natural resources affected by livestock production while at the same time maintaining a focus on strengthening food security and alleviating poverty.

About the Scientific Committee on Problems of the Environment

The Scientific Committee on Problems of the Environment (SCOPE) was established in 1969 as a nongovernmental organization by the International Council for Science (ICSU). Through its program of scientific assessments, SCOPE brings together natural and social scientists and civil society experts and practitioners from around the world to look beyond the horizons of disciplines and institutions at emerging cross-cutting environmental challenges and opportunities. Its assessments have guided and shaped many of today's global environmental programs and actively foster science-policy dialogue and decision processes.

About the Woods Institute for the Environment at Stanford University

The Woods Institute for the Environment at Stanford University harnesses the expertise and imagination of leading academics and decision makers to create practical solutions for people and the planet. The institute is pioneering innovative approaches to meet the environmental challenges of the twenty-first century by sponsoring interdisciplinary research; infusing science into policies and practices of the business, government, and not-for-profit communities; and developing environmental leaders for today and the future.

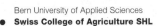
Bern University of Applied Sciences
Swiss College of Agriculture SHL

Volume 1
Drivers, Consequences, and Responses

Volume 2
Experiences and Regional Perspectives

Livestock in a Changing Landscape

Volume 1

Drivers, Consequences, and Responses

Edited by

Henning Steinfeld　　　Harold A. Mooney　　　Fritz Schneider　　　Laurie E. Neville

ISLANDPRESS

Washington | Covelo | London

Island Press is a trademark of The Center for Resource Economics.

Cataloging-in-Publication Data is available from the Library of
Congress.

ISBN (paper) 978-1-59726-671-0
ISBN (cloth) 978-1-59726-670-3

Printed on recycled, acid-free paper

Manufactured in the United States of America

10   9   8   7   6   5   4   3   2   1

Contents

Preface

This volume, and its companion featuring international case studies, is the result of an integrated collaboration involving a large community of experts from around the world in the planning, writing, and review of information and ideas. The goal of this effort was to achieve an integrated view of the global livestock sector, its drivers, consequences, and responses to issues of concerns, alongside with current practices and possible future scenarios. This effort sought to understand the opportunities enhancing the positive trends of this fundamental endeavor as well as providing insights into the negative impacts on environmental, health, and social aspects. In order to achieve this overview, many academic, technical, and socioeconomic experts engaged to consider the multiple dimensions of the livestock industry, working together and learning from one another, to produce these volumes.

This effort would not have been possible without the tireless support of many organizations, including the United Nations Food and Agricultural Organization (FAO); International Livestock Research Institute (ILRI); FAO Livestock, Environment and Development Initiative (LEAD); the Scientific Committee on Problems of the Environment (SCOPE); the Swiss College of Agriculture (SHL) of the Bern University of Applied Services; the French Agricultural Research Centre for International Development (CIRAD); and the Woods Institute for the Environment at Stanford University.

The FAO headquarters in Rome and the FAO Regional Office for Asia and the Pacific in Bangkok and the Center for Environmental Science and Policy at Stanford University hosted planning and final consultation meetings. We sincerely thank everyone, whose valuable expertise was appreciated; in particular, Linda Starke, Carolyn Opio, Anne Schram, and Charlotte Arthur, who provided key editorial support. Claudia Ciarlantini provided graphic support throughout the final preparations. In addition, the critical review process involved many individuals who contributed their time, expertise, and insight. The editors are grateful for the dedication and patience shown by all who were involved.

Executive Summary

The main technical volume and the case study volume take a detailed, comprehensive, and integrated look at the drivers of the global livestock sector; at the social, environmental, and health consequences of livestock production; and at the variety of responses to opportunities and threats associated with the sector. What emerges is indeed a picture of "Livestock in a Changing Landscape," where economic, regulatory, and environmental contexts are changing rapidly, as are the modes of production. Benefits and costs are distributed rather unevenly, and a variety of diseases and environmental threats are the source of growing concern.

Throughout history, livestock have been kept for a variety of purposes, with the almost exclusive focus on food use of livestock in modern agricultural systems being a relatively recent development. But in many developing countries, livestock are still a critical support to the livelihoods of people who live in or near poverty, and it is here that nonfood uses remain predominant. These include the use of animals for work and as a source of fertilizer (manure), as a means to store wealth, and as a buffer to hedge against the vagaries of nature and other emergencies. Livestock are often the only way to use marginal land or residues and waste material from food and agriculture. Livestock, or symbols of them, also play an important role in religious and cultural lives.

The nonfood uses of livestock, however, are in decline and are being replaced by modern substitutes. Animal draft power is replaced by machinery, and organic farm manure by synthetic fertilizer. Insurance companies and banks replace the risk management and payment functions of livestock. The many purposes for which livestock are kept are vanishing and being replaced by an almost exclusive focus on generating food for humans—meat, eggs, and milk. Hides and fiber still play a role, but these pale in comparison to food uses.

Over recent decades there has been a demand-led rapid expansion of production and consumption of animal products, the so-called livestock revolution. Demand for meat, for example, is projected to double between 2000 and 2050. Population and income growth coupled with urbanization has driven this demand. On the supply side, the livestock revolution has been fed by inexpensive, often subsidized grains, cheap fuel, and rapid technological change. This is particularly evident in poultry, pork, and dairy production. The rapid development of the livestock sector has occurred in a global environment that has been favorable to capital and market liberalization and to rapid technology flows. The recent decades have also been a period of neglect with regard to the environmental and livelihood consequences of livestock production. The response to changing disease patterns and public health threats has been slow and inadequate in many places. Similarly, the need for ever-growing amounts of animal-derived food in the diets of affluent consumers is increasingly questioned on health and environmental grounds.

In industrial countries, however, demand for livestock products is stagnant or even in gradual decline, particularly among the educated and wealthy, where concerns about health aspects, environmental issues, and animal welfare have become more widespread. Some saturation has even started to occur in developing countries, and recent price increases in livestock products may further dampen demand. Yet in most of sub-Saharan Africa and parts of South Asia, the livestock revolution has yet to occur. Overall, however, demand for livestock products may soften, and the recent economic downturn may limit expenditures, particularly for expensive livestock products such as beef. Competition for land is ever more acute, and prices for feed have reversed their decade-long declining trend. Other critical inputs to livestock production, such as water, energy, and labor, are also becoming more expensive. A still more daunting challenge is the fact that land-based livestock production is particularly exposed to the vagaries of climate change.

Increases in livestock production have come from increases in animal numbers and in yields per animal. The latter has been particularly important in dairy, pork, and poultry production and has been facilitated by the rapid spread of advanced production technologies and greater use of grains and oil crops in animal feeds.

Livestock production is practiced in many different forms. Like agriculture as a whole, two rather disparate systems exist side by side: in one case, livestock are kept in traditional production systems in support of livelihoods and household food security; at the same time, commercially intensive livestock production and associated food chains support the global food supply system. The latter provide jobs and income to producers and others in the processing, distribution, and marketing chains and in associated support services.

Social Issues

Livestock have an overwhelming importance for many people in developing countries; close to a billion poor people derive at least some part of their livelihood from livestock in the absence of viable economic alternatives. Livestock production does not require ownership of land or formal education, can be done with little initial investment, and can be transformed into money as and when required. As part of a livelihood strategy, livestock not only provide food, energy, and plant nutrients; they also have an asset function, in that they hedge against risks and play numerous sociocultural roles. The poor are mainly found in small mixed farming systems (less than two hectares) and in pastoral, dryland, or mountain areas. Projections indicate that small farms will continue to be a prominent feature in rural areas in the next decades.

Many poor producers have not benefited from livestock sector growth, however, as shown in China or India, where in the context of rapidly increasing demand for livestock products, large-scale industrial production units have grown rapidly, displacing, at least partially, the smallholder production system. Similarly, there is a risk of overuse of natural resources, leading to long-term environmental damage. On the contrary, many producers have been marginalized and excluded from growing markets. The exclusion of small producers from growing markets is often the result of heightened market barriers in the form of sanitary and other quality standards and of unfavorable economies of scale.

Soaring demand for animal products and competitive pressures in the sector have led to the emergence and rapid growth of commercialized industrial production. Here, large-scale operations with sophisticated technologies are based on internationally sourced feed and cater to the rapidly growing markets for poultry, pork, and milk. The traditional middle-level mixed family farms, while important in many places, are often relegated to the informal market and gradually squeezed out as formal market chains gain hold. In some areas, however, small producers have the potential to contribute in a sustainable way to increased production, provided the constraints of high transaction costs and product quality can be addressed.

At the same time, emerging industrial production benefits from considerable economies of scale, and labor requirements decline dramatically with growing intensification. The trend to reduce human–animal contact for health reasons further reduces employment opportunities. As a consequence, the modern livestock sector provides dramatically fewer people with income and employment than the extensive traditional sector did. This is only in part compensated by employment in the agroindustries associated with the livestock sector, such as feed mills and other input suppliers, slaughterhouses, dairy plants, and processing and retailing facilities.

Another feature of the changing landscape is that the locus of production is shifting. Because of infrastructure weaknesses in many developing countries, growth in the livestock sector is often limited to the outskirts of major consumption centers. Both intensive livestock production and associated agroindustries tend to be located here; as a result, employment and income opportunities move away from rural areas where traditional livestock raising is located. The dramatic growth in livestock production has therefore not led to broad-based rural growth and has not been tapped by the majority of smallholder producers. However, rapid intensification and industrialization have helped urban consumers as prices declined and accessibility increased.

Effective responses to these social issues have been largely absent. Although there have been many efforts to upgrade smallholder practices and, to a certain extent, the institutional framework in which they operate, this has had very little perceptible impact. Rather, many producers, particularly in rapidly developing economies, have abandoned livestock farming as an activity. As long as growing secondary and tertiary sectors can absorb excess labor, this does not necessarily need to be deplored.

Evidence shows that smallholder production can remain competitive where market barriers can be overcome and transaction costs can be reduced. This requires different institutional formats, such as cooperatives, that manage input supplies and marketing as well as the provision of knowledge. Elsewhere, contract farming, particularly in pig and poultry production, has had some success in allowing producers to remain in business, even though these farmers were seldom poor. One example of this approach is Nestlé, a large company that demonstrates how global markets can work for a large number of smallholders. The provision of credit to small producers has also proved effective, at least in part.

Environment and Natural Resources

Livestock affect the global climate, water resources, and biodiversity in major ways. Livestock occupy over one-fourth of the terrestrial surface of the planet, on pasture and grazing lands, of which a significant part is degraded. Expansion of pasture occurs in Latin America at the expense of forests. Concentrate feed demand occupies about one-third of total arable land. Pasture use and the production of feed are associated with pollution, habitat destruction, and greenhouse gas emissions. Livestock are also an important contributor to water pollution, particularly in areas of high animal densities. Both extensive and intensive forms of production contribute to environmental degradation and destruction.

Livestock, through associated land use and land use change, feed production, and digestive processes and waste, affect global biogeochemistry in major ways, particularly the carbon and nitrogen cycles. A large part of this alteration manifests itself in livestock's contribution to global greenhouse gas emissions in the form of carbon dioxide, methane, and nitrous oxides.

Extensive production is practiced by many poor producers who use low-cost or no-cost feed in the form of natural grasslands, crop residues, and other waste materials. However, a large part of the world's pastureland is degraded—releasing carbon dioxide, negatively affecting water cycles, altering vegetation growth and composition, and generally affecting biodiversity. Forest conversion to pasture and crops has important consequences for climate change and biodiversity. Pastoralists, in particular, are threatened as they lose access to traditional grazing areas as the threat of climate change increases, especially for Africa.

The environmental problems of intensive production are also associated with the production of concentrate feed and the disposal of animal waste. Feed production usually requires intensively used arable land and a concomitant use of water, fertilizer, pesticides, fossil fuels, and other inputs, affecting the environment in diverse ways. Even if increased feed production has mostly been achieved through intensification, the expansion of the area dedicated to production of crops such as soybean is now a major driver of deforestation in the Amazon region.

Because only a third of the nutrients fed to animals are absorbed, animal waste is a leading factor in the pollution of land and water resources, as observed in case studies in China, India, the United States, and Denmark. Total phosphorus excretions are estimated to be seven to nine times greater than that of humans, and livestock excreta contain more nutrients than are found in the inorganic fertilizer used annually. Through growing feed crops and managing manure, the livestock sector also emits nitrous oxides (a particularly potent greenhouse gas) and methane.

Policy makers have largely ignored environmental issues related to livestock, often because of the large role that livestock play in sustaining livelihoods and rural life. Some industrial countries have made sustained efforts, especially since the 1980s, to control waste management, with a particular focus on water pollution. Some countries, such as Denmark, have successfully reduced nitrogen leaching from animal operations. However, climate change issues related to livestock remain largely unaddressed.

Human and Animal Health Issues

Human health is affected by the livestock sector through the impact of animal source food products on human nutrition and through diseases and harmful substances that can be passed on by livestock and livestock products.

With regard to nutrition, animal source food products can be of great benefit for people who suffer from undernourishment and for those who need a diet higher in fats and protein, such as children and pregnant women. Consequently, livestock products play a large role in efforts that target improving nutrition in poor and middle-income countries—although other sources of protein and essential micronutrients might also be available at lower environmental cost. At the same time, livestock products can contribute to unbalanced diets, leading to obesity and unhealthy physical conditions. These meat products are often singled out when consumed in excessive quantities as causes or contributing factors to a variety of noninfectious diseases, including cardiovascular disease, diabetes, and certain types of cancer. Increasingly, this is being addressed by public education programs, but until now these have had little measurable impact.

Animal diseases have the potential to adversely affect people by reducing the quantity and quality of food and other livestock products. Transboundary animal diseases tend to have the most serious consequences. Many diseases are zoonotic, with the ability to be transmitted from animals to humans, threatening both human and animal well-being. Zoonotic disease outbreaks often take the form of grand-scale emergencies, requiring rapid action to prevent food supply systems and markets from collapsing.

Changing ecology, the increased mobility of people and movement of goods, and the shifting and often reduced attention by veterinary services have led to the emergence of new diseases, such as highly pathogenic avian influenza (HPAI), and to the reemergence of traditional ones such as tuberculosis. Emerging diseases are closely linked to changes in the production environment and livestock sector structure.

The focus of health concerns differs, depending on whether diseases are considered from a poverty/livelihood point of view or from a global food supply perspective. Certain diseases, such as foot-and-mouth disease, may have minor implications for smallholder production, but their presence excludes entire countries from international trade in livestock products. The control of diseases therefore tends to have far-reaching and often contrasting impacts on constituents in the livestock sector; for example, movement restrictions and sanitary controls may effectively exclude smallholders from markets and may deprive them of livestock as a livelihood option.

The coexistence of functional modern operations alongside traditional operations, in addition to environmental factors, contributes to the emergence and spread of diseases. HPAI has flourished in particular where backyard systems interact with wildlife and are connected via market channels to production systems of medium and high intensity. This explosive mix has its fuse in the structure of livestock sectors, and sector heterogeneity often results from sector protection.

The Challenges

The challenges posed by the livestock sector cannot possibly be solved by a single string of actions, and any actions require an integrated effort by a wide variety of stakeholders. These need to address the root causes in areas where the impact of livestock is negative. They also need to be realistic and take into account the livelihood and socioeconomic dimensions. For example, although in some quarters reduced intake is touted as the most effective way to address negative impacts, most people are shifting, or would likely shift if their incomes allowed, toward consuming more livestock products. Accepting individual food choices, and considering social and economic realities, several principal considerations emerge.

First, given the planet's finite land and other resources, there is a continuing and growing need for further efficiency gains in resource use of livestock production through price corrections for inputs and the replacement of current suboptimal production with advanced production methods. There appears to be little alternative to intensification in meeting the bulk of growing demand for animal products. Although niche products from extensive systems may be of importance in some markets, the bulk of animal protein supply will need to come from intensified forms of production in order to reduce substantially the requirements for natural resource use, such as nutrients, water, and land. The trends toward larger scales of production are determined by economies of scale and scope and are to a large extent inevitable. The current trend toward monogastric production and crop-based animal agriculture will likely continue and possibly accelerate if competition for land accentuates. Ruminants and roughage-based production will continue to decrease in importance; and though not mutually exclusive, this will need to balance the production of animal products with the provisioning of environmental services, including carbon sequestration, water resources, and biodiversity. Particularly in marginal areas, this balance is critical because the value of environmental services will predictably grow sharply, limiting the use of these areas for livestock production.

Second, livestock are a suitable tool for poverty reduction and economic growth in poor countries and areas that are not fully exposed to globalized food markets.

Smallholder dairy production and certain types of cooperatives and contract farming provide opportunities for smallholder livestock production. In rapidly growing and developed areas, economies of scale and market barriers will continue to push smallholders out of production, and alternative livelihoods need to be sought in other sectors. Intensification and efficiency gains from economies of scale mean that fewer and fewer people will depend on livestock production for their livelihood. Policies need to support the transition process so as to avoid social hardship and prevent rapid loss of livelihoods in cases where alternatives cannot be provided. There is more promise for growth linkages in associated agrofood industries than in primary production. Notable exceptions are dairy production in favored environments and, to some extent, poultry. In these cases, there is scope for smallholder development and a reasonable chance to compete, at least for some time to come. For the most part, however, livestock production is overwhelmingly important in sustaining livelihoods but less in providing a pathway out of poverty. Social safety nets and exit options are required for those left behind.

Third, food chains are getting longer and more complex in response to challenges and opportunities from globalization. As a consequence, food safety and quality requirements need to be applied to food chains, and sanitary requirements will be linked to them. The specific location of production and its disease status will likely become a secondary issue. Similarly, private standards are likely to be predominant over public ones. As a consequence, animal health and food safety policies will need to be applied to segments of food chains rather than territories. Public policies must foster international collaboration, upgraded monitoring of disease threats, and early reaction.

The changing landscape for the livestock sector as determined by social, economic, and biophysical aspects requires the urgent attention of policy makers, producers, and consumers alike. The institutional void and the systemic failures apparent in widespread environmental damage, social exclusion, and threats to human health need to be addressed with a sense of urgency. Only then can the livestock sector, with its vast diversity of people, critical issues, and constraints, move forward on a responsible development path.

Introduction

The domestication of plants and animals and the development of agriculture was a huge leap forward for human societies, leading to sedentary communities and some insurance against the vagaries of nature. Over the thousands of years since these innovations there have been continual improvements in the utility of domesticated organisms for humans through selection, and more recently though genetic engineering, as well as equally impressive improvements in cultural techniques and technological innovations, such as the industrial production of nitrogen fertilizers. These developments have been so successful that they have freed large segments of society to pursue endeavors other than seeking food and clothing and hence to produce other kinds of goods and services. They have fueled an increasing human population, the accumulation of per capita wealth, and an increase in individual consumption of natural resources. Animal production systems, from the most ancient to the industrial, involve a large fraction of society and most certainly a large fraction of the natural resources needed to support these systems.

As recently documented in numerous publications, these agricultural advances and their inadvertent consequences are casting a long shadow over the environment and human societies. This book concentrates on one large and important sector of agriculture: livestock. It takes a comprehensive view of livestock—considering industrial or so-called landless animal production as well as extensive grazing systems used on a large fraction of Earth's land surface. No matter where and how livestock are produced, they have an impact on the environment cumulatively, and increasingly in very indirect and complex ways. As the web of interaction among places distant in space becomes more complex due to globalization, humankind must be aware of the connective threads in order to be alert to inadvertent negative consequences or just plain overlooked adverse impacts to human welfare.

It is becoming more evident that livestock production is having far-reaching impacts not only on the environment directly but also on many social and human health endeavors and concerns. Massive social shifts are occurring related to livestock because of the altering and blending of diverse cultural viewpoints as well as the complex and rapidly changing global economic drivers that have dramatically altered how animal protein is produced and the resources required. Globalization has brought many benefits, but it has also opened floodgates that allow the rapid movement of pests and diseases.

What we attempt in this volume is to examine the many pieces of this complicated endeavor in order to determine the drivers of the many dramatic changes being seen, their consequences—historically, at present, and possibly in the future—and what can be done to have more favorable outcomes in the livestock industry, given the growing understanding that a simple fix for any one dimension of the problem may well exacerbate an additional adverse effect. Many little changes can make a big difference, but changes made in an integrated and considered manner, although difficult, may well have even greater benefits.

This volume examines the drivers of change in the livestock industry in the context of trends in agriculture as a whole. The analysis attempts to explain the revolution occurring in the livestock sector in terms of production methods and the increasing demands for animal protein and to give some texture to the differences in these developments among regions of the world.

The characterization of the changing drivers sets the stage for an analysis of the environmental impacts of previous and current livestock production methods. In this section we focus on an array of production systems—from traditional extensive grazing approaches to the very prevalent mixed farming systems and the increasingly important industrial methods.

This analysis takes in part a biogeochemical approach regarding carbon, nitrogen, and water in order to determine the role of livestock production in the large-scale alterations that are occurring to the basic functioning of Earth, including the fundamental climate system. Various chapters also look at these processes and impacts at the local level, giving a sense of how small-scale impacts cascade into global impacts. The analyses tease out how adverse effects are generated, thus laying the foundation for subsequent chapters on how to alleviate unintended consequences. The complexities of these interactions, both locally and globally, were clearly demonstrated during the preparation of this volume, when fossil fuel costs spiked at a very high level—stimulating biofuel production and leading to the "fuel versus food" debates, actions, and reactions.

The biological diversity of the landscape represents the building blocks of ecosystems and hence of the services that they provide to society. An important part of this assessment is an analysis of the impact of animal production—favorable as well as unfavorable—on the basic biodiversity of landscapes and the consequences of these changes on the delivery of services in addition to the production of animal protein and the sustenance of livelihoods.

The global scale of animal production systems is huge. The rapidly changing nature of these systems, especially increasing intensification and globalization, is playing out in complex ways across the globe. Social systems are being altered with both negative and positive consequences in differing parts of the world. Impacts are being seen on systems with very long traditions that have actually been shaped over generations by primary dependence on livestock for many of life's necessities. The loss of these traditions is yet another element of globalization and the biotic and social homogenization that is resulting. What the future will bring, and how quickly, are discussed here. In addition, the way the changing nature of the livestock sector is variably affecting the many stakeholders—providing and eliminating markets, especially for small landholders—is analyzed.

The impact of animal products in human nutrition is highly variable across the globe; people in some developing nations are struggling to obtain sufficient protein for optimal human growth and development, whereas in other regions people are consuming more than is healthy. This volume discusses these trends, along with another side of human health—the threats of animal-borne diseases. The authors look at how various animal production systems, along with global trade, protective measures, and various management approaches, can affect human health. These will be growing and increasingly complex issues for consideration in the years ahead.

Finally, and the central objective of the whole effort, we conclude with a section on what institutional and social options are available and vital when considering bringing animal production systems toward more environmentally and socially sustainable practices. This analysis also considers not only the projection of the massive changes and trends that are already in place but also some of the surprises that may be in store as the many other human endeavors compete for increasingly limited yet vital shared resources.

This book is the result of the collaboration of many scientists and experts from around the world who engaged in a mutual learning process to try to piece together the complex story of the "changing landscape" of livestock production systems globally. A companion volume provides a more in-depth view of national and regional issues. Together, these two volumes provide insights on the range of positive and negative consequences of livestock production on the environment, on social systems, and on human health, at both global and local scales, as well as the various approaches that could potentially alleviate the negative impacts while taking into account the growing needs of society.

Part I
Drivers of Change
Drivers: Perspectives on Change

Fritz Schneider

The world's population is predicted to increase by 1% annually between 2000 and 2030; increasing from 6.1 billon in 2000 to 8.2 billion in 2030, with a somewhat slower growth rate to 9.0 billion by 2050. During the same time, the demand for livestock products will increase rapidly and disproportionately. This large increase is due to a number of major drivers of change: population increase, demographic shifts (urbanization, a higher percentage of women in the workforce), rising incomes, technology in food chains, and the liberalization of trade and capital. In these contexts, the rapid increase of intensive and confined livestock production and the land and livelihood needs of extensive production systems will be crucial challenges. Furthermore, the livestock sector will emerge as a very significant contributor to environmental problems at every level—from local to global—including land degradation, climate change, air pollution, water shortage, water pollution, and loss of biodiversity.

Part I introduces this two-volume set by putting the drivers of change in global agriculture and livestock systems into context with recent developments. Over decades, global agriculture has met the demand for food and nonfood products, as evidenced by growth in agricultural output and long-term decline in commodity prices. The sharp and sudden increase in commodity prices in late 2007 and 2008 was followed by an equally sudden decrease in early 2009 to the levels of 2007. Critical factors are briefly analyzed, and the interdependence between these massive price shocks (in both directions) and the major drivers of change in global agriculture and livestock systems are illustrated. The spike in food prices has forced millions of people into poverty and hunger, showing that such fluctuations dramatically and negatively influence the weakest sections of the human populations worldwide but mainly in developing countries.

The chapters in Part I on drivers of change offer an in-depth analysis and discussion of major driving forces of identified global regional and local changes and set the stage for Parts II and III on consequences and responses, respectively. In Chapter 1, the most important driving forces are illustrated, and the factors influencing these drivers and their interactions are analyzed. The chapter

elaborates on trends in consumption, retail and supply chains, food production systems, and changes in trade patterns, mainly from the macroeconomic perspective.

In Chapter 2, the authors highlight the following facts: the share of all animal products in human diets is continuing to increase in developing countries, income growth is a major driver of increasing consumption, urbanization of populations accelerates consumption, and population growth and structure have impacts on total consumption. Global animal production is shifting from industrial to developing regions, and technological improvements will make livestock production more efficient and effective. The chapter also details health, nutrition, and food safety concerns and describes the trends in international trade of livestock and livestock products.

Chapter 3 provides a comprehensive analysis of drivers such as land, water, fossil fuels, climate, and climate change, illustrating their effects on livestock and feed production, and looks at the competition for these resources. The authors analyze productivity trends and briefly discuss trends and progress in production technology. Policies and institutions as well as the regulatory framework and incentives are important driving factors. The authors put these drivers into perspective in the context of economic, social, and ecological dimensions. The resulting structural changes within the production systems are described and discussed. In conclusion, the authors discuss the intensification of production and processing in order to leverage economies of scale and to strictly apply the "polluter pays, provider gets" principles.

In Chapter 4, the authors suggest that location matters when it comes to environmental, social, and health issues of growing livestock production and consumption. They explain that, on the one hand, land use intensification and clustering of operations is mainly associated with pig, poultry, and dairy production and, on the other hand, expansion and land marginalization are mainly associated with ruminant meat production in specific but large regions. This chapter concludes that the three major factors determining the transition are cheap transport costs, input prices, and shifting relocation of demand for animal products.

1

Drivers of Change in Global Agriculture and Livestock Systems*

Prabhu Pingali and Ellen McCullough

Main Messages

- **Over the past four decades, agricultural growth has contributed to poverty reduction and economic growth in the developing world, while also allowing food supplies to keep pace with growing demand.**
- **Economic, demographic, and technological factors have been responsible for the transformation of food systems.** These drivers include income growth, urbanization and rising female employment, technological change, and the liberalization of trade and capital.
- **Due to rising incomes and falling food prices over the last 40 years, consumers' diets in developing countries have diversified out of staples and toward higher-value foods, such as fruits and vegetables, meat and dairy, and foods consumed in processed forms.** This dietary shift has fueled and been fueled by major changes in the food retail sector, most notably the rapid *spread of supermarkets throughout the developing world.*
- **From retailers to wholesalers, processors, and input providers, major organizational changes have taken place throughout the food chain.** Growing concentration is apparent at all levels, and a greater emphasis is placed on forward and backward linkages between food chain players.
- **With major shifts in the types of products demanded and the terms by which they are procured, *small-scale farmers are facing difficulty* complying with product standards and meeting buyers' requirements.**
- **A major challenge across food systems is to increase farmers' productivity with respect to factors of production, so that farmers can stay afloat despite rising input costs.**

- **Global meat production has tripled in the past three decades and should double its present level by 2030.** The livestock sector is marked by intensification and a shift from pasture-based ruminant species to feed-dependent monogastric species.
- **The global value of traded agricultural goods has grown tenfold since the 1960s, with composition shifting toward higher-value and more-processed products.** Many farmers in developing countries have diversified into higher-value crops in response.

Introduction

Over the past four decades, global agriculture has met the demand for food and nonfood products, as evidenced by growth in agricultural output and a long-term decline in commodity prices. Though prices for major commodities spiked sharply between the fall of 2007 and 2008, they returned to their pre-spike levels by early 2009 (FAO 2008a). Some evidence suggests the world may be experiencing a reversal of the sustained decline in commodity prices due to structural shifts in demand, such as rising demand for food and animal feed in emerging economies and for biofuels stock. However, the speed at which prices rose and fell indicate that the acute price crisis cannot be attributed to changing demand alone. Critical factors in the food price spike were supply shocks, especially drought in important export-oriented bread baskets, and commodity speculation, which may have been partially fueled by expectations of demand for biofuels stock. Reactionary policy measures, such as banning exports and grain hoarding, exacerbated the problem. The food crisis forced an estimated 100 million people into poverty and 70 million people into hunger (FAO 2008b, Ivanic and Martin 2008). Low income food deficit countries, such as Haiti, were hit hard by rising food import bills. Over the medium to long term,

* The views expressed in this chapter are those of the authors and should not be attributed to the Bill & Melinda Gates Foundation.

the United Nations Food and Agriculture Organization (FAO) and the Organization for Economic Cooperation and Development (OECD) predict that prices for major food commodities will increase some relative to 2000 levels, but much less dramatically than what was experienced in 2008 (OECD/FAO 2008). One important outcome of the food crisis was the unprecedented media and political attention paid to the agricultural sector, which resulted in substantial financial commitments to the sector. The attention faded as prices began to fall and as the financial crisis began to unfold.

Agricultural growth has contributed to improvements in food security, poverty reduction, and overall economic growth in much of the developing world. The success in increasing agricultural production has not, however, been shared uniformly across regions and countries. Many of the least developed countries, particularly in sub-Saharan Africa, and marginal production environments across the developing world continue to experience low or stagnant agricultural productivity, rising food deficits, and high levels of hunger and poverty. Economic development is almost always accompanied by a falling share of agriculture in GDP. Globally, the share of agriculture in total GDP has fallen from 9% in the early 1970s to 4% in recent years (World Bank 2006a). This number is considerably higher in developing countries, although it is also on the decline. Countries can be characterized by the extent to which agricultural growth has contributed to economic growth, and the extent to which the poor people in a given country depend on agriculture for income (World Bank 2008).

Agricultural economies are those in which growth in agriculture is a large contributor to GDP growth and where the majority of poor people are found in rural areas and are largely concentrated in sub-Saharan Africa. Failed states and areas of conflict are almost always marked by low per capita GDP and a high share of agriculture in the economy. Transforming economies are also characterized by rural poverty, but their economic growth no longer results predominantly from agricultural growth. Most transforming economies are found in Asia. Finally, urbanized economies are those in which the majority of poor people live in urban areas and agriculture is not a major source of economic growth. These urbanized economies are concentrated in Latin America. In agricultural economies, agricultural growth is a means of achieving both economic growth and poverty reduction. In transforming economies, agricultural growth is essential for poverty reduction but not necessarily for economic growth. In urbanized economies, there are still many opportunities for agricultural growth to contribute to poverty reduction, and there is also a compelling need to manage large labor flows out of the agricultural sector (World Bank 2008).

The process of agricultural development also occurs amidst a major organizational transformation of food systems that was well under way by the 1990s. These changes have been led by consumption trends but were reinforced by transformation in the retail sector as well as innovations in production, processing, and distribution technologies.

Drivers of Change

Four important driving forces in agriculture are together responsible for major global shifts in consumption, marketing, production, and trade: rising incomes, demographic shifts, technology in food chains, and the liberalization of trade and capital. It is not possible to tease out each driver's individual effects, so a brief introduction of the drivers will be followed by a more detailed discussion of their collective influence on food systems.

Per capita incomes have risen substantially in many parts of the developing world over the past few decades. In developing countries, per capita income growth averaged around 1% per year in the 1980s and 1990s but jumped to 3.7% between 2001 and 2005 (World Bank 2006b). East Asia has led the world with sustained per capita growth of 6% per year in real terms since the 1980s. In South Asia, growth rates have been consistently positive since the 1980s although not as spectacular. Eastern Europe and Central Asia experienced economic decline in the 1990s but have since obtained per capita growth rates of 5% per year. Latin America and sub-Saharan Africa have also experienced negative growth rates, which reversed themselves in the 1990s in Latin America and since 2000 in sub-Saharan Africa. Income growth is closely linked with higher expenditure on food items and with diet diversification out of staples (known as Bennett's law). The effect of per capita income growth on food consumption is most profound for poorer consumers who spend a large portion of their budget on food (Engel's law).

Both urbanization and rising female employment have contributed to rising incomes for many families in developing countries. Urban dwellers outnumbered rural populations for the first time in 2007 (UN 2006). Female employment has at least kept pace with population growth in developing countries since 1980 (World Bank 2006a). Female employment rates have risen substantially in Latin America, East Asia, the Middle East, and North Africa since the 1980s. As wages increase, urban consumers are willing to pay for more convenience, which frees up their time for income-earning activities or leisure. This results in a growing demand for more processed foods with shorter preparation times. Higher rates of female participation in the work force have been linked to greater demand for processed foods (Pingali 2007, Popkin 1999, Regmi and Dyck 2001).

Technological innovation in agribusiness has contributed to major organizational change in food distribution, processing, and production. Firms have responded to variability in consumer demand by developing advanced planning systems that use quantitative modeling tools (Kumar 2001). They have then used communication tools to improve the efficiency of coordination between actors along the supply chain to shorten the response time to demand fluctuations. The Universal Product Code (UPC) emerged in the 1970s from a retail industry–led initiative to standardize a system for identifying products and managing inventories (King and Venturini 2005). Since then, other initiatives for standardizing data transfer along supply chains and across the industry have followed. While innovations in information and communications technology have allowed supply chains to become more responsive, innovations in processing and transport have made products more suitable for global supply chains. Packaging innovations throughout the second half of the twentieth century continued to extend food products' shelf lives (Welch and Mitchell 2000). Meanwhile, a downward trend in transportation costs and widespread availability of atmosphere-controlled storage infrastructure has made it cost effective to transport products over longer distances. Raw materials have been engineered to meet processing standards and improve shelf life through conventional breeding, and, more recently, genetic engineering.

"Globalization" is marked by liberalization of trade and foreign direct investment in retail and in agribusiness. Trade has matched, but not outpaced, worldwide growth in food consumption. However, trade has shifted toward higher-value and more processed products and away from bulk commodities (Regmi and Dyck 2001). Foreign direct investment (FDI) in agriculture and the food industry grew substantially in Latin America and Asia between the mid-1980s and mid-1990s, although investment remained very low in sub-Saharan Africa (FAO 2004). In Asia, FDI in the food industry nearly tripled, from $750 million to $2.1 billion between 1988 and 1997. During that same period, food industry investment exploded in Latin America, from around $200 million to $3.3 billion. (Since 1997, it has been difficult to track FDI in the food industry due to changes in data reporting by the UN Conference on Trade and Development.)

Trends

Consumption

The gains in world average food consumption reflect predominantly those of the developing countries, given that the industrial countries already had fairly high levels of per capita food consumption in the mid-1960s. The overall progress of developing countries has been decisively influenced by significant gains made in East Asia. Historical trends toward increased food consumption per capita as a world average and particularly in developing countries will likely continue in the near future, but at a slower rate than in the past as more and more countries approach medium to high per capita income levels. Real food prices have declined over the last 40 years, which gives even consumers whose income levels have not risen access to improved diets.

Income growth and urbanization in particular are feeding dietary diversification in many parts of the world. Per capita meat consumption in developing countries tripled between 1970 and 2002, while milk consumption increased by 50% (Steinfeld and Chilonda 2005). Dietary changes are most striking in Asia, where diets are shifting away from staples and increasingly toward livestock products, fruits and vegetables, sugar, and oils (Pingali 2007). Diets in Latin America have not changed as drastically, although meat consumption has risen in recent years. In sub-Saharan Africa, perhaps the most striking change was a rise in sugar consumption during the 1960s and 1970s (FAO 2006).

Retail and Supply Chain

Growth in the number and size of large urban centers creates opportunities for establishment of large supermarket chains, which attract foreign investments and advertising from global corporations (Pingali 2006). The proliferation of supermarkets in developing countries is one of the most widely cited elements of food system transformation. Structural transformation of the retail sector took off in central Europe, South America, and east Asia outside China in the early 1990s. The share of food retail sales by supermarkets grew from around 10% to 50 to 60% in these regions. By the mid- to late 1990s, the shares of food retail sales in Central America and Southeast Asia accounted for by supermarkets reached 30 to 50%. Starting in the late 1990s and early 2000s, substantial structural changes were taking place in eastern Europe, South Asia, and parts of Africa. Here supermarkets' share approached 5 to 10% in less than a decade, and it is growing rapidly (Reardon and Stamoulis 2006).

Organizational changes in food retail are felt throughout the food chain due to the interconnectedness of retail, distribution, packaging and processing, production, and input provision. Growing concentration is taking place at all levels, particularly in the retail and processing sectors, and private sector standards for food quality and safety are proliferating. Increasingly, exchange is arranged through the use of contracts. More large-scale retailers and manufacturers are relying on specialized procurement channels and dedicated wholesalers. Food is increasingly being "pulled" into formal sector retail outlets such as supermarkets rather than grown for sale in local markets.

The changes in agrifood systems pose particular risks for small-scale farmers, traders, processors, wholesale markets, and retailers, who are operating in a new game with new rules. There is a much greater degree of integration between producers and the output market, with a strong emphasis on standards in relation to quality and safety. For the small farmer there will be short-term difficulties to meet agroindustry standards and contractual requirements. Small processors increasingly will have to compete with larger-scale food manufacturers that can benefit from economies of scale in processing technologies. Traders and marketers in local markets are being squeezed by the growing importance of specialized procurement practices and certified products. Contracts often exclude small farmers either explicitly or in practice because of the difficulty of compliance, the scale of investment required, or the degree of management experience and sophistication needed to interact with buyers. It has been shown that small farms are increasingly becoming marginalized and that large farms are consolidating by relying on a casual, hired labor force that is predominantly female (Kritzinger et al., 2004, Reardon and Berdegué 2002).

Food Production Systems

Even though the global value of agricultural production per capita has had a yearly growth of 0.4% per year since 1971, not all regions have followed the same trend (World Bank 2006a). In sub-Saharan Africa per capita food production has not seen a sustained increase over the last four decades. South Asia has had a small increase, while East Asia and the Pacific have increased agricultural value added per capita by almost 200% over the last 45 years. We also find sharp heterogeneity not only between regions but also between countries, within regions, and even within countries. Countries with a high incidence of poverty and food insecurity are invariably those that have experienced poor agricultural performance over the past four decades.

Traditional agricultural systems are found where agriculture has a major role in a nation's economy and labor force. Typically, the systems consist of smallholder farmers producing staple crops for subsistence purposes. In modernizing agricultural systems, agriculture accounts for a smaller share of the GDP and employs a smaller portion of the labor force. In these systems, farmers grow food staples and cash crops, typically for national markets. In industrialized agricultural systems, agriculture's importance in a nation's economy and labor force are minor. Farmers produce a highly differentiated set of crops for national and international markets. Economies of scale are particularly important in industrialized agricultural systems.

The composition of the global agricultural production portfolio has changed considerably over the last 40 years. Production rates of cereals, oil crops, sugar, horticulture, eggs, and meat have increased faster than population growth rates since 1961, while production of pulses, roots, and tubers (which are important for the poor in many agriculture-based countries) has declined relative to total population. Cereals production grew rapidly during the 1960s and 1970s but has slowed since then. The vegetable oil sector is the most rapidly expanding, fueled by the growth of food consumption and imports in developing countries. Increasing demand for oil crops for nonfood uses is also a major contributor to growth in the subsector, as is the availability of ample land suitable for expanding oil crop production.

Total crop production growth was generally positive in the 40 years from 1961 to 2000, although the same cannot be said in per capita terms. Production growth per person was close to zero for much of the last 40 years in most regions, with the clear exception of South and Southeast Asia (FAO 2006). Area expansion has not been a major source of crop production growth in recent decades, except in sub-Saharan Africa. Further opportunities for area expansion are virtually exhausted except in parts of South America, Southeast Asia, and West Africa, where rural population densities are still low. Increases in cropping intensity may offer a partial solution for this constraint despite high rural population densities. This will require an expansion of irrigation, for which there is still some scope in regions where irrigation infrastructure is underdeveloped (Carruthers et al., 1997).

Yield growth is another way of achieving production growth while economizing on land. As expected, countries with high population densities tend to have high yields and cropping intensity and vice versa. A notable group of densely populated countries, including China, India, Indonesia, and Brazil, has achieved moderate to high labor productivity in addition to high land productivity. Growth in input use, such as fertilizer, tractors, and irrigation, can contribute to yield growth. Use of these inputs is leveling off worldwide, however, and is particularly low in Africa (FAO 2006).

Falling commodity prices and rising input prices have squeezed farmers' profits from both ends. Technology has kept farmers afloat through rising factor productivity. To date, the best prospect for raising global crop production remains the path of productivity. Investments in agricultural research and development are crucial for sustaining growth in total factor productivity. In the past, technological advances have led to reduced production costs and allowed for production to outstrip rising demand even as prices have fallen. However, opportunities for growth in total factor productivity could be adversely affected by land degradation and water resource constraints (Pingali and Heisey 1999).

Livestock Production

Globally, livestock production is the world's largest user of land and accounts for almost 40% of the total value of agricultural production. In industrial countries the share is more than half. In developing countries, where it accounts for one third, its share is rising quickly (Bruinsma 2003).

Production of meat and eggs has expanded considerably over the last 40 years and will probably continue to grow faster than the size of the population due to diversification of diets and a higher demand for meat as incomes grow. The slow growth in the milk sector is expected to accelerate, mainly because of growing demand in developing countries. Global meat production tripled from 47 million tons to 139 million tons between 1980 and 2002 (Steinfeld and Chilonda 2005). Although the pace of growth is slowing down, meat production is expected to double by 2030 to meet rising demand (Bruinsma 2003).

Until recently, a large proportion of livestock in developing countries was not kept for food but for draft power and manure and as capital assets, to be disposed of only in times of emergency. Livestock were an integral part of agricultural systems, distributed among many owners and raised close to feedstocks. Now, growing demand for meat products is being met increasingly through industrial systems, where meat production is no longer tied to a local land base for feed inputs or to supply animal power or manure for crop production (Naylor et al., 2005).

Roughly one third of Earth's terrestrial ice-free surface is devoted to livestock production, while an area one fourth that size is devoted to production of crops that are directly consumed by people (Steinfeld et al., 2006). The most land-intensive aspect of livestock production is grazing, which typically occurs on less productive land not suitable for cropping. As livestock production shifts to more intensive systems, it will place more pressure on arable land for the production of animal feed. A coinciding shift in the composition of livestock production from ruminants to monogastrics alleviates pressure to increase rangeland but exacerbates the need for more cropland to be devoted to feed production.

Satisfying the growing demand for animal food products while at the same time sustaining productive assets of the natural resource base—soil, water, air, and biodiversity—and coping with climate change and vulnerability is one of the major challenges facing world agriculture today.

Changing Trade Patterns
The total value of agricultural exports has increased tenfold since the early 1960s, while the share of agricultural trade in the total value of traded goods has followed a downward trend, from 24% to less than 10% during the same period. And during this period, the least developed countries have shifted from being net exporters to being major net importers of agricultural commodities. Imports and exports have accounted for less than 30% of domestic cereal supply in all regions save Latin America between 1960 and 2005. The relative importance of cereal imports as a percentage of domestic supply is increasing in sub-Saharan Africa (FAO 2006).

Removal of trade barriers has called into question the competitiveness of many production systems. With increased prevalence of trade and lower transportation costs, many farmers are competing on a larger scale than ever before. New technologies are making competitors more productive, and the introduction of synthetic alternatives to some commodities has led to price glut, as has the appearance of new major producers of other commodities. In many industrial countries, subsidies to support producer incomes and promote exports have contributed to pushing down world prices for many agricultural products grown in temperate zones, reducing the export earnings of developing countries that produce the same commodities, such as cotton, sugar, and rice.

Producers in some developing countries, particularly more prosperous ones, have responded to declining price trends by shifting production into higher-value, export-oriented sectors. Excluding the least developed countries, developing countries have more than doubled the share of horticultural, meat, and dairy products in their exports while reducing the share of tea, coffee, and raw materials. In contrast, the least developed countries have increased their dependence on tea, coffee, and raw materials despite sharp declines in their prices. Consequently, real agricultural export earnings of the least developed countries fell by more than 30% over the last two decades.

Conclusions
Agriculture's share in the global economy is falling, even as production shifts toward higher-value subsectors like livestock. There is considerable heterogeneity between and within regions and countries in the speed and extent of transformation in agricultural systems. For the least developed countries, familiar development problems persist even as new challenges emerge. These challenges result from changing diets, retail transformations, and major organizational changes in procurement, processing, and agroindustry. One of the most striking manifestations of food system transformation is a rise in the consumption of livestock products in developing countries. Meeting this demand in a way that is socially desirable and environmentally sustainable is a major challenge facing agriculture today.

References

Bruinsma, J. 2003. *World agriculture: Towards 2015/2030*. Rome: Food and Agriculture Organization.

Carruthers, I., M. W. Rosegrant, and D. Seckler. 1997. Irrigation and food security in the 21st century. *Irrigation and Drainage Systems* 11 (2): 83–101.

FAO (Food and Agriculture Organization). 2004. *The state of food insecurity in the world*. Rome: FAO.

FAO (Food and Agriculture Organization). 2006. FAO statistical databases. http://faostat.external.fao.org.

FAO (Food and Agriculture Organization). 2008a. *Food outlook: Global market analysis November 2008. Global Information and Early Warning System on Food and Agriculture.* Rome: FAO.

FAO (Food and Agriculture Organization). 2008b. *The state of food insecurity in the world: High food prices and food security—threats and opportunities.* Rome: FAO.

Ivanic, M. and W. Martin 2008. Implications of higher global food prices for poverty in low income countries. Policy Research Working Paper 4594. Washington, DC: World Bank.

King, R. P., and L. Venturini. 2005. Demand for quality drives changes in food supply chains. In *New directions in global food markets*, ed. A. Regmi and M. Gehlhar. Washington, DC: Economic Research Service, U.S. Department of Agriculture.

Kritzinger, A., S. Barrientos, and H. Rossouw. 2004. Global production and flexible employment in South African horticulture: Experiences of contract workers in fruit exports. *Sociologia Ruralis* 44 (1): 17–39.

Kumar, K. 2001. Technology for supporting supply chain management introduction. *Communications of the ACM* 44 (6): 58–61.

Naylor, R., H. Steinfeld, W. Falcon, J. Galloway, V. Smil, E. Bradford, J. Alder, and H. Mooney. 2005. Agriculture: Losing the links between livestock and land. *Science* 310:1621–1622.

OECD/FAO. 2008. *OECD-FAO agricultural outlook 2008–2017.* Paris: OECD.

Pingali, P. L. 2007. Westernization of Asian diets and the transformation of food systems: Implications for research and policy. *Food Policy* 32 (3): 281–298.

Pingali, P. L., and P. W. Heisey. 1999. *Cereal crop productivity in developing countries.* Mexico: International Maize and Wheat Improvement Center (CIMMYT).

Popkin, B. 1999. Urbanization, lifestyle changes and the nutrition transition. *World Development* 27 (11): 1905–1916.

Reardon, T., and J. A. Berdegué. 2002. The rapid rise of supermarkets in Latin America: Challenges and opportunities for development. *Development Policy Review* 20 (4): 371–388.

Reardon, T., and K. Stamoulis. 2006. Impacts of agrifood market transformation during globalization on the poor's rural nonfarm employment: Lessons for rural business development programs. Plenary paper presented at the 2006 meeting of the International Association of Agricultural Economists, Queensland, Australia, August 12–18.

Regmi, A., and J. Dyck. 2001. Effects of urbanization on global food demand. In *Changing structures of global food consumption and trade*, ed. A. Regmi. Washington, DC: Economic Research Service, U.S. Department of Agriculture.

Steinfeld, H., and P. Chilonda. 2005. *Old players, new players: Livestock report 2006.* Rome: Food and Agriculture Organization.

Steinfeld H., P. Gerber, T. Wassenaar, V. Castel, M. Rosales, and C. de Haan. 2006. *Livestock's long shadow: Environmental issues and options.* Rome: Food and Agriculture Organization.

UN (United Nations). Population Division of the Department of Economic and Social Affairs. 2006. *World urbanization prospects: The 2005 revision.* New York: United Nations.

Welch, R. W., and P. C. Mitchell. 2000. Food processing: A century of change. *British Medical Bulletin* 56 (1): 1–17.

World Bank. 2006a. *World development indicators.* Washington, DC: World Bank.

World Bank. 2006b. *Global economic prospects: Economic implications of remittances and migration.* Washington, DC: World Bank.

World Bank. 2008. *World development report 2008: Agriculture for development.* Washington, DC: World Bank.

2

Trends in Consumption, Production, and Trade in Livestock and Livestock Products

Allan Rae and Rudy Nayga

Main Messages

- **The share of all animal products in human diets continues to increase in the developing world.** The contribution of animal products to total calorie intakes per capita is much lower in developing countries on average than in the industrial world, but the trend is an increasing one in developing countries compared with the declining trend elsewhere. Despite this convergence between per capita livestock product consumption in different regions, consumption levels in developing countries are on average still well below those in other countries. The largest increases in consumption of livestock products per person have been in Asia, especially China and Southeast Asia, and in Brazil and Chile in South America. In sub-Saharan Africa the situation is different, with declines in per person consumption of meats and eggs, as well as static egg consumption, during the past decade. Concerns over obesity and other health risks can be related to livestock product consumption patterns. While this tends to be thought of as an industrial-country phenomenon, it is also arising in emerging higher-income regions of developing countries.

- **Income growth is a major driver of increasing consumption.** Average growth rates in real per capita incomes have been highest in regions that also tend to have the highest growth in livestock product consumption. In addition to income growth rates, the extent to which consumption levels respond to increasing incomes is important in determining the final response. Not only do developing countries display higher growth in per capita incomes than other countries, they also have higher food expenditure elasticities for livestock products. Both of these factors contribute to the observed rapid rates of growth of animal products consumption.

- **Urbanization of populations also drives growth in consumption.** Rapid urbanization of populations is a common feature in many countries, and in developing countries the average urbanization growth rate is more than four times that of industrial countries. Higher incomes may be earned in urban settings, greater shopping opportunities and food choices often abound, and well-developed refrigerated supply chains facilitate the distribution of meats and dairy products.

- **The population growth and structure impacts on total consumption combine with population growth to drive total livestock product consumption.** While growth in total populations has slowed in the developing world, it remains much more rapid than in industrial countries. Changes in the structure of populations can also be important, such as greater involvement of women in labor markets and changing proportions of young people, the elderly, and minority ethnic groups in populations.

- **Global animal production is shifting from industrial to developing regions.** Growth in the production of livestock products in the developing world has, on average, been as rapid as the growth in consumption and in most cases has been outstripping growth in the human population. Production growth in the industrial world has been stagnant, and developing countries now produce more meat than the industrial world does. For milk this is not the case, but the gap between production in various regions is narrowing. Among developing regions, production growth was more rapid where consumption was increasing the most, with the fastest production growth occurring in China. The slowest production growth occurred in sub-Saharan Africa. However, the growth and intensification of animal production have also raised concerns over environmental pollution, such as degradation of surface and groundwater and emission of greenhouse gases into the atmosphere that can threaten public health.

- **Important drivers of animal production growth include growth in the number of animals farmed and increases in yields per animal and overall productivity.** These trends have, in turn, been driven in part by market signals and profit expectations, greater availability and lower costs of purchased feedstuffs, and the development and adoption of new animal production technologies. Governments have also driven livestock production in many cases, through directives, policy pronouncements, and other nonmarket incentives. For several developing countries, total factor productivity growth in livestock production has been healthy, at rates of up to 10% per year. Also, such growth in many cases was higher in the 1990s than during the previous decade.

- **The ongoing trend away from smallholder and backyard livestock production to larger-scale and commercial production systems continues to encourage greater use of grains and oil crops in animal feeds.** Over the last decade, global growth in the use of cereals as feedstuffs was faster than growth in total cereals production. However, the fastest growth has occurred in the use of soybeans as animal feeds, with total usage in the developing world rising more than six-fold over the last 20 years.

- **Developing countries include both exporters and importers of livestock products.** The industrial countries as a group are net exporters of meats, dairy products, and eggs. Net exporters among other regions and countries include Latin America (beef, poultry), India (beef), and East and Southeast Asia (poultry and eggs), and for each of these regions net exports have increased over the past decade. Major net importers in the developing world include East and Southeast Asia (beef, pork), West Asia and North Africa (WANA), sub-Saharan Africa (meats), and all developing regions studied with the exception of India for dairy products. The major share of livestock product exports from industrial countries is sold to other such countries. Similarly, for nonindustrial countries the major share of their animal products exports is consigned to other nonindustrial countries. This reflects not only growth in import demand in the developing world but almost certainly difficulties in satisfying the food safety standards of industrial countries.

- **Food safety and health issues occasionally result in restrictions on trade in livestock products and live animals, based on standards developed through international agencies.** While this system contributes to well-functioning global markets that may ensure safe foods for consumers and protect human and animal health without erecting overly restrictive trade barriers, a key concern is the ability of emerging developing-country exporters to meet the standards of industrial-country markets or to implement such procedures themselves to protect their own human and animal populations. Animal welfare and the management of live animals in transit are also matters of public concern, and improved welfare practices are required. Further technical assistance is necessary to enable developing countries to fully realize the advantages of participating in international markets for livestock products.

- **Some countries have met increasing demands for livestock products from their own production, using either domestically produced or imported feedstuffs.** At what point might countries, especially when facing increasing pressures from land, water, and environmental constraints, resort to imports of final animal products rather than of feedstuffs? Countries of Northeast Asia have switched from importing maize to meat imports, whereas WANA's imports of meats have been relatively steady for some time, while maize imports continued to grow. China is different again: over the last 20 years it has been a net exporter of meat at the same time as maize net exports have shown a rising trend. Further research is required to help determine when, if ever, China will emerge as a persistent importer of grains for animal feedstuffs or of animal products.

Demand for Livestock Products

Major Trends in Per Capita Consumption

The past two decades have witnessed a steady increase in the share of animal products (including fish) in human diets in the developing world. (See Figure 2.1.) While the contribution of animal products to total calorie intakes per capita is much lower in developing countries on average than in industrial ones—14% versus 26% in 2002—the trend has been an increasing one in developing countries compared with a declining trend elsewhere. (See Box 2.1.) This shows that consumers in the developing world have been increasing their consumption of animal products more rapidly than they have other foods such as traditional cereals or root crops. These trends are mirrored in the consumption data for the three animal products in Table 2.1. In each case, consumption levels per person are substantially higher in industrial countries, but they are largely static. Actual consumption volumes may be less than those provided by the food balance sheet method, since the latter are estimates of food availability that do not include allowances for food losses within households or nonhuman uses of food supplies. The food balance sheet data have been standardized in that processed foods have been converted back to their primary commodity equivalents. In contrast, over the last decade per capita consumption of meat, milk, and eggs has grown each year by 3%, 2%, and 5%, respectively, in developing countries. (Unless stated otherwise, annual rates of growth are computed as compound growth rates between a pair of numbers.) These growth rates, however, mask considerable regional differences in the rate of change of animal product consumption and its role in human diets. In this chapter, unless stated otherwise, data are a three-year average centered on the quoted year, sourced from the food balance sheets of the FAOSTAT database (the latest available food balance sheet at time of writing is 2003).

The transformation of livestock food demand in the 1980s was covered comprehensively by Delgado et al. (1999). As a result, this chapter concentrates on developments since then. As in the earlier period, the most dramatic developments in livestock food consumption have taken place in Asia. (See Table 2.2.) In China, per capita consumption of milk and eggs doubled between 1992 and 2002, while that of meat increased by more than 70% (Fuller et al., 2004). East and Southeast Asia showed the next most rapid increase in meat and milk consumption, and India the next fastest growth in egg consumption per capita. Meat consumption per capita also increased in India, WANA, and Latin America; consumption of milk in India, Other South Asia, Latin America, WANA, and sub-Saharan Africa rose, and that of eggs increased in all regions. In sub-Saharan Africa, per capita meat consumption was lower in 2002 than a decade earlier.

Several researchers have raised concerns over official estimates of the level of meat consumption and production in China (e.g., see Delgado et al., 1999, Ke 1997, Longworth et al., 2001, Ma et al., 2004, Wang et al., 2004). Similar problems might also exist in other developing countries, and a degree of caution is warranted in using these data. It has been suspected by some that incentives exist for local Chinese officials to overreport livestock numbers and production. When consumption data are subsequently obtained from food balance sheets, based primarily on production and net trade data, they too may be biased upward. Hence

Table 2.1. Global per capita consumption of meat, milk, and eggs (kg).

	Industrial countries			Developing countries		
	1982	1992	2002	1982	1992	2002
Meat	75	79	79	15	21	29
Milk	194	191	201	35	39	48
Eggs	14	13	13	3	5	8

Source: FAO 2006a.

actual consumption growth rates could have been somewhat lower than those in Table 2.1. Using China's official household surveys as a check on consumption levels is also problematic since they do not adequately incorporate food purchased outside the home—likely to be important in the case of animal products. Ma et al. (2004) found a mismatch between the official demand and supply series that had worsened since the mid-1980s, and also between livestock output and feed availability data. These problems remained even after official data were adjusted following the first agricultural census in 1996. They adjusted pork production downward so that annual growth over 1987 to 1998 was 4.45% compared with 6.57% using unadjusted official data. For poultry, this discrepancy was 11.62% (adjusted) versus 14.74% (official data). For beef data, Longworth et al. (2001) describe the type of incentives that have led to exaggerated data reporting. In addition to animal numbers being overreported, slaughter numbers are not collected at abattoirs but at the point of sale, giving rise to inaccuracies when animals are traded more than once prior to slaughter.

Comparing the average regional growth rates in per capita consumption between 1992 and 2002 with those in the previous decade shows that milk consumption

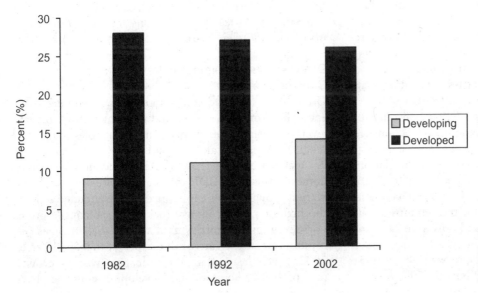

Figure 2.1. Contribution of animal products to total caloric intake per capita.
Source: FAO 2006a.

Box 2.1. Changing Consumption Patterns

Food consumption patterns change over time along with advances in a country's development. For many countries, the trend is declining consumption of traditional staples, such as rice and root crops, and increasing consumption of other foods. The latter could initially involve increased consumption of nontraditional cereals (such as wheat products) followed by increasing consumption of high-protein foods such as animal products.

East and Southeast Asia provide a good example. Between 1962 and 1982, per capita consumption of rice, wheat, and animal products increased, while that of starchy roots declined. Over the following 20 years, rice consumption leveled off while that of both wheat and animal products continued to increase and starchy roots consumption further declined. (See Box Figure 2.1 in the color well.)

Over the past 20 years, per capita consumption of all animal products (including fish) has increased in all regions shown in the table with the exception of sub-Saharan Africa and the industrial world in aggregate.

Animal products also increased their share of total calorie intake for all regions except sub-Saharan Africa and the industrial world. The most dramatic increase is observed in China, where animal products, which contributed 8% of per capita calorie intake in 1981–83, contributed 21% in 2001–03.

Per capita consumption (kg/yr) of all animal products (including fish)*

	Animal products (kcals/cap/day)		Animal products as % total food (kcals/cap/day)	
	1981–83	2001–03	1981–83	2001–03
China	186	618	8	21
India	133	198	6	8
E/SE Asia	160	254	7	9
S-S Africa	159	141	8	6
Latin America	455	559	17	20
Developing world	207	361	9	14
Industrial world	906	871	28	26
World	388	470	15	17

*Animal products include meat, milk, eggs, fish, and other animal/fish fats and oils.

increased at a faster rate during the most recent decade in all developing regions with the exception of India, where growth was still a relatively high 2% per year. Growth rates in meat consumption per capita during the 1990s were more rapid than in the 1980s for Latin America, WANA (which showed no growth in the 1980s), and sub-Saharan Africa, where the rate of decline in the 1990s was less than during the earlier decade. Per capita consumption of eggs grew substantially more quickly in WANA during the 1990s than in the 1980s, but the rate of growth was less than in the 1980s in most other regions. In contrast to these developments in the world's developing regions, per capita consumption of these livestock products in industrial countries was rather stagnant or declined during the 1980s and 1990s.

Despite the much more rapid growth in consumption in the developing world than in higher-income countries, major discrepancies remain between per capita consumption levels across both groups of countries and among and within developing countries themselves. By 2002, meat consumption per capita in developing countries averaged only 36% of that in industrial countries, and comparable percentages for milk and eggs were 24% and 59%, respectively. Latin Americans consume per person quantities of meat and milk that are well above those elsewhere in the developing world but still 50% to 75%

of industrial-country averages. As in some other regions, there are large divergences within Latin America. For example, per capita meat consumption for 2002 averaged 85 kg in Argentina and 80 kg in Brazil, compared with the Latin American average of 59 kg. India and Other South Asia showed the lowest per capita consumption for meat but high levels of milk consumption compared with other developing countries. Consumption per capita of all three livestock products was relatively low in sub-Saharan Africa. Some reasons for these differences in consumption levels are considered in the next section.

Determinants of Changes in Per Capita Consumption

Growth rates in food consumption per person are influenced by economic, social, and cultural factors. Income levels and relative prices play important roles, but so do religious, lifestyle, and technological influences. Table 2.3 includes average growth rates in incomes (real gross domestic product [GDP] per capita). Regions with the highest growth in per capita incomes also tended to have the highest growth in livestock product consumption. For developing countries as a whole, real incomes per capita grew at the rate of 3.9% per year from 1992 to 2002, much faster than the industrial world's growth rate of 0.5%. So, too, did developing countries have

Table 2.2. Per capita meat, milk, and egg consumption by region (kg).

Region	Meat			Milk			Eggs		
	1992	2002	Annual growth (%)	1992	2002	Annual growth (%)	1992	2002	Annual growth (%)
China	31	53	5.6	7	14	7.7	8	18	7.9
India	5	5	1.0	55	66	2.0	1	2	4.1
Other South Asia	8	8	−0.1	62	79	2.4	1	2	1.4
East/Southeast Asia	17	22	2.4	12	15	2.4	4	5	1.5
WANA	20	22	1.1	77	79	0.3	4	5	2.0
Sub-Saharan Africa	12	11	−0.1	29	30	0.3	1	1	−0.7
Latin America	48	59	2.1	98	106	0.8	8	8	0.6
Developing world	21	29	3.3	39	48	2.2	5	8	5.0
Industrial world	79	79	0.1	191	201	0.5	13	13	−0.2
World	34	39	1.3	75	80	0.7	7	9	2.6

Regions are defined as in FAOSTAT at the time of writing. Other South Asia is South Asia less India; China includes Mainland China and Taiwan Province; WANA (West Asia–North Africa) is Near East in Asia, Northwestern Africa, and Near East in Africa.

Source: FAO 2006a.

much higher annual growth in consumption of livestock products. Real GDP per capita in China grew at an annual rate of over 8% during this period, and China also displayed by far the highest rate of growth in per capita consumption of meat, milk, and eggs. The lowest rates of GDP per capita growth were observed for Other South Asia and sub-Saharan Africa: both these regions showed the lowest (and negative) growth in per capita meat consumption levels, and the latter region showed the lowest rate of growth in per capita consumption of milk and eggs.

In addition to the growth rate of per capita incomes, the extent to which consumption levels respond to increasing incomes is important in determining the final response. Expenditure elasticities measure the percentage change in demand in response to a 1% change in household expenditure. A consistent set of elasticities, estimated from the 1996 International Comparison Project data for 114 countries (Searle et al., 2003), is summarized in Table 2.4. Across country groups and food subgroups, demand is more responsive to changes in income in low-income countries than in high-income countries. For each income group of countries, expenditure elasticities are highest for the beverages-tobacco subgroup, but the next highest elasticities are for animal products (fish, dairy, and meat). For example, the expenditure elasticity for meat ranged from 0.86 in Tanzania (the lowest-income country in the sample) to 0.81 in Kenya, 0.77 in Pakistan, 0.73 in Indonesia, 0.66 in Brazil, 0.55 in Argentina, 0.31 in Japan, and 0.11 in the United States. Thus, not only do developing countries display higher growth in per capita incomes, they also have higher food expenditure elasticities. Both these factors combine to contribute to the observed rapid rates of growth in animal products consumption.

Rapid urbanization of the population is a common feature in many developing countries. Table 2.3 shows urbanization growth rates of up to almost 5% per year, and in all cases shown (with the possible exception of WANA) these rates are well above the growth rate of the overall population. For all developing countries, the average urbanization rate is over 3% per year, more than four times that in industrial countries.

Why might relatively rapid urbanization of a country's population encourage higher per capita consumption of foods such as livestock products? First, people are commonly attracted to urban areas by the expectation of earning higher incomes, and if this is realized they have the means to purchase increased quantities of livestock products. As a result, levels of urbanization and per capita incomes tend to be strongly correlated over time, and it may be difficult to tease out the separate impact of these factors on consumption levels. Rae (1998) attempted to do so and found consumption of livestock products to be particularly influenced by rates of urbanization in several Asian countries. Second, urban occupations may be more sedentary than jobs in rural areas, and hence there may be less need for energy-rich diets such as those dominated by cereals and starchy roots. Third, there are almost always far greater shopping opportunities and greater food choices in urban areas, including away-from-home consumption opportunities and modern supermarkets with well-developed refrigerated supply chains for the distribution of meats and dairy products, along with the associated advertising effort. Finally, urban households may be more likely to own refrigerators, allowing meats and dairy products to be stored for longer than would otherwise be the case.

Table 2.3. Income and population growth rates (% per year).

Region	Population (1992–2002)	Urban population (1992–2002)	Real GDP per capita (1990–2001)
China	0.2	3.5	8.2
India	1.8	2.6	3.7
Other South Asia	2.3	3.5	2.4
East/Southeast Asia	1.5	3.4	3.5
WANA	2.1	2.9	3.6
Sub-Saharan Africa	2.6	4.7	2.4
Latin America	1.6	2.2	2.9
Developing world	1.7	3.1	3.9
Industrial world	0.4	0.7	0.5
World	1.4	2.2	3.2

Sources: Population data from FAO 2006a; GDP data from FAO 2004.

Cultural and religious factors can also contribute to differences in livestock product consumption levels. Eggs and dairy products are acceptable livestock products in countries where vegetarianism is common. Pork is the major meat consumed in East Asian countries and China but is excluded from the diet of a large share of the world's population, including Muslim communities. For religious and cultural reasons, South Asia has lower meat consumption levels than income alone would suggest (for example, beef in India). Concerns over obesity and other health risks, particularly in industrial countries, have encouraged increased demand for low-fat foods, including low-fat milk and dairy products and lean, or white, meats. Lactose intolerance can limit consumption of milk in regions such as East Asia.

Table 2.5 shows that between 1992 and 2002 total meat consumption in industrial countries remained static, but the composition of that consumption changed, with a swing away from beef and toward poultry meat. In developing countries, total meat consumption per capita increased as did consumption of each meat type, although the most rapid increase was in consumption of poultry meat. Between 1991 and 1993 and 2001 and 2003, for example, per capita consumption of poultry meat in China increased from 4.6 kg to 10.5 kg, while in Central and South America this consumption rose from 13.8 kg to 23.9 kg. Over the same period, total consumption of poultry meat in the developing world grew by over 8% per year, compared with annual growth of just under 5% for pig meat and less than 4% for bovine meat.

Changes in Total Consumption

Observed changes in total consumption of livestock products depend on the growth rate of the total population, in addition to changes in consumption per person. Population growth rates from 1992 to 2002 are given in Table 2.3. By comparing these with the 1970–95 growth rates in Table 4 in Delgado et al., 1999, it is apparent that population growth rates in the developing world have slowed over recent years. China's population, for example, expanded at the rate of 1.6% per year during 1970–95, but by only 0.2% per year since 1992. For developing countries as a whole, these growth rates were 2.1% (1970–95) and 1.7% (1992–2002). A similar slowdown in population growth also occurred in the industrial world. Thus population growth would appear to have had a smaller impact in explaining total growth in consumption during the past decade than was the case earlier.

Population growth rates still exceeded 2% in WANA, sub-Saharan Africa, and Other South Asia during 1992–2002. So while sub-Saharan Africa showed low or negative growth in per capita consumption of livestock products during this period, total consumption increased at annual rates of 2.4% or higher. (See Table 2.6.) Annual growth in per capita consumption of milk in Other South Asia and of eggs in WANA, when combined with the rate of population growth, resulted in total consumption in these cases increasing by 4.4% or more. China shows the most rapid growth in total consumption of the livestock products in Table 2.6, primarily due to the rapid growth in per capita consumption driven by high growth rates of income and urbanization. Total consumption of meats, milk, and eggs in the industrial world increased during 1992–2002 by less than 1% per year. As a result, total meat consumption in developing countries, which amounted to 20% of global

Table 2.4. Average expenditure elasticities for food sub-categories.

Country groups	Beverages, tobacco	Breads, cereals	Meat	Fish	Dairy	Fats, oils	Fruits, vegetables	Other foods
Low income	1.25	0.53	0.78	0.91	0.86	0.55	0.64	0.78
Middle income	0.84	0.37	0.64	0.72	0.69	0.40	0.51	0.64
High income	0.44	0.17	0.36	0.39	0.38	0.20	0.28	0.36

Source: Searle et al., 2003.

Table 2.5. Per capita consumption (kg) of meat types.

	Industrial countries		Developing countries	
	1992	2002	1992	2002
Beef and buffalo	26	22	5	6
Sheep and goat	3	2	2	2
Pigmeat	29	29	9	12
Poultry meat	20	25	5	8
All ruminant	28	24	7	8
All nonruminant	49	54	14	20
Total four meats	77	78	20	28

The "total four meats" column excludes "meat, other," as this is not disaggregated to original meat type in FAOSTAT.

Source: FAO 2006a.

consumption in 1992, had increased its share to 57% by 2002. The increases in the developing-country share of global consumption were from 40% to 46% for milk and from 54% to 68% for eggs.

In addition to the growth of the overall population, changes in the structure of the total population can also affect food consumption levels and patterns. For example, the structure of diets can differ between younger and older people, by gender, and between ethnic groups. In such cases differences in population growth rates between such groups within society can affect total consumption levels. It is not uncommon for milk consumption to be higher among the very young than among teenagers and mature adults. Demory-Luce et al. (2004), for example, sampled a group of consumers in childhood and again when they were young adults, and found that milk consumption was greater in childhood and that the decrease over time was greater for males. They also found that young adulthood consumption was greater for poultry, beef, and seafood.

The growth of ethnic minorities, especially in urban areas, can have important impacts on total food consumption. Jamal (1998) reported a study of food consumption within a sample of British-Pakistanis in Bradford, England. After migration, it was found that the members of the first generation tended to consume their own foods, cooked in the traditional ways, and were reluctant to try British food regularly. The next generation, however, while continuing to consume traditional food at home, was increasingly consuming British foods outside the home. Therefore the traditional diet that such migration brings with it may become increasingly diluted over subsequent generations. At the same time, such ethnic foods may become popular among the wider population, as shown by the popularity of Asian foods among non-Asians in many European countries.

Other demographic trends, especially in but not restricted to industrial countries, are also important in influencing livestock product consumption. The number of single-person households is rapidly increasing due to declining birth rates, a higher average age of marriage, rising divorce rates, and increasing longevity. An increasing proportion of women, especially in the industrial world and in urban centers of the developing world, are now engaged in away-from-home paid employment. As a result, they have less time for cooking. Many households in a variety of countries have both spouses working outside the home. Because more women are in the labor market, an increasing number of families now enjoy double incomes and greater purchasing power. This allows them to buy more value-added food items of quality and has increased the demand for value-added or convenience meat and dairy products not only in the at-home market

Table 2.6. Total consumption of meat, milk and egg consumption by region (in million tons).

Region	Meat			Milk			Eggs		
	1992	2002	Annual growth (%)	1992	2002	Annual growth (%)	1992	2002	Annual growth (%)
China	37	69	6.5	8	18	8.6	9.7	22.8	8.9
India	4	5	2.7	48	70	3.8	1.1	1.9	5.6
Other South Asia	2	3	2.2	17	27	4.8	0.3	0.5	5.2
East/Southeast Asia	9	13	3.9	6	9	4.0	2.3	3.1	3.0
WANA	7	10	3.4	27	35	2.5	1.5	2.3	4.4
Sub-Saharan Africa	6	7	2.4	14	19	2.9	0.7	0.9	2.5
Latin America	22	31	3.8	44	57	2.5	3.4	4.3	2.4
Developing world	86	138	4.9	161	229	3.6	19.0	35.7	6.5
Industrial world	100	105	0.5	242	266	0.9	16.5	16.9	0.2
World	186	243	2.7	403	495	2.1	35.5	52.6	4.0

Source: FAO 2006a.

but also in the away-from-home and prepared foods markets (Nayga 1996).

Health, Nutrition, and Food Safety Concerns

Consumer preferences have changed due to increasing concerns for health and nutrition. Some of the more important issues are introduced here but are discussed in more detail in Chapters 11 and 12. In affluent Western countries, food saturation has largely occurred, and concerns have turned increasingly to a reduction in the consumption of fat and calories—with more grains, fruits, and vegetables being consumed (Kinsey 1992). Obesity, for instance, is a growing concern worldwide, including in more affluent regions of developing countries (Loureiro and Nayga 2005). Obesity and overweight rates are increasing in several countries, particularly in countries that belong to the Organization for Economic Cooperation and Development (OECD), where the percentage of overweight and obese people in the total population has increased from around 30% in 1985 to over 50% in 2002. According to the World Health Organization, more people are now suffering from overweight-related problems than from malnutrition. Globally, more than 1 billion adults are overweight and at least 300 million of them are clinically obese, while 800 million suffer from malnutrition.

Health-related concerns have had an influence on consumers' decisions to reduce consumption of harmful ingredients (e.g., fats, salt) or to increase consumption of healthful components in their diets. Consequently, functional foods have become increasingly popular in recent years. Defined as foods or food components that may provide a health benefit beyond basic nutrition, functional foods are widely believed to offer consumers an increased ability to reduce the risk of certain diseases or health problems. Many food companies are now developing food products with functional or health-related attributes.

In addition to obesity, nutrition, and health concerns, consumer tastes and preferences have changed due to concerns about food safety. The Centers for Disease Control and Prevention in the United States estimated that each year 76 million people there get sick, more than 300,000 are hospitalized, and 5,000 people die from foodborne illness. Although the developments related to the adoption of Hazard Analysis Critical Control Point and other safety standards may have helped reduce the incidence of foodborne illness, infections with *Salmonella*, *E. coli* O157:H7, *Campylobacter*, and *Listeria* remain and are alarmingly an ever-present phenomenon. Consequently, the use of technologies such as food irradiation is now gaining popularity, at least in the United States. Realizing the benefits of food irradiation in meats, however, will depend on consumers' acceptance of the technology. The results of a number of studies (e.g., Nayga et al., 2006) suggest that many individuals are willing to pay for irradiated meat products once they are informed about the nature of food irradiation technology and its ability to reduce the risk of foodborne illness.

Other issues influencing consumer tastes and preferences include consumer concerns about the use of hormones in animal production, about genetically modified organisms, and about animal welfare and the trend toward consumption of organically produced foods. While these issues are more predominant in industrial areas like the United States, Japan, and Europe, they will also become increasingly important for the livestock industry in developing countries as incomes increase.

Production of Livestock Products and Feed

Major Trends in Livestock Production

Worldwide, total production of meat, milk, and eggs increased at annual rates of 2.7%, 1.3%, and 4.1% in the 1992–2002 period. (See Tables 2.7–2.10.) As with consumption of livestock products, production growth was by far more rapid in developing countries than in the industrial world. For the latter region, production of meat and eggs grew annually at 0.5% or less, while milk production actually fell during the decade. By contrast, meat, milk, and egg output grew annually at between 4% and 6.5% in the developing world. The developing world's share of global production of all products increased between 1992 and 2002, and by the latter year had reached 56% (meat), 42% (milk), and 68% (eggs). Thus the 1980s trend noted by Delgado et al. (1999) of world animal production—with its attendant benefits and costs—shifting from industrial to developing countries continued through the 1990s.

Among the regions, production growth was more rapid in the same areas where consumption was increasing the most. For all products, annual growth was the fastest in China (with the same caveats about the data as noted earlier), with rates of between 6.5% (meat) and 8.9% (eggs). Based on these data, China now accounts for 28% of global meat production (up from 20% 10 years earlier) and for 42% of global egg output (compared with 27% in 1992). As for meat consumption, the next most rapid meat production growth occurred in East/Southeast Asia, Latin America, and WANA. The second highest annual growth rate in both consumption and production of milk was found in Other South Asia (4.8%), and India held this position for eggs. India now has a 14% share of global milk production, compared with an 11% share in 1992. Again almost mirroring consumption trends, sub-Saharan Africa showed the lowest annual growth rates for milk and eggs and the second lowest for meat. Among the different meat types, all showed positive annual growth rates in the developing world, with that for poultry being the most rapid at 7.6%. (See Table 6.7.) Poultry production also grew more rapidly than that of other meats in the industrial

Table 2.7. Production and trends for livestock products.

	Per capita production		Total production		Annual growth of total production
	1992	2002	1992	2002	1992–2002
	(kg)		(million MT)		(%)
Developing world					
Beef and buffalo	5	6	22	31	3.2
Pigmeat	9	12	37	57	4.4
Poultry meat	5	8	19	40	7.6
Sheep and goat	1	2	6	9	3.6
All meat	21	28	87	139	4.8
Milk	41	51	170	251	4
Eggs	5	8	22	41	6.5
Industrial world					
Beef and buffalo	26	23	33	30	−1.1
Pigmeat	29	29	37	38	0.3
Poultry meat	21	26	26	35	2.8
Sheep and goat	3	3	4	3	−2
All meat	81	81	103	108	0.5
Milk	283	266	360	352	−0.2
Eggs	15	14	18	19	0.2

Source: FAO 2006a.

world, but at the much slower rate of 2.8%. Total production of beef and sheep meat actually declined in the industrial world between 1992 and 2002.

Tables 2.7–2.10 also show livestock production volumes per capita in order to assess whether output growth is exceeding that of the total human population. This is the case for all regions shown for milk production, for all except sub-Saharan Africa for eggs, and for all except India, Other South Asia, and sub-Saharan Africa in the case of meat. For those exceptions, production per

capita was about the same in 2002 as it was in 1992. For the developing world as a whole, therefore, output per person increased during this period for meat, milk, and eggs. In contrast, milk and egg production per capita declined and that of meat remained static in the industrial world. By comparing these data with those of per capita consumption (Table 2.2), some gaps are apparent that would be accounted for by imports or exports. Per capita meat consumption was somewhat higher than that of meat production in 2002 in East/Southeast

Table 2.8. Regional meat production.

Region	Per capita production		Total production		Annual growth of total production
	1992	2002	1992	2002	1992–2002
	(kg)		(million MT)		(%)
China	31	52	37	68	6.3
India	5	5	4	6	3
Other South Asia	8	8	2	3	2.2
East/Southeast Asia	17	21	9	13	3.6
WANA	17	20	6	9	3.6
Sub-Saharan Africa	11	11	6	7	2.4
Latin America	50	64	23	34	4
Developing world	21	28	87	139	4.8
Industrial world	81	81	103	108	0.5
World	35	40	189	247	2.7

Source: FAO 2006a.

Asia and WANA, whereas Latin America had a surplus of production over consumption. (See Table 2.11.) For milk, the deficit regions in 2002 were sub-Saharan Africa and especially East/Southeast Asia, suggesting the latter is a major importer of dairy products (see following sections). Milk-surplus regions are India and the rest of South Asia, Latin America, and the industrial world.

Some Determinants of Production Growth

Production of meat, milk, or eggs is the result of animal numbers times yield per animal. Obviously, growth in either numbers or yields can drive output growth. Relative price and profit expectations (sometimes distorted through government assistance policies) can be a major driver of changes in the number of animals farmed and the choice of production techniques, while the many factors (including profit expectations) that influence technology development and adoption also influence yield and productivity developments. Government support of livestock production, including the subsidization of feedstuffs and other inputs as well as environmental and other regulations are discussed in Chapter 3. Rapid demand growth in some regions, especially for white meats and milk, has resulted in upward pressure on prices that in turn encourages increased production, the development and adoption of better technologies, and the intensification of livestock production, including increased use of grains and concentrates in animal diets. Increased availability of such feedstuffs, improvements in their quality, and reduced purchased feed costs relative to meat or milk prices have further encouraged such trends in many cases.

Tables 2.12 and 2.13 present growth rates for the past decade in both animal numbers and total production to indicate the relative importance of the former as

a driver of production growth. In most regions shown, the annual growth in milk production exceeded that in the number of cows milked, indicating that both yield growth and increases in animal numbers have been important drivers of milk production. (See Table 2.12.) The one exception is sub-Saharan Africa, where production grew at a somewhat slower rate than did cow numbers, implying negative growth in average yield per cow. For egg production, however, and the three meats (see Table 2.13), the growth rates in animal numbers have been similar to or greater than the growth rates of total egg or meat production. This suggests that for eggs and meats, growth in animal numbers was the major driver of production growth between 1992 and 2002. The few exceptions included Latin America and WANA in the production of eggs, beef, poultry meat, and pig meat.

The growth rates in yields per animal in the 1992–2002 period and actual average yields in 2002 are shown in Table 2.14. First, there is considerable variation in animal yields across regions, reflecting differences in production systems, input use, technologies used, scale of production, and environmental conditions. This variation is greatest for milk production per cow, where it ranges from 325 kg per head in sub-Saharan Africa to 2,320 kg in China and an average of 4,644 kg in the industrial world. Explanations would include wide variation in farm size and feeding regimes, differences in the range of cattle breeds that are used in milk production worldwide, and the poorer performance of the higher-yielding breeds in the developing and tropical countries. The lowest variability in animal performance is found for poultry, both meat and eggs, and in pig meat. In each of these cases, the difference between average yields in developing and industrial countries is much less than in the case of milk. An important explanation would be

Table 2.9. Regional milk production.

Region	Per capita production		Total production		Annual growth of total production
	1992	2002	1992	2002	1992–2002
	(kg)		(million MT)		(%)
China	7	14	8	18	8.5
India	64	83	56	87	4.4
Other South Asia	71	91	19	31	4.8
East/Southeast Asia	7	9	4	5	3.6
WANA	75	80	27	36	2.8
Sub-Saharan Africa	27	28	13	18	2.7
Latin America	99	116	45	62	3.2
Developing world	41	51	170	251	4
Industrial world	283	266	360	352	–0.2
World	97	97	529	603	1.3

Source: FAO 2006a.

Table 2.10. Regional egg production.

Region	Per capita production 1992 (kg)	Per capita production 2002 (kg)	Total production 1992 (million MT)	Total production 2002 (million MT)	Annual growth of total production 1992–2002 (%)
China	9	19.3	10.7	25.1	8.9
India	1.4	2.1	1.3	2.2	5.9
Other South Asia	1.5	1.8	0.4	0.6	4.4
East/Southeast Asia	5.3	6.3	2.8	3.8	3.2
WANA	4.8	6	1.8	2.7	4.4
Sub-Saharan Africa	1.7	1.6	0.9	1	2
Latin America	8.7	10	4	5.4	3
Developing world	5.2	8.3	21.7	40.8	6.5
Industrial world	14.5	14.2	18.4	18.8	0.2
World	7.4	9.6	40.1	59.6	4.1

Source: FAO 2006a.

the relative ease with which nonruminant production technologies can be transferred and adopted around the world, because they are less dependent on environmental conditions or land supplies.

Given the previous finding that growth in animal numbers has generally been more dominant than yield growth as a driver of output growth, the yield growth rates in Table 2.14 are, as expected, rather low. Apart from milk, and with few exceptions, animal yield annual growth rates have been less than 1%, and they sometimes have been negative. However, trends toward earlier slaughter of livestock and lighter carcass weights in some regions will have a negative impact on yield growth rates. Milk yields have shown faster growth, with all but one of the regions shown in Table 2.14 having annual growth rates of 1.5% or higher, with China the highest at 4.7% per year. For milk, cow yields have been growing

almost as rapidly in the developing world as in industrial countries—2.0% compared with 2.3%.

Meat, milk, or egg production per animal is a partial measure of productivity, since only one input—the animal—has been considered. It has been shown that such a partial measure is biased (Nin et al., 2003), as discussed in Box 2.2. Higher yields might, for example, be the result of increased feed use per animal and not necessarily an increase in the productivity of the production system as a whole. A better index is the total factor productivity (TFP) measure, which compares growth in output with that of the aggregate of all production inputs. Therefore the productivity performance of livestock, particularly in the developing world, may not be quite as bleak as just described. Box 2.2 contains some results from a recent study of TFP growth in China's livestock sector in the 1990s (Rae et al., 2006). Apart from milk, TFP grew at

Table 2.11. Comparison of per capita production and consumption levels, 2002 (kg).

Region	Meat Production	Meat Consumption	Milk Production	Milk Consumption	Eggs Production	Eggs Consumption
China	52	53	14	14	19.3	18
India	5	5	83	66	2.1	2
Other South Asia	8	8	91	79	1.8	2
East/Southeast Asia	21	22	9	15	6.3	5
WANA	20	22	80	79	6	5
Sub-Saharan Africa	11	11	28	30	1.6	1
Latin America	64	59	116	106	10	8
Developing world	28	29	51	48	8.3	8
Industrial world	81	79	266	201	14.2	13

Source: FAO 2006a.

Table 2.12. Annual growth rates (%) of animal numbers and milk and egg production, 1992–2002.

Region	Eggs		Milk	
	Laying hens	Production	Cows milked	Production
China	8.9	8.9	5.7	8.5
India	5.4	5.9	1.6	4.4
Other South Asia	4.9	4.4	3.2	4.8
East/Southeast Asia	3.3	3.2	1.7	3.6
WANA	2.3	4.4	1.2	2.8
Sub-Saharan Africa	2.4	2	2.9	2.7
Latin America	1.5	3	0.9	3.2
Developing world	5.3	6.5	1.9	4
Industrial world	–0.7	0.2	–1.9	–0.2
World	3.5	4.1	1.1	1.3

Source: FAO 2006a.

rates between 3.7% and 5.4% for hogs, beef, and eggs. These are healthy productivity gains and are considerably higher than the yield growth rates in Table 2.14. For milk in China, TFP grew at an annual rate of 1.3% in the 1990s—much less than the growth in yields per cow.[1]

Another comprehensive set of TFP growth estimates for livestock (and crops) is that of Jones and Arnade (2003). They used Food and Agriculture Organization (FAO) data for 27 countries in 1961–99 and published TFP results by decade. For aggregate livestock production in 1991–99, TFP annual growth varies between –6.5% (Argentina) to 10.8% (China) and 11.2% (India) for the 12 developing countries in their sample. The comparable range for industrial countries was –2.5% (Poland) to 11.9% (France). TFP growth was faster in the 1990s than in the previous decade for 9 of the 12 countries. From these studies, it can be concluded that in at least some developing countries, livestock productivity growth has in recent times been substantially higher than growth in yields might suggest.

Trends in Feed Production and Use

Consumption of nonruminant meats in particular is increasing rapidly—this applies to both poultry and pig meat in developing countries and to poultry meat in industrial countries. These trends have also been shown to apply to the production of nonruminant meat. But small-scale backyard production of pigs and poultry (and other types of livestock as well) is declining in many regions

1. The yield data in Table 2.14 are derived from official data sources, so they may also be biased due to overreporting of output relative to that of animal numbers. Rae et al. (2006) recomputed China's TFP growth using the unadjusted official data and found that for milk, eggs, and beef the use of official data gave overestimates of TFP growth. From this it could be inferred that, at least for these products in China, the overreporting problem became more severe during the 1990s.

and making an ever smaller contribution to total production. These systems, which typically use a wide range of household waste food and forages as feedstuffs, are being replaced by more-intensive or large-scale industrial nonruminant production systems that rely on cereals and processed concentrate feeds for their livestock.

Consequently, an ever increasing proportion of global cereals production is being diverted into livestock feed and away from direct human consumption. Table 2.15 (which updates Table 15 in Delgado et al., 1999) provides an update on cereal use as feed in the 1990s. Global use of cereals as feed increased by 0.9% annually between 1992 and 2002, but with negligible growth in industrial countries, up from Delgado et al.'s global estimate of 0.7% between 1982 and 1994. During 1992–2002, global growth in cereal use as feed was faster than growth in total cereals production, the reverse of the situation in 1982–94. In the developing world, cereal use as feed increased by 2.9% annually between 1992 and 2002, compared with 4.2% in 1982–94. Decreased use of feedgrains in the transition economies since around 1990 contributed to this slowdown in the average growth rate.

Developing countries, where both livestock consumption and production growth have been concentrated, increased their share of global cereal feed use from 28% in 1992 to 35% in 2002. Growth in the use of cereals as livestock feed in 1992–2002 has been fastest in India and Other South Asia, although the actual volumes involved are small compared with total global or developing region feed use. This is faster than Delgado et al. recorded for 1982–94, but they found faster growth in feed use in East and Southeast Asia during the 1980s than these figures show for the 1990s. Note that while the use of cereals as feeds increased by 2.9% per year in the developing world during 1992–2002 (Table 2.15), total production of meats in these countries grew at the

Table 2.13. Annual growth rates (%) of animal numbers slaughtered and meat production, 1992–2002.

Region	Beef and buffalo		Pig meat		Poultry meat	
	Slaughtered	Production	Slaughtered	Production	Slaughtered	Production
China	13	11.8	4.6	4.8	9.3	9.6
India	1.3	1.4	0.8	0.8	10.8	11.2
Other						
South Asia	1.6	2.3	4.2	3.3	5.9	5.8
East/Southeast Asia	0.8	0.9	2.7	3.9	4.5	4.7
WANA	0.9	2.3	2	2.9	5.1	5.8
Sub-Saharan Africa	2.3	2.1	3.1	3.2	2.8	2.9
Latin America	1.7	2.1	0.2	2.2	7	7.9
Developing world	3.2	3.2	3.8	4.4	6.9	7.6
Industrial world	−1.8	−1.1	−0.2	0.3	1.8	2.8
World	1	0.8	2.1	2.5	4.5	5

Source: FAO 2006a.

faster rate of 4.8%. This suggests productivity gains in terms of increased feed efficiency, but it also reflects the growing share of poultry in overall meat production, since poultry requires less grain per unit meat production than does pork or beef (as described later). Also contributing to this outcome would be the relative shift of world livestock production from regions that use intensive grain feeding to developing countries (Table 2.8) that show lower grain:meat ratios on average.

The official data in FAOSTAT, as shown in Table 2.15, produce a rate of growth in cereals as feed use of only 1.7% for China, compared with Delgado et al.'s rate of 5.8% during 1982–94. This appears low, given the rate of growth of livestock production in China. Two other China data sources suggest a faster growth rate.

Ma et al. (2004) adjusted the official China feed use data using their adjusted livestock numbers, and their published graph shows total feed use (including sweet potatoes) increasing from around 100 million MT in 1993 to about 125 million MT by 1999, for an annual growth rate of 3.8%. U.S. Department of Agriculture (USDA) data show total cereal feed use in China of 69 million MT in 1991–93 and 105 million MT in 2001–03, for an annual growth rate of 4.4%. While total use in the earlier period is similar to that in Table 2.15, USDA's estimate for the latter time period is substantially higher than FAOSTAT's. Both of these alternative growth estimates are slower than the rate estimated by Delgado et al. for the 1980s, but it would appear that the growth of cereal feed demand in China in the 1990s could have

Table 2.14. Partial productivity (2002) and annual growth rates (1991–2005) in livestock production.

Region	Beef and buffalo		Poultry meat		Pig meat		Milk		Eggs	
	Yield/ head (kg)	Growth (%)	Yield/ head (kg)	Growth (%)	Yield/ head (kg)	Growth (%)	Yield/ head (kg)	Growth (%)	Yield/ head (kg)	Growth (%)
China	134	−0.6	1.5	0.2	77	0.1	2320	4.7	12.5	0.2
India	118	0.1	1.0	0.2	35	−0.0	959	2.1	11.6	0.3
Other South Asia	127	0.6	1.0	0.4	34	−0.9	798	1.5	3.8	−0.3
East/Southeast Asia	189	0.2	1.1	0.1	65	0.9	1374	1.7	6.4	−0.0
WANA	156	1.3	1.2	0.6	74	0.7	1164	1.5	8.9	1.8
Sub-Saharan Africa	130	−0.1	1.0	0.1	47	−0.0	325	−0.2	2.8	−0.4
Latin America	205	0.2	1.6	1.0	74	1.8	1371	1.8	9.8	1.3
Developing world	161	−0.0	1.4	0.7	74	0.5	1017	2.0	9.8	1.1
Industrial world	259	0.7	1.8	0.9	87	0.5	4644	2.3	14.9	0.9
World	198	−0.2	1.5	0.5	78	0.3	2170	0.6	11.0	0.5

Partial productivity is production per head of livestock, a 2001–03 average. Growth rates computed by regression from 1991–2005 data. The column "Milk" "refers to cow's milk only, so is not comparable to the milk total production data in Tables 2.7 and 2.9.

Source: FAO 2006a.

Box 2.2. Livestock Productivity Growth

Commonly, livestock productivity growth has been measured by examining trends in production per animal, such as milk production per cow or average cattle carcass weights. Such a measure is a partial factor productivity (PFP) index, since it includes only one input in its construction. But production of meat, milk, or eggs from a livestock production enterprise involves the use of many inputs, such as labor, feed, and nonlivestock capital, in addition to the animal itself.

A more complete measurement of productivity is afforded by the total factor productivity index (TFP) that reflects the growth in output that is not accounted for by the growth in total inputs. TFP measures for livestock have not been common, since data on use of the various inputs in livestock enterprises have not been available or because different types of livestock are raised on the same farm, making allocation of inputs across livestock types problematic.

Nin et al. (2003) use a method that is said to overcome this problem and that compares PFP and TFP growth rates for aggregate livestock production in several countries. They conclude that PFP is a biased measure of TFP and that the former tended to overestimate TFP growth in many developing countries but underestimated it in Europe.

Rae et al. (2006) used provincial enterprise-level livestock data for China and argued that it could be used to measure TFP in hogs, milk, beef, and egg enterprises without a serious input allocation problem. This study also made adjustments to official data in an effort to overcome the suspected overreporting in official livestock statistics. During the 1990s they found TFP growth rates of 3–5% for hogs and eggs and 4% for beef. Also, the TFP

growth rate for hogs had slowed down over the 1990s compared with its growth in the previous decade.

This study also separated growth in TFP into growth in technical change (TC, upward shifts in the production frontier) and technical efficiency (TE, the rate at which farms are catching up to the frontier). This revealed that growth in TC occurred in all livestock sectors at rates between 3% and 6%. However, the rates of catch-up had been very slow or even negative. Although TFP growth for milk was much slower than for the other enterprises at 0.5–1%, growth in TC was much faster at over 6%. This was not unexpected, given that the milk sector in China has recently been undergoing very rapid growth with many new entrants, producer experimentation, and inevitable mistakes and with some slow adopters of new technologies (Fuller et al., 2004).

Technical efficiency (TE), technical change (TC), and total factor productivity (TFP) in showing growth for China in the 1990s

	TE	TC	TFP
Hogs—Backyard households	1.0	2.7	3.7
Hogs—Specialized Households	−0.7	6.1	5.4
Hogs—Commercial	−0.4	4.8	4.4
Eggs—Specialized Households	0.3	3.5	3.8
Eggs—Commercial	1.4	3.4	4.8
Beef—Rural Households	0.0	4.4	4.4
Milk—Specialized Households	−6.1	6.6	0.5
Milk—Commercial	−3.3	4.6	1.3

Table 2.15. Trends in the use of cereals as feed, 1992–2002.

Region	Annual growth rates (%)		Total cereal use as feed (million MT)	
	Total cereal production	Total cereal use as feed	1992	2002
China	−0.2	1.7	70.7	83.9
India	1.4	7.8	3.5	7.4
Other South Asia	2.7	5.5	0.7	1.2
East/Southeast Asia	2.2	3.5	19.8	27.9
WANA	1.0	3.3	28.8	39.9
Sub-Saharan Africa	2.3	2.1	3.4	4.2
Latin America	3.0	3.7	49.3	71.0
Developing world	1.2	2.9	176	235
Industrial world	0.2	0.0	445	447
World	0.7	0.9	621	682

Total cereals are as defined in FAOSTAT, and excluding use for beer.

Source: FAO 2006a.

been somewhat higher than the estimate given in Table 2.15.

The composition of animal feedstuffs has also undergone change. Table 2.16 shows that maize has been the dominant grains feedstuff for developing countries, and its usage has been increasing at a faster rate than for other coarse grains. But the fastest growth rate has been for feedstuffs derived from soybeans; between 1982 and 2002 the total quantity fed to animals increased over sixfold, and it now surpasses cottonseed products in terms of the total quantity used as a source of protein for animal feeds.

Trends in International Trade in Livestock and Livestock Products

Meats, Dairy Products, and Eggs

Between 1991–93 and 2002–04, the value of total world exports of agricultural products rose from $342 billion to $524 billion. In both these periods global exports of meats, dairy products, and eggs accounted for around 17% of total agricultural exports, rising from $61 billion to almost $87 billion (FAO 2006a). Reflecting the trends in consumption and production of livestock commodities, the composition of livestock products trade changed, with the share of poultry meats rising from 9% in 1991–93 to 14% in 2002–04. The share of bovine meats declined during this period, from 26% to 21%, while the shares of dairy products (38%) and pig meats (20%) showed little change.

The volumes of global exports of livestock products are given in Table 2.17. The average growth rate between 1992 and 2003 was highest for poultry, at nearly 10% per year, followed by pig meat at 6%. Growth in the total exports of bovine meat was considerably slower at not quite 2% per year, while the volume of ovine (sheep and goat) meats traded globally declined marginally over that period. Quantities of dairy products (in their equivalent liquid milk volumes) and eggs also increased over this period, by 3% and 2%, respectively.

Which countries and regions tend to be exporters, and which importers, of these livestock products? Especially when products are defined at a very aggregate level, as they are here, a country will likely be both an exporter and an importer. Hence net exports are measured as the difference between export and import volumes and are shown in Table 2.18 (meats) and Table 2.19 (dairy and eggs). The industrial countries in total are net exporters of meats, milk products, and eggs, implying that all other countries are in the aggregate net importers. But among industrial countries Japan stands out as the world's leading meat importer, with total net imports of 2.4 million tonnes in 2002–04. Thus net exports of other industrial countries were in excess of 6 million tonnes in 2002–04. Mention should also be made of the United States, which switched from a net importer to a net exporter of meat during the 1990s, driven by greatly increased poultry exports.

Of the developing regions shown in these tables, Latin America has become a major net exporter of beef and poultry (primarily due to rapid export growth in Brazil), but a substantial net importer of dairy products. India's net exports are rather close to zero for all livestock products, reflecting that country's self-sufficiency policies. Between 1992 and 2003, India increased its net exports for meats (especially bovine meat) and eggs and switched from being a net importer to a net exporter of dairy products. China remained a net exporter of meats in 2003, but at a much smaller volume than in 1992, reflecting the rapid growth in domestic consumption.

China's net imports of dairy products increased nearly threefold between 1992 and 2003, again reflecting rapid domestic market growth and the domestic industry's struggle to match that growth (Yang et al., 2004). Sub-Saharan Africa and WANA were net importers of all livestock products in both 1992 and 2003. Further, for all cases except that of bovine meats in sub-Saharan Africa, the volumes of net imports increased over that time period. In East and Southeast Asia the situation is rather similar, with net imports of dairy products and bovine

Table 2.16. The changing composition of feedstuffs in the developing world.

Feedstuff	Total utilization (million metric tons)			% total change 1992–2002	% total change 1982–2002
	1982	1992	2002		
Cereals	120.6	176.1	235.1	33.5	94.9
Maize	76.0	120.0	177.4	47.8	133.4
Other coarse grains	32.0	37.8	40.5	7.1	26.6
Starchy roots	62.4	93.3	114.2	22.4	83.0
Oilcrops	4.1	5.8	9.9	70.7	141.5
Soybeans	0.7	1.7	4.4	158.8	528.6
Cottonseed	2.9	3.2	3.0	−6.3	3.4

Source: FAO 2006a.

Table 2.17. Global trade in meats, dairy, and eggs.

Product	1992	2003	Annual growth (%)
	(million metric tons)		
Meat			
Bovine	6.5	7.9	1.8
Ovine	0.9	0.9	–0.1
Pig	4.4	8.6	6.2
Poultry	3.5	9.3	9.9
Total meats	15.7	27.5	5.3
Dairy products	55.2	78.4	3.2
Eggs in shell	0.8	1.0	2.0

For 1992, as in earlier sections, data are three-year averages centered on the indicated year. The latest trade data in FAOSTAT at the time of writing was 2004. "Total meats" also includes meats other than those above. Dairy products also include milk equivalents.

Source: FAO 2006a.

and pig meats increasing between 1992 and 2003. Only in the poultry sector is this region a net exporter, and increasingly so, of both eggs and meat.

Table 2.20 uses data from 2004 to examine meat and dairy trade flows (by value) between industrial and nonindustrial countries. First, the total value of meat and dairy exports from the industrial countries is much higher than that from other countries. Second, the majority of the industrial countries' exports are destined to other industrial country markets (but including the considerable intra-EU trade)—80% by value in the case of meats and 75% for dairy products. Only for poultry

meat exports was a higher proportion (40%) of exports sold to developing countries, partly reflecting the rapid demand growth for this product in developing countries. For the nonindustrial countries, the picture is quite different. Here, the major share of exports of meats and dairy products—56% and 72%, respectively—were exported to other nonindustrialized countries. While this may reflect faster demand growth in such countries, it almost certainly reflects difficulties experienced by exporters from such countries in meeting the food safety standards of industrial countries.

Trade in Live Animals

A response to increased demands for livestock products in some regions has been the importation of live animals. These may be breeding stock used to establish high-productivity livestock industries in the importing regions, perhaps through breeding programs aimed at improving native breeds, or they may be for finishing and slaughter in the importing region. For example, while some exporters supplying Muslim markets are able to slaughter livestock according to importer requirements, there may be strong cultural preferences in such markets for freshly slaughtered meat. Table 2.21 shows net exports of live animals for the same regions as in other tables, for the periods 1992–94 and 2002–04.

For all livestock types shown except cattle, the tendency is for industrial countries to be net exporters and the developing world the net importers, along with declining net exports from industrial countries. For cattle, the developing countries of Latin America (such as Mexico) are major net exporters, and the net importing regions include WANA and (increasingly) East and Southeast Asia. China switched from a net exporter to importer of cattle between 1993 and 2003; while these

Table 2.18. Net exports of meats (thousand metric tons).

Region	Bovine		Pig meat		Poultry meat		Total meats	
	1992	2003	1992	2003	1992	2003	1992	2003
China	101	–50	450	213	9	–48	582	87
India	88	317	0	1	0	3	97	332
Other South Asia	–0	1	–0	–0	–1	–6	–2	–3
East/Southeast Asia	–269	–610	–24	–247	106	251	–203	–641
WANA	–419	–427	–4	–11	–402	–823	–959	–1381
Sub-Saharan Africa	–93	–31	–28	–43	–103	–366	–229	–457
Latin America	429	1,202	–108	235	103	1418	443	2862
Industrial	944	525	110	962	737	2396	1968	4017

For China, the average unit value of poultry meat exports is greater than that of its poultry imports. Consequently, in value terms China is a net exporter of poultry meat, averaging $379 million for 2002–04, when it was also a net importer in terms of volume. The "Industrial" region is similar to the industrial countries as defined in the FAOSTAT.

Source: FAO 2006a.

Table 2.19. Net exports of dairy products and eggs.

Region	Dairy products 1992	Dairy products 2003	Eggs in shell 1992	Eggs in shell 2003
China	−1009	−2749	35	88
India	−25	123	6	35
Other South Asia	−720	−920	−3	−2
East/Southeast Asia	−3875	−6581	4	37
WANA	−6013	−7175	−37	−51
Sub-Saharan Africa	−1757	−2137	−6	−21
Latin America	−5073	−3320	−14	−20
Industrialized	19547	24388	120	12

Dairy products include milk equivalent. Values are expressed in thousand metric tons.

Source: FAO 2006a.

data do not differentiate between beef and dairy cattle, China has, for example, recently increased imports of cattle to resource its rapidly growing dairy sector. Other major trading countries are Australia and North America (as exporters) and the United States as an importer, and there is also considerable trade within the European Union (EU) in cattle, as in other types of livestock.

For live pigs, some EU countries, North America, and China are major exporters, although the latter's net exports have declined since 1993. Important importing countries include Mexico and land-scarce Asia, such as South Korea, Singapore, and Hong Kong (net exports of East and Southeast Asia fell to almost zero in 2003; see

Table 2.21). Australia is the world's principal exporter of live sheep and goats, with major import destinations being the Middle Eastern countries of the WANA region. The net volume of the live sheep and goat trade has declined over the past decade, however. Trade in live poultry is dominated by some EU countries, the United States, Malaysia, and China as major exporters. Other parts of Asia, such as Hong Kong and Singapore, in addition to EU countries and the United States, are poultry importers.

As with livestock products, trade in live animals is subject to animal health standards, such as those of the World Organisation for Animal Health (OIE, from the International Office of Epizootics in French) under the World Trade Organization's Sanitary and Phytosanitary (SPS) Agreement, as well as to private standards that may be required by buyers such as supermarkets, so as to control against unacceptable risks to animal and human health. Periodic animal disease outbreaks can disrupt international trade in live animals. Outbreaks of avian influenza have had an obvious impact on the movement of live poultry, as do outbreaks of foot-and-mouth disease on trade in cloven-footed animals. Bovine spongiform encephalitis (BSE, or mad cow disease) is another problem and was the reason China banned live cattle imports from North America in 2003, for example. Apart from the above health issues, another public concern regarding trade in live animals is animal welfare and the management of livestock when in transit. The OIE is also working with exporters in this area so as to improve animal welfare practices.

Table 2.20. Livestock product trade flows between industrial and nonindustrial countries, 2004.

Source of exports	Commodity	Exports to industrial countries (% total)	Exports to nonindustrial countries (% total)	Total exports (billion $)
Industrial countries				
	Beef	85	15	13.9
	Pig meat	86	14	13.9
	Poultry meat	60	40	6.8
	All meats	80	20	43.9
	Dairy	75	25	37.6
Nonindustrial countries				
	Beef	46	54	4.8
	Pig meat	27	73	2.2
	Poultry meat	51	49	4.1
	All meats	44	56	12.1
	Dairy	28	72	5.4

The definition of *industrial countries* is the same as that used in Tables 2.18 and 2.19.

Source: Comtrade.

Table 2.21. Net exports of live animals.

	Cattle ('000 head)		Pigs ('000 head)		Sheep & goats ('000 head)		Chickens (million head)	
	1993	2003	1993	2003	1993	2003	1993	2003
China	149	−9	2,782	1,915	272	50	41	32
India	−7	−5	−2	−6	−84	163	0	2
Other South Asia	19	27	1	11	31	−84	−1	−2
East/Southeast Asia	−133	−458	227	5	16	−0	1	−6
WANA	−707	−573	2	23	−11,015	−6,756	−64	−18
Sub-Saharan Africa	267	33	−23	−3	3,584	3,944	−13	−17
Latin America	1,046	1,142	−136	−206	−759	−41	−12	−4
Developing world	475	104	43	−175	−8,698	−4,201	−90	−47
Industrial world	−172	412	492	350	8,210	4,874	97	61

Source: FAO 2006a.

Trade in Feedstuffs

Grain use for livestock feed has been growing faster than total grain production in most regions. This implies declines in other uses of grain, such as for direct human consumption, or an increase in grain imports or possibly both. Such trends are part of the supply-side response to growing demands for livestock products in many countries and perhaps part of national self-sufficiency objectives. Increased grain imports have been especially notable in land-scarce economies such as those of Northeast Asia, where resource endowments do not encourage either large-scale cropping or pasture-based livestock production. During 2002–04, for example, global imports of maize averaged 86.8 million tonnes, of which 30.5 million tonnes (35%) were imported by Japan, South Korea, and Taiwan Province of China. Adding in WANA's imports of 14.8 mmt, these regions in total accounted for over half of the global maize trade during that period. The principal maize exporter is the United States (46.6 million tonnes in 2002–04). South America (15.1 million tonnes) is also a major maize exporter, but this region also imported 14.8 million tonnes of maize in 2002–04. (See Table 2.22.)

It is of interest to note from Table 2.22 that China remained a net exporter of maize in both periods, and during this time was a net importer of maize only during 1995 and 1996 following lower domestic production. This is despite widespread concern over China's potential to switch to become a major net importer of maize. Indeed, various projections of China's total grains trade have been made (as summarized in Fan and Agcaoili-Sombilla 1997) that projected net imports of between 11 million and 63 million tonnes in 2000, and between 14 million and 108 million tonnes in 2005. China's actual grain net imports in these two years were 3.1 mmt (2000) and 6.2 mmt (2005) (Comtrade data, as 2005 data were not available on FAOSTAT at the time of writing).

Although China's net exports of maize do fluctuate considerably from year to year—for example, net exports fell to 6 mmt in 2001 but reached 16.4 mmt in 2003—China has so far been able to meet its domestic demands for livestock maize feed largely from domestic sources. How long this will continue is not clear. Economic reforms in China have increased farmers' flexibility to switch land use away from grains to other crops, and rapid urbanization is also reducing the total area of cropland. Between 1991 and 2004, the total area of sown cropland in mainland China fell from 112.3 million ha to 101.6 million ha (9.5%). Over the same period, the area sown to grain crops fell by nearly 16%. Nevertheless, the area sown in maize actually increased by more than 17%, from 21.6 million ha in 1991 to over 25 million ha by 2004.

While China has remained a net exporter of maize despite rapid development of its livestock sector, the same cannot be said of soybeans. Between 2002 and 2006 the value of China's agricultural imports increased

Table 2.22. Net exports of maize in million metric tonnes.

Region	1992–94	2002–04
China and Taiwan	4.6	5.1
India	0.0	0.6
Other South Asia	−0.0	−0.4
East/Southeast Asia	−9.1	−13.0
WANA	−6.7	−14.5
Sub-Saharan Africa	−2.6	−1.5
Latin America	−1.9	0.3
Developing world	−15.7	−23.5
Industrial world	16.5	23.8
Of which:		
United States	39.5	46.3
Japan	−16.4	−16.7

Source: FAO 2006a.

by $20 billion, of which 25% was higher soybean imports driven in part by livestock feed demand. China is now the world's largest importer of this commodity, with the principal suppliers being the United States and Brazil.

Meeting Growing Demands: Increased Imports of Feedstuffs or Final Product?

Some countries, such as those in Northeast Asia, have met increased demands for livestock products from their domestic animals and from increasing imports of feedgrains. Others, such as China, have largely relied on domestic animal and grain resources. Especially in countries increasingly facing land, water, environmental, or other constraints to livestock production, at what point might imports of feedgrain be replaced by imports of meat products? As illustrations, Figures 2.2 and 2.3 compare maize net imports with net imports of meats measured in grain equivalents during 1975–2004 for selected countries. These grain equivalents are based on typical quantities of grain required to produce a kilogram of meat; the following coefficients are used to convert meat quantities to grain equivalents: beef 10, pork 4, and poultry 2. The grain-equivalent of imported dairy products is omitted from this analysis.

The top part of Figure 2.2 gives aggregate data for Japan, South Korea, and Taiwan Province, while Mainland China data are shown in the figure's lower half. In Northeast Asia, net imports of maize rose at an annual rate of 7.9% from 1975 until 1985, before slowing to an annual growth rate of 1.7% between 1986 and 2000 as domestic meat production met tightening constraints to further expansion. In comparison, net meat imports (in grain equivalents) rose at an annual rate of 8.1% from 1975 to 1985 and then increased to 10.9% per year between 1986 and 2000. Since 2000, imports of both

maize and meats have stabilized somewhat in response to the slowdown of demand growth in this region.

The situation in Mainland China over the same period has been quite different. Since 1985, net exports of maize have fluctuated considerably, but around a rising trend. At the same time, China has also been a net exporter of meats. It is not at all clear whether, and if so when, China will emerge as a persistent importer of grains for livestock feeding or whether further in the future China will eventually import grains embodied in meat products.

Yet another pattern is evident for WANA (also a major net meat importer) in Figure 2.3. Between 1975 and 1985, net imports of both meats and maize were increasing. Since then, meat net imports steadied while those of maize continued to grow at a time when domestic meats (and maize) production continued to increase. Only in the last few years have net meat imports shown a noticeable increase, with perhaps a leveling off of maize net imports.

Tariffs and Other Barriers to Trade in Livestock Products

The trade flows just discussed can be distorted by various barriers to trade. These include import tariffs, nontariff barriers, and export subsidies. These barriers can be severe in regard to livestock products. The bound tariffs of the Uruguay Round Agreement on Agriculture (URAA) averaged 62% globally (Gibson et al., 2001), but average tariffs for dairy products, meats, and eggs are above that. Using these averages, tariff protection among livestock products is highest for dairy. The average bound rate in OECD countries is 116%, compared with 74% in non-OECD countries. For meats, depending on the type and degree of processing, average bound tariffs vary from 82 to 106% in the OECD and between 68

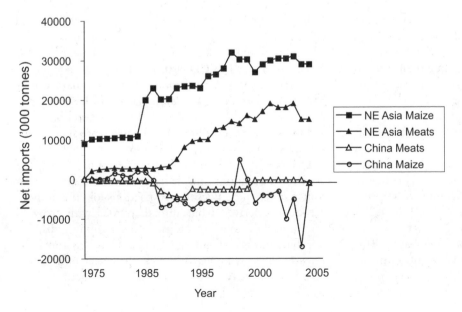

Figure 2.2. Net imports of maize and meats (grain equivalent).
Source: FAO 2006a.

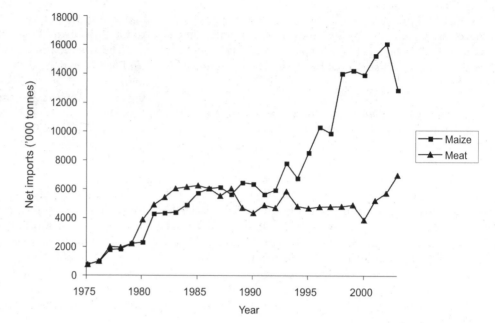

Figure 2.3. Net imports of maize and meats WANA (grain equivalent).
Source: FAO 2006a.

and 75% elsewhere. For eggs, average tariffs are 74% (OECD) and 66% in non-OECD countries. Tariff rates actually applied by a country can be lower than these bound rates, although this tends be observed in developing rather than industrial countries.

These high tariffs contribute to very high nominal protection (the ratio of producer price to world price, as a percentage) of livestock production, especially within the OECD. For example, the nominal protection coefficients for dairy in 2004 (OECD 2005) indicated average producer prices 38% and 58% above world prices in the EU25 and United States, respectively, and more than double world prices in Japan, Norway, South Korea, and Switzerland. Beef production was also highly protected in all these countries except the United States. Where high nominal protection is combined with the sometimes very low or zero tariffs on imports of feedstuffs, very high levels of effective protection can result. Such protection encourages a higher level of livestock production than would otherwise be the case, resulting in either reduced import demand or higher export supplies, and hence the distortion of international trade flows.

Quantitative restrictions also impede the international flow of livestock products. Tariff-rate-quotas (TRQs) were included in the URAA in cases where tariffs of previous quantitative restrictions would have provided little if any growth in market access. This device combines a lower tariff on the quota volume of imports and a higher tariff on above-quota imports. Of the 1,371 TRQs (WTO 2000), 31% relate to either meats or dairy products. For example, 12 OECD countries have TRQs on at least some dairy products, and 10 countries in the OECD have TRQs for fresh and/or frozen beef, pork, and/or poultry (Gibson et al., 2001). The above-quota

tariffs can be very high—for example, in 2000 they averaged more than 140% for beef and butter in the EU, over 100% and 300% for butter in the United States and Canada, respectively, and almost 680% for butter in Japan (OECD 2002).

Where high tariffs and other forms of producer assistance encourage high-cost export surpluses, export subsidies can be used to encourage the export of such supplies, up to limits specified in the URAA. The EU has accounted for around 90% of global spending on agricultural export subsidies, although its absolute level of spending has declined substantially since the mid-1990s. Dairy product exports are on average the most heavily subsidized of all commodities and accounted for around 35% or more of total spending during the 1990s. Export subsidy rates (the ratio of the per unit export subsidy to the world price, as a percentage) in 1995–97 were as high as 173% for pig meat and 118% for butter in the EU, and between 39% and 58% for dairy products from the United States (OECD 2001).

In recent years several food safety issues have arisen that involved trade in live animals or products derived from them (Mathews et al., 2003). These include bans on imports of certain animal and meat products from countries where foot-and-mouth disease is endemic (such as the trade from South to North America in beef), trade restrictions on poultry meats because of food safety risks from *Salmonella* infection, restrictions on the import of products derived from animals treated with growth-promoting hormones (such as the long-running dispute over beef exports from the United States to the EU), concerns over whether livestock drug use may erode the effectiveness of some antibiotics for human use, restrictions on trade in dairy products manufactured from non-pasteurized milk, and trade bans due to mad cow disease.

The latter, for example, became a human health issue in 1996 when a connection was announced between BSE and a variant of Creutzfeldt-Jakob disease in humans. This is a major food safety concern due to uncertainty over how the disease may be transferred to humans and the difficulties in identification and cure. It seriously affected cattle and beef exports from the United Kingdom in 1988, 1996, and 2000, and also led to bans by some countries, including China and Japan on such imports from North America from 2003 until recently. Such disease outbreaks, food safety concerns, and risks result in many countries either banning imports altogether or imposing other measures. such as tighter product standards, to prevent the introduction of such diseases through trade. In response to BSE, for example, international trade has in some cases been restricted to certain cuts or parts of a slaughtered animal.

These trade regulations are determined through rules and procedures established in the SPS Agreement of the Uruguay Round. It obligates World Trade Organization member countries to ensure that SPS regulations have scientific justification, do not discriminate between nations, are not applied as a disguised form of trade protection, and are not more trade-restrictive than is necessary to provide the chosen level of health protection. In an effort to harmonize such regulations and make them transparent, countries base their SPS measures on international standards, such as those recommended by CODEX (food standards, including those for animal products) and those of the OIE regarding animal health issues. Countries may impose higher standards, but they must be supported by scientific justification and risk assessments. Disease-free regions can also be recognized within national borders, such as regions of some South American countries that have been declared free of foot-and-mouth disease from time to time.

While this system contributes to well-functioning global markets that may ensure safe foods for consumers without erecting overly restrictive trade barriers, a key concern is the ability of emerging developing-country exporters to meet the standards of industrial-country markets or to implement such procedures themselves to protect their own human and animal populations. Technical assistance has been directed at this problem, and further assistance will be required to allow developing countries to fully realize advantages from participation in international markets for livestock products (Josling et al., 2004).

Projections of Consumption, Production, and Trade

Meat consumption per person in the developing world is expected to grow less rapidly than in the recent past (this section is taken from the projections in FAO 2006b). The rapid consumption growth observed previously, especially in China but also other countries such as some in South America, is unlikely to continue at the same rate, since relatively high consumption rates have already been reached in some instances and since population growth is slowing. This may not be the outcome in China, however, if government efforts to raise rural incomes closer to those of the urban population are successful, since rural consumption levels are much lower than in urban regions. While poultry meat consumption in India is expected to continue its recent rapid growth, it is from a relatively low base level and may not have the same impact on meat consumption growth at the global level as the expansion in China did.

Thus aggregate meat consumption in the developing world may grow during the next 30 years just half as rapidly as in the recent past. (See Table 2.23.) Meat consumption in the industrial countries is already high, at around 80 kg per person. Near-saturation levels of consumption, combined with health and obesity concerns, suggest little future growth in these countries. However, some substitution among meat types is likely to continue, with increased poultry meat consumption at the expense of beef and pork. Overall meat consumption in industrial countries is projected to grow by 10% by 2030 and by another 4% during the following two decades.

In contrast, no slowdown in dairy consumption growth in developing countries is foreseen. Consumption per person is generally still relatively low, and the potential for further growth from rising incomes and urbanization is still present in many countries. Total consumption of dairy products is projected to increase by almost 50% by 2030 and then by another 16% by 2050. For industrial countries, these increases are much lower, at 4% and 2%, respectively.

Despite the projected slowdown in meat consumption growth, expansion of international trade in meat is likely to continue. A projected feature of this trade is

Table 2.23. Projections of meat and dairy consumption (kg/person/year).

Commodity	1989/91	1999/2001	2030	2050
World				
Meats	33.0	37.4	47	52
Dairy products	76.9	78.3	92	100
Developing				
Meats	18.2	26.7	38	44
Dairy products	38.1	45.2	67	78
Industrial				
Meats	84.3	90.2	99	103
Dairy products	211.2	214	223	227

Source: FAO 2006b.

Table 2.24. Projected consumption and production growth rates (% growth per year).

Commodity	1981–2001	1991–2001	1999/01–2030	2030–50
World				
Production:				
Poultry meat	5.2	5.3	2.5	1.5
Total meat	2.9	2.7	1.7	1.0
Dairy products	0.8	1.1	1.4	0.9
Aggregate livestock	2.1	2.1	1.6	0.9
Cereal feed demand	0.6	0.8	1.6	0.8
Consumption:				
Poultry meat	5.1	5.1	2.6	1.5
Total meat	2.8	2.7	1.7	1.0
Dairy products	0.8	1.0	1.4	0.9
Developing				
Production:				
Poultry meat	8.4	8.2	3.4	1.8
Total meat	5.6	5.1	2.4	1.3
Dairy products	3.7	4.0	2.5	1.4
Consumption:				
Poultry meat	8.2	8.2	3.4	1.8
Total meat	5.6	5.3	2.4	1.3
Dairy products	3.4	3.7	2.5	1.3

Source: FAO 2006b.

that more of the meat imports may be supplied by developing country exporters, suppliers such as Brazil have the potential to further increase production. Trade in dairy products will also continue to expand, especially into the major deficit regions of East/Southeast Asia and WANA.

While feedgrain use grew at a slower rate than meat production over the past decade (Tables 2.8 and 2.15), this is not expected to continue. Instead, by 2050 total feedgrain use is projected to grow at the same annual rate as meat production, and to grow faster up to 2030 than beyond (Table 2.24). One reason is the structural shift occurring in many developing countries, from livestock production under "backyard" conditions that use various crop residues and food wastes for feed to modern commercial systems that are more reliant on grains and oil crops for feedstuffs and that could raise average grain:meat ratios. Developing countries accounted for 34% of global use of cereals for livestock feed in 2001–03 but could see this share increase to 52% in 2030 and 56% in 2050.

These projections were made before the extent of global biofuels developments was fully appreciated. Projecting feed demands from the livestock sector is now complicated by these new uncertainties in energy markets. Traditional feedstocks for animals, such as maize and soybeans, are also major feedstocks for ethanol and biodiesel production. Such emerging demands are putting upward pressure on grain and oil crop prices. Should these prices continue to strengthen, and as increased supplies of by-products suitable for livestock diets emerge from the biofuels industry, changes in feed formulation practices may occur, involving a shift away from maize and soybean and toward locally available feedstuffs and biofuel by-products. China, for example, has responded to recently escalating grain prices with new export constraints on maize and some other crops and is targeting alternative crops such as sweet sorghum for its bioethanol program. Simpson et al. (2007) suggest that such a use of sweet sorghum will add substantially to China's feedstuff production through the by-production of crushed stover and wet or dried distillers grains, and they project that use of these products could reduce China's imports of protein feed crops such as soybeans by more than 20% by 2030.

References

Delgado, C., M. Rosegrant, H. Steinfeld, S. Ehui, and C. Courbois. 1999. *Livestock to 2020: The next food revolution. Food, agriculture and the environment.* Discussion Paper 28. Washington, DC: International Food Policy Research Institute.

Demory-Luce, D., M. Morales, T. Nicklas, T. Baranowski, I. Zaker, and G. Berenson. 2004. Changes in food group consumption patterns from childhood to young adulthood: The Bogalusa heart study. *Journal of the American Dietetic Association* 104:1684–1691.

Fan, S., and M. Agcaoili-Sombilla. 1997. Why projection on China's future food supply and demand differ. *Australian Journal of Agricultural and Resource Economics* 41:169–190.

FAO (Food and Agriculture Organization). 2004. *The state of food and agriculture 2003–04.* Rome: FAO.

FAO (Food and Agriculture Organization). 2006a. FAOSTAT statistical databases. http://faostat.external.fao.org.

FAO (Food and Agriculture Organization). 2006b. *World agriculture: Toward 2030/2050, Prospects for food, nutrition, agriculture and the major commodity groups.* Interim report. Rome: FAO.

Fuller, F. H., J. C. Beghin, and S. Rozelle. 2004. *Urban demand for dairy products in China: Evidence from new survey data.* Working Paper 04-WP 380. Center for Agricultural and Rural Development. Ames, IA: Iowa State University.

Gibson, P., J. Wainio, D. Whitley, and M. Bohan. 2001. *Profiles of tariffs in global agricultural markets.* Agricultural Economic Report No. 796. Washington, DC: Economic Research Service, U.S. Department of Agriculture.

Jamal, A. 1998. Food consumption among ethnic minorities: The case of British-Pakistanis in Bradford, UK. *British Food Journal* 100:221–227.

Jones, K., and C. Arnade. 2003. A joint livestock–crop multifactor relative productivity approach. Paper presented at Southern Agricultural Economics Association Annual Meeting, Mobile, Alabama, 1–5 February.

Josling, T., D. Roberts, and D. Orden. 2004. *Food regulation and trade: Toward a safe and open food system.* Washington, DC: Institute for International Economics.

Ke, B. 1997. *Recent development in the livestock sector of China and changes in livestock and feed relationships.* Animal and Health Division (LPT2). Unpublished. Rome: Food and Agriculture Organization.

Kinsey, J. D. 1992. Seven trends driving U.S. food demands. *Choices, The Magazine of Food, Farm, and Resource Issues.* Third quarter.

Longworth, J. W., C. G. Brown, and S. A. Waldron. 2001. *Beef in China: Agribusiness opportunities and challenges.* St. Lucia: University of Queensland Press.

Loureiro, M., and R. M. Nayga, Jr. 2005. International dimensions of obesity and overweight related problems: An economics perspective. *American Journal of Agricultural Economics* 87:1147–1153.

Ma, H., J. Huang, and S. Rozelle. 2004. Reassessing China's livestock statistics: Analyzing the discrepancies and creating new data series. *Economic Development and Cultural Change* 52:117–131.

Mathews, K. H., J. Bernstein, and J. C. Buzby. 2003. International trade of meat/poultry products and food safety issues. In *International trade and food safety: Economic theory and case studies,* ed. J. Buzby. Washington, DC: Economic Research Service, U.S. Department of Agriculture.

Nayga, R. M., Jr. 1996. Wife's labor force participation and family expenditures for prepared food, food prepared at home, and food away from home. *Agricultural and Resource Economics Review* 25:179–186.

Nayga, R. M., Jr., R. Woodward, and W. Aiew. 2006. Willingness to pay for reduced risk of foodborne illness: A non-hypothetical field experiment. *Canadian Journal of Agricultural Economics* 54:461–475.

Nin, A., C. Arndt, T. W. Hertel, and P. V. Preckel. 2003. Bridging the gap between partial and total productivity measures using directional distance functions. *American Journal of Agricultural Economics* 85:928–942.

OECD (Organisation for Economic Co-operation and Development). 2001. *The Uruguay Round Agreement on Agriculture: An evaluation of its implementation in OECD countries.* Paris: OECD.

OECD (Organisation for Economic Co-operation and Development). 2002. *Agriculture and trade liberalisation: Extending the Uruguay Round Agreement.* Paris: OECD.

OECD (Organisation for Economic Co-operation and Development). 2005. *Agricultural policies in OECD countries: Monitoring and evaluation.* Paris: OECD.

Rae, A. N. 1998. The effects of expenditure growth and urbanisation on food consumption in East Asia: A note on animal products. *Agricultural Economics* 18:291–299.

Rae, A. N., H. Ma, J. Huang, and S. Rozelle. 2006. Livestock in China: Commodity-specific total factor productivity decomposition using new panel data. *American Journal of Agricultural Economics* 88:680–695.

Searle, J., A. Regmi, and J. Bernstein. 2003. *International evidence on food consumption patterns.* Technical Bulletin number 1904. Washington, DC: Economic Research Service, U.S. Department of Agriculture.

Simpson, J. R., M. J. Wang, M. S. Xiao, and Z. S. Cai. 2007. China looks to sorghum for ethanol. *Feedstuffs* 79: Issue 36 (Part 1) and Issue 37 (Part 2).

Wang, J-M, Z. Y. Zhou, and J. Yang. 2004. How much animal products do the Chinese consume? Empirical evidence from household surveys. *Agribusiness Review* 12: Paper 4.

WTO (World Trade Organization). 2000. *Tariff and other quotas: Background paper prepared by the Secretariat.* Geneva: WTO, 23 May.

Yang, J., T. G. MacAulay, and W. Shen. 2004. The dairy industry in China: An analysis of supply, demand, and policy issues. Paper presented at the Australian Agricultural and Resource Economics Society 48th Annual Conference, Melbourne, February 11–13.

3

Structural Change in the Livestock Sector

Cees de Haan, Pierre Gerber, and Carolyn Opio

Main Messages

- **The livestock sector is the most important global land user.** Increasing pressure from a growing world population and from various sectors (urbanization, tourism, biofuels) will increase the competition for land, in particular the more extensively used grasslands.
- **Competition for land, water, and fossil fuels, along with climate change, will be key driving forces shaping future livestock production systems.** Their effects can be attenuated or exacerbated by the policy framework. In the past, erroneous subsidies on inputs (feed) and distorted pricing policies for land, water, and environmental services have made the sector a major contributor to global environmental degradation.
- **The demand for feed grains, which has been rather stagnant for a few decades due to technological changes that affected feed conversion efficiency and shifts in inefficient subsidy policies, will expand to meet the continuous growth in demand for meat and milk.** Because efficiency gains similar to those in the past are unlikely, feed grain area expansion is likely to occur, which in turn can lead to land degradation, biodiversity loss from deforestation, and conflicts in the developing world.
- **Technological changes in livestock genetics, nutrition, and health that have led to increased factor productivity have greatly attenuated the need for more feed.**
- **Public and private-sector food safety and quality standards have gained major importance in the developing world not only for exporting countries but also to meet the demands of the growing middle class in regions such as South and East Asia and Latin America.**
- **Livestock production systems showed an increase in farm size because of economies of scale in about half of the world due to population pressure and demand.**

A reduction in farm size has occurred for the rest of the systems.
- **Many production systems have shifted from being grassland-based to mixed farming and, above all, to intensive production in landless systems, which has shown a strong increase in the poultry and pig sectors.**

The Impact of Natural Endowments on Production

Resource availability in terms of land, water, and fossil fuels is a main determinant of the global capacity to meet the increasing demand for livestock feed and thereby constitute major drivers shaping the livestock sector.

Land

Livestock production is strongly linked to land, as the sector uses 3.9 billion hectares—500 million hectares are for feed crops (33% of the cropland), generally intensively managed; 1.4 billion hectares are for pasture with relatively high productivity; and the remaining 2.0 billion hectares are extensive pastures with relatively low productivity (Steinfeld et al., 2006a).

Access to land and its resources is becoming an acute issue and an increasing source of competition. Disputes over access to land have caused civil unrest and even wars in the past, and resource-related conflicts are on the increase, especially in dryland areas. For example, Westing et al. (2001) report disputes over access to renewable resources—in particular, land and water—as one of the principal pathways in which environmental issues lead to armed conflicts. This may be the result of a reduced supply of resources (because of depletion or degradation), distribution inequities, or a combination of these factors.

Two concepts are central to the use of land for a given purpose: profit per unit of land and opportunity cost. Profit per unit of land (i.e., the surplus of revenue generated over expenses incurred for a particular period of time) describes the potential reason for an operator to engage in a particular use of the land. Profit thus generally depends on the biophysical characteristics of the land and on its price, as well as on factors such as accessibility to markets, inputs, and services. Opportunity cost—the cost of doing a particular activity instead of doing something else—compares the social costs related to two or more types of land use. Opportunity cost thus includes not only private costs to production but also direct and indirect costs borne by society, such as loss of ecosystem services. For example, the impossibility of using land for recreational purposes would be part of the opportunity cost of cropping a particular area.

In a situation of land surplus and low appreciation of nonmarketable ecosystem services, decisions on land use are predominantly driven by a calculation of profit per unit of land, which is usually based on tradable goods and services. As a result, nonmarketable benefits are often lost or degraded. This approach is still predominant. However, environmental and social services provided by ecosystems are receiving increasing recognition. A case in point is the growing recognition of a whole range of services provided by forests, a type of land use generally antagonistic to agricultural use.

Until the 1960s, livestock production growth was mainly based on an increase in livestock populations and in land dedicated to producing animal feed, including fodder, grass, grain, or crop residues. As a result, the conversion of natural habitats to pastures and cropland grew rapidly. (See Figure 3.1.) More land was converted to crops in the 30 years between 1950 and 1980 than in the 150 years between 1700 and 1850 (MA 2005). In particular, there was a dramatic acceleration of conversion of natural ecosystems to pasture and cropland after the 1850s (Goldewijk and Battjes 1997). Recent trends, however, show a tendency toward land use intensification.

There are major regional differences in production. Table 3.1 presents regional trends over the past four decades for three classes of land use: arable land, pasture, and forest. In North Africa, Asia, and Latin America and the Caribbean, land use for agriculture—both arable land and pasture—is expanding. The area under agriculture is high in Asia and Latin America and is progressing at substantial rates, with even accelerating expansion in Asia. Agriculture in these two regions is expanding, mostly at the expense of forest cover (Steinfeld et al., 2006a). In contrast, North Africa has seen cropland, pasture, and forestry expanding at only modest rates, with low shares of total land area covered by arable land.

Oceania and sub-Saharan Africa have limited arable land (less than 7% of total land) and vast pastureland (35–50% of total land). Expansion of arable land has been substantial in Oceania and is accelerating in sub-Saharan Africa. There is a net reduction of forested lands in both regions. Local studies have also documented replacement of pasture by cropland.

Western and eastern Europe and North America show a net decrease in agricultural land over the last four decades, often tied to either stagnation or an increase in forested land. These trends have occurred in the context of a high share of land dedicated to crops: 37.7% in eastern Europe, 21% in western Europe, and 11.8% in North America.

The Baltic and Commonwealth of Independent States (CIS) show an entirely different pattern, with decreasing land dedicated to crops and increasing land dedicated to pasture. This trend is explained by the structural and ownership changes that occurred in this region in the 1990s.

Pastureland and Grazing

Grasslands—terrestrial ecosystems dominated by herbaceous and shrub vegetation and maintained by fire, grazing, drought, or freezing temperatures—as a natural land cover are estimated to currently occupy between 29% and 40% of total emerged land (FAO 2005). Figure 3.2 in the color well shows the wide diffusion of grasslands

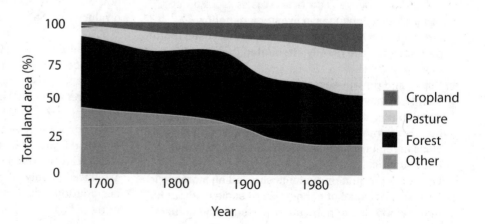

Figure 3.1. Estimated changes in land use from 1700 to 1995.
Source: Goldewijk and Battjes 1997.

Table 3.1. Regional trends in land use for arable land, pasture, and forest, 1961–2001.

	Arable land			Pasture			Forest		
	Annual growth rate (%)		Share of total land in 2001 (%)	Annual growth rate (%)		Share of total land in 2001 (%)	Annual growth rate (%)		Share of total land in 2000 (%)
	1961–1991	1991–2001		1961–1991	1991–2001		1961–1991	1990–2000	
Asia	0.4	0.5	17.8	0.9	1.0	32.9	−0.3	−0.1	20.5
Oceania	1.3	0.8	6.2	−0.1	−0.3	49.4	0.0	−0.1	24.5
Baltic states and CIS	−0.2	−0.8	9.4	0.3	0.1	15.0	n.d.	0.0	38.3
Eastern Europe	−0.3	−0.4	37.7	0.1	−0.5	17.1	0.2	0.1	30.7
Western Europe	−0.4	−0.4	21.0	−0.5	−0.2	16.6	0.4	0.4	36.0
North Africa	0.4	0.3	4.1	0.0	0.2	12.3	0.6	1.7	1.8
Sub-Saharan Africa	0.6	0.9	6.7	0.0	−0.1	34.7	−0.1	−0.5	27.0
North America	0.1	−0.5	11.8	−0.3	−0.2	13.3	0.0	0.0	32.6
Latin America and Caribbean	1.1	0.9	7.4	0.6	0.3	30.5	−0.1	−0.3	47.0
Developed countries	0.0	−0.5	11.2	−0.1	0.1	21.8	0.1		
Developing countries	0.5	0.6	10.4	0.5	0.3	30.1	−0.1		
World	0.3	0.1	10.8	0.3	0.2	26.6	0.0	−0.1	30.5

Source: FAO 2006a, except for forest growth rate 1990–2000 and share of total land in 2000, from FAO 2005.

across the world. Except for bare areas (dry or cold deserts) and dense forest, grasslands are present to some extent in all regions. They are dominant in Oceania (58% of the total area, with 63% in Australia), whereas their spread is relatively limited in North Africa and the Near East (14%) and in South Asia (15%). In terms of nominal extension, four regions have approximately 7 million km² of grassland or more: North America, sub-Saharan Africa, CIS, and Latin America and the Caribbean. Permanent pastures are estimated to cover about 34.8 million km², or 26% of total land area (FAO 2006a).

Grasslands have expanded by a factor of six since the beginning of 1800 and now occupy nearly all land that can be grazed and has no other demands on it (Asner et al., 2004). But given current trends of conversion of grasslands to cropping, there will be little room for expansion of the areas under grassland; to the contrary, the trend is expected to be continued conversion of pasture into cropland due primarily to the growing demand for feed crops (Steinfeld et al., 2006a). Significant changes in land use with direct relevance to the global environmental landscape include the following:

- *Conversion of pasture to arable land, urban areas, and forest.* While global long-term data on the rate of conversion are not available, some specific examples illustrate the general trends. For example, of the original North American tall grass prairies,

only about 9% still remain as grassland; the rest has been converted to agricultural crops (71%), urban development (19%), and forests (1%). Similarly, 21% of the South American *cerrado* woodland and savannas remains in pasture; the rest has been converted to cropland (71%), urban areas (5%), or forests (3%) (White et al., 2000). Agricultural crops in South America, particularly Brazil, are focused on the cultivation of soybeans.

The encroachment of arable farming in the dry-season grazing areas is particularly critical for migratory livestock (pastoral) systems. The conversion of such sites of high potential in the arid and semiarid rangelands removes a vital part of the pastoral extensive grazing system from the annual cycle, thereby undermining the viability of the entire system, and it is therefore a major source of conflict in the Sahel and in some South Asian countries (Scoones 1994).

- *Land degradation in arid and semiarid zones.* Extensive grazing systems of arid lands are often associated with overgrazing, leading to soil degradation caused by erosion and nutrient depletion. Population pressure, arable farming encroachment, and government policies to settle mobile pastoral groups are also major causes of soil degradation. Moreover, the communal ownership of most rangelands (as opposed to individual livestock ownership) in the arid and semiarid areas in sub-Saharan Africa and South and Central Asia constrains potential investments in improvements because the investment of one individual would be captured by other members of the society. This issue of the "tragedy of the commons" has become more critical as the social cohesion of pastoral societies has eroded, and with it the traditional discipline in the management of common lands.

However, livestock is only one of the many causes of land degradation. Agricultural cropping in marginal areas and firewood harvesting also contribute to land degradation. The most irreversible degradation occurs around settlement areas, water points, and livestock travel routes. The degree and reversibility of land degradation in the arid and semi-arid rangelands of the tropics are still major subjects of debate. The views of the United Nations Environment Programme (Middelton and Thomas 1997) and Dregne and Chou (1994) and the statement that 70% of arid land is degraded has been challenged by the "nonequilibrium" thought of contemporary range ecologists (Scoones 1994), who stress rangeland dynamics and the resilience of annual vegetation in apparently degraded areas. The latter is also highlighted by the work of Tucker et al. (1985) from the U.S. National Oceanic and Atmospheric Administration, who point to an "expanding and contracting Sahara." Finally, the resilience of the arid rangelands as presented by Breman and de Wit (1983) stresses the importance of nutrient depletion from these areas and the high efficiency of the use of the areas. (See Box 3.1.)

- *Conversion of forest to pastureland.* In Asia and Central Africa, legal and illegal timber extraction are the main reason for deforestation. In Latin America, conversion of tropical forest to cattle ranching is still important. Livestock ranching there has been associated for a long time with the deforestation of the Amazon. This negative association has gained increased significance in recent years as the rate of deforestation has increased and livestock's role in the process has become more pronounced. In the Amazon, deforestation had accelerated from about 18,000 km^2 per year in 1990–2000 to about 25,000 km^2 per year more recently. Amazonian deforestation was estimated at 0.6% between 2000 and 2005 (FAO 2007). Assisted by its comparative advantage of being free of foot-and-mouth disease and bovine spongiform encephalitis (BSE) or "mad cow disease" in some Brazilian states, beef exports soared. Over the period 1996–2004, the total export value of beef increased tenfold from $190 million to $1.9 billion, making Brazil the world's largest beef exporter (World Bank 2009). Combined with an impressive change in productivity (see Box 3.2) beef cattle ranching, which in the beginning of the 1990s was only profitable if subsidized and part of a land use sequence (including cropping)—became financially attractive in its own right and a major force in continued deforestation.

Arable Land and Feed Grain Production

Some 670 million tons of cereals are used for livestock, representing about 35% of the total world cereal use (Steinfeld et al., 2006a). Figure 3.3 in the color well shows the estimated global extent of grain production for animal feed. The reliance on different species has led to a wide diversity of feed grain crops. Corn is the predominant feed grain in Brazil and the United States, while wheat and barley are dominant in Europe and Canada. Southeast Asia relied on similar proportions of wheat until the early 1990s; since then, it has progressively shifted to corn. In contrast, Figure 3.4 in the color well shows the geographical concentration of soybean production, with eight countries supplying more than 95% of the world's consumption.

In spite of the strong increase in livestock numbers and production, the amount of grain used for livestock feed is down by 40% over the last two decades (FAO 2006b). This is due to a shift in the amount of grain required to produce a kilogram of meat. The decline reflects an overall increase in feed conversion efficiency, accompanied by a shift to a monogastric species such as poultry, which require less grain per kilogram of meat produced than beef or swine. In addition, there was a strong decline in feed use in the inefficient former Soviet economies following the transition from planned economies, as well as a change in subsidy policies in the European Union. Most of these trends are not likely to continue in the future; however, it is still expected that

---

Box 3.1. Arid Rangelands of Africa

The sub-Saharan arid rangelands are often depicted as bare wasteland, with some low-productivity cattle maintained by pastoralists for prestige reasons. However, more-detailed studies showed that they were used quite efficiently, certainly in terms of production per unit of land and per unit of fossil fuel. Analysis showed production for the various types of rangelands in Mali that ranged from 0.6 to 3.2 kg animal protein per hectare per year. This took into account the multiple uses of the pastoral herds, including milk production. This is substantially higher than the average 0.4 kg of animal protein per ha per year obtained in North American or Australian ranches under similar rainfall conditions. While part of this higher production might be due to resource depletion because of overgrazing, long-term trends show a stable meat and milk production of the Sahelian rangelands, thus pointing also to high productivity.

Source: Breman and de Wit 1983, de Haan et al., 2001.

Box 3.2. Productivity Improvement as a Major Push Factor for Deforestation in the Amazon Areas

Livestock–environment linkages have been synonymous over the last decade with deforestation of the rain forest in the Amazon. The term *the hamburger connection* stems from that perspective. Until the mid-1990s, however, cattle ranching was only profitable in most Amazon regions when it was part of a land use sequence that started with harvesting the most valuable trees, followed by cropping. Cattle grazing on the remaining natural vegetation was the final part of the land use sequence before the land was left fallow. However, over time, major technology improvements in cattle ranching within that environment contributed to improved profitability, as shown below.

Key technical parameters in the beef industry in the Amazon area of Brazil, 1985 and 2003

Parameter	1985	2003
Region	Northwest Brazil	Rondonia
Carrying capacity (AU /ha)	0.2–1	0.91
Fertility rate (%)	50–60	88
Calf mortality (%)	15–20	3
Daily weight gain	0.30 kg	0.45 kg

Source: World Bank 1991, Margulis 2004.

the total feed requirement will increase by about 1.6% per year until 2030 to about 1.1 billion tons (FAO 2006b). These requirements would need to be met by an increase in productivity per hectare in addition to an increase in cultivated area.

- *Increased past productivity of feed grains.* Impressive gains have been achieved over the last few decades in feed grain productivity, as shown in Table 3.2 for the main feed grains.
- *Expansion of feed-producing areas.* It is unclear how many more productivity gains can be achieved, as a yield limit seems to have been met for several grains. Assuming, therefore, that the demand for feed grains will reach 1.1 billion tons by 2030 and assuming some modest gains in productivity, an increase in cultivation area might be needed. This increase will have to compete with other emerging types of land

Table 3.2. Yields (kg/ha) of major feed grains, 1980–2005.

Feedgrain	1980	2005
Corn	3,154	4,754
Soybean	1,600	2,344
Wheat	1,855	2,901
Barley	1,998	2,401

Source: FAO 2006a.

use, such as the production of biomass for energy purposes (described later in the chapter). In any case, most of the increase in land area for feed grain production would have to come from the conversion of pasture into cropland, with space still available in the Southern Cone of the Americas and the arid savannas of Africa. The challenge will be to make this conversion in an ecologically sound fashion.

In this respect, one of the most debated new areas for the production of animal feed is the expansion of soybean cultivation in western Amazonia. The prohibition against using animal by-products in livestock feed because of the BSE risk has generated a strongly increased demand for plant protein, especially high-quality soybean meal for livestock feed. This, combined with the development of cultivars that are better adapted to the humid tropical environments of Amazonia, has led to a strong expansion of soybean cultivation in that region. While some soybean cultivation degraded cattle rangeland, other expansion comes from the direct conversion of savanna and forest into cropland (Steinfeld et al., 2006a). The use of new crop husbandry techniques, such as conservation tillage, facilitates direct conversion and may improve the long-term sustainability of agricultural soil management.

Water

Water availability will also be a determining factor shaping feed availability. Agriculture accounts for 70% of total water withdrawals, and the total water needs for agriculture are expected to increase an additional 14% by 2030. Withdrawals for agriculture in developing countries can be as high as 85% (World Bank 2004). Most the future demand for water is expected to occur in the developing world. These trends will exacerbate already severe water shortages, particularly in Africa and in West and South Asia. Water is less of a problem in the grazing and rainfed mixed systems, as most feed grains are produced in rainfed areas. However, competition for water in some regions will be a major factor affecting the use of irrigation water for feed or food production.

Steinfeld et al. (2006a) estimated that the production of barley, maize, wheat, and soybean crops requires about 10% of total water used for plant production in irrigated areas (that is, the total amount evapotranspired) globally. If it is assumed that those crops represent about 75% of the total feed requirement, then the share of water evapotranspired for feed production amounts to 13.3% of the total global amounts in irrigated areas. With requirements for alternative water uses also increasing, attention will have to be paid to improving irrigation efficiency, which was estimated by Bruinsma (2003) to be currently at about 38%. Investments in improved canal linings and equipment for precision irrigation, for example, would significantly improve efficiency.

However, as shown in many parts of the world, adjusting incentives and the institutional framework has to be the main instrument for improving efficiency (World Bank 2004). This can be implemented by the introduction of full-cost recovery pricing for water and energy.

Fossil Fuels

Fossil fuel needs will affect livestock production in many areas, but probably most significantly in the production of nitrogen fertilizers. Urea- and ammonia-based fertilizers have been major drivers of the increase in feed grain productivity, especially in the high-energy feed grain crops. About 97% of nitrogen fertilizers are derived from synthetically produced ammonia, which in turn is produced using natural gas, although other energy sources such as coal can be used, as in China (Steinfeld et al., 2006a).

The animal production sector is a major consumer of synthetic fertilizers. Steinfeld et al. (2006a) estimate that the share of synthetic nitrogen fertilizers used to produce feed (grains, oilseeds, roots, and tubers) varies from 70% in the United Kingdom to 16% in China. The United States (51%), Germany (62%), and France (52%) also use a major share of their fertilizer supplies to produce livestock feed.

The recent rise in energy prices affects the price of synthetic fertilizer. In the United States, for example, which has seen one of the largest increases in natural gas prices, the price of urea increased from about $120 per short ton in 2002 to about $300 per short ton in October 2005 (FAO 2005). This might offer greater opportunities for organic fertilizers, such as manures and nitrogen-fixing cover crops. This, in turn, could also change the types of feed produced, with the possible expansion of leguminous crops in mixed farming systems. The fossil fuel constraint is leading to an increase in the land area under cultivation for biofuels, which competes with feed grains for land.

Climate Change: Likely Impacts and Livestock's Contribution

Climate change, and in particular increased weather variability, will also drive the shape of the future industry. It has been estimated that crop yields could decline by one-fifth in many developing countries (Fischer et al., 2001). Water availability, particularly in the subtropics, is expected to diminish, and an increase in interannual seasonal variation (including the frequency and intensity of extreme events, such as droughts and floods) will also have a large effect on the poor. This is because their crops, livestock, homes, food stores, and livelihoods are at risk from floods and droughts, and they have few resources to carry them through difficult periods.

The effects will be most severe for feed grain production, mixed farming systems, and pastoral systems in sub-Saharan Africa and South Asia, where production systems are rainfed and thus highly vulnerable to drought. The frequency of droughts and flooding is expected to continue increasing, and reduced rainfall, higher temperatures, and changing pest populations will negatively affect food production in the tropics. On the other hand, some areas now not suitable for cropping or grazing, such as northern Canada and Russia, might come into production. Climate change might also increase the incidence of diseases, in particular vector-borne diseases such as trypanosomiasis, in many regions (Weihe and Mertens 1991, Kurukulasuriya et al., 2006).

While global climate change has important implications for future production patterns, the livestock sector itself is contributing to the changes in global climate. Agriculture currently accounts for about 30% of the global emission of greenhouse gases, with 60% coming from the livestock sector if carbon dioxide losses because of land conversion are included (Steinfeld et al., 2006a). A major challenge will be to adjust production systems so that they adapt to the changing physical environment while adopting practices that accentuate the positive environmental externalities and mitigate the adverse effects of agricultural practices on greenhouse gas emissions.

Competition for Land and Feed Resources

The livestock sector competes with other sectors for a number of resources—competition that has been spurred by either resource scarcity or the need for alternatives. This section looks at some of the challenges and emerging issues that the livestock sector is facing with regard to resource access. It focuses on issues related to feed resources and land use, analyzing how the dynamics within other sectors are affecting the livestock sector. In addition, it looks at the increasing use of livestock feed inputs in the production of biofuels, aquaculture, and other industrial purposes.

Biofuels

High world oil prices, growing demand for energy and energy security, and concerns about climate change coupled with the pressure on countries to meet greenhouse abatement targets all have been key factors driving the increasing interest in renewable energy sources, especially biofuels. Biofuels are transportation fuels derived from plant matter (OECD 2006). Two types currently account for the bulk of renewable fuels: bioethanol is made primarily from sugar (cane in Brazil and beet in the European Union) and starch crops (corn in the United States); biodiesel is produced from vegetable oilseeds such as rapeseed, sunflower, and soybean.

Biofuel production has expanded at a remarkable pace over the last decade. Global production of ethanol has increased nearly fourfold between 2000 and 2008, while production of biodiesel started from a much smaller base of 0.9 billion liters and reached some 12 billion liters in 2008. (See Figure 3.5.) Biofuel production has been supported by government programs either

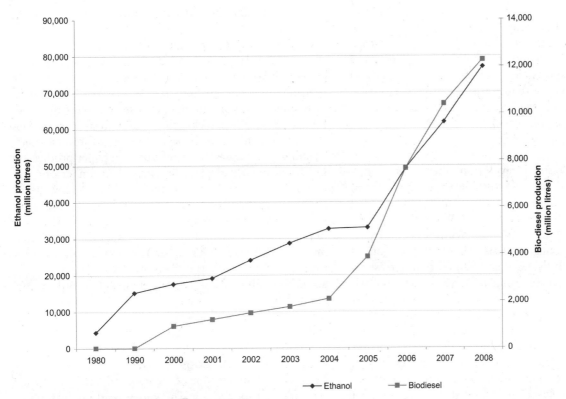

Figure 3.5. Global ethanol and biodiesel production.
Source: OECD–FAO 2008, Brown 2006.

through the provision of market incentives (tax concessions, vehicle taxes, and subsidies) or by market regulations (OECD 2006). The latter include explicit supply mandates, fuel blending standards, and bans on chemical ingredients.

There are significant trade-offs involved in the massive expansion of biofuel production. Chief among them is the increased demand for feedstock, which has led to higher prices for feed grains. Despite the recent decline in international commodity prices, world prices of commodities such as grains and vegetable oils have since risen more than 60% above 2006 levels (FAO 2009). For example, U.S. corn prices rose from $2.50 a bushel in 2005 to $4.05 a bushel in 2008 as growing demand for biofuel competed with the livestock sector for grain supplies (USDA–NASS 2009). In Europe, the price of rapeseed increased from about 180 euros a ton in 2000 to approximately 480 euros a ton in 2008 (European Commission 2008).

The rapid increase in demand for and production of biofuels, particularly bioethanol from maize and sugarcane, has had a number of effects on grain supply-and-demand systems. Expanded production of ethanol from maize, in particular, has increased total demand for maize and shifted land area away from production of maize for food and feed, stimulating increased prices for maize (IFPRI 2008). In all major producing countries there has been an increase in the crop acreage of biofuel crops.

The United States increased land in maize (corn) production in 2007 by 23%, which led to a 16% decline in soybean area and contributed to a 75% rise in soybean prices between April 2007 and April 2008 (World Bank 2008). In Brazil, the land area dedicated to sugarcane production increased by 34% between 1996 and 2006 (USDA–FAS 2006). Germany, the leading biodiesel producer in the European Union, saw its rapeseed area increase from 900,000 hectares in 1997 to 1.5 million hectares in 2007 (USDA–FAS 2009).

The biofuel process produces substantial amounts of by-products. These include dried distillers grains (DDGs), a primary feed product from dry-mill ethanol; gluten meal from wet-milling; and oilseed meal, a by-product in the production of biodiesel. Bagasse is a by-product of ethanol production from sugarcane. These products can be used partly to compensate for the loss of feed crops that go into the biofuel production process, although DDGs are more suitable for dairy cattle feeding, whereas bagasse, originally used as an animal feed, is now being diverted to generate energy on the sugar mills or used as a raw material for paper production.

The further expansion of biofuels is not clear. One estimate is that by 2030 some 15% of the liquid fuel needs will be met by biofuels, which will have major impacts on future land requirements and the price of feed grains (OECD 2006). For example, it is estimated that having biofuels meet 10% of gasoline and diesel needs would require 43% of the cropland in the United States and 38% in the European Union (IEA 2005). While new technologies such as cellulosic biomass feedstock and new processing technologies might reduce costs and land requirements, it is clear that recent high fossil fuel prices and the need to reduce greenhouse gas emission are strong drivers for increased biomass production and hence for the shape of livestock feed production.

Fishmeal

Parallel to the livestock revolution, there has been a strong expansion in demand for high-value aquacultural products, which increased at an average compounded rate of about 9.2% a year since 1970 (World Bank 2006). This fast-expanding high-value aquaculture sector based on carnivorous species (such as salmon and shrimp) is more dependent on fishmeal and fish oil than terrestrial animals and is therefore using an increasing share of the rather inflexible supply of fishmeal. For example, the share of fishmeal used by aquaculture grew from 8% in 1988 to about 35% in 2000 (Delgado et al., 2003a) and then to 45% in 2005 (World Bank 2006). Yet the prohibition against using meat offal in livestock feeds because of the BSE risk is forcing the livestock sector, in particular in East Asia, to substitute fishmeal for soybean meal as an alternative source of plant protein. Major research efforts are under way, for example, in China, to substitute fishmeal for soybean meal in aquaculture as well (see Box 3.3), putting more pressure on land in those areas that is ideally suited for soybean cultivation, such as the tropical savanna and rainforest areas.

Box 3.3. Long-Term Potential of Soybean Meal Use in Aquaculture

China's fast-growing aquaculture sector totally dominates world aquaculture. In 2002, China accounted for more than two-thirds of global aquacultural output. Aquaculture in China has expanded from 1 million tons in 1981 to 28 million tons in 2002.

Traditionally, China relied heavily on fishmeal as a source of animal feed. But with fish becoming scarce, China has turned to soybean meal as an alternative. The U.S. Department of Agriculture reports a strong import demand for soybeans over the past two decades, attributing this increasing use partly to the increase in aquaculture production. According to the American Soybean Association, soybean meal use by the Chinese aquaculture industry as feed is estimated to exceed 185 million bushels a year, the equivalent of about 5 million metric tons. To date, freshwater fish rations are now roughly one-third soybean meal, substantially higher than the 18–20% soybean meal content in livestock and poultry feeds.

Source: Brown 2004, 142–144; USDA-FAS 2005; U.S. Soybean Export Council 2006.

Agricultural Residues and By-products for Industrial Uses

Crop residues and by-products such as cereal straws, corn stover, sorghum stalks, and sugarcane bagasse, once of value only as livestock feed, now face competing uses as fuel and industrial raw materials. With a decrease in wood availability and the resulting increase in the cost of wood, the use of nonwood fiber such as agricultural crop residues is gaining significance (Hurter 1998). Crop residues such as straw and bagasse have been used in paper production in China, India, Pakistan, Brazil, and a number of other countries (Bowyer and Stockmann 2001). Today, global paper production from crop residues is on the rise, with the percentage of pulp capacity accounted for by nonwood fiber globally now close to 12%, compared with an estimated 6.7% in the 1970s. Wheat straw currently accounts for over 40% of nonwood fibers and bagasse and bamboo account for about 25% (Bowyer et al., 2004). As the large majority of these crop residues are used for ruminant feeding, these alternative uses will put more pressure on land resources and, in particular, on grasslands.

Production Technology

Technology has been one of the main drivers of increased total factor productivity. The widespread application of advanced breeding and selection methods, reproductive techniques, feeding technology, and sophisticated animal health control measures has spurred impressive productivity gains in most parts of the world. In breeding, new selection methods that disentangle genetics from feeding and management, such as the Best Linear Unbiased Prediction (BLUP), greatly improved the accuracy of the predictions of the breeding value of an animal. In reproductive techniques, advances such as artificial insemination, embryo transfer, sexing and cloning, and the identification of genetic markers have greatly accelerated the rate of multiplication of the best genotypes. This is particularly important with species with a short generation interval, such as pigs and poultry. In feeding, a much better balanced diet, including the optimum addition of amino acids and microminerals, greatly increased the feed conversion (the amount of feed required for one kg of growth). Finally, the development of improved vaccines and especially the development of pathogen-free stables reduced losses in mortality and morbidity and again increased efficiency. The synergy of the application of these technologies increased production per animal as well as the efficiency of production. (See Table 3.3.)

Policies and Institutions

Policies and institutions will be key drivers affecting the shape of the future feed and livestock industry, with incentives, regulations, and institutions as the main instruments. Public livestock policies can be seen as forces that add to the drivers described earlier and that influence changes in the sector with the aim of achieving an

Table 3.3. Comparison between some past and current key performance indicators for livestock production.

Item	Past performance and year	Current performance and year	Source
Feed conversion egg production (kg feed/kg egg), United States	2.96 (1960)	2.01 (2001)	Arthur and Albers 2003
Feed conversion broiler meat (kg feed/kg live weight gain), United States	1.92 (1957)	1.62 (2001)	Havenstein et al., 2003
Feed conversion pigs (kg feed/kg live weight gain), United States	3.34 (1980)	2.72 (2001)	de Haan et al., 2001
Milk production (kg/cow/year), OECD countries	4,226 (1980)	6,350 (2005)	FAO 2006a

identified set of societal objectives. Policies are designed and adjusted, taking into account the state of markets, available technologies and natural resources, and the current status of the sector. From this standpoint, public policies are both drivers of and responses to changes in the livestock sector. At any point in time, policies that are in existence and enforced are drivers of change, while policies in preparation are part of the public response to changes. This chapter considers only the main regulatory and policy frameworks that influence production costs and thereby shape different production systems. Other more macroeconomic policies are outside the scope of this chapter.

Incentives for Production

The livestock sector is one of the most heavily subsidized agricultural sectors, which has had major effects on the environment and on social equity. For example, in the European Union 70% of farmers' incomes in the beef and veal sector and about 40% in the dairy sector come from subsidies—significantly above the average in all other sectors except sugar (OECD 2005). In the U.S. dairy sector, about 40% of income comes from subsidies and protection. Most of these subsidies promote increased production, and only recently have some of the payments been directed to environmental sustainability. But in 2004, only 10% of the 40 billion euro support under the Common Agricultural Policy (CAP) was budgeted for environmental objectives. In the 2007–10 CAP budget, the share allocated to the environment increased to one-third (Henke and Storti 2004).

According to the World Trade Organization, feed subsidies reached 9.3 billion euros in the European Union in 2001–02 and $6.5 billion in the United States in 2002 (Berthelot 2005). Earlier preferential tariffs for cereal substitutes were one of the driving forces in the large increase in the pig and poultry sector in the Netherlands (de Haan et al., 1997). The current process of gradual reduction of these subsidies and a shift in Europe to a system of single farm payment, combined with changes in cereal subsidy policies, is expected to have a significant impact on land use as cereal prices increase (Bascou et al., 2006). This, in turn, might reduce the incentives for intensive beef production and create greater

demand elsewhere. In general, removal of production incentives would increase the price of feed and, ultimately, livestock products. It might also shift the balance away from feed grain–based nonruminant production to ruminant-based grazing systems. (See Chapter 2.)

Another major area of distorted incentives that affects feed production is the pricing of water and land. Van Beers and de Moor (2001) estimated that water subsidies in 1994–98 totaled $45 billion per year because of serious underpricing of this resource, leading to inefficiencies and waste. Subsidies on fuel exacerbate such waste, as they lead to excessive exhaustion of groundwater. Indiscriminate and subsidized water development by the public sector in arid rangelands has led to severe land degradation and poor sustainability of the infrastructure. However, water pricing is a very sensitive policy issue everywhere, as increases will probably mean that water would be directed to other uses.

Land pricing also affects land use, although less severely. As discussed, well-disciplined management of communal land by social cohesive groups has been replaced by a "free for all" on underpriced land. Moreover, as Steinfeld et al. (2006a) point out, current land pricing does not include the costs of degradation-linked externalities, such as the loss of biodiversity, carbon dioxide emissions from degrading land, and the contribution to water pollution. Including such costs in the price of land, according to Steinfeld et al. (2006a), would shift a large part of the world's arid and semiarid rangeland into providers of environmental services rather than grazing areas for livestock.

Financial incentives (subsidies and penalties) to mitigate negative environmental impacts of livestock are also an important driver of the future shape of the livestock industry. While still limited, as described earlier, subsidies on fixed assets such as assistance in the construction of manure storage facilities and on-farm or livestock retirements are being increasingly used, in particular in countries that belong to the Organization for Economic Cooperation and Development (OECD). Environmental taxes are levied on inputs, such as fertilizer, that are a potential source of environmental damage, and on phosphorus levels in feed (World Bank 2005).

Sanitary and Food Safety Regulations

The growing mobility of people and the increasing liberalization of agricultural and food trade increase the risk of the spread of diseases due to foodborne pathogens. The significant rise in the length of the food processing chain over the last few decades increases the risk of chemical residues in food. Because of their perishable nature, animal products are particularly risky. Food scares such as "mad cow disease," *E. coli* and salmonella outbreaks, and dioxin contamination in meat, to mention just a few, have heightened consumer concerns about the safety of animal products. As a result, consumers are increasingly willing to pay more for "safer" foods. This increased demand for safe and high-quality food products is manifested by the proliferation of public and private sanitary and phytosanitary-related product and process requirements. The requirement for systems based on Hazard Analysis and Critical Control Points is becoming a common feature for public food safety regulations in industrial countries, including for imported food products.

The emergence or reemergence of livestock related-diseases (zoonoses) such as severe acute respiratory syndrome (SARS), the Nipah virus, BSE, and more recently the highly pathogenic avian influenza has also increased consumer concerns. In addition, direct disease threats to the livestock populations of importers, such as foot-and-mouth disease, which have caused several billion dollars in economic losses (World Bank 2005), have also led to stricter requirements. As a result, importing countries require that livestock products meet an array of sanitary standards. Consumers, retailers, governments, and civil society now all have voices in the design and implementation of standards defining the specification of products (such as quality, size, residual chemicals, level of microbial contamination) and their production processes (such as labor standards and environmental impacts).

Meeting these standards requires access to specialized services in the area of animal health and animal husbandry. A major issue here is economies of scale, which act against smallholder farmers.

Environmental Policies

Stricter environmental policies are also major drivers of change in livestock production and processing systems. First, environmental regulations have often focused on the easily observable effects, such as odor from pig farming, rather than on the more substantial environmental effects. In addition, fewer policy options and hence regulations exist for nonpoint than for point-source pollution. The main regulatory measures are as follows (World Bank 2005):

- *Discharge standards* are designed to address end-of-the pipe emissions of pollutants from intensive production units and from the processing industry. To date, discharge standards are the most frequent

instrument, especially in the developing world. They are often derived from pollution control measures used in the industrial sector. They are thus often not adapted to the agricultural sector, and their enforcement requires substantial monitoring capacities.

- *Land/livestock balances* are of special importance in the case of intensive production systems. This is the approach underpinning the nitrate directive of the European Union, addressing both nonpoint and point source pollution sources. By setting restrictions on the quantity (and timing) of nitrogen that can be applied to the land, the directive indirectly forces a better geographical distribution of production units and a substitution between organic and chemical fertilizers. Steinfeld et al. (1997) and Gerber (2006) propose that the *zoning/spatial planning* of livestock production would further allow the designation of nutrient (nitrate)-sensitive areas, the restriction of animal numbers in a certain area, and the imposition of regional balances for nitrogen and phosphorus. Zoning has been greatly facilitated by advancements in geographic information systems, which increase the understanding of spatial and temporal relationship among landscapes, waste management, and environmental outcomes.

Little is known about the overall costs and benefits of these instruments, although there is adequate information on the financial costs and benefits of individual investments in environmental mitigation of livestock. However, these cost/benefit estimates generally lack an adequate estimate of the costs and benefits of managing the environmental externalities, although there seems to be some indication that prevention is better than mitigation. For example, figures from the Netherlands show that over the 1998–2002 period, nutrient loading decreased by about 0.2 kg phosphorus and 0.8 kg nitrogen per euro spent on a set of preventive measures (such as the introduction of mineral accounting, correcting land/livestock balances, and reducing nitrogen and phosphorus content in the feed), while removing those quantities from surface water would have cost about three times as much (RIVM 2004).

Institutions

Key institutional issues are the level and sector in which the regulations are set and implemented. There is a continuous struggle between Ministries of Agriculture, Environment, and Health on which institution should set the standards. Leaving it to the Ministry of Agriculture ensures that the interests of the production chain are duly represented, but it runs the risk that the regulations are overly biased toward farming interests. Moreover, there is an issue of central versus local monitoring and enforcement of the standards.

Resulting Structural Changes

Scale and Size of Production and Processing Units

The drivers described earlier caused farm size to increase in about half of the world, while it shrank in the other half of the world. Economies of scale in production, particularly in meeting the stricter food safety and environmental standards, and the low marginal returns to labor drive the process of increased size and scale in OECD countries. For example, in the United States the number of farms declined from a peak of 6.8 million in 1935 to about 2.2 million in 2002, whereas the average farm size increased from about 80 ha to 220 ha in that same period (USDA–ERS 2006). Similarly, in the livestock sector—and in particular in the pig and poultry sector—farm size increased significantly. Between 1989 and 2002, the share of the value of pig meat production from farms with sales of $500,000 increased from 14% to 64%, and for poultry meat the share went from 40% to 68% in the same period (McDonald et al., 2006). The same trend can be observed in Europe, where the number of farms declined from 7.9 million in 1989 to 6.7 million in 2000 (Hill 2006).

Similar trends can be observed in Australia and New Zealand and in the pig and poultry industry in middle-income countries, such as Thailand and Brazil. In Thailand, for instance, the number of small pig farms fell from 1.3 million in 1978 to 420,000 in 1998 (World Bank 2005), which could be interpreted as a significant restructuring with possible important social implications. On the other hand, in the Philippines, which also witnessed a strong increase in the industrial sector, the smallholder system of about 3 million farmers maintains itself in parallel with the industrial sector (Delgado et al., 2003b). Differences in alternative employment opportunities between the Thai and Philippine economies might explain these different trends.

In most developing countries, however, population pressure has led to an increase in the number of farm holdings and a subsequent decrease in farm size. In India, the number of farm holdings increased from 70.5 million in 1970–71 to 97.7 million in 1985–86 and then to 105.3 million in 1990–91, with a major shift to landless and marginal farm holdings (AERC 2005). In some countries, this has led to an almost complete implosion of the system. In the district of Gikongoro in Rwanda, for example, population density increased from 100 to 287 over the 1948–91 period, and as a result farm size became smaller, the capacity to grow forage was reduced, and one in every two households had to give up livestock. Soil fertility and the nutritional quality of the population's food intake declined, which in turn contributed to the collapse of Rwanda's society in the early 1990s (de Haan et al., 1997).

Scope and System of Production

Livestock production is determined by the physical (climate, soils, and infrastructure) and the biological environment (plant biomass production and livestock species composition), by economic and social conditions (prices, population pressure and markets, human skills, and access to technology and other services), and by policies (on land tenure, trade, and subsidies). These conditions together generate so-called production systems—that is, production units (herds, farms, ranches, but also regions) with a similar structure and environments—that can be expected to produce on similar production functions. Seré and Steinfeld (1996), who established the standard work on livestock production systems, distinguished at the highest level two groups of farming systems: those solely based on animal production, where less than 10% of the total value comes from nonlivestock farming activities, and those where livestock rearing is associated with cropping in mixed farming systems, with more than 10% of the total value of production coming from nonlivestock farming activities.

Table 3.4. Global livestock population and production in different systems (average for 2001 to 2003).

Type of animal/animal product	Livestock production systems			
	Grazing	Rainfed mixed	Irrigated mixed	Industrial
Livestock population (million head)				
Cattle and buffaloes	406	641	450	29.0
Sheep and goats	590	632	546	9.0
Animal product (million tonnes)				
Total beef	14.6	29.3	12.9	3.9
Total mutton	3.8	4.0	4.0	0.1
Total pork	0.8	12.5	29.1	52.8
Total poultry meat	1.2	8.0	11.7	52.8
Total milk	71.5	319.2	203.7	0.0
Total eggs	0.5	5.6	17.1	35.7

Source: Based on FAO 2006a and on calculations in Groenewold 2005.

Table 3.5. Livestock population and production in different production systems in developing countries (averages 2001 to 2003).

Type of animal/animal product	Livestock production systems			
	Grazing	Rainfed mixed	Irrigated mixed	Industrial
Livestock population (million head)				
Cattle and buffaloes	342	444	416	1
Sheep and goats	405	500	474	9
Animal product (million tonnes)				
Total beef	9.8	11.5	9.4	0.2
Total mutton	2.3	2.7	3.4	0.1
Total pork	0.6	3.2	26.6	26.6
Total poultry meat	0.8	3.6	9.7	25.2
Total milk	43.8	69.2	130.8	0.0
Total eggs	0.4	2.4	15.6	21.6

Source: Based on FAO 2006a and on calculations in Groenewold 2005.

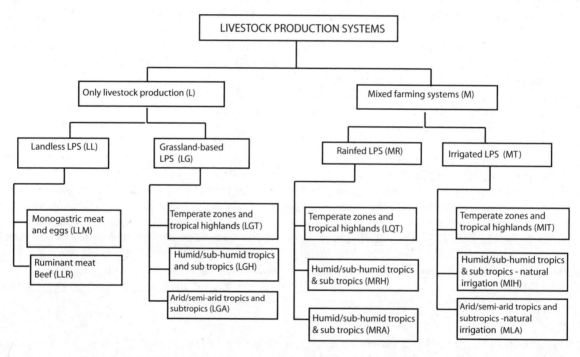

Figure 3.6. Classification of world livestock production systems. Operational considerations related to the number of different systems to be handled throughout the whole Livestock Environment Study led to the decision to limit the classification to 10 systems by retaining only the first three classification criteria. The complete classification structure is outlined. In addition, the landless system group was split into landless ruminant and landless monogastric systems. *Source*: Seré and Steinfeld 1995.

LGA livestock grassland arid semiarid tropics and subtropics
LGH livestock grassland humid/subhumid tropics
LGP length of growing period
LGT livestock grassland temperate and tropical highlands
MIA mixed farm irrigated arid semiarid tropics and subtropics
MIH mixed farm irrigated humid/subhumid tropics

MRA mixed farm rainfed arid semiarid tropics and subtropics
MRH mixed farm rainfed humid/subhumid tropics
MRT mixed farm rainfed temperate and tropical highlands
LLM livestock landless monogastric
LLR livestock landless ruminant
LPS livestock production system

Within systems that are solely based on animal production, a further distinction can be made between landless production systems, which buy at least 90% of their feed from other enterprises, and grassland-based systems. In the mixed farming systems, the availability of irrigation is a major determinant, and a first distinction is made by the absence or presence of irrigation. This leads to the main production systems presented in Figure 3.6 and to global and developing-country numbers of animals and production as provided in Tables 3.4 and 3.5. Figure 3.7 in the color well presents the geographical distribution of the main production systems.

In livestock grazing systems, there is a further distinction into three groups (Steinfeld et al., 2006):

- *Extensive grazing systems*, which cover most of the dry areas of the tropics and continental climates of central Asia, North America, and western and southern Asia, in areas of low population density. They are characterized by ruminants (cattle, sheep, goats, camels) grazing mainly on grasses and other herbaceous feed, often on communal or open areas, often in a mobile fashion. The main products of this system include about 7% of the world's global beef production, about 12% of sheep and goat meat production, and 5% of the global milk supply.
- *Intensive grazing systems*, which cover most temperate climate zones of Europe, North and South America, and increasingly the humid tropics. They are characterized by cattle (dairy and beef), in areas of medium to high population density, based on high-quality grassland and fodder production, mostly with individual landownership. They contribute about 17% of the world beef and veal supply, about the same share of the sheep and goat meat supply,

and 7% of the global milk supply as their main outputs.

- *Intensive landless systems* are found around the urban conglomerates of East and Southeast Asia and Latin America or near the main feed-producing or feed-importing (ports) areas in Europe and North America, often consisting of a single species (beef cattle, pigs, or poultry) fed on grain and industrial by-products brought in from outside the farm. They produce about 72% of the world's poultry and 55% of the pork.

Mixed farming systems can be divided into rainfed and irrigated mixed systems:

- *Rainfed mixed farming systems* occur in temperate climates of Europe and the Americas and in subhumid climates of tropical Africa and Latin America. They are characterized by individual (often family) ownership, often with more than one species. Globally about 48% of beef production, 53% of milk production, and 33% of mutton production originates from this production system (Steinfeld et al., 2006a).
- *Irrigated mixed farming systems* prevail in East and South Asia, mostly in areas with higher population density. They are an important contributor to most animal products, with about one third of the world's pork, mutton, and milk production and about one fifth of the beef production.

Together the two mixed farming systems are without doubt the most important systems worldwide, as they produce globally about 90% of the milk supply, about 70% of sheep and goat meat and beef, 43% of pork,

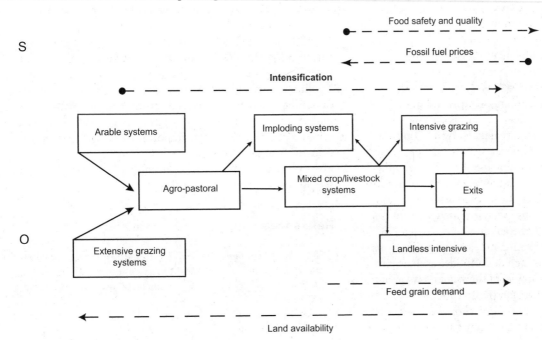

Figure 3.8. Schematic presentation of development pathways of main livestock production systems and some main drivers. *Source*: Adapted from de Haan et al., 1997.

25% of poultry meat, and practically all the buffalo meat in global production (Steinfeld et al., 2006a).

A schematized development path of the main production systems is presented in Figure 3.8, as changes in demand, consumer preferences, resource endowment, and technology cause a production system sequence to develop that moves from extensive crop and livestock systems through various levels of integration in mixed farming system to single commodity grazing or, even more important, to landless pig, poultry, or beef feedlot systems. The drivers described earlier also cause a differential growth of the different production systems. Data over the period 1981–91 (de Haan et al., 1997) showed 4.3% per year growth in meat production for the landless systems, compared with 2.2% per year for the mixed farming systems and 0.7% for the grazing systems. A more recent study covering the period 1992–2002 (Groenewold 2005) made the following observations:

- *Grazing systems have remained rather stable.* Over the decade 1992–2002, the total area under pasture increased by about 3%, mainly in South America in the Amazon area, where recent technology improvements have made the conversion of forest into pastureland for cattle ranching financially quite attractive, as noted earlier. This increase was partially offset by a reduction of the pasture area in the OECD, where pasture has been converted into forest areas as a result of subsidy policies and stricter environmental regulations. There was also a slight decrease in pasture area in sub-Saharan Africa as a result of increasing land pressure and conversion of rangeland into arable farming. For the future, little increase in productivity can be expected from the arid and semiarid regions. To the contrary, Steinfeld et al. (2006a) expect that arid and semiarid rangelands will be devoted increasingly to the provision of environmental services, such as carbon sequestration and biodiversity conservation.
- *Arable land area in mixed farming systems has increased.* The increase of 5% is a result of the conversion of pasture into arable farmland in the humid lowland and tropical highlands of Asia and Latin America, counterbalanced by a strong reduction in the area cultivated in east European and Central Asian countries following the collapse of the agricultural economy of planned economies there. Also notable is the strong increase in livestock populations and areas of fodder crops in the irrigated areas of Asia for milk production. The total area under irrigation increased by 20% in 1992–2002, and a significant part of that expansion now includes a fodder crop as part of the rotation, for example, in the winter season in the Punjab of India and Pakistan (Dost 2002).
- *Intensive, landless systems have shown the biggest increase.* As noted, there has been a major shift in the

meat sector toward monogastric (pigs and poultry) production. Total pork output rose by 24 million tons (30%) at the world level, and this increase can be found almost entirely in the increased output for Asia (20 million tons of additional production). Most regional groupings show increases in pig meat production, while for the east European and Central Asian region there is a drop of about 30%. The production of total poultry meat grew by about 75%, a much stronger expansion than for pork. Comparing growth between regions, there has been an extremely strong expansion in Asia, with about 150% production growth or 9% a year. All regions except the former Soviet countries show positive growth of between 2% and 10%. The production of table eggs grew by about 40% worldwide. Asia more than doubled its egg production in the period and produced about 50% of the world total. The landless system accounts for two-thirds of production and was growing by about 4% per year from 1992 to 2002 (Steinfeld et al., 2006b). While some of the booming growth in East Asia might level off, with the comparative advantage of such systems the consumer demand will make intensive production still the most likely major future growth area.

Conclusions

The major structural changes taking place in the livestock sector involve increases in enterprise size and the scope of production, accompanied by major shifts in production systems. Food safety and environmental regulations, which carry important economies of scale, push farm size up. In addition, the sector will face increasing competition for key natural resources of land and water for grazing and feed grain production as well as for fossil fuels, in particular for nitrogen fertilizer production. Within such an environment, the key strategies for the sector would initially be to intensify production and processing to capture economies of scale and reduce resource input levels per unit of product. A second strategy would be to apply "polluter pays and provider gets" principles to ensure sustainability, which will require strong control institutions and innovation in the area of payment for environmental services. Last, it would be important to avoid concentration by bringing livestock density into balance with the absorptive capacity of the surrounding land. Other chapters in this volume focus on this last aspect.

References

AERC (Agro-Economic Research Centre). 2005. *Consolidated report on holdings.* Department of Agriculture and Cooperation, Ministry of Agriculture, Government of India. Visvabharati: AERC.

Arthur, J. A., and G. A. A. Albers. 2003. Industrial perspective on problems and issues associated with poultry breeding. In

Poultry genetics, breeding and biotechnology, ed. W. M. Muir and S. E. Aggrey, 1–12. Wallingford, UK: CABI Publishing.

Asner, G. P., A. J. Elmore, L. P. Olander, R. E. Martin, and A. T. Harris. 2004. Grazing systems, ecosystem responses, and global change. *Annual Review of Environment and Resources* 29:261–299.

Bascou, P., P. Londero, and W. Munch. 2006. Policy reform and adjustment in the European Union: Changes in the common agricultural policy and environment. In *Policy reform and adjustment in the agricultural sectors of developed countries*, ed. D. Blandford and B. Hill. Wallingford, UK: CABI Publishing.

Berthelot, J. 2005. *The empty promise and perilous game of the European Commission to slash its agricultural supports*. Solidarité, NGO Position Paper. Geneva: World Trade Organization.

Bowyer, J., and V. E. Stockmann. 2001. Agricultural residues: An exciting bio-based raw material for the global panels industry. *Forest Products Journal* 51:10–21.

Bowyer, J., J. Howe, K. Fernholz, and P. Guillery. 2004. Paper from agricultural residues.

Breman, H., and C. T. de Wit. 1983. Rangeland productivity and exploitation in the Sahel. *Science* 221:1341–1347.

Brown, L. R. 2004. *Outgrowing the earth*. New York: W. W. Norton and Company.

Brown, L. R. 2006. Supermarkets and service stations now competing for grain. *Eco-Economy Update*. July 13. Washington, DC: Earth Policy Institute.

Bruinsma, J., ed. 2003. *World agriculture: Towards 2015/30, An FAO perspective*. Rome and London: Food and Agriculture Organization and Earthscan.

de Haan, C., H. Steinfeld, and H. Blackburn. 1997. *Livestock and the environment: Finding a balance*. Report of a Study Coordinated by the Food and Agriculture Organization of the United Nations, the United States Agency for International Development, and the World Bank. Brussels: European Commission Directorate-General for Development.

de Haan, C., T. Schillhorn van Veen, A. Brandenburg, F. Le Gall, J. Gauthier, R. Mearns, and M. Simeon. 2001. *Livestock development: Implications for poverty reduction, the environment, and global food security. Directions in development*. Washington, DC: World Bank.

Delgado, C. L., N. Wada, M. W. Rosegrant, S. Meijer, and M. Ahmed. 2003a. *Fish to 2020: Supply and demand in changing global markets*. Washington, DC, and Penang, Malaysia: International Food Policy Research Institute and WorldFish Center.

Delgado, C. L., C. A. Narrod, and M. M. Tiongco. 2003b. *Policy, technical, and environmental determinants and implications of the scaling-up of livestock production in four fast-growing developing countries: A synthesis*. Rome: Food and Agriculture Organization.

Dost, M. 2002. *Fodder production for peri-urban dairies in Pakistan*. Rome: Food and Agriculture Organization.

Dregne, H. E., and N. T. Chou. 1994. Global desertification dimensions and costs. In *Degradation and restoration of arid lands*, ed. H. E. Dregne. Lubbock, TX: Texas Technical University.

European Commission. 2008. Update on recent price developments in EU-27 agriculture and retail food. Directorate General for Agriculture and Rural Development, European Commission.

FAO (Food and Agriculture Organization). 2005. *Global forest resources assessment*. Forestry Paper No. 147. Rome: FAO.

FAO (Food and Agriculture Organization). 2006a. FAOSTAT statistical databases. http://www.faostat.external.fao.org.

FAO (Food and Agriculture Organization). 2006b. *World agriculture: Towards 2030/2050, Prospects for food, nutrition, agriculture and the major commodity groups*. Interim report. Rome: FAO.

FAO (Food and Agriculture Organization). 2007. *State of the world's forests, 2007*. Rome: FAO.

FAO (Food and Agriculture Organization). 2009. FAOSTAT statistical databases.

FAO (Food and Agriculture Organization). 2009. Food Outlook: Global Market Analysis, June 2009. Available at ftp://ftp.fao.org/docrep/fao/011/ai482e/ai482e00.pdf

Fischer, G., Shah, M., van Velthuisen, H., and Nachtergaele, F. O. 2001. *Global agro-ecological assessment for agriculture in the 21st century*. IIASA, Vienna, Austria, and FAO, Rome, Italy.

Fulton, L., and T. Howes. 2004. *Biofuels for transport: An international perspective*. Office of Energy Efficiency, Technology and R&D. Paris: International Energy Agency.

Gerber, P. 2006. Putting pigs in their place: Environmental politics for intensive livestock production in rapidly growing economies, with reference to pig farming in central Thailand. Doctoral thesis in Agricultural Economics. Zurich: Swiss Federal Institute of Technology.

Gerber, P., P. Chilonda, G. Franceschini, and H. Menzi. 2005. Geographical determinants and environmental implications of livestock production intensification in Asia. *Bioresource Technology* 96:263–276.

Goldewijk, K., and J. J. Battjes. 1997. *A hundred-year database for integrated environmental assessments*. Bilthoven, Netherlands: National Institute of Public Health and the Environment.

Groenewold, J. 2005 (unpublished). *Classification and characterization of world livestock production systems*. Rome: Food and Agriculture Organization.

Havenstein, G. B., P. R. Ferket, and M. A. Qureshi. 2003. Growth, livability, and feed conversion of 1957 versus 2001 broilers when fed representative 1957 and 2001 broiler diets. *Poultry Science* 82:1500–1508.

Henke, R. and D. Storti. 2004. Cap reform and EU enlargement: Effects on the second pillars endowments. Proceedings of the 87th EAAE Seminar. Assessing rural development policies of the CAP. Vienna, 21–23 April.

Hill, B. 2006. Structural change in European agriculture. In *Policy reform and adjustment in the agricultural sectors of developed countries*, ed. D. Blandford and B. Hill. Wallingford, UK: CABI Publishing.

Hurter, R.W. 1998. Will nonwoods become an important fiber resource for North America? World Wood Summit, Chicago.

IATP (Institute for Agriculture and Trade Policy). 2007. *The consequences of corn*. Ag Observatory. Paris: IATP.

IEA (International Energy Agency). 2005. Biofuels for transport: An international perspective. International Energy Agency, Paris.

IFPRI (International Food Policy Research Institute). 2006. Bioenergy in Europe: Experiences and prospects. In *Bioenergy and agriculture: promises and challenges*. Focus 14, Brief 9. Washington DC: IFPRI.

IFPRI (International Food Policy Research Institute). 2008. Biofuels and grain prices: Impacts and policy responses. International Food Policy Research Institute.

Kurukulasuriya, P., R. Mendelsohn, R. Hassan, J. Benhin, T. Deressa, M. Diop, H. M. Eid, et al. 2006. Will African agriculture survive climate change? *World Bank Economic Review* 20:367–388.

MA (Millennium Ecosystem Assessment). 2005. *Ecosystems and human well-being: Synthesis*. Washington, DC: Island Press.

Margulis, S. 2004. *Causes of deforestation in the Brazilian Amazon.* Working Paper 22. Washington, DC: World Bank.

McDonald, J. M., R. Hoppe, and D. Banker. 2006. Structural change in US agriculture. In *Policy reform and adjustment in the agricultural sectors of developed countries,* ed. D. Blandford and B. Hill. Wallingford, UK: CABI Publishing.

Middleton, N., and D. Thomas, eds. 1997. *World atlas of desertification.* Second edition, United Nations Environment Programme. London: Hodder Arnold Publications.

OECD (Organisation for Economic Co-operation and Development). 2005. *Agricultural policies in OECD countries: At a glance.* Paris: OECD.

OECD (Organisation for Economic Co-operation and Development). 2006. *Agricultural market impacts of future growth in the production of biofuels.* Working party on Agricultural Policies and Markets, Directorate for Food, Agriculture and Fisheries. Paris: OECD.

OECD-FAO (Organisation for Economic Co-operation and Development–Food and Agriculture Organization). 2008. *OECD–FAO Agricultural Outlook 2008–2007.* Paris: OECD.

RIVM (Rijks Instituut voor Milieu). 2004. *Minerals better adjusted: Fact finding study on the effectiveness of manure act.* Bilthoven, Netherlands: RIVM.

Scoones, I. 1994. *Living with uncertainty: New directions in pastoral development in Africa.* London: Intermediate Technology Publications.

Seré, C., and H. Steinfeld. 1996. *World livestock production systems: Current status, issues and trends.* FAO Animal Production and Health Paper 127. Rome: Food and Agriculture Organization.

Steinfeld, H., de Haan, C., and H. Blackburn. 1997. *Livestock and the environment: Issues and Options.* Brussels: European Commission Directorate General for Development.

Steinfeld, H., P. Gerber, T. Wassenaar, V. Castel, M. Rosales, and C. de Haan. 2006a. *Livestock's long shadow: Environmental issues and options.* Rome: Food and Agriculture Organization.

Steinfeld, H., T. Wassenaar, and S. Jutzi. 2006b. Livestock production systems in developing countries: Status, drivers, trends. *Revue Scientifique et Technique de l'Office International des Epizooties* 2:505–516.

Tucker, C. J., C. L. Vanpraet, M. J. Sharman, and. G. van Ittersum. 1985. Satellite remote sensing of total herbaceous biomass production in the Sahel, 1980–1984. *Remote Sensing of the Environment* 17:233–249.

USDA–ERS. 2006. Structure and finances of US farms: 2005 family farm report. USDA–ERS Economic Information Bulletin No. 12. Resource and Rural Economics Division, Economic Research Service, U.S. Department of Agriculture

USDA–FAS. 2004. *Biofuels in Germany: Prospects and limitations.* U.S. Department of Agriculture, Foreign Agricultural Service. GAIN Report GM4048. Washington, DC: USDA.

U.S. Soybean Export Council. 2006. Aquaculture marketing brochure: Aquaculture innovation in China.

USDA–FAS. 2005. *Oilseeds: World markets and trade—China's protein meal consumption rises led by soybean meal.* GAIN Report. Washington, DC: USDA.

USDA–FAS. 2006. *Brazil: Sugar.* GAIN Report. Washington, DC: USDA.

USDA–NASS. 2009. Agricultural prices. National Agricultural Statistics Service, USDA. Available at http://usda.mannlib.cornell.edu/MannUsda/viewDocumentInfo.do?documentID=1002

Van Beers, C., and A. de Moor. 2001. *Public subsidies and policy failures: How subsidies distort the natural environment, equity and trade, and how to reform them.* Northampton, MA: Edward Elgar Publisher.

Weihe, W. H., and R. Mertens. 1991. Human well-being, diseases and climate. In *Climate change: Science, impacts and policy,* ed. J. Jager and H. L. Ferguson. Proceedings of the Second World Climate Conference. Cambridge, UK: Cambridge University Press.

Westing, A. H., W. Fox, and M. Renner. 2001. *Environmental degradation as both consequence and cause of armed conflict.* Working paper prepared for Nobel Peace Laureate Forum participants by PREPCOM subcommittee on Environmental Degradation.

White, R. P., S. Murray, and M. Rohweder. 2000. *Pilot analysis of global ecosystems: Grassland ecosystems.* Washington, DC: World Resources Institute.

World Bank. 1991. *Brazil: Key policy issues in the livestock sector—towards a framework for efficient and sustainable growth.* Agricultural Operations Division, Report no 8570-BR Washington DC: World Bank.

World Bank. 2004. *Water resources sector strategy: Strategic directions for World Bank engagement.* Washington, LOGOBFH_SHL_MKPOS_24mm_e.tif

World Bank. 2005. *Managing the livestock revolution.* Report No 32725-GLB. Washington, DC: World Bank.

World Bank. 2006. *Aquaculture: Changing the face of the waters.* Washington, DC: World Bank.

World Bank. 2008. A note on rising food prices. Policy Research Working Paper No. 4682. Washington, DC: World Bank.

World Bank. 2009. *Minding the stock: Bringing public policy to bear on livestock sector development.* Report No. 44010-GLB. Washington, DC: World Bank.

4

Livestock in Geographical Transition

Pierre Gerber, Tim Robinson, Tom Wassenaar, and Henning Steinfeld

Main Messages

- **Livestock is a major user of land resources,** predominantly for fodder and concentrate feed production. On a global scale, however, meat and milk productions are growing faster than the pasture and feed crop areas.

- **"Location Matters"** when it pertains to environmental, social, and health issues. The spatial patterns of livestock production result from geographic gradients in production and distribution costs. They have implications on management practices and thus on the level of impacts of the systems (e.g., availability or absence of land in the vicinity of the animal production unit to recycle manure), as well as on the resulting implications on natural resources, human health, or social issues (e.g., presence of human settlements close to a manure discharge point).

- **While historically livestock value chains were relatively simple, the producer is now linked to the consumer via an increasingly complex network** of processors, transporters, marketers, and other agents—often over vast distances.

- **Animal food chains become longer,** involving more steps in the production chain and stakeholders who tend to relocate where operating costs are minimized, regardless of increased transportation distances.

- **Land use changes are characterized by a dichotomous trend:** land use intensification and clustering of operations are mainly associated with pig, poultry, and dairy production, while expansion (particularly in Latin America) and marginalization (in sub-Saharan Africa and Asia) are mainly associated with ruminant meat production.

- **Three major factors determine the transition:** cheap transport (although this is changing with increasing energy prices), differential resource prices, and relocation of the demand for animal products.

- **Livestock is moving from a "default land user"** strategy (i.e., as the unique way to harness biomass from marginal lands, residues, and interstitial areas) **to an "active land user" strategy**—competing with other sectors for the establishment of feed crops, intensive pasture, and production units.

Introduction

The first three chapters introduced the main drivers of change in the livestock sector and described changes from economic, production system, and institutional points of view. The livestock sector is growing and undergoing structural changes in order to meet surging demand from growing urban populations. Chapter 3 showed how the sector is changing in terms of species and products, level of intensification, size of operations, and degree of integration. This chapter assesses the geographical dimension of structural change, especially the changing geography of commodity chains and how these are being reorganized and relocated. It does not attempt to provide a comprehensive description of the sector's global geography but focuses on the geographical implications of structural change. We refer to this process as the livestock sector's "geographical transition."

A further objective of the chapter is to link earlier descriptions of how the sector is changing with assessments of the environmental, social, and public health impacts of such changes, which will be discussed in later chapters. Indeed, as shall be seen, "location matters" when it comes to environmental, social, and health issues.

Livestock's Geographical Transition

Relocation, Clustering, and Increasing Reliance on Transport

Historically, livestock value chains were relatively simple and short connections between the producer and the consumer. Production systems developed to link demand and resource availability within limited geographical areas.

Livestock have indeed traditionally been kept in close proximity to human settlements or in transhumant production systems. Monogastric species such as pigs and poultry have tended to be closely associated with human populations and raised in household backyards. Monogastric species depend on humans for their feed (e.g., household waste and crop residues and by-products) and for protection from predation. Ruminant species such as cattle, buffaloes, camels, sheep, and goats have traditionally been less closely tied to human settlements. This is largely because in traditional production systems they depend on grazing resources.

Currently, however, the producer is usually linked to the consumer via a much more complex network of processors, transporters, marketers, and other agents—often over vast distances.

Relocation of Feed and Animal Production

A first aspect of livestock's geographical transition, which is taking place on a global scale, is the emergence of new export-oriented zones and countries. Steinfeld and Chilonda (2006) have described the recent rise in production and trade of animal feed and livestock products in some major developing countries. These generally derive their competitive advantage either from cheap inputs, mainly labor and land (e.g., Brazil, China, India, and Argentina), or from good access to capital and technologies (e.g., Thailand and Brazil). Furthermore, they generally enjoy a favorable investment climate, strong transport infrastructure, and public policies that favor the livestock sector.

On the feed side, an example of rapid change in trade is the case of soybean. Soybean is widely consumed in the livestock industry but is supplied by only a few countries. The United States is traditionally the largest exporter of soybeans (29 million tons annually), followed by Brazil (17 million tons annually). Soybean exports from Uruguay and Argentina have soared over the past decades, however, making these countries major world exporters. This has been possible by taking advantage of the high availability of land, converting pasture into crops, and transferring technology from Brazil. Argentina has furthermore developed a strong milling capacity and now accounts for a large share of the soybean produced regionally.

Relocation of production is less pronounced for maize. While maize supply in Africa and Central America is predominantly dedicated to human consumption, a large proportion of the supply in Asia, Europe, and North America meets the demand for livestock feed. Overall, about 65% of maize production is dedicated to animal feeding. North and South America are the most significant maize-exporting regions. North America (the United States) has consolidated its number one position as exporting region, chiefly exporting to Central America (9.2 million tons in 2002). (Figures for 2002 in this section are averages for 2001 to 2003.) Exports from

South America have, however, increased dramatically in recent years (mainly from Argentina, Brazil, and Chile), supplying the Asian and European markets. China is another recent major maize exporter (ranking fourth in the world in 2005), mainly supplying Asian countries. The development of corn-based ethanol production will probably further change this overall picture in the short term, as the agroenergy sector diverts part of the maize supply from animal feeding and drives prices up. This is already observed in the United States. Midterm effects are less clear, however, as maize production will increase to match demand and cellulose-based biofuel processing technologies become available. (See Chapter 3.)

The global supply of poultry meat is dominated by three highly export-oriented regions. The two traditional ones—North America (3.5 million tons in 2007) and Europe (3.7 million tons)—and more recently, South America (3.7 million tons). New major exporters include Brazil, Argentina, Chile, Thailand, and China. Brazil is responsible for most exports. By combining relatively low feed grain and labor costs and by benefiting from ever-increasing economies of scale, Brazil's production costs for whole eviscerated chickens are kept lower than those of any major other supplier. In Asia, Thailand is a major exporter that has built its comparative advantage on cheap labor and advanced production and processing technology. In contrast, relocation of pig production is marginal, with the global supply being dominated by Europe and North America. Brazil is the recent exporter entering the top 10 ranking of largest exporting countries.

Oceania and South America are taking advantage of abundant land resources to supply the growing demand for beef and position themselves as world export leaders. North America is the main market for beef from Oceania (903,000 tons in 2002), but Asian imports have increased considerably in recent years. South American exports are mainly destined for Europe (390,000 tons in 2002) and Asia (270,000 tons in 2002), and volumes to both markets have increased greatly over the past 15 years. Europe and North America still make significant contributions to global bovine meat supply, based on more intensive production systems, especially in the United States, where large-scale feedlot production is widely practiced.

Clustering

Clustering occurs when there is a geographic concentration of production units. This gives rise to groups of interconnected producers, feed mills, slaughterhouses, and processing units, with the potential to generate collective actions in livestock production. The main drivers of clustering are the economies of agglomeration—the benefits that individual units obtain when they locate close to one another. Economies of agglomeration are related to economies of scale and of scope and to network effects (Fujita and Thisse 2004), all of which act to lower transaction costs. Basically, the more related units are clustered together, the lower the unit cost of production and the greater the market that individual units can sell into.

In the livestock sector, lower production costs are achieved through competition among suppliers of inputs (e.g., feed mills, veterinary and other services) and specialization and division of labor among producers (e.g., breeding operations, fattening operations, and contract farming). Minimization of transport costs, which are high for animal products, further drives the clustering of production units around slaughterhouses and dairy plants. This is amplified by the economies of scale in the postharvest sector, leading to large-scale plants that attract important numbers of producers.

The subnational trend toward clustering of landless production systems is ongoing in industrial as well as developing economies. An analysis of hen populations at the Municipio level in Brazil, for example, shows an increasing concentration from 1992 to 2001. (See Figure 4.1.)

In 1992, 5% of Brazil's area hosted 78% of the chicken population, while the same areas were home to 85% of the population in 2001. The concentration of pig production has also occurred over the same period, though not to the same extent. In 1992, 45% of the Brazilian pig population was found on 5% of the land area, but in 2001 this had risen to 56% of the population. Similar patterns have been observed in Thailand and France; for example, Figure 4.2 shows the concentration of pig production in France between 1989 and 2001.

One factor limiting the clustering of livestock production is the control of animal disease: to limit disease introduction, new large farms tend to locate away from other production units, be they large farms or small-scale units. This is the case for large pig operations in Southeast Asia, for instance, which locate in remote areas to prevent disease outbreaks despite increased operating costs.

Increasing Reliance on Transport
The livestock sector has taken full advantage of recent developments in transportation, which include improved

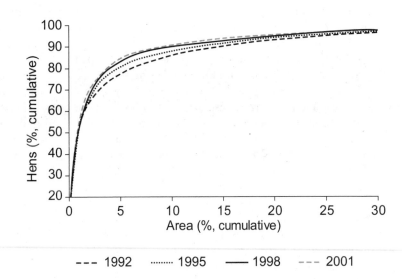

Figure 4.1. Changes in geographical concentration of poultry in Brazil from 1992 to 2001.
Source: Steinfeld et al., 2006.

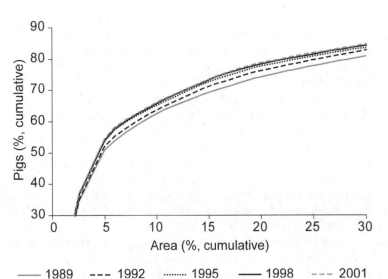

Figure 4.2. Changes in geographical concentration of pigs in France from 1989 to 2001.
Source: Steinfeld et al., 2006.

transport logistics, packaging, cold chains, and performance of transport vessels. Large-scale shipment of primary crop products and the consolidation of long-distance cold chains have played important roles in changing the shape of the livestock sector: transport of the sector's commodities has become technically possible and economically affordable.

Animal products are increasingly traded on international markets. This trend has been particularly strong for poultry meat, in which the traded proportion went from 6.5% in 1982 to 13.1% in 2002. Trade in poultry meat has overtaken trade in beef over the past 15 years, with the volume soaring from about 2 million tons in 1987 to 9 million tons in 2002, compared with 4.8 million and 7.5 million tons, respectively, for beef. With the exception of eastern Europe, all regions analyzed have become increasingly involved in trade in livestock and livestock products. In 2001–03, more than 12% of global bovine meat, poultry meat, and milk and more than 8% of pig meat was traded. These are significant increases over the amounts traded 20 years earlier. (See Table 4.1.) In addition to technical advances, the growth in trade was fostered by trade liberalization and harmonization of standards.

Trade in animal products is, however, less developed than trade in animal feedstuff in nominal terms—but also in relative terms for soymeal. About a quarter of global soybean production is traded as soymeal as a result of the concentration of production in a few countries. Cereals are traded in a lesser proportion but over much higher volumes. The difference between animal and feedstuff trade is mainly explained by the price difference of transport: the shipment of animal products relies on relatively expensive cold chains, while feedstuff can be transported by large bulk carriers (as described in the section on developments in transport). In most cases it is cheaper to transport feed and convert it into animal proteins close to the final consumer than to rear animals close to feed production areas and transport the animal products.

This general picture shown by global statistics is, however, susceptible to changes as cold chains and processing develop. The Brazilian model, for example, relies on the relocation of animal production close to feed mills and the export of animal products. Such a strategy also allows shipment of different cuts to the markets where they are most favored (e.g., chicken breasts go to the United States, legs to Europe, and wings to Asia) and thus a maximization of income per animal. Cargill, an international provider of feed and animal foodstuff, reported that the costs associated with transportation of both feed and animal products no longer represent a major constraint to the development of global food systems (Harlan, pers. comm., 2007).

Despite the increasing share of animal products entering international trade, most of the production is transported and consumed within national boundaries.

Subnational relocation of animal food chains and the development of specialized production areas result in localized supply/demand imbalances that need to be reconciled by transportation. To assess such transportation needs, supply and demand for selected animal feedstuff and products were compared globally at a 100 km resolution. Figures 4.3 and 4.4 in the color well display estimated spatial trends in feed surplus/deficits for pig and poultry, providing evidence of the sector's feed base activities concentration and its dissociation from animal production activities. Subnational imbalances (defined as more than 100 kg of cereals over/undersupply per square km) dominate for cereal, whereas countries' soybean import/export strategies translate into nationwide imbalances.

Figure 4.5 in the color well shows an estimate of the consumption/production balance of poultry meat. In most areas production levels are quite similar to those of consumption. A balanced situation (defined as ± 100 kg of meat per square km) is generally found in mixed and land-based systems. (See Chapter 3.) Areas with large surpluses are associated with a recent boom of industrial systems, whereas areas of deficit usually coincide with high population densities and rapidly expanding urban areas. The high levels of poultry exports from North and South America translate here in a dominance of red pixels (large surpluses) in these two regions, especially in export-oriented production zones.

The same analysis was conducted for pig meat. Figure 4.6 in the color well shows a similar pattern of surpluses in areas of industrial production. The geographical patterns of consumption/production balance differ, however, between the two livestock types in the level of entanglement between areas of surplus and deficit. With poultry, areas of surplus are generally more scattered among areas of deficit; with pig meat there tend to be large, solid blocks of surplus and deficit.

Figure 4.7 in the color well shows the important transportation associated with beef chains, at both national and international levels. Most beef is produced from roughage that, contrary to cereals and soybean, is expensive to transport. Production thus has to take place close to grass production, in areas where the opportunity cost of land is low and far from consumers. The surpluses area (red pixels) that is larger than for the other two commodities also indicates the relative lower land use intensity associated with beef production.

Intensification and Abandonment

The livestock sector uses land predominantly for feed production in the form of either pasture or arable land planted to feed crops. Previous studies (Steinfeld et al., 2006) have demonstrated the vastness of the area on Earth dedicated to the livestock sector. Today the livestock sector uses more than 3.9 billion hectares—about 30% of the world's entire land surface. The intensity with which this land is used varies greatly, however. It

Table 4.1. Global traded volumes versus production volume and global production volume, selected products

Product	Traded volumes vs. production (%)		Production (million metric tonnes)
	1981–83	2005–2007	2005–2007
Bovine meat	9.4	14.6	62.6
Pig meat	5.2	10.6	99.6
Poultry meat	6.5	13.1	84.7
Milk equivalent	8.9	13.5	664.1
Soybean	24.3*	27.0	217.7
Coarse grain	14.1	13.1	1028.8

*Soymeal trade over soybean production.

Data based on three-year average.

Source: FAO 2009.

is estimated that 500 million hectares are cultivated for feed crops (33% of all cropland), 1.4 billion hectares are high-productivity pastureland, and the remaining 2 billion hectares are extensive pastures with relatively low productivity.

While livestock production has expanded dramatically over past centuries to occupy such vast areas, recent trends show a tendency toward land use intensification. Most of the increase in demand for cereal feed has been met by intensification of cropland use rather than by an expansion of production area. The total supply of cereals increased by 46% from 1980 to 2004, while the area dedicated to cereal production shrank by 5.2%. In contrast, the growing soybean output has largely been achieved through the expansion of planted areas (multiplied by 7.5 over the past four decades), although yield increase has recently played a relatively more important role.

The crop intensification process, mostly driven by increasing pressure on land resources, has been aided by several technological gains, including the development of multiple cropping, the reduction of fallow periods, and the selection of higher yielding varieties (Pingali and Heisey 1999).

Where climatic, economic, and institutional conditions are favorable and where land is scarce, pasture production also becomes intensified. Such conditions are typically encountered in Europe, North America, Japan, and South Korea. In Europe, meat and dairy production units rely to a large extent on temporary pastures (leys) and the cultivation of forage crops for fresh and conserved feed. Intensification of pasture production on fertile land and degradation of marginal land are likely to continue (Asner et al., 2004). Developments in the subhumid areas of Africa, especially West Africa (Sumberg 2003), suggest a trend toward intensification—with crop–livestock integration on fertile soils with good accessibility and a progressive marginalization or even abandonment of the more remote areas. On the animal production side, improved feed conversion ratios (themselves achieved through improved breeding, feeding, and animal health), reduced feed waste, and the optimized use of carcasses have all contributed to increasing the overall output:feed ratio.

As a result of intensification on both the crop and the animal side, Figure 4.8a shows that, while the output per unit of land of meat and milk has strongly increased globally over the past four decades, there has not been a concomitant increase in land area dedicated to pasture and feed production. But there are regional variations of this global trend. In Europe, for example (Figure 4.8b), and more generally in countries that belong to the Organization for Economic Cooperation and Development (OECD), growth in meat and milk production has actually been accompanied by a reduction in pasture and feed crop areas. Part of this reduction may be explained by feed imports, in particular from South America, and indeed Figure 4.8c shows a continued increase in area under feed crops there. The area increased especially rapidly in the 1970s and late 1990s when industrial and developing countries successively engaged in livestock industrialization and started importing protein feed. A general pattern of intensification in livestock production is demonstrated in East and Southeast Asia (See Figure 4.8d). In these regions, livestock production has grown dramatically since the 1980s; however, areas under feedcrops and pastures have not increased at all. These differences in growth rates resulted not only from importation of feed resources but also from a dramatic intensification of the livestock industry, involving a shift to poultry, breed improvement, animal health protection, and enhanced animal husbandry.

In regions where land use intensification has been particularly strong and the demand for agricultural products relatively stable, the area under agricultural use

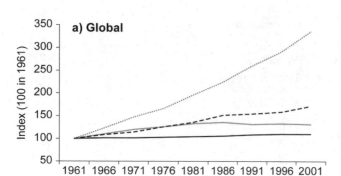

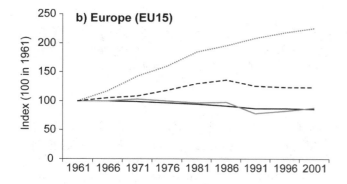

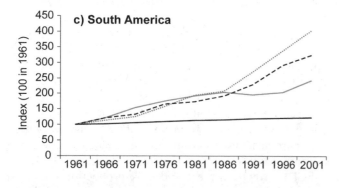

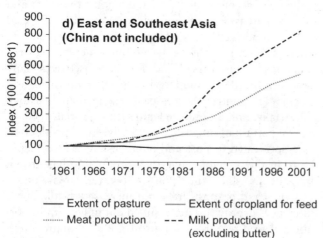

Figure 4.8. Trends in land use area for livestock production, compared to local supply of meat and milk (figures indexed to 100 in 1961).

Source: Steinfeld et al., 2006.

has actually started to shrink. Since 1950, for example, the extent of grazing land has decreased by 20% in the United States. The process of turning land that was previously used for agriculture back into forest has been named the "forest transition," which describes a historical relationship between economic development and forest cover as it occurred over the past two centuries in Europe and North America (Walker 1993). Land abandonment has tended to occur particularly in remote areas with poor soil characteristics, while more productive land with good accessibility has remained in production and intensified. As abandoned land has reverted to its natural vegetation cover, this has led to net reforestation in parts of Europe and North America since the end of the nineteenth century (Rudel et al., 2002). Forest transition is an ongoing development in Europe and North Africa, and Rudel et al. (2002) have even found evidence of similar patterns in some tropical forests. Areas of net forest gain are observed in the United States, southern Brazil, Europe, China, and Japan.

The human population is expected to grow by about 50%, reaching 9.2 billion people by 2050 (United Nations 2008). And the demand for animal products is expected to double by 2050 (FAO 2008). On the one hand, even if these projections may have to be revised downward, animal production can still be expected to grow and thus further increase land demand by the sector. On the other hand, continuous intensification—also fostered by the development of agrofuels, as mentioned earlier—will reduce the area of land used per unit of output. The balance of these two trends will ultimately determine the direction of the total area used for livestock. Based on past experience, it seems likely that the total land requirements of the sector will increase for some time and then begin to decline, with grazing area first starting to shrink, followed by a reduction of land required for feed crop production. In the mid- to long term, the cellulose-based energy sector may even outcompete livestock for access to forage and thus accelerate the reduction of pastureland.

Expansion and Mobility

Expansion of land under arable agriculture is likely to continue to be a contributing factor in increasing feed crop production, although less important than intensification. However, the potential to expand cropland in order to produce more grains and soybean varies widely around the world—it is quite limited in South and Southeast Asia (Pingali and Heisey 1999), but it exists in most of other continents, especially in Latin America and Africa. The projected contribution of arable land expansion to crop production from 1997 to 2030 is estimated to be 33% in Latin America and the Caribbean, 27% in sub-Saharan Africa, 6% in South Asia, and 5% in East Asia (FAO 2003).

Exploring options for pasture expansion, Asner et al. (2004) suggest that the expansion of grazing systems into marginal areas is already limited by climatic and edaphic factors. Any significant increase in grassland could therefore only take place in areas with high agroecological potential at the expense of cropland (highly improbable) or through the conversion of forests to pasture, as is currently happening in much of the humid tropics. Pastures are, however, on the increase in Latin America and in Africa, where the process of land colonization is still ongoing. The pace of pasture expansion into the forest will depend predominantly on macro- and microlevel policies in these areas. An opportunity for pasture expansion also exists in transition countries, where large expanses of grassland have been abandoned but could be recolonized at relatively limited environmental cost; such areas may, however, also be targeted by bioenergy companies for biomass production, and livestock producers could eventually lose out. In OECD countries, total pasture area will most likely decline as pastureland is converted to cropland, urban/industrial areas, and conserved or recreational areas. In the European Union this trend is constrained by public policies limiting conversion.

Pastures are also increasingly losing ground to cropping on most fertile soils and in prosperous climates. The trend is already prevalent in a number of places, in particular in sub-Saharan Africa and Asia, fueled by an increasing demand for grain. Recent trends in grain prices are likely to accelerate the process. Similarly, urbanized areas are likely to continue to encroach into pastureland, especially in areas with booming population and income growth. Encroachment is particularly harmful to pasture-based systems as it takes away the most productive land. Moreover, in arid and semiarid pastoral areas encroachment can have an impact that far exceeds that of the actual area encroached on, since migration routes are often cut off, isolating pastoralists from dry-season water and grazing resources (Boutrais 2007). Apart from the productivity losses this causes, it can also give rise to land degradation, since animals are increasingly concentrated on limited dry-season resources. Thus relatively small encroachments can wreak havoc on pastoral ecosystems, which can sometimes lead to violent conflict, as exemplified in Niger, Tanzania, and other countries (Mbonile 2005), where the troubles are largely attributed to encroachment of pasture resources for cropping and conflict over access to water resources.

Climate change is predicted to alter grassland-based systems and to have a greater impact on natural grasslands than on crops: in the latter, growing conditions can be manipulated more readily (e.g., through irrigation or wind protection), while the former are situated in more marginal areas where current climatic constraints are often already high. In dry areas the outcome is projected to be dramatic. Results from a case study in Mali (Butt et al., 2004) indicate that climate change will reduce forage yields by between 16% and 25% by 2030, while the yield for the most affected crops would decrease by 9%

to 17%. In contrast to these dry areas, it is predicted that pastures located in cold areas will show increased yields, benefiting from rising temperatures (FAO and IIASA 2000).

The trends observed for extensive systems are more complex than those for intensive production, and so is the nature of the geographical transition. In a nutshell, we observe the expansion into productive land with low opportunity cost in Latin America, the loss of most fertile pastures and the limitation of mobility in sub-Saharan Africa and Asia, and the recovery of abandoned grassland in transition economies.

Determinants of the Geographical Transition

The livestock geographical transition has two major aspects: one related to the transformation of animal food chains, the other to the way the livestock sector uses the land. Animal food chains become longer, involving more steps and stakeholders who tend to relocate where operating costs are minimized, regardless of the increased transportation distance this may imply. Land use changes are characterized by a dichotomous trend: intensification and a clustering of operations are mainly associated with pig, poultry, and dairy production, while expansion (in Latin America) and marginalization (in sub-Saharan Africa) are mainly associated with ruminant meat production.

Three major factors determine this transition: cheap transport, differential in resource prices, and relocation of the demand for animal products. The redistribution of production chains relates first and foremost to transportation costs and to the price differential between animal products and feedstuff transportation costs. First, relatively low private costs of transportation (although this may be changing, as discussed later) allow for a geographical decoupling of the consumer from production and thus can meet increasing urban demand. Low transport costs also allow for the relocation of each step of the animal food chain, from feed production to slaughtering and processing, in order to minimize production costs. With the cost of connecting each segment being limited, other factors play a greater role in determining the best location for each component of the chain, including the cost of natural resources, labor, energy, livestock services, taxes, and compliance with local regulations. This is especially the case for the feed industry, benefiting from particularly low transport costs. In other terms, cheap transport and the globalization of animal food chains mean that agricultural land use opportunities compete increasingly across continents, and food chains can virtually compete with others across the globe. Some constraints, however, limit what would be a purely economically driven distribution of feed and animal production: social and nontariff barriers to trade, disease control policies, and quality standards mold the final pattern of production.

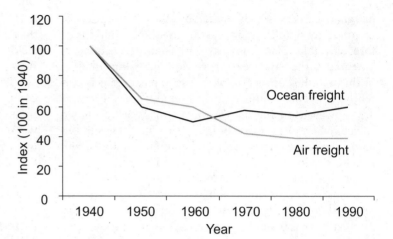

Figure 4.9. Decrease in the cost of transport and communication by mode, 1940–1990. Values expressed as indices.
Source: Molle 2003, Global Economic Institutions. London: Routledge, 2003.

Developments in Transport

Long-term trends show significant decreases in freight costs (see Figure 4.9), which have been achieved largely through improvements in technology and equipment, better logistics (freight optimization, automation), increased safety, and the substantial economies of scale associated with overall increases in the volume of shipped goods.

It is particularly the development of large bulk carriers—merchant ships used to transport unpackaged loads such as cereals—that has drawn feedstuff transportation costs down. Bulk carriers began to appear in the mid-nineteenth century and have steadily grown in sophistication. Today, over 5,500 bulk carriers make up a third of the world's merchant fleet. Cereals represent about 10% of the yearly volume transported by bulk carriers, over an average distance of 11,000 kilometers. The size of bulk carriers has also increased substantially, with large vessels (capacity greater than 60,000 tons) accounting for 1% of the fleet in the 1970s but 52% in 1996 (Duron 1999). As a result, marine transportation has become the cheapest modality of transportation over long distances. (See Figure 4.10.) In the context of poor infrastructures,

however, transporting the grain from international ports to the final purchaser substantially increases overall transportation costs. For example, in 2007, it cost on average $45 to transport a ton of grain by ship from any U.S. port to Mombasa, Kenya, but $1,500 to transport the same weight from Mombasa to Mbara, Uganda.

Transportation along cold chains is even more costly. Although data on this are relatively scarce, a transcontinental shipment of frozen food seems to cost on average about four times as much as transporting the same weight with unrefrigerated bulk carriers. Steep reduction of cold transportation has been achieved over the past through technology and logistics development, cheap energy, and economies of scale: in 2002, an estimated $1,200 billion worth of food was transported by a fleet of 400,000 refrigerated containers (Rodrigue et al., 2006).

If we assume feed conversion ratios of 2:1 for poultry, 3:1 for pork, and 7:1 for beef (feedlot), it appears preferable to transport frozen beef than grain, but more profitable to transport grain than chicken meat. The situation is more balanced for pork. Many other factors influence the balance, however, such as the synergies

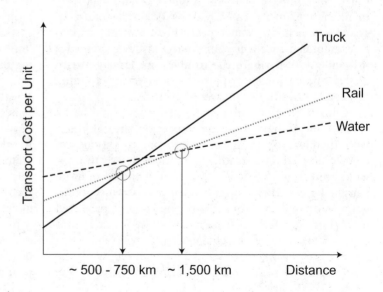

Figure 4.10. Relationships among transport cost, distance, and modal choice.
Source: Sirikijpanichkul and Ferreira 2006.

that can be developed locally with other sectors, allowing reductions in transportation costs, or the possibility of maximizing income from selling animal products by selectively exporting cuts or dairy products to the countries where prices are highest.

In recent years, however, the long-term trend toward decreasing transport costs has been reversed. Increasing fuel prices, labor costs, and the cost of insurance (due to increasing insecurity) have pulled freight prices up. The congestion of ports and, in consequence, the longer freight routes have further drawn prices up. For example, the cost of shipping one ton of frozen beef from Oakland, California, USA, to Osaka, Japan, rose from $166 in 1997 to $208 in 2003 (USDA–AMS 2006). Such changes have already contributed to modification of transportation patterns, with some countries choosing to import agricultural products from close suppliers in order to minimize transportation costs.

Resource Price Differential

In general terms, production systems develop in order to meet the demand for animal products most efficiently, and production chains become organized and positioned in order to minimize production and delivery costs. For a given technological context, production costs vary with input costs. Feed is the major input to livestock production, followed by labor, energy, water, and services. Input costs vary substantially from place to place within countries as well as across continents. Access to technology and know-how is also unevenly distributed, as is the ability to respond to changing environments and to market shocks. There are also multiple policy, institutional, and cultural patterns that further affect production costs, access to technologies, and transaction costs. The combination of these factors shapes the economic landscape within which commodity chains become distributed to minimize their overall cost, in a context of cheap (although recently increasing) and safe transportation.

Natural Resources

Livestock production is based on land for the production of feed, either in the form of cropland (concentrate feed) or pasture. Today, the livestock sector is a major land user globally. Access to land is becoming an acute issue and an increasing source of competition among the various stakeholders. Access to land has been the motivation for conflicts and wars throughout history, and in some areas resource-related conflicts are increasing. Westing et al. (2001), for example, report disputes over access to renewable resources, including land, to be one of the principal pathways by which environmental issues lead to armed conflicts. This may result from reduced supply (because of depletion or degradation), distribution inequities (such as encroachment of communal pastureland for crop farming), or a combination of these factors. Increasing land prices also reflect increasing competition for land, as seen, for example, in Europe, the United States, and Asia.

Water availability is a sine qua non of livestock production, which varies widely with location, though the price of water rarely reflects the scarcity of the resource or the pollution issues associated with its use (Steinfeld et al., 2006). Water availability thus determines the area suitable for animal production, but it does not substantially alter production costs.

Labor, Energy, and Services

The share of labor in total production costs depends not only on the level of wages but also on labor productivity, which is to a large extent determined by the production technology used. Figure 4.11 shows how labor costs can vary strongly between countries; wages in Mexico, China, and Brazil, for example, are on a different scale than those in Canada and the United States. Even within the eurozone wages differ greatly: gross annual average salaries for agricultural workers in the Netherlands and France are four times and two times greater than those in Spain (EUROSTAT database 2003). Despite the fact that production units compensate for high wages with automation and improved labor productivity, such large differences in labor costs remain prominent in overall production costs structure, as shown in Box 4.1.

In an increasingly open world, access to technology is heavily dependent on access to capital. The poultry sector, for example, shows remarkable standardization of technology worldwide among "high-tech" operations; access to production technology is mostly dependent on producers' ability to afford it. Gerber (2006) documented

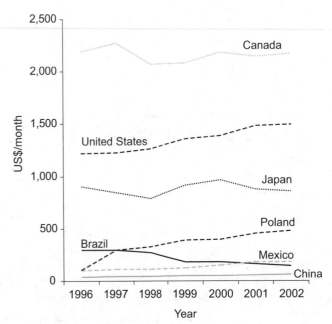

Figure 4.11. Cost of agricultural labor in selected countries (dollars per month).
Source: LABORSTAT–ILO 2008.

Box 4.1. Compared Pig Production Costs

Hoste and Backus (2003) have compared pig production costs in six major production countries: the Netherlands, Brazil, Canada, China, Poland, and the United States. The analysis was based on typical farm types that are not necessarily representative of pig production in the country but yet are those likely to compete in international markets and have some influence on national competitiveness. (Data availability was also a criterion for selecting farm types.) Farm sizes varied between 500 sows (Brazil and the Netherlands) and 800 sows (United States).

The analysis explored how differences in production technology and slaughter weight combine with differences in input costs to give a contrasting pattern of production costs. For example, in China low wages compensate for a highly labor intensive production, resulting in a limited contribution of labor to production costs. In the Netherlands, relatively high feed costs are compensated for by high feed conversion ratios.

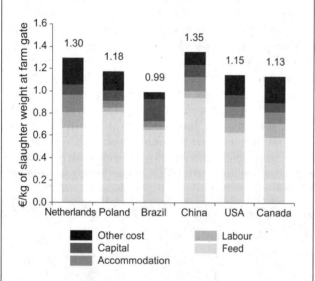

Box Figure 4.1. Compared pig production costs. Captial, accommodation, labor, and feed costs indicated for six major production countries: Netherlands, Poland, Brazil, China, United States, and Canada.

the positive impact of access to technology on the financial and environmental performance of pig farmers in Thailand. Access to capital is clearly unevenly distributed globally, being relatively good in OECD countries and near the growing urban centers of some developing countries but still very limited in much of the developing world.

Intensive livestock operations require access to specialized services, such as in animal health. The costs of drugs and veterinary services vary greatly but are estimated at about 1% of overall production costs in finishing pig operations, for example, and 2% in birth-to-finish operations in Thailand (Gerber 2006). Despite these relatively low costs, access to such services is a prerequisite for intensive operations, which simply cannot operate where such services are not available.

Effects of Policies

Livestock public policies are forces that can add to the drivers described earlier to influence changes in the sector with the aim of achieving particular societal objectives. Public policies can be both drivers of and responses to changes in the livestock sector. The FAO Pro-Poor Livestock Policy Initiative has developed a framework by which policies that are relevant to the livestock sector can be broadly categorized into three groups:

- *Policies addressing the vulnerability of small and poor producers.* These include the institutional and regulatory frameworks regarding the ownership and access to land, forage/feed, and water resources, particularly important for mobile forms of livestock production such as pastoralism and also for the capacities of livestock producers to respond appropriately to shocks and emergencies such as disease, drought, and flooding.
- *Policies creating the conditions for livestock sector development and growth.* These are largely involved in promoting access to input and output markets and are particularly relevant in rapidly growing economies. Included in this category are policies controlling access to secondary inputs, marketing policies, policies on the control of certifiable diseases, and policies determining infrastructural developments. Also important are regulations on foreign direct investment, labor policies, and in general all policies that affect the investment climate of a country.
- *Policies sustaining growth in industrializing and industrialized contexts.* At the more sophisticated end of the spectrum are trade policies (particularly those that regulate livestock markets), policies on food safety and quality, environmental policies, and those related to investment in research. Many of these have a direct impact on competitiveness and access to national and international markets. Included in this category are the incentive frameworks that further shape countries' relative competitiveness, such as decoupled farm subsidies, nontariff trade barriers, and legislation on intellectual property rights.

The influence of these public policies on the structure and distribution of value chains may be direct (in which the underlying legislation may prohibit the development of some activities in certain areas) or, more often, indirect, by altering the economic landscape in particular ways—for example, by providing tax incentives to develop livestock production in certain areas.

Relocation of Demand

More Consumers, in New Countries

On average, per capita consumption of animal-derived foods is most dynamic among lower- and middle-income groups, in areas that are experiencing strong economic growth. (See Chapter 2.) These populations are mostly found within the rapidly growing economies of Southeast Asia, coastal provinces of China, parts of India and Brazil, and the coastal urban areas of West Africa.

Table 4.2 provides an overview of the important changes that have occurred in the average diets of people in various world regions. People in industrial countries derive more than 40% of their dietary protein intake from food of livestock origin (these figures do not include fish and other seafood), and little change occurred in this situation between 1980 and 2002.

Changes have been most dramatic in Asia, where the total protein supply from livestock increased by 140% over this 22-year period. Because of mere size and strong income growth, China has been a main driver of change in Asia. In China, per capita consumption of animal source foods increased at yearly rates of 5.6%, 8.0%, and 8.1%, respectively, for meat, milk, and eggs in the last decade (FAO 2008). Growth in consumption is poised to continue, although at a slower pace. India also contributes increasingly to the changes in Asia as it is entering a dietary transition, with higher per capita intake of animal products, especially poultry.

Latin America, where per capita animal protein intake rose by 31% between 1980 and 2002, is the second biggest contributing region to consumption growth in the developing world. Per capita consumption also increased in industrial countries, although at a slower pace. Here, livestock's contribution to overall protein intake has been stable, and demand has shifted from red to white meat. In contrast, there has been a slight decline in consumption in sub-Saharan Africa, reflecting economic stagnation and a decline in available income.

Human consumption of poultry meat, pork, and beef is displayed in Figure 4.12 in the color well. The geographical distribution of the demand for animal-derived foods broadly follows that of human population. (See Figures 4.12 and 4.13 in the color well.) The pattern of geographical demand varies among species, however; pork consumption, in particular, is strongly influenced by religious and cultural factors. Patterns of poultry and beef consumption maps are similar, except for some areas such as Argentina and central and eastern Africa, where beef consumption is relatively high compared with overall consumption. Figure 4.12 illustrates the concentration of consumption in urban centers, where population growth, increasing purchasing power, and changing consumption patterns coincide to fuel a growing demand. Rural areas characterized by high consumption levels are found in China, India, OECD countries, and East Africa, fostered by population growth and increasing wealth.

Urbanization

Demand for livestock products increasingly concentrates in urban centers. Over the past two decades urbanization has been increasing in all regions of the world, and by 2006 the urban population was estimated to equal that of rural areas (FAO 2003, 2008). This trend has been especially strong in East Asia and the Pacific, coupled with a strong increase in gross domestic product (GDP) per capita. Urbanization has also been the main driver of demand in sub-Saharan Africa, although decreasing GDP per capita has undermined its effect on overall demand growth. The production and sale of food and manufactured goods in the region are increasingly driven by urban markets. In Latin America and OECD countries, the economic drive of urban markets is well established,

Table 4.2. Livestock and total dietary protein supply, 1980 and 2002.

Region	Total protein intake (g/person)		Share of protein intake supplied by livestock products (%)	
	1980	2002	1980	2002
Sub-Saharan Africa	53.9	55.1	19.3	16.9
Near East	76.3	80.5	23.9	22.5
Latin America	69.8	77	39.4	44.3
Developing Asia	53.4	68.9	13.1	23.5
Industrial countries	95.8	106.4	53.0	52.7
World	66.9	75.3	29.9	32.3

Source: FAO 2008.

with 77% of the population living in urban areas. There, too, urbanization continues, albeit at a slower pace.

The trend toward increasing urbanization is predicted to continue over the coming two decades, especially in regions with low existing levels, bridging the gap with those already well advanced on that path. (See Table 4.3.) By 2030 the Middle East and North Africa will have joined Latin America and OECD counties in the group of regions with more than 80% of the population living in urban areas. East Asia and the Pacific will come next, with more than 60% of the population living in towns and cities. In China, where the total volumes of meat consumed in urban and rural areas are currently similar, it is estimated that by 2030 urban demand will be more than twice that of the rural areas.

With high rates of consumption, rapid growth rates, and a shift toward animal-derived foods, urban centers increasingly drive the sector. This will have significant implications. First, a higher share of the production will have to be transported to the consumer, and over greater distances. Second, as urban areas grow so does the food production area on which they depend, although probably not in a linear relationship, as land use intensification of production also occurs. Demand growth in urban areas also fosters the internationalization of animal food chains: Urban centers are well served by transport infrastructure and often located in coastal areas, and they are thus able to import animal products at relatively low cost. These centers are often supplied by modern retailers that are generally connected to international providers. This results in the increased competitiveness of remote and local suppliers.

While urbanization will be a major driving force, rural demand for animal products will continue. This is especially significant in densely populated areas and in regions characterized by high GDP per capita. In Southeast Asia, especially eastern China, the coincidence of high rural population densities, relatively high income per capita, and a taste for animal products results in substantial rural demand. Similar trends are observed in South America and central Europe.

The demand for livestock products in rural areas can potentially be met through local production. This becomes difficult, however, when land is in short supply and insufficient for the production of feeds. This is the case in eastern China, for example, which compensates for land shortage by importing feedstuff. Importation of feed is, however, often limited by a poor transportation infrastructure, as in much of Africa, which increases delivery costs further.

Regional Patterns

Livestock's geographical transition is part of the structural changes that the sector is undergoing, which are happening at different speeds and with different features around the world. This section looks at regional patterns of the geographical transition and the links between urban markets and local production of livestock.

As transport infrastructure is a core determinant of the geographical transition, we have estimated human population, livestock populations, and cereal production at different distances, expressed as travel time, from urban centers (defined in this case as having at least 250,000 inhabitants). The study was conducted for

Table 4.3. Regional trends in urbanization rate and GDP per capita, 1980–2030.

Region	Present GDP/ cap. (USD)	Present Urban pop. (%)	Past trends GDP/cap. growth rate (%)		Past trends Urbanization* growth rate (%)		Future trends GDP/ cap. growth rate (%)	Future trends Urbanization* growth rate (%)
	2005	2005	1980–90	1990–2000	1980–90	1990–2000	2000–30	2000–30
East Asia and Pacific	1,355	41	5.8	7.1	2.9	2.5	5.3	1.7
Latin America and Caribbean	4,044	77	−0.9	1.7	0.9	0.6	2.3	0.4
Middle East and North Africa	1,780	57	0.0	1.8	1.0	0.7	2.4	1.6
South Asia	566	28	3.4	3.2	1.1	0.9	4.7	1.4
Sub-Saharan Africa	569	35	−1.0	−0.3	1.8	1.6	1.6	1.5
OECD	29,251	77	2.5	1.8	0.3	0.4	1.3	0.3

*Share of population living in urban areas.

Sources: World Bank 2006, *World Development Indicators*, for GDP per capita, urbanization, and total population (past and present data); FAO 2003 for projections; FAO 2008 for data on population projections.

a set of regions and countries of similar size and representing different stages in urbanization, agriculture, and transport development. Travel time was estimated by combining data on transport infrastructure, land use type, and topography using specialized routines within a geographic information system (Nelson 2007). The livestock density coverage used for the analysis was based on linear multivariate models produced at FAO (Wint and Robinson 2007), which include human population as one of the predicting variables. This may include an artifact in the results, and especially in the relation between human population and poultry population. This problem is, however, limited by the fact that the animal population regressions were used to redistribute census data within administrative boundaries that are smaller than the regions/countries used for the analysis.

Figure 4.14 presents the results of the analysis. At low levels of urbanization, the market is not dominated by urban demand, and the distribution curves of poultry, cattle, and pigs are relatively flat. This is the case, for example, in India (Figure 4.14a). Livestock populations do not appear to be concentrated in periurban areas except for a slight increase in poultry densities within two hours of urban areas, which equates to an average distance of 85 km. Livestock densities even tend to increase in areas more than seven hours away from urban areas (distance of 160 km), with some cereal production increase.

Where urbanization and economic growth translate into "bulk" demand for animal food products, specialized operators emerge in the vicinity of urban centers. Livestock products are among the most perishable goods, and their preservation as they pass along the value chain requires appropriate transportation, high levels of hygiene, chilling, and other forms of conservation. Where such infrastructure is not in place, livestock-derived foods have to be produced close to where they are consumed. Figure 4.14b illustrates this pattern for poultry in sub-Saharan Africa. Relatively poor transport and marketing infrastructures combined with strong demand from urban areas result in increasing poultry densities as the travel time to urban areas decreases. Conversely, cattle densities do not respond as strongly to travel time to urban centers in sub-Saharan Africa: densities dip in the immediate periphery of urban areas, then increase to a fairly constant level at travel times between 1 and 4.5 hours, after which they gradually tail off. This pattern may reflect the generally poor connection of extensive livestock production systems to urban markets, while the dwindling population close to urban areas is possibly caused by competition for land and the relatively low returns to extensive grazing.

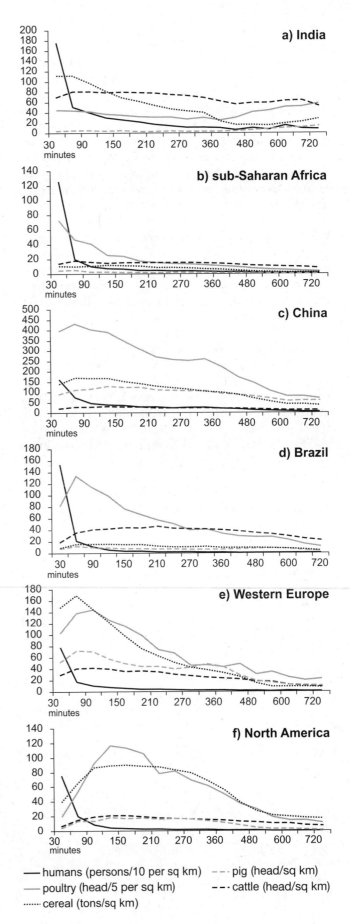

Figure 4.14. (a–f). Average densities of people, livestock species, and cereal production. Depicted at a series of travel times from urban centers in selected regions.

Where marketing and transport infrastructure are sufficiently developed, it becomes technically feasible and economically viable to keep livestock at greater distances from demand centers. Livestock production then shifts away from urban areas, driven by a series of factors such as lower land and labor prices, better access to feed, and in some cases by policy measures such as lower environmental standards or tax incentives in remote areas. This occurs, for example, in China and Brazil (see Figures 4.14c and 4.14d), where poultry density peaks in areas about one hour away from the urban centers (a distance of 50–55 km). Poultry densities in China are much higher than those calculated for the other regions, which may explain the limited decrease at the immediate periphery of the urban areas in China, despite relatively poor market infrastructure. Similar patterns are seen for pig production, though peaks occur farther away from the urban areas. In Brazil, the increases (high peaks) observed for poultry densities just outside the urban areas may be caused by the large export market. Production for export requires access to transport infrastructure and labor, thus drawing it toward the urban centers. Cattle densities in Brazil display a similar but more pronounced pattern to that of sub-Saharan Africa: densities are low in the direct vicinity of urban areas, gradually increase as travel time increases, and plateau between about three and five hours from the urban areas before tailing off—a probable result of the export-oriented nature of the production, requiring access to transport infrastructure.

Illustrating these trends over time, Figure 4.15 shows how livestock production has relocated away from urban areas in Thailand. Poultry densities in areas less than 100 km from Bangkok decreased between 1992 and 2000, with the largest decrease (40%) in the areas close to the city (less than 50 km). Beyond 100 km from urban areas, densities increased. In this particular case, the geographical shift was induced by tax incentives and improvement of transport infrastructure. Increasing land scarcity in the vicinity of Bangkok has further contributed to this geographical shift.

In OECD countries, where industrialization of livestock production occurred between the 1950s and 1970s, specialized areas became established in rural areas where there was a surplus of cereal supply. Here, livestock were initially produced as a means of diversification and value addition for existing farmers. In Europe, pig and poultry production clusters of this type include Brittany, the Po valley, western Denmark, and the Flanders. The geography of these clusters became further influenced by the increasing use of cheap, imported feed. Those well connected to ports gained in strength (e.g., Brittany, western Denmark, and Flanders in Belgium) and new production areas appeared in the vicinity of major ports (e.g., lower Saxony in Germany, the Netherlands, and Catalonia in Spain). The recent increase of feed prices does not seem to have affected the comparative advantage of these import-dependent areas, given the globalized nature of grain and soy prices.

This trend is illustrated by Figure 4.14e for western Europe and Figure 4.14f for North America. Both figures indicate a spatial correlation between cereal production and livestock production. In particular, poultry and cereal production follow similar distribution patterns. They grow as travel time to urban areas increases, reaching a maximum between an hour and an hour and a half away in western Europe (average of 60–80 km) and between an hour and a half and four hours in North America (average of 80–185 km) and then decreasing further away from urban centers. The greater distance in North America may be explained by greater availability of space and considerably lower costs of road transport. Pig densities follow a similar pattern, although the differences in densities at different travel times are much less. In western Europe there is a smaller peak in poultry and pig densities located about five hours away from urban centers (average of 190 km). These may relate to the formation of remote production clusters in environmentally less sensitive areas. Cattle densities also reach their highest values where cereal production is highest, although peaks are less pronounced than those for monogastric

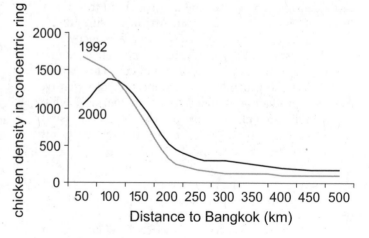

Figure 4.15. Changes in periurban geographical concentration of poultry in the vicinity of Bangkok in 2000.
Source: Gerber 2005.

species. This geographical pattern differs from those observed for India, Brazil, and sub-Saharan Africa and is probably caused by a greater share of the animal population kept in intensive systems (e.g., feedlots and dairy farms), relying on grain and imported feedstuff. These bands of relatively high cattle densities around urban centers may also be the relics of the historical "dairy belts" that used to supply city dwellers.

Conclusions

Today the livestock sector is a major user of land resources, and its land requirements are expected to grow, driven by increasing global demand for livestock products. As this sector develops, not only do land requirements grow, but the intensity of production increases and the geography of livestock production changes.

Feed crops and cultivated pastures are becoming intensified in areas with well-developed transport infrastructure, strong institutions, and high agroecological potential, while extensive pastures are tending to expand into natural habitats with low opportunity costs. Even as requirements for livestock feed and pasture continue to grow, it is likely that the bulk of the growth in land area turned over to both pasture and feed crops has already occurred. This occurs through strong intensification of the livestock sector on a global scale, with a shift from ruminant species to monogastric species. The recent development of agrofuels could play a critical role in these developing patterns.

The geographical pattern of production also changes. Production and consumption can no longer coincide geographically because consumption is located increasingly in rapidly growing urban centers—far removed from the primary natural resources required to raise livestock. Taking advantage of better and cheaper transport facilities, the sector meets demand through increasingly longer and complex food chains. Each specialized segment of the animal food chain tends to relocate to minimize its own costs. Production of grazing stock (ruminant livestock) remains linked to the feed resources because of the high transport cost of roughage. Products, however, are shipped to consumption centers through longer cold chains. In contrast, production of monogastric species is less constrained geographically.

With transport infrastructures developing, the shipment of concentrated feed is becoming increasingly cheap relative to other production costs. With increasingly high levels of processing, transportation of animal products to points of consumption or export is also becoming cheaper. Production and processing units are therefore freer to move where the policy environment (e.g., environmental standards, tax regime, and labor standards), access to resources and services, and disease constraints minimize production costs. In the absence of a well-developed transportation infrastructure, however, production units are forced to concentrate close to consumption centers.

In essence, livestock is moving from a "default land user" strategy (i.e., as the unique way to harness biomass from marginal lands, residues, and interstitial areas) to an "active land user" strategy—competing with other sectors for the establishment of feed crops, intensive pasture, and production units.

References

Asner, G. P., A. J. Elmore, L. P. Olander, R. E. Martin, and A. T. Harris. 2004. Grazing systems, ecosystem responses, and global change. *Annual Review of Environment and Resources* 29:261–299.

Boutrais, J. 2007. Crises écologiques et mobilités pastorals au Sahel: les Peuls du Dallol Bosso (Niger). *Sécheresse* 18 (1) : 5–12

Bruinsma, J., ed. 2003. *World agriculture: Towards 2015/2030, an FAO perspective.* Rome and London: Food and Agriculture Organization and Earthscan.

Butt, T. A., B. A. McCarl, J. Angerer, P. R. Dyke, and J. W. Stuth. 2004. Food security implication of climate change in developing countries: Findings from a case study in Mali. College Station: Texas A&M University.

Duron, B. S. 1999. Le transport maritime des céréales. Université de droit, d'économie et des sciences. Aix-Marseille.

EUROSTAT database. http://epp.eurostat.ec.europa.eu/portal/page/portal/eurostat/home/ (accessed May 15, 2007).

FAO (Food and Agriculture Organization). 2003. *World Agriculture: Towards 2015/2030.* Rome: FAO.

FAO (Food and Agriculture Organization). 2006b. FAO statistical databases. http://faostat.fao.org/default.aspx.

FAO (Food and Agriculture Organization). 2008. FAOSTAT statistical databases. http://www.faostat.external.fao.org.

Fujita, M., and J. F. Thisse. 2004. *Economics of agglomeration: Cities, industrial location, and regional growth.* Cambridge: Cambridge University Press.

Galloway, J. N., M. Burke, G. E. Bradford, R. Naylor, W. Falcon, A. K. Chapagain, J. C. Gaskell, E. McCullough, H. A. Mooney, K. L. L. Oleson, H. Steinfeld, T. Wassenaar, and V. Smil. 2007. International Trade in Meat: The Tip of the Pork Chop. *AMBIO* 36 (8): 622–629.

Gerber, P. 2006. *Putting pigs in their place:* Environmental policies for intensive livestock production in rapidly growing economies, with reference to pig farming in Central Thailand. Doctoral thesis in Agricultural Economics. Swiss Federal Institute of Technology. Zurich.

Hoste, R., and G. B. C. Backus. 2003. *Global Pig production costs—costs of pig production in Brazil, Canada, China, Poland and the USA compared with those in the Netherlands.* Agricultural Economics Research Institute. Wageningen: LEI.

LABORSTAT–ILO (International Labour Organization). 2008. ILO Database on Labour Statistics. Available at http://laborsta.ilo.org

LandScan Project. 2005. Oak Ridge National Laboratory. http://www.ornl.gov/sci/landscan.

Mbonile, M. J. 2005. Migration and intensification of water conflicts in the Pangani Basin, Tanzania. *Habitat International* 29:41–67.

Molle, W. 2003. *Global economic institutions*. New York: Routledge, 2003.

Nelson, A. 2007. Global accessibility model for travel time estimates to populated places of 250,000 persons or more. Joint Research Centre of the European Commission. ISPRA.[6]

Pingali, P. L., and P. W. Heisey. 1999. Cereal crop productivity in developing countries: Past trends and future prospects. Mexico: International Maize and Wheat Improvement Centre.

Rodrigue, J. P., C. Comtois, and B. Slack. 2006. *The geography of transport systems*. New York: Routledge.

Rudel, T. K., D. Bates, and R. Machinguiashi. 2002. A tropical forest transition? Agricultural changes, outmigration and secondary forests in the Ecuadorian Amazon. *Annals of the Association of American Geographers* 92:87–102.

Schnittker, J. 1997. *The history, trade, and environmental consequences of soybean production in the United States*. Report to the World Wide Fund for Nature. 110 pp.

Sirikijpanichkul A., and L. Ferreira, 2006. Contestable freight trends and implication for governments. Working Paper 1. School of Urban Development. Faculty of Built Environment and Engineering. Queensland University of Technology (QUT).

Steinfeld, H., and P. Chilonda. 2006. Old players, new players. In *The 2006 Livestock Report*, ed. A. McLeod. Rome: Food and Agriculture Organization.

Steinfeld, H., P. Gerber, T. Wassenaar, V. Castel, M. Rosales, and C. de Haan. 2006. *Livestock's long shadow: Environmental issues and options*. Rome: Food and Agriculture Organization.

Sumberg, J. 2003. Toward a disaggregated view of crop–livestock integration in Western Africa. *Land Use Policy* 20:253–264.

United Nations. 2003. *World population prospects: The 2002 revision*. New York: United Nations.

United Nations. 2008. *World population prospects: The 2008 revision*. New York: United Nations.

USDA–Agricultural Marketing Service (AMS)–Agricultural Transportation–Ocean Rate Bulletin Database. http://www.ams.usda.gov/AMSv1.0/ams.fetchTemplateData.do?startIndex=1&template=TemplateV&navID=AgriculturalTransportation&leftNav=AgriculturalTransportation&page=ATORBArchive&description=Ocean%20Rate%20Bulletin%20Database&acct=oceanfrtrtbltn>http://www.ams.usda.gov/AMSv1.0/ams.fetchTemplateData.do?startIndex=1&template=TemplateV&navID=AgriculturalTransportation&leftNav=AgriculturalTransportation&page=ATORBArchive&description=Ocean%20Rate%20Bulletin%20Database&acct=oceanfrtrtbltn (accessed May 2007).

Walker, R. 1993. Deforestation and economic development. *Canadian Journal of Regional Science* 16:481–497.

Westing, A. H., W. Fox, and M. Renner. 2001. Environmental degradation as both consequence and cause of armed conflict. Working paper prepared for Nobel peace laureate forum participants by PREPCOM Subcommittee on Environmental Degradation.

Wint, W., and T. Robinson. 2007. *Gridded livestock of the world*. Rome: Food and Agriculture Organization.

World Bank. 2006. World Development Indicators 2006. http://go.worldbank.org/RVW6YTLQH0 (accessed May 16, 2007).

Part II
Consequences of Livestock Production
Environmental, Health, and Social Consequences of Livestock Production

Harold A. Mooney

Part II assesses the many dimensions of the environmental, health, and social consequences of the livestock industry. It is important that those engaged in this endeavor, which in fact is most of the world's people, are aware of these complexities and interactions in order to make rational decisions in everyday life. All the consequences are considerable, and the scale is both spatial and temporal—extending from local to global and from daily to decadal and beyond.

Animal production systems have a major impact on global biogeochemistry—with important consequences on many dimensions of the environment, from pollution of waterways and the atmosphere to contributions to global warming and competition between water for nature versus animal protein. Chapter 5 looks at the carbon cycle; one of the major factors in carbon dioxide emissions is the conversion of forestland to pasture and crops in tropical areas. Desertification induced by livestock is another, but much smaller, contributor to climate change. Nitrous oxide emissions are mainly linked to the cultivation of grains for livestock feed and to livestock waste management practices, while methane production is inherent to the digestive process of cattle.

Another issue receiving increasing attention is the adverse environmental impacts of the vast amounts of fertilizers being used to fuel food production systems. Chapter 6 takes an integrated view of the pathways of nitrogen in the environment with a focus on animal production systems—from the positive impact on increased production to subsequent negative impacts of unintended losses to the waterways and the atmosphere. Chapter 7 then looks at total water use by animal production systems through a detailed accounting of the impacts of this activity on both the water use of irrigated crops used for feed as well as water use resulting from nonirrigated crop or pasture evapotranspiration in areas also used for animal feed.

A comprehensive framework is developed in Chapter 8 for analyzing the impacts of both intensive and extensive animal production systems on Earth's biological diversity, including impacts on all of the major ecosystem types, from waterbodies to forests and drylands. A more integrated assessment of the multiple impacts of livestock production on biodiversity, and the consequences of the losses of ecosystem services that stem from these, is a vital step for mitigating the impacts.

Chapter 9 (as well as others) deals with pollution mitigation approaches, principally for intensive production systems. The serious pollution problems associated with intensive livestock production systems have technological solutions, mostly centered on recycling of energy and nutrients. Then Chapter 10 focuses on the environmental impacts of extensive systems, describing their positive and negative impacts. The authors note that the solutions to these problems are highly geographically explicit and need the involvement of many stakeholders.

Both animal and human disease risks related to animal production systems are shifting, and in many cases increasing, due to a variety of changing drivers, including industrialization and clustering. As the complexities of the food chain grow, with the international trade of meat products, so do the difficulties of tracking sources of potential contaminants and diseases. And in order to avoid many of these problems, drugs are used in animal production systems that may actually exacerbate human health risks. These animal and human health issues are discussed in Chapter 11.

Chapter 12 addresses another important dimension of human health: the human diet involving animal products. Too much animal-based protein is not good for human diets, while too little is a problem for those on a protein-starved diet, as happens in many developing countries.

The transition in the livestock industry is also having enormous social consequences on a variety of stakeholders, including small producers, communities with large industrial facilities, and pastoralists. Chapters 13 and 14 focus on how small stakeholders are being disadvantaged generally by economic drivers. Since small animal production operations represent a way out of poverty in many parts of the world, policies that alleviate their disadvantages in the marketplace are now receiving increasing attention.

Smallholders without land are pastoralists, for whom

poverty issues and cultural identity are important. The livestock transition is playing a role in further marginalizing their way of life. Chapter 15 explores these issues in depth and points to practices that can help maintain this unique way of life, including the use of pastoralism in managing grazing lands and many policies that could serve as a lifeline to pastoralists.

The livestock industry is so vast and pervasive that the chapters on social issues could only scratch the surface of this important area of human concern. But they do assess the complexity and depth of these issues—especially in this time of rapid change in norms of practice combined with changes in population and climate.

5

Livestock and the Global Carbon Cycle

Gregory P. Asner and Steven R. Archer

Main Messages

- **The element carbon (C) is the basis for all life on Earth.** Global terrestrial net primary production—the net amount of C taken up by plants and photosynthetic microorganisms—is about 57 billion metric tons (petagrams or Pg) per year.

- **Global livestock production directly appropriates about 2 Pg C, or 3%, of global net primary production each year,** with this carbon mostly allocated from the 24 Pg C fixed in grazing lands and agricultural systems.

- **The most profound biospheric impact of livestock production on the carbon cycle is a growing set of worldwide ecological degradation syndromes,** including increasing rates of deforestation, woody vegetation encroachment on grazing lands, and desertification.

- **Livestock production results in a wide range of collateral carbon flows, including carbon losses to the atmosphere via tropical deforestation for pasture and croplands** (~1.2 Pg C yr^{-1}), C losses via desertification (~0.2 Pg C yr^{-1}), C sequestration via woody vegetation encroachment (~0.3 Pg C yr^{-1}), and C losses via methane emissions from livestock (~2.1 Pg yr^{-1} in carbon dioxide [CO_2]-equivalents). These carbon-cycle impacts far exceed those of intensive livestock production.

- **Livestock production causes total emissions of ~3.2 Pg CO_2-equivalents yr^{-1},** not including nitrogen-oxide compounds.

- **Carbon losses associated with grazing systems could be decreased, and sequestration could be enhanced,** through proactive management that maintains vegetation cover and soil carbon stores and through more explicitly integrating climatic fluctuations into planning of livestock production systems, such as where, when, and in what density to place grazing animals on the land with respect to drought events.

Introduction

Livestock production, defined here as any operation designed to manage the growth of animals for consumption, including for meat, milk, and any other major animal products, is a dominant form of land use throughout the world. There are more than 1.7 billion animals present in livestock production systems on Earth (see Chapter 2). Both extensive livestock production systems (range- or wildland-fed) and intensive ones (fed on crops and forage or managed pasture) are increasing in geographic area and output through time (Klein Goldewijk and Battjes 1997; see also Chapter 2). Currently, extensive grazing occupies more than 34 million square kilometers or 27% of the global land surface, making it the single most extensive form of land use on the planet (Asner et al., 2004; see also Chapter 3). Given that extensive grazing systems have increased more than 600% in geographic extent (from about 5.3 million km²) during the last three centuries (Klein Goldewijk and Battjes 1997), their contributions to total livestock production and the global carbon (C) cycle are substantial.

Intensification is defined here as an increase in labor or capital inputs on cultivated land, or on a combination of cultivated and grazing land, in order to increase the value of output per hectare (sensu Tiffen and Mortimore 1994). Intensification is now responsible for the rapid growth in meat production in many regions of the world (see Chapter 2). Industrial meat production operations provided 37% of global meat production in 1991–93, increasing to 43% by 1996 (Steinfeld et al., 1997), and it continues to grow today (see Chapter 2). Intensification requires a subsidy of primary production via agricultural crops (soya, corn, etc.) and forages (domesticated pastures, alfalfa, etc.), supplemented with inputs from extensive systems (e.g., native grasslands providing hay). Intensification has historically been most prominent in developed countries, which use about two-thirds of the global crop cereal production destined for livestock feed (Steinfeld et al., 2006). More recently, intensification has been ramping up in China, Southeast Asia, and South America (see Chapter 2). Today, about 4.7 million km² of arable land are required to generate feed for livestock (see Chapter 3).

This chapter looks at the major carbon flows associated with both extensive grazing systems and intensive livestock production. The effects of extensive grazing systems are geographically widespread (>34 million km²) and are thus likely the larger contributor to changes in the global carbon cycle caused by livestock production. Extensive grazing lands primarily occur in arid, semiarid, subhumid, and humid regions (hereafter "rangelands," for expediency; Holecheck et al., 2003, Asner et al., 2004), where soils, climate, and topography are not conducive to intensive agriculture or commercial forestry. However, there is also a rapid increase in machinery, irrigation, and fertilizer subsidies going to cropland products destined for supporting livestock systems in rangelands (see Chapter 2). Whereas outputs of intensive production have major impacts on C and nutrient flows in countless localized areas throughout the world (e.g., feedlots), the crop and cereal system inputs that support intensive operations also have broad impacts on the carbon cycle.

Meat, egg, and milk production appropriates about 2 billion metric tons (petagrams or Pg) of carbon from the 57 Pg fixed by terrestrial plants each year via net primary production (NPP). Although only 3% of global NPP is used directly for extensive and intensive livestock production, the processes involved lead to a series of collateral carbon gains and losses. Collateral carbon impacts are often reported as ecological degradation syndromes, mostly in extensive grazing systems, but also in some cropland systems supplying feed to intensive livestock operations. These syndromes can generally be grouped into three categories: deforestation, woody encroachment, and desertification. For example, tropical deforestation for pasturelands and croplands results in a flux of carbon from the land to the atmosphere of 0.9–1.7 Pg C each year (Achard et al., 2002, Defries et al., 2002, Canadell et al., 2007). The uncertainty in the deforestation flux is large, drawn from combinations of satellite, field, and modeling results having different geographic and temporal resolution.

Grazing, with associated decreases in fire frequency, can result in woody vegetation (brush) proliferation in subhumid regions (Archer 1994, Archer et al., 1995, van Auken 2000), potentially sequestering carbon at a rate of 0.3 Pg yr⁻¹ globally, but at the expense of decreasing forage availability and land area for grazing animals, while potentially reducing stream flow and groundwater recharge and altering wildlife habitat and biological diversity. This carbon sink estimate is also highly uncertain,

calculated here simply as the product of the estimated land area undergoing the woody encroachment and the carbon stock changes reported in the literature (Scifres 1985, Scholes and Archer 1997, Houghton et al., 1999, Geesing et al., 2000, Bovey 2001, Archer et al., 2001, Henry et al., 2002, Asner et al., 2003a, Hibbard et al., 2003, Dale et al., 2005, Hughes et al., 2006, Knapp et al., 2008). At a local to landscape scale, however, woody encroachment triggers a series of management responses aimed at reducing woody plant cover via fire, mechanical, and chemical methods, all significantly affecting the carbon cycle at the ecosystem level. When desertification occurs in response to grazing and climate variability, carbon stocks decrease and become fragmented on the landscape as wind and water erosion are accelerated, with an estimated net loss of 0.1–0.3 Pg C yr⁻¹ (Lal 2004, Valentin et al., 2005, Okin et al., 2006, Steinfeld et al., 2006).

All of these reported syndromes are associated with changes in vegetation cover, which fundamentally affects the exchange of energy, water, and carbon between our biosphere and atmosphere (e.g., Eastman et al., 2001). Methane emissions from livestock average 0.1 Pg C yr⁻¹ (enteric fermentation and manure) but with a disproportionately negative greenhouse impact on our atmosphere compared with carbon dioxide that results in an effective 2.1 Pg carbon dioxide (CO_2)-equivalents released each year (Steinfeld et al., 2006). The following sections describe in more detail how these degradation syndromes alter the flux of carbon within the biosphere and between the biosphere and atmosphere.

Primary Production to Feed Livestock

Contemporary satellite and modeling studies estimate that global terrestrial NPP is approximately 57 Pg C yr⁻¹ (Field et al., 1998, Imhoff et al., 2004). Of this total, approximately 4 Pg C are fixed via primary production each year into crops and 20 Pg C into grazable lands (savanna, grassland, desert) (Sabine et al., 2004). (See Figure 5.1.) From these source lands, roughly 2 Pg C of crop plus herbaceous NPP are destined for livestock production (Imhoff et al., 2004). Accounting for waste, seed production (for future crops), allocation to roots, and other ancillary flows, Imhoff et al. (2004) estimates that only 0.45 Pg C (22.5%) of the crops and cereals grown for livestock are actually consumed by the animals; consumption efficiencies of aboveground plant production by rangeland herbivores are on the order of 20% to 50%. Of the net primary production consumed by rangeland livestock, the conversion to secondary production is on the order of 2%, depending upon the type of animal, quality of the forage, and environmental stresses (Heitschmidt and Stuth 1991). (See Figure 5.1.) Thus, in terms of amount of carbon fixed from the atmosphere to the biosphere, livestock production is a highly inefficient process.

In extensive grazing systems, secondary production and the flow of carbon from plants to animals can be enhanced via mixed herbivore grazing systems; many pastoral societies have evolved using such systems (see Chapter 10). In industrial countries, however, extensive livestock production systems have traditionally focused on single herbivores to the exclusion of others. That approach has been steadily exported to less developed nations, often resulting in socioeconomic problems and resource degradation (see Chapters 3 and 10).

In areas where long-term records and data exist, stocking rates have declined markedly since the late 1800s and early 1900s, particularly in arid and semiarid regions. (See Figure 5.2.) Principles of rangeland livestock management did not begin to emerge until the 1940s, and adoption of the basic tenets of managing livestock grazing in wildlands lagged by decades. Thus declines in livestock numbers during this period were generally not the result of better grazing management. Although definitive studies from the early 1900s are lacking, the decline in livestock numbers during this period is widely regarded to reflect reductions in the carrying capacity known to accompany shifts in vegetation composition to plants that are less productive, less palatable, and/or less nutritious and, in many cases, the encroachment of shrubs that are unpalatable or have high levels of secondary compounds (see review by Curtin et al., 2002). These grazing-induced changes were often accompanied by accelerated rates of wind and water erosion (Thurow 1991). The resultant loss of soil fertility and water-holding capacity constitutes a positive feedback that reinforces change in species composition and loss of ground cover and forage production.

These historic changes affect current plant production potential and soil organic carbon stocks. Widespread recognition of the adverse impacts of unregulated livestock grazing in the North American West resulted in the creation of numerous "experimental ranges" in the early 1900s and spawned the fields of range science and range management, which sought to develop science-based approaches for managing grazing so as to minimize degradation and promote sustainable livestock production (e.g., Stoddart et al., 1955, Fredrickson et al., 1998, Holecheck et al., 2003). Upon implementation of such approaches, primary production and soil organic carbon stocks may rebound. We mention this because modern assessments of grazing effects on plant production and soil organic carbon pools should be considered within this historical context. That is to say, substantial livestock grazing–induced changes in the carbon cycle may have occurred well before the advent of post-1950s studies; and comparisons between currently grazed sites and sites with a "long-term" history of protection from grazing (e.g., 50–75 years) may not represent the full potential impacts.

Global Carbon Fluxes Associated with Livestock Production

Intensive Systems

A recent report derived estimates of carbon fluxes associated with intensive livestock production systems (Steinfeld et al., 2006). We summarize the fluxes for CO_2 and methane (CH_4) in terms of carbon dioxide equivalents in Table 5.1. Intensive production systems generate a suite of carbon fluxes from feed crop production and transport, livestock transport and processing, and cropland soils. Nitrogen is a critically important element used to fertilize feed crops (see Chapter 6), and more than 95% of nitrogen fertilizers are derived from ammonia produced using natural gas. This process results in an estimated 0.04 Pg of CO_2 released to the atmosphere (see Table 5.1). Vehicle use on farms for the transport of feed and livestock produces 0.09 Pg CO_2, whereas off-farm

Figure 5.1. Carbon flows from our atmosphere to vegetation in global croplands and major ecosystems supporting livestock production. Although most carbon is ultimately returned to the atmosphere, the indirect effects of livestock production—deforestation, desertification, and woody vegetation encroachment—have profound impacts on biospheric carbon storage.

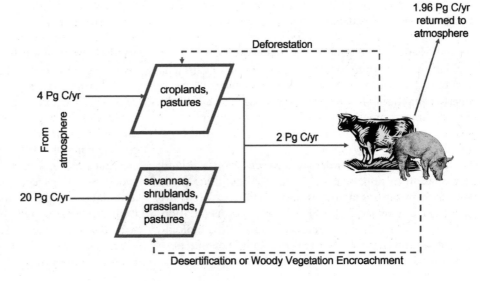

Figure 5.2. Reductions in livestock numbers on arid and semiarid rangelands [animal unit (au) years (y) or months (m)]. (A) Cattle, horses, sheep, and goats on federal lands in western USA (from Holecheck et al., 2003); (B) general trends in Australia (adapted from Ash et al., 1997); (C) cattle on desert grasslands in southern Arizona, USA (based on data from McClaran et al., 2002 and Santa Rita Experimental Range Archives, University of Arizona); and (D) sheep numbers at the Sonora, Texas, USA, Research Station (from Huston et al., 1994).
Source: Reproduced with permission from *Rangeland Journal* 19(2): 123–144 (A. J. Ash, J. G. McIvor, J. J. Mott, and M. H. Andrew, 1997). Copyright The Australian Rangeland Society. Published by CSIRO PUBLISHING, Melbourne, Australia.

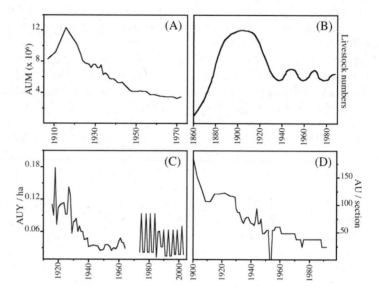

transport emits a relatively small amount (on the order of 0.001 Pg). Cropland use generates an additional 0.03 Pg CO_2, partitioned among till and no-till systems, and animal processing results in about 0.03 Pg CO_2 emitted to the atmosphere (see Table 5.1). All of these CO_2 emissions from intensive systems sum to roughly 0.2 Pg per year.

Extensive Grazing Systems and Ecological Change

Extensive grazing systems dominate many dryland landscapes of the world (see Chapter 3). The bioclimatically marginal nature of many grazing lands in arid and semiarid regions plays a key role in two of the regional ecological syndromes widely reported throughout the literature—desertification and woody plant encroachment. Concomitantly, the marginal biogeochemical characteristics of humid tropical soils play a role in the consequences of deforestation, the third regional syndrome inherent to managed grazing systems. (See Figure 5.3.) Not all extensive grazing operations result in the development of these syndromes, but an increasing number are being reported over time. Thus we define and discuss each of them with respect to the carbon cycle.

Problems defining desertification, woody encroachment, and deforestation with respect to livestock production practices are highlighted in Asner et al. (2004). These three ecological syndromes are often linked to livestock production systems throughout the world (Figure 5.4).

The United Nations defines *desertification* as "land degradation in arid, semi-arid and sub-humid areas resulting from various factors, including climatic variations and human activities" (UNEP 1994). Land degradation is further defined as the reduction or loss of the biological or economic productivity of drylands. For our purposes here, we refer to desertification physiognomically,

in terms of its expression on the landscape, as the large-scale replacement of herbaceous cover by either xerophytic shrub cover and bare soil or just bare soil. Desertification has occurred in many arid regions of the world (e.g., U.S. Southwest, Australia, Africa, Argentina), reportedly as a result of land use, especially grazing, operating against a backdrop of periodically stressful climatic conditions (e.g., drought) (UNEP 1994).

In contrast, we define woody vegetation encroachment as the widespread proliferation of woody canopies in grasslands and savannas of semiarid and subhumid regions, where it may be accompanied by slight or substantial losses of herbaceous cover and production. In both desertification and woody plant encroachment, palatable herbaceous vegetation is replaced by highly unpalatable woody vegetation. Woody encroachment has occurred in conjunction with large-scale grazing and concomitant reductions in fire (Archer 1994, Archer et al., 1995, van Auken 2000).

Deforestation is defined operationally here as the replacement of forest cover with pasture or cropland systems. Clearing for cropland and pasture has been historically common in temperate forests and is a more recent and ongoing phenomena that continues in the humid tropics in response to ever-increasing population pressures. Degradation associated with tropical deforestation is largely propelled by the fact that pastures are created on nutrient-poor soils that cannot sustain long-term grazing operations (Geist and Lambin 2001).

Some changes in ecosystem structure affecting the C cycle, including vegetation productivity, cover, and the abundance of species and life-forms, are directly attributable to grazing, whereas other changes are associated with the preparation of land for grazing. In grasslands, savannas, and shrublands, the effects of livestock production come primarily from grazing itself, whereas in

Table 5.1. Major fluxes of carbon associated with intensive and extensive livestock production systems.

Category	Extensive systems	Intensive systems
CO$_2$ from Production Operations		
N fertilizer for feed crops	—	0.04
On-farm fuel for feed transport	—	0.06
On-farm fuel for livestock transport and related	0.03	0.03
Cultivated soils, w/tillage	—	0.02
Cultivated soils, w/o tillage	—	0.01
Animal processing	—	0.03
Off-farm transport	0.001	0.001
CO$_2$ from Ecological Syndromes		
Desertification	0.2[$]	—
Tropical Deforestation*	1.2 (1.7[¥])	See note[§]
Woody vegetation (brush) encroachment*	−0.3	—
CH$_4$ from Livestock		
Enteric fermentation*	1.5	0.2
Manure management	0.2	0.2
Total CO$_2$ emissions	1.1	0.2
Total CH$_4$ emissions	1.7	0.4
Total emissions in CO$_2$ equivalents	3.2[£]	

All values are given in Pg (= 10^{15} g) of CO$_2$ equivalents (Steinfeld et al., 2006). Categories marked with (*) were estimated and/or updated by the authors.

[$] Average from Lal (2001) and Steinfeld et al. (2006).

[¥] Estimates of tropical deforestation rates for pasture range from 1.0 Pg C yr^{-1} (Houghton et al. 2005) to 1.5 Pg C yr^{-1} (Canadell et al., 2007), and for total global deforestation for pasture and feed crops may be as high as 1.7 Pg C yr^{-1} (Steinfeld et al., 2006).

[§] A portion of deforestation is driven by conversion to croplands that may support intensive livestock production (e.g., Morton et al. 2005). This fraction is included in our estimate of carbon flux from deforestation under extensive systems.

[£] This estimate excludes CO$_2$ equivalents of nitrous oxide (N$_2$O) emissions, covered by Galloway et al. (Chapter 6, this volume).

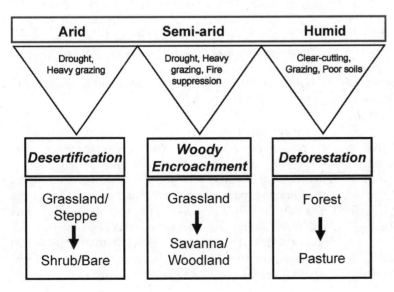

Figure 5.3. Three commonly reported regional ecological syndromes resulting from and/or affected by livestock production.
Source: Archer et al., 2001.

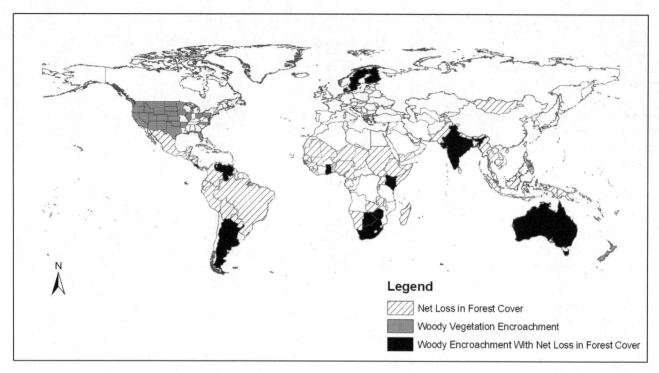

Figure 5.4. Global distribution of reported deforestation and woody encroachment associated with livestock production systems. Highlighted regions do not include nonmanaged grazing reports, such as from woody cover changes thought to be occurring in arctic regions as a result of climate change.
Source: Asner and Martin 2004.

forests and woodlands, the largest impacts of grazing are from the conversion of the system to herbaceous cover before grazing occurs. A cascade of change in associated ecosystem processes such as wind and water erosion, water drainage, disturbance types and frequency (e.g., fire), species invasion, and the structural and biochemical characteristics of soils accompany these ecosystem structural changes. All of these effects in some way alter the carbon cycle.

The effects of livestock grazing on carbon stocks in grassland, shrublands, and savannas are varied. Milchunas and Lauenroth (1993) analyzed a worldwide, 236-site data set compiled from studies that compared species composition, aboveground net primary production (ANPP), root biomass, and soil nutrients of grazed versus protected, ungrazed sites. They found the following:

- Differences in ANPP between grazed and ungrazed sites were greatest in regions where the evolutionary history of grazing was minimal and least in regions where plants had evolved with grazing herbivores, and they were greatest in areas with high ANPP, high levels of utilization, and long histories of livestock grazing.
- Differences in ANPP between grazed and ungrazed treatments were more sensitive to varying ecosystem-

environmental variables than to varying grazing variables. Within levels not considered to be abusive "overgrazing," the geographical location where grazing occurs appeared to be more important than how many animals were grazed or how intensively an area was grazed.
- Counter to the commonly held view that grazing affects root systems negatively, there was no relationship between the difference in ANPP with grazing and the difference in root mass. Root mass, soil organic matter, and soil nitrogen all displayed both positive and negative values in response to grazing.
- Grazing was a factor in the conversion of grasslands to shrublands.

The inconsistent responses to grazing in plant functional group composition and C storage highlighted in this meta-analysis reinforces the notion that climate and climatic variability may drive the C cycle to a greater extent than grazing, particularly in arid and semiarid areas. As a result, carbon losses associated with grazing systems are more likely to be decreased and opportunities for sequestration more likely to be realized with proactive management that explicitly integrates climate and climatic fluctuation into livestock stocking rate decisions (number and kind of animals, season of use, patterns of distribution, flexible strategies for de-stocking, etc.) and grazing management policy (e.g., Gillson and Hoffman 2007).

Desertification Syndrome

Published global figures on the extent of degradation of Earth's drylands range from 10% (Adeel and Safriel 2005) to 70% (Dregne and Chou 1994), with other estimates intermediate to these extremes (Oldeman and van Lynden 1997, Eswaran et al., 2001). Reasons for these varied estimates are discussed in Lal (2004) and Adeel and Safriel (2005).

In dryland regions where desertification has taken place, degradation has led to decreased productivity or vegetative cover, bringing with it a change in C stocks and C cycling. (See Figure 5.5 in the color well.) The primary ecosystem response to desertification is an increasingly heterogeneous distribution of vegetation, and the same pattern is true for carbon (Schlesinger et al., 1990). Lost vegetative cover can cause a reduction in aboveground C stocks and a decline in NPP, but there is significant variability by vegetation type and across topographic and edaphic gradients (Huenneke et al., 2002). With increasing fragmentation of vegetation cover, wind and water erosion increase, which further alters the landscape, making recovery from grazing more difficult.

In some cases, there is no net change in ecosystem C pools, as losses from herbaceous areas are compensated for by accumulations accompanying shrub encroachment (Schlesinger and Pilmanis 1998). In other cases, grazing-induced desertification may result in relatively little change in shrub cover, while causing declines in soil organic carbon and nitrogen storage of 25–80% (Asner et al., 2003b). Wind and water erosion move C and nutrients around the landscape and off-site, perpetuating a decrease in C storage averaging 30% (see Figure 5.5; Lal 2001), and, in extreme situations, causing an increasingly dune-like landscape to develop (Gallardo and Schlesinger 1992).

These patterns of fragmentation and lost productivity have been made clear at the regional scale via the perspective afforded by new aircraft sensors, which show that the fundamental interactions between the vegetation (phenology, carbon, and water) and climate variability are changed following desertification (Asner and Heidebrecht 2005).

Woody Encroachment Syndrome

In semiarid and subhumid regions historically supporting grasslands and savannas, woody plant encroachment has been widespread over the past hundred years. This proliferation of woody plants generally reduces the quality of land for animal production because the encroaching shrubs and arborescents may be of low palatability or nutritive value; may provide habitat for livestock pests, parasites, and predators; and may reduce forage production (Scholes and Archer 1997). However, in contrast to desertification, where ecosystem ANPP may decline or where production by encroaching xerophytic shrubs may barely compensate for losses of herbaceous production, woody plant proliferation in semiarid and subhumid zones can result in large increases in aboveground NPP and carbon storage (Archer et al., 2001).

A continental-scale study including eight sites in North America found shrub encroachment caused a slight decline in ANPP at arid sites (< 300 mm mean annual precipitation), but it increased ANPP as a linear function of mean annual precipitation between 300 and 1065 mm (Knapp et al., 2008). In contrast, grassland ANPP peaked and became asymptotic at ~600 mm annual rainfall. Thus increases in aboveground NPP of up to 1400 kg C ha^{-1} have been observed (Geesing et al., 2000, Hibbard et al., 2003, Hughes et al., 2006). (See Figure 5.5.) These enhanced productivities translate into increases in the aboveground carbon pool that can range from 300 to 44,000 kg C ha^{-1} in less than 60 years of woody encroachment (Asner et al., 2003a). Although these net accumulations in biomass represent C sequestration, they may be accompanied by increases in nonmethane hydrocarbon emissions and soil CO_2 fluxes and (potentially) adverse effects on aquifer recharge and stream flow (Dale et al., 2005). When the dominant woody species is a nitrogen-fixer, N accumulation can be 9–40 kg N ha^{-1} yr^{-1} greater in the woody areas than in the grasslands (Geesing et al., 2000, Hughes et al., 2006), thus augmenting a key resource that, along with water, typically co-limits ecosystem productivity. Unfortunately, much of this added productivity is not available to or useful for livestock production.

A substantial majority of the carbon in grassland and savanna ecosystems is belowground. However, it is not clear how grazing, climate, and woody plant encroachment interact to affect gains and losses from these large pools. Despite consistent increases in aboveground carbon storage with woody vegetation encroachment, the trends in soil organic carbon (SOC) are highly variable. Measuring soil carbon to 3 m depth at six sites spanning a mean annual precipitation (MAP) range of 200 to 1,100 mm, Jackson et al. (2002) found a linear, inverse relationship between changes in SOC subsequent to woody plant encroachment and MAP. They surmised that a decline in SOC in wetter ecosystems was enough to offset aboveground gains from woody encroachment, resulting in no net change in ecosystem carbon storage. However, a broader survey of the literature suggests changes in SOC following shrub encroachment may not be a simple function of annual rainfall, because woody plant encroachment has resulted in substantial increases in SOC pools in a variety of high rainfall sites, and increases in some low rainfall areas are substantially less than expected. (See Figure 5.6.)

Indeed, a broad survey of the published literature indicates woody plant encroachment can have positive, neutral, or negative effects on SOC pools (Asner and Martin 2004, Wessman et al., 2004). Figure 5.7 shows that, although both positive and negative carbon storage responses to woody encroachment have been reported (black bars), a much larger proportion of reported soil C responses are positive. Where historical grazing and woody plant encroachment effects on SOC have been explicitly taken into account, it appears that losses of SOC associated with grazing of grasslands can be recovered subsequent to woody plant encroachment; and that SOC in the resulting savanna-like ecosystem can be substantially greater than that of the original grasslands (Archer et al., 2001, Hibbard et al., 2003).

Despite the uncertainties in soil carbon responses to woody encroachment, the net effects at the global scale may be large. Houghton et al. (1999) reported that woody encroachment causes ~0.12 Pg C yr^{-1} to be sequestered in grazing systems of the United States. Given that roughly one-third of the woody encroachment reports globally have come from the United States (Archer et al., 2001, Archer 2006), we estimate that 0.3 Pg C (or more) may be stored each year in grazing systems due to woody plant proliferation. (See Table 5.1.) However, these net changes do not account for interannual variation in carbon stocks resulting from brush management

efforts (Scifres 1985, Bovey 2001, Henry et al., 2002). Therefore, the gross annual flux of woody encroachment and brush management remains highly uncertain.

Tropical Deforestation Syndrome

Deforestation for pastures occurs across a wide range of climatic conditions. We focus here on both dry and wet tropical forest conversion, where further conversion is most likely with the global expansion of grazing systems (Asner et al., 2004, Foley et al., 2005). Tropical deforestation has been driven primarily by the need for livestock pasture (Geist and Lambin 2001). Recently, however, deforestation for croplands associated with livestock intensification has become another contributing force of land-cover change. Morton et al. (2006) estimate that 17% of the deforestation in the Brazilian Amazon was driven by cropland expansion between 2001 and 2004, mainly for soya destined for livestock production systems.

In the conversion of a tropical dry forest to pasture, 50–90% of aboveground C can be lost in combustion associated with repeated burning (Kauffman et al., 1993). Water erosion and soil loss can be significant for a few months after a fire (Gimeno-Garcia et al., 2000), and concentrations of dissolved mobile nutrients (e.g., nitrate) can be substantially elevated in overland flow (Belillus and Roda 1993). Fire has little direct, immediate effect

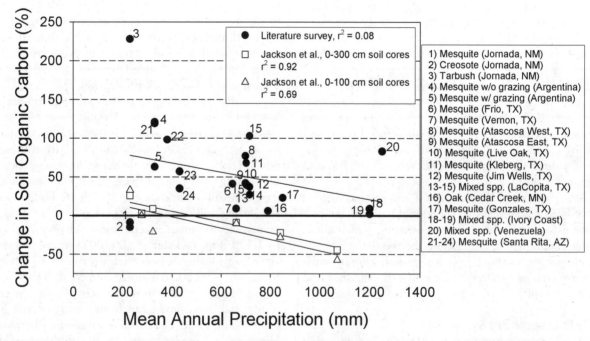

Figure 5.6. Changes in soil organic carbon (SOC) following woody vegetation encroachment versus mean annual precipitation. Open squares and triangles are the results presented by Jackson et al. (2002) for soil samples integrated over 0–300 cm and 0–100 cm, respectively. Changes in SOC occur primarily in the upper 30 cm; thus the depth of soil sampling has little effect on patterns shown. Closed circles show our newly compiled results from published studies.
Source: Numbers on symbols/legend entries indicate references: (1–3) Schlesinger and Pilmanis 1998; (4–5) Asner et al., 2003b; (6, 8–12, 17) Geesing et al., 2000; (7) Hughes et al., 2006; (13–15) Boutton et al., 1998; (16) Tilman et al., 2000; (18–19) Mordelet et al., 1993; (20) San Jose et al., 1998; (21–24) Wheeler et al., 2007

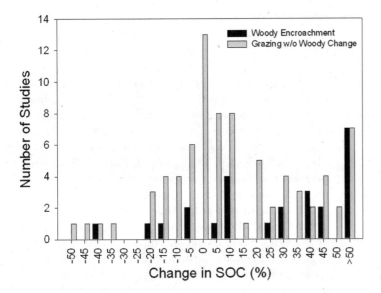

Figure 5.7. Studies that indicate soil organic carbon (SOC) responses to grazing and to grazing with woody encroachment.
Source: Asner and Martin 2004.

on total soil organic carbon (Emmerich 1999). Humid tropical forests are often found on highly weathered soils (Ultisols, Oxisols) rich in available nitrogen and poor in both phosphorus and base cations (e.g., Ca, Mg) (Sanchez 1976). Losses of these nutrients during and after forest clearing are thus important to future pasture productivity and C storage. (See Figure 5.5.) Studies show that the initial burning of slashed primary forest results in combustion of ~50% of biomass, or up to 88 Mg carbon per hectare (Guild et al., 1998) (See Figure 5.5.) A detailed accounting of changes in soil carbon that considered the effects of soil compaction and management practices suggested an average loss of 12 tons C ha^{-1} in tropical lands maintained as pasture (Fearnside and Barbosa 1998).

Fire is often used as a management tool to remove woody regrowth and weeds and to renew nutrient availability in pastures (Kauffman et al., 1998). In the Brazilian Amazon, repeated burning of cattle pasture consumes up to 46% of aboveground biomass (slash, grass, litter), with about 14 Mg C per hectare lost to the atmosphere (Guild et al., 1998). With repeated fires over a six-year period, more than 1,900 kg N ha^{-1} is lost, which is equivalent to ~90% of the aboveground pool in mature tropical forest. As a result, repeated burning can lead to N-limitation of NPP and carbon storage, even in previously nitrogen-rich tropical systems (Markewitz et al., 2004).

With decades of livestock production and frequent burning, soil P and cations (Ca, Mg, K) often decline as well (Fernandes et al., 1997). The mechanism for these nutrient losses is uncertain: some could be lost through combustion, some in erosion of ash, and perhaps some to deep soils through the leaching of compacted anaerobic soils (Townsend et al., 2002). Nonetheless, plant-available P declines and forage NPP and C accumulation in pasture soils often becomes P-limited. As a result, P

fertilization (~50 kg ha^{-1} every 5–10 years) has become a common practice in some regions such as the eastern Brazilian Amazon (Gehring et al., 1999, Davidson et al., 2004). Overall, pastures accumulate and cycle fewer nutrients than the forest and redistribute cations from trees to soils. Thus pastures have lost most of the carbon that was stored in aboveground forest biomass. (See Figure 5.5.)

Methane

Livestock operations affect the atmosphere through the direct emission of methane gas from ruminants and from livestock excreta. Ruminant livestock can produce 250–500 L of CH_4 per day (Johnson and Johnson 1995). An estimate of CH_4 from ruminant animals and animal waste is ~90 Tg CH_4, nearly a fifth of the total global emissions (IPCC 2001). However, CH_4 has a global warming potential 23 times that of CO_2. Therefore livestock produce fluxes of nearly 0.4 Pg and 1.7 Pg of "CO_2-equivalents" from intensive and extensive production systems, respectively. (See Table 5.1.) Extensive systems may dominate CH_4 emissions because they contain the majority of large livestock (e.g., cattle) and because dietary supplements provided in intensively managed settings can reduce ruminant CH_4 production (Duxbury 1994).

In total, global livestock production emits roughly 2.1 Pg CO_2-equivalent of methane to our atmosphere each year, which far exceeds any other single source of greenhouse gas resulting from livestock production. By comparison, combined methane emissions from rice paddies (53 Tg yr^{-1}) and natural wetlands (92 Tg yr^{-1}) contribute ~3.5 Pg CO_2-equivalents (Cao et al., 1998).

Numerous factors influence methane emissions from cattle, including level of feed intake, type of carbohydrate in the diet, feed processing, and alterations in ruminant microbiota (Johnson and Johnson 1995). Manipulation of these factors can reduce methane emissions

(Duxbury 1994), but options for CH_4 reductions in per-animal emissions from rangeland livestock are probably limited. From a mitigation standpoint, it is important to note that the atmospheric lifetime of CH_4 is approximately 9.6 years, whereas that of CO_2 is on the order of one to two centuries (IPCC 2001). Thus reductions in CH_4 concentrations could have a more immediate impact on radiative forcing than would reductions in CO_2 concentrations.

Summary

The major carbon fluxes associated with the three ecological degradation syndromes are summarized in Table 5.1. Desertification of arid grasslands is thought to cause a 0.1–0.3 Pg C loss to the atmosphere annually (Lal 2001, Steinfeld et al., 2006). Woody encroachment into semiarid and mesic grazing lands may sequester carbon in vegetation and soils by up to 0.3 Pg C per year. Deforestation for mainly pastureland in the humid tropics released approximately 1.2 Pg C yr^{-1}. Combined, these ecological degradation syndromes result in a net annual loss of about 1.1 Pg C per year, which far exceeds the nonmethane losses associated with intensive production systems, including feed croplands. Taking into account the net effects of these syndromes and methane emissions, the greenhouse warming impacts of extensive grazing systems are enormous—roughly 3.2 Pg CO_2-equivalents per year. The impacts of extensive grazing on the atmosphere thus far exceed those of intensive livestock production. These results stand in stark contrast to those of the global nitrogen cycle, which has been deeply affected by fertilizer production and intensive grazing systems (see Chapter 6).

Mitigation and Avoidance of Negative Impacts

Current livestock production practices mediate a range of carbon fluxes associated with processing, transport, feed production, methane emissions, deforestation, desertification, and woody encroachment. These impacts could be mitigated if both extensive and intensive grazing operations were planned with respect to background environmental conditions such as soils, topography, and climate variability. The single largest carbon loss caused by livestock production—deforestation for pasturelands—can be reduced if intensification were to take place under the appropriate environmental conditions (Vlek et al., 2004, Moutinho and Schwartzman 2005). This might involve concentrating high-production feed crop systems on fertile soils in regions of moderate rainfall in the tropics rather than on highly weathered, low-nutrient soils or in high rainfall areas. In addition, fertilizer subsidies and fire management techniques could be used to slow anthropogenically mediated soil degradation (infertility) trends in tropical Oxisols and Ultisols, thereby reducing demand for new pasture.

Mitigation in grasslands and savannas might center on relaxing or removing grazing from marginal areas and instead concentrating it in regions where ecosystem resilience is greater or where resistance to degradation is greater. This is being advocated by many environmental groups for the western United States, but it remains controversial. In some cases, livestock incomes could be replaced or augmented by income from ecotourism, recreation, hunting, etc. However, there are downsides to this as well (du Toit et al., 2009). Furthermore, while this may be a viable option in nations with available capital, it may not be realistic in poorer countries.

Desertification control and restoration of degraded soils and ecosystems would improve soil quality, increase the pool of carbon in soil and biomass, and induce formation of secondary carbonates leading to a reduction of carbon emissions to the atmosphere; in addition, they may also have numerous ancillary benefits (Lal 2001). Desertification control and soil restoration require maintaining and promoting vegetative cover, ideally with palatable species. This is difficult to do when rainfall is low and erratic and if livestock grazing pressure is not relaxed. However, relaxation of grazing represents a significant loss of revenue in commercial livestock enterprises and may affect human health and well-being in pastoral societies.

Practices aimed at promoting restoration (e.g., seeding, water harvesting and spreading, irrigation, fertilization) may also have substantial hidden carbon costs. In the past, and even currently, exotic forage grasses have been introduced with the intent of establishing ground cover and improving livestock production. Unfortunately, many of these plants can create more problems than they solve, becoming highly invasive and sometimes promoting increased disturbance via fire and erosion (Mack et al., 2000, Williams and Baruch 2000, D'Antonio et al., 2004, Franklin et al., 2006). Given the high cost and low probability of success of dryland restoration efforts, and since rates of carbon sequestration in drylands are inherently low, it is likely far more cost-effective to conserve existing carbon pools than to recover them after they are lost.

Regions prone to desertification are characterized by a high degree of interannual rainfall variability. Many of the problems in arid/semiarid regions stem from the fact that livestock management systems are not agile enough to respond to fluctuations in precipitation (Behnke et al., 1993, Ellis and Galvin 1994, Gillson and Hoffman 2007). As a result, stocking rates are kept high for too long into drought periods, and this tends to lead to degradation (Archer and Stokes 2000). Policies or coordination of regional management that would promote stocking agility and "transhumance" (e.g., moving animals to areas where there has been rain rather than confining them to certain properties or areas regardless of environmental conditions) might result in lower rates of land degradation (carbon loss) caused by grazing during drought periods. Examples of community-based adaptive management illustrate the potential of coalitions of ranchers, agencies, scientists, and environmentalists to conserve ecosystem structure and function and, in so doing, protect a matrix of communal and publicly and privately owned lands (e.g., Curtin et al., 2002). This approach to conservation is modeled after UNESCOs biosphere reserves program, wherein diverse assemblages of landscapes and cultures are protected within a matrix of different landownerships rather than a single government-regulated entity.

In drylands, sustaining ecosystem processes in the face of climatic variability requires a sound foundation of monitoring and research as well as a good working relationship between people and organizations with diverse goals and interests. This collaboration between ranching, research, and conservation communities demonstrates that these groups, working together for mutual benefit, can reach scientific and conservation goals unobtainable by any one group on its own (Curtin 2002, Herrick et al., 2006).

Regions prone to woody plant encroachment present a novel series of dilemmas and challenges to mitigation. The proliferation of woody plants in ecosystems grazed by livestock typically invokes land management actions to minimize the cover using chemical (herbicides), mechanical, or prescribed burning brush management techniques. In the 1950s–1970s, the goal of widespread, indiscriminate woody plant eradication was to increase forage production for livestock or to enhance aquifer recharge and stream flow. However, the rising cost of fossil fuels, coupled with short-lived treatment effects, has made large-scale mechanical and chemical treatments of woody plants economically tenuous or unrealistic. In addition, the recognition that herbicides can have deleterious environmental effects and that woody plants provide important habitat for wildlife has led to more progressive and selective approaches (e.g., Scifres et al., 1988). Numerous studies have challenged the traditional perspectives on the effects of woody plants on the hydrological cycle in drylands, but this topic remains highly controversial (Belsky 1996, Wilcox et al., 2005).

The recent realization that woody plant proliferation may substantially promote ecosystem carbon stocks in the geographically extensive drylands may trigger new land-use drivers as industries seek opportunities to acquire and accumulate carbon credits to offset their CO_2 emissions. Woody plant proliferation in grasslands and savannas may therefore shift from being an economic liability in the context of extensive livestock production systems to a source of income in a carbon sequestration context. If so, perverse incentives may result, as land management may shift to promoting rather than deterring woody plant encroachment. Thus potential benefits associated with carbon sequestration should be carefully weighed against costs in the form of increases in nitrogen and nonmethane hydrocarbon emissions (Guenther et al., 1999, Martin and Asner 2005); reductions in livestock production, stream flow, and groundwater recharge (e.g., Jackson et al., 2005, Wilcox et al., 2005); the extirpation of plants and animals characteristic of grasslands and savannas (loss of biological diversity); and, indeed, the local or regional extinction of grassland and savanna ecosystems (Archer et al., 2004). Quantification of these trade-offs is a challenge currently facing the scientific community and will be necessary in order to appropriately evaluate the impacts of various management and policy scenarios.

References

Achard, F., H. D. Eva, H. J. Stibig, P. Mayaux, J. Gallego, T. Richards, and J. P. Malingreau. 2002. Determination of deforestation rates of the world's humid tropical forests. *Science* 297:999–1002.

Adeel, Z., and U. Safriel. 2005. Dryland systems. In *Ecosystems and human well-being: Current state and trends, volume 1*, ed. R. Hassan, R. Scholes, and N. Ash, 623–664. Washington, DC: Island Press.

Archer, S. 1994. Woody plant encroachment into southwestern grasslands and savannas: Rates, patterns, and proximate causes. In *Ecological implications of livestock herbivory in the West*, ed. M. Vavra, W. A. Laycock, and R. D. Pieper, 13–68. Denver: Society for Range Management.

Archer, S. 2006. Proliferation of woody plants in grasslands and savannas: A bibliography. Tucson: University of Arizona. http://ag.arizona.edu/research/archer/research/biblio1.html.

Archer, S., and C. J. Stokes. 2000. Stress, disturbance and change in rangeland ecosystems. In *Rangeland desertification*, ed. O. Arnalds and S. Archer, 17–38. Dordrecht, Netherlands: Kluwer Academic Publishers.

Archer S., D. S. Schimel, and E. A. Holland. 1995. Mechanisms of shrubland expansion: Land use, climate or CO_2. *Climatic Change* 29:91–99.

Archer, S., T. W. Boutton, and K. A. Hibbard. 2001. Trees in grasslands: Biogeochemical consequences of woody plant expansion. In *Global biogeochemical cycles in the climate system*, ed. E. D. Schulze, 115–138. San Diego: Academic Press.

Archer, S., T. W. Boutton, and C. R. McMurtry. 2004. Carbon and nitrogen accumulation in a savanna landscape: Field and modeling perspectives. In *Global environmental change in the ocean and on land*, ed. M. Shiyomi and H. Kawahata, 359–373. Tokyo: Terra Scientific Publishing.

Ash, A. J., J. G. McIvor, J. J. Mott, and M. H. Andrew. 1997. Building grass castles: Integrating ecology and management of Australia's tropical tallgrass rangelands. *Rangeland Journal* 19:123–144.

Asner, G. P., and K. B. Heidebrecht. 2005. Desertification alters ecosystem–climate interactions. *Global Change Biology* 11:182–194.

Asner, G. P., and R. E. Martin. 2004. Biogeochemistry of desertification and woody encroachment in grazing systems. In

Ecosystems and land use change, ed. R. Defries, G. P. Asner, and R. A. Houghton, 99–116. Washington, DC: American Geophysical Union.

Asner, G. P., S. Archer, R. F. Hughes, R. J. Ansley, and C. A. Wessman. 2003a. Net changes in regional woody vegetation cover and carbon storage in Texas Drylands, 1937–1999. *Global Change Biology* 9:316–335.

Asner, G. P., C. E. Borghi, and R. A. Ojeda. 2003b. Desertification in central Argentina: Changes in ecosystem carbon and nitrogen from imaging spectroscopy. *Ecological Applications* 13:629–648.

Asner, G. P., A. Elmore, L. Olander, R. E. Martin, and A. T. Harris. 2004. Grazing systems, ecosystem responses and global change. *Annual Reviews of Environment and Resources* 29:261–299.

Behnke, R. H., Jr., I. Scoones, and C. Kerven, eds. 1993. *Range ecology at disequilibrium: New models of natural variability and pastoral adaptation in African savannas*. London: Overseas Development Institute.

Belillus, C. M., and F. Roda. 1993. The effects of fire on water quality, dissolved nutrient losses, and the export of particulate matter from dry heathland catchments. *Journal of Hydrology* 150:1–17.

Belsky, A. J. 1996. Western juniper expansion: Is it a threat to arid northwestern ecosystems? *Journal of Range Management* 49:53–59.

Boutton, T. W., S. R. Archer, A. J. Midwood, S. F. Zitzer, and R. Bol. 1998. Delta C-13 values of soil organic carbon and their use in documenting vegetation change in a subtropical savanna ecosystem. *Geoderma* 82:5–41.

Bovey, R. W. 2001. *Woody plants and woody plant management: Ecology, safety, and environmental impact*. New York: Marcel Dekker, Inc.

Canadell, J. G., C. Le Quéré, M. R. Raupach, C. B. Field, E. T. Buitenhuis, P Ciais, T. J. Conway, N. P. Gillett, R. A. Houghton, and G. Marland. 2007. Contributions to accelerating atmospheric CO_2 growth from economic activity, carbon intensity, and efficiency of natural sinks. *Proceedings of the National Academy of Sciences* 104:18866–18870.

Cao, M., K. Gregson, and S. Marshall.1998. Global methane emission from wetlands and its sensitivity to climate change: Effects of organic material amendment, soil properties and temperature. *Atmospheric Environment* 32:3293–3299.

Curtin, C. G. 2002. Integration of science and community-based conservation in the Mexico/ U.S. borderlands. *Conservation Biology* 16:880–886.

Curtin, C. G., N. F. Sayre, and B. D. Lane. 2002. Transformations of the Chihuahuan borderlands: Grazing, fragmentation, and biodiversity conservation in desert grasslands. *Environmental Science and Policy* 5:55–68.

Dale, V., S. Archer, M. Chang, and D. Ojima. 2005. Ecological impacts and mitigation strategies for rural land management. *Ecological Applications* 15:1879–1892.

D'Antonio, C. M., N. E. Jackson, C. C. Horvitz, and R. Hedberg. 2004. Invasive plants in wildland ecosystems: Merging the study of invasion processes with management needs. *Frontiers in Ecology and the Environment* 2:513–521.

Davidson, E. A., C. J. R. Carvalho, I. C. G. Vieira, R. O. Figueiredo, P. Moutinho, F. Y. Ishido, M.T.P. Santos, J. B. Guerrero, K. Kalif, and R. T. Saba. 2004. Nutrient limitation of biomass growth in a tropical secondary forest: Early results of a nitrogen

and phosphorus amendment experiment. *Ecological Applications* 14:S150–S163.

DeFries, R. S., R. A. Houghton, M. C. Hansen, C. B. Field, D. Skole, and J. Townshend. 2002. Carbon emissions from tropical deforestation and regrowth based on satellite observations for the 1980s and 1990s. *Proceedings of the National Academy of Sciences* 99:14256–14261.

Dregne, H. E., and N. T. Chou. 1994. Global desertification dimensions and costs. In *Degradation and restoration of arid lands*, ed. H. E. Dregne, 4–34. Lubbock: Texas Technical University.

du Toit, J., R. Kock, and J. Deutsch. 2009. *Rangelands or wildlands? Livestock and wildlife in semi-arid ecosystems*. Oxford: Blackwell Publishing Ltd.

Duxbury, J. M. 1994. The significance of agricultural sources of greenhouse gases. *Fertilizer Research* 38:151–163.

Eastman, J. L., M. B. Coughenour, and R. A. Pielke, Jr. 2001. Does grazing affect regional climate? *Journal of Hydrometeorology* 2:243–253.

Ellis, J., and K. A. Galvin. 1994. Climate patterns and land-use practices in the dry zones of Africa. *BioScience* 44:340–349.

Emmerich, W. E. 1999. Nutrient dynamics of rangeland burns in Southeast Arizona. *Journal of Range Management* 52:606–614.

Eswaran, H., R. Lal, and P. F. Reich. 2001. Land degradation: An overview. In *Responses to land degradation*. Proceedings of 2nd International Conference on Land Degradation and Desertification, ed. E. M. Bridges, I. D. Hannam, L. R. Odeman, F. W. T. Pening de Vries, S. J. Scherr, and S. Sompatpanit. Khon Kaen, Thailand. New Delhi: Oxford Press.

Fearnside, P. M., and R. I. Barbosa. 1998. Soil carbon changes from conversion of forest to pasture in Brazilian Amazonia. *Forest Ecology and Management* 108:147–166.

Fernandes, E. C. M., Y. Biot, C. Castilla, A. C. Canto, and J. C. Mantos. 1997. The impact of selective logging and forest conservation for subsistence agriculture and pastures on terrestrial nutrient dynamics in the Amazon. *Ciencia e Cultura* 49:37–47.

Field, C. B., M. J. Behrenfeld, J. T. Randerson, and P. Falkowski. 1998. Primary production of the biosphere: Integrating terrestrial and oceanic components. *Science* 281:237–240.

Foley, J. A., R. DeFries, G. P. Asner, C. Barford, G. Bonan, S. R. Carpenter, F. S. Chapin, et al. 2005. Global consequences of land use. *Science* 309:570–574.

Franklin, K. A., K. Lyons, P. L. Nagler, E. P. Lampkin, P. Glenn, F. Molina-Freaner, F. Markow, and A. R. Huete. 2006. Buffelgrass (*Pennisetum ciliare*) land conversion and productivity in the plains of Sonora, Mexico. *Biological Conservation* 127:62–71.

Fredrickson, E., K. M. Havstad, and R. Estell. 1998. Perspectives on desertification: South-western United States. *Journal of Arid Environments* 39:191–207.

Gallardo, A., and W. H. Schlesinger. 1992. Carbon and nitrogen limitations of soil microbial biomass in desert ecosystems. *Biogeochemistry* 18:1–17.

Geesing, D., P. Felker, and R. L. Bingham. 2000. Influence of mesquite (*Prosopis glandulosa*) on soil nitrogen and carbon development: Implications for global carbon sequestration. *Journal of Arid Environments* 46:157–180.

Gehring, C., M. Denich, M. Kanashiro, and P. L. G. Vlek. 1999. Response of secondary vegetation in Eastern Amazonia to relaxed nutrient availability constraints. *Biogeochemistry* 45:223–241.

Geist, H. J., and E. F. Lambin. 2001. *What drives tropical deforestation?* Rep. LUCC Report Series No. 4. Louvain-la-Neuve, Belgium: Land Use and Cover Change International Project Office.

Gillson, L., and M. T. Hoffman. 2007. Rangeland ecology in a changing world. *Science* 315:53–54.

Gimeno-Garcia, E., V. Andreu, and J. Rubio. 2000. Changes in organic matter, nitrogen, phosphorus, and cations in soil as a result of fire and water erosion in a Mediterranean landscape. *European Journal of Soil Science* 51:201–210.

Guenther, A., S. Archer, J. Greenberg, P. Harley, D. Helmig, L. Klinger, L. Vierling, M. Wildermuth, P. Zimmerman, and S. Zitzer. 1999. Biogenic hydrocarbon emissions and landcover/climate change in a subtropical savanna. *Physics and Chemistry of the Earth.* 24:659–667.

Guild, L. S., J. B. Kauffman, L. J. Ellingson, D. L. Cummings, and E. A. Castro. 1998. Dynamics associated with total aboveground biomass, C, nutrient pools, and biomass burning of primary forest and pasture in Rondonia, Brazil during SCAR-B. *Journal of Geophysical Research-Atmospheres* 103:32091-32100.

Heitschmidt, R. K., and J. W. Stuth, eds. 1991. *Grazing management: An ecological perspective.* Portland, OR: Timber Press.

Henry, B. K., T. Danaher, G. M. McKeon, and W. H. Burrows. 2002. A review of the potential role of greenhouse gas abatement in native vegetation management in Queensland's rangelands. *Rangeland Journal* 24:112–132.

Herrick, J., B. T. Bestelmeyer, S. Archer, A. J. Tugel, and J. R. Brown. 2006. An integrated framework for science-based arid land management. *Journal of Arid Environments* 65:319–335.

Hibbard, K., D. Schimel, S. Archer, D. Ojima, and W. Parton. 2003. Grassland to woodland transitions: Integrating changes in landscape structure and biogeochemistry. *Ecological Applications* 13:911–926.

Holecheck, J. L., R. D. Pieper, and C. H. Herbel. 2003. *Range management: Principles and practices.* 5th ed. London: Prentice-Hall.

Houghton, R. A., J. L. Jacklers, and K. T. Lawrence. 1999. The U.S. carbon budget: Contributions from land-use change. *Science* 285:574–578.

Huenneke, L. F., J. P. Anderson, M. Remmenga, and W. H. Schlesinger. 2002. Desertification alters patterns of aboveground net primary production in Chihuahuan ecosystems. *Global Change Biology* 8:247–264.

Hughes, R. F., S. Archer, G. P. Asner, C. A. Wessman, C. McMurtry, J. Nelson, and R. J. Ansley. 2006. Changes in aboveground primary production and carbon and nitrogen pools accompanying woody plant encroachment in a temperate savanna. *Global Change Biology* 12:1733–1747.

Huston, E., C. A. Taylor, and E. Straka. 1994. *Effects of juniper on livestock production.* Technical Report 94-2. College Station: Texas A&M University.

Imhoff, M. L., L. Bounoua, T. Ricketts, C. Loucks, R. Harriss, and W. T. Lawrence. 2004. Global patterns in human consumption of net primary production. *Nature* 429:870–873.

IPCC (Intergovernmental Panel on Climate Change). 2001. *Climate Change 2001: The scientific basis.* Cambridge: Cambridge University Press.

Jackson, R. B., J. L. Banner, E. G. Jobbagy, W. T. Pockman, and D. H. Wall. 2002. Ecosystem carbon loss with woody plant invasion of grasslands. *Nature* 418:623–626.

Jackson, R. B., E. G. Jobbágy, R. Avissar, S. B. Roy, D. J. Barrett, C. W. Cook, K. A. Farley, D. C. le Maitre, B. A. McCarl, and B. C. Murray. 2005. Trading water for carbon with biological carbon sequestration. *Science* 310:1944–1947.

Johnson, K. A., and D. E. Johnson. 1995. Methane emissions from cattle. *Journal of Animal Science* 73:2483–2492.

Kauffman, J. B., R. L. Sanford, D. L. Cummings, D. L. Salcedo, and I. H. Sampaio. 1993. Biomass and nutrient dynamics associated with slash fires in neotropical dry forests. *Ecology* 74:140–151.

Kauffman, J. B., D. L. Cummings, and D. E. Ward. 1998. Fire in the Brazilian Amazon, II: Biomass, nutrient pools and losses in cattle pastures. *Oecologia* 113:415–427.

Klein Goldewijk, C. G. M., and J. J. Battjes. 1997. A hundred year (1890–1990) database for integrated environmental assessments (HYDE, version 1.1). Rep. 422514002. Bilthoven, Netherlands: National Institute of Public Health and the Environment.

Knapp, A. K., J. M. Briggs, S. L. Collins, S. R. Archer, S. Bret-Harte, M. S. Ewers, B. E. Peters, D. P. Young, D. R. Shaver, and G. R. Pendall. 2008. Shrub encroachment in grasslands: Shifts in life form dominance mediates continental scale patterns of C inputs. *Global Change Biology* 14:615–623.

Lal, R. 2001. Potential of desertification control to sequester carbon and mitigate the greenhouse effect. *Climatic Change* 51:35–72.

Lal, R. 2004. Global extent of soil degradation and desertification. In *Soil degradation in the United States*, ed. R. Lal, T. M. Sobecki, T. Iivari, and J. M. Kimble, 23–29. Boca Raton, FL: Lewis Publishers, CRC Press.

Mack, R. N., D. Simberloff, W. M. Lonsdale, H. Evan, M. Clout, and F. Bazzaz. 2000. Biotic invasions: Causes, epidemiology, global consequences and control. *Ecological Applications* 10:689–710.

Markewitz, D., E. Davidson, P. Moutinho, D. Nepstad. 2004. Nutrient loss and redistribution after forest clearing on a highly weathered soil in Amazonia. *Ecological Applications* 14:S177–S199.

Martin, R. E., and G. P. Asner. 2005. Regional estimate of nitric oxide emissions following woody encroachment: linking imaging spectroscopy and field studies. *Ecosystems* 8:33–47.

McClaran, M. P., D. L. Angell, and C. Wissler. 2002. *Santa Rita experimental range digital database user's guide.* RMRS-GTR-100. USDA Forest Service, Rocky Mountain Research Station.

Milchunas, D. G., and W. K. Lauenroth. 1993. Quantitative effects of grazing on vegetation and soils over a global range of environments. *Ecological Monographs* 63:327–366.

Mordelet, P., L. Abbadie, and J.-C. Menaut. 1993. Effects of tree clumps on soil characteristics in a humid savanna of West Africa. *Plant and Soil* 153:103–111.

Morton, D. C., R. S. DeFries, Y. E. Shimabukuro, L. O. Anderson, E. Aria, F. del Bon Espirito-Santo, R. Frietas, and J. Morisette. 2006. Cropland expansion changes deforestation dynamics in the southern Brazilian Amazon. *Proceedings of the National Academy of Sciences* 103:14637–14641.

Moutinho, P., and S. Schwartzman. 2005. *Tropical deforestation and climate change.* Belem, Brazil: Amazon Institute for Environmental Research.

Okin, G. S., D. A. Gillette, and J. E. Herrick. 2006. Multiscale controls on and consequences of aeolian processes in landscape change in arid and semiarid environments. *Journal of Arid Environments* 65:253–275.

Oldeman, L. R, and G. W. J. van Lynden. 1997. Revisiting the glasod methodology. In *Methods for assessment of soil degradation*, ed. R. Lal, W. H. Blum, C. Valentine, and B. A. Stewart, 423–440. Boca Raton, FL: CRC Press.

Sabine, C. L., M. Heimann, P. Artaxo, D. C. E. Bakker, C. T. A. Chen, C. B. Field, N. Gruber et al. 2004. Current status and past trends of the global carbon cycle. In *The global carbon cycle*, ed. C. B. Field and M. R. Raupach, 17–44. Washington, DC: Island Press.

Sanchez, P. A. 1976. *Properties and management of soils in the tropics*. New York: John Wiley and Sons.

San Jose, J. J., R. A. Montes, and M. R. Farinas. 1998. Carbon stocks and fluxes in a temporal scaling from a savanna to a semi-deciduous forest. *Forest Ecology and Management* 105:251–262.

Schlesinger, W. H., and A. M. Pilmanis. 1998. Plant–soil interactions in deserts. *Biogeochemistry* 42:169–187.

Schlesinger, W. H., J. F. Reynolds, G. L. Cunningham, L. F. Huenneke, and W. M. Jarrell. 1990. Biological feedbacks in global desertification. *Science* 247:1043–1048.

Scholes, R. J., and Archer S. R. 1997. Tree-grass interactions in savannas. *Annual Review of Ecology and Systematics* 28:517–544.

Scifres, C. J. 1985. *Integrated brush management systems for South Texas: Development and implementation*. College Station: Texas Agricultural Experiment Station.

Scifres, C. J., W. T. Hamilton, B. H. Koerth, R. C. Flinn, and R. A. Crane. 1988. Bionomics of patterned herbicide application for wildlife habitat enhancement. *Journal of Range Management* 41:317–321.

Steinfeld H., C. de Haan, and H. Blackburn. 1997. Options to address livestock–environment interactions. *World Animal Review* 88:15–20.

Steinfeld, H., P. Gerber, T. Wassenaar, V. Castel, M. Rosales, and C. de Haan. 2006. *Livestock's long shadow: Environmental issues and options*. Rome: FAO.

Stoddart, L. A., A. D. Smith, and T. W. Box. 1955. *Range management*. New York: McGraw-Hill.

Thurow, T. L. 1991. Hydrology and erosion. In *Grazing management: An ecological perspective*, ed. R. K. Heitschmidt and J. W. Stuth, 141–160. Portland, OR: Timber Press.

Tiffen, M., and M. Mortimore. 1994. *Environment, population growth, and productivity in Kenya: A case study of Machakos district*. Drylands Issue Paper 47. London: International Institute for Environment and Development.

Tilman, D., P. Reich, H. Phillips, M. Menton, A. Patel, E. Vos, D. Peterson, and J. Knops. 2000. Fire suppression and ecosystem carbon storage. *Ecology* 81:2680–2685.

Townsend, A. R., G. P. Asner, C. C. Cleveland, M. Lefer, and M. M. C. Bustamante. 2002. Unexpected changes in soil phosphorus dynamics following forest-to-pasture conversion in the humid tropics. *Journal of Geophysical Research* 107:8067–8076.

UNEP (United Nations Environment Programme). 1994. UN Earth Summit Convention on Desertification. Presented at UN Conference on Environment and Development, Rio de Janeiro, Brazil, 1992.

Valentin, C., J. Poesen, and Y. Li. 2005. Gully erosion: Impacts, factors and control. *Catena* 63:132–153.

van Auken, O. W. 2000. Shrub invasions of North American semiarid grasslands. *Annual Review of Ecology and Systematics* 31:197–215.

Vlek, P. L. G., G. Rodríguez-Kuhl, and R. Sommer. 2004. Energy use and CO_2 production in tropical agriculture and means and strategies for reduction or mitigation. *Environment, Development and Sustainability* 6 (1–2): 213–233.

Wessman, C. A., S. Archer, L. C. Johnson, and G. P. Asner. 2004. Woodland expansion in the US grasslands: Assessing land-cover change and biogeochemical impacts. In *Land change science: Observing, monitoring and understanding trajectories of change on the earth's surface*, ed. G. Gutman, A. C. Janetos, C. O. Justice, E. F. Moran, J. F. Mustard, R. R. Rindfuss, D. Skole, B. L. Turner II, M. A. Cochrane, 185–208. Heidelberg: Springer.

Wheeler, C. W., S. Archer, G. P. Asner, and C. R. McMurtry. 2007. Climate and edaphic controls on soil carbon-nitrogen response to woody plant encroachment in a desert grassland. *Ecological Applications* 17:1911–1928.

Wilcox, B. P., M. K. Owens, R. W. Knight, and R. K. Lyons. 2005. Do woody plants affect stream flow on semi-arid karst rangelands? *Ecological Applications* 151:127–136.

Williams, D. G., and Z. Baruch. 2000. African grass invasion in the Americas: Ecosystem consequences and the role of ecophysiology. *Biological Invasions* 2:123–140.

6

The Impact of Animal Production Systems on the Nitrogen Cycle

*James Galloway, Frank Dentener, Marshall Burke, Egon Dumont, A. F. Bouwman,
Richard A. Kohn, Harold A. Mooney, Sybil Seitzinger, and Carolien Kroeze*

Main Messages

- **Nitrogen (N) is in short supply.** All organisms require nitrogen for life, but natural sources do not provide enough nitrogen to grow the plants and animals needed to sustain the world's population.
- **Human-created usable nitrogen can be used to grow plants and animals.** The development of the Haber-Bosch process in the early nineteenth century was fundamental to society's ability to grow food. Without this additional source of nitrogen, 48% of the world's population could not be fed today.
- **Most of the nitrogen added to fields or fed to an animal is lost to the environment.** Plants and animals are inefficient in their uptake and incorporation of applied nitrogen. As a result, a large proportion of the nitrogen used to grow plants and animals is released to the environment.
- **Released nitrogen has a cascading series of negative effects on ecosystems and humans.** One atom of nitrogen released to the environment will contribute to most environmental problems of today and can remain active in the environment for years to decades.
- **Industrial animal production systems (IAPSs) are among the largest contributors of nitrogen to the environment**. A large proportion of the nitrous oxide (N_2O) and ammonia (NH_3) emitted to the atmosphere are from IAPS. N_2O contributes to global warming and stratospheric ozone depletion. NH_3 enters the nitrogen cascade and can contribute photochemical smog, acid deposition, ecosystems fertilization, and coastal eutrophication. NOx is also lost from IAP systems, and can contribute to tropospheric ozone increases. In addition, much of the nitrogen lost to coastal ecosystems is from IAPS. The nitrogen first contributes to eutrophication and then, if converted to N_2O, contributes to global warming and stratospheric ozone depletion.

- **There are regions (hot spots) where nitrogen losses to the environment from IAP systems dominate the nitrogen cycle.** Atmospheric and riverine losses to the environment from IAP systems are especially pronounced in very large regions of Asia, western Europe, and smaller parts of North America, Latin America, and Africa.
- **IAP systems result in large-scale transport of nitrogen exchange among continents due to international trade in fertilizer, feed, and meat.** With the significant growth of the international transport of nitrogen-containing commodities, nitrogen from one region can be transported thousands of kilometers, used as fertilizer or feed or for another purpose, and then be lost to the environment, contributing to the global impact of nitrogen.

Introduction

We are surrounded by nitrogen, which is fortunate as we and all organisms require it for life. Unfortunately, most of the nitrogen is in the molecule N_2, which cannot be directly used by 99% of the world's species. Living organisms can only use this unreactive nitrogen if the triple bond holding the two nitrogen atoms together is broken. Once that happens, the nitrogen atoms become "reactive." This reactive nitrogen (Nr) is defined as all biologically active, chemically reactive, and radiatively active nitrogen compounds in the atmosphere and biosphere of Earth. Thus Nr includes inorganic reduced forms of nitrogen (e.g., NH_3 and NH_4^+), inorganic oxidized forms (e.g., NO_x, HNO_3, N_2O, and NO_3^-), and organic compounds (e.g., urea, amines, and proteins).

There are two natural processes, biological nitrogen fixation (BNF) and lightning, that have the energy resources needed to create reactive nitrogen; on a global basis, the former process is much more important than the latter. But these natural processes do not produce enough Nr to supply the N needed to feed the world's peoples. Initial solutions to this problem were either to plant crops (e.g., legumes) that promoted BNF or to mine existing sources of Nr (nitrate deposits or guano) and ship them to food-producing regions. These solutions worked until the end of the nineteenth century when, once again, there was not enough Nr to produce food for the world's peoples (Smil 2001). Near the beginning of the twentieth century, an industrial process was developed to convert N_2 to NH_3. Now, at the beginning of the twenty-first century, the Haber-Bosch process—the nitrogen fixation reaction of N_2 with H_2, over a substrate, to produce ammonia—supplies the majority of the Nr used in food production.

The good news is that the Haber-Bosch process can be an almost unlimited source of nitrogen not only to sustain current food production (though there are large regions that do not get enough nitrogen), but also to sustain future increases in food production driven by growth in both population and per capita food consumption. The only limitations on nitrogen produced by the Haber-Bosch process are a source of energy to drive the process and a feedstock to generate the H_2. Currently, most facilities use natural gas for both.

The unfortunate reality, however, is that all of the nitrogen used in support of the production of food is ultimately released to the environment. Some is lost immediately after fertilizer application, some is lost during crop and animal production and harvest, and some is lost during the food preparation process. And, of course, all the Nr that is ingested by humans is released to the environment by excretion or death. This would not be a problem if there were not serious impacts of the released nitrogen on both ecosystem and human health.

In summary, while energy production and industrial Nr uses account for some of the Nr released to the environment, food production in general accounts for the bulk of the remainder, with animal production accounting for the most.

As discussed in other chapters of this volume, meat production has increased by a factor of ~25 during the last 200 years (see also Steinfeld et al., 2006, Steinfeld and Wassenaar 2007). Currently about 243 million metric tons of meat is produced globally per year. Assuming that protein is 16% of the meat, and that N is 16% of the protein, this equates to ~6 Tg N embedded in the meat—all of which is eventually lost to the environment. In addition, as shown later in this chapter, for every Tg N in meat, substantially more N is released to the environment during the meat production process.

This chapter has the following objectives:

- Determine the N cycle of animal production systems in the context of the N cycle for food production and the N global cycle (1860 and current)

- Assess the degree that IAP systems contribute to the distribution of Nr via commerce and the loss of Nr to rivers and the atmosphere
- Review the consequences of the disruption at local, regional, and global scales
- Propose an overall management strategy to diminish the losses of Nr to the environment from IAP systems.

In keeping with the structure of this volume, our definition of IAP systems includes both intensive and mixed livestock production systems and does not include extensive systems that use predominantly noncommercial inputs. An important characteristic of the IAP systems is that meat production is increasingly spatially decoupled from feed production, and thus the large amounts of manure generated cannot be used to grow feed (NRC 2000, Steinfeld et al., 2006).

This chapter focuses on nitrogen in the context of IAP systems. There are, of course, many other nutrients involved in the production of meat. Notable among these is phosphorus (P), both from the magnitude of its use as a fertilizer and from the fact that it is often a limiting nutrient in natural ecosystems. In light of this importance, Box 6.1 provides a brief overview of P. Clearly a much more thorough analysis would be appropriate, not only for its own sake but also due to linkages with both the N and carbon (C) cycles.

Setting the Scene

Global N Cycle, Past and Present

Based on a recent estimate of N cycles in 1860 and the early 1990s, Figures 6.1 and 6.2 illustrate the impacts that human Nr creation has had on the global cycles of nitrogen. In 1860, the most important impacts of human action were from cultivation-induced biological nitrogen fixation (creating new Nr) and from biomass burning and animal husbandry (mobilizing existing Nr). Rates from all three of these activities were small relative to natural biological nitrogen fixation. The one exception is that even in 1860 food production in general resulted in NH_3 emissions to the atmosphere (primarily from animal waste) that were about the same magnitude as natural emissions from soils.

By the early 1990s, the creation of Nr by human activities and the subsequent movement of Nr through environmental systems on a global basis were either larger than natural processes or of the same magnitude. Arguably, even with the associated uncertainties (e.g., BNF) (Cleveland et al., 1999), human actions have altered nitrogen's cycle the most of any major elemental cycle.

There are several important points to be noted from the comparison of the 1860 N budget to that of the early 1990s:

Box 6.1. Phosphorus Utilization for Animal Production in Global Trade

The total global trade in phosphorus is very large, amounting to 10 Tg, which is a large fraction of the 16.4 Tg consumed globally in fertilizer (IFA undated, FAO 2006). Based on data from the Food and Agriculture Organization and the International Fertilizer Industry Association, 48% of the fertilizer shipped is used to grow animal feed (3.8 Tg), with the other 52% going to food or other uses. Of the phosphorus shipped in crops, 57% is used as feed (1.2 Pg). This gives a total of 5.1 Tg of the 10 Tg traded phosphorus being used for animal production. In addition, since the phosphorus in grains is not all directly available to pigs and chickens, more readily available phosphorus is added to their rations, increasing the amount of phosphorus total used by livestock (5% of phosphorus mined) (IFA undated).

The global spread of phosphorus from where it is mined (at present, principally the United States, Africa, and China) to the localities around the world where it is consumed brings special environmental problems. The phosphorus not taken up by crops or deposited through manure accumulates in the soil, and often in not readily available forms. The accumulation rate globally has been estimated at 8 Tg per year, principally to agricultural soils, which includes the amounts from weathering and natural erosion (Bennett et al., 2001). This then poses a very large "time bomb" of potential release to the ecosystems most sensitive to phosphorus additions—freshwater systems, where it can stimulate algal blooms. Whereas nitrogen moves quickly from the soil to the atmosphere and to water bodies, the low solubility of phosphorus compounds and their adsorption to soil particles means that movement across the agricultural landscape to water bodies is slow. Most of the movement is by particulate phosphorus, not by dissolved phosphorus, and hence is by erosion and thus is most prevalent in cultivated soils.

Source: FAO 2006, IFA undated.

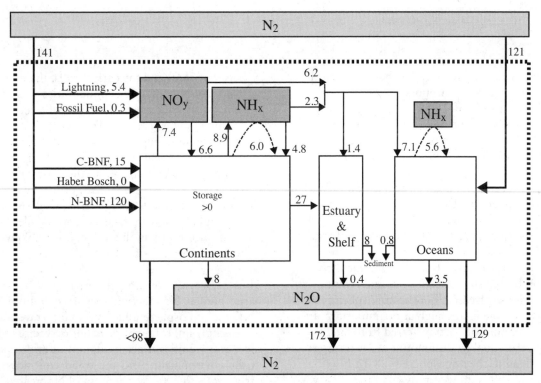

Figure 6.1. Components of the global nitrogen cycle for 1860, Tg N yr⁻¹. All shaded boxes represent reservoirs of nitrogen species in the atmosphere. Creation of Nr is depicted with bold arrows from the N₂ reservoir to the Nr reservoir (depicted by the dotted box). N-BNF is biological nitrogen fixation within natural ecosystems; C-BNF is biological nitrogen fixation within agroecosystems. Denitrification creation of N₂ from Nr within the dotted box is also shown with bold arrows. All arrows that do not leave the dotted box represent inter-reservoir exchanges of Nr. The dashed arrows within the dotted box associated with NHx represent natural emissions of NH₃ that are re-deposited on fast time-scales to the oceans and continents (Galloway et al., 2004).

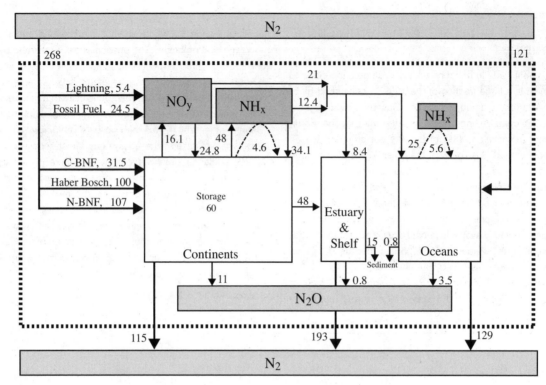

Figure 6.2. Components of the global nitrogen cycle for early 1990s, Tg N yr⁻¹. Figure format is as described for Figure 6.1.

- Human activities create more Nr than do natural terrestrial activities. Anthropogenic Nr creation increased from ~15 to ~160 Tg N yr⁻¹, primarily due to increases in food and energy production; natural Nr creation by terrestrial BNF decreased from ~120 to ~107 Tg N yr⁻¹, primarily due to land use changes.

- As more Nr is created, a larger proportion is emitted to the atmosphere. In 1860, 34 Tg N yr⁻¹ were emitted (24% of the total Nr created). By the early 1990s, 104 Tg N yr⁻¹ were emitted (~40% of total Nr created).

- As Nr creation increased from 1860 to the early 1990s, both atmospheric emissions and riverine concentrations increased, but the atmospheric emissions became increasingly more important. In 1860, the amount of Nr transferred from land to the coast was equal to the amount of Nr transferred from the land to the atmosphere (~28 Tg N yr⁻¹). By the early 1990s, however, twice as much Nr was emitted to the atmosphere (~100 Tg N yr⁻¹) than was transferred to coastal waters (48 Tg N yr⁻¹).

- As discussed in more detail later, commodity transport of Nr (i.e., fertilizer, grain, and meat) is important. In the early 1990s, the most important interregional transport process of Nr was export of N-containing commodities, primarily N fertilizer;

between 1860 and the early 1990s, the increase in export of N-containing commodities was about the same as the increase in discharge of riverine N to coastal systems (i.e., ~30 Tg N yr⁻¹).

- Nr is accumulating in environmental systems. It is estimated that of the ~160 Tg N yr⁻¹ of anthropogenic Nr created in the early 1990s, about 100 Tg N yr⁻¹ was denitrified, primarily in soil/water systems; most of the remainder, ~60 Tg N yr⁻¹, accumulated in terrestrial systems. While there is great uncertainty in this value, it is unlikely that the accumulation rate is zero.

Thus between 1860 and the early 1990s, on a global basis, human Nr creation surpassed natural terrestrial BNF, the atmosphere and commodity transport became increasingly important as distribution vectors for Nr, and there was significant accumulation of anthropogenic Nr in the environment. And the trend continues: from the early 1990s to 2005, the amount of Nr created by human activities increased from ~160 Tg N yr⁻¹ to ~190 Tg N yr⁻¹ (Galloway et al., 2008).

N Flows into and Losses from Agroecosystems

Of special relevance here is that animal production used much of the reactive N created to sustain food production. In 2004, 243 million metric tons of meat were produced:

170 million metric tons from nonruminants (poultry and pigs) and 73 million metric tons from ruminants (beef, sheep, and goats). Over 60% of the nonruminants were produced by industrial operations (pork >50%; poultry >70%), while less than 5% of the ruminant production was from industrialized facilities (Table 6.1), suggesting that N wastes associated with raising nonruminant production are typically much more concentrated geographically than those associated with ruminants.

Furthermore, Table 6.1 shows that nonruminants account for roughly 80% of concentrate feed inputs to animal agriculture globally, the production of which is typically associated with large losses of N to agroecosystems. Cattle, on the other hand, are typically grazed on unfertilized pasture or fed with forages or by-products that require little N-fertilizer input to produce. Thus, despite the relative overall inefficiency with which ruminants convert feed to meat and the larger total feed use by the ruminant sector, the geographical concentration and dependence on concentrate feeds for nonruminant production suggest that this type of production has the more important impacts on the global nitrogen cycle. It is important to note that this chapter focuses on animal production systems and does not consider milk and egg production systems. These systems are also important players in the human impact on the N cycle.

The actual use of nitrogen for animal production in general in the context of total food production is illustrated in Figure 6.3 in the color well.

Globally, croplands receive 170 Tg N yr^{-1}. Of this amount, 49 Tg N yr^{-1} are converted to food for people (one third) and animals (two thirds), while the rest is lost to the environment or recycled back into the system. Of the 33 Tg N yr^{-1} that is fed to animals, only 5 Tg N yr^{-1} enters a human mouth; the rest is lost to the environment or recycled back. Another way of saying this is that of the 170 Tg N yr^{-1} that enters cropland, ~70% is lost to the environment, ~20% is fed to animals, and ~10% is fed to humans. Of the 20% fed to animals, 17% is lost to the environment and 3% is fed directly to humans as an animal product—overall, most of the nitrogen applied to croplands is used to sustain animals; however, a much greater proportion of that N is lost to the environment without entering a human mouth.

The possible fates of the 70% lost to the environment (~120 Tg N yr^{-1}) are emission to the atmosphere (NH_3, N_2O), losses to waters (e.g., nitrate), storage in biomass or soils (e.g., organic N), or conversion to N_2 via denitrification. The allocation to one fate or another depends on a multitude of factors, often determined by the type of animal production system and the characteristics of the region. While the losses can be determined for any specific system, a major unknown in the ultimate fate of the lost nitrogen is whether it is stored as reactive nitrogen or converted back to N_2, thus removing it from the system.

Figure 6.3 is meant to depict the overall N flow in the global agroecosystem. There will be substantial regional variability in most of the terms. One example of

Table 6.1. Global meat production and estimated feed use, 2002 (million metric tons, meat and crop).

	Ruminants		Nonruminants		
	Beef	Sheep and goat	Pork	Poultry	Total
Production (carcass weight)	61	12	95	75	243
Production, industrialized (%)	7	1	57	72	0.45
Estimated concentrate feed inputs					
Cereals	87	7	221	121	436
Oilseed meals	36	2	91	50	179
Roots and tubers	—	—	85	14	99
Subtotal	123	9	397	185	714

	Ruminants	Nonruminants	Total
Estimated nonconcentrate feed inputs			
Forage (arable)	90	—	90
Forage (nonarable)	813	—	813
By-products	75	75	150
Crop residues	350	—	350
Subtotal	1328	75	1403
Total feed inputs	1460	657	2117

such a regional analysis is that of pork production in the United States, as discussed in the next section.

N Flows and Losses during Meat Production in Industrial Animal Production Systems

Between the application of fertilizer to a cropland used to grow feed and the consumption of an animal product, there are numerous steps where N can be lost from the production system and released to the environment. To illustrate these losses, we use the example of pork production in the United States and follow nitrogen from its application to a field to its uptake by a feed crop, its harvest, its uptake by an animal, and the conversion of the animal to a carcass and a product, and ultimately to a consumed pork product.

The flows of reactive N for a typical industrial swine operation are shown in Figure 6.4 in the color well. For every 100 mass units of N fixed by the Haber-Bosch process and legume crop fixation, humans will consume only 17%. The remaining 83% is lost to air or water resources or denitrified back to N_2 at various stages of production. Nitrogen for swine production derives from the application of fertilizer for cereal grain (e.g., corn) or nitrogen fixation from soybeans. These crops take up N already in the soil as well as use N from fertilizer or fixation. When crops are harvested, the roots remain to decompose and supply N for subsequent crops, and often crop vegetative mass (stover) is returned to the field. In a given year, a typical crop may acquire less than half its N from fertilizer or fixation, with the remainder coming from the soil and biological processes from earlier years. However, the apparent crop N uptake may be 80–90% and represents the N in the crop as a percentage of what was applied (or assumed fixed) that year. It is the net uptake, year over year, considering the temporary losses to soil that are recovered in subsequent years.

Only 35% of the N fed to hogs is captured in animal growth, with the remaining 65% excreted to manure. Manure is generally applied to crops but used at reduced efficiency compared with chemical fertilizer. Organic N in manure is mobilized for crop growth over time, but the timing is difficult to predict, adding to uncertainty that is met with additional N application. Harvested animals provide meat for human consumption and by-products that are used for pet foods or other animal feeds. For purposes of these calculations, these were assumed to be recycled for swine production.

The flows of N for industrial swine production are similar to what would be expected for poultry meat production (CAST 1999). In contrast, dairy and beef cattle consume larger quantities of by-products and whole-plant material (forages) that cannot be consumed directly by humans. Thus the efficiency of feed N utilization is lower but the conversion of N in crop to feed is higher, and more N is recycled from other food production systems (CAST 1999). With the increasing use of corn ethanol production for energy, the by-product of fermentation containing much of the N is distillers grain, which is used by cattle.

In contrast, plant-derived foods consumed directly by humans vary greatly in the amount of N lost to the environment in their production. Humans consume cereal and legume grains that can be produced with little N loss to the environment, but some industrial vegetable and fruit production uses high levels of fertilizer application relative to the N in harvested and consumed food. For example, a typical recommendation for N fertilization of corn grain is 170 kg/ha (Maryland Cooperative Extension 2000), while typical yields would be 85 kg N per ha (Maryland Department of Agriculture 1998). Thus the apparent conversion of N to grain would be about 50% (85/170) of applied N. Typical fertilizer application for lettuce would be 80 kg/ha (Maryland Cooperative Extension 2000) with typical yields of 15 kg N per ha (Maryland Department of Agriculture 1998). The apparent conversion of applied N to food would be 19% (15/80).

These show greater efficiency of N capture by animal production than reported previously (Smil 1999, 2001). In the earlier analysis, the effect of N recycling within the system was not considered, and older data were used to estimate the efficiency of feed N utilization by animals.

N Losses to the Environment

Nitrogen is lost to the environment via emissions to the atmosphere, discharge to surface and groundwaters, and retention in soils. On a global basis, in the early 1990s, of the 268 Tg N yr^{-1} of reactive N that was created on continents (~60% from human activities), 115 Tg N yr^{-1} was emitted to the atmosphere as N_2 and removed from active interaction with people and ecosystems. Of the remaining 153 Tg N yr^{-1}, 48 Tg N yr^{-1} were discharged to coastal ecosystems via rivers, and ~40 Tg N yr^{-1}, 48 Tg N yr^{-1}, and 12 Tg N yr^{-1} were emitted to the atmosphere as NOx, NH_3, and N_2O, respectively.

A number of processes contributed to these losses of reactive N to the environment—natural and anthropogenic (energy production, crop production, and animal production, including IAPSs). The remainder of this section presents spatially defined maps of N losses to coastal systems and of NO, NH_3, and N_2O emissions to the atmosphere due to all processes and due to IAPSs alone. With this format, we can then discuss the role of IAPSs in altering the nitrogen cycle.

As will be seen, there is substantial regional variability in the losses of N to the environment, with hot spots occurring in regions with dense population of animals commonly associated with IAPSs. There are thus local impacts in these hot spots, as well as impacts thousands of kilometers downwind and downstream of the hot spots.

The losses of nitrogen to the environment are not evenly distributed over the continents. There are hot spots due to both energy and food production. For IAPSs, the hot spots can be inputs to coastal ecosystems downstream of regions of intense feed production or animal production or they can be N deposition to terrestrial and aquatic ecosystems downwind of regions with large ammonia emissions. There are also hot spots of N_2O emission, but due to the long-lived nature of N_2O, the emissions are well distributed through the atmosphere.

Unless otherwise noted, the uncertainties in the fluxes are noted in the supporting references.

NH_3 Losses to the Atmosphere and Subsequent Deposition

In 1860, the three largest sources of NH_3 to the global atmosphere were soils, oceans, and domestic animals, all at ~6 Tg N yr^{-1} (Galloway et al., 2004). By 2000, while the first two remained the same, emissions from domestic animals had increased to 33.3 Tg N yr^{-1}, with emissions from IAPS being by far the largest component at 29.6 Tg N yr^{-1}.

Figure 6.5a in the color well shows the global 1° × 1° distribution of total annual NH_3 emissions, consisting of emissions from intensive and extensive agriculture; natural emissions from land and oceans; emissions related to the burning of agricultural waste, biofuel, savannas, and forests; and emissions related to domestic and industrial processes. While there are hot spots of NH_3 emissions in most regions, they are especially prevalent in South and East Asia, where emissions > 5 g N m^{-2} yr^{-1} are common (in contrast to background emission rates on the order of 0.1 g N m^{-2} yr^{-1}).

To illustrate the influence of IAPS on global and regional NH_3 emissions, we distinguish between volatilization from grazing animals, animal houses and manure storage systems, and manure spreading in cropland and grassland. The volumes of manure in each system are taken from Bouwman et al., 2006a. Emission factors for grazing animals and animal houses and manure storage systems are from Bouwman et al., 1997. Volatilization from spreading of animal manure and fertilizer is calculated with the empirical model presented by Bouwman et al., 2002a, based on crop type, manure or fertilizer application mode, soil cation exchange capacity, soil pH, and climate. Fertilizer type is used as a model factor for calculating NH_3 volatilization from fertilizers. All manure applied to cropland is presumed to be incorporated, while manure is applied to grassland by broadcasting.

The results are that total NH_3 emissions from ~4,900 Mha of agricultural land are 33.3 Tg N per year with a contribution of 29.6 Tg NH_3-N per year from the 2,000 Mha covered by IAPSs and only 3.6 Tg NH_3-N per year for pastoral systems. (See Figure 6.5b in the color well.) Note that the emissions from IAPSs and pastoral systems both include emissions from fertilizer use, manure spreading, grazing, and animal houses and manure storage. However, in pastoral systems emissions from grazing animals are dominant.

While hot spots of total NH_3 emission from IAPSs are found mostly in Asia, on a relative basis there are large regions of the world where NH_3 emissions from IAPSs are >80–90% of total NH_3 emissions. In most of central North America, southern Latin America, western and central Europe, southern Asia, and southern Australia, emissions from IAPSs account for >80% of the total NH_3 emissions to the atmosphere. (See Figure 6.5c in the color well.)

Once emitted to the atmosphere, NH_3 can react with acidic aerosol to form ammonium. Ammonium aerosol can travel hundreds of kilometers from its point of emission. Thus an assessment of the impact of IAPS NH_3 emissions on ecosystems has to be expressed in terms of where it is deposited, how much N is deposited, and what fraction came from IAPSs. Figure 6.6a in the color well gives the deposition of global total N (NOy and NHx) deposition calculated using the TM5 model (Krol et al., 2005) on a global 3° × 2° resolution using meteorology for the year 2000. Given that background N deposition to terrestrial systems is on the order of 100 g N m^{-2} yr^{-1} (Galloway et al., 2004), it is interesting to note that very large regions of most continents receive at least 10 times that amount. (See Figure 6.6a.)

The portion of total N deposition that is directly attributable to IAPS sources is illustrated in Figure 6.6b in the color well. Not surprisingly, N deposition in East and South Asia (especially), western Europe, and the middle of the United States is heavily influenced, if not controlled, by IAPSs. Relative to other sources of NHx in deposition, there are large regions of all continents except Africa and Antarctica where IAPSs account for >70% of the total N deposition. (See Figure 6.6c in the color well.) In other words, ammonia emissions from IAPSs contribute >70% of total NHx deposition in large regions of the United States, Europe, and Asia. Relative to total (NOy and NHx) deposition, NHx deposition attributable to IAPSs is still >70% for regions of South Asia and South America (Figure 6.6d in the color well).

N_2O Losses to the Atmosphere

Calculations for N_2O emission from soils under natural vegetation are based on the model developed by Bouwman et al. (1993). The calculations are based on a regression model for the natural N_2O soil emissions, and fluxes vary with climate, net primary production, and soil conditions. This regression model is modified from Kreileman and Bouwman (1994) on the basis of a more extensive set of measurement data covering a wider range of ecosystems.

Land cover data are from the IMAGE 2.4 model (Bouwman et al., 2006b). Data on N fertilizer and animal manure inputs (Bouwman et al., 2005) are used to calculate emissions from agricultural fields of N_2O and for each 0.5° × 0.5° grid cell. For direct N_2O emissions from fertilizer and animal manure application, we use residual maximum likelihood (REML) models. REML is a statistical approach that is particularly appropriate for analyzing unbalanced data sets with missing data. The REML model for N_2O is based on 846 series of measurements in agricultural fields (Bouwman et al., 2002b). The calculations differ slightly from the original results because here we used updates for the spatial datasets for soil organic carbon content, soil drainage, and soil pH (Batjes 2002).

The model is based on environmental factors (climate, soil organic C content, soil texture, drainage, and soil pH), management-related factors (N application rate per fertilizer type, type of crop, with major differences between grass, legumes, and other annual crops), and factors related to the measurements (length of measurement period and frequency of measurements). The factors used for calculating NO emissions include the N application rate per fertilizer type, soil organic C content, and soil drainage. The advantage of these N_2O and NO models is that they compute total emissions instead of the methods for calculating the anthropogenic fertilizer-induced emissions of N_2O (as proposed by IPCC 1997). In addition, since they are based on a larger data set, the uncertainty is much smaller than in estimates for the fertilizer-induced emission of IPCC (1997).

N_2O emissions from animal manure housing and storage systems and from nitrogen excreted during grazing are based on the corresponding emission factors from IPCC (1997) and estimates of animal storage as presented by Bouwman et al. (2006a).

The global emissions from soils under natural vegetation for the year 2000 amount to 4.6 Tg N_2O-N per year, while total emissions from ~4,900 Mha of agricultural land are 6.0 Tg N_2O-N per year, with a contribution of 4.1 Tg N_2O-N per year from the 2,000 Mha covered by IAPSs. The emissions from the IAPSs include 2.9 Tg N_2O-N per year from fertilized cropland, 0.2 Tg N_2O-N per year from fertilized grassland, 0.5 Tg N_2O-N per year from housing and storage systems, and 0.5 Tg N_2O-N per year from grazing. (See Figure 6.7 in the color well.) This reflects the intensive use of nitrogen fertilizers and the cycling of nitrogen in the intensive livestock production system.

The spatial patterns of the emissions in the IAPSs clearly reflect the regions with high fertilizer consumption levels, such as western Europe, China, India, and North America. High emission rates are also seen in parts of Argentina and Brazil, where large-scale cultivation of soybean occurs.

The contribution of pastoral systems to agriculture of 1.9 Tg N_2O-N per year is also considerable, reflecting the vast area covered by pastoral grazing (~2,800 Mha) and the important contribution of the so-called background emissions. On the basis of the revised IPCC methodology, indirect N_2O emissions from the global agricultural system amount to only about 0.1 Tg N_2O-N per year.

Riverine Fluxes of Dissolved Inorganic Nitrogen

The amount of nitrogen exported by rivers to coastal waters is to a large extent determined by human activities on the land. We used the NEWS-DIN (nutrient export from watersheds; dissolved inorganic nitrogen) model developed by Dumont et al. (2005) to calculate export of DIN by rivers that is attributed to inputs from manure. NEWS-DIN is a spatially explicit model for predicting DIN export by rivers to coastal waters as a function of watershed characteristics (e.g., population density and agricultural activities) and river characteristics (e.g., runoff, human water extraction, and location of dams). The model uses the $0.5° \times 0.5°$ STN30 global river network database by Vörösmarty et al. (2000a, 2000b) as a basis for the calculations. NEWS-DIN can be used not only to quantify DIN export rates by rivers but also to estimate the contributions of different land-based sources of N to total DIN export rates.

The NEWS-DIN model only includes dissolved inorganic N, not dissolved organic N (DON) or particulate N (PN), for which models exist as well (Beusen et al., 2005, Harrison et al., 2005). Global estimates indicate that DIN accounts for about one-third of the total N export to coastal seas, while DON and PN account for two thirds (Seitzinger et al., 2005). Even though DON and PN are not directly biologically available, they may be mineralized over time; therefore we may be underestimating the impact of animal production on coastal ecosystems if we focus only on DIN. This underestimation is, however, probably small. NEWS model results indicate that, globally, only 11% of global DON inputs to coastal waters can be attributed to anthropogenic diffuse sources, including animal production (Seitzinger et al., 2005). For PN it is as yet not possible to estimate the impact of agriculture on river export. It can be argued, however, that most anthropogenic PN fluxes are not associated with industrialized animal production systems. We believe that using one of the available global TN models (Bouwman et al., 2005, Boyer et al., 2006) for our study would not have resulted in better estimates of the impact of animal production systems on Nr inputs to coastal waters.

A number of sources contribute to riverine DIN export: agricultural N inputs (mainly synthetic fertilizer and manure), atmospheric N deposition, biological N_2 fixation, and sewage effluents. NEWS-DIN results indicate that global DIN export by rivers amounts to 24.8 Tg N y^{-1}. (See Figure 6.8a in the color well.) Animal manure contributes 4.5 Tg N y^{-1} (almost 20% of total DIN export) to this total. This estimate includes all manure excretion in animal houses as well as during grazing. Industrialized animal production contributes 3.0 Tg N y^{-1} (about 12% of total DIN export) (see Figure 6.8b in the color well), with the largest fluxes attributable to IAPSs in Asia. Manure produced in animal houses is often collected and used as fertilizer in arable fields or on meadows. In the soils, crops take up part of the manure N. The remainder is lost, in part through leaching and runoff. During N transport from soils via groundwater or surface water to coastal seas, a considerable part is denitrified or accumulates in sediments. The NEWS-DIN estimates take this denitrification into account.

The share of manure N in DIN export differs regionally. (See Figure 6.8c in the color well.) It is interesting to note that hot spots can be found on most continents. Relatively large contributions of manure to total DIN export rates are calculated for areas in western and central Europe, South and East Asia, the East and Gulf Coasts of the United States, and parts of Latin and South America. In addition, parts of Eastern Australia and New Zealand show up as hot spots.

International Trade of N-Commodities (Fertilizer, Grain, and Meat)

Figure 6.9 in the color well shows N contained in continental trade of meat, feed, and fertilizer associated with industrialized animal production. It was created by aggregating reported bilateral (i.e., country-to-country) trade data for fertilizer, feed, and meat to the continent level, using data from the FAO (2006). Data are from the year 2004. For fertilizer trade, reported data record total shipments of fertilizer (not N) and do not distinguish end use (e.g., agriculture versus other industrial uses). To determine the amount of shipped N in fertilizer going into industrialized animal production, we multiply reported fertilizer trade by the percent N in each type of fertilizer (derived from EFMA 2006), by the estimated percent going to agriculture, and finally by the percent of agricultural production being used as feed (assumed to be the same ratio as feed use relative to total crop production, derived from FAO 2006).

Meat shipments include live and processed meat products from all ruminants and monogastrics, dairy, and egg products. We assume that all traded meat derives from industrialized production. Feed shipments include all cereals, oilseeds, roots and tubers, and forage, corrected for the percentage of crop used as feed in the importing country (based on data derived from FAO 2006). (Feed data for many legumes and oilseeds are opaque in FAO data, due to intermediate and unrecorded processing stages. We assume 80% of soybeans, 50% of rapeseed and related meals, and 50% of potato are used as animal feed.) N contents for meat and feed commodities are taken from Grote et al. (2005) and NRC (1994).

Half of fertilizer N traded is used to produce feed to sustain IAPSs, with major flows going to North America from Latin America and the Middle East and with intraregional flows within North America, Europe, and Asia. (See Figure 6.9a in the color well.) For feed, 72% of the grain N that is traded internationally is used to produce animals in IAPSs (Figure 6.9b in the color well), and a surprisingly large amount of the meat produced

(10%) in IAPSs is traded internationally (Figure 6.9c in the color well).

The total volume of N contained in traded fertilizer used to produce animal feed is roughly 1.7 times the amount of N contained in traded industrial animal feed and almost 20 times the amount contained in traded meat. This relative scaling makes sense for a number of reasons. The Haber-Bosch process used to create most N-rich fertilizer depends on having a ready source of hydrogen (usually derived from natural gas), which many countries wishing to grow crops do not have locally in abundance. Large exporters of N fertilizer are thus typically rich in natural gas, such as Russia (to the European Union) and Trinidad and Tobago (to the United States). Furthermore, most fertilizer is much easier to store and ship than meat, requiring no refrigeration (with the exception of ammonia, which is shipped under pressure as a liquid).

The historical absence of cold-storage technology has also encouraged trade in feed rather than meat, although technological advances over the last few decades have helped to expand meat trade at roughly twice the speed of feed trade. International trade in animal products has also been hobbled historically by relatively high tariff barriers (particularly for sectors such as dairy) and nontariff barriers (e.g., sanitation standards). These barriers have proved slow to come down (OECD 2002). The relative size of N-related feed trade relative to meat has also been fueled in recent years by explosive growth in the international trade of soy. Brazil, in particular, has become a major exporter of soybeans and soymeal in recent years, now rivaling the traditional U.S. dominance in that category. Due to the importance of soy in many industrialized animal diets and to the N-richness of soy, trade in feed accounts for roughly 70% of total crop-related N trade. And summing across all categories, industrialized meat-related products as a whole (feed + meat + fertilizer for feed) account for 23.6 Tg of trade in N, or an amazing 58% of the total agriculture-related international trade in N.

Shipments of N-rich agricultural goods could in some cases add to nutrient depletion or enrichment issues in exporting or importing countries (e.g., Grote et al., 2005). For fertilizer trade, it is unlikely that export of fertilizer will deplete local N sources, since the lion's share of exported N is fixed from abundant available atmospheric nitrogen. Export of feed could also deplete soil N in exporting countries if replacement rates from applied fertilizer do not match what is being extracted and shipped in the feed. However, most soils in large feed-producing areas are considered to be in steady state with respect to N levels, given typically high rates of fertilizer application (e.g., Cassman et al., 2002).

Issues differ for importers of N-rich agricultural goods. Since global trade in nitrogen fertilizer is just under half of total global production, fertilizer trade is clearly crucial for sustaining food production systems around the world. However, abundant fertilizer import and application to crops leads to serious N runoff problems in many parts of the world, given the relative inefficiency with which most major feed crops use applied N (e.g., Cassman et al., 2002). Import of feed into industrial livestock production systems can also indirectly add to local problems of N surplus, given the relative inefficiency with which various livestock convert protein and the associated point source N losses from confined feeding operations. Import of meat for consumption is of the least concern from an environmental perspective, since N wasted during human consumption is typically captured by urban water purification systems and does not enter the environment in a damaging way.

Consequences of N from Meat Production

There are positive and negative consequences of the human alteration of the nitrogen cycle. On the positive side, at the end of the twentieth century there was a growing realization that supplies of naturally occurring nitrogen-containing materials were not sufficient to sustain a growing population and that new nitrogen sources were needed. The Haber-Bosch process provided this new nitrogen to such a degree that 48% of the world's population is fed today because of it (Smil 2001, Erisman et al., 2008).

There are two major classes of negative aspects. The first is that there are still several regions of the world (e.g., sub-Saharan Africa) that are not benefiting from the Haber-Bosch process and do not have enough Nr to sustain food demands. The second is that for those regions that do use commercial N for food production, most of the N is lost to the environment without being incorporated into food, and the N that is incorporated is also lost to the environment.

For the same reasons that N is necessary to produce food (i.e., required for protein synthesis), N is also an important component of ecosystems and is often limiting in their productivity. Given the "leakage" of Nr from the food production process, this often results in unwanted productivity in terrestrial and aquatic ecosystems, especially marine ones.

In addition to the impacts of too much N on ecosystems, there are a number of other negative consequences of excess Nr in environmental reservoirs, including acid deposition, tropospheric ozone, stratospheric ozone, and the greenhouse effect. The combination of the multitude of effects, coupled with the dynamic nature of the nitrogen biogeochemical cycle, has led to the concept of the "nitrogen cascade" (Galloway et al., 2003), which, simply stated, is that once a molecule of Nr is created by energy or food production, it will move through the environment contributing to numerous effects, in sequence. The nature of the effect, and its magnitude, depends on the sensitivity of the environmental system to increases in Nr concentrations and the residence time of the Nr.

As the cascade progresses, the origin of Nr becomes unimportant. The only way to eliminate Nr accumulation and stop the cascade is to convert Nr back to nonreactive N_2.

More specifically, and as presented in Galloway et al. (2003) but updated here, these releases have the following negative consequences:

- Increases in Nr lead to production of tropospheric ozone and aerosols that induce serious respiratory illness, cancer, and cardiac disease in humans (Pope et al., 1995, Wolfe and Patz 2002, Townsend et al., 2003).
- Forest and grassland productivity increase and then decrease wherever atmospheric Nr deposition increases significantly and critical thresholds are exceeded; Nr additions probably also decrease biodiversity in many natural habitats (Aber et al., 1995).
- Reactive N is responsible, together with sulfur (S), for acidification and loss of biodiversity in lakes and streams in many regions of the world (Vitousek et al., 1997).
- Reactive N is responsible for eutrophication, hypoxia, loss of biodiversity, and habitat degradation in coastal ecosystems. It is now considered the biggest pollution problem in coastal waters (Howarth et al., 2000, NRC 2000).
- Reactive N contributes to global climate change and stratospheric ozone depletion, both of which have impacts on human and ecosystem health (e.g., Cowling et al., 1998).

At the simplest level, the contributions of industrial animal production to these effects can be attributed to the proportion of the Nr that is mobilized for animal production. In that regard, Steinfeld et al. (2006) estimate that ~25% of the N fertilizer produced is used to grow feed, oilcakes (mostly soy), and products other than oilcakes (mostly bran) to sustain animal production, mostly industrial. By extension, on a global basis, it can be said that industrial animal production accounts for about 25% of the negative consequences of Nr lost to the environment.

However, that analysis is too simplistic for two reasons—there is tremendous regional variability in both the intensity and the absolute amounts of IAPSs, so impacts have to be examined on a regional basis with a special focus on the hot spots (Figures 6.5–6.9 in the color well), and Nr losses from IAPS do not contribute equally to all environmental and human health issues.

The developing world currently accounts for just over half of the world's poultry, pig, and cattle production, a share expected to grow to around 70% by 2050 (Steinfeld et al., 2006). The developing world thus bears much of the burden of the environmental impacts of Nr loss, and this burden will only increase. Much of the pig and poultry production is via IAPSs. China alone accounts for a quarter of the poultry and half of the pig production. At the other end of the scale, West Asia and North Africa, and sub-Saharan Africa account for 15% of poultry production and 2% of pig production, respectively (Steinfeld et al., 2006). While these regions are less affected by the losses of Nr to the environment, they are also reaping less of the benefits from meat production.

Figure 6.10 in the color well illustrates conceptually where Nr is lost to the environment and the environmental issues associated with the initial losses. Figures 6.5–6.8 illustrate the hot spots of the losses to coastal systems and to the atmosphere. The primary losses to the environment from agroecosystems in general and IAPS specifically are, in order of approximate decreasing magnitude at the global scale, atmospheric NH_3 emissions (volatilization from manure), NO_3^- losses to surface waters and groundwaters (leaching from fertilized fields, manure), atmospheric NOx emissions (fossil fuel combustion in support of IAPSs, emissions from soils), and atmospheric N_2O emissions (from soils). On a subglobal level, generally the greater the level of IAPSs, the greater will be the losses to the atmosphere and waters.

As discussed earlier, these losses can be quantified at the global and regional level. For N_2O, of the 4.5 Tg N emitted to the atmosphere, 1.8 Tg (40%) is due to IAPSs, with regional hot spots occurring in western Europe and South Asia. Given the long residence time of N_2O, the effects of its loss to the atmosphere (greenhouse effect, stratospheric ozone depletion) are not limited to the region of emission—it is a true global consequence of N losses from IAPSs.

On the global scale, 24.8 Tg N of dissolved inorganic nitrogen (mostly NO_3^-) was discharged to coastal ecosystems. Approximately an equal amount was denitrified during the transit of Nr from its point of transfer from the terrestrial system to the stream–river continuum that leads to the coast (Seitzinger et al., 2006). Some of the Nr denitrified is emitted to the atmosphere and contributes to the N_2O emissions discussed in the previous paragraph. Of the DIN that reaches the coast, only about 12% was due to manure generated by IAPSs. While at the global scale this is a small amount, there are hot spots of loss on the regional scale, and numerous studies have shown that Nr losses to coastal ecosystems in those regions have caused eutrophic conditions, leading to hypoxia and anoxia and accompanying ecosystem impacts (NRC 2003).

Integrated View of Animal Protein Production

As summarized in Figure 6.11, N that is fixed for agricultural production either ends up in food products or is lost to the environment. Improving the efficiency of N utilization within the production system is the only way to reduce losses without decreasing food production.

It is difficult to quantify the individual flows of NH_3 and NO_x from agriculture to water and air. Most of these flows are detrimental to the environment, however, and continue to be passed from one reservoir to another once they leave the farm. Therefore it may be more beneficial to aim at reducing the total N loss from the agricultural system rather than reducing only one or another form of loss (NRC 2003). It is also more reliable to quantify total N losses from quantities that are routinely measured and documented on farms, namely the N inputs such as fertilizer and outputs such as animal products. Taking a systems approach enables us to aim at reducing all N losses from the farm, a goal that is more easily quantified as inputs minus production outputs.

The systems approach also applies to optimizing the entire agricultural system rather than focusing on specific farms or processes (NRC 2003). For an individual farm, inputs might include feeds from other farms and outputs might include excess manure. Importing feeds or exporting manure might decrease the farm balance and make it appear more benign to the environment because some N losses will occur on the farm producing the crop or using the manure. Changes on that farm may or may not decrease the total N losses from the agricultural system, including the other farms participating in the process. Nonetheless, it may be good for the environment to outsource crop production to farms that are more efficient at it.

The global demand for specific food products will be one of the most difficult aspects of agricultural production to control. With increasing demand, greater N losses to the environment can be expected unless there is an improved efficiency of utilization of N inputs. Domestic animal and crop species have been selected for economic efficiency, which largely has meant high levels of production per unit of time and space. The ample availability of fixed N has decreased the benefit for high efficiencies of N utilization. When considering environmental costs of N utilization, this equation may differ in the future.

Conclusions

At the beginning of the nineteenth century, global meat consumption was about 10 million metric tons (~10 kg per capita^{-1}) and population was 1.6 billion people. At the beginning of the twenty-first century, population had grown to 6.4 billion people, global meat production had increased to 243 million metric tons, and per capita rates had increased to ~40 kg yr^{-1} on average and to ~80 kg yr^{-1} in developed countries. This increase has a large impact on the release of nitrogen to environmental systems. This impact is magnified because almost half of the meat is produced in IAPSs.

A large fraction (for some regions, >80%) of the N_2O and NH_3 emitted to the atmosphere and Nr injected into coastal ecosystems are from IAPSs. The Nr lost to the environment contributes to most environmental issues

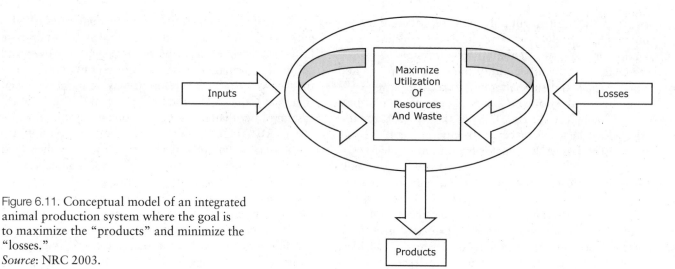

Figure 6.11. Conceptual model of an integrated animal production system where the goal is to maximize the "products" and minimize the "losses."
Source: NRC 2003.

of the day, and the Nr remains in the environment for long periods of time. A particularly vexing question is— of the N released from IAPSs, how much is stored as Nr in both aquatic and terrestrial ecosystems versus being denitrified to N_2? And if it is denitrified, how long has the Nr stayed active before denitrification occurs?

Given that Nr will be created in ever increasing amounts due to IAPSs, the challenge to society is that if N management is not enhanced, environmental damages will be magnified as Nr accumulates in environmental systems. The even greater challenge is that the increasing injection of Nr into environmental systems has profound effects on other biogeochemical cycles, most notably P and C. For many regions of the world, the increasing amount of Nr results in a spinning up of the P and C biogeochemical cycles, which will have additional consequences for both humans and ecosystems.

Acknowledgments

Thanks to Roberta Martin, Henning Steinfeld, and Robert Howarth for helpful comments. We are grateful also to the workshop discussants for their very helpful comments. A very special thank you to Laurie Neville, who with patience and humor helped bring this paper to closure.

References

Aber, J. D., A. Magill, S. G. McNulty. R. D. Boone, K. J. Nadelhoffer, M. Downs, and R. Hallett. 1995. Forest biogeochemistry and primary production altered by nitrogen saturation. *Water, Air, and Soil Pollution* 85:1665–1670.

Batjes, N. H. 2002. Revised soil parameter estimates for the soil types of the world. *Soil Use and Management* 18:232–235.

Bennett, E. A., S. R. Carpenter, and N. F. Caraco. 2001. Human impact on erodable phosphorus and eutrophication: A global perspective. *BioScience* 51:227–234.

Beusen, A. H. W., A. L. M. Dekkers, A. F. Bouwman, W. Ludwig, and J. Harrison. 2005. Estimation of global river transport of sediments and associated particulate C, N, and P. *Global Biogeochemical Cycles* 19:GB4S05.

Bouwman, A. F., I. Fung. E. Matthews, and J. John. 1993. Global analysis of the potential for N_2O production in natural soils. *Global Biogeochemical Cycles* 7:557–597.

Bouwman, A. F., D. S. Lee, W. A. H. Asman. F. J. Dentener, K. W. Van der Hoek, and J. G. J. Olivier. 1997. A global high-resolution emission inventory for ammonia. *Global Biogeochemical Cycles* 11:561–587.

Bouwman, A. F., L. J. M. Boumans, and N. H. Batjes. 2002a. Estimation of global NH_3 volatilization loss from synthetic fertilizers and animal manure applied to arable lands and grasslands. *Global Biogeochemical Cycles* 16:1024.

Bouwman, A. F., L. J. M. Boumans, and N. H. Batjes. 2002b. Modeling global annual N_2O and NO emissions from fertilized fields. *Global Biogeochemical Cycles* 16:1080.

Bouwman, A. F., G. van Drecht, and K. W. van der Hoek. 2005. Nitrogen surface balances in intensive agricultural production systems in different world regions for the period 1970–2030. *Pedosphere* 15:137–155.

Bouwman, A. F., A. H. W. Beusen, K. W. van Der Hoek, W. A. H. Asman, and G. van Drecht. 2006a. Global Inventory of Ammonia Emissions from Global Livestock Production and Fertilizer Use. In *Agricultural air quality*, ed. V. P. Aneja, J. Blunden, P. A. Roelle, W. H. Schlesinger, R. Knighton, D. Niyogi, W. Gilliam, G. Jennings, and C. S. Duke, 180–185. Potomac, MD: Ecological Society of America.

Bouwman, A. F., T. Kram, and K. Klein-Goldewijk. 2006b. *Integrated modeling of global environmental change: An overview of image 2.4.* Bilthoven: Netherlands Environmental Assessment Agency.

Boyer, E. W., R. W. Howarth, J. N. Galloway, F. J. Dentener, P. A. Green, and C. J. Vörösmarty. 2006. Riverine nitrogen export from the continents to the coasts. *Global Biogeochemical Cycles* 20:GB1S91.

Cassman, K. G., A. Dobermann, and D. T. Walters. 2002. Agroecosystems, nitrogen-use efficiency, and nitrogen management. *Ambio* 3:132–140.

CAST (Council for Agricultural Science and Technology). 1999.

Animal agriculture and global food supply. Task Force Report Number 135. Ames, IA: Council for Agricultural Science and Technology.

Cleveland, C. C., A. R. Townsend, D. S. Schimel, H. Fisher, R. W. Howarth, L. O. Hedin, S. S. Perakis, et al. 1999. Global patterns of terrestrial biological nitrogen (N₂) fixation in natural systems. *Global Biogeochemical Cycles* 13:623–645.

Cowling, E. B., J. W Erisman, S. M. Smeulders, S. C. Holman, and B. M. Nicholson. 1998. Optimizing air quality management in Europe and North America: Justification for integrated management of both oxidized and reduced forms of nitrogen. *Environmental Pollution* 102:599–608.

Dumont, E., J. A. Harrison, C. Kroeze, E. J. Bakker, and S. P. Seitzinger. 2005. Global distribution and sources of DIN export to the coastal zone: Results from a spatially explicit, global model. *Global Biogeochemical Cycles* 19:GB4S02.

EFMA (European Fertilizer Manufacturers Association). 2006. Fertilizers: An introduction. http://www.efma.org.

Erisman, J. W, M. S. Sutton, J. N. Galloway, Z. Klimont, W. Winiwarter. 2008. A century of ammonia synthesis. *Nature Geosciences* 1:1–4; doi:10.1038/ngeo325.

FAO (Food and Agriculture Organization). 2006. FAOSTAT statistical databases. http://faostat.external.fao.org.

Galloway, J. N., J. D. Aber, J. W. Erisman, S. P. Seitzinger, R. W. Howarth, E. B. Cowling, and B. J. Cosby. 2003. The nitrogen cascade. *Bioscience* 53:341–356.

Galloway, J. N., F. J. Dentener, D. G. Capone, E. W. Boyer, R. W. Howarth, S. P. Seitzinger, G. P. Asner et al. 2004. Nitrogen cycles: Past, present and future. *Biogeochemistry* 70:153–226.

Galloway, J. N., A. R. Townsend, J. W. Erisman, M. Bekunda, Z. Cai, J. R. Freney, L. A. Martinelli, S. P. Seitzinger, and M. A. Sutton. 2008. Transformation of the nitrogen cycle: Recent trends, questions and potential solutions. *Science* 320:889–892.

Grote, U., E. Craswell, and P. Vlek. 2005. Nutrient flows in international trade: Ecology and policy issues. *Environmental Science and Policy* 8:439–451.

Harrison, J. A., N. Caraco, and S. P. Seitzinger. 2005. Global patterns and sources of dissolved organic matter export to the coastal zone: Results from a spatially explicit, global model. *Global Biogeochemical Cycles* 19:GB4S04.

Howarth, R. W., D. Anderson, J. Cloern, C. Elfring, C. Hopkinson, B. Lapointe, T. Malone, et al. 2000. Nutrient pollution of coastal rivers, bays, and seas. *Issues in Ecology* 7:1–15.

IFA (International Fertilizer Industry Association). IFADATA statistical databases. http://www.fertilizer.org/ifa/Home-Page/STATISTICS.

IPCC (Intergovernmental Panel on Climate Change). 1997. *Guidelines for national greenhouse gas inventories*. Paris: IPCC and Organisation for Economic Co-operation and Development.

Kantor, L. S., K. Lipton, A. Manchester, and V. Oliveira. 1997. Estimating and addressing America's food losses. USDA, Economic Research Service. *Food Review* 20:1–12.

Kreileman, G. J. J., and A. F. Bouwman. 1994. Computing land use emissions of greenhouse gases. *Water, Air, and Soil Pollution* 76:231–258.

Krol, M., S. Houweling, B. Bregman. M. van den Broek, A. Segers, P. van Velthoven, W. Peters, F. Dentener, and P. Bergamaschi. 2005. The two-way nested global chemistry transport zoom model TM5: Algorithm and application. *Atmospheric Chemistry and Physics* 4:3975–4018.

Maryland Cooperative Extension. 2000. *Commercial vegetable production recommendations*. Extension Publication EB-236. College Park: Maryland Cooperative Extension.

Maryland Department of Agriculture. 1998. *Agriculture in Maryland: Summary*. Annapolis: Maryland Department of Agriculture.

NRC (National Research Council). 1994. *Nutrient requirements of poultry*. 9th ed. Washington, DC: National Academy Press.

NRC (National Research Council). 2000. *Clean coastal waters: Understanding and reducing the effects of nutrient pollution*. Washington, DC: National Academy Press.

NRC (National Research Council). 2003. *Air emissions from animal feeding operations: Current knowledge, future needs*. Washington, DC: National Academies Press.

OECD (Organization for Economic Co-operation and Development). 2002. *Agricultural policies in OECD countries*. Paris: OECD.

Pope, C. A. III, M. J. Thun, M. M. Namboodiri, D. W. Dockery, J. S. Evans, F. E. Speizer, and C. W. Heath Jr. 1995. Particulate air pollution as a predictor of mortality in a prospective study of US adults. *American Journal of Respiratory and Critical Care Medicine* 151:669–674.

Seitzinger, S. P., J. A. Harrison, E. Dumont, A. H. W. Beusen, A. F. Bouwman. 2005. Sources and delivery of carbon, nitrogen, and phosphorus to the coastal zone: An overview of global Nutrient Export from Watersheds (NEWS) models and their application. *Global Biogeochemical Cycles* 19:GB4S01.

Seitzinger, S. P., J. Harrison, J. Bohlke, A. Bouwman, R. Lowrance, B. Peterson, C. Tobias, and G. van Drecht. 2006. Denitrification across landscapes and waterscapes: A synthesis. *Ecological Applications* 16:2064–2090.

Smil, V. 1999. Nitrogen in crop production: An account of global flows. *Global Biogeochemical. Cycles* 13:647–662.

Smil, V. 2001. *Enriching the earth*. Cambridge, MA: The MIT Press.

Smil, V. 2002. Nitrogen and food production: Proteins for human diets. *Ambio* 31:126–131.

Steinfeld, H., and T. Wassenaar. 2007. The role of livestock production in carbon and nitrogen cycles. *Annual Review of Environment and Resources* 32:271–294.

Steinfeld, H., P. Gerber, T. Wassenaar, V. Castel, M. Rosales, and C. de Haan. 2006. *Livestock's long shadow: Environmental issues and options*. Rome: Food and Agriculture Organization.

Townsend, A. R., R. H. Howarth, F. A. Bazzaz, M. S. Booth, C. C. Cleveland, S. K. Collinge, A. P. Dobson, et al. 2003. Human health effects of a changing global nitrogen cycle. *Frontiers in Ecology and Environment* 1:240–246.

Vitousek, P. M., R. W. Howarth, G. E. Likens, P. A. Matson, D. W. Schindler, W. H. Schlesinger, and D. G. Tilman. 1997. Human alteration of the global nitrogen cycle: Causes and consequences. *Issues in Ecology* 1:1–17.

Vörösmarty, C. J., B. M. Feteke, M. Meybeck, and R. Lammers. 2000a. Geomorphometric attributes of the global system of rivers at 30-minute spatial resolution. *Journal of Hydrology* 237:17–39.

Vörösmarty, C. J., B. M. Feteke, M. Meybeck, and R. Lammers. 2000b. A simulated topological network representing the global system of rivers at 30-minute spatial resolution (STN-30). *Global Biogeochemical Cycles* 14:599–621.

Wolfe, A., and J. A. Patz. 2002. Nitrogen and human health: Direct and in-direct impacts. *Ambio* 31:120–125.

7

Water-Mediated Ecological Consequences of Intensification and Expansion of Livestock Production

Lisa Deutsch, Malin Falkenmark, Line Gordon, Johan Rockström, and Carl Folke

Main Messages

- **Water needed for animal feed is by far the dominant freshwater resource challenge in the livestock sector, affecting both blue and green water flows in the hydrological cycle.** Two types of water are involved in feed production: "green water" (e.g., naturally infiltrated rain) attached to soil particles and accessible to the roots, is the basis for rainfed production; and "blue water" (e.g., liquid water from rivers and aquifers) is also used for irrigated feed production.

- **Feed production affects water flows in three main ways:** (1) blue water withdrawal (e.g., to irrigate feed crops) affects the availability of blue water downstream; (2) land cover change (e.g., deforestation for soybean production) affects water partitioning and green water use; and (3) changes in land use management (e.g., use of machinery resulting in soil compaction) influence infiltration and water-holding capacity of the soil and therefore blue/green water partitioning at the land surface.

- **Although water is a foundation of functioning ecosystems, lack of comprehensive understanding of this relationship has led to inadequate management.** Current water use estimates of the livestock sector are gross oversimplifications for three reasons: most often one simple figure is used, although there are large differences in total water use between different production systems (and it is particularly important to account for grazing properly); interlinkages between all the processes in the entire hydrological cycle are neglected; and water use figures need to be tied to ecosystem changes—both to understand ecosystem consequences of actions and to allocate water costs correctly.

- **Agricultural activities for feed production influence ecosystem water determinants and therefore disturb ecosystem production and functioning.** The roles that freshwater plays in the landscape as the bloodstream of the biosphere mean that changes in livestock production systems will impact on blue and green water flows, alter ecosystem processes and functions, and ultimately be reflected in the ecosystem services provided. Terrestrial ecosystems depend on green water and aquatic ecosystems on blue water.

- **A five-part ecohydrological framework provides a way to analyze ecosystem effects of hydrological alterations due to livestock production changes.** Livestock activities drive agricultural production system changes, which affect blue/green water partitioning processes; changes in partitioning processes are then reflected in alterations of ecosystem water determinants, and finally in consequences for ecosystem functioning and services.

- **Livestock production involves consumptive use of huge amounts of freshwater for animal feed, whether based on pasture or crops.** In today's livestock production, almost 20% of green water flow, or ~11,900 km^3/yr is used (some 1400 for feed, 10,300 for grazing, and almost 300 for irrigation). Most of the water used in livestock production is green water; blue water plays a minor role.

- **Since food production will have to expand to feed a growing population and meet nutritional improvements as well as new dietary preferences, considerable changes may be expected in blue as well as green water use.** We estimate that past deforestation of lands now used for grazing has already involved large changes in green water flows—with a global reduction of 2200 km^3/yr. By 2050 we may expect a 50% increase of total cropland area, mainly for feed production, which can further decrease green water if it results from deforestation. However, we also expect a 10% increase in total green water flow for livestock production, around 1,000 km^3/yr, mainly from increased production of feed by 2050 in the developing countries. Significant changes, whether decreases or increases, in water use are likely to generate alterations in ecosystem processes, resulting in changes in ecosystem functions and ultimately in the ecosystem services provided.

- **In view of such expected disturbances of ecosystem services in the next four to five decades due to livestock production expansion, strategies and policies have to be developed to facilitate knowledge-based trade-offs between water for meeting increasing demands in terms of human animal protein supply and water for secured ecosystem production and functioning.** To secure ecosystem protection, future water management and research need to expand their blue water focus. A shift in thinking is essential so that green water is properly addressed, and integrated water resources management is extended to incorporate land use.

Water Use in Livestock Production: Problems and Estimation Methods

Livestock production is potentially a huge water user, and global scenarios of future water use show that a shift in diets toward more animal products will be an important driver of increased water scarcity globally (Molden 2007, Rockström et al., 2007; see also Chapter 12). Estimates indicate that, on average, plant-based foods consume approximately 0.5 m^3 of freshwater for every 1,000 kcal, compared with 4 m^3/1,000 kcal for animal-based foods (Falkenmark and Rockström 2004). This is important, since humans already use around 50% of accessible blue water (Postel et al., 1996).

Water Involved in Livestock Production

In livestock production, water is used for meeting animals' drinking needs, servicing (e.g., to wash and cool animals, clean production facilities, and dispose of wastes), processing livestock products (e.g., slaughter, tanning), and producing animal feeds. The annual global water requirements for the first three uses are estimated to be relatively small—no more than 20 km^3 or less than 1% of total annual human freshwater withdrawals (Oki and Kanae 2006, Steinfeld et al., 2006). The main effects on freshwater resources and ecosystem functioning associated with drinking, servicing, and processing are related to pollution (see Chapter 9).

Producing a growing amount of feed for animals, on the other hand, constitutes a very large freshwater use. No economic sector consumes as much freshwater as agriculture, which accounts for 70% of withdrawals of runoff water (so-called blue water) from rivers, lakes, and groundwater resources and 86% of total green and blue water use (Hoekstra and Chapagain 2007, 2008). Grazing uses the soil moisture resource from infiltrated rainfall in the root zone, which becomes evapotranspiration (ET) (green water) from the world's terrestrial

ecosystems (Rockström et al., 1999, Falkenmark and Rockström 2004). Thus, in terms of freshwater use, the water flow needed to sustain plant growth for animal feed is by far the dominant freshwater resource challenge in the livestock sector, affecting both blue and green water flows in the hydrological cycle.

Basically, animals consume crops and/or graze. The four most widely used feed crops are barley, maize, wheat, and soybean (FAO 2007). Ruminants (cattle, sheep) require more fiber in their diets than monogastrics (pigs, poultry) and thus also consume cultivated fodder crops such as alfalfa and hay. The proportion of cropland to grazing land needed will be determined by the particular animal involved (monogastric or ruminant), the location of production, and the production system that is chosen (e.g., intensive can be purely crop based). (See Figure 7.1 in the color well.) In relation to water, animal feed uses freshwater withdrawals (blue water) on irrigated croplands and naturally infiltrated rainwater stored in soil moisture (green water) for crops and grazing lands in rainfed areas.

Feed production affects water flows in three main ways: blue water withdrawal (e.g., to irrigate feed crops, affecting the availability of blue water downstream); land cover change (e.g., when rain forest is converted to soybean production, which affects water partitioning and green water use), and altered water partitioning due to changes in land use management (e.g., runoff increases with soil compaction). Various other effects such as chemical (nutrients, pesticides, organic matter, salinity), physical (suspended sediment load), and bacterial pollution are discussed in Chapters 5, 6, and 9.

Ecosystems depend on rainwater, groundwater, surface water, and water vapor (Falkenmark 2003). Although water is a foundation of functioning ecosystems, bias in study and research has led to a lack of comprehensive understanding of this relationship and hence inadequate management (Falkenmark and Folke 2002). First, traditional freshwater estimations, management, and policy have focused on the liquid water part (the blue runoff flows) of the hydrological cycle and on aquatic ecosystems (planning and allocating stable runoff resources in rivers, lakes, and aquifers) (Falkenmark and Folke 2003). This is despite the fact that blue water on average only accounts for some 40% of the global terrestrial hydrological cycle. The rest of the cycle, reflected in vapor or green water flows (evaporation and transpiration flows), has multiple functions in the human life-support system, such as sustaining all biomass growth. Second, conventional resource management has addressed land and water separately (Falkenmark and Lundqvist 1997). A deeper understanding of the interactions between terrestrial systems and freshwater flows is particularly important in light of present widespread land cover changes (Gordon et al., 2005, Lambin and Geist 2006). Third, only recently has it been recognized that calculations should not focus solely on flows used by humans but should also include water needed for ecosystem function in terrestrial and aquatic systems (Jansson et al., 1999, Rockström 2003, Smakhtin et al., 2004). Thus water must be managed wisely to maintain both agricultural productivity and desired ecosystem services.

To understand water-related effects of feed production, it is important to distinguish between the different water flows in the hydrological cycle that are associated with livestock production. Rainfall is partitioned into blue and green water flows. (See Figure 7.2 in the color well.) Falkenmark coined the term *green water resource* in the mid-1990s to describe the stock of naturally infiltrated rainwater in the root zone (soil moisture) and *green water flow* for the vapor flow in terms of evapotranspiration (Falkenmark 1995).

Blue water flows are liquid water fluxes, such as groundwater flow, base flow in rivers, and surface runoff, while blue water resources consist of the liquid water stored in aquifers, lakes, wetlands, and dams. Blue water flows are formed in two partitioning points. The first, on the land surface, determines the partitioning between surface runoff, direct interception flow (evaporation from ponded water and canopy surfaces), and rainfall infiltration. Surface runoff is a blue flow of "fast" runoff, which forms stream flows in rills and gullies and ultimately feeds lowland areas and rivers. Infiltrated rainfall forms the green water resource in the root zone. A part of this resource forms groundwater flow in the lower partitioning point, while the remainder returns to the atmosphere as green water flow. The groundwater flow (a blue water resource) passes underground through the landscape and surfaces as baseflow to water bodies such as rivers, lakes, and estuaries (a blue water flow). The aggregate of the two horizontal flows, surface and subsurface runoff, forms the blue water resource (aquifers and lakes) and blue water flow (groundwater recharge and river flow).

The green water flow consists of a nonproductive flow (evaporation, E) and a productive flow (transpiration, T), which directly contributes to biomass growth. As will be emphasized in this chapter, there are good opportunities to improve the productivity of water use by shifting vapor from nonproductive E-flow to productive T-flow, thereby not affecting blue water generation. This is critical, as green and blue water flows interact spatially and temporally. There is generally a linear relationship between productive green water flow and biomass growth, which means that more pasture or feed production per unit land area is related to increased green water use, which may reduce blue water generation.

The blue water resource, when used for irrigation, partly undergoes a blue to green redirection and forms evapotranspiration (a green water flow) from the irrigated cropland, while a significant portion normally forms a return flow back to water bodies (a blue water flow). Finally, the amount of precipitation in an area is dependent on vapor from water bodies and oceans and on the green water from terrestrial vegetation in downwind areas.

Therefore, for livestock production, which depends strongly on pastures and feed production, the dominant use of freshwater is not blue water flows, but rather green water flows—that is, infiltrated rainfall forming soil moisture in the root zone, which flows back to the atmosphere as vapor.

Previous Estimates of Water Use by the Livestock Sector

Global Water Estimates

Global evapotranspiration from croplands is estimated at 5,400 km³/year (Rockström et al., 1999) (Figure 7.3 in the color well). It is estimated that nearly 15% (Steinfeld et al., 2006) of the evapotranspiration from irrigated croplands can be attributed to livestock production: 300 km³ of the 1,800 km³ total annual irrigated ET (Postel et al., 1996). Irrigation influences water withdrawals directly but then flows back to the atmosphere as ET. Of the remaining 3,600 km³ ET from rainfed croplands, approximately one-third (1,400 km³/year) is attributable to livestock (Steinfeld et al., 2006). Further, grasslands constitute a major freshwater-consuming biome. An estimated 68% of grassland ET is associated with permanent pastures (livestock grazing, which amounts to an estimated 10,300 km³/year). Thus we estimate that green water flows linked to livestock production from rainfed feed crops, grazing lands, and irrigation-fed feed crops altogether amount to 11,900 km³/year or just less than 20% of the total global green water flow from the continents. These figures do not include forests and wetlands as there are no estimates available at present. As noted earlier, blue water flows used by livestock for drinking, servicing, and processing are tiny—estimated at almost 20 km³/year (Steinfeld et al., 2006). Thus most of the water used is green water.

To give an overview of water flows, Figure 7.3 presents estimates associated with livestock production. Annually, an average of approximately 111,000 km³ of water precipitates on land surfaces globally (Oki and Kanae 2006). Of this total, 61,900 km³ (Rockström et al., 1999) evapotranspires and approximately 49,100 km³ remains available for blue water withdrawal. According to the compilation of data in Figure 7.3, livestock's total share of these global hydrological flows is estimated as equal to about 11,900 km³ or approximately 10% of global water flows.

Despite the importance of livestock and the animal-based foods they provide in determining global freshwater use, the estimates of actual water use by livestock vary tremendously and are often based on overly simplistic generalizations (particularly for intensive livestock systems, where feed is predominantly assumed to originate from crops, and mainly from irrigated crops, which rarely is the case) (see, e.g., Pimentel et al., 1997). Major advances have been made over the past years in revising water use estimates in light of different sources of feed and fodder (de Fraiture et al., 2007).

But estimates of livestock water use remain overly blue-water-oriented as they do not include green water resources from different land use types. What has been lacking from earlier studies are estimates and discussions of the water-related consequences of land cover changes and management practices related to livestock production and how these changes have affected rainwater partitioning processes and therefore caused changes in both green and blue water flows. A critical factor is the analysis of multiple ecological functions provided from water use on pasturelands and the opportunity costs, in both social and ecological terms, of using the water currently consumed on pastureland for other uses.

Estimates for Different Livestock Production Systems

Several studies have approximated the amounts of water used to produce a variety of animal products and animal feeds. Estimated water use varies according to animal production systems, as noted earlier (Figure 7.1), as well as different methods used to calculate and allocate water use, as clearly reflected in the ranges of estimates of water use. (See Table 7.1.)

The many different methods used to calculate water use for animal products and feed have resulted in very different estimates. For example, as seen in Table 7.1, water use in beef cattle production varies from 8,999 to 200,000 L/kg—so by a factor of 20. A major difference between the lowest estimate (Chapagain and Hoekstra 2003) and the highest one (Pimentel et al., 1997) is that the first is for an intensive crop-based feed system where animals grow faster and have shorter life spans, thus increasing the measured water efficiency of production. In contrast, the highest estimate allocates the full ET of all vegetation in rangelands to grazing—that is, it assumes that all vegetation is used by livestock and it allocates no water use to other purposes or to multiple ecological functions in nature. Since water use for feed crops also depends on specific crop water requirements, calculations for extensive livestock production (grazing animals) assume significantly more water use because grasslands and noncultivated fodder crops have lower yields and are less energy-rich and protein-rich feed sources, thus requiring greater areas where livestock products are allocated the

Table 7.1. Estimates of water use for animal products and feeds.

	Pimentel[1,2]	Hoekstra[3]	Chapagain & Hoekstra[4]	Hoekstra & Chapagain[5,6]	SIWI IFPRI IUCN IWMI[7]	*Range of different studies*
Beef cattle:						
Gradient from intensive grain fed grazing	100,000[1]– 200,000	16,000	8,999–13,824		15,000–70,000	8,999–200,000
Dairy cows			85,955–159,523			85,955–159,523
Beef meat			22,000	15,500		15,500–22,000
Milk		900	800	990		800–990
Cheese		5,300	4,600	3,190		3,190–5,300
Poultry (broiler[1])	3,500[1]	2,800	1,028–7,702		3,500–5,700	1,028–7,702
Laying hens			7,241–38,519			7,241–38,519
Eggs		4,700	2,000	3,340		2,000–4,700
Pork meat	6,000[2]	5,900	3,500	4,860		3,500–6,000
Pigs			2,415–4,344[m]			2,415–4,344
Sheep	51,000[2]		6,074–6,435			6,074–51,000
Sheep meat			9,000	6,140		9,000
Soybean	2,000[1]	2,300		1,790	1,100–2,000	1,100–2,300
Maize–Corn	1,400[1]	450		910	1,000–1,800	450–1,800
Alfalfa	900[1]			130[6]	900–2,000	900–2,000
Potatoes (dry wt)	500[1]	160		255[6]	500–1,500	160–1,500
Wheat	900[1]	1,200		1,330	900–2,000	900–2,000

Liters/kg = m³/ton

[m] = mixed animal production system.

Sources: [1](Pimentel et al., 1997); [2](Pimentel et al., 2004); [3](Hoekstra 2003); [4](Chapagain and Hoekstra 2003); [5](Hoekstra and Chapagain 2007); [6](Hoekstra and Chapagain 2008); [7](SIWI et al., 2005).

full water debt. Thus, in order to discuss the differences in livestock water use it is key to account properly for grazing.

Accounting for Grazing

One of the main reasons for the huge differences in the numbers associated with livestock production is that it is difficult to account for grazing. Basically there seem to be three different approaches. The first one argues that since pasturelands would still use water, even without grazing, and that in many cases there is no other likely use of the areas other than grazing, the alternative water cost is zero, and thus water use attributed to livestock in these grazed areas should be zero (Peden et al., 2007, Steinfeld et al., 2006). The second approach allocates the full ET of rangelands to grazing (Pimentel et al., 1997). This results in extensive livestock production getting significantly higher water use than intensive livestock production. The third approach only accounts for the ET for the actual portion of the vegetation consumed by livestock in grasslands (Postel et al., 1996, Falkenmark and Rockström 2004, de Fraiture et al., 2007).

The first two totally contrasting approaches are actually both partially right. There are three issues that should be addressed with regard to allocation of the water to be accounted for in livestock production: multifunctionality of ecosystems, degradation of ecosystems, and the full hydrological effects associated with livestock. The first approach of "no allocation" is correct for ecologically sustainable systems (no land degradation) where grazing is but one of several ecosystem services generated from this landscape. And the approach of "allocating the full water burden for livestock" is correct if grazing is substantially degrading the ecosystem and thus taking place at the expense of other ecosystem services. But there is also a need to consider effects on the entire hydrological cycle, since grazing can also reduce overall water use if it is coupled with deforestation.

Multifunctionality of Ecosystems

Ecosystems are multifunctional systems in that they can generate a whole bundle of ecosystem services at the same time, including provisioning services such as food and fiber, regulating services such as pollination and erosion control, and cultural services such as spiritual values and the potential for recreation (Foley et al., 2005, MA 2005). Depending on management practices, the relative abundance of different services in a specific ecosystem can change. For example, a crop field managed for optimal crop production with little consideration of other services can be expected to have high levels of provisioning services but low levels of regulating and cultural services. A more multifunctional cropland may have less crop yield but higher values of several other services (Foley et al., 2005). Sometimes, synergistic effects between yields and other functions can be found. For example, a

review by Pretty et al. (2006) of investments in resource-conserving agriculture across almost 200 projects in developing countries showed that it is possible to increase yields while enhancing other ecosystem services such as carbon sequestration and pest control.

Similarly, grazed grasslands also generate a full bundle of ecosystem services. If grazing were not practiced, the bundle of services generated would change. Depending on how grazing is managed, the level of other services can be high or low. For example, the reduction of grazing in Sweden has led to losses of cultural ecosystem services associated with an open landscape and biodiversity as low-diverse forests replace meadows (Eriksson et al., 2002, Pykälä 2000). However, since it is so difficult to quantify the specific amount of water that goes only to sustaining grazing, the third approach is not accurate. When grazing sustains or enhances other ecosystem services, the alternative water costs should be zero.

Degradation of Ecosystems

It is estimated that 20% of global pastures and rangelands have been degraded (Steinfeld et al., 2006). In these systems, grazing has reduced the bundle of ecosystem services generated in the system. For example, soil losses due to erosion, bush encroachment, and sedimentation in runoff water causing siltation of downstream systems has reduced grass productivity and resulted in a loss of biodiversity. Thus grazing activities dominate the behavior of the system at the expense of other services and can be allocated the full cost of the entire system.

Such contrasting approaches to allocating costs can result in dramatic differences in estimates of the total water allocated to livestock. This section presents generic approximations of annual water consumption for three main beef-producing livestock systems and considers the impacts of whether pasturelands are attributed to the production system. (See Table 7.2.) If a grazing system is assumed to be sustainable and none of the cost to beef production is allocated to it, then the total water use is 12,000 L/kg meat. This is less than half of the water used if the full burden of grazing is allocated for a degraded system (30,300 L/kg meat).

Full Hydrological Effects: Green Water Flow Changes from Land Conversion to Grazing Lands

This chapter focuses attention on water use from feed crops and grazed grasslands in terms of the amount of green water that leaves land for the atmosphere. In many places around the world grazing takes place in deforested areas. Since forests generally generate higher green water flows than grasslands and pastures do, water use can in fact be reduced by grazing. Gordon et al. (2005) quantified human-induced change in green water flows from deforestation and irrigation at the global scale by comparing actual vegetation cover (Goldewijk 2001) with potential vegetation cover (Ramankutty and Foley 1999).

Table 7.2. Estimates of water use for beef meat in three different production systems.

Feed mix	Different beef production systems							
	INTENSIVE % of feed allocation — Liter water/ kg beef meat	Feed mix %	MIXED % of feed allocation — Liter water/ kg beef meat	Feed mix %	GRAZING zero allocation — Liter water/ kg beef meat	Feed mix %	GRAZING full allocation — Liter water/ kg beef meat	Feed mix %
Crops	32,800	50	20,800	30	7,400	10	7,400	10
Fodder	20,400	50	4,200	30	4,600	10	4,600	10
Pasture	0	0	13,000	40	0	80	18,300	80
TOTAL	53,200		38,000		12,000		30,300	

In our estimates, we allocate the use of the grazing system according to how the pasturelands contribute to the production system. In intensive and mixed systems the allocated amount is determined by the percentage of feed contributed by pastures. In grazing systems, there is zero allocation when grazing sustains or enhances other ecosystem services. However, when grazing activities dominate the behavior of the system at the expense of other services they are allocated the full cost of the entire system.

(Potential vegetation is the vegetation cover assumed to exist in the absence of human activities and is mainly driven by climate.) Figure 7.4 in the color well uses the data from Gordon et al. (2005) to estimate changes in green water flows, but only those associated with grazing (in the vegetation class "marginal croplands, lands used for grazing"). The green areas in Figure 7.4 show no net change, while the orange to red areas show reductions of green water flows. This shows a total global reduction of green water flows of approximately 2,200 km³. This is thus a first rough estimate of how much and where conversion to grazing lands has reduced green water flows to the atmosphere. It agrees with previous results (Gordon et al., 2003) that estimated that the entire change of available blue water for Australia during the last 200 years could be attributed to deforestation.

Need to Link Water Estimates to Ecosystem Changes
An overarching synthesis of the consequences of hydrological alterations related to livestock production for ecosystem goods and services is sorely needed. Such a synthesis is urgent, considering the strong driving forces, including increasing pressure on land and water for agricultural production (Molden 2007) and projected changes in diets, market forces, and trade regimes are all related to livestock production. Increasing demand for animal products is predicted and is expected to be met by further intensification of livestock production, both the grazing-related and the intensive crop-based parts of feed production (FAO 2003). Currently, as previously stated, feed production is responsible for almost 20% of both the blue water use in irrigation and the total vapor flow

from continental lands. In coming decades, Asner et al. (2004) do not anticipate further expansion of marginal lands for pastures, but in fact pasture areas are predicted to intensify and actually lose area to croplands (FAO 2003, Steinfeld et al., 2006). A recent estimate has suggested that by 2050 almost 1,000 additional km³/yr of green water flows will be required for grain feed production for livestock (Rockström et al., 2007).

In conclusion, present water use estimates for livestock production have improved, for example, with the inclusion of green water in water use numbers. But what do the present estimates really say? Current water use estimates of the livestock sector are gross oversimplifications for three reasons. First, most often one simple figure is presented, although there are large differences in total water use between different production systems (this in addition to the large variation between the climate and agriculture practices among different regions, which also affects water usage for feed). Since such large amounts of water are involved in livestock production, several estimates of different production systems need to be made. For example, an estimated 12,000 to 30,300 liters of water are used per kg beef produced in grazing systems, and 53,200 L/kg in intensive systems. It is a gross oversimplification to use one figure to represent the variety of livestock production systems, and it is particularly important to account for grazing properly. Sustainable use of grazing lands allows the areas to provide other ecosystem services as well and thus should not be allocated the full water cost. Second, water use must be considered in relation to all the processes in the entire hydrological cycle, since they are interlinked. Third, water use figures need to be tied to ecosystem changes—both for the sake of understanding ecosystem consequences of actions and to correctly allocate water costs. Scenarios may use water use estimations to discuss trade-offs, but they are insufficient to make management decisions—mainly because production should be tied to the most fundamental factor of production: functioning ecosystems. Therefore, both for the sake of understanding ecosystem consequences of actions and to correctly allocate water costs, this chapter moves from estimates of water use to water flows to ecosystems and the services that ecosystems provide to humans.

Water-Mediated Changes of Increased Livestock Production on Ecosystem Functioning

The Ecohydrological Framework as a Conceptual Model

Because water flows are of fundamental importance to functioning ecosystems, it is imperative to analyze possible challenges to ecosystems and to the goods and services that they provide that are associated with intensified livestock production. This section explores the relationship of livestock production to freshwater resources and ecosystem functioning. In order to link the discussion of livestock-driven activities to changes in ecosystem services and increase understanding of the role of water in dynamic landscapes and ecosystems, an ecohydrological framework is developed. The basic building blocks of the framework are shown in Figure 7.5 in the color well.

Developing the Matrix

We developed a matrix for analyzing ecosystem effects of hydrological alterations due to livestock production changes based on a five-part ecohydrological framework. The method used is an adaptation of the matrix approach developed by Falkenmark (1989) for global change–induced consequences for water-related phenomena. (See Figure 7.6 in the color well.) Progressing through the five parts in four steps A–E, livestock activities (A) drive agricultural production system (agrosystem) changes (B), which affect water partitioning processes (C); the resultant effects of changes in processes are then reflected in alterations of ecosystem water determinants (D), and finally in consequences for ecosystem services (E). The matrix is marked only for the examples described here in the text. In the text the elements of the first part of the step are in bold text and those in the second part of the step are in italics.

Step A–B: Livestock-Driven Activities and Related Agrosystem Changes

The first section of the matrix shows step A–B, linking livestock-driven activities and agrosystem changes. Three general livestock-driven activities (A) result in agrosystem changes (B). As stated earlier, in livestock

production water is used for drinking, servicing, and processing, for production of animal feeds, and for grazing. The resulting agrosystem changes are related to blue water withdrawals for irrigated feed crops, water trade-offs from land cover change (e.g., deforestation to grow feed crops), and the impact on water use of changes in land use management practices (e.g., runoff increases with soil compaction).

As described at the start of this chapter, blue *water withdrawal* for **drinking, servicing, and processing** livestock is minor compared with use for feed production. Therefore the effects of this activity are not explored here, although in some cases it certainly has significant local impacts.

Feed production, on the other hand, affects all three agrosystem changes. It may involve *water withdrawal* when feed crops are irrigated. In the last half-century, irrigated areas have doubled and water withdrawal amounts have tripled (Molden 2007).

Livestock feed production is the dominant way in which *land cover is changed* today (Ramankutty and Foley 1999, Asner et al., 2004). Today, crops and pasture areas are equal to 40% of Earth's land surface, and predictions are that further land conversion and intensification of agriculture will occur (Hill et al., 2006, Ragauskas et al., 2006). Land cover is defined "by the attributes of the Earth's land surface and immediate subsurface, including biota, soil, topography, surface and groundwater, and human (mainly built-up) structures" (Lambin and Geist 2006). It is estimated that animal feed crop production presently appropriates one-third of total cropland production (Asner et al., 2004, Steinfeld et al., 2006). Since the 1950s, livestock production has steadily intensified. Presently, there is a relatively larger increase in the production of monogastrics, particularly chickens, which are more land-efficient in terms of areas (Gerber et al., 2005, Steinfeld et al., 2006). Croplands used by livestock have increased (see Chapter 4), and use of feed concentrates has risen due to increased crop yields and decreased postharvest losses (Asner et al., 2004, Steinfeld et al., 2006).

Hydrologically relevant *land cover conversions* related to the intensification of livestock production include the following:

- Conversion of forests, natural grasslands, or grazing lands to feed crops (such as coarse grains and soybeans), which can either raise (in the grazing to crop conversion) or lower (in the forest to crop conversion) ET used for livestock production.
- Drainage of wetlands for feed croplands.

Feed production sometimes involves only altered *land use management practices* that do not necessarily result in land use conversions but that have indirect impacts on hydrological phenomena, such as the following:

- Changes to other crops—for example, from rice or wheat to soybeans—with differing crop water requirements.
- Increased use of heavy machinery, compacting soils and affecting infiltration and runoff.
- Use of different tillage practices—conventional, conservation, or no tillage—which affects infiltration, drainage, runoff, and soil water-holding capacity.
- Increased use of pesticides, which can reduce key soil organisms and affect infiltration and soil water holding capacity.
- Changes in crop rotation, intercropping, and continuous cropping practices, which can potentially influence soil nutrients status and affect plant water uptake.

As previously mentioned, increasing **grazing** areas for animals remains a key driving force behind land use and land cover changes (Fearnside 2001, Wassenaar et al., 2006). The most significant *land cover change* that indirectly influences the hydrological balance is the conversion of forests and natural grasslands to grazing lands. There is also an opposite trend, though on a much smaller scale, in Europe. Some extensification in the form of increased use of meadows and pastures (i.e., a conversion back from domestic croplands) is actually taking place to maintain biodiversity and open cultural landscapes. However, this is only possible through the supplement of imported protein feed crops.

The following are some changes in *land use management* practices associated with grazing that affect water use:

- *Increasing grazing pressure*—which can compact soils and affect infiltration and runoff; remove vegetation affecting ET levels; and increase erosion, which removes organic matter and affects water holding capacity in soils.
- *Direct modification of vegetation by livestock*, which can influence infiltration patterns and competition in the root zone between, for example, grasses and woody species.

Step B–C: Changes in Water Partitioning Processes Resulting from Agrosystem Changes
The second section of the matrix shows step B–C, where the three types of agrosystem changes (B) just discussed are linked to resulting hydrological alterations of five water partitioning–related processes (C) in ecosystems: vegetation uptake/evaporation (moisture in soil), infiltration/overland flow, groundwater recharge/groundwater flow, stream flow generation, and green water flow (moisture in atmosphere).

The fact that water withdrawals influence stream flow (C.4) is probably obvious. However, the modifications caused by land use change and altered land management practices in the water partitioning processes of incoming rainfall are much more complex, as illustrated in Figure 7.2.

Although livestock practices leading to **water withdrawal** may be relatively small at the global scale, they are not insignificant. Irrigation for feed currently amounts to almost 300 km^3 (Figure 7.3). Water use involved in irrigation, for all agricultural purposes not just feed production, is a redirection of blue to *green water flows*. As stated earlier, this water withdrawal obviously influences *stream flow*. Unfortunately, there are many examples of stream flow depletion, shrinking lake volumes, and river depletion in the irrigated regions of the world (Falkenmark et al., 2007). However, presently, livestock's portion of this depletion is unknown.

Land cover change can affect all the water partitioning processes through biophysical determinants. Changing the vegetative cover can have significant effects on the amount, location, and timing of *vegetation uptake processes*. The most common conversions include forests to grazing lands, forests to feed crops, grasslands to croplands, and wetlands drained for croplands. Land cover change may influence the *infiltration* of rainwater. If infiltration decreases, runoff may increase, raising flood flows (i.e., *streamflow generation*). If water infiltrates and is not taken up by vegetation it contributes to *groundwater recharge*. In Australia, when lands were deforested to make grazing lands for sheep, the rainwater that no longer evapotranspired from trees in the flatlands, percolated downward and raised groundwater levels (bringing grave salinization problems) (Gordon et al., 2003). *Green water flow* depends on transpiration from vegetation and evaporation from soils and water bodies. When a forest is removed and replaced with grazing areas or with nonirrigated crop fields, the amount of transpiration from vegetation is most often reduced. This may also have effects on atmospheric vapor flow and moisture recycling and therefore on precipitation elsewhere, a feedback that has been noted in the Amazon (Oyama and Nobre 2003, Salati and Vose 1984, Savenije 1995).

Land use management practices in feed-producing systems will directly influence water flow partitioning. Changes in management practices will alter the biomass productivity of the feed or fodder production system. As shown by Rockström (2003), while there is a relatively linear relationship between plant transpiration and biomass growth, the relationship between evapotranspiration (i.e., total *green water flow*) and biomass growth is highly nonlinear, particularly at low productivity levels (<4–5 t/ha). This is because *evaporation flows* constitute a large portion of green water flows at lower canopy densities, while transpiration flows dominate in high-yielding systems. Management practices, including tillage, fertilization, crop management, and soil and water management, will affect yields and thereby water use. For example, farmers producing feed cereal crops with poor management systems yielding in the range of 2–4 t/ha of grain dry matter will consume 1,500–3,000 m³/t, while a farmer adopting efficient management systems (e.g., no-tillage systems with good fertilization management) will typically produce in the range of 5–10 t/ha, which will consume at a level of 1,000 m³/t. This has a major impact on green to blue water trade-offs.

Step C–D: Effects on Ecosystem Water Determinants Resulting from Changes in Water Partitioning Processes
The third section of the matrix shows step C–D, linking changes in water partitioning processes to ecosystem water determinants. This next section analyzes how the five mentioned changes in water partitioning processes (C) affect a set of water determinants of ecosystems (D). The four water determinants of ecosystems selected for attention are green water availability (soil moisture), groundwater flow/water table, stream flow/flow regime, and precipitation pattern. Water determinants are the specific environmental conditions of particular importance to a certain type of ecosystem. For example, stream flow/flow regime (D.3) is decisive for the habitat of aquatic ecosystems and inundation-dependent wetlands.

First, **vegetation uptake/evaporation** is the portion of the green water flow that is tied to vegetation uptake in the root zone in the soil. Alterations in vegetation uptake can influence *groundwater flow* if all water is taken up by vegetation and nothing percolates down. For example, the replacement of endogenous cool-season natural vegetation with a warm-season feed crop, like corn or soybean, results in a vegetation demand that is not adapted for local precipitation patterns. This results in vegetation taking up all soil moisture (*green water availability*) during the summer and reduces groundwater recharge. This will also influence the *stream flow regime* in terms of reduced base flow in rivers fed by groundwater outflow. This is of relevance in rivers used for irrigating crop fields (e.g., the Mississippi River in the United States). Vegetation uptake can also influence *precipitation* patterns if there are large changes in the amounts of water taken up in the root zone and thus in the amounts of ET released into the atmosphere. Presently, the Amazon is the best-known example of the effects of vegetation uptake changes associated with the removal of forests and replacement with grasslands. This deforestation broke the regional moisture feedback cycle, transpiration was greatly reduced, and in effect only vapor flow from the ocean remained, thus precipitation was greatly reduced (Oyama and Nobre 2003).

Second, alterations in the ratio of the amounts of water that infiltrate into the soil instead of running off (**infiltration/overland flow**) will directly influence *green water* availability in the soil.

Third, alterations in **groundwater recharge** naturally influence the water table. Wherever percolation increases, *groundwater flows* are increased and *water tables* rise.

Fourth, alterations in **stream flow generation** may affect *stream flow regimes*. Regime changes are related to variation over the course of a year, such as peak and base flows. As stated earlier, stream flow depletion is a widespread phenomenon in rivers with large-scale irrigation, including the Yellow, Indus, Nile, Aral Sea tributaries, Ganges, Murray-Darling, Chao Phraya, Incomati, Rio Grande, and Pangani (Falkenmark et al., 2007). Although agriculture is a major driver of alterations, the portion attributable solely to livestock—while certainly smaller—is not yet determined.

And fifth, **green water flow** is the atmospheric portion of the green water flow and is tied to the moisture feedback phenomenon. (The portion of vegetation uptake in the soil was described earlier.) Large alterations in evapotranspiration, which contributes to atmospheric vapor flow, may influence atmospheric vapor flux and therefore also *precipitation* patterns in downwind regions.

Step D–E: Effects on Ecosystem Services Resulting from Altered Water Determinants
The fourth section of the matrix shows step D–E, linking changes in ecosystem water determinants to ecosystem services. Analysis of the consequences for ecosystems requires consideration of the biological changes resulting from the physical effects of altered water determinants. Ecosystems and ultimately the services they provide depend on a variety of environmental conditions and hydrological phenomena that vary according to the particular type of ecosystem.

The four water determinants (D) can be coupled to ecosystem services (E) as follows:

- **Green water availability** (D.1) determines the soil moisture available for *terrestrial ecosystems*.
- **Groundwater flow/water table** (D.2) determines the water accessible for deep roots of *terrestrial ecosystems*.
- **Stream flow/flow regime** (D.3) is decisive for the habitat of *aquatic ecosystems* and for inundation-dependent *wetlands*.
- The **precipitation** pattern (D.4) is a defining variable behind the hydroclimate in an area and therefore the ecosystems developing there.

In *terrestrial ecosystems*, plant production, both of crops and in natural ecosystems, depends on the green water in the soil because this is where plant roots have access to water. Decreased infiltration leading to reduced green water in soils and ultimately reduced plant production can often be linked to poor management of grazing and croplands. Some terrestrial systems are also particularly vulnerable to changes in the water table. In savanna areas there is very limited lateral groundwater flow, and the areas depend on exceptionally deep-rooted trees and bushes that tap the water table or its capillary fringe directly. Removal of woody vegetation for grazing in these systems can create vulnerability to waterlogging and salinization if the groundwater rises, as happened in Australia after extensive deforestation (Gordon et al., 2003).

Further, grass and crop production in terrestrial agro-ecosystems may be vulnerable to altered streamflow regimes. For example, decreased flood flow can result in a reduction of deposition of nutrients on inundated floodplains used for grazing and feed crops, while changes in precipitation patterns influence the typical terrestrial ecosystems that can develop or exist in a region. The most well-known example tied to livestock production is from the Amazon, where up to 60% of precipitation comes from moisture recycling. Deforestation for expansion of grazing areas and soybean cultivation for livestock have altered moisture feedback, reducing precipitation, and may ultimately lead to savannization of the region. Likewise, the conversion of grasslands to croplands and the resulting increased vapor flows due to irrigation can also alter precipitation patterns, including increasing thunderstorm activities (Falkenmark et al., 2007).

Aquatic ecosystems, such as stream, riverine, and downstream lake habitats, depend on the characteristics of the stream flow, both in terms of average quantity, seasonality (flood flow, dry season flow), variability between and within years, and quality. Reduction in total stream flow, base flow, and flood flow can lead to fragmentation and loss of habitats, changes in aquatic communities, loss of species, and thus severe degradation of downstream ecosystem services (including drinking water) as well as loss of local livelihood options (such as fishing and tourism) along rivers or in downstream lakes (Finlayson et al., 2005, Vörösmarty et al., 2005). Decreased flood flow can also result in less sedimentation and deposition of nutrients on floodplains and reduced flows and nutrient deposition in parts of the coastal zone (Finlayson et al., 2005). Instead, the sediments are deposited in other areas as flows decrease and sediments are carried shorter distances.

Effects on the quantity of the outflow of freshwater to coastal areas is also important to other aquatic ecosystems, especially lagoon ecosystems dependent on slightly brackish water generated by the mixing of fresh groundwater discharge and limited seawater incursion at exceptionally high sea tides. Today, there are many cases where streamflow depletion is evident as an effect of large-scale irrigation in tropical and subtropical regions. However, the specific portion of this that can be tied to livestock production alone is poorly understood and remains to be assessed. Riparian ecosystems in humid regions (which include a variety of ecosystems along upper reaches of a river system) can be affected by groundwater flow alterations if they are fed partly by perennial groundwater discharge and partly by intermittent groundwater flow.

Wetland disappearance is most often due to intentional drainage or infilling for croplands. However, wetland degradation can be coupled to water determinants, depending on wetland type or hydrological context. A wide range of wetland types are determined by their location within the catchment and connectivity with groundwater and the downstream river system (Bullock and Acreman 2003). Examples range from simple surface depressions to groundwater-dependent mountain bogs and flood-dependent inland river floodplains.

Groundwater-dependent wetlands are vulnerable to alterations in groundwater flow coupled with, for example, the lowering of the water table for irrigation. Inundation-dependent wetlands are vulnerable to alterations in stream flow. Base flows can be important for sustaining wet springs in the landscape for local wildlife and small wetlands (potholes) as stopping areas for birds during migrations or in dry seasons. Again, the portion of wetland loss due to hydrological alterations attributable solely to livestock has yet to be determined, as studies with this focus have not yet been done. However, wetlands have been associated with a diversity of hydrological functions, and thus losses are reflected in a number of desirable ecosystem services, such as bird habitats, coastal protection during storms, groundwater recharge, reducing and/or delaying floods, and reduction of lake and coastal eutrophication and plant biodiversity.

Ecosystem Disturbances Generated by Increased Livestock Production

This chapter has shown that there are numerous ways in which agricultural land and water-related activities may influence ecosystem determinants and therefore disturb ecosystem production and functioning. The roles that freshwater plays in the landscape as it circulates as the bloodstream of the biosphere mean that changes in livestock production systems will be reflected in alterations in ecosystem processes and functions and ultimately in the ecosystem services provided. The chapter has developed a knowledge-based matrix for analyzing the impact of livestock production in terms of ecosystem alterations, shown in Figure 7.6.

There has been concern about emerging blue water scarcity and how much additional water will be required for increased feed production. Yet the most recent assessment suggests that the blue water component related to livestock is probably small, currently some 300 km^3/year (Steinfeld et al., 2006). Moreover, it is probable that future blue water use for feed irrigation will remain limited (Molden 2007). However, water-mediated ecological consequences originate less from blue water use for irrigation of feed production and livestock drinking, servicing, and processing than from green water use and altered land use and land management practices that change rainwater partitioning at the ground surface.

Increased feed production basically implies increased green water use, with possible consequences for run-off production and therefore stream flow. Land cover changes can increase green water flows if crops replace grasslands or drastically decrease flows when forests are removed for grazing. Just as changes in land management practices, such as increased animal densities or use of heavy machinery, can affect water processes, in these cases increasing runoff and reducing infiltration.

Thus increased livestock production will affect green water as well as blue. Since food production will have to expand to feed a growing population and meet nutritional improvements as well as new preferences when urbanization expands further, considerable increases may be expected in blue as well as green water use, with consequent effects on ecosystems. Significant changes in water determinants are likely to generate alterations in ecosystem processes, resulting in changes in ecosystem functions and ultimately in the ecosystem services provided.

The provision of livestock goods and services is based on functioning ecosystems. Ecosystems are complex, adaptive systems that are characterized by non-linearity (Levin 1999), multiple states (Scheffer and Carpenter 2003), threshold effects, and time-lags (Jackson 2001) between the onset of resource extraction and consequent changes in ecological communities and, thus, variability and limited predictability (Folke et al., 2004). An increasing number of studies show the existence of thresholds in ecosystems and provide examples of unintended ecosystem regime shifts (Gunderson and Pritchard 2002, Scheffer et al., 2001, Scheffer and Carpenter 2003, Walker and Meyers 2004). These shifts are associated with changes in the ecosystem services provided by these systems to humans and often have negative consequences for human societies. Whether or not the hydrological changes discussed here will trigger a rapid regime shift is still speculative (Gordon et al., 2008). However, increasing studies concerning the savannization of the Amazon, changes in the African and Asian monsoons (Fu 2003), and rainfall changes in West Africa (Zheng and Eltathir 1998) show potentially dramatic changes that may be irreversible. Therefore, it is urgent that humans take responsibility for stewardship (Falkenmark 2003) and actively manage for the uncertainty faced today in terms of how ecosystems will respond in the future to anticipated intensification of livestock production (Steffen et al., 2004).

Conclusions: Future Choices and Trade-offs

The relationship of livestock production to freshwater resources and ecosystem functioning needs further research. This chapter provides a framework to increase understanding of the role of water in dynamic landscapes and ecosystems and uses this framework for livestock production. Although this linear, simplified approach does not capture the dynamic feedbacks in social–ecological systems, it provides a more systematic way to examine what changes in livestock production systems can mean for ecosystem functioning and the ecosystem services provided.

In its scenario chapter (de Fraiture et al., 2007), the Comprehensive Assessment estimated that the production of cereals for feed will rise from 645 Mton/year by 2000 to 1,010 Mton/year by 2050, corresponding to a production increase of 55%. Such production increases will involve considerable agrosystem changes. The water-related changes caused by these agrosystem changes have been estimated by Rockström et al. (2007), based on the water required to produce the recommended 3,000 kcal/day (FAO 2003), of which 20% is animal protein, for 92 developing countries. Calculations based on water productivity data in the literature (see Rockström et al., 1999), estimate that 4 m^3 of water is used per 1,000 kcal consumed. Linking the different categories of changes in that study with the agrosystem changes considered in this chapter, it can be tentatively stated that the scales of water flows involved related to anticipated agrosystem changes will be of three principal types:

- *Increased blue water withdrawal.* Irrigated feed production is estimated to rise from approximately 300 km^3/year to some 500 km^3/year by 2050 (derived from Rockström et al., 2007)—that is, almost doubling. This would lead to further stream flow depletion and would influence the habitats of aquatic ecosystems in the rivers concerned.
- *Land cover change.* Agricultural production areas can expand by some 450 Mha, or by 50%, by 2050. These will certainly involve some land cover changes. On the one hand, land cover changes that increase canopy on feed-producing croplands can increase ET. On the other hand, if the opening of new grazing lands involves the transformation of forests to grasslands and marginal croplands, this implies an overall reduction of ET by some 2,200 km^3/year (Gordon et al., 2005) and an equivalent increase of runoff. Thus there are large changes in the hydrological cycle associated with land cover changes.
- *Altered land management practices.* Practices can be changed to improve water productivity by finding ways to turn evaporation into transpiration. This will depend on improved infiltration and plant water uptake and can involve some 1,500 km^3/year of water that now evaporates but could be brought into the soil. This is an amount of the same

order of magnitude as the water consumed for feed production in 2000 (de Fraiture et al., 2007). Land management practices may also be expected to improve rain capture through rainwater harvesting, reducing overland flow, and are estimated to provide some additional 900 km³/year for feed grain production (Rockström et al., 2007). This corresponds to an approximate 60% increase in the current water use for feed production mentioned earlier.

These projections of water quantities involved in future feed-production-based livestock production suggest that the water implications may be quite considerable. The many ways in which ecosystem services may be disturbed by livestock production expansion in the next four to five decades will therefore require the development of strategies and policies for knowledge-based trade-offs between water for meeting increasing demands in terms of human animal protein supply and water for secured ecosystem production and functioning.

Most of the water involved in meeting the water requirements in livestock production is currently green water influenced by changes in biomass production, land cover, and land management practices. The alterations are by no means limited in size, since they involve green water use increases in the order of 50% or more. Green water flow figures are much larger than blue water flows; in fact, estimated changes of green water tend to be one order of magnitude larger than for blue water. This shows how important the green water perspective is for estimating the ecosystem impacts of an expanding livestock production and provides further support for the call to reconnect livestock to the land (Naylor et al., 2005).

What the world faces is therefore a future full of choices and trade-offs, as well as a future of uncertainty and potentially rapid change (Folke et al., 2004, Kates et al., 2001). To secure ecosystem protection, future water management needs to expand its blue water focus—that is, where aquatic ecosystem protection is addressed by directions of minimum stream flow to be left in the river (so-called environmental flows). A shift in thinking will be essential so that green water is properly addressed, and integrated water resources management is extended to incorporate land use. One of the largest challenges we face today is that we, through our changes, have substantially altered the ecological controls for many of Earth's freshwater systems (Gordon et al., 2005, Meybeck 2003). The magnitude of the effects on ecosystem production and functioning and the risk for passing thresholds will have to be urgently evaluated.

Three main research challenges emerge from this synthesis:

- Better data are needed from studies that can illustrate how water determinants are linked to more specific ecosystem services.

- Although there are several global models of land–water interactions, these are generally too coarse to include the interactions of water determinants and ecosystem services. More specifically, while they include land cover characteristics they cannot deal with changes in management practices (such as rotational grazing, buffer-zones along rivers, or fine-scale nutrient management) in agricultural land. It is therefore currently not possible, at the global scale, to model the effects on ecosystem services of changes in ongoing land use practices, although this may be the area where most potential lies in terms of securing ecosystem services and avoiding trade-offs.

- Finally, this discussion has mainly used a linear cause–effect analysis of changes in water determinants on ecosystem services. In reality, these relations include complex feedbacks that often lead to nonlinear responses in the behavior of ecosystems as well as of hydrological systems. Researchers must do more to include these feedbacks in their analyses and case studies.

Acknowledgments

Thanks to Tove Gordon at tovedesign.se for figures 1, 5 and 6, Robert Kautsky at azote.com for figures 2–3, Garry Peterson for assistance with figure 4. We also greatly appreciate the comments of Jennie Barron, Patrick Fox, Arjen Hoekstra, Jakob Kijne, and Don Peden on earlier drafts of the chapter.

References

Asner, G. P., A. J. Elmore, L. P. Olander, R. E. Martin, and A. T. Harris. 2004. Grazing systems, ecosystem responses, and global change. *Annual Review of Environment and Resources* 29:261–299.

Bullock, A., and M. Acreman. 2003. The role of wetlands in the hydrological cycle. *Hydrology and Earth System Sciences* 7:358–389.

Chapagain, A. K., and A. Y. Hoekstra. 2003. *Virtual water flows between nations in relation to trade in livestock and livestock products.* Report No. 13. Delft, Netherlands: UNESCO–IHE.

de Fraiture, C., D. Wichelns, J. Rockström, E. Kemp-Benedict, N. Eriyagama, L. Gordon, M. A. Hanjra, J. Hoogeveen, A. Huber-Lee, and L. Karlberg. 2007. Looking ahead to 2050: Scenarios of alternative investment approaches. In *Water for food, water for life: A comprehensive assessment of water management in agriculture*, ed. D. Molden, 91–145. London: Earthscan.

Eriksson, O., S. A. O. Cousins, and H. H. Bruun. 2002. Land use history and fragmentation of traditionally managed grasslands in Scandinavia. *Journal of Vegetation Science* 13:743–748.

Falkenmark, M. 1989. Global-change-induced disturbances of water-related phenomena: The European perspective. In Collaborative Paper CP-89-1. Laxenburg, Austria: International Institute for Applied Systems Analysis.

Falkenmark, M. 1995. Land–water linkages: A synopsis, land and water integration and river basin management. Land and Water Bulletin, 15–16. Rome: Food and Agriculture Organization.

Falkenmark, M. 2003. Freshwater as shared between society and ecosystems: From divided approaches to integrated challenges. *Philosophical Transactions of the Royal Society B: Biological Sciences* 358:2037–2049.

Falkenmark, M., and C. Folke. 2002. The ethics of socio-ecohydrological catchment management. *Hydrology and Earth System Sciences* 6:1–9.

Falkenmark, M., and C. Folke. 2003. Introduction. *Philosophical Transactions of the Royal Society B: Biological Sciences* 358:1917.

Falkenmark, M., and J. Lundqvist. 1997. *Comprehensive assessment of the freshwater resources of the world: World freshwater problems call for a new realism*. Stockholm: Stockholm Environment Institute.

Falkenmark, M., and J. Rockström. 2004. *Balancing water for humans and nature: The new approach in ecohydrology*. London: Earthscan.

Falkenmark, M., and J. Rockström. 2006. The new blue and green water paradigm: Breaking new ground for water resources planning and management. *Journal of Water Resources Planning and Management* 132:129–132.

Falkenmark, M., C. M. Finlayson, L. Gordon, E. Bennett, T. M. Chiuta, D. Coates, N. Ghosh, et al. 2007. Agriculture, water and ecosystems: Avoiding the costs of going too far. In *Water for food, water for life: A comprehensive assessment of water management in agriculture*, ed. D. Molden, 233–277. London: Earthscan.

FAO (Food and Agriculture Organization). 2003. *World agriculture: Towards 2015/2030. An FAO perspective*. Rome: Food and Agriculture Organization.

FAO (Food and Agriculture Organization). 2007. FAOSTAT statistical databases. http://faostat.external.fao.org.

Fearnside, P. M. 2001. Soybean cultivation as a threat to the environment in Brazil. *Environmental Conservation* 28:23–38.

Finlayson, C. M., R. D'Cruz, N. Davidson, J. Alder, S. Cork, R. d. Groot, C. Lévêque, et al. 2005. *Ecosystems and human well-being: Wetlands and water synthesis*. Washington DC: Island Press.

Foley, J. A., R. DeFries, G. P. Asner, C. Barford, G. Bonan, S. R. Carpenter, F. S. Chapin, et al. 2005. Global Consequences of Land Use. *Science* 309:570–574.

Folke, C., S. Carpenter, B. Walker, M. Scheffer, T. Elmqvist, L. Gunderson, and C. Holling. 2004. Regime shifts, resilience, and biodiversity in ecosystem management. *Annual Review of Ecology, Evolution & Systematics* 35:557–581.

Fu, C. 2003. Potential impacts of human-induced land cover change on East Asian Monsoon. *Global and Planetary Change* 37:219–229.

Gerber, P., P. Chilonda, G. Franceschini, and H. Menzi. 2005. Geographical determinants and environmental implications of livestock production intensification in Asia. *Bioresource Technology* 96:263.

Goldewijk, K. 2001. Estimating global land use change over the past 300 years: The HYDE database. *Global Biogeochemical Cycles* 15:417–433.

Gordon, L., M. Dunlop, and B. Foran. 2003. Land cover change and water vapour flows: Learning from Australia. *Philosophical Transactions of the Royal Society B: Biological Sciences* 358:1973.

Gordon, L. J., W. Steffen, B. F. Jonsson, C. Folke, M. Falkenmark, and A. Johannessen. 2005. Human modification of global water vapor flows from the land surface. *Proceedings of the National Academy of Sciences* 102:7612–7617.

Gordon, L. J., G. D. Peterson, and E. Bennett. 2008. Agricultural modifications of hydrological flows create ecological surprises. *Trends in Ecology and Evolution* 23:211–219.

Gunderson, L. H., and L. E. Pritchard. 2002. *Resilience and the behavior of large-scale ecosystems*. Washington, DC: Island Press.

Hill, J., E. Nelson, D. Tilman, S. Polasky, and D. Tiffany. 2006. From the cover: Environmental, economic, and energetic costs and benefits of biodiesel and ethanol biofuels. *Proceedings of the National Academy of Sciences* 103:11206–11210.

Hoekstra, A. 2003. Virtual water trade between nations: A global mechanism affecting regional water systems. IGBP Global Change Newsletter 54:23–24. Available at http://www.igbp.kva.se/documents/resources/NL_54.pdf

Hoekstra, A. Y., and A. K. Chapagain. 2007. Water footprints of nations: Water use by people as a function of their consumption pattern. *Water International* 21:35–48.

Hoekstra, A. Y., and A. K. Chapagain. 2008. *Globalization of water: Sharing the planet's freshwater resources*. Oxford, UK: Blackwell Publishing.

Jackson, J. B. C. 2001. What was natural in the coastal oceans? *Proceedings of the National Academy of Sciences* 98:5411–5418.

Jansson, Å., C. Folke, J. Rockström, and L. Gordon. 1999. Linking freshwater flows and ecosystem services appropriated by people: The case of the Baltic Sea drainage basin. *Ecosystems* 2:351–366.

Kates, R. W., W. C. Clark, R. Corell, J. M. Hall, C. C. Jaeger, I. Lowe, J. J. McCarthy, et al. 2001. Environment and development: Sustainability science. *Science* 292:641–642.

Lambin, E. F., and H. J. E. Geist. 2006. *Land-use and land-cover change: Local processes and global impacts*. Berlin: Springer.

Levin, S. A. 1999. *Fragile dominion: Complexity and the commons*. Reading: Perseus Books.

MA (Millennium Ecosystem Assessment). 2005. *Ecosystems and human well-being*. Washington, DC: Island Press.

Meybeck M. 2003. Global analysis of river systems: From earth system controls to anthropocene syndromes. *Philosophical Transactions of the Royal Society B: Biological Sciences* 358:1935.

Molden, D. 2007. *Water for food, water for life: A comprehensive assessment of water management in agriculture*. London: Earthscan.

Naylor, R., H. Steinfeld, W. Falcon, J. Galloway, V. Smil, E. Bradford, J. Alder, and H. Mooney. 2005. Agriculture: Losing the links between livestock and land. *Science* 310:1621–1622.

Oki, T., and S. Kanae. 2006. Global hydrological cycles and world water resources. *Science* 313:1068-1072.

Oyama, M. D., and C. A. Nobre. 2003. A new climate–vegetation equilibrium state for tropical South America. *Geophysical Research Letters* 30:2199.

Peden, D., G. Tadesse, A. K. Misra, F. A. Ahmed, A. Astatke, W. Ayalneh, M. Herrero, et al. 2007. Water and livestock for human development. In *Water for food, water for life: A comprehensive assessment of water management in agriculture*, ed. D. Molden, 485–514. London: Earthscan.

Pimentel, D., Berger, B., Filiberto, D., Newton, M., Wolfe, B., Karabinakis, E., Clark, S., Poon, E., Abbett, E., Mandagopal, S. 2004. Water resources: Agricultural and environmental issues. *BioScience* 54:909–918.

Pimentel, D., J. Houser, E. Preiss, O. White, H. Fang, L. Mesnik, T. Barsky, S. Tariche, J. Schreck, and S. Alpert. 1997. Water resources: Agriculture, the environment, and society. *BioScience* 47 (2): 97–106.

Postel, S., G. Daily, and P. Ehrlich. 1996. Human appropriation of renewable fresh water. *Science* 271:785–788.

Pretty, J., A. D. Noble, D. Bossio, J. Dixon, R. E. Hine, F. W. T. Penning de Vries, and J. I. L. Morison. 2006. Resource-conserving agriculture increases yields in developing countries. *Environmental Science and Technology* 40:1114–1119.

Pykälä, J. 2000. Mitigating human effects on European biodiversity through traditional animal husbandry. *Conservation Biology* 14:705–712.

Ragauskas, A. J., C. K. Williams, B. H. Davison, G. Britovsek, J. Cairney, C. A. Eckert, W. J. Frederick, Jr., et al. 2006. The path forward for biofuels and biomaterials. *Science* 311:484–489.

Ramankutty, N., and J. A. Foley. 1999. Estimating historical changes in land cover: North American croplands from 1850 to 1992. GCTE/LUCC research article. *Global Ecology and Biogeography* 8:381–396.

Rockström, J. 2003. Water for food and nature in drought-prone tropics: Vapour shift in rain-fed agriculture. *Philosophical Transactions of the Royal Society B: Biological Sciences* 358:1997.

Rockström, J., L. Gordon, C. Folke, M. Falkenmark, and M. Engwall. 1999. Linkages among water vapor flows, food production and terrestrial ecosystem services. *Conservation Ecology* 3:5.

Rockström, J., M. Lannerstad, and M. Falkenmark. 2007. Assessing the water challenge of a new green revolution in developing countries. *Proceedings of the National Academy of Sciences* 104:6253–6260.

Salati, E., and P. Vose. 1984. Amazon Basin: A system in equilibrium. *Science* 225:129–138.

Savenije, H. 1995. New definitions for moisture recycling and the relationship with land-use changes in the Sahel. *Journal of Hydrology* 167:57–78.

Scheffer, M., and S. R. Carpenter. 2003. Catastrophic regime shifts in ecosystems: Linking theory to observation. *Trends in Ecology and Evolution* 18:648.

Scheffer, M., S. Carpenter, J. A. Foley, C. Folke, and B. Walker. 2001. Catastrophic shifts in ecosystems. *Nature* 413:591–596.

SIWI, IFPRI, IUCN, IWMI. 2005. *Let it rein: The new water paradigm for global water security.* Final report to CSD-13. Stockholm: Stockholm International Water Institute. Available at http://www.gwsp.org/downloads/SIWI_report05.pdf

Smakhtin, V., C. Revenga, and P. Döll. 2004. A pilot global assessment of environmental water requirements and scarcity. *Water International* 29:307–317.

Steffen, W., A. Sanderson, P. Tyson, J. Jäger, P. Matson, B. Moore III, F. Oldfield, et al. 2004. *Executive summary: Global change and the Earth system: A planet under pressure.* Stockholm: International Geosphere-Biosphere Programme.

Steinfeld, H., P. Gerber, T. Wassenaar, V. Castel, M. Rosales, and C. de Haan. 2006. *Livestock's long shadow: Environmental issues and options.* Rome: Food and Agriculture Organization.

Vörösmarty, C. J., C. Lévêque, C. Revenga, R. Bos, C. Caudill, J. Chilton, E. M. Douglas, et al. 2005. Chapter 7: Fresh water. In *Ecosystems and human well-being: Current state and trends, volume 1,* ed. R. Hassan, R. Scholes, and N. Ash, 165–208. Washington, DC: Island Press.

Walker, B. H., and J. A. Meyers. 2004. Thresholds in ecological and social-ecological systems: A developing database. *Ecological Society* 9:3.

Wassenaar, T., P. Gerber, H. Verburg, M. Rosales, M. Ibrahim, and H. Steinfeld. 2006. Projecting land use changes in the Neotropics: The geography of pasture expansion into forest. *Global Environmental Change* 17:86–104.

Zheng, X., and E. Eltathir. 1998. The role of vegetation in the dynamics of West African monsoons. *Journal of Climate* 11:2078–2096.

8

Global Livestock Impacts on Biodiversity

Robin S. Reid, Claire Bedelian, Mohammed Y. Said, Russell L. Kruska, Rogerio M. Mauricio,
Vincent Castel, Jennifer Olson, and Philip K. Thornton

Main Messages

- **Livestock are having widespread direct and indirect impacts on the foundation of all life**—biodiversity, which signifies the number and diversity of genes, species, populations, and ecosystems. These impacts affect every square kilometer of Earth—on land, in the sea, and throughout the atmosphere.

- **The overall driving causes of biodiversity loss through livestock are similar to those for other aspects of the environment and include the increasing demand and consumption of milk, meat, and eggs,** which leads to a greater need to grow crops and harvest fish to feed livestock. These causes are getting harder to monitor because of long food chains, where impacts occur far from consumers.

- **The impacts of livestock on biodiversity are principally negative, although there are some positive impacts as well.** The effect of livestock on biodiversity depends on the magnitude (or exposure) of livestock impacts, how sensitive biodiversity is to livestock, and how biodiversity responds to the impacts.

- **The negative impacts of livestock on biodiversity include heavier grazing impacts on plants and animals when livestock populations expand;** biodiversity loss from forests as pastures and croplands for feed expand in the tropics (often driven by long-distance trade in feeds); emissions of greenhouse gases that cause climate change and then affect biodiversity; diseases spread by livestock to wildlife; simplification of landscapes through intensification; competition of livestock with wildlife; pollution of watercourses with nutrients, drugs, and sediments, with related effects on aquatic biodiversity; native biodiversity loss through competition with nonnative feed plants; and overfishing to create fishmeal for livestock.

- **The positive impacts include increasing efficiency of production,** where fewer natural resources are used for each kilogram of milk, meat, or eggs produced; increased species diversity in moderately grazed pastures; and pastoral land use protecting wildlife biodiversity in savanna landscapes.

- **While livestock have many direct impacts on biodiversity through trampling, grazing, and defecation, the bigger impacts appear to be indirect**—through deforestation to create pastures; emissions of methane and other greenhouse gases; the growing feed trade; and the pollution of streams, rivers, lakes, and oceans.

- **Effects on marine systems are multiple and unexpected,** through fish harvest for fishmeal, coral loss through climate change, introduction of marine invasion species, and probably dust-transmitting pathogens reaching coral reefs.

- **The impacts occur together in ecosystems in what might be called syndromes.** In the extensive dryland syndrome, on the wetter fringes of the vast drylands, rangelands are contracting to make way for cropping and settlement, with significant impacts on biodiversity. In a second syndrome, affecting dryland key resources like towns, markets, riverine areas, and wetlands, livestock grazing is heavy, wildlife are all but excluded, and only plants tolerant of heavy grazing thrive. In the more productive forests, the deforestation/reforestation syndrome means pastures signify either massive biodiversity loss in tropical rain forests or valuable sources of biodiversity in abandoned pastures that are being reforested in the temperate regions. In drier forests and woodlands used more intensively, the simplified intensive system syndrome is evident: where invasive species are pervasive, wildlife are gone, nutrient pollution is common, and biodiversity is simplified and homogeneous.

- **One considerable gap in understanding includes whole ecosystems or whole region assessments of**

the total impacts of livestock on biodiversity, based on useful tools like life cycle approaches. Another gap is the estimations of what will happen to biodiversity in the future as the demand for meat, milk, and eggs continues to grow. This suggests that one next step is a set of scenarios, based on the information in this assessment, that show different possible alternatives futures for biodiversity, driven by the multifaceted impacts of livestock.

Introduction

The diversity of life on Earth, or biodiversity, is the foundation for its environment and thus the foundation for the survival, health, wealth, and happiness of human society (MA 2005a). This diversity contains all the genes that humans have manipulated throughout history to create more productive and adapted crops, livestock, trees, and fish (Diamond 2002), which are the basis of the human food chain. More important than these very visible services that biodiversity provides to human society are the less obvious ways that biodiversity (and other factors) creates a viable and stable planet for life on Earth as compared to nearby planets. For example, the great diversity of genes, populations, species, and ecosystems on land and in the sea allows the environment to recover more quickly from stress like droughts, floods, or pollution; in turn, it allows human societies to survive, adapt, and be resilient to change, buffering human society from both "natural" and "human-produced" environmental stress (Folke et al., 2004). This diversity can thus be thought of as the safety net for life on Earth by creating system redundancy and dampening variation in critical ecosystem services (see, e.g., Srivastava and Vellend 2005). And livestock are a major force in shaping Earth's native biodiversity and contributing to its domesticated biodiversity (Steinfeld et al., 2006).

It is common to define biodiversity in a cascade of scales from small to large as the diversity of genes, populations, species, and ecosystems on the planet. Biodiversity can also refer to the diversity of functional groups (species with similar functions in the environment, such as pollinators), endemic species (those found only in a limited area and nowhere else), and life forms (plants differing in structure, such as vines, leafy herbs, and trees) (Dirzo and Raven 2003). Biodiversity usually does not refer to the diversity of human ethnic groups or societies. But the term does include the diversity of domesticated plants and animals, like crop varieties and livestock breeds and species (MA 2005a). Because of the large scale of human impacts on native species and the millions of native species in existence (most of which are still unknown to science), the impacts of people on the few thousand domesticated species usually receives less attention than impacts on native species (Sala et al., 2000, MA 2005a). This is the case in this chapter as well.

Biodiversity at the species level is in double jeopardy today because there is limited ability to keep track of species loss, although it is known to be pervasive. Keeping track requires knowing what species exist. But scientists have only a vague idea how many exist: there are between 3 million and 30 million species, of which only 2 million are now known to science (May 1990, MA 2005a). By tracking known species, Pimm and colleagues (1995) estimated that biodiversity loss today is 100–1,000 times faster than before humans started affecting ecosystems and that the loss may increase tenfold by the end of the twenty-first century.

What is the role of people and livestock in extinguishing Earth's biological diversity at unprecedented rates? People eliminate or reduce the diversity of genes, species, populations, and ecosystems by changing habitats of native species to other uses, introducing diseases through the spread of pathogens, introducing nonnative plant and animal species, overharvesting plants and animals, and causing the climate to change (MA 2005a). Acid rain, the deposition of nitrogen, and increased atmospheric carbon dioxide (CO_2) are also important (Sala et al., 2000). There is strong agreement that degradation and loss of habitat are the most important causes (Chapin et al., 2000, Sala et al., 2000, MA 2005b), with climate change a close second (Thomas et al., 2004). As described in this chapter, livestock are part of all these more general causes of biodiversity change, and they have a special role in pollution (Steinfeld et al., 2006).

Broad Trends in Livestock Related to Changes in Biodiversity

Overall, global trends in human society cause changes in the livestock sector and, in turn, these changes initiate local and global environmental changes that affect biodiversity directly. (See Figure 8.1 in the color well.) Some trends magnify the impacts of livestock on biodiversity while others lessen it. And the importance of these trends is not the same everywhere, differing on land and in the water and by the type and intensity of livestock production. There are also important follow-on effects of biodiversity maintenance or loss on other ecosystem services, and feedbacks to human society, but these, although very important, will not be covered in this chapter.

The most important trend is the rising demand for and consumption of livestock products, particularly in the developing world. Already the global livestock sector provides livelihoods for 1.3 billion people, constitutes 40% of total agricultural output, and is growing faster than any other part of agriculture (Steinfeld et al., 2006). As consumer incomes rise, people eat more animal foods and fewer staples, like cereals and root crops (Bennett 1941). This can also happen without a rise in income when rural populations move to cities (Leppman 1999). Production of meat and eggs is rising faster than human population, although milk output has not risen as fast (see Chapter 1). These trends are strongest in the developing world, and, particularly in Asia, will continue in the future (Rosegrant et al., 1999). Add rising meat and milk consumption to human population growth (Chapter 2), and the stage is set for the consumption of meat and milk to nearly double from 2002 to 2030 in the developing world, while rising less than 20% in industrial countries (Steinfeld et al., 2006). Globally, this means that the production of livestock must become far more efficient over the next two decades just to maintain the level of livestock impacts on the environment seen today, much less diminish those impacts.

It is not only the production of animals that is rising rapidly but also the production of the food livestock eat. Already, more than a third (37%) of the cereals farmers grow in their fields are fed to livestock (WRI 2005). Pulses (legumes), starchy roots (potatoes, yams, cassava), and oilseeds are also fed to livestock, as well as fishmeal and the by-products of producing human food (e.g., oilcakes, the byproducts of oil production) (Bouwman and Booij 1998). The fastest growth is in soybeans, which increased sixfold in the last 20 years (see Chapter 2). This means that livestock are responsible not only for animal-related impacts on biodiversity but also the impacts of the production of plants to feed livestock, as described later.

Unfortunately, the impacts of livestock on biodiversity are often invisible to consumers because the producers are so far away (Deutsch and Folke 2005). Food chains are becoming longer and more complicated in the developing world (Reardon et al., 2003), as food travels farther from farm to plate and passes through more hands along this chain. Because of the strong movement of rural populations into urban centers, many consumers live far from farms even within their own countries. And as trade in agriculture expanded 1,000% since 1960 (see Chapter 1), these chains spread horizontally across the entire globe in a spider web of complicated connections.

For example, demand for soybeans from Brazil in both Europe and China has jumped in the last decade, as Europeans partially shift away from livestock by-product feeds to soy feeds in response to "mad cow" disease and as Chinese pig and poultry production grows (Naylor et al., 2005, Nepstad et al., 2006). In Sweden, almost 80% of the feed to support livestock in 1999 came from abroad, often from far away, dominated by soybean cake from Brazil and palm oil seeds from Malaysia and Indonesia (Deutsch 2004, Deutsch and Folke 2005). Food often travels through a few, centralized supermarket chains far away before consumers buy it (Reardon et al., 2003). In addition, new concerns over food safety and the demands of the supermarket chains mean that farm products often need to meet a limited set of standards that simplify and homogenize the products themselves, the

types of animals that produce them, and the farms and landscapes that support livestock (see Chapter 1).

Other important trends for biodiversity are changes in the type of meat people eat. Worldwide, consumers are eating more meat from pigs and poultry, which are monogastrics and emit little methane, and less meat from cows, sheep, goats, and camels, which are ruminants and emit significant methane (see Chapter 1). Monogastrics not only emit less methane, they appear to emit as much or less nitrous oxide, another greenhouse gas, as ruminants do (Steinfeld et al., 2006). Thus this shift may reduce greenhouse gas emissions, slow climate change, and slow impacts on biodiversity. In addition, pigs and poultry are more often produced in intensified systems in the Americas, Asia, and Europe (Gerber et al., 2005; see also Chapter 4), thus this dietary change will shift the impacts of people through livestock away from more extensive systems to the more intensive ones. In these systems, especially if they are industrial, farmers and industries produce meat with little tie to the local land base (Naylor et al., 2005), immediately losing the nutrient recycling that occurs in tightly coupled crop–livestock systems. This means that pollution from livestock will affect biodiversity more and more in the future.

On the positive side, farmers now produce livestock more efficiently: milk, meat, and eggs take less water and feed to produce than they did in the past. In relation to biodiversity, this trend may "save land for nature" (Tilman et al., 2002) and slow the loss of biodiversity caused by livestock. Communications and transport are also becoming more efficient (see Chapter 1), which may reduce the environmental impacts of the livestock sector.

Effect of Livestock Production and Consumption on Genes, Populations, Species, and Ecosystems

Most concerns about livestock and biodiversity surround visible impacts directly connected to livestock, such as farmers felling trees to clear rain forest for pasture, or goats walking through bare soil on a livestock trail in the Sahel. The connection between these events and livestock are obvious and strong. There has been more scientific work in specific places on how livestock herds affect plants and soils, as described later, but it is now apparent that some of the most important impacts on biodiversity may be indirect or even largely invisible and global. On the surface, for example, clearing rain forest or savanna to cultivate soybeans seems unconnected to livestock, yet many of these soybeans become livestock feed (Nepstad et al., 2006). Another indirect impact is climate change, of which an estimated 18% can be attributed to the livestock sector (Steinfeld et al., 2006). As difficult as it is to see these indirect impacts, the picture becomes even less clear when livestock impacts combine with other common human and environment-driven causes of biodiversity change. The complexity of the impacts means there can never be a complete picture of livestock impacts on biodiversity.

People, through livestock, can have negative, neutral, and possibly positive impacts on the diversity of genes, populations, species, and ecosystems, as Figure 8.1 indicated. Negative impacts are rife where large numbers of livestock graze and trample the land heavily or where soils, plants, animals, or whole ecosystems (e.g., rain forest) are particularly sensitive to the way people manage the land to produce livestock (e.g., land clearing, intensification). But there appear to be some surprisingly positive impacts of livestock on biodiversity, especially where livestock grazing prevents people from using land in ways that are less compatible with biodiversity.

What determines, generally, how much people affect biodiversity through livestock? One way to consider this is to borrow a framework used to measure vulnerability of human populations to droughts and other stresses that focuses on the exposure, sensitivity, and resilience of biodiversity to livestock (e.g., Turner et al., 2003). Exposure refers to direct and indirect actions of livestock themselves, including the duration, magnitude, and frequency of their impacts on biodiversity (species, ecosystems). Sensitivity is an inherent character of a species or ecosystem itself and measures how likely biodiversity is to change once exposed to livestock (e.g., slow or fast species loss). Resilience refers to how well biodiversity copes with or responds to livestock impacts.

For example, clearing tropical rain forest for pasture has such strong impacts on biodiversity because the exposure is high (removal of the forest), the forest supports many and very sensitive species, and forest species are less able to cope with (or are less resilient to) clearing because of the massive change of the structure of their habitat. On the other hand, savannas that receive significant rainfall in Africa have supported livestock for thousands of years, and wild grazers for millions of years, and thus their sensitivity to livestock grazing is reasonably low. It may seem simple to predict the impacts of livestock with a framework like this, but as indicated below, different types of species and ecosystems react differently to livestock grazing, with several different responses in a single plant community at a single time (Stoddart et al., 1975, Vesk and Westoby 2001).

How much of Earth's biodiversity is likely influenced by people through livestock? All species and ecosystems, whether they are in protected areas, in Antarctica, or in the sea, are affected by livestock through climate change. This probably includes the majority of the 30 million or so species on the planet. Livestock are a threat to biodiversity in about a third of Earth's biomes and two-thirds of Earth's biodiversity hotspots (Steinfeld et al., 2006). They are a major cause of biodiversity loss in Latin American rain forests and the Brazilian *cerrado*, and probably in some of Earth's marine dead zones, through nitrogen pollution, as in the Gulf of Mexico. So far, livestock do not appear to be a major direct cause of biodiversity loss in most forest systems other than in Latin America, in very cold regions, or in the deep sea.

One way to visualize likely direct impacts of livestock on biodiversity is to map how many livestock (cattle, sheep, and goats) are close to parks and reserves around the world. (See Figure 8.2 in the color well.) Around forested parks and reserves, people often need to clear forest to create pasture for livestock, and these pastures often do not support as many native species as the intact forest does. The pasture often serves as a "sink" for forest biodiversity, with fewer and fewer native species in pastures farther and farther from the forest edge (e.g., Pulliam 1988, Ricketts et al., 2001). Around grassland parks and reserves, the impacts of livestock are not nearly as strong, but some species groups, like large mammals, can suffer from conflict with livestock for forage and water (Prins 2000). More generally, savanna parks surrounded by more people (who also keep livestock) lost more ungulate, carnivore, and primate species in the last 30 years than those surrounded by fewer people (Brashares et al., 2001).

In a map of densities of livestock in close proximity to parks and reserves, livestock are not particularly abundant around many protected areas, particularly those in the cold lands in the north. But in much of the tropics, India, China, eastern North America, Europe, and New Zealand, many livestock live close to protected areas and are a potential threat to the biodiversity these areas protect. Of particular concern are the "at risk" protected areas within biodiversity hotspots: eastern Africa, particularly the Horn and Rwenzori range; Madagascar; along the west coast of India; the Himalayas; southern Europe and eastern Asia; the Amazon and southern Brazil; northwestern Colombia, Ecuador, and Venezuela; most countries of Central America; and southern Mexico. Even protected areas that are not in hot spots are important for some rare and endangered species, such as those in the eastern United States and central Europe, India, China, and New Zealand.

Most of the remainder of this chapter describes eight different major impacts that people have on biodiversity through livestock, followed by a discussion of the relative importance of these impacts. Each section considers the balance of the negative, neutral, and positive impacts as far as they are known at this time.

Impact 1: Broad-Scale Habitat Loss and Fragmentation through Livestock Production

The growing demand for livestock globally is pushing four general trends or processes of habitat change at broad scales that all affect biodiversity, both negatively and positively. These four trends are different in wetter forests or woodlands, where rain is sufficient to support crop cultivation, versus drier savannas, steppe, or deserts, where crop cultivation is limited to areas with irrigation or local groundwater. The first trend is that of expansion (often called extensification) of the area grazed or the land ploughed for feed crop production. In dryland systems, herders extend livestock grazing into places previously inaccessible to them, perhaps because of newly built water points. In wetter forests or woodlands, farmers and herders clear land to grow more pasture or cultivated feeds. Even though there are few places on Earth where new land is available to support this expansion (Rosegrant et al., 2001), agricultural land is still expanding in more than two-thirds of the world's countries (MA 2005a). The second trend is intensification, where farmers and herders concentrate more resources (labor, capital, natural resources) from within and outside farms to produce more livestock or feed crops on the same unit area of land (e.g., Boserup 1965). If this increase in intensity of inputs per unit land is also accompanied by an increase in productivity (e.g., more kg of meat produced per mouthful of feed), then total environmental impacts may decline. For feed crop cultivation, farmers add more plant inputs like fertilizer, water, and pesticides. The third trend is contraction of rangelands, the mirror opposite of expansion of grazing, where herders become farmers or incoming migrants cultivate or settle in rangelands or governments establish protected areas in rangelands, thus shrinking the area grazed. The fourth trend is abandonment, where farmers and herders leave former grazing areas in pastures to regenerate into the pregrazing vegetation (such as a forest).

Deforestation: Pasture Expansion in Forests and Woodlands

Probably the best known example of extensification caused by human demand for livestock products is deforestation in the wet tropics. Before the 1990s, farmers and ranchers cleared rain forest and woodland, sometimes to secure ownership, and they grazed cattle on the new pastures (Schmink and Wood 1992, Fearnside 1993, Hecht 1993). But more recently, as noted earlier, farmers started to clear forest to plant soybeans to feed livestock (Hecht 2005), and this is now outstripping the clearing of forest for pasture (Morton et al., 2006, Nepstad et al., 2006). Although forest loss is common across the tropics and subtropics in Latin America, Asia, and Africa, it is only in Latin America that livestock are clearly the primary cause (Lambin et al., 2003). Deforestation occurs not only in the Amazon but across Latin America in other biodiverse ecosystems, such as the forested uplands of the seasonally flooded Pantanal and the *cerrado* savanna woodlands (Seidl et al., 2001).

Tropical deforestation is so devastating because rain forests support more species than any other biome on Earth. Covering only 7% of Earth's surface, these forests may contain up to 50% of all species in existence (Myers 1988). Many of these are limited only to a forest in a particular region or to a particular area within a large forest. For example, 700 species might be found in just 10 hectares—the same number as found in all of North America (Wilson 1988). Thus rain forest systems are particularly vulnerable (or sensitive) to clearing because the number of species per unit area is very high and the distribution of those species is often so limited. Wilson (1988) estimated that 10,000 species are lost per year in rain forests because of deforestation.

Compounding this situation, as mentioned earlier, rain forest environments are very sensitive to change. For example, removing even a single tree can affect lizard assemblages in the Amazonian rain forest, favoring predators that may strongly affect populations of smaller prey species (Vitt et al., 1998). As clearing continues, smaller, isolated forest patches support fewer bird species than large patches that are close to other patches (Ferraz et al., 2007). Most species that have adapted to the cooler shady conditions of the forest floor are completely unsuited to life in a newly cleared cow pasture. For example, a study of raptors in French Guiana indicated that patches of up to 300,000 ha may be needed to ensure the representation of all raptor species in a tropical rain forest (Thiollay 1989).

Recently ecologists have discovered powerful connections between small forest openings and larger forest loss. Farmers and ranchers often use fire to clear and maintain forest openings. Unfortunately, these small openings dry out the surrounding forest, making it more susceptible to second and third fires, and each successive fire is more severe (Cochrane et al., 1999). With each fire, the forest opening enlarges, drying out more fuel at the edge, making the forest more susceptible to burning again. Forest patches burned only three times can lose up to 96% of their forest species, replaced by common exotic weeds that outcompete regenerating forest tree seedlings (Cochrane and Schulze 1998).

Intensification and Native Biodiversity

Herders and farmers often intensify their production of livestock by adding more inputs per unit land through improving feed and watering, using more productive livestock breeds, controlling livestock diseases better, and using other means mentioned earlier. In more extensive dryland systems, intensification creates a land use, commercial ranching, where herders build fences and water points (discussed further later), which fragments grasslands and savannas into smaller areas (Huenneke and Noble 1996, Geist and Lambin 2004, Hobbs et al., 2007). Here, fragmentation can cut migration routes of wildlife entirely (Boone and Hobbs 2004), but more subtly it can reduce the diversity of grazing-sensitive, native plant species, even far from waterpoints (Landsberg et al., 2003). As ranching intensifies further, commercial ranchers will also often grow improved pasture species, some of which are native plants but others of which are exotics. Within most dryland landscapes, irrigated cultivation—which is an even more intensive use—is possible in small pockets along rivers, where groundwater collects and where wells and boreholes allow (Scoones 1991).

Some of these irrigated crops may be used to feed livestock, but more often farmers grow human food here. In the most intensified case, farmers and settlers around towns and cities in drylands bring in most feed and water from outside their farms to livestock. The most intensified systems are industrial feedlots or poultry operations where there is no direct connection between land and livestock production (Naylor et al., 2005). When comparing these ways of producing livestock, a simplistic view is that native biodiversity generally declines with intensification locally, where the direct impacts of intensification are largely negative. However, if this production is more efficient (and productivity increases), it may "save land for nature" by reducing the amount of land needed to produce livestock.

In areas where livestock and crop production are not limited by rainfall (which were often originally forests or woodlands), livestock production can become much more intensive than in most drylands, following a similar general sequence. Here, "natural" ecosystems, which are often forests because of rainfall, are first cleared as farming, and grazing expands (extensification); native biodiversity declines, as in the deforestation case described earlier. As land becomes more scarce, farmers and grazers begin to intensify production, including linking crops and livestock in the same system, or mixed farming (McIntire et al., 1992). At a small scale, farmers in these linked and moderately intensified systems cycle nutrients from livestock to crops (Tarawali et al., 2004) and usually maintain a diversity of vegetation patches on their farms, like woodlots, hedgerows, and home gardens (Altieri 1999). These highly diverse landscapes are very attractive to many plants and smaller animals, like butterflies, birds, small mammals, reptiles, amphibians, and insects (Wilson et al., 1997, Altieri 1999, Daily et al., 2001, Ricketts et al., 2001, Boutin et al., 2002). But mixed farms also attract invasive species of plants, small mammals, and insects, particularly where grazing and cultivation are heaviest (e.g., Vitousek et al., 1997b). Farmers and settlers usually exterminate all the dangerous and damaging animals that kill livestock and people or damage crops (Woodroffe and Ginsberg 1998, Hoare 1999, Naughton-Treves and Treves 2005). Thus the large carnivores and herbivores disappear at this stage, along with native plants sensitive to human use. But other smaller species can persist even in heavily used landscapes (e.g., Maestas et al., 2003).

These integrated systems, which can support significant biodiversity, become disconnected in the next stage of specialization (Naylor et al., 2005). Here, livestock farmers often specialize in a particular species, such as dairy cattle, pigs, or poultry. These enterprises range from small-scale dairies and pig/poultry farms to large-scale agribusiness owned by multinational corporations. These "industrial" and "smallholder landless" systems are found in the eastern half of the United States, Europe, southern Brazil, Ecuador, central Mexico, China, the Near East, and Southeast Asia (Steinfeld et al., 2006). In the process of intensification, farming landscapes lose their previous complexity, many species disappear (Altieri 1999, White and Kerr 2007), and heavy nutrient loads from concentrated manure pollute waterways and damage aquatic species (Rouse et al., 1999). The diversity that forms the foundation of Earth's life-support system is much diminished as these specialized systems spread, as landscapes become more homogeneous, as insects and animals that control pests decline, and as weeds become more widespread (Matson et al., 1997, McKinney and Lockwood 1999, Smart et al., 2006).

Not only are intensive systems themselves poor in biodiversity, but they "export" their biodiversity impact to other ecosystems around the world because they need to import concentrated feed, water, and a wide range of drugs for livestock (de Haan et al., 1997). It is difficult to trace fully the more complex and less visible impacts of livestock on biodiversity, but attempts to do so show they are significant and far-reaching (Deutsch and Folke 2005, Steinfeld et al., 2006). Probably the biggest impact exported from these farms is the clearance of land, the addition of fertilizers and pesticides, and the transport needed to produce feed for livestock (covered in detail later). But everything else associated with these systems usually comes from outside: the transport for human labor to run these operations, the construction materials to build feedlots and slaughterhouses, the veterinary and pharmaceutical industry that supplies drugs, and many other elements.

Intensification and Livestock Breed Diversity

While the major biodiversity losses during intensification concern "wild" species, livestock breed diversity often also plummets during intensification (Matson et al., 1997). Domestication of livestock began 12,000 years ago and has evolved through the selective breeding efforts of farmers to produce several thousand domestic animal populations (Hoffmann and Scherf 2006). But today the increased demand and growth in consumption of livestock products relies on increasing a few livestock breeds that are highly productive. Of the 30–40 mammalian and bird species recognized as domesticated, fewer than 14 account for 90% of global livestock production (UNEP 1995). And this limited diversity of livestock often depends on a limited number of crops for feed. Data collected on domestic animal diversity in 170 countries identified 6,379 livestock breeds, of which 740 were recorded as extinct and 1,335 are classified at a high risk of loss (Scherf 2000). These figures represent a 13% increase in the number of breeds recorded at risk from 1993 to 2000 (Scherf 2000). Less productive but genetically valuable breeds are continually threatened with extinction. For example, scholars estimate that 1 of the 44 Eurasian cattle breeds may be lost by 2050 (Bennewitz et al., 2006).

A reduction in breed diversity may result in a reduced resistance to diseases, as pathogens can spread more easily when their hosts are uniform and abundant (Tilman et al., 1999). For example, only a few local breeds of cattle in west and central Africa are resistant to trypanosomes, a parasitic disease transmitted by the tsetse fly (Murray et al., 1990). Loss of this diversity would extinguish the information about disease resistance locked in the genes of these few breeds.

In the case of highly pathogenic avian influenza (HPAI or bird flu) H5N1, the severity of outbreaks may have been exacerbated by lack of diversity in disease resistance in poultry to new strains (Tilman et al., 1999). Concerns of some governments over the role of smallholder backyard flocks of poultry in the persistence of the virus could favor exotic breeds in intensive production systems as resource-poor smallholder producers, with their largely indigenous breeds, are forced to abandon poultry keeping (Otte et al., 2006). However, results have shown that backyard flocks are at significantly lower risk of HPAI infection than commercial-scale operations (Otte et al., 2006). This may suggest, contrary to conventional wisdom, that small operations are sometimes an important safety net for poultry production in the face of bird flu.

Breeds in Europe and North America are more at risk than those in Asia and Africa (Scherf 2000), although data for developing countries are few. In the industrial world, intensive production systems have led to a reduction in the use of indigenous breeds, and several rare breeds are maintained by hobby farmers for conservation (Hoffmann and Scherf 2006). In the developing world, indigenous breeds are central to the livelihoods of poor people (IDLGroup 2002). For example, herders in extensive pastoral production systems of Africa have developed well-adapted breeds that cope well with drought, fodder scarcity, climatic extremes, and diseases (Mathias et al., 2005). A quarter of global livestock diversity is adapted to drylands, but fortunately the loss of livestock genetic diversity is lower here than in wetter production systems (FAO 2006b). However, the globalization and extension of livestock markets has encouraged the replacement of indigenous breeds with exotic ones, eroding the stock of local genetic resources (Tisdell 2003). This suggests that dryland herders play an important role in maintaining not only native diversity but also livestock diversity.

Indirect and Positive Impacts of Intensification

More positively, if intensive systems are more efficient than extensive systems (e.g., more meat or milk is produced for each amount of natural resource used), although this is not always the case, the increasing efficiency can sometimes slow the need to clear land for pasture/feed crops or heavily graze other areas. For example, Swedish farmers rarely use low-yielding pastures for livestock grazing today, so the extent of livestock grazing has contracted significantly since the 1960s (Deutsch and Folke 2005). Indeed, without increases in efficiency, the world would need far more land to produce livestock today, likely with far greater biodiversity loss.

As farming becomes more efficient, for example, less tropical rain forest may need to be cut to produce the same amount of beef (Angelsen and Kaimowitz 2001). However, it is not clear that intensification of agriculture actually saves forests from being cleared (Hecht 1989, Angelsen and Kaimowitz 2001, Geist and Lambin 2002). As feed crop production or livestock grazing becomes more profitable, it can attract new migrants and accelerate deforestation for more pasture and feed crops. And there are limits to efficiency gains; for cereal crops, the big gains in production caused by adding nitrogen fertilizer have slowed considerably since the 1970s, although this is more true for the industrial than the developing world (Tilman et al., 2002).

It is tempting to assume that intensification will take pressure off the land and biodiversity, but the reality is that it is simply not clear if this is the case. The crux of the matter is whether this increase in intensity is accompanied by an increase in amount of livestock produced per amount of natural resource used. What is needed is a full accounting of the total environmental impacts of producing each kilogram of meat, milk, and eggs—comparing extensive systems to mixed crop–livestock systems to more-intensified systems in different contexts.

A comparison of the greenhouse gas emissions associated with beef production, for instance, shows that for each kilogram of beef produced, greenhouse emissions are almost double in an intensive U.S. feedlot than in an extensive Sahelian pastoral system of West Africa (Subak 1999). Even though Sahelian cattle emit twice as much methane as American cattle, the latter are fed feedlot corn, which is fertilized and transported by truck, resulting in double the emissions overall. This means that, for greenhouse gas emissions at least, the extensive Sahelian system is more efficient than the intensive American feedlot, and thus the intensive production is more environmentally damaging. This type of assessment needs to be done for the full environmental impacts—water, biodiversity, soils, pollution, atmospheric dust, and so on—of production systems at different intensities.

Impact 2: Livestock Grazing as "Biodiversity Protection"

Farmers, herders, settlers, and governments around the world are converting rangelands into other uses like cropland (Campbell et al., 2000), irrigated land (Blench 2000), exurban development (Hansen et al., 2002), and protected areas (Brockington et al., 2006). Where livestock grazing slows such conversions, livestock have a new role in what could be called "biodiversity protection." This happens in four different contexts.

First, in the dryland tropics, crop farmers and herding farmers (Campbell et al., 2000) extend farms and plough rangelands for crops, a land use that is generally less compatible with both plant and animal diversity than extensive livestock grazing. In East African savannas, for example, a strong cause of this change is the shift from communal to private landownership and then the sale of some grazing land to farmers or settlers, resulting in a patchwork of land uses that make fluid movement of livestock and wildlife through landscapes difficult (Reid et al., 2004). And this trend of loss of biodiverse rangelands in these savannas is set to continue, because the financial returns to landholders from livestock and wildlife together do not compete with the financial returns from crop cultivation in most areas (Norton-Griffiths et al., forthcoming).

How extensive might this replacement of rangelands by croplands be, given human population growth and climate change by 2050? We used the estimates of human population growth and greenhouse gas emissions from one of the scenarios ("Global Orchestration") in the Millennium Ecosystem Assessment (Carpenter et al., 2005) to develop human population and climate change surfaces (as projected by the Hadley CM3 climate model) to 2050. Using the extent of cropland and rangeland from the GLC 2000 (JRL 2005), we then assumed that rangeland shifted into cropland wherever there were more than 20 people km² and more than 60 days of growing period per year. In this estimate (see Figure 8.3 in the color well), almost all of the rangeland that shifts into cropland by 2050 is in Africa, with scattered patches on other continents. This suggests that the phenomenon of "biodiversity protection" may be uniquely an African one.

The second example is in the drylands of western North America, where some livestock ranchland is attractive to wealthy urban dwellers who want to move out of congested areas or build second homes, in "exurban" development (Hansen et al., 2002, 2005). In other rangelands, as in Australia, pastoral ranches near towns split up into small private holdings to absorb growing urban expansion (Stokes et al., 2007). New settlers build roads, houses, and fences; bring in exotic species like hobby livestock and pets; and generally fragment these former extensive ranches. Largeholder ranches in Colorado support more native carnivores, fewer nonnative plants, and more ground nesting birds, while new exurban homesteads have more domestic cats and dogs, more nonnative and weedy plants, and more birds that people attract (Maestas et al., 2003).

In a slightly different case, the contraction of

extensive grazing in Europe is allowing forest that has been cleared for farming and grazing centuries ago to be reestablished. Industrialization and depopulation of rural areas is a long-term trend in parts of Europe (Laiolo et al., 2004); as a result, traditional farmland practices and agricultural lands in remote areas with poorer soils and harsher climatic conditions have been abandoned. Left unmanaged, these pastures revert back to trees and shrubs, and ultimately to forest. Low-intensity farmland systems, characterized by low nutrient inputs and low outputs, contain more than 50% of Europe's most valued habitats (Bignal and McCracken 1996) and are now the focus of numerous agroenvironment and nature conservation schemes. Europe faces a polarization of agriculture practices—toward intensification in one direction and abandonment of farms in the other—which threatens many seminatural habitats like livestock pastures (Baldock et al., 1996). Out of the 195 habitat types considered of European importance due to their high biodiversity value (in the Habitats Directive of the European Union, 1992), 65 are threatened by intensification of grazing and 26 are threatened by abandonment of rural activities (Ostermann 1998).

Across Europe, livestock grazing plays an important role in maintaining valuable habitat, and virtually all the remaining grasslands of high conservation value are associated with low-intensity livestock systems, where livestock graze in extensive pastureland (Bignal and McCracken 1996). In the pastures, livestock grazing creates and maintains structural heterogeneity, which in turn affects the number of plants and animals in pastures (Rook et al 2004). Livestock grazing can increase grass species richness in hay meadows (Smith and Rushton 1994), mesic grasslands (Pykälä 2005), heathlands (Bokdam and Gleichman 2000), and dry calcareous grasslands (Barbaro et al., 2001).

The last context is the contraction of rangelands to establish protected areas, which is pronounced in Africa. Here, protected areas often maintain biodiversity better than adjacent pastoral lands (Prins 1992), but sometimes protected areas have no or unknown effects on biodiversity (Ward et al., 1998, Brockington and Homewood 2001). In a few cases, removal of livestock grazing to create protected areas can have a negative effect on wildlife diversity, particularly in the wet season (Reid et al., 2001). When protected areas prevent cropland or heavy livestock grazing, they are very important for biodiversity protection. But outside protected areas, pastoral grazing may be important for maintaining the diversity of large wildlife, a pattern just emerging in Africa. Here, the removal of grazing and burning by herders, without its replacement by grazing and burning by conservationists, can leave grasslands long and rank, which are unattractive to small and medium-sized wildlife (poor-quality food, risky in terms of predators) (Western and Gichohi

1993). For example, in the Mara of Kenya, wildlife avoid the long-grass areas of the reserve and cluster instead around adjacent pastoral settlements (Reid et al., 2003).

Impact 3: Effects of Grazing and Trampling by Livestock on Species Diversity and the Balance of Trees and Grass

Negative Impacts of Livestock Grazing

While extensive livestock grazing in drylands is more compatible with biodiversity than crop farming and exurban development are, it still affects the diversity of plants and animals at local scales and the diversity of ecosystems at broader scales. One long-running controversy among experts on grazing is whether livestock do or do not cause changes in grazing ecosystems, including changes in biodiversity (Illius and O'Connor 1999, Sullivan and Rohde 2002, Briske et al., 2003, Vetter 2005). Where rainfall is very variable, human populations are low, and livestock movement is unfettered (as in many drylands in Africa and Central Asia), recurrent droughts force herders to move livestock frequently or often cause major die-offs of livestock so that permanent damage to rangeland vegetation is rare (Ellis and Swift 1988, Fernandez-Gimenez and Allen-Diaz 1999, Sullivan and Rohde 2002, Vetter 2005). In Mongolia, for example, livestock had little effect on the number and composition of plant species in dry desert steppe ecosystems but had strong effects in wetter steppe and mountain steppe ecosystems (Fernandez-Gimenez and Allen-Diaz 1999).

Also important is the long-term history of grazing in rangelands (Mack and Thompson 1982); those that evolved with wild grazers often change little when people bring in livestock, like the shortgrass prairie of the United States, which had a history of bison grazing (Milchunas et al., 1988, Laurenroth et al., 1999) or the savannas of East Africa (McNaughton 1979). If there is abundant rainfall in these systems, livestock grazing actually increases the number of plant species, as described at the end of this section.

But livestock can strongly change and diminish native biodiversity where people and their livestock concentrate and graze heavily year-round in rangelands or where rangelands are particularly sensitive to grazing because it is a new phenomenon there. As people concentrate around natural or artificial water sources, or as they build towns, livestock populations remain high even during drought. This also happens where ranchers purchase feed to maintain livestock during dry periods if they do not move livestock frequently to allow the vegetation to recover from grazing (Ellis and Swift 1988). This continual heavy grazing (and the concentration of people) can shift the composition of grasses and leafy herbs from those palatable for livestock to those that are less palatable (Hiernaux 1998).

Even more important for biodiversity, heavy and sedentary grazing can discourage herbaceous plants like grasses and leafy herbs in favor of woody plants like shrubs and trees (Asner et al., 2004, Vetter 2005, Rohde et al., 2006), often called woody encroachment. Encroachment is a global phenomenon (Skarpe 1991, Archer 1994, Van Auken 2000, Asner et al., 2004), but it does not happen everywhere (Witt et al., 2006). This shift to woody plants is particularly acute in arid systems not used to grazing, with a history of light grazing by wildlife followed by heavy livestock grazing (Milchunas et al., 1988, Pieper 1999). For example, in the southwest United States and northern Mexico in the Mohave, Chihuahuan, and Great Basin areas, light wildlife grazing was followed by heavy livestock grazing in the late 1800s, causing extensive woody encroachment, shifts in plant diversity, and loss of valuable nitrogen-fixing soil crusts (Laurenroth et al., 1999, Pieper 1999, Belnap 2003). Of particular concern are critical habitats like ribbons of vegetation along rivers. In the western United States, livestock grazing removes grazing-sensitive, native plants along rivers and also diminishes valuable fish populations (such as trout and salmon) (Belsky et al., 1999).

At a local scale, livestock also encourage the growth of nitrogen-tolerant plants and drive wildlife away from water points. Around water points, livestock trample vegetation and deposit nitrogen, phosphorus, and other nutrients in manure (Andrew 1988). Generally, this encourages the establishment of plants that are nonnative, short-lived, and grow well in the abundant plant nutrients provided by the manure (Landsberg et al., 2003, Brooks et al., 2006, Todd 2006). Fortunately, these heavier grazed places are often limited to within 100–200 m of water points (Landsberg et al., 2003, Brooks et al., 2006).

Livestock can also drive wildlife away from water points during the day (de Leeuw et al., 2001), although wildlife can return at night to water. In the case of the rare Grevy's zebra in northern Kenya, this shift to night grazing because of livestock is more dangerous for mothers and foals, thus likely reducing population sizes (Williams 1998). The extent of these impacts across landscapes depends on the density of water points and settlements. In Turkana, Kenya, for example, 95% of the landscape is more than 5 km from a settlement or major water point and thus lightly grazed (Reid and Ellis 1995). But in other landscapes, these impacts can be heavy (see, e.g., Sinclair and Fryxell 1985).

Livestock grazing can also exclude or reduce wildlife populations by changing the vegetation, competing with wildlife for food, or passing diseases (which is discussed later). For example, sheep and goat grazing and herding push the endangered Tibetan argali (*Ovis ammon*) into less productive habitat in Ladakh, India (Namgail et al., 2007). Heavy grazing by horses, cattle, or sheep reduces the number and diversity of land snails in a nutrient-poor montane grassland in Switzerland (Boschi and Baur 2007). In the shortgrass steppe of the western United States, heavy livestock grazing strongly diminishes the number of macroarthropods and lagomorphs (Milchunas et al., 1998). In southwestern Kenya, some species, like elephant, lion, and rhino, avoid herding settlements, probably because people chase them away to protect themselves and their livestock (Reid et al., 2003). Herders also kill wildlife directly (Prins 1992). Among the Barabaig of Tanzania, anyone who kills a lion, elephant, rhino, or buffalo is rewarded with gifts of livestock (Klima 1970). Many conflicts occur when herders and livestock farmers move into former wildlife habitat. For example, large cats, like jaguars and pumas, switch to killing livestock as farmers clear Amazonian forests for ranching, thus putting carnivores at danger because of human encroachment on their habitat (Michalski 2006).

But are these impacts of livestock and herders completely irreversible or at least hard to reverse? If livestock cause the extinction of a species, this is irreversible. There is no evidence of livestock causing the extinction of an entire species, but it is difficult to know this for sure, because so many changes are unmeasured. More common are changes that take decades to reverse. For example, in a dry Karoo savanna in South Africa, it took 10 years for perennial plants to recover after sheep and goats were removed and replaced by wild grazers (Kraaij and Milton 2006). Other wetter savannas can remain in a bushland state when heavily grazed by cattle, and it requires either great expense or 30–40 years to shift back to grassland (Westoby et al., 1989, Scholes and Archer 1997, Anderies et al., 2002, Wiegand et al., 2006).

Potential Positive Impacts of Livestock Grazing

Grazing by livestock can also maintain or boost the diversity of native plants and animals in wetter grazing systems if grazing is not too heavy. Here, there are few species where there is no or low grazing, many species where there is moderate grazing, and few again where grazing becomes heavy (Milchunas et al., 1988). Research from lakes to savannas is beginning to show that moderate grazing by species as different as limpets and wildebeest particularly increases plant diversity in productive ecosystems (those that are wet and rich in nutrients) (Bakker 1989, Hodgson and Illius 1996, Olff and Ritchie 1998, Harrison et al., 2003) and has no effect or decreases diversity in nonproductive ecosystems (dry and poor in nutrients) (Proulx and Mazumder 1998). The reason this happens is that productive ecosystems often have a few species that garner all the resources, and grazing prevents them from doing so by removing more of the few dominant species. This applies not only

to wild grazers but to livestock too (Bakker 1989, Hodgson and Illius 1996, Olff and Ritchie 1998, Harrison et al., 2003). But there is an important caution here: even if plant species rise with grazing, the additional species may be common plants or nonnatives (Belsky et al., 1999). And when grazing is too intense, diversity can plummet, encouraging the invasion of weeds (Milchunas et al., 1988, Hobbs and Huenneke 1992).

Herders and their livestock may have positive impacts on wildlife too. For example, wildlife prefer to graze on grass atop abandoned livestock corrals in central Kenya rather than in places far from old corrals (Young et al., 1995). Turkana pastoralists inadvertently plant most of the Acacia trees in their arid, Kenyan landscape by feeding their livestock seed pods from Acacia trees (Reid and Ellis 1995). Large trees are like small ecosystems; they support many more other species than small trees do. In this case, large trees are not only larger, but longer lived and structurally more complex than small trees (Dean et al., 1999).

Impact 4: Livestock-Generated Pollution

Global and Production System Impacts

At the global level, pollution is one of the main causes of biodiversity loss in terrestrial, freshwater, and marine ecosystems (MA 2005a); livestock are largely responsible for this pollution. Livestock deposit and redistribute nitrogen (N), phosphorus (P), and other nutrients directly through manure and indirectly through the fertilizer used on feed crops and pastures (see Chapter 6). Manure, carcasses, and fertilizers also contribute other pollutants like heavy metals, drug residues, and sediments. These livestock-generated pollutants, when they exceed critical thresholds, affect biodiversity directly by killing or weakening plant and animal species and indirectly by degrading habitats. They often concentrate close to where livestock deposit them, but when these pollutants enter watercourses or the atmosphere, they can affect distant lands and waters.

While livestock recycle nutrients all over the world, these nutrients become overabundant principally in intensified livestock systems. But this can also happen in extensive grazing systems, where manure and fertilizer inputs are usually low per unit area and where livestock concentrate manure around settlements and water points, as described earlier. This concentration of nutrients in particular places probably "mines" nutrients from the wider landscape where livestock graze (Turner 1998), but the impacts of this mining on biodiversity are unknown. Farmers in mixed crop–livestock systems often recycle livestock wastes on crops (Tarawali et al., 2001), which is generally the case in the developing world. But crops grown on large commercial farms to produce concentrate feed for livestock, more common in the industrial world, use significant amounts of fertilizer, which pollutes surface and groundwater with N and P—both close to farmers' fields and far downstream (Galloway et al., 2003).

The role of livestock in crop fertilizer use is large; in the United States, about 60% of the crops produced are used to feed livestock (Howarth et al., 2002). Furthermore, manure and fertilized soils emit ammonia and nitrogen oxide to the atmosphere that then fertilize ecosystems through nitrogen deposition in rainfall (Tilman et al., 2002). Ammonia from livestock in intensive systems, for example, is responsible for more than 70% of the nitrogen redeposited from the atmosphere in the United States, Europe, and Asia (see Chapter 6).

Nutrient Pollution, Particularly Nitrogen

An overabundance of nutrients from livestock and feed crops first affects biodiversity on land and in the atmosphere. Addition of nitrogen and phosphorus through manure and fertilizer increases nutrients in the soil, which generally decreases the diversity of plants, as nitrogen-loving plants outcompete nitrogen-sensitive ones (Tilman 1987, Vitousek et al., 1997a, Haddad et al., 2000). Deposition of nitrogen from the atmosphere, some originating from the ammonia applied to grow livestock feed and then emitted to the atmosphere, can cause loss of plant species in sensitive, acid grasslands in Britain and across Europe (Stevens et al., 2004) and even affects plants inside protected areas (Bank et al., 2006).

Fertilizer and manure runoff flow into local water systems, polluting streams and rivers but also affecting lakes and oceans downstream on a larger scale. Globally, about 20% of the nitrogen produced by all sources flows into inland waters, with 80% flowing to the sea; flows are greatest in Asia, followed by Europe, the former Soviet Union, and North America (Boyer et al., 2006). The livestock sector is responsible for about a third of the N and P discharges into watercourses in the United States (Carpenter et al., 1998, Steinfeld et al., 2006). When excessive, these nutrients can damage aquatic species directly (Boyer and Grue 1995, Camargo et al., 2005). For example, about 20% of the watersheds bordering the Great Lakes in the United States contained enough nitrates to kill or cause developmental anomalies in amphibians, insects, and fish (Rouse et al., 1999). Marine pollutants often stimulate the growth of bacteria that invade coral tissues and make them more prone to disease (Porter et al., 1999). Conversely, in oak savannas of California, livestock grazing on grass reduces the amount of nitrates that flows into local streams and keeps nitrate concentrations below levels where it is considered a water pollutant (Jackson et al., 2006).

Not only can nutrients from livestock affect biodiversity directly but they can promote the growth of algae and higher plants at the base of the food chain,

productivity that then cascades through the other organisms in the ecosystem (Carpenter et al., 1999). Eutrophication also creates large algal blooms, which upon decay can deplete dissolved oxygen periodically or permanently, killing many other aquatic organisms (Rabalais 2002). For example, pig waste from industrial production systems is the main source of nutrient loading affecting water systems in China (Guangdong Province), Thailand, and Vietnam, contributing from 14% to 72% of the nitrogen and 61% to 94% of the phosphorus (FAO 2004). Pig waste undoubtedly caused part of the massive algal bloom in 1998 that killed more than 80% of the fish in a 100 km^2 area of the South China Sea, a waterbody particularly rich in aquatic biodiversity (Groombridge and Jenkins 1998). Since 2002, giant jelly fish have proliferated in this region because of the increasing availability of zooplankton, now reaching and affecting fisheries along the Japanese coast (Kawahara et al., 2006).

In its extreme, eutrophication can create "dead zones" in freshwater and marine ecosystems—places with little oxygen—and loss of seabed vegetation, benthic invertebrates, and bottom-dwelling demersal fish (Belsky et al., 1999, Diaz 2001, Rabalais et al., 2002). While livestock are not the only cause of hypoxia, runoff from manure and fertilizer is partly responsible for dead zones. Hypoxia caused the collapse of the Norway lobster fishery in Kattegat between Sweden and Denmark, the collapse of the cockle fishery in Sommone Bay in France, and the death of invertebrates at the base of the food chain in many other coastal and inland areas (Diaz 2001). All of the major dead zones are off the coasts of North America, Europe, Japan, and New Zealand and in the inland waters of the Black, Baltic, and Caspian seas (Diaz 2001). These dead zones are growing as nutrient inputs increase (Rabalais 2002).

Drug and Chemical Residues, Heavy Metals

Livestock tissues and excreta—fertilizers used on feed crops and in slaughterhouses and tanneries—affect biodiversity by introducing heavy metals, antibiotics, hormones, insecticides, and pesticides into the environment. For example, residues of the drug diclofenac, concentrated in livestock carcasses, killed 95% of the vulture population in South Asia in the 1990s (Oaks et al., 2004). Hormones from livestock can affect the neurological and endocrine development in wildlife, including cases of reproductive tract changes and cancers (Soto et al., 2004). Wastewater from tanneries, with high concentrations of chromium and hydrogen sulfides, may also affect morbidity and mortality of aquatic species (Galiana-Aleixandre et al., 2005). The role of livestock is significant here; the sector consumed 37% of the agricultural pesticides used worldwide in 2001, uses half of the antibiotics produced annually in the United States, and produces 5.5

million tons of raw hides (Harrison and Lederberg 1998, USDA 2001).

Sedimentation of Water Sources

The livestock sector can contribute strongly to soil erosion through heavy grazing, crop cultivation, and deforestation. In the United States, for example, the livestock sector causes 55% of the erosion from agricultural land (USDA 2001, FAO 2006c, Steinfeld et al., 2006). In western rangelands, livestock graze heavily on riverine vegetation, increasing sedimentation in streams that reduces important fish species (Belsky et al., 1999). This erosion particularly affects freshwater species and coral reefs along coastlines. For example, high sediment loads are the main threat to imperiled freshwater fauna in the eastern United States (Richter et al., 1997).

Impact 5: Emerging Livestock Diseases

Disease emergence today is largely caused by human environmental change, driven by globalization of agriculture, climate change, commerce, and human travel (Vitousek et al., 1997b, Daszak et al., 2001). Many human diseases originated from livestock, evolving to attack humans as people became more crowded in cities. Livestock are a major contributor to disease emergence, and sometimes these diseases affect biodiversity. For example, viral rinderpest, brought to Africa by imported cattle in the late nineteenth century, spread across Africa and killed most of the continent's cattle and also many of the ruminant wildlife, such as buffalo, giraffe, and eland (Ford 1971). Bovine tuberculosis, a disease whose natural host is cattle, is now a major disease in buffalo in Africa, possums in New Zealand, white-tailed deer in north central United States, and badgers in the United Kingdom and Ireland (Bengis et al., 2002).

Livestock pass diseases to wildlife through "pathogen pollution" and "spillover" processes that often act at the same time (Daszak et al., 2000). Rinderpest is an example of pathogen pollution, where a previously unknown disease in wildlife was introduced by people, through cattle, and caused significant sickness and death of wildlife. Spillovers occur when domesticated animals outnumber wildlife and continually infect a local wildlife population with a shared disease until the wildlife population goes extinct. For example, canine distemper spilled over from the abundant populations of domestic dogs west of Serengeti National Park into lion populations inside the park in the early 1990s, causing significant mortality (Cleaveland et al., 2000).

One of the most prominent emerging diseases, which appears to move between poultry and wild birds (and also to people), is the highly pathogenic avian influenza (HPAI) H5N1 virus, or H5N1 bird flu. The disease was first identified in domesticated geese in southern China in 1996 (Sims et al., 2005) and in humans in Hong Kong in 1997 (Jong et al., 1997). Thus the first outbreak of this strain of bird flu probably occurred in livestock (domesticated geese), but this is still uncertain. In late 2002 H5N1 caused deaths among migratory birds and resident waterfowl in two Hong Kong parks (Ellis et al., 2004, Sturm-Ramirez et al., 2004). This was the first time an outbreak of lethal avian influenza had been reported in wild aquatic birds since 1961 (Becker 1966). H5N1 avian influenza has spread rapidly, with outbreaks in poultry, wild birds, and other mammalian species in almost 60 countries and with 312 human cases, including 190 deaths (OIE 2007, WHO 2007b). More than 200 million poultry have been killed by the virus or culled to prevent its spread (IISD 2006). The geographical spread and numbers of infected flocks have been unprecedented in the history of avian flu outbreaks (Sims et al., 2005).

During 2005, H5N1 outbreaks spread north and then westward from China to Mongolia, Russia, Kazakhstan, and Europe, and in 2006 into Africa. The spread of H5N1 in Asia and Africa was due to introductions by poultry and wild birds, whereas the spread to European countries was more consistent with the movements of wild birds (Kilpatrick et al., 2006), although this interpretation is debated (Butler 2006, Feare and Yasué 2006, Fergus et al., 2006, Normile 2006, Birdlife 2007). An introduction into the Americas, where the H5N1 strain is yet to be detected, would mostly likely be through the trade of poultry rather than migrating birds (Kilpatrick et al., 2006).

Regardless of the role of wild birds in the transmission of H5N1 bird flu, the potential threat to wild bird diversity is high, particularly for birds found in inland wetlands and along coastlines. (See Box 8.1.) The H5N1 virus has infected more than 50 species of wild birds (Olsen et al., 2006) in at least 12 of the 27 avian orders, with most frequent infections reported in ducks, swans, and geese (order Anseriformes, family Anatidae) and in gulls and terns (order Charardriiformes) (Gilbert et al., 2006). It is estimated that 84% of all bird species could be at risk from H5N1 (Roberton et al., 2006). Pheasants, grouse, partridges, and turkeys (order Galliformes) have the most threatened or near threatened species at risk (39.4% of species at risk), followed by the parrots and cockatoos (order Psittaciformes, 34% at risk) and by medium to large waterbirds (order Pelecaniformes, 33.9% at risk) (Roberton et al., 2006). In addition, there is the risk that many wild birds will die through culling to prevent further outbreaks and spread of disease, even though there is no scientific evidence to support culling (Olsen et al., 2006). Moreover, proposals to drain wetlands as a way of deterring migratory birds would harm much wider biodiversity than only wild birds and probably do little to control the spread of the disease (Birdlife 2005).

Avian flu threatens biodiversity well beyond wild and domestic birds (see Figure 8.4), with some evidence of deaths in a wide range of mammalian species already.

Box 8.1. Poultry Density, Wild Birds, and Avian Flu

One way to visualize potential threats of avian flu to bird diversity in the future is to map the density of poultry and their proximity to wetlands, which shows places where chickens and wetland birds are close to each other around the world. (See figure for Box 8.1 in the color well.) Such a map does not suggest there is or will be contact or transmission of avian flu between poultry and wild birds, only that there is the potential for this to occur. Avian flu is probably, but not conclusively, transmitted between poultry and wild birds, and many of these wild birds are associated with wetlands.

To create this map, we took actual and predicted poultry density data from the Gridded Livestock of the World database (GLW) (FAO 2007) and global lakes and wetlands from the Global Lakes and Wetlands Database (GLWD) (Lehner and Döll 2004), along with outbreak data from (WHO 2007a). All categories of water and wetlands were included in the analysis with the exception of the intermittent wetland/lake delineation. The remaining areas were then expanded to include all areas within a 3 km distance of the edge of the wetland using a buffering algorithm. Poultry den-

sity was then extracted for only the water/wetland areas and the buffered zones. No outbreaks have been reported on the American continents, so these have been excluded.

Where does the map suggest chickens and wetland birds are close to each other, and does this have any association with actual outbreaks of avian flu in poultry and wild birds? Or, from a biodiversity perspective, where might wetland birds be in danger of contracting avian flu from poultry or being culled because they are close to poultry? Poultry densities are high (more than 200/km²) and within 3 km of wetlands in eastern China, Southeast and South Asia, the Philippines, Iran, Turkey, central Europe, northwestern South America, across Central America, eastern and north central United States, and northern New Zealand. In Africa, this occurs in Uganda, Nigeria, and parts of the Nile valley. These match the outbreaks of avian flu in poultry in much of Asia and Africa, but not in the Americas and New Zealand. There are also places where wetland birds may be at lower risk, even though there have been outbreaks in poultry in Scotland and western China.

Two tigers and two leopards were found dead with the H5N1 virus in a Thailand zoo in December 2003 (Keawcharoen et al., 2004). Zookeepers fed these carnivores chicken carcasses from a local slaughterhouse. In another Thailand zoo, the virus was transmitted between tigers, and 147 out of 441 tigers were either killed or euthanized (Thanawongnuwech et al., 2005). The potential

role of cats in the epidemiology of the virus is still largely unknown; however, domestic cats have been found dead due to H5N1 in Asia and Europe (FAO 2006a, Kuiken et al., 2006, Yingst et al., 2006, FAO 2007). In June 2005, infections occurred in a captively bred population of Owston's civet, a globally threatened viverrid, in a Vietnamese national park (Roberton et al., 2006). Here,

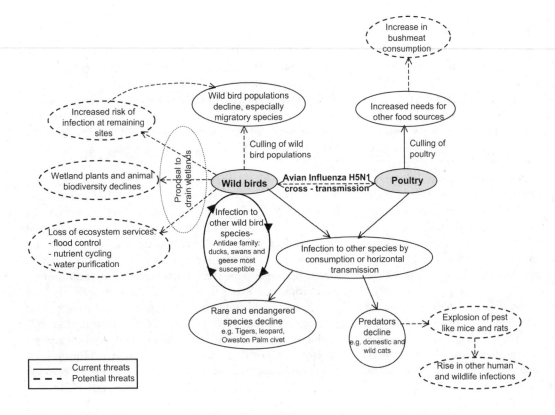

Figure 8.4. Potential avian flu transmission zones between wetland birds and poultry, and avian flu outbreaks, 2003–07.

park officials did not feed the civets dead poultry, so the source of infection remains unclear. In Germany, in early 2006, H5N1 was found in a stone marten, a Mustelidae, presumed to have acquired its infection after feeding on a diseased bird (Kuiken et al., 2006).

Given the broad spectrum of the virus, it is possible that wild or domesticated mammals including pinnipeds, mustelids, and other fur-bearing animals will become infected by contacting infected animals (FAO 2006a). Mammals at risk of eating infected dead birds include carnivores, primates, rodents, marsupial carnivores, Virginia opossums, and some bats (Roberton et al 2006). This strain of bird flu may most strongly affect those species already endangered by people, like tigers and leopards (Roberton et al., 2006). Of the mammalian species listed as threatened or near threatened in the 2004 IUCN Red List, 37.2% of the carnivores and 58.8% of the primates were at risk (Roberton et al., 2006). And if people cull chickens, they may switch to consuming bushmeat, which is already overharvested in many places (CBD 2006).

Impact 6: Invasive Alien Plant Species and Native Biodiversity

People have purposely and accidentally carried species from one place to another for millennia (Traveset and Richardson, 2006). Most transported species do not flourish in their new environments (Mack et al., 2000) and go unnoticed in ecosystems and people's livelihoods. Others become the foundation of human food chains far from where they originated, like maize in Africa, wheat in North America, or sheep in Australia. About 1% of these successful newcomers become aggressive invasives or pests (Williamson 1996), spreading rapidly and replacing natives (Stromberg et al., 2007).

Altogether, over time, these newcomers are numerous; for example, 45% of pasture species in the United States are nonnatives (Pimentel et al., 2005). The invasive species can spread in favor of natives because they lack native predators, or they do well in new habitats disturbed by people, or they are highly adaptable and outcompete natives (Pimentel et al., 2005). This can damage the native diversity of plants, sometimes causing loss of local plant populations (Gurevitch and Padilla 2004), often affecting animals and ecosystems in significant ways (Brooks et al., 2004, MacDougall and Turkington 2005, Traveset and Richardson 2006). This may influence the future evolution of species and their ecosystems over the long term (Mooney and Cleland 2001, Strayer et al., 2006).

Livestock themselves are invasive species in some parts of the world (Steinfeld et al., 2006). People domesticated livestock about 10,000 years ago from species native to Asia and northern Africa (Clutton-Brock 1981, Hanotte et al., 2002, Marshall and Hildebrand 2002). Over time, people then brought livestock to other continents, such as Australia and the Americas, that had no previous history of livestock grazing (Steinfeld et al., 2006). Thus many of the impacts of livestock described in this assessment are also those of a nonnative mammal introduced into new environments to produce food.

Intentionally or accidentally, people also brought nonnative plant species for many uses, some to feed livestock. In South America, for example, African grasses were imported for grazing and now dominate many pastures (Pivello et al., 1999). In South Africa, about 12% of the species are aliens, with over 60% imported from Europe (Milton 2004). Of the 580 nonnative grasses brought into Britain, a full 430 (74%) may have arrived in wool (Milton 2004). In addition, livestock grazing itself can encourage establishment of nonnative plants if they are present in pastures and more tolerant of grazing than some native plants.

These alien species, some imported to feed livestock, are one of the major threats to native biodiversity globally. About 42% of the 958 native species listed as threatened and endangered in the United States and up to 80% of those in other countries globally are threatened by competition or predation by nonnatives (Wilcove et al., 1998, Pimentel et al., 2005). Many grasses introduced for pasture are biologically adapted to spread rapidly because they have abundant and persistent small seeds, an ability to survive under stressful environmental conditions, and a tolerance for burning and heavy grazing (D'Antonio and Vitousek 1992, Milton 2004). This makes grasses good competitors in many of the environments where people choose to herd livestock. For example, the grasses that now dominate many South American pastures evolved in Africa, where there is strong pressure to adapt to a wide range of wild grazers (Klink 1996, Pivello et al., 1999). Similarly, the tussock grass *Nasella*, introduced from South America to South Africa, replaces natives in pastures there, reducing their productivity (Milton 2004). The net effect of this superior competitive ability is often a more simple pasture with fewer species, and thus less genetic variability to use to respond to new and different conditions in the future.

In the United States, at least 50,000 species have been introduced by people for various needs, including livestock and forage species (Pimentel et al., 2005). A stark example of nonnatives replacing natives occurred after settlers brought the European grasses *Avena* and *Bromus* and other annual grasses to central Californian grasslands in the nineteenth century. Now these nonnatives dominate more than 9.2 million hectares of these grasslands, with only rare pockets of the native perennial bunch grasses left (D'Antonio and Vitousek 1992, Malmstrom et al., 2006). The success of the annual grasses was not their superior competitive abilities; they relied on heavy livestock grazing and drought to provide the conditions for their spread (Seabloom et al., 2003).

Perhaps the most striking example of invasive pasture grasses damaging native plant diversity is found in the Brazilian *cerrado*. The *cerrado* is one of the most diverse savannas on Earth and covers 200 million ha (24% of Brazil's territory) (Brossard and Barcellos 2005). However, only 43% of this biome is now preserved, principally in fragments surrounded by other land uses (Brossard and Barcellos 2005), like soybeans and pasture for livestock. African grasses (*Melinis minutiflora* Beauv., *Hyparrhenia ruffa* [Nees] Stapf, *Andropogon gayanus* Kunth, *Brachiaria decumbens* Staf and *Panicum maximum* Jacq) were introduced for pasture by the Portuguese in 1600 and later by Brazilian researchers (Rocha 1991). These grasses are invasive, highly competitive, and resistant to fire, and they also grow better in areas that were originally used for cattle ranching (Pivello et al., 1999). They are now widespread throughout the *cerrado*, with 80% of the area covered by *Brachiaria* alone, a species even found inside protected areas (Pivello et al., 1999). Ironically, the pastures created with nonnative grasses to feed livestock are degrading over the long term, which then lowers beef and milk production (Brossad and Barcellos 2005).

On the Arabian Peninsula, heavy grazing in some places has encouraged growth of unpalatable and poisonous species, which suggests that plants sensitive to grazing are being lost. Some 3,500 plant species here are adapted to extreme climatic conditions characterized by low rainfall, high evaporation rates, and high temperatures (Peacock et al., 2003). Since the 1960s the population of camels, sheep, and goats has risen sharply with improved veterinary care and supplemental feeding; higher populations encouraged the spread of nonnative plants through heavy grazing. And in some areas, farmers grow two exotic plants, alfalfa (*Medicago sativa*) and Rhodes grass (*Chloris gayana*), for supplemental feed, but under heavy irrigation (Peacock et al., 2003).

In Australia, decades of introduction of plant species to feed livestock provided dismal performance: only 5% were useful, and of the remaining useless species, 13% became weeds (Lonsdale 1994). In the future, a full 81% will likely become weeds. For example, *Prosopis*, a nonnative tree used for fodder, shade, and other purposes, was introduced from the Americas to Australia in the 1930s (Klinken et al., 2006). In some areas, this species is turning native grassland and shrubland into dense thorn forest, removing most of the value of the land for cattle grazing. This genus is now considered one of the 20 most significant weeds in Australia (Klinken et al., 2006).

Some managers attempt to slow the spread of nonnative grasses by having livestock graze them. For example, an African exotic grass, *Melinis minutiflora* Beauv., slows tree regeneration and fuels more fire in gallery forests of Brazil (Hoffmann et al., 2004). But light grazing on this grass by livestock can improve regeneration of trees and reduce fuel loads (Posada et al., 2000).

There are clear trade-offs between introducing more-productive grasses for livestock, protecting native biodiversity, and producing other ecosystem services. For example, a pasture dominated by African grasses mixed with native legumes in South America allows ranchers to produce 25 times more beef than on a pasture dominated by native grasses alone (Fisher et al., 1994). In addition, the exotic deep-rooted African grasses sequester 25% more carbon in the soil than native grasses do, although this is not true elsewhere (Wilsey and Polley 2006). In such cases, the double benefit of food production and carbon sequestration will encourage ranchers to plant nonnative species unless they receive strong incentives to preserve natives through biodiversity payment schemes (Pagiola et al., 2004) that make raising cattle on ranches with native pasture much more profitable.

Impact 7: Feed Production and Trade

Human need for livestock is directly responsible for the environmental (and biodiversity) costs and benefits of 37% of the cropland used to grow cereals, because farmers grow these cereals to feed livestock (WRI 2005). The proportion of cropland dedicated to feed livestock varies strongly from one region to another—from 50% to 60% in Europe and the Americas to 21% in Asia and only 8% in Africa (WRI 2005). The biodiversity costs of these croplands include the loss of ecosystems that they replace, their contribution to climate change, and the pollution from fertilizers they use, as described earlier. This means, for example, that livestock are doubly responsible for the aquatic dead zones around the world, not only directly through livestock urine and manure but also indirectly through nitrates, pesticides, and other chemicals that are used on croplands grown to feed livestock and that run off into waterways.

As mentioned earlier, growth and extension of the global trades in livestock feeds means that livestock produced in one part of the world often depend on ecosystems far away, where the impacts on biodiversity are invisible to those who consume livestock (Lebel et al., 2002). As feed demand rose in the last decade, deforestation expanded rapidly in the Amazon (Nepstad et al., 2006). Feed made from soybean cake is also produced in the Brazilian *cerrado*, which, as described, is a highly diverse savanna largely converted to soybeans and cattle ranching, with remaining fragments largely invaded by exotic forage plants. The actual movement of feed from one part of the world to another, usually by ship, consumes fossil fuels that contribute to climate change that affects biodiversity. Ship ballast is also the principal way that exotic marine species (plankton, fish) invade new waters globally (Carlton 1985, Ruiz et al., 1997).

Livestock also affect the sea through the fishmeal they eat. The world's largest users of fishmeal are the poultry and swine industries (Pike 1998, Naylor et al., 2000). For example, in 1994 Swedish farmers used fishmeal to feed chickens (55% of their feed), pigs (44%), and cattle (1%) (Deutsch and Folke 2005). Fishmeal for Swedish animals is mostly from the North Sea, consisting of sprat and herring. About 12% of the total land and marine areas used to produce feed for Swedish livestock are marine (Deutsch and Folke 2005). Global fisheries are already seriously overused, so any dependence of livestock on fishmeal only exacerbates loss of this resource.

Impact 8: Livestock-Generated Greenhouse Gas Emissions and Climate Change

The livestock sector, including feed production and transport, contributes substantially to greenhouse gas emissions globally; this was recently estimated as 18% of all greenhouse gas emissions as measured in CO_2 equivalents and 80% of all the emissions from agriculture (Steinfeld et al., 2006). Livestock may also account for 9% of global carbon dioxide emissions through deforestation for pasture and feed crops and land degradation (Steinfeld et al., 2006), 65% of human-related nitrous oxide, 35–40% of all human-produced methane, and 64% of ammonia, which contributes significantly to acid rain. Based on Steinfeld et al. (2006), we estimate that Asian livestock emit more methane than any other region—a full third of the world total—while livestock in Africa emit 14% of the world's methane from enteric fermentation and only 3% of that from manure, with emissions from the Americas and Europe falling in between. Considering carbon emissions from livestock in CO_2 equivalents, 34% comes from deforestation for pastures and feed crops in South America, 25% from enteric fermentation, 12% from direct manure application/deposition, 8.7% from indirect manure emission, and 5.2% from manure management (Steinfeld et al., 2006).

Climate change, partially caused by livestock, is already altering species distributions and population sizes (Parmesan and Yohe 2003, Root et al., 2003, Thomas et al., 2004), the timing of reproduction or migration events (Ogutu and Owen-Smith 2006, UNEP 2006), and the frequency of pest and disease outbreaks (MA 2005a). Parmesan and Yohe (2003), in a meta-analysis of more than 1,700 species, suggest that global warming is responsible for the observed movement of species toward the poles at a rate of 6.1 km per decade and the advancement of spring by 2.3 days per decade, which affects the seasonal movement of species. Root et al. (2003) also report consistent temperature-related shifts in the distributions of breeding species, ranging from mollusks to mammals and from grasses to trees.

Contributions to climate change resulting from livestock production are also partially responsible for an increase in the severity of diseases (Harvell et al., 2002, IPCC 2002) and the melting of icecaps. Both of these changes affect biodiversity. As described earlier, infectious diseases can threaten biodiversity by catalyzing declines and accelerating extinctions of wild animal populations

(Harvell et al., 2002). Pathogens were involved in the recent declines in Australian and Central American frogs, Hawaiian forest birds, African wild dogs and lions, and black-footed ferrets (Harvell et al., 2002). In the Arctic, the melting of major parts of the polar icecaps removes habitat for wide-ranging species like the polar bear (Robinson et al., 2005). Glacier melting in the Alps has accelerated since 1980, and 10–20% of glacier ice was lost in less than two decades (Haeberli and Beniston 1998). In tropical Africa, the glaciers on Mt. Kilimanjaro lost 80% of their area during the last century; despite persisting for over 10,000 years, they are likely to disappear by 2020 (Thompson et al., 2002).

Particularly important is the multiple impacts of livestock on marine ecosystems, including through climate change. Elevated sea surface temperatures "bleach" highly diverse and productive coral reefs, often killing coral (MA 2005a, UNEP 2006). Coral reefs are extremely important for biodiversity, providing a home to over 25% of all marine life (Moberg and Folke 1999). Some scholars estimate that many coral ecosystems will disappear in a few decades (Williams et al., 1999). In addition, global fisheries are in major decline, and livestock contribute to all of the major causes of that decline: overfishing, climate change, pollution, and introduction of marine invasive species (Sala and Knowlton 2006). In addition, dust from heavily grazed parts of the Sahel may be linked to coral diseases in the Caribbean (Shinn et al., 2000).

By 2050 climate change will likely cause the extinction of a substantial number of the endemic species currently being analyzed globally (Thomas et al., 2004). Climate change is projected to most strongly affect species with intrinsically low population numbers, those inhabiting restricted or patchy areas, and those that have limited climatic ranges (Sala et al., 2000, Thomas et al., 2004, UNEP 2006, IPCC 2007). So far, global climate change has caused the extinction of at least one endemic vertebrate species, the golden toad, from the forests of Costa Rica (Pounds et al., 2006, UNEP 2006). In the future, Thomas et al. (2004) predict that between 15% and 37% of all species on Earth could be threatened with extinction as a result of climate change.

Summary of the Relative Importance of Livestock's Different Effects on Biodiversity

The impacts of livestock on biodiversity are pervasive, but they differ in kind, magnitude, and geographic spread (Table 8.1). If specific locations of particular problems can be identified, the way to make livestock production more compatible with biodiversity conservation will become more apparent. For example, we calculate from Steinfeld et al. (2006) that 90% of the effect of livestock on global warming comes from methane from livestock, methane and nitrous oxide from manure, and CO_2 from deforestation. If allocated regionally (Steinfeld

Table 8.1. Relative importance of geographic spread and level of impact of different processes affecting biodiversity through need for livestock by continent

Process that affects biodiversity	Geographic spread	Impact on biodiversity	Africa	Asia	Australia	Europe	Latin America	North America
Expansion of pasture and feed crops[2]	Regional	High	** feed crops	—	—	—	*** Pasture and feed crops	—
Feed demand and trade[3]	Regional	Medium to high	*	*** (eastern)	*	*** (central)	***	*** (eastern)
Invasion of nonnative species[4]	Global	Medium to high	*	**	**	***	***	***
Heavy grazing[2]	W	L/M	**	**	**	*	*	**
Manure and fertilizer pollution[3]	Regional	High	*	***	*	***	**	**
Intensification of feeds and livestock[4]	Regional/global	Medium	**	***	*	***	**	***
Disease emergence[4]	Regional	Low to high	**	***	*	*	**	*
GHG emissions caused by livestock[5]	Global	Low/high	**	***	*	**	*** From deforestation and methane	**
Contraction of rangelands/pastures[6]	Global	Low	***	*	*	—	—	**
Abandonment of pastures[4]	Regional	Low/medium	—	*	—	**	—	**
Increased human well-being[4]	Regional	Medium to high (?)	***	**	*	*	**	*

Processes in left-hand column are ordered by their estimated relative impact on biodiversity from high to low[1]:

— = no impacts, * = low impacts, *** = high impacts in each region.

[1] Estimate based on the review of literature in this chapter and in MA 2005c, Figure 4.16, p. 843.

[2] Based on the review of literature in this chapter and in Chapter 5.

[3] Based in the review of literature in this chapter and in Chapter 6.

[4] Based on the review of literature in this chapter.

[5] Based on our calculations from Steinfeld et al., 2006.

[6] Based on Figure 8.2.

et al., 2006), these sources are concentrated first in Latin America and China, followed closely by sub-Saharan Africa, India, and western Europe, and then by North America and the rest of Asia. For other impacts of livestock on biodiversity, precise locations are difficult to assess; they are generally as shown in Table 8.1, based on the best available information.

Per hectare, the heaviest impacts are probably those places where people modify the environment the most and the ecosystem modified is particularly fragile. This certainly occurs when farmers and ranchers clear complex rain forests to create open pastures and fields to grow feed crops. Deforestation for livestock also releases large amounts of CO_2, which has a very important global impact on biodiversity. Another consideration is the anthropogenic impact through livestock production on marine coral reefs through climate change. The review suggests that people, through livestock, have unexpectedly large impacts on biodiversity in rivers, lakes, and the sea more broadly by removing riverine vegetation, polluting water, and contributing to ocean dead zones.

In summary, it appears that different suites of impacts occur together in particular systems, depending on the inherent productive potential and the current intensity of the livestock systems. (See Figure 8.5.) These four suites of impacts, or syndromes, simplify the complexity of the real world, where different systems often exist side by side, but they serve to illustrate the interrelated nature of impacts of livestock on biodiversity. The first one is the extensive dryland syndrome. Here, on the wetter fringes of the vast drylands, rangelands are contracting to make way for cropping and settlement, with significant impacts on biodiversity. In some areas, exurban development has also diminished biodiversity. But when grazing is profitable enough to slow these developments, livestock can protect biodiversity in these landscapes. Where grazing is moderate to heavy, invasive species are important in some areas. In many cases there is significant room for native wildlife and plants to share the landscapes with livestock, as long as they are not overexploited. Bushlands and shrublands are replacing grasslands in many of these lands (Asner et al., 2004). Diseases can be an issue, but because of the wide open places, spread is usually slow. Greenhouse gas emissions are important only because these lands are the most extensive on Earth and the only place where livestock are the central focus of people's livelihoods.

The second, perhaps least extensive, syndrome is the one that affects dryland key resources. These include towns, markets, riverine areas, and wetlands where unusually high numbers of people and livestock gather. Here, livestock use is intensive, grazing is heavy, and wildlife are all but excluded. Generally only plants tolerant of heavy grazing thrive. Landscapes are highly fragmented. People bring in nonnative species for food, fodder, and fuel. Pollution by greenhouse gas emissions and nutrients is concentrated but limited because these places are limited in extent. This syndrome can be found all across the world in specific areas within drylands.

In the more productive, wetter parts of the world, there are two very different situations. In forests, the deforestation/reforestation syndrome is evident, where pastures signify either massive biodiversity loss in tropical rain forests or valuable sources of biodiversity in abandoned pastures being reforested in temperate regions. In this case, deforestation is a significant source of greenhouse gases, and slowing the abandonment of pastures slows carbon sequestration. In forests and woodlands used more intensively, the simplified intensive system

syndrome is found. Invasive species are pervasive, wildlife are absent, nutrient pollution is common, and biodiversity is simplified and homogeneous from place to place. In many ways, this system replaces native ecosystems, with pronounced and negative implications for biodiversity.

While it is impossible to get a global and accurate measure of all the impacts of livestock on biodiversity, it is clear that the effects are felt more in some places than others, often through several effects simultaneously, and that the negatives appear to outweigh the positives. One big gap in understanding includes whole ecosystems or whole region assessments of the total impacts of livestock on biodiversity, using useful tools like life cycle approaches. Another big gap is estimations of what will happen to biodiversity in the future, as the demand for meat, milk, and eggs continues to grow. This suggests that one next step is a set of scenarios, based on the information in this assessment, that show different alternative futures for biodiversity, driven by the multifaceted impacts of livestock.

Conclusions

Many of the needed responses by citizens, civil society, governments, and the private sector to the impacts of livestock on biodiversity are similar to the responses needed to reduce livestock's impacts on carbon, nitrogen, and water, and thus are described elsewhere in this assessment. See Chapter 9 for manure management to reduce nutrient pollution and greenhouse gas emissions, and Chapter 6 for whole system nitrogen management. Chapter 5 refers to some measures to reduce deforestation, sequester carbon, restore degraded soils,

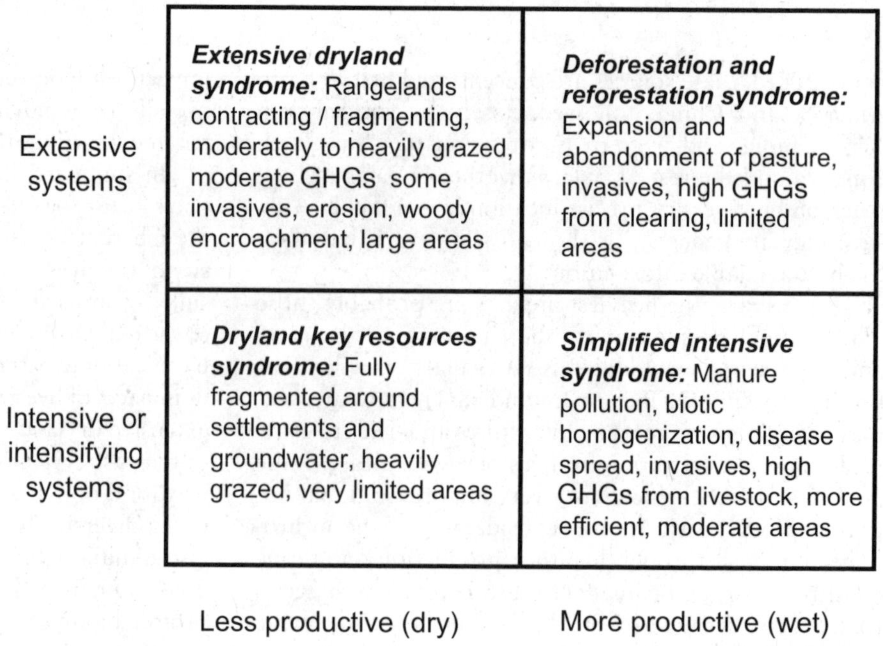

Extensive systems	**Extensive dryland syndrome:** Rangelands contracting / fragmenting, moderately to heavily grazed, moderate GHGs some invasives, erosion, woody encroachment, large areas	**Deforestation and reforestation syndrome:** Expansion and abandonment of pasture, invasives, high GHGs from clearing, limited areas
Intensive or intensifying systems	**Dryland key resources syndrome:** Fully fragmented around settlements and groundwater, heavily grazed, very limited areas	**Simplified intensive syndrome:** Manure pollution, biotic homogenization, disease spread, invasives, high GHGs from livestock, more efficient, moderate areas
	Less productive (dry)	More productive (wet)

Figure 8.5. Syndromes of livestock impacts on biodiversity distinguished by intensity of the livestock production system and the inherent productivity of the land.

and manage woody encroachment. See Chapter 7 for responses to livestock impacts on water resources. This conclusion covers responses to intensification, invasive species, diseases, and contraction of rangelands—topics not covered elsewhere in this volume.

From a biodiversity perspective, management of intensification needs to focus on improving incentives for farmers and pastoralists to maintain landscape diversity (horizontally and vertically) as they seek to make farms more efficient. Incentives can include subsidies, green labeling, taxes, easements, and other devices. (See Chapter 16 for more details.) As indicated earlier, maintaining a diversity of vegetation patches like hedgerows, woodlots, and other features is very important for preserving diversity on farming landscapes where farmers grow livestock and their feed. Sometimes this results in elusive win-wins, such as pastures with more species producing more hay in southern England (Bullock et al., 2007). Low tillage and conservation agriculture approaches need to be encouraged through incentives (Gregory et al., 2005).

For industrial systems, there is a need for regulations and incentives to encourage those who produce livestock and feed to internalize pollution costs and to control pollution from their facilities (Naylor et al., 2005). To conserve livestock breeds, government policies need to favor the conservation of their local animal genetic resources by commercial and subsistence farmers. In intensifying rangelands, rotating grazing and resting grasslands at different times will maintain a diversity of landscape conditions for native animal and plant species. Particularly important is conservation of biodiverse areas of these landscapes like wetlands and riparian corridors (Belsky et al., 1999), as well as protection of water sources like rivers, streams, lakes, and coastal areas.

For invasive species, prevention is far better than a cure. Eradication of established invasive species is very costly; early prevention is less costly and more successful (Mooney and Hobbs 2000). But often preventing exotic plant seeds from establishing is not the main issue; rather, it is land use and climate that are more important in preventing the spread of invasives (Corbin and D'Antonio 2004). In some cases, livestock themselves can help control weeds (Oba et al., 2000), but in others livestock encourage weed spread; understanding which situation applies and where is important. Successful control of invasives depends more on the persistence of the control efforts over long periods of time than on the technique of the control itself (Mack et al., 2000). Unfortunately, the problem of invasives is likely to accelerate, given the growth of global trade and the movement of feeds and livestock.

Avian flu presents a particular case where widespread culling could strongly affect wild bird diversity. Efforts to prevent poultry and wild bird populations from mixing are needed, and enhanced monitoring and surveillance in both domesticated and wild populations, particularly around wetlands and lakes, should be put in place to prevent spread of the virus. Culling should be done only if there is a clear connection between wild bird disease and domestic disease. Firm policies need to tackle the illegal trade of poultry, wild birds, and other wildlife. To reduce further outbreaks in threatened biodiversity there needs to be enhanced biosecurity, particularly in protected areas, captive breeding locations, and wildlife trade (Roberton et al., 2006). Of course, if there is a global human pandemic of bird flu, then many of these more cautious approaches will be abandoned.

If livestock grazing is generally more compatible with high levels of biodiversity than crop farming and exurban development, then strong incentives, both positive and negative, are needed to ensure higher returns to pastoral lifestyles. A whole suite of positive incentives are now being tried, like schemes for ecosystem or biodiversity payments, conservation easements, local conservancies, and public–private investment partnerships. For example, in Kitengela, Kenya, local landowners receive payments to keep the land open so that livestock and wildlife can continue to move freely, and payments double the incomes of the poorest households in the dry season (Kristjanson et al., 2002, Reid et al., 2007). Schemes to pay premiums for "conservation beef" from herds that are managed in ways that are compatible with biodiversity need to be expanded. The best schemes not only provide positive monetary incentives, but also devolve real power to local communities to manage biodiversity in ways that bring them economic returns. In drier rangelands, herders need to be able to maintain mobility but also have access to schools, clinics, and shops: services that move with herding families are sorely needed. In fragmented landscapes, support is needed for institutions that allow landowners to "stitch" the landscape back together through reciprocal grazing and watering rights.

One cornerstone to these responses is better linking of local and nonlocal knowledge with action on the ground (Cash et al., 2003). When community members, policy makers, and scientists work together to produce knowledge in a collaborative way, action is much more likely to be at least partially based on evidence rather than wholly based on politics.

Finally, one of the major ways to reduce livestock's impact on biodiversity is to consume fewer livestock products, particularly meat. This applies only to people who have adequate access to protein and micronutrients and can "afford" to limit meat consumption (see Chapter 12). This might also imply a shift from ruminant meat to pig and poultry meat in order to reduce greenhouse gas emissions and grazing impacts, as long as pollution from industrial pig and poultry production is carefully controlled. This can be done by individual action and also through organized incentives to encourage more-sustainable meat production and consumption.

References

Altieri, M. A. 1999. The ecological role of biodiversity in agroecosystems. *Agriculture Ecosystems and Environment* 74:19–31.

Anderies, J. M., M. A. Janssen, and B. H. Walker. 2002. Grazing management, resilience, and the dynamics of a fire-driven rangeland system. *Ecosystems* 5:23–44.

Andrew, M. H. 1988. Grazing impact in relation to livestock watering points. *Trends in Ecology and Evolution* 3:336–339.

Angelsen, A., and D. Kaimowitz. 2001. *Agricultural technologies and tropical deforestation.* Wallingford, UK: CAB International.

Archer, S. A. 1994. Woody plant encroachment into southwestern grasslands and savannas: Rates, patterns and proximate causes. In *Ecological implications of livestock herbivory*, ed. M. Vavra, W. Laycock, and R. Pieper, 13–68. Denver: Society for Range Management.

Asner, G. P., A. J. Elmore, L. P. Olander, R. E. Martin, and A. T. Harris. 2004. Grazing systems, ecosystem responses, and global change. *Annual Review of Environment and Resources* 29:261–299.

Bakker, J. P. 1989. *Nature management by cutting and grazing.* Dordrecht: Kluwer

Baldock, D., G. Beaufoy, F. Brouwer, and F. Godeschalk. 1996. Farming at the margins: Abandonment or redeployment of agricultural land in Europe. London: IEEP, and The Hague: LEI-DLO.

Bank, M. S., J. B. Crocker, S. Davis, D. K. Brotherton, R. Cook, J. Behler, and B. Connery. 2006. Population decline of northern dusky salamanders at Acadia National Park, Maine, USA. *Biological Conservation* 130:230–238.

Barbaro, L., T. Dutoit, and P. Cozic. 2001. A six year experimental restoration of biodiversity by shrub-clearing and grazing in calcareous grassland of French Prealps. *Biodiversity and Conservation* 10:119–135.

Becker, W. B. 1966. The isolation and classification of Tern virus: Influenza A-Tern South Africa–1961. *Journal of Hygiene* (London) 64:309–320.

Belnap, J. 2003. The world at your feet: Desert biological soil crusts. *Frontiers in Ecology and the Environment* 1:181–189.

Belsky, A. J., A. Matzke, and S. Uselman. 1999. Survey of livestock influences on stream and riparian ecosystems in the western United States. *Journal of Soil and Water Conservation* 54:419–431.

Bengis, R. G., R. A. Kock, and J. Fischer. 2002. Infectious animal diseases: The wildlife/livestock interface. *Scientific and Technical Review* (Office International de Epizooties) 21:53–65.

Bennett, M. K. 1941. International contrasts in food consumption. *Geographical Review* 31:365–376.

Bennewitz, J., J. Kantanen, I. Tapio, M. H. Li, E. Kalm, J. Vilkki, I. Ammosov, et al. 2006. Estimation of breed contributions to present and future genetic diversity of 44 North Eurasian cattle breeds using core set diversity measures. *Genetics Selection Evolution* 38:201–220.

Bignal, E. M., and D. I. McCracken. 1996. Low-intensity farming systems in the conservation of the countryside. *Journal of Applied Ecology* 33:413–424.

Birdlife. 2005. Waterbird culls and draining of wetlands could worsen spread of avian influenza: 20 October 2005. http://www.birdlife.org/news/news/2005/10/avian_flu_bird_culls.html (date accessed 15 June 2008).

Birdlife. 2007. BirdLife International Statement of Avian Influenza: 9th February 2007. http://www.birdlife.org/action/science/species/avian_flu/ (date accessed 15 June 2008).

Blench, R. 2000. "You can't go home again," extensive pastoral livestock systems: issues and options for the future. London: ODI/FAO.

Bokdam, J., and J. M. Gleichman. 2000. Effects of grazing by free-ranging cattle on vegetation dynamics in a continental north-west European heathland. *Journal of Applied Ecology* 37:415–431.

Boone, R. B., and N. T. Hobbs. 2004. Lines around fragments: Effects of fencing on large herbivores. *African Journal of Range and Forage Science* 21 (3): 147–158.

Boschi, C., and B. Baur. 2007. The effect of horse, cattle and sheep grazing on the diversity and abundance of land snails in nutrient-poor calcareous grasslands. *Basic and Applied Ecology* 8:55–65.

Boserup, E. 1965. *The conditions of agricultural growth.* London: Allen and Unwin.

Boutin, C., B. Jobin, L. Belanger, and L. Choiniere. 2002. Plant diversity in three types of hedgerows adjacent to cropfields. *Biodiversity and Conservation* 11:1–25.

Bouwman, A. F., and H. Booij. 1998. Global use and trade of feedstuffs and consequences for the nitrogen cycle. *Nutrient Cycling in Agroecosystems* 52:261–267.

Boyer, R., and C. E. Grue. 1995. The need for water quality criteria for frogs. *Environmental Health Perspectives* 103:352–357.

Boyer, E. W., R. W. Howarth, J. N. Galloway, F. J. Dentener, P. A. Green, and C. J. Vörösmarty. 2006. Riverine nitrogen export from the continents to the coasts. *Global Biogeochemical Cycles* 20:GBIS91.

Brashares, J. S., P. Arcese, and M. K. Sam. 2001. Human demography and reserve size predict wildlife extinction in West Africa. *Proceedings of the Royal Society of London Series B-Biological Sciences* 268:2473–2478.

Briske, D. D., S. D. Fuhlendorf, and F. E. Smeins. 2003. Vegetation dynamics on rangelands: a critique of the current paradigms. *Journal of Applied Ecology* 40:601–614.

Brockington, D., and K. Homewood. 2001. Degradation debates and data deficiencies: The Mkomazi Game Reserve, Tanzania. *Africa* 71:449–480.

Brockington, D., J. Igoe, and K. Schmidt-Soltau. 2006. Conservation, human rights, and poverty reduction. *Conservation Biology* 20:250–252.

Brooks, M. L., C. M. D'Antonio, D. M. Richardson, J. B. Grace, J. E. Keeley, J. M. DiTomaso, R. J. Hobbs, M. Pellant, and D. Pyke. 2004. Effects of invasive alien plants on fire regimes. *BioScience* 54:677–688.

Brooks, M. L., J. R. Matchett, and K. H. Berry. 2006. Effects of livestock watering sites on alien and native plants in the Mojave Desert, USA. *Journal of Arid Environments* 67:125–147.

Brossad, M., and A. O. Barcellos. 2005. Conversão do Cerrado em pastagens cultivadas e funiconamento de latossolos. *Cadernos de Ciência and Tecnologia* 22:153–168.

Bullock, J. M., R. F. Pywell, and K. J. Walker. 2007. Long-term enhancement of agricultural production by restoration of biodiversity. *Journal of Applied Ecology* 44:6–12.

Butler, D. 2006. Blogger reveals China's migratory goose farms near site of flu outbreak. *Nature* 441:263.

Camargo, J. A., A. Alonso, and A. Salamanca. 2005. Nitrate toxicity to aquatic animals: A review with new data for freshwater invertebrates. *Chemosphere* 58:1255–1267.

Campbell, D. J., H. Gichohi, A. Mwangi, and L. Chege. 2000. Land use conflict in Kajiado District, Kenya. *Land Use Policy* 17:337–348.

Carlton, J. T. 1985. Trans-oceanic and interoceanic dispersal of coastal marine organisms: The biology of ballast water. *Oceanography and Marine Biology* 23:313–371.

Carpenter, S. R., N. F. Caraco, D. L. Correll, R. W. Howarth, A. N. Sharpley, and V. N. Smith. 1998. Nonpoint pollution of surface waters with phosphorus and nitrogen. *Ecological Applications* 8:559–568.

Carpenter, S. R., D. Ludwig, and W. A. Brock. 1999. Management of eutrophication for lakes subject to potentially irreversible change. *Ecological Applications* 9:751–771.

Carpenter, S. R., P. L. Pingali, E. M. Bennett, and M. B. Zurek, eds. 2005. *Ecosystems and human well-being: Scenarios, volume 2.* Washington, DC: Island Press.

Cash, D. W., W. C. Clark, F. Alcock, N. M. Dickson, N. Eckley, D. H. Guston, J. Jager, et al. 2003. Knowledge systems for sustainable development. *Proceedings of the National Academy of Sciences* 100: 8086–8091.

CBD. 2006. Press Release: Avian Flu May Prove Big Threat to Biological Diversity. Convention of Biological Diversity. Available at http://www.unep.org/Documents.Multilingual/Default.asp?DocumentID=471&ArticleID=5235&l=en

Chapin, F. S., E. S. Zavaleta, V. T. Eviner, R. L. Naylor, P. M. Vitousek, H. L. Reynolds, D. U. Hooper, et al. 2000. Consequences of changing biodiversity. *Nature* 405:234–242.

Cleaveland, S., M. G. J. Appel, W. S. K. Chalmers, C. Chillingworth, M. Kaare, and C. Dye. 2000. Serological and demographic evidence for domestic dogs as a source of canine distemper virus infection for Serengeti wildlife. *Veterinary Microbiology* 72:217–227.

Clutton-Brock, J. 1981. *Domesticated animals from early times.* Austin: University of Texas Press.

Cochrane, M. A., and M. D. Schulze. 1998. Forest fires in the Brazilian Amazon. *Conservation Biology* 12:948–950.

Cochrane, M. A., A. Alencar, M. D. Schulze, C. M. Souza, D. C. Nepstad, P. Lefebvre, and E. A. Davidson. 1999. Positive feedbacks in the fire dynamic of closed canopy tropical forests. *Science* 284:1832–1835.

Corbin, J. D., and C. M. D'Antonio. 2004. Competition between native perennial and exotic annual grasses: Implications for an historical invasion. *Ecology* 85:1273–1283.

Daily, G. C., P. R. Ehrlich, and G. A. Sanchez-Azofeifa. 2001. Countryside biogeography: Use of human-dominated habitats by the avifauna of southern Costa Rica. *Ecological Applications* 11:1–13.

D'Antonio, C. M., and P. M. Vitousek. 1992. Biological invasions by exotic grasses, the grass/fire cycle, and global change. *Annual Review of Ecology and Systematics* 23:63–87.

Daszak, P., A. A. Cunningham, and A. D. Hyatt. 2000. Emerging infectious diseases of wildlife: Threats to biodiversity and human health. *Science* 287:443–449.

Daszak, P., A. A. Cunningham, and A. D. Hyatt. 2001. Anthropogenic environmental change and the emergence of infectious diseases in wildlife. *Acta Tropica* 78:103–116.

de Haan, C., H. Steinfeld, and H. Blackburn. 1997. *Livestock and the environment: Finding a balance.* Fressingfield, UK: WRENmedia,

de Leeuw, J., M. N. Waweru, O. O. Okello, M. Maloba, P. Nguru, M. Y. Said, H. M. Aligula et al. 2001. Distribution and diversity of wildlife in northern Kenya in relation to livestock and permanent water points. *Biological Conservation* 100:297–306.

Dean, W. R. J., S. J. Milton, and F. Jeltsch. 1999. Large trees, fertile islands, and birds in arid savanna. *Journal of Arid Environments* 41:61–78.

Deutsch, L. 2004. *Global trade, food production and ecosystem support: Making the interactions visible.* Stockholm: Stockholm University.

Deutsch, L., and C. Folke. 2005. Ecosystem subsidies to Swedish food consumption from 1962 to 1994. *Ecosystems* 8:512–528.

Diamond, J. 2002. Evolution, consequences and future of plant and animal domestication. *Nature* 418:700–707.

Diaz, R. J. 2001. Overview of hypoxia around the world. *Journal of Environmental Quality* 30:275–281.

Dirzo, R., and P. Raven. 2003. Global state of biodiversity and loss. *Annual Review of Environment and Resources* 28:137–167.

Ellis, J., and D. M. Swift. 1988. Stability of African pastoral ecosystems: Alternative paradigms and implications for development. *Journal of Range Management* 41:450–459.

Ellis, T. M., R. B. Bousfield, L. A. Bissett, K. C. Dyrting, G. S. M. Luk, S. T. Tsim, K. Sturm-Ramirez, et al. 2004. Investigation of outbreaks of highly pathogenic H5N1 avian influenza in waterfowl and wild birds in Hong Kong in late 2002. *Avian Patholology* 33:492–505.

FAO (Food and Agriculture Organization). 2004. Payment for environmental services in watersheds. Regional Forum, 9–12 June 2003, Arequipa, Peru. Rome: FAO.

FAO (Food and Agriculture Organization). 2006a. *Animal health special report: H5N1 in cats.* Animal Production and Health Division. Rome: FAO.

FAO (Food and Agriculture Organization). 2006b. *Breed diversity in dryland ecosystems.* Commission on Genetic Resources for Food and Agriculture. Rome: FAO.

FAO (Food and Agriculture Organization). 2006c. FAO statistical databases. http://faostat.external.fao.org.

FAO (Food and Agriculture Organization). 2007. FAO Newsroom: Avian Influenza in cats should be closely monitored: 8 February. Rome: FAO.

Feare, C. J., and M. Yasué. 2006. Asymptomatic infection with highly pathogenic avian influenza H5N1 in wild birds: How sound is the evidence? *Virology Journal* 3:96.

Fearnside, P. M. 1993. Deforestation in Brazilian Amazonia: The effect of population and land tenure. *Ambio* 22:537–545.

Fergus, R., M. Fry, W. B. Karesh, P. P. Marra, S. Newman, and E. Paul. 2006. Migratory birds and avian flu. *Science* 312:845.

Fernandez-Gimenez, M. E., and B. Allen-Diaz. 1999. Testing a non-equilibrium model of rangeland vegetation dynamics in Mongolia. *Journal of Applied Ecology* 6:871–885.

Ferraz, G., J. D. Nichols, J. E. Hines, P. C. Stouffer, R. O. Bierregaard, and T. E. Lovejoy. 2007. A large-scale deforestation experiment: Effects of patch area and isolation on Amazon birds. *Science* 315:238–241.

Fisher, M. J., I. M. Rao, M. A. Ayarza, C. E. Lascano, J. I. Sanz, R. J. Thomas, and R. R. Vera. 1994. Carbon storage by introduced deep-rooted grasses in the South American savannas. *Nature* 371:236–238.

Folke, C., S. Carpenter, B. Walker, M. Scheffer, T. Elmqvist, L. Gunderson, and C. S. Holling. 2004. Regime shifts, resilience, and biodiversity in ecosystem management. *Annual Review of Ecology, Evolution and Systematics* 35:557–581.

Ford, J. 1971. *The Role of Trypanosomiases in African Ecology.* Oxford, UK: Clarendon Press.

Galiana-Aleixandre, M. V., A. Iborra-Clar, B. Bes-Pia, J. A. Mendoza-Roca, B. Cuartas-Uribe, and M. I. Iborra-Clar. 2005. Nanofiltration for sulfate removal and water reuse of the pickling and tanning processes in a tannery. *Desalinisation* 179:307–313.

Galloway, J. N., J. D. Aber, J. W. Erisman, S. P. Seitzinger, R. W. Howarth, E. B. Cowling, and B. J. Cosby. 2003. The nitrogen cascade. *BioScience* 53:341–356.

Geist, H. J., and E. Lambin. 2002. Proximate causes and underlying driving forces of tropical deforestation. *BioScience* 52:143–149.

Geist, H. J., and E. Lambin. 2004. Dynamic causal patterns of desertification. *BioScience* 54:817–829.

Gerber, P., P. G. Chilonda, G. Franceschini, and H. Menzi. 2005. Geographical determinants and environmental implications of livestock production intensification in Asia. *Bioresource Technology* 96:263–276.

Gilbert, M., X. Xiao, J. Domenech, J. Lubroth, V. Martin, and J. Slingenbergh. 2006. Anatidae migration in the Western Palearctic and spread of highly pathogenic avian influenza H5N1 virus. *Emerging Infectious Diseases* 12:1650–1656.

Gregory, M. M., K. L. Shea, and E. B. Bakko. 2005. Comparing agroecosystems: Effects of cropping and tillage patterns on soil, water, energy use and productivity. *Renewable Agriculture and Food Systems* 20:81–90.

Groombridge, B., and M. Jenkins. 1998. *Freshwater biodiversity: A preliminary global assessment.* Cambridge, UK: World Conservation Monitoring Centre.

Gurevitch, J., and D. K. Padilla. 2004. Are invasive species a major cause of extinctions? *Trends in Ecology and Evolution* 19:470–474.

Haddad, N. M., J. Haarstad, and D. Tilman. 2000. The effects of long-term nitrogen loading on grassland insect communities. *Oecologia* 124:73–84.

Haeberli, W., and M. Beniston. 1998. Climate change and its impacts on glaciers and permafrost in the Alps. *Ambio* 27:258–265.

Hanotte, O., D. G. Bradley, J. W. Ochieng, Y. Verjee, E. W. Hill, and J. E. O. Rege. 2002. African pastoralism: Genetic imprints of origins and migrations. *Science* 296:336–339.

Hansen, A. J., R. Rasker, B. Maxwell, J. J. Rotella, J. D. Johnson, A. W. Parmenter, L. Langner et al. 2002. Ecological causes and consequences of demographic change in the new west. *BioScience* 52:151–162.

Hansen, A. J., R. L. Knight, J. M. Marzluff, S. Powell, K. Brown, P. H. Gude, and A. Jones. 2005. Effects of exurban development on biodiversity: Patterns, mechanisms, and research needs. *Ecological Applications* 15:1893–1905.

Harrison, P. F., and J. Lederberg. 1998. *Antimicrobial resistance, issues and options. Forum on emerging infections.* Institute of Medicine. Washington, DC: National Academy Press.

Harrison, S., B. D. Inouye, and H. D. Safford. 2003. Ecological heterogeneity in the effects of grazing and fire on grassland diversity. *Conservation Biology* 17:837–845.

Harvell, C. D., C. E. Mitchell, J. R. Ward, S. Altizer, A. P. Dobson, R. S. Ostfeld, and M. D. Samuel. 2002. Climate warming and disease risks for terrestrial and marine biota. *Science* 296:2158–2162.

Hecht, S. 1989. The sacred cow in the green hell: Livestock and forest conversion in the Brazilian Amazon. *Ecologist* 19:229–234.

Hecht, S. 1993. The logic of livestock and deforestation in Amazonia. *BioScience* 43:687–695.

Hecht, S. 2005. Soybeans, development and conservation on the Amazon frontier. *Development and Change* 36:375–404.

Hiernaux, P. 1998. Effects of grazing on plant species composition and spatial distribution in rangelands of the Sahel. *Plant Ecology* 138:191–202.

Hoare, R. E. 1999. Determinants of human–elephant conflict in a land-use mosaic. *Journal of Applied Ecology* 36:689–700.

Hobbs, R. J., and L. F. Huenneke. 1992. Disturbance, diversity, and invasion: Implications for conservation. *Conservation Biology* 6:324–337.

Hobbs, N. T., R. S. Reid, K. A. Galvin, and J. E. Ellis. 2007. Fragmentation of arid and semi-arid ecosystems: Implications for people and animals. In *Fragmentation in semi-arid and arid landscapes: Consequences for human and natural systems*, ed. K. A. Galvin, R. S. Reid, R. H. Behnke, and N. T. Hobbs, 25–44. Dordrecht, Netherlands: Springer.

Hodgson, J. G., and A. W. Illius. 1996. *The ecology and management of grazing systems.* Wallingford, UK: CAB International.

Hoffmann, I., and B. Scherf. 2006. Animal genetic resources—time to worry? Rome: Food and Agriculture Organization.

Hoffmann, W. A., V. Lucatelli, F. J. Silva, I. N. C. Azeuedo, M. D. Marinho, A. M. S. Albuquerque, A. D. Lopes, and S. P. Moreira. 2004. Impact of the invasive alien grass *Melinis minutiflora* at the savanna–forest ecotone in the Brazilian cerrado. *Diversity and Distributions* 10:99–103.

Howarth, R. W., E. W. Boyer, W. J. Pabich, and J. N. Galloway. 2002. Nitrogen use in the United States from 1961–2000 and potential future trends. *Ambio* 31:88–96.

Huenneke, L. F., and I. Noble. 1996. Ecosystem function of biodiversity in arid ecosystems. In *Functional roles of biodiversity: A global perspective*. ed. H. A. Mooney, J. H. Cushman, E. Medina, O. E. Sala, and E. D. Schulze, 99–128. New York: John Wiley & Sons.

IDLGroup. 2002. *Poverty and livestock breed diversity—the way forward for DFID.* Draft consultation document. Bristol, UK: The IDL Group.

IISD (International Institute for Sustainable Development). 2006. A summary report of the scientific seminar on avian influenza, the environment and migratory birds. *Avian Influenza and WildBirds Bulletin* 123:1–8.

Illius, A. W., and T. G. O'Connor. 1999. On the relevance of non-equilibrium concepts to arid and semiarid grazing systems. *Ecological Applications* 9:798–813.

IPCC (Intergovernmental Panel on Climate Change). 2002. *Climate change and biodiversity.* IPCC Technical Paper V. Geneva: IPCC.

IPCC (Intergovernmental Panel on Climate Change). 2007. *Climate change 2007: The physical science basis.* Contribution of Working Group I to the Fourth Assessment Report of the Intergovernmental Panel on Climate Change. Cambridge: Cambridge University Press.

Jackson, R. D., B. Allen-Diaz, L. G. Oates, and K. W. Tate. 2006. Spring-water nitrate increased with removal of livestock grazing in a California oak savanna. *Ecosystems* 9:254–267.

Jong, J. C. d., E. C. J. Class, A. D. M. E. Osterhause, R. G. Webster, and W. L. Lim. 1997. A pandemic warning? *Nature* 389:554.

JRL (Joint Research Laboratory). 2005. *GLC 2000 (Global Land Cover) data layer.* Ispra, Italy: JRL.

Kawahara, M., S. Uye, K. Ohtsu, and H. Iizumi. 2006. Unusual population explosion of the giant jellyfish *Nemopilema nomurai* (Scyphozoa: Rhizostomeae) in East Asian waters. *Marine Ecology-Progress Series* 307:161–173.

Keawcharoen, J., K. Oraveerakul, T. Kuiken, T. Fouchier, A. Amonsin, S. Payungporn, S. Nopponpanth, et al. 2004. Avian influenza H5N1 in tigers and leopards. *Emerging Infectious Diseases* 10:2189–2191.

Kilpatrick, A. M., A. A. Chmura, D. W. Gibbons, R. C. Fleischer, P. P. Marra, and P. Daszak. 2006. Predicting the global spread of H5N1 avian influenza. *Proceedings of the National Academy of Sciences* 103:19368–19373.

Klima, G. 1970. *The Barabaig: East African Cattleherders*. Long Grove, IL: Waveland Press.

Klink, C. A. 1996. Competition between the African grass *Andropogon gayanus* Kunth. and the native cerrado grass *Schizachyrium tenerum* Nees. *Revista Brasileira de Botânica São Paulo* 19:11–15.

Klinken, R. D., J. Graham, and L. K. Flack. 2006. Population ecology of hybrid mesquite (*Prosopis* species) in Western Australia: How does it differ from native range invasions and what are the implications for impacts and management? *Biological Invasions* 8:727–741.

Kraaij, T., and S. J. Milton. 2006. Vegetation changes (1995–2004) in semi-arid Karoo shrubland, South Africa: Effects of rainfall, wild herbivores and change in land use. *Journal of Arid Environments* 64:174–192.

Kristjanson, P. M., M. Radeny, D. Nkedianye, R. L. Kruska, R. S. Reid, H. Gichohi, F. Atieno et al., eds. 2002. *Valuing alternative land-use options in the Kitengela wildlife dispersal area of Kenya*. Nairobi: International Livestock Research Institute.

Kuiken, T., A. D. M. E. Osterhaus, and P. Roeder. 2006. Feline friend or potential foe? *Nature* 440:741–742.

Laiolo, P., F. Dondero, E. Ciliento, and A. Rolando. 2004. Consequences of pastoral abandonment for the structure and diversity of the alpine avifauna. *Journal of Applied Ecology* 41:294–304.

Lambin, E. F., H. J. Geist, and E. Lepers. 2003. Dynamics of land-use and land-cover change in tropical regions. *Annual Review of Environment and Resources* 28:205–241.

Landsberg, J., C. D. James, S. R. Morton, W. J. Muller, and J. Stol. 2003. Abundance and composition of plant species along grazing gradients in Australian rangelands. *Journal of Applied Ecology* 40:1008–10024.

Laurenroth, W. K., D. G. Milchunas, J. L. Dodd, R. H. Hart, R. K. Heitschmidt, and L. R. Rittenhouse. 1999. Effect of grazing on ecosystems of the Great Plains. In *Ecological implications of livestock herbivory in the West*, ed. M. Vavra, W. A. Laycock, and R. D. Pieper, 69–100. Denver: Society of Range Management.

Lebel, L., N. H. Tri, A. Saengnoree, S. Pasong, U. Buatama, and L. K. Thoa. 2002. Industrial transformation and shrimp aquaculture in Thailand and Vietnam: Pathways to ecological, social and economic sustainability? *Ambio* 31:311–323.

Lehner, B., and P. Döll. 2004. Development and validation of a global database of lakes, reservoirs and wetlands. *Journal of Hydrology* 296:1–22.

Leppman, E. J. 1999. Urban–rural contrasts in diet: The case of China. *Urban Geography* 20:567–579.

Lonsdale, W. M. 1994. Inviting trouble: Introduced pasture species in northern Australia. *Journal of Ecology* 19:345–354.

MA (Millennium Ecosystem Assessment). 2005a. Biodiversity. In *Ecosystems and human well-being: Current state and trends, volume 1*, ed. R. Hassan, R. Scholes, and N. Ash, 77–122. Washington, DC: Island Press.

MA (Millennium Ecosystem Assessment). 2005b. Biodiversity across scenarios. In *Ecosystems and human well-being:*

Scenarios, volume 2, ed. S. R. Carpenter, P. L. Pingali, E. M. Bennett, and M. B. Zurek, 375–408. Washington, DC: Island Press.

MA (Millennium Ecosystem Assessment). 2005c. *Ecosystems and human well-being: Current state and trends, volume 1*. Washington, DC: Island Press.

MacDougall, A. S., and R. Turkington. 2005. Are invasive species the drivers or passengers of change in degraded ecosystems? *Ecology* 86:42–55.

Mack, R. N., and J. N. Thompson. 1982. Evolution in steppe with few large, hooved mammals. *American Naturalist* 119:157–173.

Mack, R. N., D. Simberloff, W. M. Lonsdale, H. Evans, M. Clout, and F. A. Bazzaz. 2000. Biotic invasions: Causes, epidemiology, global consequences, and control. *Ecological Applications* 10:689–710.

Maestas, J. D., R. L. Knight, and W. C. Gilgert. 2003. Biodiversity across a rural land-use gradient. *Conservation Biology* 17:1425–1434.

Malmstrom, C. M., C. J. Stoner, S. Brandenburg, and L. A. Newton. 2006. Virus infection and grazing exert counteracting influences on survivorship of native bunchgrass seedling competing with invasive exotics. *Journal of Ecology* 94:264–275.

Marshall, F., and E. Hildebrand. 2002. Cattle before crops: The beginnings of food production in Africa. *Journal of World Prehistory* 16:99–143.

Mathias, E., I. Köhler-Rollefson, and J. Wanyama. 2005. Pastoralists, local breeds and the fight for livestock keepers' rights. Prepared for PENHA, Pastoralism in the Horn of Africa: Surviving against all odds. 15th Anniversary Conference.

Matson, P., W. J. Parton, A. G. Power, and M. J. Swift. 1997. Agricultural intensification and ecosystem properties. *Science* 277:504–509.

May, R. M. 1990. How many species? *Philosophical Transactions of the Royal Society of London Series B-Biological Sciences* 330:293–304.

McIntire, J., D. Bourzat, and P. Pingali. 1992. *Crop–livestock interaction in sub-Saharan Africa*. Washington, DC: World Bank.

McKinney, M. L., and J. L. Lockwood. 1999. Biotic homogenization: A few winners replacing many losers in the next mass extinction. *Trends in Ecology and Evolution* 14:450–453.

McNaughton, S. J. 1979. Grazing as an optimization process: Grass–ungulate relationships in the Serengeti. *American Naturalist* 113:691–703.

Michalski, F. 2006. Human–wildlife conflicts in a fragmented Amazonian forest landscape: Determinants of large felid depredation on livestock. *Animal Conservation* 9:179.

Milchunas, D. G., O. E. Sala, and W. K. Lauenroth. 1988. A generalized model of the effects of grazing by large herbivores on grassland community structure. *American Naturalist* 132:87–106.

Milchunas, D. G., W. K. Lauenroth, and I. C. Burke. 1998. Livestock grazing: Animal and plant biodiversity of shortgrass steppe and the relationship to ecosystem function. *Oikos* 83:65–74.

Milton, S. J. 2004. Grasses as invasive alien plants in South Africa. *South African Journal of Science* 100:69–75.

Moberg, F., and C. Folke. 1999. Ecological goods and services of coral reef ecosystems. *Ecological Economics* 29:215–233.

Mooney, H. A., and E. E. Cleland. 2001. The evolutionary impact of invasive species. *Proceedings of the National Academy of Sciences* 98:5446–5451.

Mooney, H. A., and R. J. Hobbs. 2000. *Invasive species in a changing world*. Washington, DC: Island Press.

Morton, D. C., R. S. DeFries, Y. E. Shimabukuro, L. O. Anderson, E. Arai, F. D. Espirito-Santo, R. Freitas, et al. 2006. Cropland expansion changes deforestation dynamics in the southern Brazilian Amazon. *Proceedings of the National Academy of Sciences* 103:14637–14641.

Murray, M., J. C. Trail, and G. D. D' Ieteren. 1990. Trypanotolerance in cattle and prospects for the control of trypanosomiasis by selective breeding. *Revue scientifique et technique* 9:369–386.

Myers, N. 1988. Tropical forest and their species. In *Biodiversity*, ed. E. O. Wilson, 28–35. Washington, DC: National Academy Press.

Namgail, T., J. L. Fox, and Y. V. Bhatnagar. 2007. Habitat shift and time budget of the Tibetan argali: The influence of livestock grazing. *Ecological Research* 22:25–31.

Naughton-Treves, L., and A. Treves. 2005. Socio-ecological factors shaping local support for wildlife: Crop-raiding by elephants and other wildlife in Africa. In *People and wildlife: Conflict or coexistence*, ed. R. Woodroffe, S. Thirgood, and A. Rabinowitz, 252–277. Cambridge: Cambridge University Press.

Naylor, R. L., R. J. Goldburg, J. H. Primavera, N. Kautsky, M. C. M. Beveridge, J. Clay, C. Folke, et al. 2000. Effect of aquaculture on world fish supplies. *Nature* 405:1017–10124.

Naylor, R., H. Steinfeld, W. Falcon, J. Galloway, V. Smil, E. Bradford, J. Alder, and H. Mooney. 2005. Losing the links between livestock and land. *Science* 310:1621–1622.

Nepstad, D. C., C. M. Stickler, and O. T. Almeida. 2006. Globalization of the Amazon soy and beef industries: Opportunities for conservation. *Conservation Biology* 20:1595–1603.

Normile, D. 2006. Avian influenza: Evidence points to migratory birds in H5N1 spread. *Science* 311:1225.

Norton-Griffiths, N., M. Y. Said, S. Serneels, D. S. Kaelo, M. B. Coughenour, R. Lamprey, D. M. Thompson, and R. S. Reid. 2008. Land use economics in the Mara area of the Serengeti ecosystem. In *Serengeti III: Impacts of people on the Serengeti ecosystem*, eds. A. R. E. Sinclair, C. Parker, S. A. R. Mduma, and J. M. Fryxell, 379–416. Chicago: Chicago University Press.

Oaks, J. L., M. Gilbert, M. Z. Virani, R. T. Watson, C. U. Meteyer, B. A. Rideout, et al. 2004. Diclofenac residues as the cause of vulture population decline in Pakistan (letter). *Nature* 427:630–633.

Oba, G., E. Post, N. C. Stenseth, and W. J. Lusigi. 2000. The role of small ruminants in arid zone environments: A review of research perspectives. *Annals of Arid Zone* 39:305–332.

Ogutu, J. O., and N. Owen-Smith. 2006. Oscillations in large mammal populations: Are they related to predation or rainfall? *African Journal of Ecology* 43:332–339.

OIE (World Organisation for Animal Health). 2007. Update on highly pathogenic avian influenza in animals (type H5 and H7). http://www.oie.int/downld/AVIAN%20INFLUENZA/A_AI-Asia.htm.

Olff, H., and M. E. Ritchie. 1998. Effects of herbivores on grassland plant diversity. *Trends in Ecology and Evolution* 13:261–265.

Olsen, B., V. J. Munster, A. Wallensten, J. Waldenström, A. D. M. E. Osterhaus, and R. A. M. Fouchier. 2006. Global patterns of influenza A virus in wild birds. *Science* 312:384–388.

Ostermann, O. P. 1998. The need for management of nature conservation sites designated under Natura 2000. *Journal of Applied Ecology* 35:968–973.

Otte, J., D. Pfeiffer, T. Tiensin, L. Price, and E. Silbergeld. 2006. *Evidence-based policy for controlling HPAI in poultry: Bio-security revisited*. Pro-Poor Livestock Policy Initiative. Baltimore, MD: Johns Hopkins School of Public Health.

Pagiola, S., P. Agostini, J. Gobbi, C. de Haan, M. Ibrahim, E. Murgueitio, E. Ramirez, M. Rosales, and J. Pablo Ruiz. 2004. *Paying for biodiversity conservation services in agricultural landscapes*. Environment Department Paper No. 96. Washington, DC: World Bank.

Parmesan, C., and G. Yohe. 2003. A globally coherent fingerprint of climate change impacts across natural systems. *Nature* 421:37–42.

Peacock, J. M., M. E. Ferguson, G. A. Alhadramis, I. R. McCann, A. H. A, A. Saleh, and R. Karnik. 2003. Conservation through utilization: A case study of the indigenous forage grasses of the Arabian Peninsula. *Journal of Arid Environments* 54:15–28.

Pieper, R. 1999. Ecological implications of livestock grazing. In *Ecological implications of livestock herbivory in the West*, ed. M. Vavra, W. A. Laycock, and R. D. Pieper, 177–211. Denver: Society of Range Management.

Pike, I. H. 1998. In S. Fraser, ed. International Aquafeed Directory. Middlesex, UK: Turret.

Pimentel, D., R. Zuniga, and D. Morrison. 2005. Update on the environmental and economic costs associated with alien-invasive species in the United States. *Ecological Economics* 53:273–288.

Pimm, S. L., G. J. Russell, J. L. Gittleman, and T. M. Brooks. 1995. The future of biodiversity. *Science* 209:347–350.

Pivello, V. R., C. N. Shida, and S. T. Meirelles. 1999. Alien grasses in Brazilian savannas: A threat to the biodiversity. *Biodiversity and Conservation* 8:1281–1294.

Porter, J. W., S. K. Lewis, and K. G. Porter. 1999. The effect of multiple stressors on the Florida Keys coral reef ecosystem: A landscape hypothesis and a physiological test. *Limnology and Oceanography* 44:941–949.

Posada, J. M., T. M. Aide, and J. Cavelier. 2000. Cattle and weedy shrubs as restoration tools of tropical montane rainforest. *Restoration Ecology* 8:370–379.

Pounds, J. A., M. R. Bustamante, L. A. Coloma, J. A. Consuegra, M. P. L. Fogden, P. N. Foster, E. L. Marca, et al. 2006. Widespread amphibian extinctions from epidemic disease driven by global warming. *Nature* 439:161–167.

Prins, H. H. T. 1992. The pastoral road to extinction: Competition between wildlife and traditional pastoralism in East Africa. *Environmental Conservation* 19:117–123.

Prins, H. H. T. 2000. Competition between wildlife and livestock in Africa. In *Wildlife conservation by sustainable use*, ed. H. H. T. Prins, J. G. Grootenhuis, and T. T. Dolan. Dordrecht, Netherlands: Kluwer Academic Publishers.

Proulx, M., and A. Mazumder. 1998. Reversal of grazing impact on plant species richness in nutrient-poor vs. nutrient-rich ecosystems. *Ecology* 79:2581–2592.

Pulliam, H. R. 1988. Sources, sinks and population regulation. *American Naturalist* 132:652–661.

Pykälä, J. 2003. Effects of restoration with cattle grazing on plant species composition and richness of semi-natural grasslands. Biodiversity and Conservation 12:2211–2226.

Rabalais, N. N. 2002. Nitrogen in aquatic ecosystems. *Ambio* 31:102–112.

Rabalais, N. N., R. E. Turner, and W. J. Wiseman. 2002. Gulf of Mexico hypoxia, aka "The dead zone." *Annual Review of Ecology and Systematics* 33:235–263.

Reardon, T., C. P. Timmer, C. B. Barrett, and J. Berdegue. 2003. The rise of supermarkets in Africa, Asia, and Latin America. *American Journal of Agricultural Economics* 85:1140–1146.

Reid, R. S., and J. E. Ellis. 1995. Livestock-mediated tree regeneration: Impacts of pastoralists on dry tropical woodlands. *Ecological Applications* 5:978–992.

Reid, R. S., M. E. Rainy, C. J. Wilson, E. Harris, R. L. Kruska, M. Waweru, S. A. Macmillan et al. 2001. *Wildlife cluster around pastoral settlements in Africa*. People, Livestock and Environment (PLE) Science Series. Nairobi: International Livestock Research Institute.

Reid, R. S., M. Rainy, J. Ogutu, R. L. Kruska, M. Nyabenge, M. McCartney, K. Kimani, et al. 2003. *Wildlife, people, and livestock in the Mara ecosystem, Kenya: The Mara count 2002*. Nairobi: International Livestock Research Institute.

Reid, R. S., P. K. Thornton, and R. L. Kruska. 2004. Loss and fragmentation of habitat for pastoral people and wildlife in East Africa: Concepts and issues. *South African Journal of Grass and Forage Science* 21:171–181.

Reid, R. S., H. Gichohi, M. Y. Said, D. Nkedianye, J. O. Ogutu, M. Kshatriya, P. Kristjanson, et al. 2007. Fragmentation of an peri-urban savanna, Athi-Kaputiei Plains, Kenya. In *Fragmentation in semi-arid and arid landscapes: Consequences for human and natural systems*, ed. K. A. Galvin, R. S. Reid, R. H. Behnke, and N. T. Hobbs, 195–224. Dordrecht, Netherlands: Springer.

Richter, B. D., D. P. Braun, M. A. Mendelson, and L. L. Master. 1997. Threats to imperiled freshwater fauna. *Conservation Biology* 11:1081–1093.

Ricketts, T. H., G. C. Daily, P. R. Ehrlich, and J. P. Fay. 2001. Countryside biogeography of moths in a fragmented landscape: Biodiversity in native and agricultural habitats. *Conservation Biology* 15:378–388.

Roberton, S. I., D. J. Bell, G. J. D. Smith, J. M. Nicholls, K. H. Chan, D. T. Nguyen, P. Q. Tran, et al. 2006. Avian influenza H5N1 in viverrids: Implications for wildlife health and conservation. *Proceedings of the Royal Society* B 273:1729–1732.

Robinson, R. A., J. A. Learmonth, A. M. Hutson, C. D. Macleod, T. H. Sparks, D. I. Leech, G. J. Pierce, M. M. Rehfisch, and H. Q. P. Crick. 2005. *Climate change and migratory species*. Research report 414. Thetford, Norfolk, UK: Bird Trust for Ornithology.

Rocha, G. L. 1991. Ecosistemas de pastagens. Piracicaba, Brazil: Fundação de Estudos Agrários Luiz de Queiroz.

Rohde, R. F., N. M. Moleele, M. Mphale, N. Allsopp, R. Chanda, M. T. Hoffman, L. Magole et al. 2006. Dynamics of grazing policy and practice: Environmental and social impacts in three communal areas of southern Africa. *Environmental Science and Policy* 9:302–316.

Rook, A. J., B. Dumont, J. Isselstein, K. Osoro, M. F. WallisDeFries, G. Parente, and J. Mills. 2004. Matching type of livestock to desired biodiversity outcomes in pastures: A review. *Biological Conservation* 119:137–150.

Root, T. L., J. T. Price, K. R. Hall, S. H. Schneider, C. Rosenzweig, and J. A. Pounds. 2003. Fingerprints of global warming on wild animals and plants. *Nature* 421:57–60.

Rosegrant, M. W., N. Leach, and R. V. Gerpacio. 1999. Alternative futures for world cereal and meat consumption. *Proceedings of the Nutrition Society* 58:219–234.

Rosegrant, M. W., M. S. Paisner, S. Meijer, and J. Witcover. 2001. *Global food projections to 2020: Emerging trends and alternative futures*. Washington, DC: International Food Policy Research Institute.

Rouse, J. D., C. A. Bishop, and J. Struger. 1999. Nitrogen pollution: An assessment of its threat to amphibian survival. *Environmental Health Perspectives* 107:799–803.

Ruiz, G. M., J. T. Carlton, E. D. Grosholz, and A. H. Hines. 1997. Global invasions of marine and estuarine habitats by non-indigenous species: Mechanisms, extent, and consequences. *American Zoologist* 37:621–632.

Sala, E., and N. Knowlton. 2006. Global marine biodiversity trends. *Annual Review of Environment and Resources* 31:93–122.

Sala, O. E., F. S. Chapin, J. J. Armesto, E. Berlow, J. Bloomfield, R. Dirzo, E. Huber-Sanwald, et al. 2000. Global biodiversity scenarios for the year 2100. *Science* 287:1770–1774.

Scherf, B. 2000. *World watch list for domestic animal diversity*. Rome: Food and Agriculture Organization.

Schmink, M., and C. H. Wood. 1992. *Contested frontiers in Amazonia*. New York: Columbia Press.

Scholes, R. J., and S. R. Archer. 1997. Tree–grass interactions in savannas. *Annual Review of Ecology and Systematics* 28:517–544.

Scoones, I. 1991. Wetlands in drylands: Key resources for agricultural and pastoral production in Africa. *Ambio* 20:366–371.

Seabloom, E. W., W. S. Harpole, O. J. Reichman, and D. Tilman. 2003. Invasion, competitive dominance, and resource use by exotic and native California grassland species. *Proceedings of the National Academy of Sciences* 100:13384–13389.

Seidl, A. F., J. D. V. de Silva, and A. S. Moraes. 2001. Cattle ranching and deforestation in the Brazilian Pantanal. *Ecological Economics* 36:413–425.

Shinn, E. A., G. W. Smith, J. M. Prospero, P. Betzer, M. L. Hayes, V. Garrison, and R. T. Barber. 2000. African dust and the demise of Caribbean coral reefs. *Geophysical Research Letters* 27:3029–3032.

Sims, L. D., J. Domenech, C. Benigno, S. Kahn, A. Kamata, J. Lubroth, V. Martin, and P. Roeder. 2005. The origins and evolution of H5N1 highly pathogenic avian influenza in Asia. *Veterinary Record* 157:159–164.

Sinclair, A. R. E., and J. M. Fryxell. 1985. The Sahel of Africa— ecology of a disaster. *Canadian Journal of Zoology-Revue Canadienne De Zoologie* 63:987–994.

Skarpe, C. 1991. Impact of grazing in savanna ecosystems. *Ambio* 20:351–356.

Smart, S. M., K. Thompson, R. H. Marrs, M. G. Le Duc, L. C. Maskell, and L. G. Firbank. 2006. Biotic homogenization and changes in species diversity across human-modified ecosystems. *Proceedings of the Royal Society B-Biological Sciences* 273:2659–2665.

Smith, R. S., and S. P. Rushton. 1994. The effects of grazing management on the vegetation of mesotrophic (meadow) grassland in northern England. *Journal of Applied Ecology* 31:13–24.

Soto, A., J. M. Calabro, N. V. Prechtl, A. Y. Yau, E. F. Orlando, A. Daxenberger, A. S. Kolok, L. J. Guillette, Jr., B. le Bizec, I. G. Lange, and C. Sonnenschein. 2004. Androgenic and estrogenic activity in water bodies receiving cattle feedlot effluent in eastern Nebraska, USA. *Environmental Health Perspectives* 112:346–352.

Srivastava, D. S., and M. Vellend. 2005. Biodiversity-ecosystem function research: Is it relevant to conservation? *Annual Review of Ecology Evolution and Systematics* 36:267–294.

Steinfeld, H., P. Gerber, T. Wassenaar, V. Castel, M. Rosales, and C. de Haan. 2006. *Livestock's Long Shadow*. Rome: Food and Agriculture Organization.

Stevens, C. J., N. B. Dise, O. Mountford, and D. J. Gowing. 2004. Impact of nitrogen deposition on the species richness of grasslands. *Science* 303:1876–1879.

Stoddart, L. A., A. D. Smith, and T. W. Box. 1975. *Range management*. New York: McGraw-Hill.

Stokes, C. J., R. R. J. McAllister, A. J. Ash, and J. E. Gross. 2007. Changing patterns of land use and land tenure in the Dalrymple Shire, Australia. In *Fragmentation of semi-arid and arid landscapes: Consequences for human and natural systems*, ed. K. A. Galvin, R. S. Reid, R. H. Behnke, and N. T. Hobbs, 93–121. Dordrecht, Netherlands: Springer.

Strayer, D. L., V. T. Eviner, J. M. Jeschke, and M. L. Pace. 2006. Understanding the long-term effects of species invasions. *Trends in Ecology and Evolution* 21:645–651.

Stromberg, J. C., S. J. Lite, R. Marler, C. Paradzick, P. B. Shafroth, D. Shorrock, J. M. White, et al. 2007. Altered stream-flow regimes and invasive plant species: The *Tamarix* case. *Global Ecology and Biogeography* 16:381–393.

Sturm-Ramirez, K. M., T. Ellis, B. Bousfield, L. Bissett, K. Dyrting, J. E. Rehg, L. Poon, et al. 2004. Re-emerging H5N1 influenza viruses in Hong Kong in 2002 are highly pathogenic to ducks. *Journal of Virology* 78:4892–4901.

Subak, S. 1999. Global environmental costs of beef production. *Ecological Economics* 30:79–91.

Sullivan, S., and R. Rohde. 2002. On non-equilibrium in arid and semi-arid grazing systems. *Journal of Biogeography* 29:1595–1618.

Tarawali, S. A., A. Larbi, S. Fernandez-Rivera, and A. Bationo. 2001. The contribution of livestock to soil fertility. In *Sustaining soil fertility in West Africa*, 281–304. Madison, WI: Soil Science Society of America and American Society of Agronomy.

Tarawali, S. A., J. D. H. Keating, J. M. Powell, P. Hiernaux, O. Lyasse, and N. Sanginga. 2004. Integrated natural resource management in West African crop–livestock systems. In *Sustainable crop–livestock production for improved livelihoods and natural resource management in West Africa*, ed. T. O. Williams, S. A. Tarawali, P. Hiernaux, and S. Fernandez-Rivera, 349–370. Ibadan, Nigeria: International Institute of Tropical Agriculture.

Thanawongnuwech, R., A. Amonsin, R. Tantilertcharoen, S. Damrongwatanapokin, A. Theamboonlers, S. Payungporn, K. Nanthapornphiphat, et al. 2005. Probable tiger-to-tiger transmission of avian influenza H5N1. *Emerging Infectious Diseases* 11:699–701.

Thiollay, J. M. 1989. Area requirements for the conservation of rain-forest raptors and game birds in French-Guyana. *Conservation Biology* 3:128–137.

Thomas, C. D., A. Cameron, R. E. Green, M. Bakkenes, L. J. Beaumont, Y. C. Collingham, B. F. N. Erasmus, et al. 2004. Extinction risk from climate change. *Nature* 427:145–148.

Thompson, L. G., E. Mosley-Thompson, M. E. Davis, K. A. Henderson, H. H. Brecher, V. S. Zagorodnov, T. A. Mashiotta, et al. 2002. Kilimanjaro ice core records: Evidence of Holocene climate change in tropical Africa. *Science* 298:589–593.

Tilman, D. 1987. Secondary succession and the pattern of plant dominance along experimental nitrogen gradients. *Ecological Monographs* 57:189–214.

Tilman, G. D., D. N. Duvick, S. B. Brush, R. J. Cook, G. C. Daily, G. M. Heal, S. Naeem, et al. 1999. *Benefits of biodiversity*. Task Force Report 133. Ames, IA: Council for Agricultural Science and Technology.

Tilman, D., K. G. Cassman, P. A. Matson, R. Naylor, and S. Polasky. 2002. Agricultural sustainability and intensive production practices. *Nature* 418:671–676.

Tisdell, C. 2003. Socioeconomic causes of loss of animal genetic diversity: Analysis and assessment. *Ecological Economics* 45:365–376.

Todd, S. W. 2006. Gradients in vegetation cover, structure and species richness of Nama-Karoo shrublands in relation to distance from livestock watering points. *Journal of Applied Ecology* 43:293–304.

Traveset, A., and D. M. Richardson. 2006. Biological invasions as disruptors of plant reproductive mutualisms. *Trends in Ecology and Evolution* 21:208–216.

Turner, B. L., R. E. Kasperson, P. A. Matson, J. J. McCarthy, R. W. Corell, L. Christensen, N. Eckley, et al. 2003. A framework for vulnerability analysis in sustainability science. *Proceedings of the National Academy of Sciences* 100:8074–8079.

Turner, M. D. 1998. Long-term effects of daily grazing orbits on nutrient availability in Sahelian West Africa: Effects of a phosphorus gradient on spatial patterns of annual grassland production. *Journal of Biogeography* 25:683–694.

UNEP (United Nations Environment Programme). 1995. *Global biodiversity assessment*. Cambridge, UK: Cambridge University Press.

UNEP (United Nations Environment Programme). 2006. *Migratory species and climate change: Impacts of a changing environment on wild animals*. Bonn: UNEP/Convention on Migratory Species Secretariat.

USDA (U.S. Department of Agriculture). 2001. *Agricultural chemical use*. Washington, DC: National Agricultural Statistic Service, USDA Economics and Statistics System.

Van Auken, O. W. 2000. Shrub invasions of North American semiarid grasslands. *Annual Review of Ecology and Systematics* 31:197–215.

Vesk, P. A., and M. Westoby. 2001. Predicting plant species' responses to grazing. *Journal of Applied Ecology* 38:897–909.

Vetter, S. 2005. Rangelands at equilibrium and non-equilibrium: Recent developments in the debate. *Journal of Arid Environments* 62:321–341.

Vitousek, P. M., J. Aber, R. Howarth, G. E. Likens, P. Matson, D. Schindler, W. Schlesinger, and G. D. Tilman. 1997a. Human alteration of the global nitrogen cycle: Causes and consequences. *Ecological Applications* 7:737–750.

Vitousek, P. M., H. A. Mooney, J. Lubchenco, and J. M. Melillo. 1997b. Human domination of Earth's ecosystems. *Science* 277:494–499.

Vitt, L. J., T. C. S. Avila-Pires, J. P. Caldwell, and V. R. L. Oliveira. 1998. The impact of individual tree harvesting on thermal environments of lizards in Amazonian rain forest. *Conservation Biology* 12:654–664.

Ward, D., B. T. Ngairorue, J. Kathena, R. Samuels, and Y. Ofran. 1998. Land degradation is not a necessary outcome of communal pastoralism in arid Namibia. *Journal of Arid Environments* 40:357–371.

Western, D., and H. Gichohi. 1993. Segregation effects and the impoverishment of savanna parks: The case for ecosystem viability analysis. *African Journal of Ecology* 31:269–281.

Westoby, M., B. Walker, and I. Noymeir. 1989. Opportunistic management for rangelands not at equilibrium. *Journal of Range Management* 42:266–274.

White, P. J. T., and J. T. Kerr. 2007. Human impacts on environment–diversity relationships: evidence for biotic homogenization from butterfly species richness patterns. *Global Ecology and Biogeography* 16:290–299.

WHO (World Health Organization). 2007a. Areas reporting confirmed occurrence of H5N1 avian influenza in poultry and wild birds since 2003. http://gamapserver.who.int/mapLibrary/Files/

Maps/Global_SubNat_H5N1inAnimalConfirmedCUMULA-TIVE_20070608.png (accessed 8 June 2007).

WHO (World Health Organization). 2007b. Avian influenza. http://www.who.int/csr/disease/avian_influenza/en (accessed March 2007).

Wiegand, K., D. Saitz, and D. Ward. 2006. A patch-dynamics approach to savanna dynamics and woody plant encroachment—insights from an arid savanna. *Perspectives in Plant Ecology Evolution and Systematics* 7:229–242.

Wilcove, D. S., D. Rothstein, J. Bubow, A. Phillips, and E. Losos. 1998. Quantifying threats to imperiled species in the United States. *BioScience* 49:607–615.

Williams, E. H. J., P. J. Bartels, and L. Bunkley-Williams. 1999. Predicted disappearance of coral-reef ramparts: A direct result of major ecological disturbances. *Global Change Biology* 5:839–845.

Williams, S. D. 1998. Grevy's zebra: Ecology in a heterogeneous environment. Ph.D thesis. London: University College.

Williamson, M. 1996. *Biological invasions*. London: Chapman and Hall.

Wilsey, B. J., and H. W. Polley. 2006. Above ground productivity and root-shoot allocation differ between native and introduced grass species. *Oecologia* 150:300–309.

Wilson, C. J., R. S. Reid, N. L. Stanton, and B. D. Perry. 1997. Ecological consequences of controlling the tsetse fly in southwestern Ethiopia: Effects of land-use on bird species diversity. *Conservation Biology* 11:435–447.

Wilson, E. O. 1988. *Biodiversity*. Washington, DC: National Academy Press.

Witt, G. B., J. Luly, and R. J. Fairfax. 2006. How the west was once: Vegetation change in south-west Queensland from 1930 to 1995. *Journal of Biogeography* 33:1585–1596.

Woodroffe, R., and J. R. Ginsberg. 1998. Edge effects and the extinction of populations inside protected areas. *Science* 280:2126–2128.

WRI (World Resources Institute). 2005. *World resources 2005—the wealth of the poor: Managing ecosystems to fight poverty*. Washington, DC: WRI.

Yingst, S. L., M. D. Saad, and S. A. Felt. 2006. Qinghai-like H5N1 from domestic cats, northern Iraq. *Emerging Infectious Diseases* 12:1295–1297.

Young, T. P., N. Partridge, and A. Macrae. 1995. Long-term glades in Acacia bushland and their edge effects in Laikipia, Kenya. *Ecological Applications* 5:97–108.

9

Impacts of Intensive Livestock Production and Manure Management on the Environment

Harald Menzi, Oene Oenema, Colin Burton, Oleg Shipin, Pierre Gerber, Tim Robinson, and Gianluca Franceschini

Main Messages

- **Intensive livestock production systems are characterized by a high stocking density, a high output of animal products per unit surface area, and a relatively large share of the milk, beef, pork, and egg and poultry production worldwide.** Cattle, pigs, and poultry are the dominant species. These systems rely extensively on imported animal feed and are known as landless systems.

- **Intensive livestock production systems are rapidly expanding, especially in East and Southeast Asia and Latin America.** This expansion is characterized by the following: an agglomeration of livestock production near urban (market) centers; a shift from ruminants to pigs and poultry; a trend toward large, landless, highly specialized farms; a shift from litter-based housing to slurry-based systems, with insufficient manure storage, poor manure management, and little account of the crop-nutritive value of livestock manure.

- **Nitrogen (N) and/or phosphorus (P) excretions by livestock per unit surface area are indicators of the environmental pressure exerted by livestock production.** Areas with a high livestock stocking density are characterized by high N and P loadings (with high N and P surpluses), while the areas that provide animal feed commonly have N and P deficits.

- **The total amounts of nutrients (N, P, and potassium [K]) in livestock excreta are as large or larger than the total amounts of N, P, and K in fertilizers used annually.** Cattle contribute about 40% to total livestock P excretion, and pigs and poultry about 20% each.

- **In many parts of the world, environmentally sound manure management is hindered by the fact that manure is considered a waste rather than a nutrient and energy source and by the lack of environmental legislations and its enforcement.**

- **Poor manure management has serious impacts on the environment.** It contributes to pollution and eutrophication of surface waters, groundwater, and coastal marine ecosystems. It contributes to air pollution through emissions of odor, ammonia, methane, and nitrous oxide, and it contributes to soil pollution through the accumulation of heavy metals. These pollution and eutrophication effects subsequently lead to loss of human health and biodiversity, to climate change and acidification, and to ecosystem degradation.

- **The risk for the environment of intensive livestock production is highest in East and Southeast Asia because of the current high livestock stocking density, the rapid expansion of intensive pig and poultry production, and the lack of legislative restrictions and enforcement.** The region also lacks a tradition of recycling liquid manure to crops and has only limited awareness of environmental concerns.

- **Improving the environmental performance of intensive livestock production systems requires an integral whole-farming systems approach.** The weakest part of the whole chain of activities in a farming system should be cured first. Three types of measures are available: management measures, technological measures, and structural measures. Management and technological measures are usually the cheapest and should be considered first.

- **Currently there is a lack of uniform performance standards in livestock production and manure management.** International agreements are needed among countries with intensive livestock production systems for deriving and implementing such standards.

- **A wide range of proven manure treatment options exist, including separation technologies, composting, and anaerobic digestion—all of which have a role in certain situations.** These can contribute to safe manure recycling to crops and to meeting special objectives like sanitation, odor control, biogas production, and manure's value as fertilizer. These options will contribute to replacing fertilizer nutrients with manure nutrients, thereby lowering the environmental impact of food production.

Introduction

This chapter focuses on intensive livestock production for the production of meat, milk, and eggs. These production systems are characterized by a high stocking density and a high output of animal products per unit of agricultural land and per unit of stock (i.e., livestock unit). This is generally achieved by high efficiency in converting animal feed into animal products. Because of their capacity to respond rapidly to growing demand, intensive livestock production systems now account for dominant shares of global pork, poultry meat, and egg production (56%, 72%, and 61%, respectively) and for a significant share of milk production (Steinfeld et al., 2006).

As described in earlier chapters, these systems are dominated by three species/animal categories: cattle, pigs, and fowl (mainly chickens), although sheep, goat, and buffaloes may also be kept in intensive production systems. Intensive livestock production systems include mixed systems and industrial systems. Mixed systems include intensive grazing systems that import part of the animal feed and systems where a significant part of the value of the total production comes from activities other than animal production. Industrial systems have livestock densities larger than 10 livestock units (500 kg liveweight per livestock unit) per hectare and they depend primarily on outside supplies of feed, energy, and other inputs, as in confined animal feeding operations. If less than 10% of the dry matter fed to animals is produced on the farm, the classification "landless" production system is often used (Kruska et al., 2003).

For maintenance of body functions and production, livestock require water, energy (carbohydrates), protein, and 18 mineral nutrients, including phosphorus (P), calcium (Ca), potassium (K), magnesium (Mg), selenium (Se), copper (Cu), and zinc (Zn). A large percentage of the substances consumed in the feed are excreted again via dung and urine—typically anywhere from 70% to over 90% of the nitrogen (N) and the mineral nutrients present in the feed, depending on animal species, feed composition, and management. A large portion of the detrimental environmental effects of intensive livestock production relate to poor management of this livestock excreta, which contain large amounts of undigested organic matter and mineral nutrients (Schröder 2005, Sims et al., 2005).

Estimates of the global amounts of nutrients excreted by livestock are highly uncertain, mainly because of poor information about the intake and composition of the feed. Sheldrick et al. (2003) estimated that global livestock excreta in 1996 contained 94 Tg N, 21 Tg P, and 67 Tg K. These amounts are larger by a factor of 1 to 3 than the amounts of fertilizer N, P, and K used annually.

Animal manure is an important source of plant nutrients, especially in organic farming and in developing countries (Rufino et al., 2006). Also, in countries with intensive agricultural production, animal manure is a major source of nutrients used in crop and forage production. In Switzerland, for example, the contribution of livestock manure to total agricultural fertilizer use is about 60% for N, 70% for P, and over 90% for K (FAL 2005 and personal estimates). In general, however, only a fraction of the nutrients excreted by livestock is properly collected and managed as manure (which can also contain bedding material); the remainder is left unmanaged as animal droppings in grazed pastures and paddocks or is simply discharged (Sheldrick et al., 2003, Rufino et al., 2006, Oenema et al., 2007). Increasing the use of nutrients from animal manure will not only reduce negative environmental impacts of livestock production, it will also help save limited resources like rock phosphates used for P fertilizer production and fossil energy used for N fertilizer production. (For an interesting focus on this, see Mikkelsen et al., 2009.)

Global Distribution of Cattle, Pigs, and Poultry

While more than 60% of milk, beef, and poultry meat production is situated in Europe and America, 60% of the world's pork is produced in Asia. Africa, the Middle East, and Oceania have only a small share of cattle, pig, and poultry production. In terms of animal numbers, Latin America leads for cattle and China for pigs and poultry. The relative distribution of the number and production of cattle, pigs, and poultry in different parts of the world according to FAOSTAT (2008) is shown in Table 9.1. Note that there are large differences between regions in the share of animal numbers and production. For example, North America has only 4% of the total number of dairy cows but supplies 16% of global milk, indicating intensive production systems. Conversely, Africa has 14% of the world's dairy cows but only 2% of global milk production, indicating extensive production systems. Differences between regions in the share of animal numbers and production are smaller in pig and poultry production.

The spatial distributions of cattle, pig, and poultry density (in kg liveweight per hectare area) are shown in Figures 9.1, 9.2, and 9.3 in the color well. Cattle density is highest in Europe and India, but parts of Brazil and central East Africa also show rather high densities. From +30 to +50 and below −30 degrees latitude, cattle are the dominant livestock species. In medium latitudes, the dominant species varies: poultry in the Middle East, parts of Africa, and western Latin America; pigs in Southeast and East Asia and parts of Latin America and Africa;

Table 9.1. Relative distribution of livestock numbers and production (% of global total) by region in 2005.

	Cattle stock total (%)	Dairy cows (%)	Other cattle (%)	Production meat (%)	Milk (%)	Pigs stock total (%)	Production meat (%)	Poultry stock total (%)
Europe (excluding former USSR)	8	14	7	16	32	17	22	10
EU, Switzerland, Norway	7	11	6	15	29	17	21	8
Americas	37	21	40	49	29	15	17	27
USA, Canada	8	4	9	22	16	8	11	13
Latin America	29	17	31	27	12	8	6	14
Asia (excluding USSR)	32	30	32	18	18	64	60	46
China	9	5	9	12	5	51	49	26
China, Thailand, Vietnam	9	5	10	12	5	55	52	28
SE-Asia	11	6	12	14	6	59	56	39
India	13	15	0	2	7	1	0	3
Former USSR	4	10	3	7	12	3	3	5
Africa	13	14	13	5	2	2	1	7
Middle East	3	9	3	1	3	0	0	4
Oceania	3	3	3	5	5	0	1	1

Source: FAO 2008.

and cattle in India, large parts of Africa, eastern Latin America, and Oceania.

The spatial distribution of cattle, pork, and poultry is related to the consumption patterns of meat, milk, and eggs; to cultural practices; and to production circumstances. At the end of the 1990s, average meat consumption per person ranged from 12 kg per year in Africa to 127 kg per year in North America. Average meat consumption in Asia was 27 kg (in India and Bangladesh, <5), in Europe 69 kg, and in Latin America 43 kg per person. The total amount of animal protein available per person per day was 12 g in Africa, 19 g in Asia, 34 g in Latin America, 58 g in Europe, and 69 g in North America (Smil 2002).

Intensive livestock production is found in all parts of the world, but especially in North America and western Europe and increasingly in East and Southeast Asia, the Middle East, and Latin America. For example, North America produces 22% of the world's beef, 11% of the pork, and 24% of the poultry meat (FAOSTAT 2008) on, respectively, 8%, 8%, and 13% of the stock of cattle, pigs, and poultry. Latin America contributes 22% of total beef and 24% of poultry meat production. The largest livestock density is found in southern Brazil and southern Mexico. The production intensity is generally relatively low, apart from dairy production in Mexico, Honduras, Puerto Rico, and Argentina and pig production in Puerto Rico.

In Europe, cattle, pig, and poultry production is rather intensive, though there are large regional differences. Total livestock density is highest in the Netherlands, Denmark, northern Germany, southern United Kingdom and Ireland, Brittany in northern France, and northern Italy. Cattle in Europe are the dominant species, and predominantly kept on a roughage diet (grassland-based), supplemented with concentrates (mixed systems). Europe produces 32% of the milk and 16% of the beef worldwide with 14% of the dairy cows and 8% of the cattle. Following considerable growth in the 1960s and 1970s, cattle numbers have been decreasing there since the 1980s, which is related to the implementation of the milk quota system in 1984 in the European Union combined with increasing milk yields per cow. Also the collapse of the centrally planned economies in central Europe contributed to the decrease in livestock numbers during the 1990s. In spite of this decrease, Europe still maintains one of the highest livestock densities worldwide. Landless livestock production is common only on strongly specialized pig and poultry farms, especially in the Netherlands and various regions in France, Italy, Spain, and Germany.

East and Southeast Asia (China, Thailand, Viet Nam, Malaysia, Philippines, South Korea, Cambodia, Laos, and Myanmar) are the main pig and poultry producers worldwide. Moreover, the number of pigs and poultry is rapidly increasing, while most of additional pigs and poultry are kept on highly intensive, landless farms around cities. India has a high cattle and buffalo density in the north and east. It is the largest milk-producing country in the world, although production efficiency is low. The remainder of Asia has a rather low livestock density. The structure of livestock production varies greatly across Asia. Typically for rapidly developing countries, traditional small and land-bound livestock production systems are often situated next to large intensive and landless pig and poultry farms. While ruminant production is still predominantly extensive, the share that is intensive is rapidly increasing for pig and poultry production.

Nutrient Elements in Livestock Excreta

A proxy indicator of the environmental pressures related to livestock production is the amount of nutrient elements in livestock excreta expressed per unit of surface area, using the approach described by Gerber et al. (2005). First, mean livestock production rates were assessed on a national level for different livestock categories, using production data in FAOSTAT (2008). Second, N and P excretions of the various livestock categories were calculated and down-scaled to a five arc-minutes pixel grid in geographic projection, using spatial modeling. Here, we use phosphorus as an indicator because it has a high relevance for eutrophication of surface waters, it usually is the first nutrient that constrains the livestock carrying capacity of land, and it is nonvolatile and strongly adsorbed to soils and therefore does not escape from manure in gaseous forms or through leaching (in contrast to nitrogen). The spatial distribution of total P excretion in kg km^{-2} is illustrated in Figure 9.4 in the color well.

Clearly, the spatial distribution of total P excretion largely follows the spatial distribution of cattle, pigs, and poultry. The highest P loads in livestock excreta are found in northeastern China, north India, and northwestern Europe. Locally high concentrations are also found in the U.S. Midwest, in southern Brazil and northern Argentina, in the Alpine parts of Europe, and in some parts of East Africa, India, eastern China, Indonesia, southern Australia, and New Zealand. However, not all hot spots of high P excretion relate to intensive cattle, pig, and poultry production. For example, in India the high P excretion density is the result of a very high density of extensive livestock production systems, especially cattle. Note that our estimate of the global P excretion of livestock is 9.5 Tg, which is lower than the estimate of Sheldrick et al. (2003) by more than a factor of 2. We differentiated the P excretion per animal according to production intensity and average liveweight, while Sheldrick et al. used the standard excretion per animal based on production levels in North America and other countries of the Organization for Economic Cooperation and

Development (OECD). Thus our excretions per animal are lower than their estimates for countries with extensive production systems and indigenous species. According to our calculations about 40% of the global livestock P excretion is from cattle and a bit more than 20% each from pigs and poultry.

Another proxy and more integrated indicator of the environmental pressures related to crop and animal production in agriculture is the P surplus of the balance of total P input via animal manure and fertilizers and the P output via harvested crop (Gerber et al., 2005). (See Box 9.1.)

To assess the potential load to the environment resulting from livestock production relative to that of the human population, we estimated the liveweight and P excretions of livestock and humans in different parts of the world. For livestock, we used the liveweight and excretion figures of Gerber et al. (2005). For humans, an average mean body weight of 50 kg was assumed for Africa and Asia and 60 kg for the rest of the world (derived from WHO 2008). Based on Jönson et al. (2004), an average annual P excretion of 0.4 kg was assumed for Africa and Asia and 0.55 kg for the rest of the world.

The global average ratio between livestock biomass and human biomass is 1.4—that is, total livestock liveweight is approximately 35% higher than total human liveweight. (See Table 9.2.) With a value of 8.2, the ratio is highest in Oceania (New Zealand 18.1). At 2.2, Central and Latin America is also clearly above the average, while Asia (1.1) and the Middle East (1.0) are below average. Over 40% of the total livestock biomass is in Asia, which also contributes more than half of the total worldwide human liveweight.

The estimated mean global ratio of livestock to human P excretion is 7.6. (See Table 9.3.) This is much higher than the liveweight ratio of livestock to humans, indicating that on average more P is excreted by livestock than by humans per unit liveweight. Evidently, livestock have a higher P excretion because their main aim is producing liveweight gain, milk, and eggs, while humans are mostly adults who only have to cover the maintenance requirements for bodily functions. Europe, America, the Middle East and Oceania are clearly above the world average; Asia is slightly below that number, and Africa clearly below the world average ratio of P excretion by livestock compared with humans. The contribution of Asia to the world P excretion is around 50% for both humans and livestock.

The ratios between livestock and human liveweight and between livestock and human P excretion are indicators of the potential environmental pressures originating from livestock manure relative to that of human wastes. The actual pollution greatly depends on how the manure and wastes are handled. Based on the following considerations, we may conclude, however, that the

Box 9.1. Example of P_2O_5 Balance in Asia and Contribution of Different Sources to the Potential Nutrient Input to the South China Sea

Gerber et al. (2005) estimated surpluses of P on nearly 24% of the agricultural land, mainly in eastern China, the Ganges basin, and around urban centers such as Bangkok, Ho Chi Minh City, and Manila. Deficits of P (i.e., P output via harvested crops exceeds the P input via animal manure and fertilizer) were found mainly in western China and southern and western India. (See figures A and B in the color well for Box 9.1.) On average, livestock manure accounted for an estimated 39% of the total P supply, while chemical fertilizer contributed 61%. Livestock was the dominant P source around urban centers and in livestock specialized areas, while chemical fertilizers were the dominant source in crop intensive areas (rice). However, various areas with a large supply of manure P also use large amounts of fertilizer P, suggesting either that farmers do not account for the P applied via manure or that the manure is not recycled to crop-

land. When applied to cropland, most if not all of the surplus P is stored in the soil, because soils have a strong P adsorption capacity. This capacity is finite, however. Moreover, P-enriched topsoils are prone to P losses via erosion and overland flow. Hence, a P surplus does indicate an increased potential of eutrophication of nearby surface waters.

Badayos and Dorado (2004) made model calculations on the relative contribution of pig wastes, domestic wastes, and runoff/nonpoint sources to the N and P loading of the South China Sea and the Gulf of Thailand from Thailand, Guangdong Province in China, and Viet Nam. (See Table.) According to this estimate, pig waste is by far the major source of P in all three countries and also a major source of N in Guangdong and Viet Nam. The contribution of pig waste is estimated to be 4, 18, and 94 times larger than that of domestic wastes in, respectively, Thailand, Viet Nam, and Guangdong.

Potential load of nitrogen and phosphorus reaching the South China Sea from pig waste, domestic wastewater, and runoff in Guangdong Province (China), Thailand, and Viet Nam.

Country or province	Nutrient	(1,000 tons)	Potential load Contribution (%)		
			Pig waste	Domestic wastewater	Runoff/nonpoint source
Guangdong, China	N	530	72	9	19
	P	220	94	1	5
Thailand	N	491	14	9	77
	P	53	61	16	23
Viet Nam	N	442	38	12	50
	P	212	92	5	3

Source: Badayos and Dorado 2004.

environmental pressures originating from livestock excreta are often larger than those originating from human excreta:

- Total P excretions by livestock are much larger than total P excretions by humans in all parts of the world. A similar relationship can be assumed for N excretions.
- Only a fraction of the nutrients in animal manure and human wastes are recycled for crop production; this fraction is larger for animal manure than for human wastes. Most of the human wastes are collected via the sewage system and partially discharged after treatment or end up in latrine pits in the ground, while animal manure is either applied to cropland or discharged into surface waters, or it ends up in the soil in various manure storage systems.

- A considerable share of the nutrients in municipal wastewater is removed in the sewerage treatment system, while such treatment is not applied to livestock waste prior to its discharge.
- About two-thirds of the P excreted by livestock is in emerging and developing countries that often lack policies or the ability to enforce environmental standards.
- In countries where municipal sewerage treatment is common and manure management is poor, the pollution potential of animal manure is much larger than that of human wastes, especially for water pollution. This is the case for many emerging and developing countries.

Good livestock and manure management greatly decreases the risk of environmental pollution by animal manure. Yet the potential threat to the environment

Table 9.2. Relative contribution of humans and livestock to global total liveweight by region, 2006.

Region	World liveweight		Ratio of livestock to human liveweight
	Humans (%)	Livestock (%)	
Europe (excl. former USSR)	10	12	1.5
United States, Canada	6	7	1.7
Central and Latin America	10	16	2.2
Asia (excl. former USSR)	53	43	1.1
Former USSR	5	5	1.4
Africa	13	12	1.2
Middle East	3	2	2.3
Oceania	1	3	8.2
World	100	100	1.4

Table 9.3. Contribution to total phosphorus excretion by humans and livestock by region in 2006.

Region	Global P excretion		Ratio of livestock to human P excretion
	Humans (%)	Livestock (%)	
Europe (excl. former USSR)	11	14	9.6
United States, Canada	6	9	11.0
Central and Latin America	11	15	10.5
Asia (excl. former USSR)	50	45	6.8
Former USSR	4	4	6.7
Africa	13	8	5.0
Middle East	3	2	13.7
Oceania	1	3	36.6
World	100	100	7.6

arising from operational management failures and malfunctioning equipment will remain; this threat is larger for very large operations with no or weak controls or enforcement of environmental regulations.

Characteristics of Intensive Pig, Poultry, and Dairy Cattle Production in Developing Countries

The considerable environmental consequences of livestock production are related to several important characteristics of intensified production: the concentration of livestock production close to urban centers; the shift from ruminants to pigs and poultry; the trend toward large, highly specialized, landless livestock farms; the shift from litter-based to slurry-based systems; and the neglect of manure storage and of the crop-nutritive value of livestock manure.

Concentration of Livestock Production Close to Urban Centers

In a globalizing world with open markets and more or less uniform prices, profitability is largely determined by the cost of production. Areas offering favorable production conditions and favorable logistics for transport,

processing, marketing, and retail have a comparative cost advantage. Intensive livestock production thus develops near urban areas following typical patterns, largely in accordance with the theories of Ricardo, von Thünen, and Sinclair (Sinclair 1967, Block and Dupuis 2001). This can lead to a high animal density locally and to a high environmental pressure associated with the animal manure. A concentration of livestock in urban areas is characteristic for countries in early stages of industrialization. With a growing economy, improving infrastructure, and rising awareness of environmental problems, livestock production is expelled from cities. (Chapter 4 discusses in more detail the driving factors for the location of livestock operations in different phases of development.)

Shift from Ruminants to Monogastrics

Within the livestock production sector, pig and poultry production have witnessed the largest increases during the last 20 years, as described in Part I of this volume (de Haan et al., 1998). This development has several explanations. First, pigs and poultry have much shorter production cycles than ruminants, while the high number of offspring per year can achieve a rapid growth and allows farms to respond quickly to fluctuating demand. Second, pig and poultry production do not necessarily rely on locally produced feed; the animals have relatively low feed conversion rates and, unlike ruminants, do not require roughage, which is voluminous and therefore relatively expensive to transport. Third, investments in land, buildings, and expertise are relatively low in pig and poultry production and are often facilitated by feed companies and/or slaughterhouses. As a result, meat production costs are lower than in beef production.

Large and Specialized Units with Little or No Land
Dictated by economics, livestock farms become more and more specialized, intensive, and larger because they have competitive cost advantages relative to diversified, extensive, small farms. (See Chapters 1–3.) However, these trends magnify the manure problem, as the amounts of manure nutrients per unit of surface area increase.

Growing Importance of Slurry-Based Systems
Most modern pig and cattle farms use little or no litter and collect the dung and the urine combined as slurry, often using large amounts of water for cleaning and cooling purposes. Slurry-based systems require less labor than litter-based systems with solid manure or solid and liquid manure collected separately. Although often illegal, slurry and slurry effluents from manure treatment processes are also easily discharged into nearby streams. Flushing water will help, following the credo "the solution to pollution is dilution." Apart from the fact that solid manure is relatively easy to heap and transport, and therefore to "commercialize," there is a century-long tradition and experience with the handling of solid manure. In contrast, proper management of slurry is often impeded by a lack of experience with this form of manure. This constraint and the smelly, liquid appearance often make specialized crop production farmers reluctant to use slurry, especially when "clean and cheap fertilizers" are available.

Neglect of the Nutritive Values of Manure Nutrients
The esteem given to manure management is related to the existence of governmental policies and is inversely related to the price and availability of chemical fertilizer. In countries where chemical fertilizer is hardly available or very expensive for farmers (e.g., Myanmar, Laos, or Cambodia), the interest in manure is higher than, for example, in China or Viet Nam, where chemical fertilizer is cheap and already used in high doses. Labor constraints and a lack of education also play a role in the lack of proper collection and storage of manure nutrients in practice (Schröder 2005, Rufino et al., 2006).

A simplified and approximate model calculation for European conditions can illustrate that the nutrient value of manure can usually not come close to manure handling costs, at least for slurry. Huijsmans et al. (2004) give average application costs of about €5 per m^3 slurry for a farm applying 1,000 m^3 per year. Further costs are incurred by the slurry storage. Assuming capital investment costs of €100 per m^3 slurry store, based on the range of €30–150 per m^3 given by Burton and Turner (2003), a storage capacity of six months, and an amortization period of 20 years, the investment costs per m^3 of slurry per year are €2.5. Including labor costs for manure collection, costs for pumping and mixing equipment, and so forth, manure handling cost would thus be around €10 per m^3. If, on the other hand, we can assume a typical nutrient content of 1.5 kg crop available N, 1 kg P_2O_5, and 2 kg K_2O per m^3 and an approximate price level in 2005 of €0.5 kg^{-1} N, €1 kg^{-1} P, and €0.4 kg^{-1} K, the nutrient value per m^3 of slurry would be around €1.5 per m^3. If the slurry is more diluted than approximately 1 part of water per part of undiluted slurry (values going up to 20:1 in Asia), the nutrient value would be even lower.

Although the rapidly growing fertilizer prices in 2008 have led to an increase in the nutrient value by a factor of 2 to 3, slurry management is economically reasonable only if the amount of mineral fertilizer is limited by legislation, availability, or price or if costs for alternative disposal of the manure are also included in the calculation.

Farm Structure and Manure Management
The effect of livestock production on the environment depends on a large number of structural and management factors, including livestock density and the link between livestock and land, the livestock operation's location, livestock management, the housing system, the manure management and storage systems, and the treatment and use of manure. Proper manure management should be an integral part of livestock production systems. However, manure management implies both cost and effort, which in many cases is deemed to bring little benefit to the farmer. Box 9.2 provides a short summary of the manure management situation in Europe that illustrates these issues.

Livestock Density and the Link between Livestock and Land
Livestock stocking density and the geographical coupling or uncoupling of livestock production units and feed production areas have a large influence on the flow of nutrients and "virtual water" (the total amount of water needed directly or indirectly for production of feed and other purposes to produce one unit of meat or milk product) in animal feed across the globe and on the way manure can be recycled (Hoekstra and Hung 2005, Naylor et al., 2005, Chapagain et al., 2006, Galloway et al., 2007, Smaling et al., 2008). Direct recycling of nutrients from manure is possible only if the amounts of manure nutrients produced on a farm or in a region

Box 9.2. Manure Management in Europe

- Manure management is highly variable across Europe due to different housing, feeding, and manure handling systems; climatic and structural conditions; traditions; and so forth.

- The discharge of manure to rivers, streams, and lakes is prohibited in most parts of Europe. To forestall serious water pollution, most countries do not even allow the discharge of wastewater or highly treated effluents from farms. Nearly all the livestock excreta and wastewater is therefore used on agricultural land (including horticulture) as fertilizer.

- The nutrient content of the manure is only partly accounted for in fertilization (corresponding reduction of mineral fertilizer use) unless this is legally required (e.g., in Denmark and Switzerland).

- Especially for slurry, the most common use is application of the manure on the farm where it is produced. Slurry is transferred between farms only in regions with a high livestock density (e.g., the Netherlands) and if there is legislation that restricts the nutrient balance surplus or the livestock density. An actual market exists for solid manure in most southern and eastern European countries.

- The proportion of liquid manure is high (>65%) in most central European countries and low in eastern Europe as well as in the United Kingdom and France.

- Liquid manure (slurry) is mainly stored in tanks, which in most countries are usually not covered. Lagoons are sometimes used in southern and eastern European countries, but they are never the major storage system.

- Most countries have guidelines about the minimal period of storage for manure. The average storage capacity for slurry is around six months (longer in Scandinavia, shorter in some southern and eastern countries). For solid manure, the storage capacity varies from 2 to 12 months.

- Most countries have guidelines on the application of manure covering timing, location, and quantity. Applications in adverse weather conditions or onto bare soil or sloping ground are generally discouraged.

- Special treatment of liquid manure to improve handling and to reduce pollution or odor or for biogas production is practiced only in certain countries or regions. This includes biogas production in Germany and Denmark and aerobic treatment in France. Composting of solid manures is common on organic farms.

Source: Menzi 2002, Menzi et al., 2002, Burton and Turner 2003.

do not surpass the nutrient requirements of the farmer's cropland or the cropland within the region at feasible transport distance. If local recycling back to land is not possible, the manure has to be transported to other areas, undergo treatment to remove surplus nutrients, or be discharged to the environment. The first two options will incur considerable costs and will therefore only be implemented if required by an enforced legislation. The third option will inevitably have strong environmental impacts on aquatic and terrestrial ecosystems.

Even if local cropland is available for manure recycling, environmental impacts can only be prevented if the amount of nutrients in the manure does not surpass the nutrient requirements of the crops. This concept of nutrient balance or livestock carrying capacity of the land (i.e., the maximum livestock density that will allow the local recycling of the manure without causing nutrient overloads) is discussed by Menzi and Gerber (2006). The need for balanced crop fertilization is also the basis for the annual application limit of 170 kg per hectare of manure N stipulated in the European Union (EU) Nitrates Directive (EC 1991).

The link between livestock and crop production may not necessarily be found within the same farm. Collaboration between specialized livestock farms and specialized crop farms is possible, even though the crop farms do not provide feed to the livestock farms. For example, 53% of the 91Gg manure P produced in the Netherlands in 2007 was applied on mixed farms where the manure was produced, 28% was transported to other farms (mostly specialized crop farms), and 18% was exported to other countries (Hoogeveen et al., 2008). The Netherlands has a high mean livestock density (~4 livestock units per ha), imports a significant fraction of animal feed from elsewhere, and has strict environmental regulations for manure management (Oenema and Berentsen 2004, Schröder et al., 2007). There is an ongoing debate about whether livestock production at such high density is sustainable in the long term.

The more specialized the livestock farm and the larger the number of animals, the more likely it is that manure cannot be recycled locally and that the nutrient cycle will be disrupted (Menzi and Gerber 2006). This leads to increasing nutrient surplus for the livestock-producing area and increasing soil nutrient depletion for the feed-producing area. With the increasing importance of international trade in feedstuffs, the problem of disrupted nutrient cycles is increasingly shifting from a local and national scale to regional and continental scales (Grote et al., 2005, Galloway et al., 2007, Smaling et al., 2008). Recycling of manure nutrients is a challenge for large livestock farms with little or no available land in a region dominated by small crop farms. For example, a landless farm with 20,000 fattening pig places requires 2,000–3,000 hectares of cropland to recycle properly the manure it produces. In Thailand, the average crop farm is 2–3 hectares, suggesting that each pig farm would need the full cooperation of approximately 1,000 neighboring farms annually.

Farm Location

Apart from the local livestock density, the location of a farm is an important factor in its potential threat to the environment. The risk is especially high if the operation is near a vulnerable habitat, whether terrestrial or aquatic ecosystems. Species-rich natural grassland, heath land, and bogs are especially sensitive to increased deposition of atmospheric N, which may originate from nearby livestock units (Dragosits et al., 2006). Restrictions on farming within zones around particularly sensitive areas are a possible solution; the restriction may include a limit on the stocking density or the need to implement measures to cut ammonia emissions (e.g., changes to building, air purification, covered manure storage, or land application of manure using low NH_3 emission techniques). Restrictions on farming have been implemented for such areas within the European Union through the Habitat Directive (EC 1996). The location of a livestock farm is also relevant for surrounding crops; the amount of manure that can be recycled can vary considerably between different crops.

Livestock Management

Both livestock diet and livestock performance (yield expressed as feed conversion efficiency: kg feed per kg liveweight growth) directly influence the excretions of the animals and thus the nutrient load on the environment. The feed conversion efficiency depends on the animal type, the genetic potential of the herd, animal housing and welfare conditions, the productivity level, and diet composition and water use (Tamminga 1996, Castillo et al., 2000, Broderick 2003, Geers and Madec 2006, Rufino et al., 2006). Commonly, the feed conversion efficiency is better in intensively managed livestock production systems than in other systems because of the better genetic potential, feed quality, housing quality, and management (Ke 2009). However, as intensive production relies heavily on imported animal feed, the risk of disrupted nutrient cycles and corresponding environmental impacts is higher.

Housing Systems

The housing systems for livestock production determine the manure type and in part also its composition. Depending on the system, farms may produce liquid manure (slurry), solid manure, or a combination of slurry and solid manure. The system chosen depends on local tradition, economic considerations such as labor costs, storage and transport costs, the availability of water and bedding, animal welfare considerations, and farm size and structure.

Solid manure usually contains bedding material (straw, wood chips) with a high organic matter content, which increases its perceived value as "soil conditioner." There is also some suggestion that overall ammonia losses are lower from farms using solid manure (Misselbrook et al., 2000). Yet, for economic reasons, new modern livestock operations commonly have slurry-based systems. Slurry does have the advantage of readily "flowing" from the animal housing system to an outside storage system. Further, the ratio of mineral N to organically bound N is higher in slurry than in solid manure, and as a consequence slurry has a higher percentage of readily available N needed for crop growth (Schröder 2005, Rufino et al., 2006).

Water Use and Water Content of Manure

Total water use in livestock production, including feed production and processing, is high; the production of 1 kg of chicken meat, pork, and beef requires on average 3.9, 4.9, and 15.5 cubic meters of water, respectively (Chapagain and Hoekstra 2004), although differences between farms and countries can be large. (See also Chapter 7.) Most of this water is used for animal feed production. However, water use on the livestock farm can be high too. Water is used not only for drinking and cleaning purposes but also for flushing and cooling the animals. For example, the amount of water used per fattening pig can vary from 5–10 liters per day to 100 liters per day, as reported by Dan et al. (2004) for southern Viet Nam. Rainwater falling on open slurry stores and rainwater from roofs, paved areas, and areas draining directly into lagoons often further increase the slurry volume. The variability of the dry matter content of slurry can therefore be very large. For example, the dry matter content of pig slurry can range from 5–10% in northern Europe to well below 1% in tropical zones where large amounts of water are used (e.g., Rattanarajcharkul et al., 2000). Correspondingly, the amount of fresh slurry produced per fattening pig can vary from 1 to well over 20 cubic meters.

Manure Storage Systems

In order to be able to recycle manure nutrients, livestock manure must be collected and properly stored. It should then be applied to cropland when the crops need the nutrients for growth and development. If applied too early or too late, the manure nutrients may not be used effectively by the crop and may be lost through leaching, runoff, and denitrification (Marschner 1985, Laegreid et al., 1999, Schröder 2005). Depending on climate and crop types, the storage capacity for livestock manure needed ranges from a few weeks to nine months. This requires a considerable investment, especially in the case of strongly diluted slurry.

Even without dilution, 1–2 cubic meters of slurry with a dry matter content of 7–8% is produced per pig fattening place per year, assuming three cycles of fattening from 25 to 100 kg. Similarly, 15–25 cubic meters of undiluted slurry is produced per dairy cow per year, depending on milk yield. With the usual addition of water, these volumes become a factor of 1.5 to 10 larger. In Europe, the capital investments in slurry storage vessels are in the range of €30–150 per cubic meter, depending on the construction material used and the coverage (Burton and Turner 2003), which translates to €30–600 per fattening pig place and €400–4,000 per dairy cow. Such investments in manure storage capacity increase the cost of production considerably.

The appropriate storage capacity depends on the maximum length of time during which manure cannot be applied to land effectively. In Europe, the required storage capacity ranges from four to nine months, depending on climate. With multiple cropping systems common in the tropics and subtropics, it may be less. The limiting factor is often the winter break of the vegetation period or the rainy season, when manure application is strongly limited by the crop nutrient uptake potential, transport and spreading possibilities, and the absorption capacity of the soil. Storages should be leak-tight. If there is no natural clay layer that prevents leaching, lagoons for liquid manure should be lined with plastic or concrete. To circumvent N losses via NH_3 volatilization, manure storage systems must be covered (EC 2003, Rotz 2004).

Manure Disposal

Landless livestock farms have three basic options for manure disposal:

- Transport the manure to crop farms that are able and willing to accept it as a source of organic matter and nutrients. If they are nearby, the cost of transport of the manure is limited. When crop farmers hesitate to accept the manure, for whatever reasons, livestock farmers may have to pay a fee to get them to do so.
- Treat the manure to obtain manure fractions that can be disposed of more easily than untreated manure. Such treatment usually includes separation of solid and liquid fractions. The solid fractions are composted and/or dried and pelleted, and then transported and applied to cropland or incinerated for energy production. The liquid fractions have to be treated further, so as to allow the controlled discharge to surface waters, or they are applied to cropland.
- Simply discharge the manure to the environment, either through dumping on cropped land (implying substantial overapplication) or via ditches into surface waters. Although this implies little or no cost to the farmer, it has huge environmental impacts.

The third option is generally prohibited in OECD countries. In emerging countries with rapidly growing intensive livestock production, all three options are still considered. However, it can be expected that uncontrolled discharge will probably not be tolerated in the future. Livestock farms will therefore need an improved manure management strategy. The most appropriate strategy depends on the farm and its location. The findings from the project "Area-Wide Integration of Specialized Crop and Livestock Activities" (see Box 9.3) suggest the following for choosing the correct management option:

- With investments in the order of magnitude of $10–15 per fattening pig place plus running costs, "full treatment" of the manure to achieve discharge standards is costly but an option under specific circumstances.
- For existing landless farms in an area with few options for recycling manure, "full treatment" of waste may be the only option. It is then an economic decision in terms of whether the costs are bearable or the farm needs to be relocated or even closed down.
- For existing landless farms in areas with crop farms, spreading manure on cropland will be cheaper than "complete treatment." On large farms, the distance to transport the manure will be a serious constraint. However, the scale of the operation may enable large farms to bear those high costs.
- For existing landless farms, it may be advantageous to collect all solid manure, as this fraction is relatively easy to transport and market.
- For new farms, it is economically and environmentally beneficial to choose the farm location according to the potential for recycling the manure in the vicinity (Gerber 2006).

Manure Treatment

Manure treatment can be used to address specific problems of manure management or to meet special aims (Burton and Turner 2003). As it is easier to transport smaller amounts of solid rather than dilute liquid manure, different separation techniques (in-house separation by regular removal of solids, sedimentation, and technical separation) are being applied to obtain a nutrient- and organic matter–rich fraction with high dry matter content. Treatment can also help sanitize the manure (e.g., by the thermophilic anaerobic or aerobic treatment or by direct pasteurization), change the characteristics of the manure (e.g., composting or drying solid manure), remove N via nitrification and denitrification cycles, and address specific emission issues (e.g., NH_3 emissions, odor). A combination of separation of liquids and solids, followed by pressing and drying the solids and (ultra-) filtration and reverse osmosis of the liquid, yields the stable and valuable fraction—but at high (energy) costs (Meunier 2007).

Numerous treatment systems exist for handling the various organic waste products from a livestock farm. For an overview of current available systems, see the review

Box 9.3. Southeast Asia Case Study

Between 2000 and 2004 the Livestock Environment and Development Initiative coordinated by FAO studied the current environmental relevance of intensive livestock production in Southeast Asia and potential measures to improve the situation. Activities focused mainly on Thailand (near Bangkok), Viet Nam, and China (see also Ke 2009), but some preliminary analysis was also done in Myanmar, Cambodia, and Laos. The conclusions of the project can be summarized as follows:

- Liquid manure (including "wastewater" and slurry) is the main environmental threat of intensive livestock production. Activities to reduce environmental impacts of manure management should therefore have a first-priority focus on liquid manure. Special attention should be given to the situation during the rainy season.
- Pig production is in most cases a much bigger environmental threat than poultry production, which mostly produces solid manure.
- Both large pig farms (typical in Thailand) and a high concentration of small pig farms (common in Viet Nam) cause environmental pollution if the liquid manure cannot be properly recycled.
- Environmental impacts per pig are lower when the excreta are partly or fully collected as solid manure.
- Environmental impacts of livestock production are becoming a serious problem, especially in areas around large urban centers. Concerted actions on a regional and national scale are urgent to meet this problem.
- In countries like Myanmar, Laos, and Cambodia where intensive specialized livestock production is just getting started, appropriate awareness raising, capacity building, and policy measures should be taken to guide development into a sustainable direction. Such activities might capitalize on the higher esteem accorded to manure due to very limited access to mineral fertilizers.

of methods used in Europe by Burton and Turner (2003). Figure 9.5 summarizes the main options shown as elements of a single process: mixing and storage, mechanical separation, decantation, composting, and ponding systems. The last major group to be included is intensive biological treatment that can be subdivided into aeration and anaerobic digestion with biogas production.

In reality, complete processes as illustrated are both rare and not usually required because manure treatment should only be considered if needed to achieve a specific aim such as the reduction of nutrient surplus or organic matter, ease of handling, or odor.

Improving Handling of Liquids
For many farms, the first consideration in manure management lies with easier handling and thus avoiding problems of collection and transport. The simple removal of coarse suspended matter such as bedding straw by mechanical separation is a practical way of achieving this. The screened liquid can then be more easily pumped with a reduced risk of pipe blockage, and land application can be done in a more controlled way. The solid material removed is often "stackable" with less than 80% water content, and it may represent 10% or more of the raw effluent. However, its removal will not reduce the nutrient load of the raw slurry by much more than this, with most of the active content being soluble or associated with finer particles. Nonetheless the screened solids can be readily composted, producing a reasonably stable product that can be easily transported. Using finer screens in the separation operation can remove a greater proportion of the suspended matter, but the solids produced tend to have higher water content. This disadvantage can be offset by the use of screw-press separators or (for a more intensive operation) decanter centrifuges, but such systems can be very costly and have a relatively low capacity.

Other than separation, the main strategy to improve handling of liquids depends on adequate mixing, which can be achieved by a wide range of equipment. Crucially, this avoids the problems of separation, which can otherwise result in the development of crust or sludge layers, both being difficult to remove from the storage unit. However, the formation of a crust may in some circumstances be desirable in reducing gaseous emissions. But once well formed, removing such a layer can be difficult. The sludge presents its own problems by reducing the effective volume of the storage unit as it accumulates. In addition, zones of methanogenic activity in aging sludge increase methane (CH_4) production.

Reducing Organic Loading
Manure from livestock can be expected to contain a large proportion of organic matter in a reactive form. Within hours of production, a high level of microbiological activity can be expected, with the production of a range of odorous by-products if anaerobic conditions prevail. If these are allowed to persist, methanogenic activity can be expected in a few days, with the production of methane along with substantial emissions of ammonia. Separation (either mechanical or physical as sedimentation) can remove part of the organic load, but this is often the less reactive material; the greater problems are associated with the smaller and more reactive molecules that tend to be more soluble. Biological treatment is most effective in this case either as the anaerobic or aerobic option. Anaerobic treatment is often preferred, as there is the clear benefit of a useful gas product, but enthusiasm is less for small operations or where there is a cheap source of electricity available.

For larger farms, anaerobic digestion can be linked with generators to produce electricity in schemes that can sometimes be self-financing, especially if the electricity

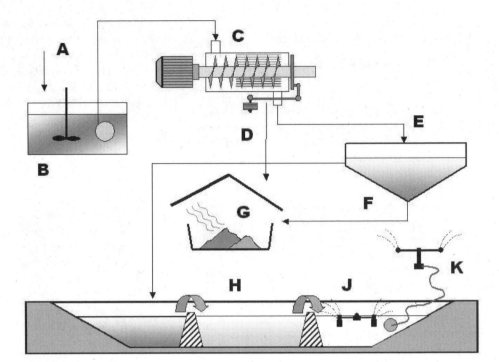

Figure 9.5. General scheme summarizing main manure treatment systems in common use.

(A) raw effluent from animal buildings; (B) mixing and separation; (C) raw feed; (D) separated solids; (E) clarified effluent; (F) settled sludge; (G) composting systems; (H) ponding system; (J) aeration; (K) irrigation of final treated wastewater.

can be used locally, such as for evaporative cooling systems. Any process that strives to maximize gas production can be expected to also achieve near-total removal of reactive organic matter and produce a final digestate that is both stable and relatively free of offensive odors. Contrary to some expectations, however, anaerobic treatment does not eliminate all pathogens. More often it only reduces their presence, especially under conditions common on farms where new slurry enters the system continuously.

Aeration systems are generally less attractive because of the running costs of the aerator, which can exceed $1 per cubic meter of liquid manure treated. On the other hand, it is both quicker and more effective than anaerobic digestion.

Ponding systems (also known as lagooning) can also enable both aerobic and anaerobic degradation of organic matter. The treatment times tend to be much longer, running into weeks or even months in some cases. This, along with the need to keep the lagoons relatively shallow to meet process needs, results in a large area requirement. Nonetheless, running costs are very low, and the low technology feature makes this an attractive option in poorer regions (Mara 2004). If retention times are sufficient, a high level of treatment can be achieved, with aerobic activity easily possible in the final lagoons by keeping the depth to no more than 1 meter. In hot climates, losses by evaporation can help reduce the quantities that need to be irrigated in dry periods, but during the rainy season the capacity of the system might be hampered by the large amounts of rainwater. The impression that this is a closed system is false, as there is

both the progressive accumulation of a sludge layer and substantial emissions of methane and ammonia.

Removing Nitrogen and/or Phosphorus

Organic nitrogen can be concentrated along with other suspended matter in sedimented sludge or (to a lesser extent) in screened solids. The proportion of nitrogen in organic form can be increased by limited biological treatment, but most of it will remain in the soluble form. Nitrogen in the form of ammonium (NH_4^+) and ammonia (NH_3) is not easily concentrated. If it is unwanted, the main option will rely on the process of nitrification and denitrification. The main final product of this process is harmless di-nitrogen (N_2) gas, but the process may be combined with the production of nitrous oxide, which reduces its environmental credibility.

Removal of phosphorus from animal manure and exporting the P-rich fraction from the farm can be an effective way to diminish the problems of excess P accumulation in manure-amended soils (e.g., Szogi and Vanotti 2009). The P-rich fraction has to be used as P fertilizer on agricultural land that needs P fertilization.

Concentrating Nutrients and Production of Organic Products

Any of the separation processes (screening, centrifuge, or settling) can be used as the first stage of a process in which most of the insoluble (and some of the soluble) components are concentrated into a fraction representing a small volume of the original manure.

The most common operation for the production of solid products to be used as organic fertilizer is clearly the versatile and well-established compost operation. However, this is strictly an aerobic process that requires a regular mixing to supply enough oxygen. It does offer a simple and readily accessible method of reworking a variety of organic materials into a useful product. For the most successful schemes, the availability of a solid substrate to give the product structure is essential; straw from animal bedding, chopped wood or twigs, or leaf/vegetable matter would be suitable. In Europe the use of the manure/bedding mix (often known as farmyard manure) is common. However, given the lack of bedding in hotter regions, straw or similar materials may need to be brought in, thus adding costs to the process. Manure sources suitable for composting include collected solid dung, screened solids, and settled sludge, although the proportion of the latter is limited due to its relatively high water content.

The main shortfall for composting schemes often comes from a poor mix of materials, which is often the result of trying to incorporate too much of the unwanted "waste" component. Control of feed materials, even animal manures, is important if the intention is to produce a high-quality product suitable for sale. A second problem can lie with the process itself, which can imply an elaborate piece of equipment to ensure a consistent and high-quality product. Simpler schemes for smaller quantities of compost do exist, but they tend to be much more labor-intensive or produce a poor and variable product.

The second main process to produce concentrated manure products is based on drying. In areas where there are long periods of dry warm weather, this option is very attractive. Solid manure intended for direct use as an organic fertilizer can be adequately dried over a week or so when spread out as a thin layer on a concrete surface. Longer storage can both improve the drying and greatly reduce the disease risk. Other than solar drying, there is the limited option of drying with applied heat in an enclosed process, but this is rarely practical even at the large scale unless there is a ready local source of waste heat. Formulated organic products (compound fertilizers) from an industrial operation would follow this line with acidification to conserve the nitrogen as ammonium salt.

Combining Separation, Ultrafiltration, and Reverse Osmosis

The attractiveness of basic separation such as screening or sedimentation as a treatment option for effluents lies with its relative simplicity and the associated low costs for installation, operation, and maintenance. The obvious drawback, as detailed by Burton (2007), is that it can only remove and concentrate components that are insoluble and, for the most part, present as relatively large particles. The exception lies with membrane processes, which in the case of reverse osmosis can even separate out salt ions. Such technology has thus received a great deal of interest, especially in the supply of drinkable water (for the removal of contaminants) and in the treatment of dilute wastewaters (Tchobanoglous et al., 2002, Shannon et al., 2008).

Not surprisingly, there has been interest in applying the same technologies to stronger effluents, including animal slurries. Pieters et al. (1999) demonstrated such a system treating sow slurry using reverse osmosis as the final stage after a pretreatment to remove suspended matter. The final permeate was indeed largely free of salts but the reported cost was high, at €7 per ton of slurry, and practical problems such as membrane cleaning remained. Preclarification is essential, and a large volume of relatively dilute concentrate for disposal will still remain. However, the factor of greatest significance is the costs of systems with enough capacity to handle typical volumes from a large livestock unit, which can be expected to be prohibitive for many farmers.

Manure Application to Cropland

A key challenge is to transport manure to the field where it can be used effectively and with minimal costs and nutrient losses. Appropriate equipment for transporting and spreading manure and applying the correct dose according to crop nutrient requirements is therefore critical (Burton and Turner 2003, Rotz 2004, Rufino et al., 2006, IFA 2007, Petersen et al., 2007). Furthermore, manure spreading must also consider the trafficability of the soil in terms of restrictions to manure application near sensitive ecosystems, watercourses, or drinking water catchments.

For manure application (particularly slurry), the appropriate technique strongly depends on the structure of the farm, the size and shape of the fields, the available mechanization, the characteristics of the liquid manure, the crop, and the topography. Important concerns include achieving even and uniform spreading and preventing losses. It is worth noting that the land spreading techniques developed in Europe (e.g., Huijsmans et al., 2004) are often not applicable in countries in Asia because of small or bounded fields, access problems, and so forth. The adaptation of spreading systems and the development of new techniques is an urgent challenge.

Nutrient applications to cropland should not surpass the crop's nutrient requirements. To achieve this, the farmer needs reliable guide values on the crop and recommendations about the manure's nutrient content (e.g., Laegreid et al., 1999, IFA 2007). Based on this information, the appropriate dose can be calculated. As the manure dose needed to meet the P requirement of the crop is often lower than that for N, K, and other nutrients, especially for pig and poultry manure, it is advisable to calculate the maximum annual manure dose on the basis of the P content. If the manure is applied in higher doses as soil conditioner once every couple of years, the nutrient input can be reduced accordingly in the successive years. For N, the full amount in the manure is not available for crops in the first year because the N contained in the organic substance is only degraded in the course of several years and because of inevitable losses. On plots regularly receiving manure, 50–70% of the total N content in slurry and 10–50% of that in solid manure are usually considered as available for the crops, depending also on the time scale considered (Schröder 2005).

Constraints to manure recycling depend on farm-specific conditions:

- A primary constraint is usually the cost. Often, good manure management with collection, storage, transport, spreading, and sometimes treatment is more costly than to buy the equivalent amount of nutrients in the form of mineral fertilizer. This is especially the case for strongly diluted slurry. Under such circumstances, economic considerations must not only consider the potential benefit from the manure but also the cost of its disposal.
- A further important constraint is the uncertain crop yield effect of the manure. Because part of the readily available N fraction for plants is lost through ammonia volatilization and the organic fraction of N is available only after degradation, it is more difficult to dose manure correctly than to use mineral N fertilizer. To avoid the risk of yield decrease, farmers therefore often tend to overdose the manure or do not reduce mineral fertilizer use accordingly after having used manure.
- In many developing countries (e.g., in East and Southeast Asia), the use of slurry as fertilizer is also impeded by the lack of experience of the farmers, who often worry that the slurry would have negative effects on their crop quality and yield. This fear is often based on a few cases where excessive doses of slurry or overflowing lagoons damaged crops. Furthermore, the farmers lack reliable recommendations about proper slurry use and often do not have appropriate technology for transport and spreading.
- In addition to the general requirement to minimize emissions, there can be site-specific constraints for manure use. For example, liquid manure should not be spread very close to watercourses or vulnerable ecosystems, and special care (reduced dosage, etc.) is required on fields with a high water table, especially when the groundwater is used as drinking water resource. Furthermore, any specific national or local legal restrictions must be respected.

Environmental Impacts of Nutrient Losses from Manure

Elements are lost from livestock manure management systems at different rates. The descending order is roughly as follows: C > N >> S > K, Na, Cl, B > P, Ca, Mg, Fe, Mn, Cu, Zn, Mo, Co, Se, Ni. This order is related to the reactivity, speciation, solubility, and fugacity of the nutrient element species and their mass fractions in manure. The double mobility of C, N, and S in soluble waterborne compounds as well as in gases makes cycling and loss pathways much faster and more complex than those of the mineral elements. Carbon is released from manure in gaseous forms (mainly carbon dioxide and methane), in dissolved forms as inorganic and organic C (ΣHCO_3, DOC), and as particulate matter (via runoff). Nitrogen is also released in gaseous forms (mainly NH_3, N_2, N_2O, NO), in dissolved forms as inorganic and organic N (NO_3^-, NH_4^+, DON), and as particulate matter (via runoff). Sulfur is mainly lost via volatilization of sulfides (H_2S) and sulfur oxides (SO_2) and the leaching of sulfate (SO_4^{2-}) and particulate matter. There are extensive reviews on gaseous emissions concerning the mechanisms and size of losses from manure management systems. These reviews also allow proper parameterization of models used to assess the emissions at farm level and at regional, national, and global levels (e.g., Bouwman et al., 1997, Sommer et al., 2006, Velthof et al., 2009).

The mineral elements can be lost via leaching in dissolved form or via runoff in particulate matter at various stages of the feed–animal–manure–soil–crop chain. (See Figure 9.6.) Leaching of effluents from ensiled feed and manure from animal houses and storage systems is negligible when the ground is sealed and the drainage water collected. High concentrations of, for example, K, Cl, and NO_3^- in wells near farm houses are often seen as evidence of leaching losses from manure heaps on top of unsealed soil (e.g., Culley and Phillips 1989, Withers et al., 1998). Potassium, sodium, chloride, and boron have high solubility in water, and the main loss pathway is via leaching. Finally, the elements P, Ca, Mg, Fe, Mn, Cu, Zn, Mo, Co, Se, and Ni have low mobility because of their low solubility and high reactivity with soil constituents. Some of these elements (Fe, Mn, Se) are redox-active and more mobile in chemically reduced forms. Some elements may form complexes with dissolved organic carbon and inorganic anions and thereby increase their mobility.

Few studies have quantified nutrient losses from animal manure storage systems via leaching and runoff. The available information suggests that N leaching losses from slurries in unsealed lagoons in EU-27 are in the range of 5% of the N in slurry when covered and in the range of 10% when uncovered. Similarly, N leaching losses from solid manure in unsealed heaps are in the range of 2% of N in the manure when covered and 5–10% when uncovered.

Current environmental concerns relate mainly to gaseous emissions of NH_3, N_2O, and CH_4 from manure management systems to the atmosphere, to the leaching of NO_3^- to groundwater, and to N (both inorganic and

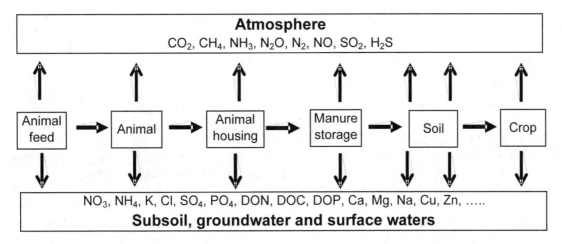

Figure 9.6. Possible loss pathways of nutrient elements from the feed–animal–manure–soil–crop chain. Losses via gaseous emissions to the atmosphere are shown in the upper half, losses via leaching and runoff of nutrient elements to the subsoil and to groundwater and surface waters are in the lower half.
Source: Oenema et al., 2007.

organically bound N species) and P (ΣPO_4, particulate matter) losses from manure management to surface waters. The excessive accumulation of Cu and Zn in soils following heavy applications of pig and poultry manures received considerable attention in Europe in the 1980s and 1990s but is now less a concern following governmental regulations on their additions in animal feed and on manure application rates (Jondreville et al., 2003; Schlegel et al., 2008). However, this is not yet the case in many countries in Asia and Latin America.

Pollution of Freshwater Ecosystems
The effects of livestock wastes on aquatic ecosystems are basically the same as for household and municipal wastes. These effects can be summarized by eutrophication and toxification, which lead subsequently to ecosystem changes. The effects follow from the presence of degradable organic matter, nutrients (mainly N and P), heavy metals (mainly Cu and Zn), and possibly pathogenic microorganisms, hormones, and antibiotics in the wastes.

Eutrophication has been among the main water quality issues for decades. It is caused by excess loads of nutrients (mainly N and P) and organic matter in surface waters. This leads to excessive growth of phytoplankton and subsequently to decreases of water clarity and biodiversity, oxygen depletion, fish kills, and odor problems, while harmful toxic algal blooms may develop more frequently (Smith 2003, Rabalais 2004). Agriculture contributes roughly 50% to the total loading of surface waters with N and P (Carpenter et al., 1998, EEA 2005), but the specific contribution of animal agriculture is unknown. Badayos and Dorado (2004) estimated that in some parts of East and Southeast Asia pig production could contribute up to 90% of the P load reaching the

South China Sea (as noted in Box 9.1). Effects of eutrophication can be exacerbated by the natural growth of toxic cyanobacteria (sometimes incorrectly referred to as blue-green algae). These microorganisms commonly occur in all waterbodies but are particularly abundant in tropical and subtropical climates and are notorious for their production of toxins that are detrimental or even deadly for humans and animals consuming water with high concentrations of cyanobacteria.

Pathogens associated with livestock and belonging to such microbial groups as viruses, bacteria, protozoa, and parasitic helminths are known to penetrate into aquifers and surface waterbodies. They present a threat to humans and animals alike. (See Chapter 11.) These threats are taken more and more seriously, particularly in light of the recent SARS and avian influenza epidemics. These epidemics, though viral by nature, demonstrate and accentuate an intrinsic potential danger of all groups of microorganisms emanating from livestock through wastes.

Pollution of Coastal Marine Ecosystems
Through the transport of nutrients via rivers and the deposition of ammonia-N originating from manure, livestock production also contributes to the eutrophication of marine coastal ecosystems (see Chapter 6; see also Seitzinger et al., 2005, 2006; Boyer et al., 2006). The flux of nutrients from land to sea has increased since the 1960s by more than three to four times (Schaffelke et al., 2005), and nutrients originating from livestock production played a prominent role in this. Eutrophication stimulates the growth of marine algae, be it microalgae (which lead to "red and brown tides") or macroalgae (seaweeds). Transfer of pathogens from livestock wastes to sea is not an uncommon occurrence and results in

temporary closures of marine aquaculture farms due to health risks resulting from elevated *coliform* bacteria counts. Furthermore, eutrophication-related impacts on sensitive estuaries, coastal zones, and coral ecosystems, which are among the most productive ecosystems on the planet, lead to loss of biodiversity, which may lead to coastal erosion (Duxbury and Duxbury 1997). This may coincide with decreased commercial harvests of marine products and necessitates rehabilitation associated with high costs. Eutrophication of marine ecosystems is a global concern (Smith 2003).

Recently, a direct relationship between eutrophication and harmful algal blooms, frequently referred to as red or brown tides, was established (Sellner et al., 2003). Similar to those in freshwater ecosystems, these microorganisms produce toxins that, through the trophic chain (via shellfish, fish, and other marine products), threaten human health directly as food supplies or indirectly through residual contact via recreational activities (contaminating beaches and coming in contact with skin). Harmful algal blooms, as well as blooms of nontoxic, "nonharmful" microalgae, eventually decompose with overconsumption of oxygen, leading to hypoxia in bottom layers of marine waters. This is particularly serious in lagoons and bays highly prone to water stratification when replenishment of oxygen from the surface is impeded and leads to fish kills (Duxbury and Duxbury 1997) and can have an impact on marine mammal survival (Landsberg 2002).

Groundwater Pollution

The leaching of nitrate to groundwater is considered a problem because when the resource is used as drinking water, the nitrate is consumed by humans and may cause methemoglobinemia (blue baby disease). Livestock excrements contribute to nitrate leaching indirectly through crops and grassland highly fertilized with manure as well as when grazing livestock are fed protein-rich diets. The nitrate is generated by the nitrification of ammonium from fertilizer and excreta (Ryden et al., 1984). The nitrate may also originate from leakages of manure storage systems (Withers et al., 1998, Oenema et al., 2007). Nitrate pollution of groundwater affects water resources available to communities, leading to health and social problems. However, there is still debate about the actual toxicity of nitrates in drinking water consumed by humans (Powlson et al., 2008).

Leaching losses resulting from leaking storage systems or fields receiving high manure doses can also lead to groundwater pollution with pathogens and residues of pharmaceutical products (mainly antibiotics).

Nutrient Accumulation in Soil Ecosystems

Plants need N, P, K, and 10 other nutrients for their growth and functioning (Marschner 1985). In crop production, nutrient uptake per crop from the soil may range from 50 to 300 kg ha^{-1} for N, from 10 to 40 kg ha^{-1} for P, and from 50 to 400 kg ha^{-1} for K. In double and triple cropping systems, uptakes of N, P, and K may be even higher. These wide ranges reflect differences between crop types and soil fertility and fertilization levels (Marschner 1985, Laegreid et al., 1999).

Theoretically, the whole nutrient requirement of the crops could be met by livestock manure. However, in practice this is usually not feasible because the ratio of the different nutrients in manure is not the same as that for crop requirements and because sometimes it is not possible to spread manure when there is a particular N demand of the crops (e.g., during a wet spring in Europe or during the rainy season in the tropics). When application rates exceed the uptake capacity of the crop, surplus nutrients accumulate in the soil or are lost to the wider environment, depending on the mobility and reactivity of the nutrients and the environmental conditions (such as rainfall and topography). Phosphorus and also zinc and copper are strongly adsorbed to the soil and therefore have low mobility and hence tend to accumulate. In contrast, nitrogen and sulfur have high mobility and tend to dissipate into the wider environment. Calcium, magnesium, and potassium take an intermediate position (Marschner 1985).

For P, the strong accumulation in the topsoil will mainly be a concern for aquatic ecosystems when the topsoil can enter surface waters via erosion (Carpenter et al., 1998). High K concentrations in soils may lead to a high K uptake by the crop, which may lead to grass tetany (or hypomagnesemia) in grazing dairy cows. Extensive studies by Kemp and others (e.g., Kemp and Geurink 1978) indicated this to be associated with Mg deficiency, due to insufficient Mg intake and/or availability, and found that Mg availability was regulated by the K and N contents of the diet (Mg content should increase proportionally to the product of K × N contents). Furthermore, high amounts of K can cause an increase of soil salinity, with negative impacts on crop yield and soil fertility. Lowering high P and K levels in soils may take many years; it requires greatly decreasing the application rates of animal manure (and fertilizers) (Koopmans et al., 2004).

Nutrient Depletion and Soil Degradation in Areas That Export Feed Products

Farming regions with a net export of crop products are faced with a depletion of soil nutrients and a decrease of soil organic matter if they cannot replenish with commercial mineral or organic fertilizers along with the crop residues (Burresh et al., 1997, Galloway et al., 2007). Even if they can fully meet the nutrient requirement of the crops with mineral fertilizer, the organic matter status of the soils will diminish. Apart from that, intensive crop production can have a range of negative impacts on the environment (Matson et al., 1997).

In many developing countries the increasing demand for animal feed for intensive livestock production (grain, soy, manioc, etc.) and the growing demand for grain for human consumption have triggered a market-oriented intensification of crop production. With the resulting shift toward cash crops, the intensification of production, and the increasing use of mineral fertilizers and pesticides, negative impacts on the environment are inevitable.

Soil Pollution with Heavy Metals and Organic Compounds

Some heavy metals are used in livestock feed in amounts surpassing the essential animal requirements. In particular, copper and zinc are often used in relatively high doses in pig production because of their growth-promoting and therapeutic effects (Eckel et al., 2008, Poulsen and Carlson 2008). Other heavy metals that are sometimes added to animal feed are chromium (Cr), cobalt (Co), and arsenic (As). Greater than 90% of the heavy metal inputs to feedstuffs are excreted by the animal. Manure, especially pig and to a lesser extent poultry manure, therefore contains relatively high concentrations of heavy metals. The Cu and Zn content commonly found in pig slurry in Europe already leads to an accumulation of these metals in the soil. Excessive manure application to soils will further increase this accumulation. Because heavy metals are not degraded or lost, it is only a question of time until a level is reached where the soil fertility is harmed and crop yields are affected. Especially in areas with fast-growing livestock production, high livestock density, and weak environmental legislation, heavy metals are increasingly gaining importance as a threat to the environment (Menzi 2008).

Pharmaceutical products used in livestock production can lead to residues of these products or their derivates in the manure (Burkhardt et al., 2004, Gilchrist et al., 2007). Furthermore, any contamination of the feed with organic compounds can lead to a contamination of the manure. Little is known about the potential threat of such residues for soil fertility (Tamis et al., 2008). A special aspect of such residues is the contribution of livestock production to antibiotic resistance in humans.

Damage to Soil Structure

Heavy equipment is often used for manure application to land. This can lead to soil compaction, with negative impacts on soil fertility, especially if inappropriate equipment is used or if manure is applied to wet soil. Furthermore, on clay soils, high sodium levels resulting from excessive slurry application can harm the soil structure (Huijsmans et al., 1998).

Emissions to the Atmosphere

Apart from the effects of N from livestock manure in groundwater and surface waters, livestock manure also contributes to volatilization of nitrous oxide and ammonia into the atmosphere and, in consequence, to increased atmospheric N deposition (Asman et al., 1998; see also Chapter 6). Atmospheric N deposition is especially a problem in vulnerable ecosystems sensitive to high N input (Galloway et al., 2003, Erisman et al., 2007). Typically, from 20% to over 50% of the N excreted by housed livestock is lost through ammonia volatilization (Oenema et al., 2007). Thus livestock production typically contributes from 70% to over 90% of total ammonia emissions (Bouwman et al., 1997). Furthermore, livestock production (especially ruminants) and manure management are an important source of methane emissions and also contribute to nitrous oxide emission, both of which are important greenhouse gases. (Aspects of carbon, methane, and nitrogen are discussed in detail in Chapters 5 and 6.)

Sanitary Considerations

Concerns over health risks from manure handling can be divided into five areas: direct transfer of zoonotic pathogens to other livestock, workers, and communities; the cause and spread of disease affecting other farm animals and wildlife; contamination of water sources; contamination of food crops; and contamination of animal products. Aspects of zoonoses are covered in Chapter 11.

The contamination of water resources with pathogens is a major concern, especially if surface water is used within the local community. Direct release of untreated farm effluents should therefore never be accepted as a satisfactory disposal route. Even the accidental release of farm wastes to water has resulted in outbreaks of serious illness in local communities, including some deaths (Guan and Holley 2003).

The health impact of manures on field crops will depend on the crop, the dose, and the method of application. The risk is particularly high if effluents are applied to the foliage of a growing crop where the leaves are consumed or on crops that are consumed raw, while they are minimal for other field crops. Practically no risk and therefore no need for special treatment exist when manure is used on plantations such as rubber, eucalyptus, oil palm, and so forth.

A special issue is manure use on paddy fields. From the aspect of the food chain, there should be no public health problem, but it is questionable if, with manual seeding and weeding of paddies, workers come into close contact with the manure over a longer time.

Contamination of animal products relates to practices within the abattoir, where animal excreta can inadvertently contaminate the meat that is produced. In addition, there are risks linked to fish products when production ponds receive quantities of manure ostensibly as "fish feed." This practice is common in parts of Asia, for example, and it should be noted that there is a greatly enhanced risk of pathogen contamination of fish subjected to such practices.

Improving the Environmental Performance of Intensive Livestock Farms

Environmental policy and measures in agriculture aim at changing farmers' behavior and thereby improving the environmental performance of farms. This can involve three types of changes: managerial, technological, and structural. Managerial changes involve investments in the "software" of the farm, in the execution of day-to-day farming practices, and are usually relatively cheap. These changes require a better integration of farmers' knowledge and scientific knowledge through interactive training, education, and demonstration activities. And they require suitable platforms for interaction between farmers, researchers, and industry. Technological changes, by comparison, involve investments in the "hardware" of the farm—the buildings, machines, and equipment. These investments are often costly and usually only made when combined with an enlargement of the farm to make the investment cost-effective or following suitable incentives (subsidies, tax reduction, or fines). Structural changes, the third type, involve the organization of agriculture—the type (land-based versus landless systems, specialized versus mixed systems), size, and location of farming systems and vertical (from suppliers to consumers) versus horizontal (cooperation among farmers) structure. They involve changes in the relative importance of production factors and resources (land, labor, capital, energy, and management) and may entail changes in ownership of farms and farmland and in the organization of farmers and the institutionalization of their organizations.

From a cost-effectiveness point of view, management-related measures must be considered first. While doing so, technological measures have to be considered. Structural measures must be considered when management and technological measures do not bring solutions.

Improving performance requires farm-specific analyses of the feed animal–animal produce–animal manure–crop production chain. The weakest part of this chain in terms of agronomic, economic, and environmental performances usually determines the best options for improvement in terms of effectiveness and efficiency. A series of consecutive steps have to be considered (Oenema and Pietrzak 2001, Oenema and Tamminga 2005). The weakest part of the chain in intensive production systems is commonly the storage and use of manure. Although the focus in this chapter is on improving animal manure management, we start the discussion with the producers of manure: livestock.

Improving Livestock Management

The amounts and composition of animal excrements are directly dependent on the number, type (species), productivity, and feeding of the animals, while the total amounts of manure to be handled strongly depends on the amounts of water and litter used in animal housing.

Thus the following aspects of livestock management are relevant for manure management:

- The higher the productivity and the more efficient the feed conversion, the lower the amount of manure produced per unit product. This is a key option for improving the economic and environmental performances of livestock farms. It requires improvement of the genetic potential of the herd, phase feeding (precision feeding), and proper housing of the animals. However, improved productivity might not diminish the load to the environment if it is achieved with commercial feed that replaces local resources.

- Protein and P concentrations in the diet should be restricted to the minimum requirement of the animals so as to limit the amount of N and P in the manure and maximize efficient resource use. This aspect is especially important for N, because a protein reduction in the diet reduces the proportion of N excreted in soluble form and will thus strongly reduce the ammonia volatilization potential.

- Concentrations of heavy metals in animal feed must be kept to the minimum requirement of the animals in order to limit heavy metal concentrations in the manure and to prevent accumulation in the soil.

Improving Manure Management

As discussed earlier, proper manure recycling and application to crops means that all livestock excreta must be collected, stored with minimal losses until the nutrients they contain are needed by the crop, transported to the field, and applied evenly and in doses corresponding to crop nutrient requirements. This implies minimizing the amount of water being collected together with the slurry, sufficient leak-tight storage capacity, and minimizing NH_3 volatilization through covered slurry stores and low-emission spreading techniques.

There is a wealth of information about improving the recycling of manure nutrients effectively in crop production through proper implementation of management and technological measures. Some of this information is derived from experiments, but some is from practice. For example, the default N fertilizer and P fertilizer values of pig slurry in Denmark are 70% and 100%, respectively, indicating that 70% of the N and 100% of the P in the pig slurry must be accounted for in the nutrient balance—in other words, they must be considered as effective as high-quality N and P fertilizers (Mikkelsen et al., 2009).

System Improvements and Structural Adjustments

A fundamental disadvantage of intensive livestock production systems is the large-scale geographic uncoupling of feed production from animal production. As discussed in the opening of this chapter and in Chapters 1–4, landless intensive livestock production systems tend to agglomerate near large markets and to import essentially all animal feed from elsewhere, often from abroad. By doing so, nutrients in soils of crop-producing areas are depleted while livestock-producing areas become enriched with nutrients. Nutrient depletion in soils can be remediated via fertilizer applications, but nutrient enrichment cannot be solved without carrying the nutrients away from the site of manure production. Basically, nutrient depletion and nutrient enrichment are naturally occurring processes, but the current scale and intensity associated with intensive livestock production in some regions is unprecedented and is likely to increase further. Are current agglomerations of intensive livestock production systems sustainable in the long term? If the answer is no, the next question is: What solutions are available?

Although the answer to the first question is indeed largely "no," this depends in part on the scale and intensity of the agglomerations and on the perception of "sustainability." Sustainability is less of an issue when the manure from intensive livestock production systems can be disposed of properly on nearby cropland compared with systems that would have to transport the manure hundreds of kilometers away and therefore dump or discharge the manure locally. Further, the term *sustainability* is value-laden; basically it deals with the quality of life and about the possibilities to maintain that quality in the future. This in turn depends on society's notions about the quality of life and its geographic distribution, as well as on a scientific understanding of the functioning of humans and natural ecosystems on Earth.

Current scientific understanding is that the disruption of the natural nutrient cycles through agglomerations of industrialized livestock production systems is not sustainable (Steinfeld et al., 2006). It leads to eutrophication and loss of biodiversity in livestock-producing areas and to soil depletion and degradation in animal feed–producing areas. Note also that easily accessible phosphorus rock reserves will be depleted in 50–100 years (Laegreid et al., 1999, Smil 2000). Further, society may not accept the local dumping and discharge of manure nutrients from industrialized livestock farms and may call for governmental policy and measures, as is the case in the European Union (Romstad et al., 1997). Restrictions on dumping and discharge of manure will increase the cost of production and will force farmers of intensive livestock production systems to search for other approaches.

What sort of innovative solutions are available? When management and technological measures are insufficient to solve the nutrient imbalance, fundamental changes in the structure of the production–consumption chain are needed. Such structural adjustments may lead to transitions of agricultural systems (Dovring 1965).

In this case, the structural adjustments have to correct the fundamental flaw associated with industrialized livestock production systems. Roughly, four lines of thought regarding structural adjustments are possible:

- High-tech manure processing, yielding marketable manure products that can be stored, transported, and applied to cropland elsewhere; however, manure processing techniques that yield marketable manure fractions are expensive. As a result, intensive livestock production systems with manure processing technology may have to be combined and integrated with intensive crop production systems, food processing, and heat and element recycling units in high-tech agroproduction parks (Annevelink et al., 2003, Smeets 2004). Because of their high-tech, knowledge-intensive, cost-effective, and environmentally friendly reuse and recycling of wastes, these systems may have a comparative advantage over conventional livestock farms that dispose of manure inappropriately. A few of these agroproduction parks are now under development in China, India, and the Netherlands; their success in practice still needs to be proved.

- Spatial zoning and planning of intensive livestock production systems to such an extent that the livestock density within a region allows the proper disposal of all untreated animal manure on nearby croplands within environmental, economic, and social constraints. Evidently, this spreading of intensive livestock production systems over larger areas goes against the trend of agglomeration to benefit from site-specific conditions (markets, logistics, marketing, extension and research services, etc.). Such spatial zoning and planning have been introduced via (tradable) pig and poultry quota per farm combined with ceilings of the maximum number of pigs and poultry per region in the Netherlands, so as to limit the further expansion of the intensive livestock pig and poultry sector (Oenema and Berentsen 2004). However, such an animal quota system may also have (economic) side effects. To be maximally effective, a quota system should be implemented when animal stocking densities in a region do not yet exceed critical limits.

- A greater structural change would occur if animal production were made land-bound again. This would connect crop production to animal production locally and regionally. The large-distance transport of animal feed would be replaced by large-distance transport of meat, while the small-distance exchange of animal feed and animal manure would allow effective recycling of nutrient elements at low cost. Evidently, achieving such a structural change would require a major transition, as the whole production–consumption chain has to be relocated. It would take generations and require strategic international agreements, as it relates to food security, entrepreneurship, stewardship, and trust.

- Even greater structural changes would be brought about through a transition in protein production and consumption, shifting from meat to plant proteins. (See Figure 9.7.) A shift from a diet with relatively large proportions of meat protein to plant-derived protein products appears to be environmentally more sustainable, technological feasible, and socially desirable (Aiking et al., 2006). It would also help release pressure on freshwater resources and decrease the energy intensity of food production. Again, such a protein transition would need to be implemented at the global level. It would more than halve the agricultural land needed and more than halve the eutrophication associated with food production (Aiking et al., 2006, Weidema et al., 2008).

Clearly, these four lines of thought encompass transitions of the entire food production (and consumption) chain. Fundamental to transitions is the change in structure and characteristics of (part of the) society, its large scale, and its large impact. These structural changes occur gradually but continuously in periods of decades to centuries, generally with distinct phases.

Governmental Policy Measures and Awareness of Stakeholders

Government legislation is an important determinant of proper manure management and thus of reducing the environmental pressure arising from intensive livestock production. Legislation may contribute to overcoming the following barriers to proper manure management:

- Manure is considered to be a waste product, especially in landless livestock production systems, when it has to be disposed of at minimal economic costs.
- Proper manure management is associated with considerable investments in capital, labor, and knowledge, which increase production costs.
- The lost value of nutrients is relatively low compared with the costs associated with proper management, and farmers therefore do not have any incentive to reduce emissions (losses).
- Livestock farmers who do produce part of the animal feed often have little incentive to use slurry properly instead of fertilizers, because slurry is voluminous and has a low and variable nutrient content, while artificial fertilizer has a high and known nutrient content and is relatively cheap in industrial countries.
- Farmers' awareness and knowledge about the long-term consequences of environmental pollution due to animal manure are still limited.

Though legislation may help even out some of these barriers, there is often a gap between the theory and practice of government policy (Romstad et al., 1997, Gunningham and Grabosky 1998). The variable and slow responses of environmental policies in agriculture have been ascribed to the huge differences in farming systems and environmental conditions in practice; to a variable interpretation of policy measures; to legislative delays and lack of enforcement; to farmers' failure to implement measures due to system constraints, perceived costs, and the learning time needed; and to antagonisms between measures due to lack of integration. Moreover, there is lack of agreement about global environmental protection at the level of the World Trade Organization. Agreements about trade liberation and environmental protection need to go hand in hand. Implementation of environmental policies at the national or regional level creates unfair competition as intensive livestock production systems increasingly produce for the global market. This is often a major reason that farmers in industrial countries are reluctant to invest in proper manure management.

The EU Nitrates Directive (EC 1991) sets an annual limit of 170 kg per hectare of manure N to be applied to

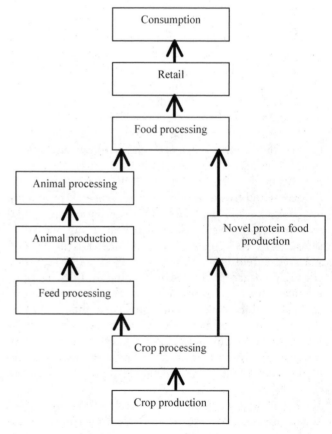

Figure 9.7. The food chain from production to consumption. On the left-hand side, via animal production to consumption; on the right-hand side, via novel protein production.
Source: After Aiking et al., 2006.

agricultural land. This is meant to reduce and prevent the pollution of groundwater and surface waters with nitrate from agricultural sources. Through its manure application limit, it couples animal production to crop production indirectly within regions. This directive has been heavily debated and has a significant influence on the further development of the intensive livestock production sector in the EU (Romstad et al., 1997, Oenema and Berentsen 2004, Mikkelson et al., 2009). It also includes various good agricultural practice guidelines for the application of manure and fertilizers (Mikkelsen et al., 2009).

Concerning manure disposal to surface waters, there are two divergent reasons in legislative policy. The first one allows manure substances to be discharged; it includes discharge standards that define the maximum content of organic substance and nutrients of liquid effluents to be discharged to watercourses. This is the common policy in many developing countries (Ke 2009). It allows the discharge of effluents that have been sufficiently treated to meet established standards. In some cases, the standards for livestock operations are derived from those of industrial operations and are applied not only for discharge to watercourses but also for the application to land. The second approach does not allow any discharge of effluents from livestock operations to watercourses (a zero discharge standard). Examples of this level of environmental protection include the EU Nitrates Directive (EC 1991) and the EU Integrated Pollution Prevention and Control Directive (EC 1996). Effluents from livestock production cannot be discharged to watercourses even if they meet the discharge standards applied to other sources of effluent (Mikkelsen et al., 2009, Powell et al., 2009).

With respect to NH_3 emissions to the atmosphere from animal manure in animal housings and storages and following application to land, several EU members follow the strict regulations in the U.N. Economic Commission for Europe's Gothenborg Protocol (UNECE 2006). Without such measures, up to 50% of the N from animal manure may be lost via NH_3 volatilization to the atmosphere during storage and following the application of manure to land (Sommer et al., 2006), while with proper implementation of the various measures, losses can be minimized to ~10% of the N in manure.

The best available techniques (BATs) associated with the EU Directive on Integrated Pollution Prevention and Control (EC 1996) applies to operations holding more than 2,000 fattening pigs, 750 sows, 30,000 laying hens, and 40,000 broilers. These farms need a permit (a license to produce) and have to use best available techniques for livestock and manure management. These BATs include measures for minimizing NH_3 emissions from animal houses and manure storages and for minimizing leaching losses.

In summary, environmental policy for intensive animal production is a relatively new subject that emerged first in the European Union in the late 1980s. The policy measures were aimed at decreasing N and P emissions to air and water. They have been developed and implemented in practice, while the understanding of the processes, cycling, and functioning of N and P in the biosphere—in air, water, soil, plants, and animals—is still limited. In part because of limited understanding of complex N and P cycles and in part because of the compartmentalization of society and especially governments, the focus in developing policy measures has been on single N species (NO_3^-, NH_3, N_2O or NO_x), on single P measures, on single sectors (households, industries, traffic, crop production, animal production), on single environmental compartments (air, water, nature, humans), or on single effects (human health, animal health, food quality, eutrophication, acidification, biodiversity loss, global warming). This multiaspect approach has contributed to a "wealth" of policies, with some interactive effects (both synergistic and antagonistic effects). As a result, there is an increasing quest for integrating environmental policy measures. Livestock farmers in the EU feel limited by the large number of regulations. Yet these regulations have helped improve the environmental performance of intensive livestock farms. (For further details on EU and Denmark, see Mikkelsen et al., 2009).

Conclusions

Projections for 2030 suggest further increases in animal numbers in the range of 30–50% relative to 2000, with the largest increases for poultry and pigs kept in intensive livestock farms (Bruinsma 2003). Increases will be relatively large in developing countries near large cities, while a continuing decrease is anticipated in some affluent countries in response to globalization of markets, to societal concerns about animal welfare and eating too much animal protein, and to environmental policy measures. Without a change in current practices, the projected increases in intensive livestock production systems will double the current environmental burden and will contribute to large-scale ecosystem degradation unless appropriate measures are taken. As discussed by Erisman et al. (2008), nitrogen has played and will play a major role in the development of agriculture and its environmental impacts.

Current trends of globalization and liberalization of agricultural markets and the economics of scale, specialization, and intensification will further contribute to the agglomeration of intensive livestock production systems near rapidly developing markets. Small, mixed farming systems will not be able to compete, and fewer people will be living in rural areas. Some people in the countryside may even eat animal products produced in urban areas, because it may be cheaper to produce meat in large facilities there than in small mixed rural farms. This may further increase the environmental pressures of intensive livestock production systems unless appropriate measures are taken.

These sobering projections are largely based on extrapolations of current trends. At the same time, there is evidence that the environmental performance of some intensive livestock production systems is high, while others can be improved greatly when effective policies are in place. These would typically include a mix of incentive and disincentive measures, combined with training and demonstration of best available techniques and management tools. Currently, there is a lack of proper incentives, standards, and control in various areas with intensive production systems. The geographical concentration and size of the intensive systems, however, tend to ease the enforcement of policies and especially to reduce the support and monitoring costs compared with more diffuse forms of production (e.g., pastoral systems).

Clearly, international agreements about animal production and animal manure management standards are needed among countries with intensive production systems. Such agreements should provide a basis for environmentally sound and socially acceptable livestock production and should also provide a level playing field to prevent unfair competition. The agreements should include regional livestock stocking density limits, expressed in livestock units per hectare or preferably in P excretion or P surplus per hectare.

Intensive livestock production systems can make a significant contribution to reducing the load of N and P in the wider environment. This requires cooperation among countries in setting standards for livestock production and manure management. It also requires collaboration among animal scientists, agronomists, technologists, economists, and ecologists to derive these standards and the best available techniques and management tools. Foremost, it requires cooperation among the stakeholders along the food chain, including consumers, retailers, the food processing industry, representatives of livestock farmers, animal feed processing companies, and technology companies. Adopting standards and improved technologies requires additional skills, labor, and investments in new technology and equipment by farmers. They need to be convinced about the expected costs and benefits. Clearly a joined effort is needed.

Capacity building on good manure management means that reliable recommendations are available and accepted. To provide these is a challenge for joint activities of scientists, extension officers, and progressive farmers and should also rely on pilot farms demonstrating the feasibility of the proposed measures.

References

Aiking, H., J. de Boer, and J. Vereijken (eds.). 2006. *Sustainable protein production and consumption: Pigs or peas?* Environment and Policy 45. Dordrecht: Springer.

Annevelink, E., A. Vink, W. G. P. Schouten, A. C. Smits, S. Hemming-Hoffmann, E. J. J. Lamaker, and P. W. G. Groot Koerkamp. 2003. Food park, a case study of an integrated sustainable agro production park system designed with Agro Innovation Framework (AIF). In EFITA 2003: Fourth European Conference of the European Federation for Information Technology in Agriculture, Food and the Environment, Debrecem, Hungary. 90–96.

Asman, W.A. H., M. A. Sutton, and J. K. Schjoerring. 1998. Ammonia: Emission, atmospheric transport and deposition. *New Phytologist* 139:27–48.

Badayos, R. B., and M. A. Dorado. 2004. Environmental baseline study (EBS): Nutrients migration to South China Sea. Internal report to FAO.

Block, D., and E. M. Dupuis. 2001. Making the country work for the city: Von Thünen's ideas in geography, agricultural economics and the sociology of agriculture. *American Journal of Economics and Sociology* 60:79–98.

Bouwman, A. F., D. S. Lee, W. A. H. Asman, F. J. Dentener, K. W. van der Hoek, and J. G. J. Olivier. 1997. A global high-resolution emission inventory for ammonia. *Global Biogeochemical Cycles* 11:561–587.

Boyer, E. W., R. W. Howarth, J. N. Galloway, F. J. Dentener, P. A. Green, and C. J. Vörösmarty. 2006. Riverine nitrogen export from the continents to the coasts. *Global Biogeochemical Cycles* 20:GB1S91.

Broderick, G. A. 2003. Effects of varying dietary protein and energy levels on the production of lactating dairy cows. *Journal of Dairy Science* 86:1370–1381.

Bruinsma, J. E. 2003. *World agriculture: Towards 2015/2030.* An FAO perspective. London: Earthscan Publications Ltd.

Burkhardt, M., K. Stoob, C. Stamm, H. Singer, and S. Mueller. 2004. Veterinary antibiotics in animal slurries—a new environmental issue in grassland research. *Grassland Science in Europe* 9:322–324.

Burresh, R. J., P. A. Sanchez, and F. Calhoun (eds.). 1997. *Replenishing soil fertility in Africa.* Madison, WI: ASA, CSSA, SSSSA.

Burton, C. H. 2007. The potential contribution of separation technologies to the management of livestock manure. *Livestock Science* 112:208–216.

Burton, C. H., and C. Turner. 2003. *Manure management—treatment strategies for sustainable agriculture.* 2nd ed. Silsoe: Silsoe Research Institute.

Carpenter, S. R., N. F. Caraco, D. L. Correll, R. W. Howarth, A. N. Sharpley, and V. H. Smith. 1998. Nonpoint pollution of surface waters with phosphorus and nitrogen. *Ecological Applications* 8:559–568.

Castillo, A. R., E. Kebreab, D. E. Beever, and J. France. 2000. A review of efficiency of nitrogen utilisation in lactating dairy cows and its relationship with environmental pollution. *Journal of Animal Feed Science* 9:1–32.

Chapagain, A. K., and A. Y. Hoekstra. 2004. *Water food prints of nations.* Value of Water Research Report Series No 16. Delft, NL: UNESCO-IHE.

Chapagain, A. K., A. Y. Hoekstra, and H. H. G. Savenije. 2006. Water saving through international trade of agricultural products. *Hydrology and Earth System Sciences* 10:455–468.

Culley, J. L. B., and P. A. Phillips. 1989. Groundwater quality beneath small-scale, unlined earthen manure storages. *Transactions of the ASAE* 32:1443–1448.

Dan, T. T., T. A. Hoa, L. Q. Hung, B. M. Tri, H. T. K. Hoa, L. T. Hien, N. N. Tri, P. Gerber, and H. Menzi. 2004. Animal waste management in Vietnam—problems and solutions. In *Sustainable organic waste management for environmental protection and food safety*. Proc. 11th Conference of the Recycling Agricultural, Municipal and Industrial Residues in Agriculture Network (RAMIRAN), Murcia, Spain, 6–9 October, ed. M. P. Bernal, R. Moral, R. Clemente, and C. Paredes, 337–340.

de Haan, C., H. Steinfeld, and H. Blackburn. 1998. *Livestock and the environment, finding a balance*. European Commission Directorate-General for Development, Food and Agricultural Organization.

Dovring, F. 1965. The transformation of European agriculture. In *Cambridge economic history of Europe*, ed. M. M. Postan and H. J. Habakkuk, 604–672. Cambridge: Cambridge University Press.

Dragosits, U., M. R. Theobald, C. J. Place, J. U. Smith, M. Sozanska, L. Brown, D. Scholefield, et al. 2006. The potential for spatial planning at the landscape level to mitigate the effects of ammonia deposition. *Environmedntal Science and Policy* 9:626–638.

Duxbury, A. C., and A. B. Duxbury. 1997. *An introduction to the world's oceans*. 5th ed. Dubuque, IA: Wm C. Brown Publishers.

EC (European Commission). 1991. Council Directive of 12 December 1991 concerning the protection of waters against pollution caused by nitrates from agricultural sources (Directive 91/676/EEC). Brussels: European Commission.

EC (European Commission). 1996. Council Directive of 24 September 1996 concerning integrated pollution prevention and control (Directive 96/61/EC). Brussels: European Commission.

EC (European Commission). 2003. Reference documents on best available techniques for intensive rearing of poultry and integrated pollution prevention and control. Brussels: European Commission.

Eckel, H., U. Roth, H. Döhler, and U. Schultheiss. 2008. Assessment and reduction of heavy metal input into agro-ecosystems. In *Trace elements in animal production systems*, ed. P. Schlegel, S. Durosoy, and A. W. Jongbloed, 33–44. Wageningen: Wageningen Academic Publishers.

EEA (European Environmental Agency). 2005. *Source apportionment of nitrogen and phosphorus inputs into the aquatic environment*. Copenhagen: European Environmental Agency.

Erisman, J. W., A. Bleeker, J. Galloway, and M. S. Sutton. 2007. Reduced nitrogen in ecology and the environment. *Environmental Pollution* 150(1): 140–149.

Erisman, J. W., M. A. Sutton, J. Galloway, Z. Klimont, and W. Winiwarter. 2008. How a century of ammonia synthesis changed the world. *Nature Geoscience* 1:636–639.

FAL. 2005. Evaluation der Ökomassnahmen–Bereich Stickstoff und Phosphor. *FAL-Schriftenreihe* 57. Federal Research Station for Agroecology and Agriculture, Zurich-Reckenholz.

FAO (Food and Agriculture Orgnaization). 2008. FAOSTAT statistical databases. http://faostat.external.fao.org.

Galloway, J. N., J. D. Aber, J. W. Erisman, S. P. Seitzinger, R. W. Howarth, E. B. Cowling, and B. J. Cosby. 2003. The nitrogen cascade. *BioScience* 53:341–356.

Galloway, J. N., M. Burke, G. E. Bradford, R. Naylor, W. Falcon, A. K. Chapagain, J. C. Gaskell, et al. 2007. International trade in meat: The tip of the pork chop. *Ambio* 36:622–629.

Geers, R., and F. Madec. 2006. *Livestock production and society*. Wageningen, Netherlands: Wageningen Academic Publishers.

Gerber, P. 2006. Putting pigs in their place, environmental policies for intensive livestock production in rapidly growing economies, with reference to pig farming in Central Thailand. PhD Thesis Swiss Federal Institute of Technology, Zurich.

Gerber, P., P. Cilonda, G. Franceschini, and H. Menzi. 2005. Geographical determinants and environmental implications of livestock production intensification in Asia. *Bioresource Technology* 96:263–276.

Gilchrist, M. J., C. Greko, D. B. Wallinga, G. W. Beran, D. G. Riley, and P. S. Thorne. 2007. The potential role of concentrated animal feeding operations in infectious disease epidemics and antibiotic resistance. *Environmental Health Perspectives* 115(2): 313–316.

Grote, U., E. Craswell, and P. Vlek. 2005. Nutrient flows in international trade: Ecology and policy issues. *Environmental Science and Policy* 8:439–451.

Guan, T. T. Y., and R. A. Holley. 2003. *Hog manure management, the environment and human health*. New York: Kluwer Academic/Plenum Publishers.

Gunningham, N., and P. Grabosky. 1998. *Smart regulation: Designing environmental policy*. Oxford: Oxford University Press.

Hoekstra, A. Y., and P. Q. Hung. 2005. Globalisation of water resources: international virtual water flows in relation to crop trade. *Global Environmental Change* 15:45–56.

Hoogeveen, M. W., H. H. Luesink, and J. N. Bosma. 2008. *Summary of 2007 monitoring data on the Dutch manure "market."* WOt-rapport 72. Wageningen: Statutory Research Tasks Unit for Nature and the Environment.

Huijsmans, J. F. M., J. G. L. Hendriks, and G. D. Vermeulen. 1998. Draught requirement of trainling-foot and shallow injection equipment for applying slurry to grassland. *Journal of Agricultural Engineering Research* 71:347–356.

Huijsmans, J. F. M., B. Verwijs, L. Rodhe, and K. Smith. 2004. Costs of emission-reducing manure application. *Bioresource Technology* 93:11–19.

IFA (International Fertilizer Industry Association). 2007. *Fertilizer best management practices: General principles, strategy for their adoption and voluntary initiatives vs regulations*. Paris: International Fertilizer Industry Association.

Jondreville, C., P. S. Revy, and J. Y. Dourmad. 2003. Dietary means to better control the environmental impact of copper and zinc by pigs from weaning to slaughter. *Livestock Production Science* 84:147–156.

Jönson, H., A. R. Stinzing, B. Vineras, and E. Salomon. 2004. *Guidelines on the use of urine and feces in crop production*. EcoSanRes Publication series, report 2004-2. Stockholm: Stockholm Environment Institute.

Ke, B. 2009. China: the East–West dichotomy In *Livestock in a changing landscape: Experiences and regional perspectives*, ed. P. Gerber, H. Mooney, J. Dijkman, S. Tarawali, and C. de Haan. Washington, DC: Island Press.

Kemp, A., and J. H. Geurink. 1978. Grassland farming and minerals in cattle. *Netherlands Journal of Agricultural Science* 26:161–169.

Koopmans, G. F., W. J. Chardon, P. A. I. Ehlert, J. Dolfing, R. A. A. Suurs, O. Oenema, and W. H. van Riemsdijk. 2004. Phosphorus availability for plant uptake in a phosphorus-enriched noncalcareous sandy soil. *Journal of Environmental Quality* 33:965–975.

Kruska, R. L., R. S. Reida, P. K. Thorntona, N. Henningerb, and P. M. Kristjansona. 2003. Mapping livestock-oriented agricultural production systems for the developing world. *Agricultural Systems* 77:39–63.

Laegreid, M., O. C. Bockman, and O. Kaarstad. 1999. *Agriculture, fertilizers and the environment.* Wallingford, UK: CAB International.

Landsberg, J. H. 2002. The effects of harmful algal blooms on aquatic organisms. *Reviews in Fisheries Science* 10(2): 113–390.

Mara, D. 2004. *Domestic wastewater treatment in developing countries.* London: Earthscan.

Marschner, H. 1985. *Mineral nutrition of higher plants.* London: Academic Press.

Matson, P. A., W. J. Parton, A. G. Power, and M. J. Swift. 1997. Agricultural intensification and ecosystem properties. *Science* 277:504–509.

Menzi, H. 2002. Manure management in Europe: Results of a recent survey. In Proc. 10th Conference of the FAO/ESCORENA network on Recycling Agricultural, Municipal and Industrial Residues in Agriculture (RAMIRAN), Strbske Pleso, Slovak Republic, May 14–18, ed. J. Venglovsky and G. Greserova, 93–102.

Menzi, H. 2008. Consideration of heavy metals in manure recycling strategies in South East Asia in the nutrient flux model NuFlux. In *Trace elements in animal production systems,* ed. P. Schlegel, S. Durosoy, and A. W. Jongbloed, 63–76. Wageningen: Wageningen Academic Publishers.

Menzi, H., and P. Gerber. 2006. Nutrient balances for improving the use-efficiency of non-renewable resources: Experiences from Switzerland and Southeast Asia. In *Function of soils for human societies and the environment,* ed. E. Frossard, W. E. H. Blum, and B. P. Warkentin, 171–181. *Special publication of the Geological Society* 266. London: The Geographical Society of London.

Menzi, H., B. Pain, and S. G. Sommer. 2002. Manure management: The European perspective. In *Global perspective in livestock waste management,* ed. H. K. Ong, I. Zulkifli, T. P. Tee, and J. B. Liang, 35–44. Malaysia: Malaysian Society of Animal Production.

Meunier, J. 2007. Membrane filtration, microfiltration, ultrafiltration, reverse osmosis. http://www.johnmeunier.com/en/files/?file=980. Retrieved 08/04/2008.

Mikkelsen, S. A., T. M. Iversen, B. H. Jacobsen, and S.S. Kjaer. 2009. Denmark–EU: The regulation of nutrient losses from intensive livestock operations. In *Livestock in a changing landscape: Experiences and regional perspectives,* ed. P. Gerber, H. Mooney, J. Dijkman, S. Tarawali, and C. de Haan. Washington, DC: Island Press.

Misselbrook, T. H., T. J. van der Weerden, B. F. Pain, S. C. Jarvis, B. J. Chambers, K. A. Smith, V. R. Phillips, and T. G. M. Demmers. 2000. Ammonia emission factors for UK agriculture. *Atmospheric Environment* 34:871–880.

Naylor, R., H. Steinfeld, W. Falcon, J. N. Galloway, V. Smil, G. E. Bradford, J. Alder, and H. Mooney. 2005. Losing the links between livestock and land. *Science* 310:1621–1622.

Oenema, O., and P. M. B. Berentsen. 2004. Manure policy and MINAS: Regulating nitrogen and phosphorus surpluses in agriculture of the Netherlands. Paris: Organisation for Economic Co-operation and Development.

Oenema, O., and S. Pietrzak. 2001. Nutrient management in food production: Achieving agronomic and environmental targets. *Ambio* 31:159–168.

Oenema, O., and S. Tamminga. 2005. Nitrogen in global animal production and management options for improving nitrogen use efficiency. *Science in China (Series C)* 48:871–887.

Oenema, O., D. A. Oudendag, and G. L. Velthof. 2007. Nutrient losses from manure management in the European Union. *Livestock Science* 112:261–272.

Petersen, S. O., S. G. Sommer, F. Béline, C. Burton, J. Dach, J. Y. Dourmad, A. Leip, et al. 2007. Recycling of livestock manure in a whole-farm perspective. *Livestock Science* 112:180–191.

Pieters, J. G., G. G. J. Neukermans, and M. B. A. Colanbeen. 1999. Farm-scale membrane filtration of sow slurry. *Journal of Agricultural and Engineering Research* 73 (Issue 4): 403–409.

Poulsen, H. D., and D. Carlson, 2008. Zinc and copper for piglets—how do high dietary levels of these minerals function? In *Trace elements in animal production systems,* ed. P. Schlegel, S. Durosoy, and A. W. Jongbloed, 151–160. Wageningen: Wageningen Academic Publishers.

Powell, J. M., P. Russelle and N. P. Martin 2009. Dairy industry trends and the environment in the USA, In *Livestock in a changing landscape: Regional perspectives,* ed. P. Gerber, H. Mooney, J. Dijkman, S. Tarawali, and C. de Haan. Washington, DC: Island Press.

Powlson, D. S., T. M. Addiscott, N. Benjamin, K. G. Cassman, T. M. de Kok, H. van Grinsven, J. L. L'hirondel, A. A. Avery, and C. van Kessel. 2008. When does nitrate become a risk for humans? *Journal of Environmental Quality* 37:291–295.

Rabalais, N. N. 2004. Eutrophication. In *The sea, volume 13, the global coastal ocean: Multiscale interdisciplinary processes,* ed. A. R. Robinson and K. Brink, 819–865. Harvard: Harvard University Press.

Rattanarajcharkul, R., W. Rucha, S. Sommer, and H. Menzi. 2000. *Area-wide integration of specialised livestock and crop production in Eastern Thailand.* In Proc. 10th International Conference of the Recycling Agricultural, Municipal and Industrial Resiudes in Agriculture Network (RAMIRAN), Gargnano, Italy, 6–9 September, ed. J. Venglovsky and G. Greserova, 22295–22300. Kosice, Slovak Republic: University of Veterinary Medicine.

Romstad, E., J. Simonsen, and A. Vatn. 1997. *Controlling mineral emissions in European agriculture: Economics, policies and the environment.* Wallingford, UK: CAB International.

Rotz, C. A. 2004. Management to reduce nitrogen losses in animal production. *Journal of Animal Sciences* 82:E119–E137.

Rufino, M. C., E. C. Rowe, R. J. Delve, and K. E. Giller. 2006. Nitrogen cycling efficiencies through resource-poor African crop–livestock systems. *Agriculture, Ecosystems and Environment* 112:261–282.

Ryden, J. C., P. R. Ball, and E. A. Garwood. 1984. Nitrate leaching in grassland. *Nature* 311:50–53.

Schaffelke, B., J. Mellors, and N. C. Duke. 2005. Water quality in the Great Barrier Reef region: Responses of mangrove, seagrass and macroalgal communities. *Marine Pollution Bulletin* 51:279–296.

Schlegel, P., S. Durosoy, and A. W. Jongbloed. 2008. *Trace elements in animal production systems.* Wageningen: Wageningen Academic Publishers.

Schröder, J. J. 2005. Revisiting the agronomic benefits of manure; a correct assessment and exploitation of its fertilizer value spares the environment. *Bioresource Technology* 96:253–261.

Schröder, J. J., H. F. M. Aarts, J. C. van Middelkoop, R. L. M. Schils, G. L. Velthof, B. Fraters, and W. J. Willems. 2007. Permissible manure and fertilizer use in dairy farming systems on sandy soils in the Netherlands to comply with the Nitrates Directive target. *European Journal of Agronomy* 27:102–114.

Seitzinger, S. P., J. A. Harrison, E. Dumont, A. H. W. Beusen, and A. F. Bouwman. 2005. Sources and delivery of carbon, nitrogen, and phosphorus to the coastal zone: An overview of global Nutrient Export from Watersheds (NEWS) models and their application. *Global Biogeochemical Cycles* 19:GB4S01.

Seitzinger, J. Harrison, J. Bohlke, A. Bouwman, R. Lowrance, B. Peterson, C. Tobias, and G. van Drecht. 2006. Denitrification across landscapes and waterscapes: A synthesis. *Ecological Applications* 16:2064–2090.

Sellner, K. G., G. J. Doucette, and G. J. Kirkpatrick. 2003. Harmful algal blooms: Causes, impacts and detection. *Journal of Industrial Microbiology and Biotechnology* 30 (7): 383–406.

Shannon, M. A., P. W. Bohn, M. Elimech, J. G. Georgiadis, B. J. Marinas, and A. M. Mayes. 2008. Science and technology for water purification in the coming decades. *Nature* 452:301–310.

Sheldrick, W. F., J. K. Syers, and J. Lingard. 2003. Contribution of livestock excreta to nutrient balances. *Nutrient Cycling in Agroecosystems* 66:119–131.

Sims, J. T., L. Bergström, B. T. Bowman, and O. Oenema. 2005. Nutrient management for intensive animal agriculture: Policies and practices for sustainability. *Soil Use and Management* 21:141–151.

Sinclair, R. 1967. Von Thünen and urban sprawl. *Annals of the Association of American Geographers* 47:72–87.

Smaling, E. A. M., R. Roscoe, J. P. Lesschen, A. E. Bouwman, and E. Communello. 2008. From forest to waste: Assessment of the Brazilian soybean chain, using nitrogen as marker. *Agriculture, Ecosystems and Environment* 128:185–197.

Smeets, P. J. A. M. 2004. Agriculture in the Northwest-European delta metropolis. In *The New Dimensions of the European Landscape*, ed. R.H.G. Jongman, 59–71. Dordrecht: Springer.

Smil, V. 2000. Phosphorus in the environment: Natural flows and human interferences. *Annual Review of Energy and the Environment* 25:53–88.

Smil, V. 2002. Eating meat: Evolution, patterns, and consequences. *Population and Development Review* 28:599–639.

Smith, V. 2003. Eutrophication of freshwater and coastal marine ecosystems—a global problem. *Environmental Science and Pollution Research* 10:126–139.

Sommer, S. G., O. Oenema, D. Chadwick, T. H. Misselbrook, R. Harrison, N. J. Hutchings, H. Menzi, G. J. Monteny, J. Q. Ni, and J. Webb. 2006. Algorithms determining ammonia emission from buildings housing cattle and pigs and from manure stores. *Agronomy Journal* 89:264–335.

Steinfeld, H., P. Gerber, T. Wassenaar, V. Castel, M. Rosales, and C. de Haan. 2006. *Livestock's long shadow: Environmental issues and options*. Rome: Food and Agriculture Organization.

Szogi, A. A., and M. B. Vanotti. 2009. Removal of phosphorus from livestock effluents. *Journal of Environmental Quality* 38:576–586.

Tamis, W. L. M., P. G. L. Klinkkamer, E. van der Meijden, G. R. de Snoo, J. A. van Veen. 2008. Potential ecological effects of veterinary medicines in the terrestrial environment. *CML Report* 178. Leiden, the Netherlands: University of Leiden (in Dutch).

Tamminga, S. 1996. A review on environmental impacts of nutritional strategies in ruminants. *Journal of Animal Science* 74:3112–3124.

Tchobanoglous, G., F. Burton, and H. D. Stensel. 2002. Wastewater engineering: Treatment and reuse. In *Advanced wastewater treatment* (4th ed.). McGraw-Hill Series in Civil and Environmental Engineering. New York: Metcalf and Eddy Inc.

UNECE (United Nations Economic Commission for Europe). 2006. *Strategies and policies for air pollution abatement, review 2006. Convention on Long-range Transboundary Air Pollution*. New York and Geneva: United Nations.

Velthof, G. L., D. A. Oudendag, H. P. Witzke, W. A. H. Asman, Z. Klimont, and O. Oenema. 2009. Integrated assessment of nitrogen emissions from agriculture in EU-27 using MITERRA-EUROPE. *Journal of Environmental Quality* 38:402–417.

Weidema, B. P., M. Wesnæs, J. Hermansen, T. Kristensen, and N. Halberg. 2008. *Environmental improvement potentials of meat and dairy products*. JRC Scientific and Technical Reports. Seville, Spain: European Commission Joint Research Centre.

WHO (World Health Organization). 2008. Mean body weights and age distributions for countries reporting 97. 5th percentile consumption. http://www.who.int/foodsafety/chem/en/acute_hazard_db2.pdf.

Withers, P. J. A., H. G. McDonald, K. A. Smith, and C. G. Chumbley. 1998. Behaviour and impact of cow slurry beneath a storage lagoon, I: Groundwater contamination 1975–1982. *Water, Air and Soil Pollution* 107:35–49.

10

Impacts of Livestock Systems on Terrestrial Ecosystems

Bernard Toutain, Alexandre Ickowicz, Céline Dutilly-Diane, Robin S. Reid, Amadou Tamsir Diop, Vijay Kumar Taneja, Annick Gibon, Didier Genin, Muhammad Ibrahim, Roy Behnke, and Andrew Ash

Main Messages

- **Extensive livestock systems (mobile pastoral, ranching, and mixed farming systems) use land areas of low production, providing food, employment, and income in a considerable number of rural societies.** Due to their great geographical extent and the significant number of livestock involved, these systems interact significantly with the environment, with a range of positive and negative consequences.

- **Under conditions of moderate grazing, many livestock systems provide the environmental services of vegetation cover, biodiversity, and carbon sequestration.** They contribute to maintaining natural resources, and even wildlife, and to preserving them from more aggressive pressures and land uses, including land clearing, agriculture, and fires. Reducing range resources and livestock grazing is particularly detrimental to small farmers and pastoralists, leading to land degradation and to a greater need for intensified livestock production.

- **Excessive livestock pressures alter ecosystems, however, leading to erosion and land degradation.** Overgrazing in general and grazing in unsuitable ecosystems (forests, swamps) have negative impacts on those ecosystems and their associated environmental services.

- **Land tenure and natural resources management policies play crucial roles in sustaining resources.** Livestock's ecological, social, and economic effects on ecosystems must be assessed in order to adapt policies that can mitigate environmental degradations. Given that each situation is highly specific, a participatory approach and decentralization give stakeholders the opportunity to adapt decisions to local situations.

- **Biophysical and socioeconomic issues associated with the management of extensive grazing lands cannot be separated.** Constraints and management problems are best addressed by viewing these pastoral environments as linked, complex socioecological systems. The challenge is to understand these complexities in a systems context and reduce them to relatively simple management responses.

Introduction

In response to the growing demand for animal products at the global level, intensive livestock production systems are addressing emerging needs (see Chapter 2), but extensive and mixed livestock systems continue to contribute significantly to meat production. (See Figure 10.1.) Extensive livestock production is mainly based on traditional inherited knowledge, especially in the developing world. But all livestock systems, whether extensive or intensive, are constantly adapting to a "changing landscape."

In spite of their apparent traditional image, extensive systems are influenced by current events. Not only do they have to cope with climatic variations, climate change, and the consequences of land degradation and desertification, they also have to evolve in reaction to changes in agriculture and markets, population growth, political events, societal changes, and shifts in food consumption habits. Considering specific pressures on resources, extensive livestock systems affect the environment directly or indirectly. So changing practices in this form of extensive livestock production modifies other trends, including pressures on resources and ecosystems.

The environmental consequences of extensive livestock production vary from one context or region to another. The observations and comments in this chapter refer mostly to seven regions of the world: sub-Saharan Africa (pastoral and mixed farming systems), West Asia and North Africa (mostly pastoral), India (mostly mixed farming and some pastoral systems), Central Asia

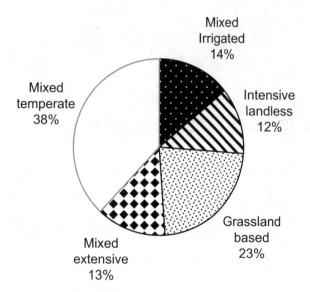

Figure 10.1. Contribution of livestock systems to the world beef/veal production. About two thirds of beef and veal is produced by extensive or mixed systems.
Source: After Seré and Steinfeld 1996.

(mostly pastoral), Latin America (ranching and some mixed farming), Australia (ranching), and Southern Europe (pastoral and silvopastoral systems).

Environmental consequences are not all negative, nor are they all unavoidable. Societies have the capacity to decide on appropriate policies and to promote good practices for sustainable development, provided they have been made aware of the interactions between livestock systems and the environment. This chapter gives particular attention to the responses already used to mitigate negative environmental consequences.

Extensive livestock production systems refer to animal production based on the use of resources available at low or very low cost and are driven by access to feed resources with minimum investment. This chapter defines extensive livestock systems as pastoral systems, ranching systems, and mixed farming systems, following Seré and Steinfeld's classification (1996).

The pastoral system is predominantly grazing, which depends on the natural productivity of rangelands. Mobility is one of its characteristics. In the nomadic system, people and livestock move according to pasture and water availability; transhumance is the seasonal movement of livestock with all or part of the human groups. Animals have the capacity to range freely and to endure temporarily harsh conditions. Often, depending on available resources, individual animal productivity is low. These systems provide animal products at low prices because the cost of land and inputs is very low. They are embedded in traditional exchange and market networks, and they contribute significantly to animal production

markets in developing countries but are not able to satisfy the high increase in demand for animal products (Boutonnet 2005).

Ranching is a sedentary grazing system with potential rotational management of pastures, as in North, Central, and South America, and sometimes large-scale mobility based on contracts, as in Australia and South Africa. This system specializes in meat or wool production. The productivity of labor is high, whereas that of land is low. It can provide the market with products at low cost but often is not able to increase production significantly.

Mixed farming systems are primarily conducted on a small scale by households and combine various sources of feed: natural pastures (rangelands), on-farm products like crop residues and forage crops, and feed bought at markets that includes by-products, grains, and other supplements. This system operates at different levels of intensification and is more or less narrowly dependent on farm crop production, but it has close interactions with the natural resource base (Tarawali et al., 2004). Generally, extensive mixed livestock systems are characterized by low external inputs. They are widely represented in most developing countries. Livestock contribute to further intensification of agriculture by providing draft power and nutrient transfers. The production increase is significant but limited by the increase of feed production and the constraints on market capacity (Boutonnet 2005).

It is worth mentioning that backyard systems that use very low cost feeds, such as domestic and crop residues, mainly raise livestock for family consumption, but they also supply niche markets. Expansion of these systems is limited by the availability of crop residues. In spite of their advantage for reducing poverty's impact and improving food security, they are likely to have a limited environmental impact (with the exception of zoonotic disease transmission) and thus are not discussed in this chapter.

The direct impact of livestock on terrestrial ecosystems varies with the type of ecosystem. See Table 10.1 for various ecosystems used in extensive livestock production. Arid regions account for 40% of the emerged land mass, with the majority of arid regions (70%) located in developing countries. Rangelands account for more than 85% of land use (MA 2005b). Livestock modify the structure, composition, and functioning of these ecosystems, resulting in specific types of landscapes.

The Livestock, Environment and Development Initiative (LEAD) of the United Nations Food and Agriculture Organization (FAO) has adopted an approach to environmental assessment that addresses major physical and biological components to analyze the interactions between environment and livestock. The main impacts of livestock on these components involve varying positive and negative impacts on ecosystems. (See Table 10.2.)

Table 10.1. Global ecosystems used by extensive livestock systems

	Tropical	Dry continental	Temperate	Cold
Rangeland	Savanna Steppe Mountainous grassland	Steppe	Grassland Prairie Steppe Mountainous grassland	
Forest	Woodland Dry and wet forests Riparian forest Bush Mangrove	Forest Bush Riparian forest	Forest Riparian forest	Tundra
Wetland	Grassland Swamp	Swamp	Grassland Swamp	
Cropland	Fallow Cultivated pasture Field after harvest	Fallow Field after harvest	Fallow Cultivated pasture Field after harvest	

The impacts of extensive livestock systems on the environment have been—and still are—important issues globally. Many political options and scientific controversies have taken place on this subject, with major consequences for extensive livestock systems (Behnke et al., 1993, de Haan et al., 1997). The lack of biological, sociological, and economic scientific data and the multiple factors that lead to the various ecological impacts make the debate a difficult one. Impacts of extensive livestock systems that consider livestock–ecosystem compatibility and the associated drivers are important to consider. (See Table 10.3.)

Table 10.2. Impacts of extensive livestock systems on the main components of the environment

Environmental component	Feature	Negative impacts	Positive impacts
Land and soil	Soil structure Soil fertility Water balance	Compaction Erosion Salinization	Soil crust breaking Fertilization
Air and atmosphere	Gaseous composition Air temperature Air transparency	Air pollution Methane emission	Contribution to carbon sequestration
Fresh water	Underground water Superficial water	Water pollution Water waste	Infiltration
Plant and vegetation	Vegetation structure Botanical composition Plant diversity	Land cover degradation Deforestation Loss of biodiversity Bush fire Weed invasion	Rangeland and landscape maintenance Seed dissemination Fire control
Animal and wildlife	Habitat Wildlife diversity Genetic diversity Animal health	Degradation of habitats Loss of biodiversity Concurrence for feed Loss of livestock genetic diversity	Rangeland and landscape maintenance

Table 10.3. Main impact of extensive livestock system given the ex-ante livestock–ecosystem compatibility and the main drivers

	Livestock system in competition with natural ecosystem (case I)	Livestock system is the comparative advantage of the ecosystem (case II)		Livestock system complementary to other land use (case III)	Livestock system in competition with other land use (case IV)
Drivers and changes					
Drivers	Population growth	Population growth / Increased animal population / Sedentarization / Decreasing animal mobility (case IIa)	Population migration / Decreased animal population / Economic liberalization (case IIb)	Population growth / Increased animal population	
Changes					
Land use change	Deforestation	Crop encroachment	Abandonment	Common rangeland converted into individual plots	Transhumant flocks accessing cropping zones previously inaccessible to animals
Management	(–) Competition on forage / (+) Intensification of the system	(–) Overgrazing / early grazing / (–) External feeding	(–) Undergrazing	(–) Overgrazing / (+) Range improvement (shrubs plantations, alley cropping, fertilization)	
Regional perspective	Tropical areas of Latin America	Sahel / Driest region of West Asia and North Africa Andes / Australia / United States	± Central Asia, southern Europe	Subhumid Sub-Saharan Africa / Marginal regions of West Asia and North Africa / India / Mexico / Tropical highlands	Regions with high crop yield potential

Impacts

	Land degradation	No land degradation		Not concerned
Land and soil	Soil and water erosion; Land cover degradation; Shrubs invasion		Soil and water erosion; Soil compaction	
Carbon sequestration	Reduction (direct impact through deforestation)	Increased	Unknown	
Biodiversity	Wildlife endangered (loss of habitat); Flora endangered	Increased/change; Local livestock breeds endangered	Local livestock breeds endangered (switch or crossing with better performing races)	Poor impact
Other environmental impact	Increased frequency and severity of dust storm	Decreased landscape maintenance; Bush fire hazard		
Other impacts (nonenvironmental)	Change in livestock composition (substitution of sheep to cows or of goats to sheep); Disease spreading		Exclusion of mobile pastoralists; Land use conflicts	Conflicts

Responses

Preserving grazing system as land use	Preventing further land use change through conservation measures or payment for environmental services	Pastoral legislation (pastoral codes); Organization of land users and professionals; Managing sedentarization process; Promoting animal mobility; Payment for environmental services; Land tenure	Agri-environmental policy; Landscape policy; Rural development policy; Payment for environmental services	Appropriate land tenure; Appropriate incentive policies
Promoting sustainable and equitable management practices	Natural resource management and sustainable intensification, for converted land			Community-based natural resource management; Integrated natural resource management; Improve pastures; Controlled grazing pressure

Drivers of Change

The pressures of livestock on natural resources and the environment depend on the number of animals grazing, and this in turn is related to how many people are raising livestock. Intensity of pressure also depends on the production system: mobility, seasonal use of resources, and intensification level of production. Social forces, objectives of production, profitability of livestock, market price for animal products, attractiveness of other sectors, and social or political capacity of investment in the sector, all play a role in the number of animals and the production systems. Policy makers have a certain capacity for directing evolution of the sector—and hence part of the responsibility for the impacts on ecosystems.

Evolution of Livestock Numbers

In the regions where extensive livestock production has significant importance, the evolution of animal numbers is driven not only by availability of pasture resources but also by population growth, by the market network and demand for animal products (which is linked to societal changes and urbanization), and by the political situation.

In developing countries, even with climatic and soil constraints, livestock numbers increase parallel with human population growth. In sub-Saharan Africa, the global number of livestock is constantly increasing; the mean annual growth rate for the last 30 years is approximately 3.5% for goats, 3% for sheep, 2% for cattle, and 1% for camels (after FAOSTAT, 1975–2005 period). But productivity levels remain low. In the agropastoral zones of West Africa, a shift from pastoral systems has led to livestock growth, resulting in increased livestock density. (See Figures 10.2 and 10.3.)

In India, the growth rate of livestock numbers has shown a declining trend: the increase in livestock was

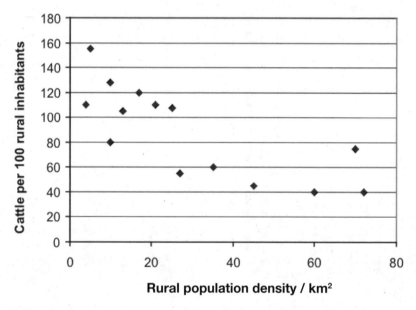

Figure 10.2. Cattle number owned in relation to rural population density in savanna regions of Senegal and Côte d'Ivoire
Source: After Landais et al., 1991.

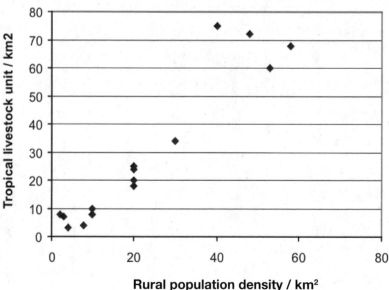

Figure 10.3. Rural population and livestock densities in savanna regions of Senegal and Côte d'Ivoire
Source: After Landais et al., 1991.

lower during 1992–2003 than in 1982–92 for all species except sheep, where a substantial increase in numbers was noted in rainfed and coastal zones (Chandel and Malhotra 2006). In rural areas, cattle, sheep, and goats decreased between 1991–92 and 2002–03 (Chacko et al., 2009).

In some regions of the pastoral steppe of North Africa, the number of sheep increased over 15 years and reached stocking rates that were five times the real carrying capacity (Boutonnet 1989).

In some countries with extensive livestock production oriented to export (Australia and North and Latin America), booming demand boosted beef production for the international market. (See Box 10.1.) In industrial countries, the shift from extensive to intensified systems accompanying the transition to a modern economy since World War II progressively decreased the number of extensive livestock while production was growing globally.

In wet tropical Latin America, livestock are attractive to smallholders for complex reasons: the use of large areas of land with little labor, the availability of livestock for household consumption or to obtain cash, lower labor costs than crops, a source of capital for cash during emergencies, flexibility of timing of sale, and ease of transport. Timber sales and grain production provide the savings to purchase cattle (Sunderlin 1997).

Changes in Livestock Systems

Producers are often forced to modify their production systems. This can take the form of a continuous and permanent adaptation to change over time as a response to the growing number of producers, changing environmental conditions, or economic and social crisis. When the environmental resources supporting extensive production are uncertain or lost, pastoral activity is often abandoned, and mixed crop–livestock systems become more attractive and secure. Small-scale operations with diversification may provide more-efficient food production and income when adverse events such as drought, zoonotic diseases, and restricted land access endanger household food security. Former pastoralists in sub-Saharan Africa shift back from mixed farming to pastoral systems when climatic and resource conditions allow it.

Adaptations occur within the pastoral system itself, as particular livestock breeds are replaced by others better adapted to the new conditions. In sub-Saharan Africa, Sahelian pastoralists adapting themselves to droughts replaced cattle with more-resilient sheep, goats, and camels. In arid zones of India, small ruminants, which were the predominant species, are being replaced by cattle and buffalo. These trends indicate adjustment of different species and numbers to available feed resources, production systems, and environment.

Production systems are changing as well in India—with food production from animals becoming more

Box 10.1. Evolution of Livestock Numbers in Australia

Extensive livestock production in Australia started in the mid-1800s, initially with sheep in the arid and semiarid areas of mainly southern Australia. Sheep numbers rose rapidly. The high stocking rates were not sustainable, and when the first major drought events occurred there was a major loss of perennial grasses and carrying capacity. Sheep numbers dropped considerably (McKeon et al., 2004). Overgrazing was compounded by an invasion of rabbits, which increased soil erosion and prevented recruitment of new plants when good rains returned. Similar scenarios of rapid expansion of sheep numbers followed by crashes in vegetation and carrying capacity were repeated across southern Australia through the first half of the twentieth century. Although livestock numbers never again reached their initial high levels, a stable and prosperous wool industry persisted throughout part of the twentieth century.

In northern Australia a combination of tick infestations, poorly adapted British breeds of cattle, nutrient deficiencies, regular droughts, and poor infrastructure limited the development of the extensive beef industry until the 1960s and 1970s. However, around this time the introduction of tropically adapted *Bos indicus* cattle, urea and phosphorus supplementation, and infrastructure development contributed to a rapid increase in cattle numbers. This increase was stalled by a slump in world beef prices in the mid-1970s, which led to livestock being retained rather than being sold at very low prices. In a number of areas, animal numbers exceeded forage supply, which resulted in loss of perennial grasses, accelerated soil erosion, and the inevitable drop in the livestock population (Ash et al., 1997).

Land settlement policies have contributed to this overstocking. There was a drive by the Australian government for people to settle in the extensive livestock zones following World Wars I and II, and this resulted in smaller properties that eventually could not sustain a livelihood without increasing livestock numbers (Passmore and Brown 1992).

important than draft and manure. This shift suggests further shrinkage of pasture-based systems, with semi-intensive/industrial production systems becoming more important. The migratory system of grazing small ruminants, which has existed for a long time, has been declining, mainly because of reduced grazing areas and pasture productivity and its shift in the society. The number of nomadic and pastoral communities has also decreased.

In West Asia and North Africa, extensive livestock farming has been the dominant form of livelihood: in the Maghreb (northwest Africa), estimates indicate that in 1880 only 45% of the population was sedentary. Nomadism was characterized by the high mobility of livestock and humans and the use of extensive territories, collectively managed on a tribal basis, with complex access rights and land use rules. This feature has changed significantly during the past century, mainly due to active public policies initiated by colonial powers and independent governments.

Bourbouze (2000) distinguishes four major changes: regression of traditional organizations and social equity, which led to individual decisions prevailing at the expense of the groups involved; drastic changes in land tenure toward either privatization or nationalization; regression of mobility and reorganization of pastoral territories; and intensification of livestock systems with widespread use of feed supplements, agricultural products, mechanization of transport, and higher integration to market. Population growth and a strong increase in meat demand also brought in new actors and new forms of livestock farming oriented toward markets and financial investments. Such changes contribute to weakening a number of pastoral systems and increasing the poverty of pastoralists.

Rangeland and forest conversion to establish cropping and grazing areas can alter ecosystems dramatically, disturbing soils and vegetation and ultimately contributing to the destruction of biodiversity (Humphries 1998). Many agricultural frontiers are moving into former rangelands or forests on all continents, changing considerably the conditions of livestock farming and its environmental impacts.

Extensive systems move toward intensification following livestock system development and when the market opportunities make this evolution profitable. In India, the drivers affecting livestock systems are mainly linked to the evolution of the Indian economy, the lack of productivity increase in extensive systems, and the change in land rights. Per capita income in India is growing along with the urban population and public awareness of nutrient-rich foods of animal origin increasing among both urban and rural consumers. This suggests that the livestock sector will have a major role to play in the growth of the agricultural economy.

Changes in Resources

For several decades the resources available for livestock in extensive systems have been changing faster than ever, as seen in two major trends: for pastures, in area and in fodder yield and quality, and for water, in the number of water points and access rights (key resources for the use of rangeland areas). Desertification and land degradation are the consequence of human activities and climate change.

In Africa, West Asia, and India, changes in land use and land tenure lead to rangeland fragmentation, reduction of livestock mobility, and transhumance, and thus local animal concentration and overgrazing, which have a direct impact on the environment (see Chapter 3). The changes in pasture access and vegetation have led

either to higher herd seasonal mobility or to long-term migrations (Petit 2000). In the first half of the twentieth century, 15% of dryland rangelands were converted to cultivated systems, and an even greater transformation took place in the second half of the century (MA 2005b).

In tropical Africa north of the equator, savanna is converted to cropland at an estimated rate of 0.15% per year, representing today 14% of the original cover (MA 2005b). In India, the change in ownership and management of common property resources from communities to *Panchayats* (village-elected representatives), redistribution of common lands to landless people, and the diversion of grazing lands to community development activities led to further reduction in areas under common property resources (Jodha 1990). Many poor farmers keeping livestock in India are landless, relying on communal lands that compete with the establishment of tree plantations for firewood and are in decline in surface area and in quality.

The annual biomass production of rangelands and pastures has been slowly decreasing in most dry zones, including western Asia and North Africa, sub-Saharan Africa, India, and far eastern Asia; other zones, like Central Asia and Australia, have been less affected. In large regions of sub-Saharan Africa, the natural environment is in decline: dry rangelands have lost trees and perennial grasses in terms of density and diversity (but much less in annual grasses), and savanna ecosystems are changed by shrub encroachment, low productive grasses, or weeds (Gaston 1981, Miehe 1991, Diouf et al., 2005, Ly et al., 2009). In India, during the last two decades the overall grazing areas declined across all regions except hill and mountain zones, where it increased (Chandel and Malhotra 2006).

The productivity of grazing lands has been low and is declining further due to inefficient management, unregulated land use, and overgrazing. As a consequence, the grazing pressure on land has increased and the quality of grazing resources has declined. For instance, in rainfed areas the present stocking rate is 1–5 adult cattle units (ACUs)/hectare against a carrying capacity of 1 ACU/hectare, while in arid zones the stocking rate is 1–4 ACUs/hectare against a carrying capacity of 0.2–0.4 ACU/hectare (Shankar an d Gupta 1992).

In North Africa, the surface area of rangelands is decreasing in all countries. Expansion of crops and fallows generated postcultivation steppes with simplified flora and fragmented rangelands. The pure steppe areas of the neighboring Jeffara region in southeast Tunisia, observed by satellite, declined by 36% between 1972 and 2001 (Hanafi et al., 2004), becoming a mosaic of crops and steppe with new forms of livestock pressure on residual rangelands. This situation has largely influenced the structure and functioning of range ecosystems.

Climate change is contributing to desertification.

Since the 1970s, a succession of dry years in sub-Saharan Africa, West Asia, and North Africa enhanced the consequences of grazing pressure on vegetation and soil. But in the Sahel, following consecutive years with good rainfalls, natural dry forests formerly damaged by drought are being reconstituted (A. Marty, pers. comm., 2008). In northern China, climate change brings temperature increase and dry and windy periods, with multiplication of erosion events: traditional rangeland management becomes no longer appropriate in a pastoral equilibrium system (Han et al., 2008). In northern Australia, climate change brought more rains: ranchers used this opportunity to intensify by improving and sowing pastures. (See Box 10.2.)

Changes in Society and Social Demand

In developing and emerging countries, rapid population increase is combined with income growth and major changes in society. Urbanization and the expansion of cities, development of the secondary and tertiary sectors, and production of wealth are modifying food habits and increasing demand for animal products, such as meat (especially the less costly meats), eggs, and milk (see Chapters 1 and 2). Extensive systems are still increasing because livestock provide crucial familial food security and income to a growing number of households; extensive livestock systems are a long-term economic support to small producers in many arid regions. However, trends indicate that younger generations are reluctant to continue rearing livestock because of low incomes. Surveys have shown that people often prefer alternative types of employment to farming, a statement confirmed among pastoralists (Prévost 2007).

In Latin America, the main drivers of deforestation are land occupation, land tenure, poverty, and product marketing (Walker and Moran 2000). Livestock production is a secondary factor. In the Amazon, large landowners are less interested in raising cattle than in securing land tenure. Under Brazilian legislation, clearing land for pastures is considered "effective use" and is the first step toward landownership (Hecht and Cockburn 1990). Deforested land also costs 5 to 10 times as much as forested land, making land clearing an attractive tool for economic gain (Mertens et al., 2002). In Central America, 50% of the population lives at the poverty level on average and 23% is extremely poor; in addition, millions of poor people who lack access to land migrate to agricultural frontiers, trying to improve their livelihoods but exerting pressure on forested areas (Kaimowitz 1996). The general pattern of this expansion starts with clearing and burning established forests, followed by some years of shifting agriculture for subsistence and surplus for market. After a few years, soils are no longer productive and farmers either convert their land to pasture, sell it to larger cattle ranchers, or leave it fallow (Smith et al., 2001).

Box 10.2. Climate Change in Australia

Climate change is an increasingly important driver. There have been strong rainfall trends in the last 30 years that deviate from the long-term historical record, and it is still not clear whether this is related to climate change or to longer-term natural variability. However, pastoralists are already adapting to these rainfall trends. Northwestern Australia, for example, has had a 30% increase in rainfall in the last 30 years, and this is contributing to some of the drive to intensify production. In relation to climate, a major change in policy occurred in 1992 when the emphasis shifted from government assistance and subsidies during droughts to a policy based on self-reliance; this led pastoralists to adapt better to and cope with climate variability.

With regard to demand for animal products, in Latin America and the Caribbean (excluding Brazil), meat consumption is historically higher than in other developing country groups and is predicted to increase further. For example, the demand for meat in Latin America is expected to increase from 26.0 million metric tons in 1997 to 44.9 million metric tons by 2020—an increase of 72% (Rosegrant et al., 2001, Hall et al., 2004). This will result in more pressure on forest reserves for pasture and feedstuff production.

In industrial countries, a growing social interest in preserving the environment has given a new focus to rangeland ecosystems. Rangelands, being considered a "natural" environment, are maintained with consideration for landscape management and biodiversity preservation (Woinarski and Fisher 2003). In southern Europe, extensive systems are strengthened by rising recognition of the environmental services they provide. (See Box 10.3.)

Changes in Public and International Policies

The public policies with impacts on rural development and the livestock sector contribute to the sector's evolution and indirectly to livestock's impact on the environment. Governments are increasingly concerned about the globalization of livestock trade (see Chapter 1), the decentralization of governance, economic development to improve livelihoods, and food security. Since the 1990s, the rising influence of environmental policies at the global level has been added to the list, as these have significantly affected livestock sector development.

Abrupt and fundamental institutional reforms have driven land use change in the rangelands of Central Asia, for example, for the last 15 years. When the Soviet Union collapsed at the end of the 1980s, Central Asian livestock producers lost both their secure Soviet export markets and the benefits they derived from Soviet noncommercial

Box 10.3. New Trends in Southern Europe

Throughout the twentieth century, changes in southern European livestock farming systems have been influenced by the traditional practices and the evolution of European society. In mountainous and hilly areas, ancient rural communities used to associate self-sufficiency with strict control of land use on both private and common lands, given the long-term perspective (Bourbouze and Gibon 1999). Rural depopulation began in the late nineteenth century with the general growth of the European economy and was further accelerated by agricultural development policy during the second half of the twentieth century (Caraveli 2000). Farm numbers dropped dramatically, and a substantial part of the agricultural area was either abandoned or subjected to urbanization (MacDonald et al., 2000). In many areas, livestock production, traditionally included in crop–livestock systems, became specialized. In place of raising various animal species and multipurpose breeds, most farmers changed to more-intensive meat or milk production systems in search of improved economic efficiency. While animal feeding in traditional systems largely relied on the use of grazing lands, supplementation with cereal and concentrate feedstuffs became increasingly important, especially in the diet of dairy flocks in Greece, Spain, and France. The progressive loss of interest in shepherding jobs was an additional factor of change in herd management practice that hampered the use of grazing lands in many countries, especially Greece and Spain (Gibon et al., 1999).

However, natural environmental constraints and the cultural traits of local rural communities have limited the intensification and farm enlargement process and the move toward food production for mass consumption. In the 1980s, growing awareness of the negative impacts of intensification on the environment contributed to restoring interest in extensive systems and traditional livestock management. Southern Europe produces a wide variety of local-specific meat and dairy products using traditional farming practices and adapted processing technologies. The development of tourism in the region and the growing interest of European consumers in high-quality and specific agricultural products contribute to sustained local rural development (Rubino et al., 2006). After some initial regulations supporting mountain agriculture in the 1980s, successive reforms of the Common Agricultural Policy (CAP) since the early 1990s increasingly acknowledged the important role of extensive livestock systems in supporting environment-friendly and sustainable rural development in the less favored areas of Europe (Caraveli 2000, Bignal and McCracken 2000).

The CAP strengthened the connection between agricultural policy and a new approach of rural development that takes the landscape into account. Landscape is regarded as "a societal requirement" having a role to play in the cultural, ecological, environmental, and social fields, favorable to economic activity (Council of Europe 2000). Both the 2003 CAP reform and the Rural Development Regulations introduced with Agenda 2000 promote agri-environment schemes as compulsory elements of rural development programs, requiring that farmers' production activities be compatible with maintenance of landscape and protection of the environment (European Union 2002). But the European farming population is aging, and the prospects for economic viability of extensive livestock farms are weak within the current conditions (Pfimlin and Perrot 2005). Extensive livestock systems will be able to survive and play their complex role only if agri-environmental measures are enforced and rural development policies are extended.

---

pricing systems. At the national level, command economies based on state purchase orders and planned production targets were either dissolved or restructured. At the local level, the state and collective farms that had encompassed almost all aspects of rural life were either reorganized or abandoned.

These changes took place at different times and varying degrees in the five Central Asian republics that emerged out of the Soviet Union. Tajikistan was embroiled in a civil war from 1992 to 1997 and began the reform process late and impoverished (Robinson et al., 2008). In Kazakhstan and Kyrgyzstan, privatization of land and livestock began in the early 1990s and was complete by the end of the decade. In Uzbekistan and Turkmenistan, rangelands and many livestock still remain state property, with pastoralists working within what amounts to reformed and renamed Soviet farms (Behnke et al., 2005).

Changes in pastoral land use reflect these national policy differences. In Kyrgyzstan and Kazakhstan, the sudden and chaotic privatization of livestock caused the loss of about three-quarters of the national herds in the mid-1990s (Schillhorn van Veen 1995). Fodder and hay production also declined markedly. With the collapse of rural livelihood systems, there were high levels of emigration from pastoral areas into towns and changes in rural settlement patterns. Gone were the collective farm structures that had once supported a dispersed settlement pattern, flock mobility, and the use of seasonal pastures.

In Kyrgyzstan, pastoral settlements became smaller, poorer, and less geographically dispersed (Farrington 2005). In Kazakhstan, rural farmsteads and wells were abandoned and destroyed, and many remote seasonal pastures were unused as flock owners retreated to larger rural settlements (Behnke 2003). Around 1999 or 2000, these downward trends were halted and then reversed in Kazakhstan, as flocks expanded for the first time in a decade and larger flock owners began to recolonize isolated farmsteads and wells (Kerven et al., 2003). For the first time since the breakup of collective farms, pastoralists also began to compete to acquire private leases from government and to occupy the most attractive key resources, such as water points, pastures, and infrastructure.

Turkmenistan represents the antithesis of Kazakhstan's radical reforms. Independent Turkmenistan operates a centralized agricultural economy, modeled on farm reforms that were being implemented in the Soviet Union in the late 1980s just before Turkmenistan became independent. Households are permitted to lease livestock from the state, and private ownership of livestock is permitted, but there is no private ownership or leasing of pastures or water points (Behnke et al., 2005). The proportion of the national flock that is private or state owned is unclear, though official statistics state that well over half of all small ruminants are now in private hands. The slow pace of agricultural reform in Turkmenistan and Uzbekistan did not lead to the catastrophic livestock losses that accompanied radical reforms in Kazakhstan and Kyrgyzstan (Kerven et al., 2003). Official statistics state that livestock are now more numerous in Turkmenistan than at any other time in the area's history. It is official policy to increase the size of the national herd, which was also an objective of Soviet livestock policy.

International Trade

Sub-Saharan Africa, West Asia, and North Africa have important extensive livestock sectors, and sub-Saharan Africa exports live animals, but they are all net importers of beef and poultry meat. (See Chapter 2.)

The economic importance of the extensive livestock sector and its direct or indirect dependence on international markets becomes clear during crisis situations. In sub-Saharan Africa, for example, imports of subsidized low-cost animal products from Europe have significantly influenced local prices and resulted in artificial competition with local products, leading to continued poverty among livestock farmers. In southern Australia, a collapse in wool prices in the early 1990s left the pastoral industry in a depressed state. Policy responses to this encouraged either property buildup or an exit from the sector, with pastoral land being converted to conservation reserves (Fargher et al., 2003).

In contrast, in northern Australia the extensive beef industry has been experiencing a boom due to increases in beef prices, the development of the live export market to Southeast Asia, and improvements in infrastructure and transport. Both corporate and large family operations are looking to take advantage of this boom by expanding: the push to acquire additional properties has forced pastoral land prices up by about 150% over the last decade (ABARE 2006). Increasing capital values are in part driving intensification of extensive properties as businesses try to maintain a reasonable return on investment. This has some environmental risks, as pastoralists may be forced to exploit the land in the short term to meet financial obligations.

Land Tenure

The law influences economic development. Extensive livestock production refers in particular to the rural and the pastoral code. In many developing countries, crop farming receives more attention than animal production, and an increasing trend toward land privatization leads to the fragmentation of pastoral lands and reduces the rangeland area and potential livestock mobility. Certain national policies still recommend sedentarization of pastoralists, in spite of risks for the environment and herders' vulnerability.

Nevertheless, many arid countries have established a pastoral code. Following the example of Guinea and Mauritania, the Sahelian countries of West Africa are revising their rural code—defining and protecting rights of pastoralists, common land, and pastoral resources. This evolution encourages pastoralists to apply good resource management practices. But their response is often delayed due to social and political restraints.

In West Asia and North Africa, changes in livestock production systems are driven by land tenure reforms and input subsidies (for seeds, orchard seedlings, fertilizers, supplementary feed, etc.), which promoted individual appropriation of land, cultivation of rangelands, and intensification of livestock farming (Abaab and Genin 2004). In Latin America, pasture establishment responds to individual strategies of private land acquisition, a mechanism often supported by governments but encouraging deforestation and pasture degradation.

Incentive Policies

Subsidies are common and powerful tools to implement national policies and achieve precise objectives. But all the consequences of offering incentives cannot be foreseen. Subsidizing agricultural inputs such as fuel or equipment (tractors) stimulates agriculture and moves agricultural frontiers but has led to crop encroachment in pastoral areas of Africa and the Middle East (Pratt et al., 1997). In Brazil and Central America, in the 1970s the livestock sector secured a disproportionate share of credit at subsidized rates and with lenient reimbursement conditions and control, leading to deforestation for livestock (Kaimowitz 1996). Where credit was available in the Brazilian Amazon, the conversion of forest to pasture was more profitable than sustainable use of already cleared land (Moran 1993, Laurance et al., 2002).

Infrastructures

Pastoral infrastructures that are linked to livestock services are essentially public facilities offered to livestock keepers to improve the conditions of production and livestock productivity. The most effective interventions involve animal health programs (public and/or private veterinary services), quality control of animal products, and markets. Other important components include water delivery for livestock (wells, boreholes, ponds), fixed transhumance routes, corral infrastructure and dipping tanks (related to vaccination campaigns and parasites control), and feed supplements (by-products, dry forage, grains, and trace minerals).

Most of these services are connected and extended through basic infrastructures, which are common with other activity sectors and essential to improve communication between people. Links to fundamental infrastructure include roads, railways, ports, airstrips, markets, and telecommunications, which facilitate services delivery like broadcasting, postal services, social services, and marketing and financial services.

Significant transformations of infrastructure occurred almost everywhere in recent decades, including in remote regions with pastoral areas. Such advances directed to extensive livestock production have resulted in some noticeable investment programs for infrastructure. Examples in sub-Saharan Africa include boreholes programs in Sahel, dipping tanks in southern countries, vaccination campaigns, and fixed transhumance routes supported by international or bilateral funding.

Improved infrastructures have two environmental consequences. While they give livestock managers the opportunity to adapt, diversify, and improve their livestock practices, they also create environmental externalities not always expected. Providing water, for instance, has changed pastoral patterns of seasonal mobility and migrations between dry and wet season grazing areas. The increased number of boreholes in dry areas of the Sahel and Central Asia has improved livestock dispersion in many rangelands and the conditions of transhumance and has reduced concentrations of animals around villages, especially during the cropping time. But it has also concentrated herds and flocks in zones with water where no alternative for mobility existed previously. As a result, areas previously undisturbed and with high levels of biodiversity are now subject to grazing pressures.

Similarly, the designation of transhumance corridors and pastoral areas in western and central Africa has reduced conflicts with farmers in some areas and led to a more controlled dispersion of livestock in cropping areas and along reserves of biodiversity. But it has also had negative impacts on forests and led to the illegal use of pastures in parks and reserves. And facilitation of transportation that provides supplementary feed and water can have appreciable technical and economic results, but the trade-off in the long term can be environmental degradation due to higher stocking rates on land that is augmented by these supplements.

Investments in road infrastructure and modern communication networks have provided tremendous support for extensive livestock modernization and isolated populations. In remote areas in arid regions, mobile and satellite phones are now used by pastoralists to take advantage of grazing schedules in available pastures. New roads permit the transport of live cattle and sheep, milk, butter, cheese, and dry meat to urban markets, providing further motivation for improving livestock productivity. Roads increase opportunities for accelerated transhumance by vehicles, but the pastures near the most frequented routes tend to become overgrazed (Valenza 1981). Throughout Latin America, roads provide access to regions that were previously inaccessible, leading to accelerated clearing of the forest for timber and subsequent crop production and pasture for livestock.

Drought Policies

To reduce drought losses and alleviate social consequences, governments of many arid countries have introduced drought management policies, through either feed subsidies, credit rescheduling, or emergency food aid. In North Africa, for example, governments decided to support the livestock sector, particularly following years of drought. For several years government subsidies on the price of grains used as feed supplements for sheep allowed fodder plantations to be established, providing temporal livestock safeguard plans in dry years (Bourbouze 2000). These interventions have succeeded in protecting producers' incomes during drought. Yet they also introduce bias in the natural regulation of the stocking rate in rangelands, leading producers to keep livestock numbers high and indirectly accelerating rangeland degradation (Boutonnet 1989, Blench 1998, Hazell et al., 2001).

Governments of dry countries have organized specific emergency action plans for pastoral populations faced with climatic crises. The plans are based on the experiences acquired during the drought periods of the recent decades, but beyond their real economic and social advantages, their efficiency for the pastoral population should be improved.

Research and International Development

Until recently, research focused mainly on improving productivity and the efficiency of grazing resources and grazing systems in response to increasing demand (Toutain and Lhoste 1999). Targeted technological innovations were inspired by intensification processes and consisted of various aspects such as pasture improvement (rehabilitation, introduction of legumes, seed production); the adoption of rotational, deferred, or cut-and-carry systems; improvement of herd/flock productivity through genetic, feeding, and sanitary development; and improvement of the efficiency of household enterprises through risk management and marketing development. In sub-Saharan Africa, veterinary research has fought with noticeable success against epizootic diseases and livestock parasites and could, for instance, eradicate rinderpest.

In the 1990s, ecologists described the negative impacts of livestock on the environment; as a the result, there was a drastic reduction of international support for livestock development, even for extensive systems. A global assessment of interactions between livestock and the environment was undertaken to evaluate the challenge (de Haan et al., 1997, Steinfeld et al., 1997). Since then, information and dialogue have been directed toward policy makers and the private sector through international initiatives (Steinfeld et al., 2006).

Impacts on Terrestrial Ecosystems

Herbivores are plant predators: by grazing and browsing in natural vegetation, they not only remove leaves but also break tree branches, reduce flower and seed production, trample plants, and can damage young seedlings. But many biological mechanisms give particular plants an advantage, increasing production when they are grazed; the mechanisms include sprouting and regrowth after cutting, multiplication of grass tillers, adoption of a creeping or prostrate shape, and expansion by seed dispersal via transportation of seeds by animals (Danthu et al., 1996). In addition, stimulation of hard seed germination for many plant species relies on the digestive tract of herbivores (Danthu et al., 1996). In many systems, reduction of dead litter on ground occurs, which promotes germination in grasses. In the savannas and grasslands of Africa, Europe, Asia, and North America, plant species and herbivores have evolved under their mutual influence, with livestock species taking the place of reduced numbers of wild herbivores. In Oceania and the savannas of South America, on the other hand, native vegetation often appears less adapted to grazing, leading to greater impacts on biodiversity and the transformation of ecosystems caused by increased pressures related to livestock due to detrimental overgrazing and the introduction of invasive species (Ibrahim et al., 2003).

Like native herbivores, livestock can be part of the ecosystem. But overstocking of grazing lands significantly affects ecosystem functioning, and livestock can transform shrublands, grasslands, and forests and compete with wild herbivores. Predators are considered a threat by herders, who respond with eradication of top predators, which in turn affects other aspects of biodiversity in a given region. Livestock's impact on soils leads to land degradation. Overgrazing reduces the function of protection of the soil surface through reduction of plant cover and contributes to the degradation of riparian forests, with detrimental effects on water flow. (See Chapter 7.) In addition, overstocking of livestock causes erosion, and in some areas bush fires used by herders have long-term consequences on soil fertility. Also, disturbances by livestock facilitate dissemination of invasive plant species in degraded environments.

Impacts on Rangelands

Extensive grazing modifies the competition and balance between plant species (Hiernaux 1998) and between herbaceous and tree/shrub covers: at the landscape level, grazed grasslands and rangelands often expand while wooded areas or ligneous plant density are reduced (except in wooded savannas, where grazing can stimulate shrub encroachment) (Zoumana et al., 1996, Ouattara and Louppe 2001). Mixed production systems combine land clearing for crop establishment with livestock grazing. This can result in potential overgrazing through the concentration of animals around fields and watering areas.

Overgrazing is often the initial cause of desertification in places where animal density is high and livestock mobility insufficient. Degradation of grazing resources leads also to uncertainty of fodder availability for livestock, low livestock productivity, and loss of pasture species; this is occurring in India (Singh et al., 2005), where the result is a reduction of rangeland biomass production except in mountainous regions (Chandel and Malhotra 2006).

In West Asia and North Africa, desertification is a major threat. Overgrazing due to extensive livestock farming is the most common cause of range degradation. In Syria, the livestock population is three times the estimated rangeland carrying capacity (Nahal 2006). The consequence is the loss of biodiversity, forage biomass, and soils. In the southern steppe of Tunisia, the perennial plant cover—an indicator of rangeland ecosystem health—has been so degraded during the last three decades that the natural vegetation has been replaced by invasive and unpalatable plant species such as *Astragalus armatus* (Jauffret and Visser 2003).

In Australia, native perennial grasses are particularly susceptible to overgrazing because they have not evolved with large herbivore numbers (Ash and McIvor 1998), which is a different situation than is found in most other pastoral regions in the world. Exotic species have been selected and seeded for pastures, as the native grassland species are less palatable and less resilient to grazing. This reduces the native biodiversity, which leads to less stable ecosystems. Invasion by alien invasive plants has led to major changes in extensively managed landscapes.

In Central America, the Amazon Basin, and the Andes region, various estimates indicate that between 30 and 60% of the area under pasture is degraded (Szott et al., 2000, Etter et al., 2006). In Guatemala, animal productivity declines linearly as the level of pasture degradation increases; average losses in income on degraded pastures compared with the potential production of well-managed pasture are estimated at 60% (Betancourt 2006).

Livestock facilitate the spread of alien plants in pastures, some of them becoming invasive in overgrazed areas (e.g., *Sida rhombifolia*, and *Chromolaena odorata* in tropical Africa, *Euphorbia esula* in North America).

Protected forests are often illegally used by herders for shelter, grass, forage trees, and shade, particularly during cropping time. Invasive shrubs are a problem in northern Australia (Grice 2004), and introduced grass and forb species brought in to increase livestock productivity are spreading. While this is positive for livestock productivity, it is also raising environmental concerns (Childs 2002).

Impacts on Vegetation

Most natural rangelands and grasslands have been grazed for long periods of time, often for centuries, without serious degradation. This has often been the case in the Old World, such as the arid regions of sub-Saharan Africa and India, the steppes of Central Asia, western Asia, and North Africa, the tundra of northern Europe, and the alpine meadows of many mountainous regions (Table 10.4). But with today's expanding human settlements and more-intensive grazing, these regions are adversely affected by livestock grazing (Conant et al., 2001). (See Box 10.4.) Vegetation studies demonstrate ecosystem sustainability and the reversibility of plant diversity; other areas show signs of heavy grazing and degradation of vegetation.

According to the paradigm of nonequilibrium in dry rangelands between annual pasture biomass and stocking rate (Ellis and Swift 1988, Behnke et al., 1993), the pastoral system itself is naturally regulated by the level of resources, a mechanism protecting the environment. But external interventions that create overstocking problems, with long-term negative environmental consequences, such as the growing feed supplementation enabled by transportation to livestock in rangelands, bolsters high stocking rates that ecosystems cannot support.

In sub-Saharan Africa, impacts of livestock on vegetation vary: in arid Sahelian regions (such as Mauritania), the desert has progressed under the double pressure of livestock and climate change, whereas large semiarid pastoral Sahelian regions from Senegal to Chad showed degradations during and after the drought years of the 1970s and 1980s, but recovery has occurred so that vegetation cover is more or less comparable to the predrought years (Toutain and Forgiarini 1996, Diouf and Lambin 2001, Diop 2007). The trend of grassland biomass recovery is linked with higher rainfall, with 76% of annual rainfall since 1994 being above normal (Anyamba and Tucker 2005). In subhumid and humid zones, heavy grazing is resulting in shrub encroachment and regression of the perennial herbaceous flora, leading to diminished pastoral value (Boudet 1991). These areas continue to produce and support pastoral activities.

In Sahelian Africa, near villages and watering areas, the grass biomass falls below initial levels of the first period of measurements in 1930–65 (Valenza 1981). But there is no certainty that this situation is irreversible and that the vegetation has changed in composition

(Diouf and Lambin 2001). In Central Asia, the steppe is degraded around settlements, but elsewhere there is no evidence of rangeland degradation (Gintzburger et al., 2005). In pastoral areas of China, the most degraded areas are the agricultural and the semipastoral rangelands (Han et al., 2008).

Moderate grazing can have positive impacts on grasslands by breaking crusted soils and distributing beneficial seeds (Miehe 1991). In the African Sahel, some studies that completely excluded livestock from an area showed the rapid reestablishment of a complex vegetation cover (Miehe 1991). In subhumid wooded savannas of western Africa, excessive grazing and poor management of grazing areas support the establishment or regrowth of natural woody shrubs. This often leads to shrub encroachment and invasive plants. However, appropriate grazing management can provide conditions that optimize competition between trees, grasses, and herbaceous growth (Rousset and Lepart 1999). Yet in India, Namibia, and other areas, livestock dung provides an alternative to fuelwood as a source of energy in some areas, which may alleviate pressure on ecosystems in terms of deforestation. Still, studies show that dung is often a substitute where fuelwood or charcoal is already scarce, and this use is variable, influenced by complexities in economy and the time spent obtaining fuel (Heltberg et al., 2000, Palmer and MacGregor 2009).

Under a temperate climate and in tropical mountainous areas, grazing contributes to the maintenance of grassland meadows and prevents natural expansion of unwanted forests, protecting the diversity of the ecosystems. This is often the case in mountainous regions of Europe (see Box 10.5) and in Madagascar (Bloesch et al., 2002).

Impacts on Forests

In tropical forest ecosystems, including dry, subhumid, and humid climates, livestock are generally considered a direct or indirect agent of degradation—directly through excessive grazing pressure on plants and indirectly when humans clear forest to establish pasture. Dry tropical forests are threatened everywhere not only by firewood harvest or land clearing for cropping but also by grazing ruminant livestock. In southern Europe, extensive systems are strengthened by rising recognition of the environmental services they provide.

In wet tropical ecosystems, deforestation has reached critical levels; for several reasons, throughout Central and Latin America deforestation is oriented to livestock ranching. Land clearing to secure land rights is the first objective, and livestock farming is usually only a subsequent activity on cleared lands.

Forest degradation and deforestation transform biologically diverse and complex tropical forests into simplified shrub and grassland with introduced flora (Harvey et al., 2005, Etter et al., 2006). The direct effects of deforestation include degradation of natural habitats, ecosystem fragmentation, and concurrent loss of landscape connectivity (Laurance et al., 2002). The remaining natural habitats are degraded by hunting, plant and animal extraction, domestic animals, and the introduction of invasive alien species. In such climates, the native vegetation is not adapted to grazing, and pastures are established with introduced, nonnative perennial forage species, which often require labor-intensive management, alter ecosystem processes within specific locations, and can change fire regimes and other natural processes (D'Antonio and Vitousek 1992, Mack et al., 2000).

Impacts on Biodiversity, Flora, and Fauna

Faunal biodiversity outside of protected areas has been rapidly declining for several decades in most tropical countries due to various human activities: agriculture and livestock expansion, hunting and poaching, disease transmission from domestic animals to wildlife, natural habitat fragmentation and ecosystems degradation, and political conflicts (Chardonnet 1995). In pastoral areas, livestock and wildlife can coexist. But the most important impacts of livestock on biodiversity may be indirect and not easy to see. (See Chapter 8.)

For most of the extensive grazing lands of the world, loss of biodiversity remains a major challenge. While the loss is not as great as in highly developed and fragmented agricultural landscapes, a greater societal responsibility seems to be placed on pastoralists to better manage and conserve biodiversity because of the structural intactness of most extensive landscapes.

Under certain conditions, cohabitation of pastoralists and wildlife can positively contribute to maintenance of rangeland ecosystems and can provide potentially favorable conditions for wildlife outside of protected areas. But inside these areas, livestock are rarely tolerated, and conflicts between wardens and herders occur (Toutain et al., 2004). (Conservation initiatives and wildlife–livestock interface in Africa, where the diversity of wildlife is threatened, are discussed in Chapter 8.) In Australia, particularly in Queensland, extensive tree clearing and sowing with introduced pasture species has greatly improved beef production, but this has come at a high cost to biodiversity (Lonsdale 1994).

Deforestation leads to the loss of native plant communities, loss of habitat and resources for wildlife, and the disruption of ecological processes such as wildlife territory expansion, plant pollination, and seed dispersal. In line with biodiversity hot spots, endemic species undergo exceptional loss of habitat, as has happened in Mesoamerica (Myers et al., 2000), Madagascar, or New Caledonia. Species that are particularly vulnerable to deforestation include those that require large contiguous forest, "forest interior" species that can only survive in intact forests, endemic species with small geographic range, and species with small population sizes that are vulnerable to extinction (Laurance et al., 1998, Harvey et al., 2006). Certain taxons are more vulnerable to fragmentation than others, including birds, large predators, primates, butterflies, and solitary wasps (Bierregaard et al., 2001, Harvey et al., 2005). Despite the concerns of conservationists, deforestation in Latin America continues to occur, even in areas strategic for the conservation of regional and global biodiversity (Wassenaar et al., 2007).

In sub-Saharan Africa, plant diversity is markedly reduced in ecosystems heavily overgrazed. But under moderate grazing pressure, there is no evidence in most of the rangelands that grazing is detrimental to the vegetation or reduces the number of species (Hiernaux 1998, Diouf et al., 2005, Wassenaar et al., 2007). Positive impacts on ecosystem integrity are obtained when animal grazing pressure is distributed throughout large areas and when there is flexibility of livestock feeding management in the case of drought (Bollig 2006).

The diversity of ecosystems represents a high value, not only environmentally but also pastorally, by offering livestock complementary feed resources. Protecting this environmental diversity may provide profitable benefits to both the environment and livestock.

The diversity of domestic animals is also a genetic capital under threat. The indigenous livestock breeds have been selected and adapted for a long time to the specific environments and are still improved in accordance to the new rearing conditions. They constitute a considerable element of sustainability for the extensive systems and contribute to reducing the risks when there are difficult sanitary, climatic, or feeding conditions. But they are threatened by the preference for European breeds, by cross-breeding with exogenous animals, and by other forces (Ly et al., 2009). A synopsis of regional specificities of the interactions between extensive livestock systems and the environment is provided in Table 10.4.

Impacts on Soil

In agropastoral zones, erosion and the loss of soil fertility have complex impacts on vegetation, the quantity and quality of water (see Chapter 7), the quality of the atmosphere, and human livelihoods (Landais et al., 1991). Soil degradation involves livestock, agriculture, and forest management. But sound agricultural practices can also sustain soil fertility: incorporation of manure and waste products, for example, contributes to recycling nutrients. (See Chapter 9.) Manure and nutrient transfers from livestock can contribute to 80% of carbon, nitrogen, and phosphorus returns to the soil (Buerkert and Hiernaux 1998, Dugué 1998, Manlay et al. 2004). In arid zones, clearing land for cropping or mining can damage soils more severely than pastoral activities and, in combination with pastoralism, can have a serious impact on land productivity. In Australia, maintaining ground cover and minimizing soil erosion are particularly important for

Table 10.4. Synopsis of regional specificities of interactions between extensive livestock systems and environment

	Tropical Africa	India	West Asia North Africa	Southern Europe	Central Asia	Australia	Latin America
Changes	Livestock number increasing	Growth rate of extensive livestock declining except sheep	Grazing resources declining in productivity and in area	Extensive systems remaining in less favored areas (LFAs)	Livestock number decrease or increase according to country	Sheep number stable (South), beef number increase (North)	Fast extension of (sown or planted) pasture lands after forest clearing
	Grazing resources influenced by climate change	Grazing resources declining in productivity and in area	Pastoral mobility declining	Shrub encroachment	Change in pastoral use. Pastoral mobility declining	Practice of modern transhumance	Easiness of rearing cattle
	Fragmentation of pastoral lands	Pastoral mobility declining	Intensification (supplementary feed, transport)	Reforestation of extensive pastures due to undergrazing		Intensification (pasture improvement by sowing)	Small farms acquired by big livestock farms
	Programs for better access to water	Intensification (supplementary feed)		Intensification (supplementary feed) and specialized livestock systems			
	Interpenetration of livestock and agriculture	Low efficiency of extensive livestock systems					
Drivers	Climate change (more arid, more variable)	Population increase, growing urbanization. Increasing demand for food of animal origin	Population increase, growing urbanization	Rural depopulation	Collapse of Soviet Union, degradation of meat export market and price	Climate change (wetter)	Increasing demand for meat
	Population increase, growing urbanization, increasing demand for food of animal origin	Change of ownership of common land	Decline of pastoral traditional organization to more individual decision	Demand for specific animal products	Rapid privatization of land and livestock in some countries	Fluctuation of wool and meat prices	Migration of poor, agricultural frontiers
	Change of land rights on common land	Provide livelihood to smallholders	Subsidies to agricultural input and feed	Decline of traditional land use control		Development of tourism	Land clearing secures land tenure
	Provide livelihood to rural people		Change of ownership of common land (state, private)	Social pressure to maintain landscape, environment and cultural heritage, development of tourism		Increase of property size after periods of decrease	Price of cleared land higher than forest
						Pasture lease holding/free holding	

Impacts of extensive systems	Extensive systems well integrated within dry environment; Degradation of dry forests and woodlands by grazing; Overgrazing, land degradation; Nutrient cycling by manure; Wildlife threatened	Degradation of grazing resources by overgrazing; Nutrient cycling by manure; Dung source of energy	Traditional systems well integrated within environment; Land degradation (desertification) by overgrazing; Fragmentation of habitats, loss of biodiversity	Systems well integrated within environment; Loss of biodiversity if overgrazing; Sustain biodiversity and landscapes; Reduce fire hazard	Most grazing lands intact, sustain natural ecosystems; Overgrazing and land degradation around settlements and wells	Most grazing lands intact, sustain natural ecosystems; Localized overgrazing and land degradation; Invasion by exotic weeds, pests, and pasture plants; Fragmentation of habitats, loss of biodiversity
						Degradation of rain forest by land clearing; Degradation of pastures by mismanagement; Fragmentation of habitats, serious loss of biodiversity; Carbon sequestration by permanent pastures
Responses	Improve pastures by sowing pasture species; Strengthen professional organizations; Adapt policies about resource access rights	Improve pastures by sowing pasture species; Intensification of livestock systems, supplemental feeding; Adapt policies about resource access rights	Facilitate livestock mobility; Link innovative technologies and organization of production; Recognize and use local knowledge and traditional range management strategies; Adapt policies about resource access rights	Promote ecosystem custodianship by landholders with agri-environmental subsidies (state or EU support)	Facilitate livestock mobility; Improve flock dispersal	Link innovative technologies and organization of production; Adapt policies about resource access rights; Promote ecosystem custodianship by landholders with public support
						Improve pastures by sowing pasture species; Promote intensification of livestock systems; Adapt policies about resource access rights to reduce deforestation; Facilitate open trade agreements

Box 10.4. Livestock Impact on Rangelands in Central Asia

The environmental impact of livestock reflects variations in reform policies. Over the last decade, Kazakhstan and Kyrgyzstan carried out unintended and economically debilitating destocking experiments, with mixed results. In Kazakhstan, there is some evidence of regeneration of some destocked seasonal pastures, while other areas show little response to the withdrawal of grazing, perhaps because they were not as damaged as the Soviet authorities had asserted (Robinson et al., 2002, Ellis and Lee 2003, Coughenour 2006). There is also evidence of increased degradation around large settlements (Alimaev 2003, Alimaev et al., 2008). The availability of grazing around these settlements was initially not a problem when flocks were small, but it has become a constraint as animal numbers rebound. In terms of environmental impact, the spatial relocation of livestock following market reforms has therefore been as important as the decline in total stock numbers.

In Turkmenistan, there is evidence of degradation in the form of reduced rain use efficiency even around wells used by one or two flocks—from 500 to 1,600 head of small stock (Gintzburger et al., 2005). Larger water points attract still more settlers, support more animals, and produce observable changes in the composition and biomass of surrounding vegetation, though it is unclear whether the vegetation changes are caused primarily by grazing or by firewood collecting (Khanchaev et al., 2003, Khanchaev 2006, Behnke et al., 2008, Coughenour et al., 2008).

Herd and flock dispersal and potential seasonal mobility limit the environmental impact of livestock. This is one of the main results of a multicountry study on pastoralism in Inner Asia, the region to the east of Central Asia and consisting of Mongolia, western China, and southern Siberia: "The highest levels of degradation were reported in districts with the lowest livestock mobility; in general, mobility indices were a better guide to reported degradation levels than were densities of livestock. This pattern corresponded with the experience of local pastoralists. At six sites, locals explicitly associated pasture degradation with practices that limited the mobility of livestock" (Sneath 1998, p. 2). Research in Kazakhstan supports these conclusions (Kerven et al., 2003, Alimaev et al., 2008).

In general, from an international perspective, livestock production is considered less damaging than the more easily substantiated environmental degradation in Central Asia—including nuclear pollution, the decimation of wild saiga (*Saiga tatarica*) populations by poachers, the dessication of the Aral Sea due to excessive water extraction for cotton production, and soil salinization caused by poor irrigation management (Glantz et al., 1993, O'Hara 1997, Milner-Guland et al., 2001).

Box 10.5. Landscape Management by Pastoralists in Southern Europe

In Europe, the environmental function of extensive livestock systems was unknown or neglected. When pressure on natural resources was high due to a high population and livestock density, the general opinion put emphasis on damages from overgrazing. Since the second half of the twentieth century, the positive role of environmental preservation has been recognized and assessed (MacDonald et al., 2000). Today traditional practices in pastoral-based systems are generally considered well integrated within the environment (Bignal and McCracken 2000). The social control of traditional communities on land use is acknowledged to have positively preserved resources over the long term (Balent and Stafford-Smith 1993). Over centuries, traditional agropastoral systems shaped the ecosystems and the landscape. The current biological and aesthetic value of the wide variety of rural landscapes in southern Europe has been created and maintained by long-established farming systems (Rubino et al. 2006).

Throughout southern Europe, it is evident that change in land management strategies at farm level and accelerated agricultural depopulation (1960–1980) and had major negative impacts on the environment (MacDonald et al., 2000, Hadgigeorgiou et al., 2005). Pressures to maintain farm incomes resulted in intensification of the most accessible and highest quality lands along with extensification or abandonment of the other areas. The reduction or collapse of traditional management practices, in particular for pastoralism, hay making, and meadow management, had a strong negative impact on the environment, causing loss of diverse grassland vegetation and associated faunal biodiversity. Land abandonment results in shrub encroachment and spontaneous reforestation, which can be detrimental to the services and the cultural and ecological value of landscapes. Such changes can also jeopardize local economies that rely increasingly on leisure activities and tourism.

trying to minimize sediments and nutrients going into rivers, especially those that feed into sensitive marine environments such as the Great Barrier Reef (O'Reagan et al., 2005).

Some agroforested landscapes are specific to mixed farming systems, and these may include hedges of native or introduced bushes to protect crops from livestock (euphorbs and cacti) and agroforests with forage trees (*Faidherbia albida*) that provide for integrated utilization of resources.

Wildfires contribute to soil and ecosystem degradation. In dry areas, pastoralists usually avoid bushfire events and may organize fire prevention measures. In most subhumid and humid zones of western Africa, burning savannas is a common practice. When applied early and controlled to meet the needs of livestock, fire management can be an efficient tool for pasture and open landscape maintenance (for Europe, as an example, see Buffière 2000). In Australia, alteration of fire regimes and overgrazing have resulted in an increase in native shrubs in many rangeland regions (Burrows et al., 2002).

Impacts on Greenhouse Gas Emissions

The reduction of woody vegetation in several grazed terrestrial ecosystems such as those in Latin America represents a loss of carbon sequestered by vegetation. When adding the corresponding quantity of carbon dioxide (CO_2) represented by the loss of forests and the global methane production of ruminants, livestock substantially contribute to carbon dioxide emissions. Grazing and mixed systems in the tropics and subtropics are the main contributors to methane emissions by ruminants, explained by the low productivity of livestock and the poor quality of feed. But from the fermentation of manure, extensive livestock systems produce much less methane than intensive mixed farming and industrial systems do (de Haan et al., 1997).

Yet under conditions of moderate grazing, many livestock systems provide the environmental service of carbon sequestration. In spite of their low carbon contained in soils and biomass, rangelands play an important role in sequestering atmospheric carbon because of their large size. And the permanent pastures established after deforestation represent carbon sequestration by underground and aboveground biomass, as well as by an increase in organic matter contained in soil (Dutilly-Diane et al., 2004). Trees naturally protected in grazed wooded rangelands and savannas, as well as trees in silvopastoral systems and planted trees in agroforestry systems, are other sinks of carbon.

The mitigation effects on greenhouse gas (GHG) emissions of different practices in cropland, livestock, and manure management are largely acknowledged in the scientific community, but the importance of grazing lands management and pasture improvement is often not taken into consideration. By reestimating the global CO_2 mitigation potential from croplands and permanent pastures from previous studies (Lal and Bruce 1999, Conant et al., 2001), Smith et al. (2008) showed that the management of permanent pastures through a combination of practices like grazing intensity, fertilization, nutrient management, fire management, and species introduction (legumes) could represent a CO_2 mitigation potential of 1360–1560 Mt CO_2~eq. yr^{-1}, which represents 70% of what would be the potential of cropland management. In Latin America, deep-rooted grasses sequester significant amounts of organic carbon deep in the soil. Fisher et al. (2002) evaluated sequestration of 100–500 Mt CO_2~eq. yr^{-1} by introduced pasture species in the savannas. In Central Asia, recent estimates show that rangelands are important in the global sequestration of carbon (Johnson et al., 1999). Within the former Soviet Union, restoring the rangelands could be equivalent to a 30% reduction in carbon emissions caused by humans (Demment 1997).

Responses to Terrestrial Impacts

Sustainable Improvement of Extensive Systems Performance

The most obvious and most common way to develop extensive systems has been upgrading their performance. The technological innovations adopted have included pasture and livestock improvement, while improving transhumance conditions have contributed to rangeland management.

Pasture Productivity Improvement

The low natural productivity of rangelands constitutes a limit to improving extensive production. Improved use of natural pastures has been proposed and tentatively applied on numerous occasions, either on controlled or private lands (see, for example, descriptions of North American or Australian rangeland management) or on communal rangelands. (See Box 10.6.)

The main pillars of improving rangeland use sustainably include the following:

- Moderate pasture use and a sound management plan that preserves plant regeneration (often through keeping animals out on a rotational system) and that regulates grazing pressure and stocking rate
- The rehabilitation of degraded areas (periodic exclusion of livestock for vegetation resting, reseeding, adapted tillage)

Box 10.6. Adoption of Environmental Practices on Australian Farms

In Australia, while the issues and drivers of land degradation are now well understood, it is still a challenge for the pastoral industry to adapt and change management practices to avoid degradation. The solution lies with government policies around land stewardship. However, it is unlikely that all land managers will achieve economic, social, and environmental outcomes without a more proactive land stewardship policy that recognizes such issues. Caring for biodiversity conservation is a responsibility of government and may be beyond the reasonable duty of landholders.

A unique feature of Australian agriculture in the last 15 years has been the Landcare movement. This was largely driven from the ground up as a landholder initiative to improve land management. The Commonwealth government quickly realized the importance of this initiative and has facilitated its implementation and growth through grants to coordinate and undertake on-ground improvements.

• The introduction and reintroduction of native (or carefully selected introduced) species in the pastoral vegetation (adapted grasses, legumes, fodder trees)

These principles can be strategically used by grassland managers when they control areas and livestock, based on their level of knowledge, labor availability, and economic considerations.

In the case of communal pasture areas, the resources are shared between users, and other regulation principles apply: the main one is the mobility of livestock—allowing herders to avoid overgrazed or poor areas and reach better pastures. The other is based on collective decisions of management (e.g., to avoid temporarily sensitive areas), reinforced by an authority recognized by all. Another tool of regulation can be payment for using the resources (water access, pasture use), which is described further later in this section. The relevance of this principle depends on the social and cultural context; it appears less applicable in most communal rangelands—at least for pastures and to a lesser extent for water (Ancey et al., 2008).

Many development projects in dry areas have involved rangeland management activities; however, most of them are usually the result of a top-down approach by remote central administrations, and they often fail because they do not take into account actual constraints and opportunities of the farmers. The concept of "holistic management" of pasture (Savory and Butterfield 1999) has been tested in various situations, with a positive impact in controlled situations but with unequal results in collective areas. With the similar objective of organizing pasture use and rational management among herders, a project based on wire fencing implemented in a communal pastoral area of the Senegalese Sahel failed, as it did not work either ecologically or socially (Thébaud et al., 1995). For rangeland rehabilitation, several technologies are applicable, but the results in the long term depend completely on the ability to control degradation factors (Toutain et al., 2006).

Livestock Productivity Improvement

Progress in animal genetics for extensive livestock depends on two contradictory requirements: an increased capacity for production through productivity improvement and a preserved hardiness and resistance to harsh and variable environmental conditions. Livestock farmers and pastoralists permanently practice genetic adaptation of their animals, crossing females with males acquired on the market or even imported. In western Africa, pastoralist migrants in savanna regions have crossed their Sahelian zebu cattle with local bulls (*Bos* taurus cattle), which are much more tolerant of trypanosomiasis. In Latin America, ranchers and *fazendeiros* create crossed herds with imported bovine breeds from the United States or Europe.

However, crossing cattle breeds is not applicable in particular environments. In sub-Saharan Africa, several experimental importations of productive breeds and crossbreeding attempts did not succeed and were soon abandoned due to disease problems and lack of resistance to insect vectors. In tropical environments, the local breeds—in spite of their limited performance—have demonstrated better resistance to stress and adaptability than Western breeds.

Pastoralists and indigenous communities play a role in the preservation of livestock genetic diversity. The majority are not concerned with the expansion of the more common commercial breeds; they practice their own selection in relation to local constraints. Producers in remote areas are not easily involved with development projects (Maier et al., 2002). Koehler-Rollefson (2002) cites the importance of acknowledging the economic potential that indigenous breeds represent for the livelihoods and well-being of herders through an economic offset that considers their commercial potential.

Intensification

The low productivity and reduction in grazing areas suggest that maintaining livestock production, particularly small ruminant production, would be difficult if the systems relied entirely on grazing resources. Supplemental feeding is critical to support higher weights and improve reproductive efficiency. Such a strategy is the beginning of intensification. The level of intensification depends on the investment capacity and the availability of supplemental feeding. But a compromise must be found between more efficient extensive systems and the environmental impacts they generate.

Market Improvement

The specificity and sometimes the special quality of the animal products produced by extensive livestock are an added value. New urban food habits and rising incomes have increased interest in safe and regional products. In Europe, extensive livestock farming is now adapting its traditions to high-quality production, on-farm services related to tourism, and the preservation of natural and cultural heritage (Rubino et al., 2006).

Globalization and open trade agreements create conditions for modernizing the livestock sector and improving competitiveness. However, international standards and regulations to control animal diseases during transboundary movements and to improve food safety and the environment can marginalize small farmers while benefiting wealthier producers (Henson and Loader 2001, Hall et al., 2004). The trade agreement between Central America and the United States, for instance, imposes health, food safety, and environmental standards on exports (Nicholson et al., 1995). In Argentina, the private standards set up by the dairy sector are much more stringent than public standards (Farina et al., 2005). Special policies must be developed to incorporate small farmers in the livestock marketing chain (Cordero 2005). For example, the milk sector in Costa Rica has organized a massive program in technical assistance and training for farmers (Pomareda 2001).

Multifunctional Resources Strategy
As far as extensive grazing systems are concerned, individuals involved in livestock production, development, and research do (or should) consider biophysical and socioeconomic issues together. The reduction of pastoral areas, the evolution of land tenure, degradation of pastoral resources, and changes in livestock systems make sense if they are understood within the global ecological, economical, and social evolution in which pastoral activity is only one of various livelihood strategies used by rural households. Solutions to constraints and management problems are best addressed by viewing these pastoral environments as linked socioecological systems that are complex in nature (Fernandez et al., 2002). Understanding this complexity is necessary in order to propose and apply simple management responses adapted to the context. In addition, land management at the landscape level requires coordination and spatial organization between all land users, not just those involved in livestock production.

Developing simple management responses requires a complete change in management philosophy and approach, but there are some good examples of how this has been achieved successfully (Landsberg et al., 1998). Joint innovative technologies and organizational aspects have been explored in West Asia and North Africa to improve animal production and sustainable range management, such as growing useful and ecologically adapted forage plants (Nefzaoui 2003) or optimizing the use of local resources (Genin et al., 2007). The Jordan's Wadi Araba project uses a participatory approach somewhat effectively to associate conservation of resources with pastoral production (Rowe 2003). Community-based goat fattening and marketing units are associated with a plan of rangeland management that alleviates grazing pressure.

Payment for Environmental Services
The concept of environmental services provided by natural ecosystems is expanding to the global arena. However, the methodology of valuation is a challenging component due to the inherent spatial and temporal variability of the natural processes involved and their interdependent and often intangible nature (Turner et al., 2003). The challenge when valuing ecosystem services is considering the ecological, economic, and societal aspects of livestock production together (Daily et al., 1998, Winkler 2006).

In some regions, payments for environmental services are already in place, focusing on specific ecological objectives. In southern Europe, societal and policy pressures on agriculture for the delivery of environmental and other public goods are steadily rising. Increasing attention is paid to the capacity of livestock systems to manage land and maintain sustainable rural area development. The future of extensive livestock farming depends on its capacity to meet such multifunctional objectives.

In the United States, the Grassland Reserve Program of the Natural Resources Conservation Service pays ranchers to protect or restore private rangeland (USDA 2003). The participants can choose between a rental agreement for a period of 10–30 years or an easement (30 years and beyond).

The idea of paying for environmental services provided by healthy rangelands is also emerging in developing countries. A program of payment for the conservation of wildlife is being promoted in East and Central Africa (ALive 2006). With the growing carbon market, new opportunities might arise through emerging carbon management programs or voluntary carbon markets (Taiyab 2005). In Central America, farmers practicing silvopastoralism are paid for the carbon sequestrated in tree plantations. The evaluation of the potential of carbon sequestration offered by rangelands is presently extended to more arid regions (FAO 2004).

Management of Mobility and Transhumance
Mobility allows pastoralists and farmers to feed livestock adequately and to distribute grazing pressure. Livestock mobility in extensive systems is at the same time a technical necessity and a warranty of good pastoral resource management. Mobility is made possible by development of judicial measures (specific rights enacted in rural and pastoral codes), political support, and development of pastoral infrastructure. In West Asia and North Africa, access to rangeland and water sources has been shared and managed by tribes of nomads under complex traditional rights. Public policies have progressively replaced traditional powers, and pastoral laws were enacted. In sub-Saharan Africa, pastoral codes were or are being established, defining the rights and duties of pastoralists and livestock herders. Guinea ratified a pastoral code in 1998, and Mauritania followed suit in 2000. The terms were discussed after consultation with existing professional organizations. The first laws are less than 10 years old, and their efficiency needs to be evaluated in light of their results.

However, most administrative authorities tend to prefer a more stable citizenry, as mobile people are a potential source of conflict, are often suspected of ignoring the law, and are easy prey for corrupt people. Livestock mobility and transhumance are at times barely socially and politically acceptable. Sometimes sedentary farmers, needing land for agriculture, contest the rights of transhumance and do not respect the corridors or the water access for herds. In western Africa, the regional institution CEDEAO (Communauté Économique des États de l'Afrique de l'Ouest) has protected transboundary transhumance between member states since 1998, but this is not always easy: for instance, Benin closed its borders to foreign herds in 1995 to avoid animal disease introduction and conflicts with local farmers, which can increase the risk of illegal grazing by Nigerian cattle in the "W" Region park, a UNESCO biosphere reserve.

Development projects aim at securing mobility. In eastern Chad, for example, demarcation of corridors for livestock and the creation of water sources were part of a livestock development project; they have helped thousands of Sahelian herders practice annual transhumance to the South over more than 800 kilometers (Barraud et al., 2001) and avoid overgrazing in the northern driest rangelands, which are not yet degraded.

Sustaining mobility includes protecting access to all areas essential to livestock, like dry-season pastures and field crops after harvest. Networks of livestock paths for transhumance and commercial herds have been marked out in western Africa to secure mobility and reduce conflicts. Burkina Faso has experimented with "pastoral zones," areas delimited and defined as pastoral reserves where field crops are not allowed.

The international project World Initiative for Sustainable Pastoralism (WISP), hosted by the International Union for Conservation of Nature and Natural Resources (IUCN) initially supported by the United Nations Environment Programme and the Global Environment Facility, aims at supporting the empowerment of pastoralists and advocates for livestock mobility in pastoral areas.

Knowledge Improvement and Transmission

Considerable scientific and technical knowledge on extensive livestock improvement is already available in reports and through resource persons. Even before the 1980s some environmental considerations were evident, and the idea of sustainability of the systems was attracting attention. But part of this information is misused or insufficiently known for several reasons: past failures, most of which can be explained by insufficient social acceptance or by excessive recurrent charges for labor and other costs; weak knowledge transmission; and insufficient adaptation to current conditions and social needs. As a consequence, greater attention is now being paid to indigenous knowledge.

Traditional Knowledge and Practices

Most rural people have valuable knowledge on the status of resources and the conditions of their renewal, which they acquired through traditional transmission and experience. Local knowledge built on centuries of such experience sometimes leads to very sound strategies of rangeland management. In the High Atlas of Morocco, for instance, the traditional *Agdal* system has proved to be ecologically and economically profitable and at the same time to improve cohesion and responsibility among the local population (Auclair et al., in press).

Agdal involves managing rangelands collectively by resting during periods critical for vegetation growth and by organizing pasture use. It also includes forested areas near the villages, which provide emergency tree forage for livestock in case of long periods of snow in the winter. Other systems, like the *Hema* (Draz 1990), inspired new proposals for institutional range management, with more or less successful results. Other varieties of extensive livestock farming like those using multispecies production contribute to secure livelihoods and to preserving range resources and a diversity of landscapes (Tichit et al., 2004).

Understanding local knowledge can strengthen traditional pastoral institutions. It forms the basis for developed partnerships between the state, these institutions, and various other stakeholders through participatory democracy for effective comanagement of rangelands (Taghi Farvar 2003). Several projects have thus been launched, such as Conservation De la Biodiversité Par la Transhumance Dans le Versant Du Sud Du Haut Atlas in Morocco, which aims at organizing traditional pastoral practices (transhumance corridors, opportunistic management of rangelands) and sustaining current patterns of living (mobile schools, local institutions).

Some creative individual pastoralists and livestock farmers are able to develop efficient innovations on their own or collectively. Such innovators, leaders, or entrepreneurs open new efficient ways of production and organization. For example, the Mauritanian company Tiviski SARL processes 15,000 liters per day of camel milk, collecting the output of hundreds of producers and selling various products, including ultra-high-temperature milk bricks.

Professional Information and Communication

Information systems and early warning systems for the livestock sector in some developing countries are tools used not only by policy makers for prompt decisions but also by farmers, pastoralists, and other stakeholders. Services provided on-line in several languages by FAO/LEAD and the LEAD Livestock and Environment Toolbox aim at transmitting available knowledge to all interested users, including policy makers and technicians.

In West Africa, FAO/SIPES (Système d'Information Pastoralisme et Environnement au Sahel) is a regional information system based on a network for pastoral societies and policy-makers that facilitates sustainable pastoral resource management, adaptation to resource variations, and animal product marketing. SIPES is established at the national level and involves eight countries, giving information on a dozen criteria or indicators selected by the stakeholders and collected by specialists—from satellite images to data collected in the field and the local markets. In Saudi Arabia, the Environment Support of Nomads project promotes remote sensing data to provide relevant technical information helping livestock owners make decisions and minimize land degradation (Al-Gain 1998).

The Famine Early Warning Systems Network provides early warning and vulnerability information on human food security and covers several developing countries. In East Africa, the Livestock Early Warning System is using monitoring systems (e.g., NDVI [Normalized Difference Vegetation Index], meteorological data) and on-ground post-effect surveys of human/animal conditions, vegetation cover, food consumption habits, and movement and market patterns associated with pastoralists. These networks do not target environmental sustainability issues, but they could indirectly contribute to better resource use. In western Africa, the AGRHYMET Regional Centre (ARC), specialized institute of the Permanent Interstate Committee for Drought Control in the Sahel (CILSS), has started monitoring Sahelian natural resources for information and is training the livestock sector and pastoralists.

Training and Information

The recent trend in training programs is to pay greater attention both to extensive livestock systems and to environmental preservation. Numerous examples in each region could illustrate this trend.

Different levels of training are addressed to the following stakeholders:

- Farmers and rural people, through active participation in research (e.g., participatory mapping, participatory discussions for management decisions, use of a multi-agent system to draw scenarios) and through participatory schools (e.g., farmer field schools [FFS])
- Technicians and extension agents, through environmental courses during their training in livestock, specialized technical journals, and the broadcasting of specific courses and interesting examples
- Scientists and policy makers, either through papers in scientific journals or through workshops, conferences, and congresses, as well as through specialized high school training.

Information addressed to rural people and public opinion is provided by articles in papers, conferences, books, and movies.

Policies, Organizations, and Governance

National and International Organizations

Globalization of exchanges and regional development have increased the political attention given to organizations focused on rural and livestock development as well as those advocating for the environment. Such organizations have been strengthened and become useful partners to national organizations and structures.

Laws and Codes

Several national legislatures consider rangeland and pastoral areas as specific entities. Most African Sahelian countries have revised their rural codes, land tenure legislation, and pastoral codes, even if further improvements are needed (Hesse 2001, Touré 2004). By assuring pastoralists of access to strategic resources and managing conflicts resulting from increased competition or interactions between agricultural and pastoral activities, the lawmakers help prevent ecosystem degradation, land fragmentation, and the destruction of rangelands, forests, and wildlife habitats. In Central America, for instance, interest is growing at the policy level in environmental laws to reduce deforestation and recover degraded pastures. In Costa Rica, new forest laws and incentives aim to increase tree resources in landscapes dominated by cattle rearing.

Land Tenure and Land Occupation

Land tenure and rights of access to rangelands must be considered with respect to pastoral needs. Open access to resources can be detrimental to ecosystem integrity, but privatization of rangelands often leads to similar tragedies due to inequitable land appropriation and cultivation in poorly adapted zones. In India, livestock farming is practiced on over 83% of agricultural land, while grassland systems encompass just 4%. Livestock farming primarily consists of smallholders on less than two hectares and the landless. More than 80% of resource-poor households depend exclusively on common property resources to provide feed for livestock.

State ownership can also lead to conflicts and mismanagement (Rae et al., 2000). In relation to common lands, however, it can potentially be a way to promote and support sustainable extensive livestock farming by central government or local regulations and application control.

For much of the Australian extensive grazing lands, either property sizes are still large enough or additional properties have been acquired to maintain economic viability. Maintaining a large spatial scale of operation also helps to buffer climatic risk, and it allows modern transhumance with herds or flocks shifted from properties that may be short in forage to other lands that have more abundant resources. A large percentage of the extensive grazing lands in Australia is leasehold rather than freehold, and governments are reflecting societal concerns about biodiversity by increasingly attaching duty-of-care responsibilities to lease conditions.

Professional Organizations and Public Governance

Most of the professional organizations in the extensive livestock sector have emerged during the last two or three decades. They have progressively increased in number of members, capacity, and power; some federations arise at national and even regional levels. They develop a dialogue with policy makers, advocate for their sector and their rights, and promote collective actions. In several developing countries, the traditional authorities, which used to supervise the collective use of natural resources, have been weakened by land nationalization, new official power authorities, and development projects. The new professional organizations now have a major role to play at the local level with the local authorities, other land users, and their own members in managing natural resources.

Public Investments and Incentives
Public, private, and nongovernmental organizations' investments in resource management infrastructure like pastoral wells, boreholes, and pond programs are sometimes suspected of being linked with environmental degradation (livestock increase, overgrazing, and water pollution). The same comments could be made about the social services addressed to the livestock sector, the veterinary campaigns against epizootics, and the road infrastructure—all of which have resulted not only in the reduction of farmers' vulnerability and improvements in their livelihoods but also in population and livestock increases.

In fact, those public investments facilitate product marketing and natural resources management by opening access to remote rangelands and securing livestock activities. Basic infrastructures crucial for making marketing and exports easy—like roads, railways, shelters, abattoirs and ports, quarantine and storage facilities—are progressively improved, given national and regional political priorities (see Maitima et al., 2009). Veterinary science and health programs have significantly reduced disease risks, minimized loss of animals (which results in global livestock productivity increase), and improved the quality of products. Animal health and product quality controls contribute to improving the competitiveness of marketed livestock products. These improvements can generate environmental externalities, but most of the projects now include environmental considerations and measures. For example, tsetse control programs—the Regional Tsetse and Trypanosomiasis Control Programme in southern Africa and the Farming in Tsetse Controlled Areas program in East Africa—have incorporated environmental components.

On the other hand, some policies promote the diversification of activities to sustain and improve the livelihoods of small livestock farmers. In eastern Africa, dairy in zero grazing units (mixed farming systems) and tourism (pastoral systems) can provide significant revenues. Decentralized processing of livestock products provides an added value to the localities.

Research
Investment in research for extensive livestock production has been weakened by decreasing confidence in obtaining commercially significant results and by environmental concerns. Changing the paradigm from production increase to the search of preserving functional integrity of ecosystems in cohabitation with these unavoidable livestock systems is providing research with new challenges. Scientists also opened the field of research to landscape management, considering livestock as one of multiple challenges: multidisciplinary programs and teams include technical, economic, social, and environmental targets. A great deal of research still needs to be done. For example, further research should evaluate the impacts of forest and rangeland fragmentation on biodiversity, quantify negative externalities of pasture degradation, and assess externalities that affect livestock farmers and the livelihoods of the rural poor.

Conclusion: Some Lessons Learned

A Diversity of Situations
A quick survey of all the extensive livestock systems around the world reveals the diversity of situations, societies, breeds, productions, difficulties, and environmental impacts and services. This chapter has tried to extract the most generic issues and solutions, but it is important to remember the variety of local realities. Each decision must be contextualized and adapted.

Today's large variety of extensive systems is the present stage of dynamic systems that are constantly influenced by changing environment, history, technical and social care, investment, and now more than ever policies. The environmental impacts of livestock also contribute to the changes, modifying in return these impacts.

The Place of Livestock in Ecosystems
Reliable extensive livestock systems have the tendency to be integrated into ecosystem functioning in a sustainable way. Basically, grazing systems represent a sustainable activity with no harmful predisposition to the environment. (See Chapter 8.) This pattern is true as long as the resource use remains compatible with the ecosystem capacity of regeneration. The great variety of impacts on ecosystems depends on the diversity of composition and functioning of ecosystems, which determines their resilience to livestock.

When levels of exploitation are too high (or too low) and surpass ecological thresholds, the environment is altered and the sustainability of livestock production becomes questionable. Such cases result in vicious circles, including the drivers (often ignoring the externalities), unfair practices, and inappropriate policies that impair natural adjustments between resource level and animal pressure. If the resilience of the natural resource base is not taken into account, livestock lead in dry areas to desertification and everywhere to depletion of ecosystem goods and services (Requier-Desjardins and Bied-Charreton 2006, MA 2005a).

Compatibility of Livestock and Ecosystems

When extensive livestock activities are observed inside their usual terrestrial ecosystems and in interaction with other ecosystem components, they can be classified by their tendency to maintain or not maintain sustainable relationships and a kind of balance. Such an analysis is convenient to assess the performance of the livestock and environmental managements and to foresee the trends of evolution and limits. Four distinct situations portray the compatibility between livestock systems and ecosystems. Natural and cultivated ecosystems have variable interactions with livestock production systems in various regions (See Table 10.5.)

In natural ecosystems, two opposite extreme situations can be distinguished: In the first case the presence of livestock leads to an important disturbance of ecosystem functioning and to noticeable ecosystem changes. Such situations can result in serious environmental degradation. In the second case livestock activity does not cause noticeable degradation, or at least it permits the harsher human impacts to be avoided or even provides beneficial effects for ecosystems. Landscape maintenance is preserved or facilitated. In this situation, livestock systems have a comparative advantage over other human activities.

In areas with an agricultural potential, correct integration of livestock and agriculture in mixed farming systems can be profitable for animal production with limited environmental externalities. In cultivated ecosystems, two opposite cases can also be distinguished. In one case, associating livestock with cropping systems, farmers not only take technical advantage of joint production but also provide some services to ecosystems, such as soil fertilization, wild plant and animal diversity maintenance, tree protection, and rural landscape maintenance. In another case, extensive livestock production is not as productive as intensified agriculture, which is preferred. The environment has the capacity for profitable and intensive land use, productive agriculture, and a more intensified livestock system than extensive grazing; in this case, ecological considerations have another meaning.

Table 10.5. Changes in classification of the main ecosystems and their predisposition to interact with extensive livestock production systems

	Natural ecosystems		Cultivated ecosystems	
	Case I	Case II	Case III	Case IV
	Livestock system in competition with natural ecosystem	Livestock system has a comparative environmental advantage to other rural activities	Livestock system is complementary to other land use	Extensive livestock system is in competition with other land use
Type of ecosystems/ resources	Forest (wet, dry) Riparian forest Mangrove Swamp Grassland on soils susceptible to trampling Wildlife habitats	Steppe Rangelands Mountainous grassland Prairie Tundra	Savanna Field after harvest Open woodland Grazed Wetland Grassland in cultivated area	Cultivated systems Cultivated pasture Fallow
Main regions	Tropical areas of Latin America Tropical coastal areas Humid sub-Saharan Africa Forests in dry lands	Driest regions of: West Asia and North Africa United States Central Asia Sub-Saharan Africa (Sahel) Andes Australia South Europe	Subhumid Sub-Saharan Africa Marginal regions of West Asia and North Africa India Mexico Tropical highlands	Regions with high crop yield potential

Human Capacity to Pilot Interactions between Livestock and the Environment

Human factors greatly influence the degree of impacts. Responsibility for the environmental impacts of livestock production must be shared by all the stakeholders of the extensive livestock sector and by consumers. Decisions and practices at all levels can affect the evolution of environmental externalities. To exercise this responsibility, stakeholders need access to appropriate information. Scientists and technicians have a responsibility to assess ecosystems influenced by livestock and to keep livestock producers and policy makers informed. The role of decision makers and livestock policies is crucial in facilitating the evolution of extensive livestock systems while preserving other ecological services.

References

Abaab, A., and D. Genin. 2004. Politiques de développement agropastoral au Maghreb: enseignements pour de nouvelles problématiques de recherche-développement. In *Environnement et sociétés rurales en mutations: approches alternatives*, ed. M. Picouet, M. Sghaier, D. Genin, H. Guillaume, and M. Elloumi, 341–358. Paris: Coll. Latitudes 23, IRD Éditions.

ABARE (Australian Bureau of Agricultural and Resource Economics). 2006. *Livestock commodities*. 13 (1): March quarter. Canberra: Government of Australia.

Al-Gain, A. 1998. Maintaining the viable use of marginal resources: the environmental support of nomads (ESON) project (Saudi Arabia). In *Drylands: Sustainable use of rangelands into the twenty-first century*, ed. V. Squires and A. Sidhamed, 437–443. Rome: International Fund for Agricultural Development.

Alimaev, I. I. 2003. Transhumant ecosystems: Fluctuations in seasonal pasture productivity. In *Prospects for pastoralism in Kazakstan and Turkmenistan: From state farms to private flocks*, ed. C. Kerven, 31–51. London: Routledge Curzon.

Alimaev, I. I., C. Kerven, A. Torekhanov, R. Behnke, K. Smailov, V. Yurchenko, Z. Sisatov, and K. Shanbaev. 2008. The impact of livestock grazing on soils and vegetation around settlements in Southeast Kazakastan. In *The socio-economic causes and consequences of desertification in Central Asia*, ed. R. Behnke, 81–112. Dordrecht, Netherlands: Springer.

ALive., 2006. Investing in maintaining mobility in pastoral systems of the arid and semi-arid regions of sub-Saharan Africa. Washington, DC: World Bank.

Ancey, V., A. Wane, A. Müller, D. André, and G. Leclerc. 2008. Payer l'eau au Ferlo Stratégies pastorales de gestion communautaire de l'eau. IRD (France). *Rev. Autrepart* 46:51–66.

Anyamba, A., and C. J. Tucker. 2005. Analysis of Sahelian vegetation dynamics using NOAA-AVHRR NDVI data from 1981–2003. *Journal of Arid Environments* 63:596–614.

Ash, A. J., and J. G. McIvor. 1998. Forage quality and feed intake responses to oversowing, tree killing and stocking rate in open eucalypt woodlands of north-east Queensland. *Journal of Agricultural Science, Cambridge* 131:211–219.

Ash, A. J., J. G. McIvor, J. J. Mott, and M. H. Andrew. 1997. Building grass castles: Integrating ecology and management of Australia's tropical tallgrass rangelands. *Rangeland Journal* 19:123–144.

Auclair, L., D. Genin, S. Hammi, and L. Kerautret. In press. Les agdals du Haut Atlas central: les effets d'un mode de gestion traditionnel sur les ressources sylvopastorales. In *Séminaire international sur le développement de la montagne marocaine*, December 2005, AMAECO, Rabat, Morocco.

Balent, G., and M. Stafford Smith. 1993. A conceptual model for evaluating the consequences of management practices on the use of pastoral resources. In *Proceedings of 4th International Rangeland Congress*, 1158–1164. Montpellier, France: CIRAD Editions.

Barraud, V., Saleh O. Mahamat, and D. Mamis. 2001. *L'élevage transhumant au Tchad Oriental*. N'Djamena, Chad: Vétérinaires Sans Frontières/SCAC Ambassade de France.

Behnke, R. H. 2003. Reconfiguring property rights and land use. In *Prospects for pastoralism in Kazakstan and Turkmenistan: From state farms to private flocks*, ed. C. Kerven, 75–107. London: Routledge Curzon.

Behnke R. H., I. Scoones, and C. Kerven. 1993. *Range ecology at disequilibrium: New models of natural variability and pastoral adaptations in African savannas*. London, Overseas Development Institute, International Institute for Environment and Development, and Commonwealth Secretariat.

Behnke, R. H., A. Jabbar, A. Budanov, and G. Davidson. 2005. The administration and practice of leasehold pastoralism in Turkmenistan. *Nomadic Peoples* 9:147–170.

Behnke, R. H., G. Davidson, A. Jabbar, and M. Coughenour. 2006. Human and natural factors that influence livestock distributions and rangeland desertification in Turkmenistan. In *The human causes and consequences of desertification*, ed. R. H. Behnke, 141–168. Dordrecht, Netherlands: Springer.

Betancourt, H. 2006. *Evaluación bioeconómica del impacto de la degradación de pasturas en fincas ganaderas de doble propósito en El Chal, Petén, Guatemala*. Tesis Mag. Sc. Turrialba, Costa Rica: CATIE.

Bierregaard, R. O., W. F. Laurance, C. Gascon, J. Benitez-Malvido, P. M. Fearnside, C. R. Fonseca, G. Ganade, et al. 2001. Principles of forest fragmentation and conservation in the Amazon. In *Lessons from Amazonia: The ecology and conservation of a fragmented forest*, ed. R. O. Bierregaard, C. Gascon, T. E. Lovejoy, and R. Mesquita, 371–385. New Haven: Yale University Press.

Bignal, E. M., and D. I. McCracken. 2000. The nature conservation value of European traditional farming systems. *Environmental Review* 8:149–171.

Blench, R. 1998. Rangeland degradation and socio-economic changes among the Bedu of Jordan: Results of the 1975 IFAD survey. In *Drylands: Sustainable use of rangelands into the twenty-first century*, ed. V. Squires and A. Sidhamed, 397–424. Rome: International Fund for Agricultural Development.

Bloesch U., A. Bosshard, P. Schachenmann, H. R. Schachenmann, and F. Klötzli. 2002. Biodiversity of the subalpine forest–grassland ecotone of the Andringitra Massif, Madagascar. In *Mountain biodiversity: A global assessment*, ed. C. H. Kömer and E. M. Spehn, 165–175. London: The Parthenon Publishing group/ CRC Press Company.

Bollig, M. 2006. *Risk management in a hazardous environment, A comparative study of two pastoral societies*. New York: Springer.

Boudet, G. 1991. *Manuel sur les pâturages tropicaux et les cultures fourragères*. Paris : La Documentation Française.

Bourbouze, A. 2000. Pastoralisme au Maghreb: La révolution silencieuse. *Fourrages* 161:3–21.

Bourbouze, A., and A. Gibon. 1999. Ressources individuelles ou ressources collectives? L'impact du statut des ressources sur la gestion des systèmes d'élevage des régions du pourtour de la Méditerranée. *Options Méditerranéennes Serie A* 32:289–309.

Boutonnet, J. P. 1989. *La spéculation ovine en Algérie, un produit clé de la céréaliculture*. Notes et documents n° 90. Institut Nationale de la Recherche Agronomique/Ecole Nationale Supérieure Agronomique de Montpellier (INRA-ENSAM).

Boutonnet, J. P. 2005. Les conditions économiques du développement des productions animales. In *Manuel de zootechnie comparée Nord Sud*, ed. A. Thewis, A. Bourbouze, R. Compère, J. M. Duplan, and J. Hardouin, 519–544. Institut National de la Recherche Agronomique/Agence Universitaire de la Francophonie (INRA/AUF).

Buerkert, A., and P. Hiernaux. 1998. Nutrients in the West African Sudano-Sahelian zone: Losses, transfers and role of external inputs. *Z. Pflanzenerärh. Bodenk* 161:365–383.

Buffière, D., ed. 2000. Brûlages dirigés. *Pastum* 51–52.

Burrows, W. H., B. K. Henry, P. V. Back, M. B. Hoffmann, L. J. Tait, E. R. Anderson, N. Menke, T. Danaher, J. O. Carter, and G. M. McKeon. 2002. Growth and carbon stock change in eucalypt woodlands in northeast Australia: Ecological and greenhouse sink implications. *Global Change Biology* 8:769–784.

Caraveli, H. 2000. A comparative analysis on intensification and extensification in Mediterranean agriculture: Dilemmas for LFAs policy. *Journal of Rural Studies* 16:231–242.

Chacko, C. T., Gopikrishna, V. Padmakumar, S. Tiwari, V. Ramesh. 2009. India: Growth, efficiency gains, and social concerns. In *Livestock in a changing landscape (vol. 2): Experiences and regional perspectives*, 55–73. Edited by P. Gerber, H. Mooney, J. Dijkman, S. Tarawali, and C. de Haan. Washington DC: Island Press.

Chandel and Malhotra. 2006. Livestock systems and their performance in poor endowment regions of India. *Agricultural Economics Research Review* (19)2.

Chardonnet, P., ed. 1995. *Faune sauvage africaine, La ressource oubliée*. Luxembourg: Office of the Official Publications of the European Community.

Childs, J. 2002. Where to for tropical pasture improvement—silver bullet, weed or. . . . *Tropical Grasslands* 36:199–201.

Conant, R. T., K. Paustian, and E. T. Elliot. 2001. Grassland management and conversion into grassland: Effects on soil carbon. *Ecological Applications* 11:343–355.

Cordero, S. P. 2005. *El Comercio internacional de carne bovina en Centro America*. San Jose, Costa Rica: SIDE.

Coughenour, M., R. Behnke, J. Lomas, and K. Price. 2008. Forage distribution, range condition, and the importance of pastoral movement in Central Asia—A remote sensing study. In *The socioeconomic causes and consequences of desertification in Central Asia*, ed. R. Behnke, 45–80. Dordrecht, Netherlands: Springer.

Council of Europe. 2000. *European landscape convention*, Florence, 20/10/2000. European Treaty Services.

Daily G., P. Dasgupta, B. Bolin. 1998. Food production, population growth, and the environment. *Science* 281:1291–1292.

Danthu, P., A. Ickowicz, D. Friot, D. Manga, and A. Sarr. 1996. Effet du passage par le tractus digestif des ruminants domestiques sur la germination des graines de légumineuses ligneuses des zones tropicales sèches. *Revue Elev. Méd. Vét. Pays Trop.* 49 (3): 235–242.

D'Antonio, C. M., and P. M. Vitousek. 1992. Biological invasions by exotic grasses, the grass/fire cycle, and global change. *Annual Review of Ecology and Systematics* 23:63–87.

de Haan, C., H. Steinfeld, and H. Blackburn. 1997. *Livestock and the environment. Finding a balance*. Washington, DC: Food and Agriculture Organization, U.S. Agency for International Development, and World Bank.

Demment, M. 1997. University of California–Davis. At www.news.ucdavis.edu/search/news_detail.lasso?id=3942.

Diop, A. T. 2007. *Dynamique écologique et évolution des pratiques dans la zone sylvo-pastorale du Sénégal: Perspectives pour un développement durable*. PhD thesis, Université Cheikh Anta Diop de Dakar (UCAD). Dakar, Senegal.

Diouf, A., and E. F. Lambin. 2001. Monitoring land-cover changes in semi-arid regions: Remote sensing data and field observations in the Ferlo, Senegal. *Journal of Arid Environments* 48:129–148.

Diouf, J. C., L. E. Akpo, A. Ickowicz, D. Lesueur, and J. L. Chotte. 2005. Dynamique des peuplements ligneux et pratiques pastorales au Sahel (Ferlo, Sénégal). Atelier 2: Agriculture et biodiversité. *Actes de la Conférence International sur la Biodiversité, Sciences et Gouvernance*, Paris, January 24–28. Muséum National d'Histoire Naturelle.

Draz, O. 1990. The Hema system in the Arabian peninsula. In *The improvement of tropical and subtropical rangelands*, 321–331. Washington DC: BOSTID, National Academy Press.

Dugué, P. 1998. Les transferts de fertilité dus à l'élevage en zone de savane. *Agriculture et Développement* 18:99–107.

Dutilly-Diane, C., N. McCarthy, F. Turkelboom, A. Bruggeman, J. Tiedeman, K. Street, and G. Serra. 2004. Could payments for environmental services improve rangeland management in Central Asia, West Asia and North Africa? *Tenth Biennial Conference of the IASCP*, Oaxaca, Mexico, 9–13, August.

Ellis, J., and R. Y. Lee. 2003. Collapse of the Kazakhstan livestock sector: A catastrophic convergence of ecological degradation, economic transition and climate change. In *Prospects for pastoralism in Kazakstan and Turkmenistan: From state farms to private flocks*, ed. C. Kerven, 52–74. London: Routledge Curzon.

Ellis, J. E., and D. M. Swift. 1988. Stability of African pastoral ecosystems: Alternative paradigms and implications for development. *Journal of Range Management* 41 (6): 458–459.

Etter, A., C. McAlpine, D. Pullar, and H. Possingham. 2006. Modelling the conversion of Colombian lowland ecosystems since 1940: Drivers, patterns and rates. *Journal of Environmental Management* 79:74–87.

European Union. 2002. *Europa, Activities of European Union. Summary of legislations. Mid-term review of the common agricultural policy*. Brussels.

FAO (Food and Agriculture Organization). 2004. Carbon sequestration in dryland soils. In *World soils resources report*. Rome: FAO.

Fargher, J. D., B. M. Howard, D. G. Burnside, and M. H. Andrew. 2003. The economy of Australian rangelands—Myth or mystery? *Rangeland Journal* 25:140–156.

Farina, E., G. Gutman, P. Lavarello, R. Nunes, and T. Reardon. 2005. Private and public milk standards in Argentina and Brazil. *Food Policy* 30:302–315.

Farrington, J. D. 2005. De-development in eastern Kyrgyzstan and persistence of semi-nomadic livestock herding. *Nomadic Peoples* 9:171–198.

Fernandez, R. J., E. R. M. Archer, A. J. Ash, H. Dowlatabadi, P. H. Y. Hiernaux, J. F. Reynolds, C. H. Vogel, B. H. Walker, and T. Wiegand. 2002. Degradation and recovery in socio-ecological systems. In *An integrated assessment of the ecological, meteorological and human dimensions of global desertification*, ed. M. Stafford Smith and J. Reynolds, 297–323. Berlin: Dahlem Press.

Fisher, M. J., I. M. Rao, M. A. Ayarza, C. E. Lascano, J. I. Sanz, R. J. Thomas, and R. R. Vera. 2002. Carbon storage by introduced deep-rooted grasses in the South American savannas. *Nature* 371:236–238.

Gaston, A. 1981. *La végétation du Tchad: Evolutions récentes sous les influences climatiques et humaines*. Thèse de doctorat ès Sciences Naturelles, Paris XII-Créteil.

Genin, D., T. Khorchani, and M. Hammadi. 2007. Improving nutritive value of a North African range grass (*Stipa tenacissima*): Effect of dung ash and urea treatment on digestion by goats. *Animal Feed Science and Technology* 136 (1–2): 1–10.

Gibon, A., J. Lasseur, E. Manrique, P. Masson, J. Pluvinage, and R. Revilla. 1999. Livestock farming systems and land management in the mountain and hill Mediterranean regions. *Options Méditerranéennes Série B* No. 27.

Gintzburger, G., S. Saïdi, and V. Soti. 2005. Rangelands of the Ravnina region in the Karakum Desert (Turkmenistan): Current conditions and utilisation. In *Final report to the DARCA project, Desertification or regeneration: Modelling the impact of market reforms on Central Asian rangelands*. Montpellier, France: Centre de Coopération Internationale en Recherche Agronomique pour le Développement/Département d'Élevage et de Médecine Vétérinaire des pays tropicaux (CIRAD-ECONAP).

Glantz, M. H., A. Z. Rubinstein, and I. Zonn. 1993. Tragedy in the Aral Sea Basin: Looking back to plan ahead. *Global Environmental Change* 3:174–198.

Grice, A. C. 2004. Weeds and the monitoring of biodiversity in Australian rangelands. *Australian Ecology* 29b:51–58.

Hadjigeorgiou, I., K. Osoro, J. P. Fragoso de Almeida, and G. Molle. 2005. Southern European grazing lands: Production, environmental and landscape management aspects. *Livestock Production Science* 96:51–59.

Hall, D., S. Ehui, and D. Delgado. 2004. The livestock revolution, food safety, and small-scale farmers: Why they matter to us all. *Journal of Agricultural and Environmental Ethics* 17:425–444.

Han, J. G., Y. J. Zhang, C. J. Wang, W. M. Bai, Y. R. Wand, G. D. Han, and L. H. Li. 2008. Rangeland degradation and restoration management in China. *Rangeland Journal* 30:233–239.

Hanafi, A., D. Genin, and A. Ouled Belgacem. 2004. Steppes et systèmes de production agropastorale dans la Jeffara tunisienne: Quelles relations dynamiques? *Options Méditerranéennes* 62:223–226.

Harvey, C. A., C. Villanueva, J. Villacís, M. Chacón, D. Muñoz, M. López, M. Ibrahim, et al. 2005. Contribution of live fences to the ecological integrity of agricultural landscapes in Central America. *Agriculture, Ecosystems and Environment* 111:200–230.

Harvey, C., A. Medina, D. Sanchez, S. Vilchez, B. Hernandez, J. C. Saenz, J. M. Maes, F. Casanoves, and F. Sinclair. 2006. Patterns of animal diversity in different forms of tree cover in agricultural landscapes. *Ecological Applications* 16:1986–1999.

Hazell, P., P. Oram, and N. Chaherli. 2001. Managing droughts in the low-rainfall areas of the Middle East and North Africa. *EPTD Discussion Paper No. 78*. Washington, DC: International Food Policy Research Institute.

Hecht, S. B., and A. Cockburn. 1990. *The fate of the forest: Developers, destroyers, and defenders of the Amazon*. New York: HarperCollins.

Heltberg, R., T. C. Arndt and N. U. Sekhar. 2000. Fuelwood consumption and forest degradation: A household model for domestic energy substitution in rural India. *Land Economics* (76) 2: 213–232.

Henson, S. and R. Loader. 2001. Barriers to agricultural exports from developing countries: The role of the sanitary and phytosanitary requirements. *World Development* 29 (1): 85–102.

Hesse C. 2001. Gestion des parcours: qui en est responsable et qui y a droit ? In: *Élevage et gestion de parcours au Sahel, implications pour le développement*, ed. E. Tielkes, E. Schlecht, and P. Hiernaux, 139–153. Beuren (Germany): Verlag Ulrich E. Grauer.

Hiernaux, P. 1998. Effects of grazing on plant species composition and spatial distribution in rangelands of the Sahel. *Plant Ecology* 138 (2): 191–202.

Humphries, D. 1998. Milk cows, migrants, and land markets: Unravelling the complexities of forest-to-pasture conversion in Northern Honduras. *Economic Development and Cultural Change* 47(1):95–124.

Ibrahim, M., L'T. Mannetje, and S. Ospina. 2003. Prospects and problems in the utilization of tropical herbaceous and woody leguminous forages. In *Matching herbivore nutrition to ecosystems biodiversity*, ed. L'T Mannetje, L. Ramírez-Avilés, C. Sandoval-Castro, and J. C. Ku-Vera. VII International Symposium on the Nutrition of Herbivores. Merida, Mexico, Universidad Autónoma de Yucatan, 19–24 October.

Jauffret, S., and M. Visser. 2003. Assigning life-history traits to plant species to better qualify arid land degradation in Presaharian Tunisia. *Journal of Arid Environments* 55:1–28.

Jodha, N. S. 1990. Rural common property resources: Contributions and crisis. *Economic and Political Weekly* 25 (26): A65–78.

Johnson, D. A., T. G. Gilmanov, N. Z. Saliendra, E. A. Laca, K. Akshalov, M. Dourikov, B. Madranov, and M. Nasyrov. 1999. *Dynamics of CO_2 flux and productivity on three major rangeland types of Central Asia: 1999 growing season*. Davis, CA: Global Livestock CRSP, University of California–Davis.

Kaimowitz, D. 1996. *Livestock and deforestation in Central America in the 1980s and 1990s: A policy perspective*. Jakarta, Indonesia: Center for International Forestry Research.

Kerven, C., I. I. Alimaev, R. Behnke, G. Davidson, L. Franchois, N. Malmakov, E. Mathijs, A. Smailov, S. Temirbekov, and I. Wright. 2003. Retraction and expansion of flock mobility in Central Asia: costs and consequences. In *Proceedings of the 7th International Rangelands Congress*, Durban, South Africa.

Khanchaev, K. 2006. Pasture conditions in Goktepe District. In *Final report to the DARCA project: Desertification or regeneration: Modelling the impact of market reforms on Central Asian rangelands*. Aberdeen, UK: Macaulay Institute.

Khanchaev, K., C. Kerven, and I. A. Wright. 2003. The limits of the land: Pasture and water conditions. In *Prospects for pastoralism in Kazakstan and Turkmenistan: From state farms to private flocks*, ed. C. Kerven. New York: Routledge Curzon.

Koehler-Rollefson, I. 2002. Why we need "livestock keepers' rights" to save livestock genetic diversity. In *Livestock diversity: Keepers' rights, shared benefits and pro-poor policies*. Documentation of a Workshop with NGOs, Herders, Scientists, and FAO, ed. J. Maier, S. Gura, I. Kohler-Rollefson, E. Mathias, and S. Anderson. German NGO Forum on Environment & Development. Bonn, Germany.

Lal, R., and J. P. Bruce. 1999. The potential of world cropland soils to sequester C and mitigate the greenhouse effect. *Environmental Science and Policy* 2:177–185.

Landais, E., P. Lhoste, and H. Guerin. 1991. Systèmes d'élevage et transferts de fertilité. In *Savanes d'Afrique, terres fertiles?* ed. Focal Coop, Ministère de la coopération et du développement, 219–270. Paris.

Landsberg, R. G., A. J. Ash, R. K. Shepherd, and G. M. McKeon. 1998. Learning from history to survive in the future: Management evolution on Trafalgar Station, north-east Queensland. *Rangeland Journal* 20:104–118.

Laurance, W. F., L. V. Ferriera, J. M. Rankin de Mérona, and S. G. W. Laurance. 1998. Rain forest fragmentation and the dynamics of Amazonian tree communities. *Ecology* 79:2032–2040.

Laurance, W. F., A. K. Albernaz, G. Schroth, P. M. Fearnside, S. Bergen, E. M. Venticinque, and C. Da Costa. 2002. Predictors of deforestation in the Brazilian Amazon. *Journal of Biogeography* 29:737–748.

Lonsdale, W. M. 1994. Inviting trouble: Introduced pastures species in Northern Australia. *Australian Journal of Ecology* 19:345–354.

Ly, C., A. Fall, and I. Okike. 2009. Livestock sector in West Africa: In need of regional strategies. In *Livestock in a changing landscape (vol. 2): Experiences and regional perspectives*, 27–54. Edited by P. Gerber, H. Mooney, J. Dijkman, S. Tarawali, and C. de Haan. Washington DC: Island Press.

MA (Millennium Ecosystem Assessment). 2005a. Dryland system. In *Ecosystems and human well-being*. Volume 1: *Current state and trends*, chapter 22. Washington, DC: Island Press.

MA (Millennium Ecosystem Assessment). 2005b. *Ecosystems and human well-being: Desertification synthesis*. Washington, DC: World Resources Institute.

MacDonald, D., J. R. Cratbee, G. Wiesinger, T. Dax, N. Stamou, J. Gutierrez Lazpita, and A. Gibon. 2000. Agricultural abandonment in mountain areas of Europe: Environmental consequences and policy response. *Journal of Environmental Management* 59:47–69.

Mack, R. N., D. Simberloff, W. M. Lonsdale, H. Evans, M. Clout, F. A. Bazzaz. 2000. Biotic invasions: Causes, epidemiology, global consequences, and control. *Ecological Applications* 10:689–710.

Maier, J., S. Gura, I. Kohler-Rollefson, E. Mathias, and S. Anderson, eds. 2002. *Livestock diversity: Keepers' rights, shared benefits and pro-poor policies*. Documentation of a Workshop with NGOs, Herders, Scientists, and FAO. Ed. J. Maier, S. Gura, I. Kohler-Rollefson, E. Mathias, and S. Anderson. German NGO/CSO Forum on Environment and Development. Rome, 13 June.

Maitima, J. M., M. Rakotoarisoa, and E. K. Kang'ethe. 2009. Horn of Africa: Responding to changing markets in a context of increased competition for resources. In *Livestock in a changing landscape (vol. 2): Experiences and regional perspectives*, 4–26. Edited by P. Gerber, H. Mooney, J. Dijkman, S. Tarawali, and C. de Haan. Washington DC: Island Press.

Manlay, R. J., A. Ickowicz, D. Masse, C. Feller, and D. Richard. 2004. Spatial carbon, nitrogen and phosphorus budget of a village of the West African savanna, II: Element flows and functioning of a mixed-farming system. *Agricultural Systems* 79 (1): 83–107.

McKeon, G., W. Hall, B. Henry, G. Stone, and I. Watson, eds. 2004. *Pasture degradation and recovery in Australia's rangelands: Learning from history*. Australia: Queensland Department of Natural Resources, Mines and Energy.

Mertens, B., R. Poccard-Chapuis, M.-G. Piketty, A.-E. Lacques, and A. Venturieri. 2002. Crossing spatial analysis and livestock economics to understand deforestation processes in the Brazilian Amazon: The case of São Félix do Xingú in South Pará. *Agricultural Economics* 27:269–294.

Miehe, S. 1991. *Inventaire et suivi de la vegetation dans les parcelles pastorales à Widou Thiengoly*. Résultats des recherches effectuées de 1988 à 1990 et evaluation globale proviso ire de l'essai de pâturage contrôlé sur une période de 10 an. Widou/Göttingen: Deutsche Gesellschaft für Technische Zusammenarbeit (GTZ).

Milner-Gilland, E. J., M. V. Kholodova, A. B. Bekenov, O. M. Bukreeva, I. A. Grachev, A. A. Lushchekina, and L. Amgalan. 2001. Dramatic declines in saiga antelope populations. *Oryx* 35:340–345.

Moran, F. 1993. Deforestation and land use in the Brazilian Amazon. *Human Ecology* 21 (1): 1–21.

Myers, N., R. A. Mittermeier, C. G. Mittermeier, G. A. B. de Fonseca, and J. Kent. 2000. Biodiversity hotspots for conservation priorities. *Nature* 403:853–858.

Nahal, I. 2006. *La désertification dans le monde*. Paris: L'Harmattan.

Nefzaoui, A. 2003. Cactus to prevent and combat desertification? In *Desert and dryland development: challenges and potential in the new millennium*, ed. J. Ryan, 261–269. Aleppo, Syria: International Center for Agricultural Research in the Dry Areas (ICARDA).

Nicholson, C. F., R. Blake, and D. Lee. 1995. Livestock, deforestation, and policy making: Intensification of cattle production systems in Central America revisited. *Journal of Dairy Science* 78:719–734.

O'Hara, S. L. 1997. Irrigation and land degradation: Implications for agriculture in Turkmenistan, central Asia. *Journal of Arid Environments* 37:165–179.

O'Reagan, P. J., J. Brodie, G. Fraser, J. J. Bushell, C. H. Holloway, J. W. Faithful, and D. Haynes. 2005. Nutrient loss and water quality under extensive grazing in the upper Burdekin river catchment, North Queensland. *Marine Pollution Bulletin* 51:37–50.

Ouattara, N'K., and D. Louppe. 2001. Influence du pâturage sur la dynamique de la végétation ligneuse en Nord Côte d'Ivoire. In *Aménagement intégré des forêts naturelles des zones tropicales sèches de l'Afrique de l'Ouest*, 221–230. Actes du séminaire international, November 16–20, 1998, Ouagadougou (Burkina Faso): Centre National de la Recherche Scientifique et Technologique (CNRST).

Palmer, C., and J. MacGregor. 2009. Fuelwood scarcity, energy substitution, and rural livelihoods in Namibia. *Environment and Development Economics*. Cambridge University Press

Passmore, J. G. I., and C. G. Brown. 1992. Property size and rangeland degradation in the Queensland mulga rangelands. *Rangeland Journal* 14:9–25.

Petit, S. 2000. *Environnement, conduite des troupeaux et usage de l'arbre chez les agropasteurs peuls de l'ouest burkinabé*. PhD Thesis, University of Orléans, France.

Pfimlin, A., and C. Perrot. 2005. Diversity of livestock farming systems in Europe and prospective impacts of the 2003 CAP reform. *56th Annual Meeting of EAAP*, Uppsala, Sweden, June 3–6.

Pomareda, C. 2001. Políticas para la competitividad del sector lácteo en Honduras. *Políticas económicas y productividad*. Tegucigalpa, Honduras.

Pratt, D. J., F. Le Gall, and C. de Haan. 1997. *Investing in pastoralism: Sustainable natural resource use in arid Africa and the Middle East*. World Bank Technical Paper No. 365. Washington, DC: World Bank.

Prévost, S. 2007. *Les nomades d'aujourd'hui: Ethnologie des éleveurs Raika en Inde*. Montreuil (France) Aux Lieux d'Etre, Institut Français de Pondichéry.

Rae, J., G. Arab, T. Nordblom, K. Jani, and G. Gintzburger. 2000. Property rights and technology adoption in rangeland management, Syria. In *Property rights, collective action, and technology adoption*, ed. R. Meinzen-Dick, P. Hazell, and B. Swallow. Washington, DC: International Food Policy Research Institute.

Requier-Desjardins, M., and M. Bied-Charreton. 2006. Evaluation des coûts économiques et sociaux de la dégradation des terres et de la désertification en Afrique. Versailles (F), C3ED/AFD.

Robinson, S., E. J. Milner-Gulland, and I. Alimaev. 2002. Rangeland degradation in Kazakhstan during the Soviet era: Re-examining the evidence. *Journal of Arid Environments* 53:419–439.

Robinson, S., I. Higginbotham, T. Guenther, and A. Germain. 2006. The process and consequences of land reform in Tajikistan: Inequality, insecurity and the new landless. Prepared for the Advanced Research Workshop on The Socio-economic Causes and Consequences of Desertification in Central Asia, Bishkek, Kyrgyzstan, May 30–June 1. In *The human causes and consequences of desertification*, ed. R. H. Behnke, L. Alibekov, and I. I. Alimaev. Dordrecht, Netherlands: Springer.

Rosegrant, M. W., M. S. Paisner, S. Meijer, and J. Witcover. 2001. *Global food projections to 2020: Emerging trends and alternative futures*. Washington, DC: International Food Policy Research Institute.

Rousset, O., and J. Lepart. 1999. Evaluer l'impact du pâturage sur le maintien des milieux ouverts. Le cas des pelouses sèches. *Fourrages* 159:223–235.

Rowe, A. 2003. Conservation and pastoral livelihood in Jordan's Wadi Araba. In *Desert and dryland development: Challenges and potential in the new millennium*, ed. J. Ryan, 225–230. Aleppo, Syria: International Center for Agricultural Research in the Dry Areas (ICARDA).

Rubino, R., L. Sepe, A. Dimitriadou, and A. Gibon. 2006. *Livestock farming systems: Product quality based on local resources leading to improved sustainability*. Wageningen, Netherlands: Wageningen Academic Publishers.

Savory A., and J. Butterfield. 1999. *Holistic management: A new framework for decision making*. Washington, DC: Island Press.

Schillhorn van Veen, T. W. 1995. The Kyrgyz sheep herders at a crossroads. *Pastoral Development Network Paper 38d*. London: Overseas Development Institute.

Seré, C., and H. Steinfeld. 1996. *World livestock production systems. Current status, issues and trends*. Animal Production and Health Paper No. 127. Rome: Food and Agriculture Organization.

Shankar, V., and J. N. Gupta. 1992. Restoration of degraded rangelands. In *Restoration of degraded lands—Concepts and strategies*, ed. J. S. Singh, 115–155. Meerut, India: Rastogi Publications.

Singh, V. K., A. Suresh, D. C. Gupta, and R. C. Jakhmola. 2005. Common property resources rural livelihood and small ruminants in India: A review. *Indian Journal of Animal Sciences* 75 (8) : 1027–1036.

Smith, J., B. Finegan, C. Sabogal, M. Gonçalvez Ferreira, G. Siles Gonzalez, P. van de Kop, and A. Diaz Barba. 2001. Management of secondary forests in colonist swidden agriculture in Peru, Brazil and Nicaragua. In *World forests, markets and policies*, ed. M. Palo, J. Uusivuori, and G. Mery, 263–278. Dordrecht, Netherlands: Kluwer Academic Publishers.

Smith, P., D. Martino, Z. Cai, D. Gwary, H. H. Janzen, P. Kumar, B. McCarl, et al. 2008. Greenhouse gas mitigation in agriculture. *Philosophical Transactions of the Royal Society, B* 363:789–813.

Sneath, D. 1998. State policy and pasture degradation in Inner Asia. *Science* 281:1147–1148.

Steinfeld, H., C. de Haan, and H. Blackburn. 1997. *Livestock-environment interactions: Issues and options*. Washington, DC: Food and Agriculture Organization, U.S. Agency for International Development, and World Bank /LEAD.

Steinfeld, H., P. Gerber, T. Wassenaar, V. Castel, M. Morales, and C. de Haan. 2006. *Livestock's long shadow: Environmental issues and options*. Rome: LEAD/Food and Agriculture Organization.

Sunderlin, W. D. 1997. Deforestation, livelihoods, and the preconditions for sutainable management in Olancho, Honduras. *Agriculture and Human Values* 14:373–386.

Szott, L., M. Ibrahim, and J. Beer. 2000. *The hamburger connection hangover: Cattle pasture land degradation and alternative land use in Central America*. Serie Técnica. Informe técnico/ CATIE No. 313. Costa Rica: Centre Agronómico Tropical de Investigación y Enseñanza (CATIE).

Taghi Farvar, M. 2003. Mobile pastoralism in West Asia: Myths, challenges and a whole set of loaded questions. *Policy Matters* 12:31–41.

Taiyab, N. 2005. *The market for voluntary carbon offsets: A new tool for sustainable development?* Gatekeeper series 121. London: International Institute for Environment and Development.

Tarawali, S., J. D. H. Keatinge, J. M. Powell, P. Hiernaux, O. Lyasse, N. Sanginga. 2004. Integrated natural resource management in West African crop-livestock systems. In *Sustainable crop-livestock production for improved livelihoods and natural resource management in West Africa*, eds. Williams, T. O., Tarawali, S., Hiernaux, P., and Fernandez-Rivera, S., 349–370. Nairobi, Kenya, International Livestock Research Institute (ILRI) and Wageningen, Netherlands, Technical Centre for Agricultural and Rural Cooperation (CTA). Proceedings of an international conference held in the International Institute for Tropical Agriculture, Ibadan, Nigeria, 19–22 November 2001.

Thébaud, B, H. Grell, and S. Miehe. 1995. Recognizing the effectiveness of traditional grazing practices: Lessons from a controlled grazing experiment in northern Senegal. Drylands Programme, Paper No. 55, IIED (International Institute for Environment and Development), London, UK.

Tichit, M., B. Hubert, L. Doyen, and D. Genin. 2004. A viability model to assess sustainability of mixed herds under climatic uncertainty. *Animal Research* 53:405–417.

Touré, O. 2004. *Impact des lois pastorales sur la gestion équitable et durable des ressources naturelles en Guinée*. Dossier No. 126, London: International Institute for Environment and Development.

Toutain, B., and G. Forgiarini. 1996. Projet Almy-Bahaim d'hydraulique pastorale dans le Tchad Oriental. Cartographie de la végétation pastorale. Montpellier, France: CIRAD-EMVT.

Toutain, B., T. Guervilly, A. Le Masson, and G. Roberge. 2006. Lessons learnt from trials to regenerate Sahelian rangelands. *Science et changements planétaires/Sécheresse*. 17(1): 72–75.

Toutain, B., and P. Lhoste. 1999. Sciences, technologies et gestion des pâturages au Sahel. In *Horizons nomades en Afrique sahélienne: Sociétés, développement et démocraties*, ed. A. Bourgeot, 377–394. Paris: Karthala.

Toutain, B., M. N. De Visscher, and D. Dulieu. 2004. Pastoralism and protected areas: Lessons learned from Western Africa. *Human Dimension of Wildlife* 9:287–295.

Turner, R. K., J. Paavola, P. Cooper, S. Farber, V. Jessamy, and S. Georgiou. 2003. Valuing nature: Lessons learned and future research direction. *Ecological Economics* 46:493–510.

Valenza, J. 1981. Surveillance continue de pâturages naturels Sahéliens Sénégalais. Résultats de 1974 à 1978. *Rev. Elev. Med. vét. Pays trop.* 34 (1): 83–100.

Walker, R., and E. Moran. 2000. Deforestation and cattle ranching in the Brazilian Amazon: External capital and household processes. *World Development* 28 (4): 683–699.

Wassenaar, T., P. Gerber, P. Rosales, M. Ibrahim, P. H. Verbung, H. Steinfeld, 2007. Projecting land use changes in the Neotropics: The geography of pasture expansion into forest. In *Global Environmental Change* 17:86–104.

Winkler, R. 2006. Valuation of ecosystem goods and services, I: An integrated dynamic approach. *Ecological Economics* 59:82–93.

Woinarski, J. C., and A. Fisher. 2003. Conservation and the maintenance of biodiversity in the rangelands. *Rangeland Journal* 25:157–171.

Zoumana, C., P. Yesso, and J. César. 1996. La production des jachères pâturées dans le Nord de la Côte d'Ivoire. In *Actes de l'atelier "Jachère, lieu de production,"* Bobo Dioulasso, 2–4 October 1996, ed. C. Floret, 113–121. Dakar: CORAF.

11

Human Health Hazards Associated with Livestock Production

*Bassirou Bonfoh, Karin Schwabenbauer, David Wallinga, Jörg Hartung, Esther Schelling,
Jakob Zinsstag, François-Xavier Meslin, Rea Tschopp, Justin Ayayi Akakpo, and Marcel Tanner*

Main Messages

- **Livestock production has long been associated with possible threats to human health** in terms of zoonotic diseases, food safety hazards from infectious agents, and antibiotic resistance in humans arising from excessive use of antibiotics in livestock.

- **More livestock are kept in towns and intensively marginalized** as a result of the steadily growing demand for livestock products, a gradually increasing livestock population, urbanization, and globalization, as well as shocks including conflict and war.

- **Human health risks particularly associated with intensive systems include occupational risks,** mainly respiratory problems arising from poor air quality in animal houses. These systems are also associated with antimicrobial resistance in human pathogens arising from the use of antibiotics to increase production and to treat livestock diseases.

- **Extensive livestock are increasingly marginalized and the specific human health risks associated with such systems include disease transmission between animals, wildlife, and people** arising from their close interactions, as well as the lack of animal and human health services that limit the ability to control diseases such as brucellosis and echinococcosis. Pressures on land and water resources, which force extensively kept livestock to come into more regular contact with each other and to reoccupy the same pasture more frequently, have reduced the ability to control livestock diseases through traditional measures of movement and separation.

- **Changes in the management of livestock contribute to promoting the emergence and reemergence of zoonoses.** In some production systems, weak capacities and infrastructure increase the human–animal–wildlife interface, with negative consequences for human and animal health.

Introduction

Livestock production has long been associated with possible threats to human health in terms of zoonotic diseases, food safety hazards from infectious agents, and antibiotic resistance in humans arising from incorrect use of antibiotics in livestock. With major changes in livestock production systems, the types of threats have changed, depending on the type of production. For example, while livestock in extensive systems may be more exposed to infections arising from casual contact with wildlife and other livestock, the risks and hazards associated with intensive production systems are more likely to arise from the large numbers and high density of animals and their contact with farm workers and traders carrying disease agents on their vehicles and clothes (Nunn and Black 2006).

Livestock producers, consumers of livestock and livestock products, and traders and processors of livestock products are at risk of contracting zoonoses, including foodborne infections and intoxications in both intensive and extensive systems. Nevertheless, in pastoral communities of Chad, in comparison to other predominant morbidity such as respiratory tract disorders, alimentary disorders, or malaria, the impact of selected zoonotic diseases on the health status of the study populations seemed to be comparatively low (Schelling et al., 2005). However, the disease burden in terms of disability-adjusted life-years (DALYs) has so far been determined for very few zoonoses, and experts expect that once underreporting is better assessed, zoonoses will start to score high from a total disease burden perspective.

This chapter examines the factors that predispose different livestock production systems to creating human health hazards from zoonotic diseases and environmental conditions and the problems that arise for humans from antibiotic resistance originating in livestock systems.

Food safety issues are addressed in more detail in the Responses section of this volume (see Chapter 17).

Features of the Livestock Sector and Implications for Human Health Hazards

The livestock sector is undergoing major changes, with a new dynamic in evidence over the last 20 years. These are described elsewhere in this publication (see Chapter 2) but include the following main developments (Steinfeld et al., 2006) all of which are taking place simultaneously:

- An increased human population that has created a fast-growing demand for animal proteins in developing economies
- Human migration toward urban areas and the associated development of informal urban livestock production in developing economies
- Changes in the global distribution of livestock production from traditional producers like the United States and the European Union to East and South Asia (China and India) and Latin America (Brazil)
- Changes in production from extensive to more intensive systems, relying on imported feed supplies combined with new management practices in order to supply a growing demand for animal products of the quality required by urban retailers
- Marginalization of pastoralists and other smallholders in less productive zones.

In addition, the last decade has seen an increased demand for and production of crop- derived biofuel to help mitigate climate change. Added to the existing competition between humans and animals for feed grains, this is putting additional pressure on land and bringing new land into agricultural use, resulting in closer contact between livestock and wildlife. At the same time, the impacts of climate change on livestock production are threefold, affecting the distribution of crop and livestock production systems, the water supply, and the distribution of disease pathogens.

Social disruption induced by war adds a further dimension to disease spread when people migrate and take their animals with them. Continuous conflict in some areas has led to massive movements of livestock and populations across country borders and contributed to unstable livestock production systems as well as to total collapse of basic services for people and livestock, including food insecurity. International regulations are not adapted to this mobility and may contribute to pushing livestock and products trade into illegality.

These various changes in the ecology of livestock systems have in different ways been associated with the resurgence of classic zoonotic diseases, such as bovine tuberculosis, as well as the emergence of new diseases such as the H5N1 highly pathogenic avian influenza virus. Some of these main features of the changing livestock sector are elaborated in this chapter.

Urbanization

People moving to urban centers in various developing regions of the world generally keep some small stock (poultry, goats, pigs) for security in the face of uncertainty. Urban livestock production is associated with high densities of people and close contact with animals, facilitating the transmission of pathogens. The lack of waste management attracts wild animals (rodents, birds) and endangers the supply of potable water.

In addition, the increased demand for food in urban and periurban centers creates opportunities for the growth of commercial production systems. As animal health systems are generally not the first priority when production is extended, appropriate disease control mechanisms are not in place, so growth often occurs in conditions with minimal regulations to preserve animal and human health and environmental quality. As a result, fast-growing commercial livestock production is often associated with poor biosecurity and inadequate waste management; processing and marketing systems that lack food safety precautions are generally missing, so there is an increased risk for food- and waterborne diseases.

Globalization

Changes in the global distribution of livestock production from traditional producing countries have been associated with long global market chains for livestock products, creating the potential for animal and human health hazards to be transferred over long distances. When livestock production is transferred to new regions, there is an additional risk that this new agroecological environment will provide an opportunity for the emergence of diseases, including zoonoses.

Intensification

Intensification refers to changes in the production systems from extensive to more intensive systems, relying on imported feed supplies and new management practices. Intensification without careful management can lead to human health hazards at livestock production sites associated with air quality and the pathogen load in the environment (as discussed later in this chapter). It can also encourage the emergence of new pathogens that may be zoonotic (e.g., H5N1). In addition, lack of infrastructure for milk and meat processing, including slaughtering, may lead to contamination of milk and meat, creating food safety hazards.

Growing and intensifying livestock sectors tend to be characterized by periods when significant portions of the population buy and sell their animal products through informal markets, which, by definition, escape regulation. There are often inadequate food processing, hygiene, and management protocols. The poor hygiene of processed livestock products is a result of the low quality of the raw products and developing countries' lack of infrastructure and capacity for risk analysis and quality control. The lack of easy and rapid diagnosis, weak governmental surveillance and control capacities, trade systems, food processing and eating habits, lack of awareness, and inappropriate information on zoonoses all further exacerbate zoonotic infection risk of the population.

Competition for Natural Resources

Where forest and rangeland are opened up to animal agriculture, or where hunting is a common source of meat or recreation, there is closer contact between wildlife, people, and domestic animals, with opportunities to exchange pathogens and amplify the load of pathogens and the pathogenicity of disease agents. When production and marketing chains for livestock products become physically longer, even crossing national boundaries, trade offers a convenient way for pathogens to move over large distances in a short time.

As people, livestock production, and wildlife become more closely associated in landscapes, pathogens have many opportunities to move among these sectors, creating public health hazards and the opportunity for the emergence and evolution of diseases in new ecological environments. Institutions dealing with consumer protection and health rarely keep pace with the changes in production systems. This may be one reason for the ongoing Highly Pathogenic Avian Influenza (HPAI) panzootic, where old forms of poultry production systems are mixed with new ones in a sector that is mainly privately organized and where poultry health systems are only accessible to large commercial producers (Schwabenbauer and Rushton 2007, Slingenbergh and Gilbert 2007).

Global warming and unusual weather conditions like heavy rainfall can lead to a new distribution of vectors and diseases, with the emergence and persistence of new pathogens or the spread of pathogens to new areas. The spread of bluetongue in northern Europe suggests that even a small shift in climatic conditions may aid the establishment of a disease agent. The bluetongue virus was known to have *Culicoides imola,* a midge insect, as a main vector. As this vector does not exist in northern Europe, the virus was only able to survive in the region when *C. obsoletus* and other native midges became more strongly involved in maintaining it. A particularly hot summer in the first year that bluetongue virus, type 8 (BTV8) was present in Europe may have facilitated this establishment.

The combination of new environments and climate change poses a challenge to human and animal health that is not yet adequately addressed by traditional health systems.

Economic Consequences of Zoonotic Disease

The impacts of zoonotic diseases are felt in the public health sector (via disease burden and costs of treatment), the private sector (including out-of-pocket expenditures and opportunity costs), the livestock sector (reduced productivity and market losses), and the environmental sector (disease prevention with failure to consider risk zones infected by zoonotic diseases like trypanosomiasis or anthrax).

The social and economic costs of zoonoses are substantial. Assessments of the burden of viral and parasitic diseases show high costs for society (Carabin et al., 2005, Knobel et al., 2005, Budke et al. 2006). The Sanitary and Phytosanitary Measures of the World Trade Organization have a direct impact on national and global economies when meat trade is reduced or displaced because of embargoes placed on export from countries with notifiable livestock diseases. When these diseases are known to be transferable to human beings (e.g., HPAI or bovine spongiform encephalopathy [BSE]), there can be an additional market shock effect when consumers decide not to consume certain livestock products.

When the diseases have an associated human pandemic risk, such as HPAI, control measures at the source can be particularly severe. In 2003 the Food and Agriculture Organization of the United Nations (FAO) reported that one-third of the global meat trade was embargoed due to livestock disease epidemics, which included Avian Influenza and BSE (Karesh et al. 2005). 44 million birds were culled in Viet Nam at an estimated cost of more than $120 million to stop the spread of avian flu (World Bank 2006).

The costs of zoonoses fall on both animal and human health systems. For example, the total cost of an average rabies postexposure prophylaxis per individual varies depending on the country, from less than US$50.00 to several hundred dollars.

Zoonoses are among the most seriously underdiagnosed of all human diseases (WHO/DFID-AHP, 2006), and therefore their economic impacts are also underestimated. Some zoonoses fall into the group of more neglected zoonoses, such as brucellosis or tuberculosis, which tend to be undetected, unreported, and uncontrolled (FAO 2005). Some information on these is available, however. The estimated global human burden of echinococcosis (tapeworm) may be as high as 1 million DALYs; this equates to a loss of $765 million annually. In Tanzania, rabies is estimated to be responsible for the loss of 42,669 DALYs per year.

Zoonoses management and control is not limited to the livestock sector. It has tremendous implications for public health, wildlife and biodiversity, international and tourism. There is evidence of the economic benefit in all sectors and the cost-effectiveness of zoonoses control (Knobler et al., 2004, Kahn 2006).

For the reduction of 52% brucellosis transmission between cattle and small ruminants through large-scale vaccination programs, the estimated intervention costs are $8.3 million with the overall benefit of $26.6 million (see Figure 11.3) (Zinsstag et al., 2003). If the costs of the intervention are shared between the sectors proportionally to the benefit to each, the public health sector should contribute to the intervention cost. If the costs of livestock brucellosis vaccination are allocated proportionally relative to all benefits, the intervention becomes profitable and cost-effective for both the agricultural and the health sectors.

Control and elimination of zoonoses is most effective with interventions applied to the animal reservoir (Zinsstag et al., 2005b). Complex and multisectoral implications of zoonoses call for a comprehensive interdisciplinary and intersectoral evaluation and implementation of control options. Further evidence has confirmed that interventions to control zoonoses, particularly in extensive production systems, are highly cost-effective when considered from a societal point of view. This could be an area where targeted interventions have an enormous potential for poverty alleviation, provided that it improves livestock production and human health.

In many developing countries, large rural populations, especially children and women, are not only at higher risk for zoonotic diseases by their close contact with livestock, they also become more vulnerable to poverty with the loss of livestock productivity and their marginalization from health social services such as systematic disease control (Perry et al., 2002).

Risk Perception

The perception of human health risk affects the way that both livestock owners and health systems deal with human health hazards of animal origin. These hazards are often overlooked when they fall between the public health and animal health systems or when, as previously suggested, their burden to both is underestimated. This is the case for diseases such as rabies, being transmitted by dogs, or Rift valley fever transmitted by infected ruminants, as well as for foodborne diseases like brucellosis or salmonellosis. This misperception of the problems often jeopardizes control programs for these diseases.

In addition, the risk perception for zoonoses often depends on the stage of economic development of a society. Where people encounter numbers of health problems, like diarrhea, malaria, or tuberculosis, and when they might rely mainly on livestock for their livelihoods, they may choose to accept the risk to themselves in return for the economic benefit derived from their animals. Consumers or retailers in richer countries, in contrast, may be risk-averse. At the same time, risk perception in

economically developed countries is not equally present: whereas consumers in Europe are very afraid of contracting BSE, the perception of risk for other foodborne diseases like campylobacteriosis or salmonellosis is generally low. This is a major challenge for risk mitigation and risk reduction strategies, and it can only be solved through a partnership between the public and private sectors.

Emerging and Reemerging Zoonoses in Extensive Livestock Production Systems

Characteristics and Dynamics of Extensive Production Systems

Extensive livestock production systems are those that make limited use of commercial inputs (Steinfeld and Mäki-Hokkonen 1995). However, they can be variously defined to include only those that use limited land—predominantly the transhumant and nomadic pastoralist systems in geographically remote rural zones where communities are "hard to reach" and live at the margins of the outreach of social services (see Figure 11.1)—or those in mixed crop–livestock farming that make little or no use of commercial feed, such as scavenging poultry that are housed only at night and roam freely during the day to find whatever feed they can.

The transhumant and nomadic systems have the following characteristics that affect the way in which diseases are spread through livestock populations and from livestock to humans:

- *Mobility*—In order to make the best use of scarce pasture and water, large animals in extensive systems are moved around, and live animals are moved and traded within and across international borders.
- *Distance from towns*—Transhumant and nomadic systems are often found in remote areas and increasingly in areas with scarce natural resources unsuitable for crops or intensive livestock. In such places, it is difficult to provide support services.

Human Health Hazards from Extensive Production Systems

Over the last few decades, rapid and major sociopolitical, economic, cultural, institutional, and ecological changes have put heavy and unprecedented pressures on many extensive livestock production systems. Two trends have implications for human disease risk.

The first is a transformation away from extensive and toward more intensive systems, with all of the animal and human health risks associated with that change. This is occurring most markedly through sedenterization of mobile livestock. Examples can be seen in pastoral zones of West Africa, the Horn of Africa, and Central Asia, where livestock keeping is often restricted to subsistence, mobile monetary capital, and social prestige. In

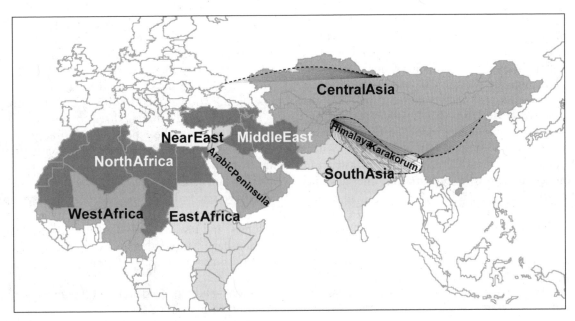

Figure 11.1. Distribution of nomadic pastoralists.
Source: Adapted from Scholz 1991.

recent decades, however, production systems have experienced important transformations, with a shift from resource-driven (access to resources) to demand-driven (access to market) actions and with dynamic actors' responses. In developing countries, former mobile populations are settling, and intensification of production occurs as they have fewer options. (See Figure 11.2.) The impacts of zoonotic diseases in intensive systems are discussed later in this chapter.

The second trend is the continued marginalization of extensive systems into more remote and resource-scarce areas, where access to markets and services is limited. Examples can be seen in traditional African pastoral societies, mobile for the most part, that have supported sizable human populations under often severe environmental stress in the past (Majok and Schwabe 1996). However, in the Sahelian zone, the equilibrium of cohabitation between pastoralists and farmers/fishers is

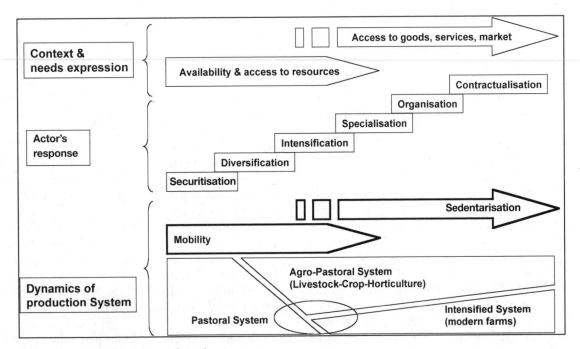

Figure 11.2. Transformation of production systems.
Source: Adapted from Djamen et al., 2005.

increasingly disturbed (Thébaud et al., 2001). The exclusion of pastoralists from more productive pastures (with higher agronomic potential) by farmers and the blocking of traditional transhumance routes lead to significant disruption of the annual transhumance cycle, increasing the ecological and economic vulnerability of pastoral systems in dryland Africa. Attempts to replace traditional systems with new production forms underestimated the efficiency of the traditional systems. Within these marginalized extensive areas, lack of infrastructure and resources increases the potential for the incidence and impact of zoonoses.

Examples of zoonotic diseases typical of extensive systems include brucellosis and anthrax, two of the more serious ones. These diseases are not exclusive to extensive systems, but they can be problematic in these cases.

Brucellosis in Livestock Extensive Production Systems

McDermott and Arimi (2002) stated that the available data on brucellosis in humans and livestock are limited. The highest incidences of brucellosis are found in pastoral production systems (e.g., 0.5% in Kyrgyzstan).

In Mongolia, after democratic reform and the shift away from dependence on the former Soviet Union in 1990, human brucellosis reemerged as a major, although preventable, important morbidity. Central Asian countries have a very high prevalence of brucellosis (Lundervold et al., 2004). The World Health Organization (WHO) raised the question of whether, from a health sector perspective, mass animal vaccinations or other control efforts such as culling are cost-effective. The economic benefits, cost-effectiveness, and distribution of benefit of mass vaccination against brucellosis in cattle and small ruminants were estimated in light of societal, public health, and agricultural sector costs and benefits by an animal–human transmission model linked to an economic assessment (Zinsstag et al., 2005b). If costs of livestock brucellosis vaccination are allocated proportionally to all benefits, the intervention may become profitable and cost-effective for both the agricultural and the public health sectors (Roth et al., 2003). (See Figure 11.3.)

Anthrax in Extensive Systems

The loss or limited use of pastures due to increased population pressure, degradation, or desertification forces communities to use lower-quality zones, such as those contaminated with anthrax spores. These are present only as a result of previous livestock cases, or perhaps wild animal cases, but the highest contamination rates come with high stocking rates. For example, Tajikistan has a hyperendemic/epidemic status for anthrax (WHO 2005). After agriculture reorganization occurred in 1993, within a few years the private sector gained ownership of 90% of livestock. Almost 2274 cases of anthrax in livestock and 1687 cases in humans were confirmed during the period from 1992 to 2003. A significant number (1139) of these human cases occurred in southeast Tajikistan, namely, Khatlon, Rasht, and Pamir (Navruzshoyeva 2005), where areas contaminated by anthrax from earlier outbreaks are used for agriculture and livestock production as land has become more scarce.

For increasingly marginalized communities, zoonoses and food safety can be expected to become growing problems since people in these communities are often suffering from malnutrition, which exacerbates the effects of diseases and causes people to eat meat they might otherwise have rejected. But one encouraging development is that intervention trials involving training, simple technologies, and institutional innovations such as self-certification have had good results in improving food safety and consumer confidence (Hatanka et al., 2005).

Livestock–Human–Wildlife Interface and Emergence of Zoonoses

The reduction of land available for free-range livestock production systems, leading to growing interaction between wildlife, livestock, and humans, increases the possibilities for transfer of disease between these populations. This interface is flexible and not well defined. It can move temporally and spatially depending on various factors, and new interfaces are constantly created within the system as populations shift with resource availability.

Examples have included the transmission of Nipah, Hendra, SARS, and Ebola viruses to humans involving intermediate amplification hosts such as pigs, horses, civets, and primates, respectively. Understanding the biology of the natural reservoir(s), amplifying host(s), modes of disease persistence, and transmission as well as identifying at-risk human populations are crucial in the control of emerging zoonoses. Vertebrate animals (including humans) are reservoirs for zoonotic infection, and disease agents (bacterial, rickettsial, viral, parasitic, and fungal) are transmitted directly or indirectly between them.

Furthermore, escalating climate change is altering the distribution and risk of vectorborne zoonoses (Harrus and Baneth 2005). Rift Valley fever is an example of major human diseases that can be expected to be influenced by climate change (Jouan et al., 1990, Digoutte and Peters, 1989). Mauritania, in proximity to the Senegal River and in particular to the Diama and Manantali dams, was considered potentially an important risk of amplification of the virus in relation to the migration of people, livestock, and wildlife. Over a thousand Rift Valley fever cases occurred in the town of Rosso alone, with an estimated 47 deaths.

The following factors influence the creation of a physical wildlife–livestock–human interaction:

- Livestock production systems prevailing in the region or in the country (e.g., extensive ranching system, pastoral, agrosylvopastoral)

- Social factors, such as population growth and movements, population displacements and migrations, changing population behavior patterns, conflicts and political instability, and food habits (e.g., consumption of bush meat and raw products)
- Environmental factors, such as land degradation, habitat destruction, deforestation or reforestation, decrease of natural resources, and climatic factors (e.g., drought, heavy rainfalls)
- Biological factors, such as present wildlife species, seasonal population migration, and wildlife breeding.

Deficiencies in Support Services to Extensive Production

In the areas where extensive production systems prevail, there are often inadequate support services (feed and water supply, veterinary services, waste management, etc.) due to the lack of infrastructure and the large distances that need to be covered to reach scattered populations of animals. Zoonoses are often overlooked because they fall in the gap between veterinary and public health. Veterinary and public health services historically have worked independently, and the lack of cooperation has contributed to inadequate attention to and ineffective control of zoonoses. In addition, the marketing of livestock products often follows the traditional pattern through live animal markets and informal butcheries.

Many countries, especially those with resource constraints, lack information on the burden and the distribution of zoonotic diseases. Risks for zoonoses are considered negligible compared with those for diseases of higher consequence because the societal consequences of zoonoses are not recognized by specific sectors. For example, outbreaks of Rift Valley fever among humans in Mauritania were mistakenly identified as yellow fever. The correct diagnosis was made only after public health services contacted livestock services, which informed them of abortions in cattle, a common occurrence with these outbreaks (Digoutte 1999).

In resource-limited and transitioning countries, many zoonoses are not controlled effectively because adequate policies and funding are lacking. However, transmission of zoonoses to humans can already be greatly reduced by health information and changes in behavior. Authorities in Kyrgyzstan, for example, have started an information campaign to reduce brucellosis transmission to small-ruminant herders by encouraging them to wear gloves for lambing and to boil milk before drinking it. Interventions in livestock should always be accompanied by mass information, education, and communication programs (Zinsstag et al., 2007). Zoonoses in livestock and humans also go undetected and unreported due to lack of appropriate diagnostic tools, such as tests that can be used in poorly equipped laboratories, and to the lack of skilled technicians.

In many developing and transition countries where extensive systems prevail, public health surveillance is not well designed and lacks the human resources and financial means to respond adequately to existing and emerging zoonoses. Reporting and control are threatened by a chronic lack of reliable and updated data specific to livestock populations and zoonoses. Insufficient systems for comprehensive disease surveillance and weak health systems in general combined with resource constraints lead to poor awareness and information on old as well as emerging diseases. This is usually due to the scale of priorities in the veterinary and public health sectors.

Where vaccines against zoonoses (e.g., anthrax) are available, the quality of livestock vaccines is not always ensured. In some instances there is a lack of independent and regular quality control in veterinary vaccine production facilities of low-income countries. In addition, due to the weak health services and the lack of resources, vaccine delivery is not always ensured. Impaired safety and efficacy of livestock vaccines have a direct impact on the income of livestock holders and, for vaccines against zoonotic diseases, also on their health.

One important outcome of the Pan African Program for the Control of Epizootics is the revitalization of the Pan African Vaccine Centre in Ethiopia, which, among other tasks, has the mission of international independent quality control of veterinary vaccines in Africa. There is an urgent need to enhance capacities in these countries and subsequently to connect various surveillance and early warning, alert, and response systems at the regional and international levels.

It is unrealistic to assume that in most countries delivery of veterinary services and food-safety regulations in extensive production systems can reach international trade standards in the short or medium term. Simpler procedures, such as the application of Hazard Analysis and Critical Control Point (FAO 2006), and zoning and compartmentalization (OIE 2009) of easy and accurate diagnostic tools and other processing techniques are commendable but must be adapted to local demands.

The Effect of Behavior and Perception on Zoonosis Control in Extensive Systems

In communities that practice extensive livestock keeping, close housing with livestock is not uncommon for social and security reasons. Aerosol transmission from animals may occur as a result of close contact with infected animals. Manipulation of abortion materials and weak lambs (e.g., brucellosis in Mongolia), eating raw and undercooked meat, and drinking unpasteurized milk (in Central Asia, West Africa, and the Horn of Africa) also exacerbate the transmission of bovine tuberculosis, Q-fever, and brucellosis. Being a camel breeder was a risk factor for human Q-fever seropositivity, since 80% of camels were seropositive (Schelling 2003). For almost all pastoral populations, three zoonoses are considered constantly endemic and are determined also by human behavior: echinococcosis, brucellosis, and rabies. Other zoonotic infections are also important, including trypanosomiasis and leishmaniasis in Africa and plague in Asia (Zinsstag et al., 2006).

Important public health diseases may not be perceived as such by a community. Pastoralist groups generally have no concept of diseases that are transmitted from animals to people (zoonoses), with the exception of anthrax. Treatment-seeking behaviors may be influenced by cultural norms—for example, the Fulani concept of *pulaaku*, meaning a high degree of self-control, results in them using health services only at an advanced stage of disease (Krönke 2001).

The environment of livestock production is changing rapidly, and developing countries have not been able to respond with improved infrastructure and new or revised initiatives supporting veterinary public health resources.

Systems in Transition

Systems in transition between extensive/low input and intensive/high input pose a particular challenge for human health. One reason, although not the only one, is that livestock and their products tend to be sold in uncontrolled and informal markets favored by low incentives for quality products, while producers and poor consumers have poor access to regulated markets or choose not to obtain them because of cost. Marketing of livestock products through informal markets often carries food safety risks (Omore et al., 2001, Bonfoh et al., 2006), exposing consumers to hazards including *Brucella*, *Mycobacterium tuberculosis* complex, *Cysticerca*, *Cryptosporidium*, *Escherichia coli*, antibiotic residues, aflatoxins, *Listeria*, *Staphylococcus*, *Streptococcus*, and *Salmonella*.

Human Health Hazards Associated with Intensive Livestock Production

Intensive livestock management generally means high-performing animals, indoor stock-keeping all year-round, high animal densities per square meter of housing, high-energy feedstuff, a high degree of mechanization and automation (e.g., in feeding, water supply, manure removal, and ventilation), a low physical labor requirement, and often a small air volume in relation to the number of animals in the housing unit. Typical examples of intensive management are found in high-yielding dairy cow herds and in the sectors of poultry and pig production. These environments can create hazards for humans, two of which are examined here: the effects of poor air quality on people working with intensive livestock and the problems of antimicrobial resistance created by incorrect use of antibiotics.

Occupational Health Risks for Humans Working in Animal Houses

Health risks associated with dust are a major hazard to humans working with intensive livestock. Numerous studies have demonstrated links between dust and human health in a number of livestock-related industries (Donham 1995). A survey of 69 full-time poultry stockmen found that 20% were exposed to levels of dust 2.5 times the figure of 10 mg/m recommended under occupational health and safety guidelines (Whyte et al., 1994). Findings such as these have led to the introduction of strict codes to protect people involved in the intensive livestock industries in several countries, including Denmark and Sweden. Guidelines have also been recommended for the Australian pig industry (Jackson and Mahon 1995).

The first reports indicating health hazards for humans working in intensive livestock production systems were published over 30 years ago (Donham et al., 1977). A number of syndromes have been recognized in such individuals. They range from an acute syndrome that develops within a few hours to days of exposure to animal sheds and that is accompanied by a variety of clinical signs ranging from lethargy, a mild febrile reaction, headaches, joint and muscle aches, and general malaise to more chronic responses. In some cases, the initial attack is so severe that the employee terminates employment within a matter of days. In general, episodes last 12 to 48 hours, with chronic fatigue and congested respiratory passages being the most common clinical signs. The condition has been referred to as organic dust toxic syndrome (ODTS) or toxic alveolitis. ODTS has been quoted as affecting from 10 to 30% of workers, depending on the type of intensive animal production and the facilities used (Donham 1995).

Exposure to dust produces a variety of clinical responses in individuals. These include occupational asthma due to sensitization to allergens in the airspace and to airborne-resistant bacteria and resistance determinants (Keokuk County Study, pers. comm., David Wallinga) chronic bronchitis, chronic airways obstructive syndrome, allergic alveolitis, and ODTS (Iversen 1999). A range of acute respiratory symptoms, described by employees following contact with their work environment but not necessarily associated with a generalized clinical syndrome, have also been documented (Brouwer et al., 1986). In several studies in North America, Sweden, and Canada, the prevalence of acute symptoms was found to be 1.5 to 2.0 times higher than chronic symptoms. In a similar study in the Netherlands, however, the prevalences were reported to be similar (Brouwer et al., 1986). The suggestion that the primary clinical problem is an obstruction of the airways is supported by various studies in which workers have been subjected to lung function tests (e.g., Haglind and Rylander 1987).

Exposure to bioaerosols has also been shown to cause a bronchoconstriction, hyperresponsiveness, and increased inflammatory cells in bronchoalveolar lavage fluids in naive subjects (Malberg and Larsson 1992). It is assumed that bronchoconstriction followed by reduced ventilation of the lungs can be caused by inhaled endotoxin (ET). Experiments using nasal lavage show that pig house dust containing different concentrations of ETs increases the inflammatory reaction of the nasal mucous membranes of humans (Nowak et al., 1994). The bronchoconstrictive effects of bioaerosols have also been demonstrated in guinea pigs (Zuskin et al., 1991) as well as stockpersons in Sweden and North America (Donham 1995).

The number of farmers and employees complaining about respiratory symptoms during and after work in animal houses has risen in recent years. The number of obstructive airway diseases caused by allergic compounds rose from about 90 in 1981 to approximately 700 in 1994; a slightly smaller increase from 8 to 50 was observed for obstructive diseases caused by chemical irritants or toxic compounds (according to the Agricultural Occupation Health Board in Lower Saxony, Germany, 1996). In a study of 1861 farmers in north Germany, about 22% of the pig farmers, 17% of the cattle farmers, and 13% of the poultry farmers displayed airway problems (Nowak 1998). (See Table 11.1.) Although the causes of the relatively high incidence of health problems associated particularly with pig farming are not yet completely understood, it seems that factors like high concentrations of air pollutants, the composition of pig house bioaerosols, insufficient ventilation, and poor system management may play a role. The results may also be biased by the fact that most pig farmers in Germany work on their own farms, which they do not easily abandon, even in the case of health problems.

Air Pollutants Found in Farm Animal Houses

Providing a safe and healthy work environment for employees is an important aspect of any industry—including animal farming (Cargill and Hartung 2001). Modern animal production systems are, however, increasingly regarded as a source of air pollutants that can be harmful for farmers. There is epidemiological evidence that the health of farmers working in animal houses may be harmed by regular exposure to air pollutants like ammonia, dust, microorganisms, and endotoxins (Donham 1987, Whyte et al. 1994, Donham 1995, Radon et al. 2002, Hartung 2005).

Further studies are required to understand the building features and animal husbandry practices that increase the concentration of airborne pollutants in buildings housing animals and to determine the key pollutants involved. The evidence collected in buildings housing pigs suggests that issues such as hygiene and stocking density (kg biomass/m) are key factors but that the composition of pollutants or bioaerosols may vary significantly from shed to shed, depending on a range of factors (Banhazi et al., 2000). These include hygiene, diet, and the type of bedding and effluent disposal system used. The composition of bioaerosols might be more important than just the concentration of airborne particles within an animal house atmosphere for the severity of specific occupational health problems.

The key pollutants recognized in the airspace of pig houses are listed in Table 11.2. Under commercial production conditions the airborne particles will contain a mixture of biological material from a range of sources, with bacteria, toxins, gases, and volatile organic compounds adsorbed to them. Hence a more descriptive term for these airborne particles is bioaerosol (Cargill and Banhazi, 1998). The typical character of bioaerosols is that they may affect living things through infectivity, allergenicity, toxicity, and pharmacological or other processes. Their sizes can range from aerodynamic diameters of 0.5 to 100 μm (Hirst 1995, Seedorf and Hartung 2002).

Dust

Dust is the most visible part of the bioaerosol. In animal houses it originates from the feed, the bedding material, and the animals themselves. A small amount enters the

Table 11.1. Frequency of workplace-related respiratory symptoms in livestock farmers/employees in Lower Saxony, Germany

Animal species		Number of persons	Share of complaints (%)
Pig	Sow	619	22.7
	Fattening	799	21.9
	Weaner	551	23.0
Cattle	Cow	1,245	17.4
	Beef	895	17.2
	Calf	1,190	17.8
Laying hen Broiler		279	14.7
		47	12.8

Source: Nowak (1998).

Table 11.2. Common air pollutants in pig houses in Denmark, Germany, The Netherlands, and the United Kingdom

Gases	Ammonia, hydrogen sulphide, carbon monoxide, carbon dioxide, 136 trace gases, osmogens
Bacteria/fungi	100 bis 1000 cfu/L air
	80% staphylococcaceae/streptocococcaceae
Dust	e.g., 10 mg/m³ inhalable dust
	organic matter approx. 90 %
Endotoxin	e.g., 2 µg/m³ in piggeries

animal house with the incoming ventilation air. The dust particles are carriers for gases, microorganisms, endotoxins, and various other substances such as skin cells and manure particles (Donham 1989). Animal house dust consists of up to 90% of organic matter (Aengst 1984).

The amount of airborne dust fluctuates greatly in the course of a day and according to the type of animal. Investigations carried out in 329 animal houses in four different European Union countries (Denmark, Germany, The Netherlands, and the United Kingdom) revealed the dust concentrations given in Table 11.3. The results represent 24 hours of mean values for inhalable and respirable dust (Takai et al., 1998). The highest dust concentrations are found in poultry housing, followed by pig and cattle.

The health effects of dust particles depend very much on the nature of the dust (organic, not organic), the compounds the particles are carrying (bacteria, toxins), and the diameter of the particles. Particles with aerodynamic diameters smaller than 5 µm can penetrate deep into the lung. The larger particles are deposited in the upper airways. High dust concentrations can irritate the mucous membranes and overload the lung clearance mechanisms.

Microorganisms

Along with dust particles, microorganisms can be transported into the respiratory system, causing infections. The composition of the respirable particles in animal houses is associated with compounds such as dried dung and urine, skin dander, and undigested feed. The majority of bacteria found in shed airspace have been identified as gram-positive organisms, with *Staphylococcus* spp. predominating. In piggeries a considerable part of the airborne bacteria can also be gut-associated or fecal bacteria (which are often gram-negative). Table 11.4 gives concentration values for bacteria and fungi measured in the air of different sow and fattener units. The total bacteria count does not differ very much between the production systems. The amount of enterobacteria in relation to the total bacteria count is distinctly lower in the sow houses (2.6 %) than in the fattener units (7.2 %). The samplings took place during the daytime; during the night, the concentrations can be lower depending on the activity of the animals and the efficiency of the ventilation system (Seedorf et al., 1998).

A considerable number of pathogens and zoonotic agents are also present in the air of intensive animal production units. Table 11.5 shows that 12 bacteria, 2 fungi, and 4 viruses identified in pig and poultry house air have zoonotic properties.

Microorganisms and endotoxins belong to the aerial pollutants that have been linked with several production diseases (Hartung 1994, Wathes 1994) and that are assumed to pose a specific risk to the health of farmers and workers in relation to allergic–toxic reactions of the respiratory tract (Donham 1990) and to the residential areas around intensive livestock enterprises. Concentrations of airborne microorganisms are particularly high in pig and poultry houses, where they can reach about 6.43 log CFU per m³ air (Clark et al., 1983, Erwerth et al. 1983, Cormier et al. 1990, Seedorf et al. 1998).

Table 11.3. Mean dust concentrations in the air of livestock housings, mg/m³, *n* = 329

Animal species	Inhalable dust	Respirable dust
Beef	0.15–1.01	0.04–0.09
Calves	0.26–0.33	0.03–0.08
Cows	0.10–1.22	0.03–0.17
Fattening pigs	1.21–2.67	0.10–0.29
Sows	0.63–3.49	0.09–0.46
Piglets	2.80–5.50	0.15–0.43
Broilers	3.83–10.4	0.42–1.14
Laying hens	0.75–8.78	0.03–1.26

Source: Takai et al. (1998).

Table 11.4. Airborne bacteria concentrations in sow and fattener units

cfu/m³	Sows	Fatteners
Total bacteria	139,972 (9)	139,500 (4)
Enterobacteriaceae	3,645 (8)	10,005 (4)
Fungi	8,940 (8)	5,417 (4)

Source: From Seedorf et al. (1998), modified in parentheses (number of samples).

Endotoxins

Endotoxins can trigger allergic reactions in the airways of susceptible humans, even in low concentrations. Where antibiotics are commonly used, dust can also contain considerable amounts of antibiotics, up to total concentrations as high as 12.5 mg/kg dust, which may have the potential to trigger antibiotic resistance problems when inhaled (Hamscher et al., 2003).

Results from Seedorf et al. (1998) reveal highest concentrations in pig and poultry barns. For inhalable ETs, mean concentrations in cattle barns range between 74 and 639 endotoxin units (EU) m^{-3} and for respirable ET, between 6 and 67 EU m^{-3}. In pig barns the respective values range between 523 and 1865 EU m^{-3} and between 74 and 189 EU m^{-3} (respirable dust). Concentrations were highest for poultry; mean values ranged between 3389 and 8604 EU m^{-3} in inhalable dust fractions and from 96 to 581 EU m^{-3} in respirable dust. These reported concentrations are far beyond an earlier proposed occupational health limit of 50 EU m^{-3}, which was never enacted.

Influence of Straw Litter

Surveys of straw-based shelters have shown that the concentrations of airborne particles and viable bacteria are generally higher than found in nonstraw production systems. In a survey by Banhazi et al. (2000), the concentrations of respirable particles and airborne viable bacteria in straw-based shelters housing pigs were around three times higher than in naturally ventilated sheds. (See Table 11.6.) Similarly, concentrations of ETs and total airborne particles were also higher in straw-based systems.

Strategies to Minimize the Risk of Air Pollutants for Employees and Animals

Several approaches aimed at protecting employees on the job are available. These include wearing protective gear, reducing exposure levels within the buildings, and eliminating pollutants at the source. Readers are directed elsewhere for a recent survey on protective gear (Anonymous 2005).

Various strategies have been recommended for reducing the concentrations of airborne pollutants in animal houses. These include management measures as well as strict hygienic rules and direct reduction techniques such as fogging sheds with oil and water (Pedersen 1998, Banhazi et al., 1999). All these methods have to be investigated carefully to see whether they have side effects on

Table 11.5. Pathogens including zoonotic agents (Z) known to be transmitted aerially in pig and poultry houses.

Bacteria

Bordetella bronchiseptica	*Haemophilus parasuis*	*Mycoplasma hyopneumoniae*
Brucella suis (**Z**)	*Haemophilus pleuropneumoniae*	*Pasteurella multocida* (**Z**)
Chlamydia psittaci (**Z**)	*Listeria monocytogenes* (**Z**)	*Pasteurella pseudotuberculosis*
Corynebacterium equi	*Leptospira pomona* (**Z**)	*Salmonella pullorum* (**Z**)
Erysipelothrix rhusiopathiae (**Z**)	*Mycobacterium avium* (**Z**)	*Salmonella typhimurium* (**Z**)
Escherichia coli (VTEC) (**Z**)	*Mycobacterium tuberculosis* (**Z**)	*Staphylococcus aureus*
Haemophilus gallinarum	*Mycoplasma gallisepticum*	*Streptococcus suis* (**Z**)

Fungi

Aspergillus flavus	*Aspergillus nidulans*	*Cryptococcus neoformans* (**Z**)
Aspergillus fumigatus	*Coccidioides immitis*	*Histoplasma farciminosum* (**Z**)
Aspergillus niger		

Viruses

African Swine Fever-like Virus (ASFV)	Avian infectious bronchitis virus (AIBV)	Newcastle Disease Virus (**Z**)
Avian Leukose Virus (ALV)	Infektiöses Laryngotracheitis Virus	Porcines respirat. Coronavirus
Avian Influenza A-Virus (**Z**)	Herpesvirus 2 (chicken)	Swine Influenza Virus (SIV)
Avian infectious bronchitis virus (AIBV)	Foot and Mouth Disease (FMD) Virus (**Z**)	Teschen/Talfan Disease Virus
	Swine Herpes Virus 1 (**Z**)	Hog cholera Virus (HCV)
	Swine Herpes-Virus 2	

Source: After Wathes (1995) and Schulz (2008).

Table 11.6. Concentrations of respirable and total particles, airborne bacteria, and endotoxins in three production systems

System	Respirable particles (mg/m³)	Viable bacteria (× 1000 cfu/m³)	Total particles (mg/m³)	Endotoxin (EU/m³)
Natural ventilation	0.23 ± 19	99 ± 67	1.65 ± 1.64	29 ± 30
Straw based	0.68 ± 67	350 ± 181	2.58 ± 2.18	89 ± 71
Mechanical ventilation	0.30 ± 29	89 ± 42	2.22 ± 1.29	42 ± 33

Source: Banhazi et al. (2000).

the animals, the environment, or the meat quality (Cargill and Hartung 2001).

Reducing air pollutants in livestock housing will provide a safer and healthier work environment for employees and a better atmosphere for the animals, resulting in improved health, welfare, and performance. A future-oriented sustainable farm animal production should enhance consumer protection, economy, and the environment in addition to addressing aspects of animal welfare and occupational health.

Antimicrobial Resistance

With the global epidemic of antimicrobial resistance, increasing numbers of human illnesses are approaching the point of being untreatable with current medicines. Two main drivers of resistance are the extent of antimicrobial use and the prevalence of resistance genes (Levy 1998).

Global resistance is affected by both human and nonhuman uses of antimicrobials (OIE/WHO/FAO 2004a). Among antimicrobials—antivirals, antiparasitics, antibacterials, and the like—public health concern mostly focuses on resistance to antibiotics, especially resistance developing and spreading among bacteria that are commensal or human pathogens. Resistance in human pathogens is largely the consequence of use in humans and in terrestrial animal agriculture (OIE/WHO/FAO 2006).

Both medical and veterinary uses of antibiotics can create selection pressure resulting in the appearance of resistant strains of bacteria. The presence of an antibiotic may kill most of the bacteria in an environment. But resistant survivors can eventually reestablish themselves, passing on their resistance genes to offspring and, often, to other species of bacteria. Resistance genes give rise to proteins that protect bacteria from antibiotics.

Bacteria are "promiscuous." Microbiologist Stuart Levy writes, "the exchange of genes is so pervasive that the entire bacterial world can be thought of as one huge multicellular organism in which the cells interchange their genes with ease" (Levy 1998). Resistance in pathogenic bacteria is important because it leads to harder-to-treat infections, with more morbidity and mortality, as well as to a greater incidence of infections. But even resistance among nonpathogens can be dangerous because these bacteria can transfer their antibiotic resistance genes to disease-causing bacteria. Hospitals, communities, the

human gut, farms, and farm animals can all be viewed as a single bacterial ecosystem. The genetic determinants of resistance are disseminated throughout, respecting neither geographic borders nor phylogenetic boundaries (Levy 1998).

With their propensity to exchange DNA, and with generation times as short as 15 to 20 minutes under some conditions, bacteria are incredibly adaptive. Gene-based antibiotic resistance therefore can evolve quickly and can spread rapidly worldwide (Aarestrup et al., 2008). It can take only months to years to go from the widespread adoption of new antibiotics to the appearance of clinically significant resistance (Walsh 2000).

Animals and humans constitute overlapping "ecological" reservoirs of resistance (Wegener 2003). Human pathogens, for example, share approximately half of the resistance genes identified in fish pathogens (OIE/WHO/FAO 2006). Thus antibiotic use in food animals can contribute to resistance developing and being disseminated in other contexts such as in human medicine (OIE/WHO/FAO 2004a). In addition, trade in food animals, animal feed, and food of animal origin is global, which means the occurrence of antibiotic resistance in one country's food supply quite possibly will pose a problem for all countries. Global initiatives across human and animal disciplines as well as across borders are therefore necessary for monitoring and controlling antibiotic resistance and overuse (Aarestrup et al., 2008).

Antibiotic Use in Animal Agriculture

Very large quantities of antibiotics are used in current food animal production (Aarestrup et al., 2008). Indeed, antibiotic usage has been integral to many intensive livestock production systems. Food animals are typically confined in large groups, often at high animal density and in inappropriate environments. Inappropriate housing conditions, management, and handling of animals are the most important factors that contribute to the development of multifactorial diseases—infectious or noninfectious ones. The common response to these modern diseases is currently the massive application of antibiotics as feed additive or as prophylaxis. Meat-producing systems are designed to use breeds that can be raised quickly to slaughter weight, often before the animals reach physical maturity. Because facilities have been specialized for raising animals only to a particular age, the animals must be frequently moved as they grow and progress to maturity. The large groups and the high densities, animals' immaturity, and the frequent movement all predispose to the outbreaks and spread of infectious disease (Wegener 2003).

With the use of antibiotics, the development of resistance is simply a question of "when," not "if" (Walsh 2000). Antibiotics used for food animals are in many cases identical to or belong to the same classes as those used in human medicine, including many considered by the WHO to be "critically important" (Aarestrup et al., 2008). From the standpoint of efficiency, the delivery of antibiotics via animal feed or drinking water might be perceived as a practical solution, especially for systems raising poultry in very large numbers (McEwen and Fedorka-Cray 2002). These routes for delivering antibiotics often expose bacteria to sublethal levels, however. All these conditions favor the selection, persistence, and spread of antibiotic-resistant bacteria capable of causing infections in both animals and people.

Antibiotics in such systems can generally be used for treating diagnosed disease or for prophylaxis or prevention of disease, including in animals not yet sick, or to promote feed efficiency and growth. Antimicrobial use, including antibiotic use, is not well tracked in many countries, including the United States. Precise figures on the amounts of antibiotics used in humans and in animals therefore are often impossible to obtain. What is more definite is that in developed countries, animal production has accounted for a high share of total antibiotic use (Steinfeld et al., 2006). How high depends on the individual country and its oversight of antibiotic use in animal agriculture.

As an example, relying on U.S. sources as diverse as the Institute of Medicine of the National Academy of Sciences, nonprofit advocacy groups, and the animal pharmaceutical industry, it is possible to derive a consistent estimate of U.S. agricultural antimicrobial use in 2001 of approximately 20 to 30 million pounds (IOM 1998, Mellon and Fondriest 2001, Animal Health Institute 2002).

THERAPEUTIC USE AND METAPHYLAXIS
Antibiotics can be used therapeutically to treat individual animals with diagnosed disease, whether in extensive or intensive animal farming systems. Also common is treatment of large groups of animals as soon as clinical symptoms appear in a few of them. This is also considered therapeutic use, or, alternatively, *metaphylaxis* (McEwen and Fedorka-Cray 2002; Wegener 2003).

Organic animal production is a positive approach aiming to keep animals healthy and prevent disease from occurring in the first place. Avoiding the use of antibiotics is a priority in that system. Certified organic production in the United States prohibits any antibiotic use.

Sick animals in these systems will still receive treatment, of course, but the treated animal gets diverted into a noncertified production system (Wells 2003).

Antibiotics deemed "critically important" to human medicine are widely used for therapeutic reasons in food animals. For example, two fluoroquinolone (FQ) antibiotics, enrofloxacin and sarafloxacin, were approved by the U.S. Food and Drug Administration (FDA) in the mid-1990s for treating respiratory disease in poultry. The FQs were delivered via a flock's drinking water supply, typically after the first birds in the flock had become sick. FDA approval came over the objections of public health scientists who warned it likely would increase reservoirs of FQ resistance in the human population. That is exactly what happened. By 2000, after seeing dramatic increases in FQ resistance among *Campylobacter* bacteria in poultry, on poultry meat, and in the human population, the FDA had asked the manufacturers to voluntarily withdraw their products from the poultry market (FDA 2000). *Campylobacter* is the leading cause of foodborne illness in the United States, and ciprofloxacin—another FQ—is one of the primary treatments for severe cases of *Campylobacter* food poisoning.

While one manufacturer immediately removed its poultry product from market, the other refused to do so (FDA 2001b). By the time an administrative judge finally ruled in FDA's favor, five years later, the agency was estimating that therapeutic FQ use in poultry had resulted in 153,580 cases of FQ-resistant *Campylobacter* infections per year in the United States alone (FDA 2001a). In contrast, in Australia, where FQs were never approved for broad-scale use in animals, there has been an absence of FQ resistance in human *Campylobacter* infections (Unicomb LE et al., 2006).

Cephalosporins are another important example of a critical human antibiotic used in food animals (FDA 2003, WHO 2007). Cephalosporins, in particular the newer injectable third- and fourth-generation ones, are often the "last line of defense" in treating severe human infections—such as bacteremia or meningitis involving bacteria of food origin like *Salmonella* or *Escherichia coli*—that are resistant to other antibiotics (Collignon and Aarestrup 2007). It is particularly concerning, therefore, to see in these same pathogens rising levels of multidrug or extended spectrum beta (ß) lactamase (ESBL) resistance, including resistance to third and fourth generation cephalosporins (Helfand et al., 2006, Rodriguez-Bano 2006).

Beta lactamases are enzymes made by bacteria to destroy one or more antibiotics. Acquiring the ability to produce beta lactamase or ESBL enzymes is one way that bacteria develop antibiotic resistance. ESBLs break down an exceptionally wide array of antibiotics. In addition to cephalosporins, ESBL-carrying bacteria can often break down aminoglycosides, FQs, tetracyclines, chloramphenicol, and sulfamethoxazole-trimethoprim. Such broad resistance makes it more likely that a physician's first choice of an antibiotic to treat infection by ESBL-producing bacteria is likely to fail, possibly delaying more effective antibiotic therapy for days.

There are only minimal restrictions on food animal use of cephalosporins in industrial countries. Fewer controls in most developing countries make it likely that third- and fourth-generation cephalosporin use is even more widespread there (Collignon and Aarestrup 2007). In the United States and many other countries, the approval of a new antibiotic for use with animals comes with a requirement that products be labeled specifying the particular purpose and animal species in which it is to be used. However, the use of that same antibiotic in other animal species for other "off label" or "extra label" purposes is legal unless explicitly prohibited.

Off-label use of critically important antibiotics like the cephalosporins is common in the United States, Australia, and many other countries and is of grave public health concern. For example, FDA records reveal that 78% of U.S. hatcheries surveyed in 2001 injected ceftiofur, a third-generation cephalosporin, into the eggs of broiler chickens just before hatching (FDA 2002, Collignon and Aarestrup 2007). The surveyed hatcheries accounted for 500 million chickens per year. The practice is suspected of selecting for ceftiofur-resistant pathogens in these birds; 9% of *Salmonella* Heidelberg isolates in the United States are now resistant to ceftiofur (Zhao et al., 2008).

Even where label restraints on cephalosporins have been attempted, as in Australia, they have been widely ignored by agriculture authorities. Agribusiness interests have also defied attempts to restrict antibiotic usage via changes in labeling (Collignon and Aarestrup 2007). In July 2008, the FDA proposed banning any extra-label use of third- and fourth-generation cephalosporins in food animals, citing the public health risks (FDA 2008a). In November 2008, the Centers for Disease Control and Prevention (CDC) supported the proposal (CDC 2008). On November 26th, however, the FDA reversed itself and revoked its proposal (FDA 2008b, Mundy and Favole 2008). Ironically, the FDA and the CDC are both public health agencies and different parts of the same U.S. federal department, the Department of Health and Human Services.

PROPHYLAXIS AND GROWTH PROMOTION

In many food animal production systems, it is routine practice to treat groups of animals with antibiotics before clinical symptoms of disease actually appear (*prophylaxis*). Prophylactic use is typical of intensive livestock systems employing production-enhancing practices that also predispose to outbreaks of disease. Antibiotics are also given to food animals to enhance growth and increase feed efficiency (known as growth promotion) (McEwen and Fedorka-Cray 2002, Wegener 2003). Definitions are not uniform across the world. The American Veterinary

Medical Association, for example, defines "therapeutic" as including antibiotics used for treatment, control, and prevention of bacterial disease (McEwen and Fedorka-Cray 2002).

Debate over how to best address antibiotic misuse and overuse in food animal production can be sidetracked by opposing claims as to the prevalence of growth promotion versus other uses. In the United States and many other countries lacking programs to collect data on antimicrobial usage, opposing claims on the degree of growth promotion are impossible to reconcile with any certainty. Based on the limited publicly available data, the most transparent estimate appears to be that nontherapeutic uses of antimicrobials (including ionophores) in food animal production in the United States could account for around 70% of total antimicrobial use, including human uses (Mellon and Fondriest 2001). The general agreement on this figure calls into question pharmaceutical industry claims that growth promotion constitutes as little as 5% of animal antimicrobial use in the United States (Animal Health Institute 2008).

From a producers' perspective, the delivery of antibiotics via feed may seem like the most efficient or practical solution. It ensures, however, that weaker, sicker animals will consume lower concentrations of antibiotics than healthy animals do (Wegener 2003). Beyond the concentrations used, another critical distinction between therapeutic and other antibiotic uses is how routinely they are used throughout an animal's life. In many intensive production systems, animals are given feed antibiotics and other antimicrobials throughout much of their lives, exposing bacteria in these animals and their farm environments to near constant selection pressure for resistance (NAS/NRC 1999, Chapman and Johnson 2002).

As with therapeutic antibiotics, many antibiotics added to animal feed belong to classes that are identical or nearly so to antibiotics used in human medicine. Widely used antibiotic growth promoters, for example, belong to classes of antibiotics sometimes developed only much later into critical human medicines. Avoparcin, a glycopeptide, was a common, approved growth promoter in Europe and Australia (but not the United States). In 1992 over 120,000 kilograms of avoparcin (10% active ingredient by weight) were used in animals in Australia (predominantly as a growth promoter), while only 68 kilograms of the human glycopeptide, vancomycin, were used (JETACAR 1999).

As human pathogens like enterococci became resistant to other antibiotics, vancomycin emerged as a critical human drug often used as an antibiotic of "last resort." Among nosocomial or hospital-acquired infections, enterococci are the second to third most important bacterial genus—vancomycin-resistant enterococci (VRE) are a dangerous scourge. VRE has caused much more human illness in Europe than in the United States, possibly due to cross-resistance

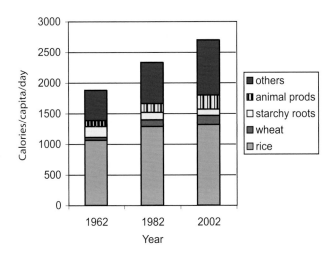

Box Figure 2.1. East and southeast Asia: changing food consumption patterns.

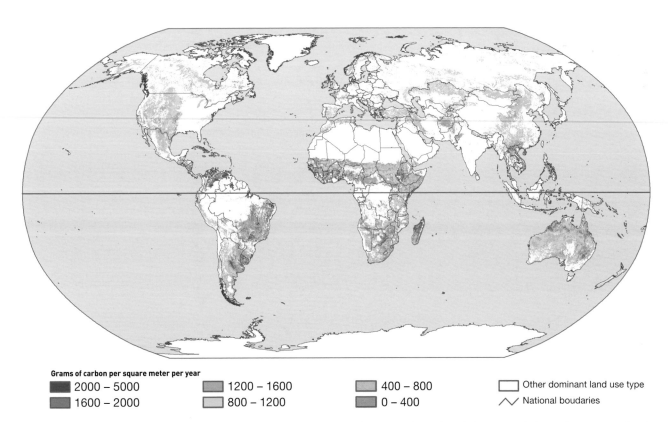

Figure 3.2. The distribution of grasslands across the world. With the exception of arid deserts and dense forest, grasslands are present to some extent in all regions.

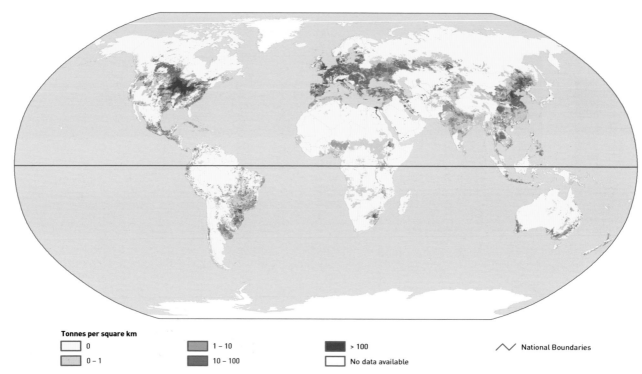

Tonnes per square km

☐ 0   ▨ 1 – 10   ▣ > 100   ⋀⋁ National Boundaries
▨ 0 – 1   ▣ 10 – 100   ☐ No data available

Figure 3.3. The estimated global extent of grain production for animal feed.
Source: Steinfeld et al., 2006.

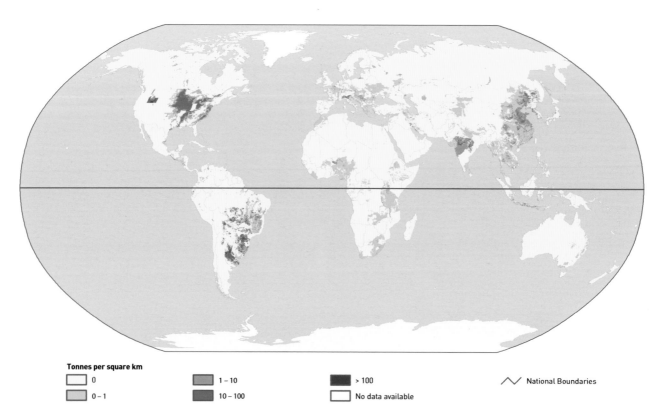

Tonnes per square km

☐ 0   ▨ 1 – 10   ▣ > 100   ⋀⋁ National Boundaries
▨ 0 – 1   ▣ 10 – 100   ☐ No data available

Figure 3.4. Geographic concentration of soybean production.
Source: Steinfeld et al., 2006.

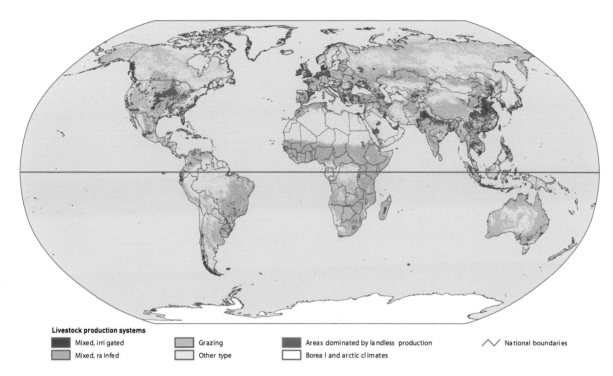

Figure 3.7. The global geographical distribution of the main livestock production systems.
Source: Steinfeld et al., 2006.

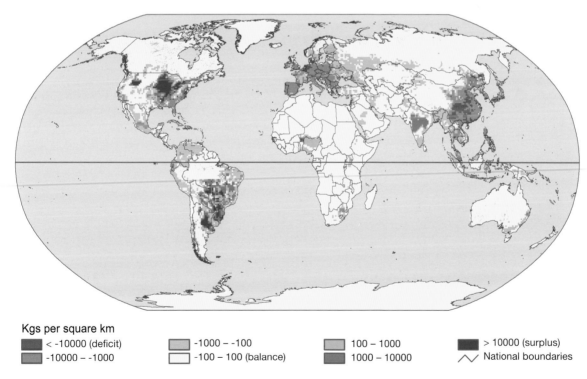

Figure 4.3. Estimated feed surplus/deficit soymeal in pig and poultry production. For each 100 × 100 km cell, the balance is calculated as the difference between the estimated soymeal production for pig and poultry feed and the soymeal consumption by pig and poultry. The soymeal production map is derived from the estimated soybean production for animal feed (map 8), removing the fraction dedicated to ruminants (Galloway et al., 2007) and applying a bean to meal weight conversion factor (Schnittker 1997). The consumption map was calculated from pig and poultry meat production maps (Steinfeld et al., 2006). National level indexes derived were first used to estimate live weight production and total feed consumption (FAO 2006b). For each country, the share of soymeal in the feed basket composition was then extrapolated from available data (Steinfeld et al., 2006). This share was finally used to calculate the soymeal consumption by pig and poultry in each cell.
Source: Steinfeld et al., 2006.

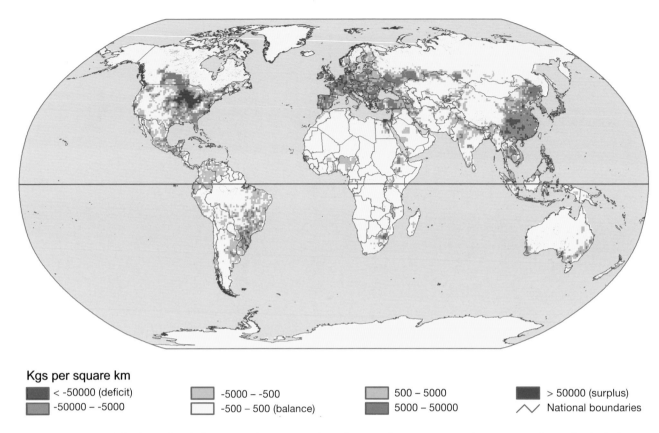

Kgs per square km

■ < -50000 (deficit)	■ -5000 – -500
■ -50000 – -5000	□ -500 – 500 (balance)

■ 500 – 5000	■ > 50000 (surplus)
■ 5000 – 50000	∧∨ National boundaries

Figure 4.4. Estimated feed surplus/deficit—cereals in pig and poultry production. For each 100 × 100 km cell, the balance is calculated as the difference between the estimated cumulated maize, wheat, and barley (MWB) production for pig and poultry feed and the MWB consumption by pig and poultry. The production map is derived from estimated MWB production for animal feed (Steinfeld et al., 2006), removing the fraction of MWB feed dedicated to ruminants (Galloway et al., 2007). The consumption map was calculated from pig and poultry meat production maps (Steinfeld et al., 2006). National level. Indexes derived were first used to estimate live weight production and total feed consumption (FAO 2006b). For each country, the share of MWB in the feed basket composition was then extrapolated from available data (Steinfeld et al., 2006). This share was finally used to calculate the MWB consumption by pig and poultry in each cell. *Source*: Steinfeld et al., 2006.

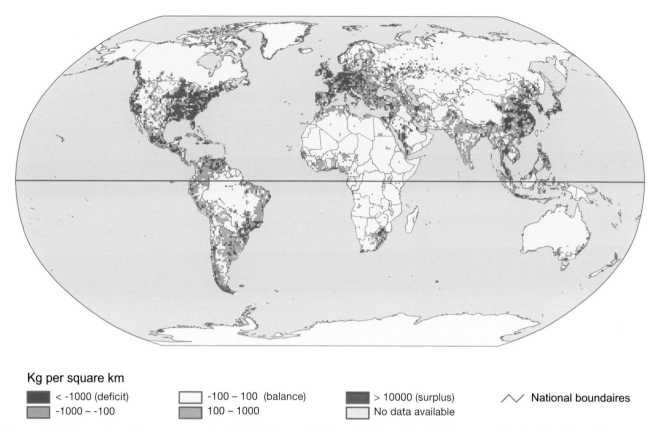

Kg per square km

■ < -1000 (deficit)	□ -100 – 100 (balance)	■ > 10000 (surplus)	∧∨ **National boundaires**
▨ -1000 – -100	▨ 100 – 1000	□ No data available	

Figure 4.5. Estimated poultry meat surplus/deficit. For each 100 × 100 km cell, the balance is calculated as the difference between estimated poultry meat production and consumption. The production map is based on national level statistics (FAO 2006b) distributed along animal densities corrected for the level of production intensity. The consumption map was calculated by distributing national level statistics (FAO 2006b) along human population. In developing countries, higher consumption levels were attributed to urban areas than to rural areas (LandScan 2005).
Source: Steinfeld et al., 2006.

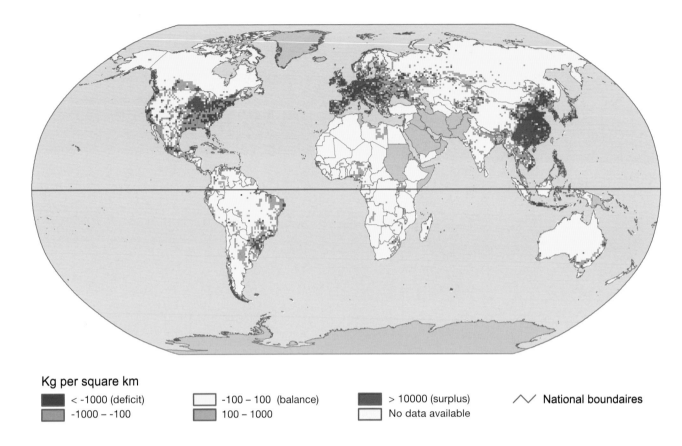

Kg per square km

■ < -1000 (deficit)	□ -100 – 100 (balance)
▧ -1000 – -100	▨ 100 – 1000

■ > 10000 (surplus)	⋀⋁ **National boundaires**
□ No data available	

Figure 4.6. Estimated pig meat surplus/deficit. For each 100 × 100 km cell, the balance is calculated as the difference between estimated pig meat production and consumption. The production map is based on national level statistics (FAO 2006b) distributed along animal densities corrected for the level of production intensity. The consumption map was calculated by distributing national level statistics (FAO 2006b) along human population. In developing countries, higher consumption levels were attributed to urban areas than to rural areas (LandScan 2005).
Source: Steinfeld et al., 2006.

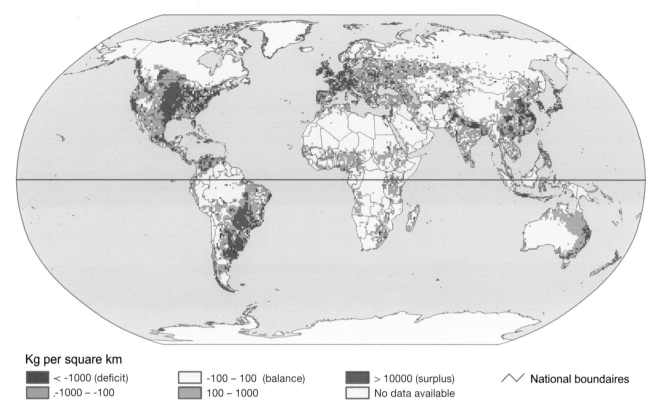

Kg per square km

■ < -1000 (deficit)	□ -100 – 100 (balance)	■ > 10000 (surplus)	⋀⋁ National boundaires
▨ .-1000 – -100	▨ 100 – 1000	□ No data available	

Figure 4.7. Estimated beef surplus/deficit. For each cell, the balance is calculated as the difference between estimated beef production and consumption. The production map is based on national level statistics (FAO 2006b) distributed along animal densities. The consumption map was calculated by distributing national level statistics (FAO 2006b) along human population. In developing countries, higher consumption levels were attributed to urban areas (LandScan 2005) than to rural areas.
Source: Steinfeld et al., 2006.

Beef (a)

Pork (b)

Poultry (c)

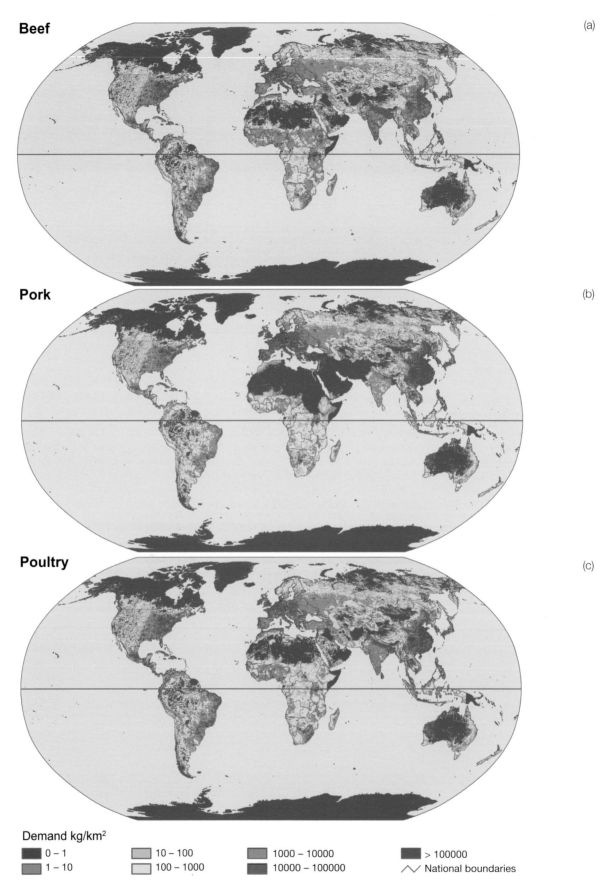

Demand kg/km²

- 0 – 1
- 1 – 10
- 10 – 100
- 100 – 1000
- 1000 – 10000
- 10000 – 100000
- > 100000
- ∧∨ National boundaries

Figure 4.12. Estimated distribution of meat consumption. The map was calculated by distributing national level statistics (FAOSTAT) along human population. In countries not belonging to the OECD, higher consumption levels of poultry (a), pork (b), and beef (c) were attributed to urban areas than to rural areas (LandScan 2005).

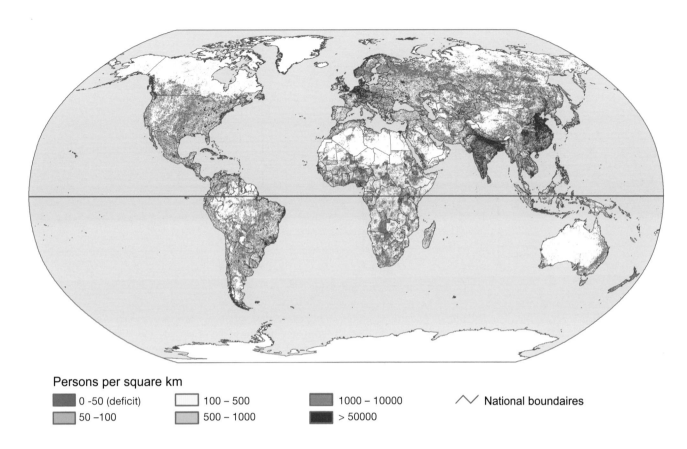

Persons per square km

■ 0 -50 (deficit)	□ 100 – 500	■ 1000 – 10000	⋀ National boundaires
■ 50 –100	■ 500 – 1000	■ > 50000	

Figure 4.13. Estimated distribution of human population.

	C Stocks Before	Change in C Flows	C Stocks After	Net C Affect
Desertification	21 ⬇ 1	Small decrease in NPP ⬇ **Desertification** Increased spatial heterogeneity of C and nutrients ⬇ Increased Erosion Losses	13 ⬇ 0.7	↓
Woody encroachment	2,100 ⬇ 16,800	700 increase in NPP ⬇ **Woody Encroachment** Increased spatial heterogeneity of C and nutrients ⬇ Erosion Losses	12,000 ? ⬇ 19,000 ?	↑
Tropical Deforestation	130,000 ⬇ 210,000	88,000 — **Fire & Conversion** — Leaching losses 14,000 — **Repeat Burning** (each burn) — Erosion Losses	3,900 ⬇ 200,000	↓

Figure 5.5. Three ecological degradation syndromes associated with livestock production systems. Values indicate mean carbon stocks (kg ha⁻¹) or fluxes (kg ha⁻¹ yr⁻¹) as reported throughout the scientific literature (from Asner et al., 2004). Net effect on C storage is depicted on far right.

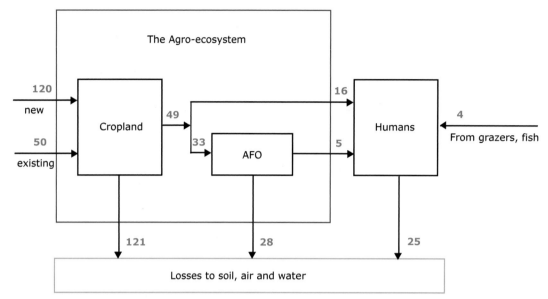

Figure 6.3. Major reactive nitrogen (Nr) flows in crop production and animal production components (AFO: animal feeding operation) of the global agroecosystem (Tg N yr^{-1}). Croplands create vegetable protein through primary production; animal production uses secondary production to create animal protein. Reactive nitrogen inputs represent new Nr (created through the Haber-Bosch process and through cultivation-induced biological nitrogen fixation) and existing Nr that is reintroduced in the form of crop residues, manure, atmospheric deposition, irrigation water, and seeds. Portions of the Nr losses to soil, air, and water are reintroduced into the cropland component of the agroecosystem (Smil 1999, 2001, 2002; Galloway et al., 2003). The uncertainties of these fluxes are generally presented in Smil 1999.

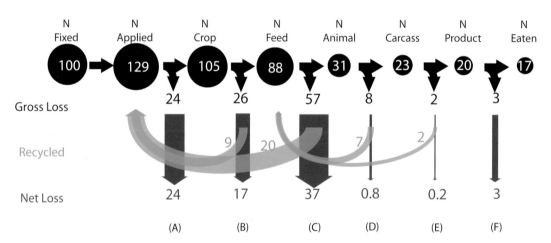

Figure 6.4. The fate of 100 mass units of nitrogen (44 mass units fixed by Haber-Bosch process and applied to agricultural field and 56 mass units produced within agricultural field by BNF in legumes) introduced to agroecosystems and used to raise pork in industrial animal production systems. Co-products include the N lost from the system and the N contained in by-products of production (plant matter, animal parts, etc.) that are recycled or used elsewhere. Red arrows are net N losses; green arrows are amounts of N recycled.

Notes:

(A) Applied N losses are weighted average of losses from grain-based feed and losses from legume-based feed, where the weights are the relative share of grain and legume protein in pig diets. Assuming a pig diet of 80% grains (8.5% crude protein) and 20% soybean meal (44% crude protein), the diet crude protein percentage would be .80(.085) + .20(.44) = 15.6%. Grains would contribute .80(.085)/.156 = 44% of the crude protein, and legumes would contribute the remaining .20(.44)/.156 = 56%. Assuming apparent N uptake of 80% for corn (Cassman 2002), 90% for soybeans, and 65% for recycled crop residues and manure, total losses are [1](.2).44 + (.1).56 + .35(29) = 24. The initial value of 129 units of N applied refers to N application, fixation, and available N from recycling.

(B) Losses for both legumes and grains are 25%, but 35% of plant matter is recycled back to the field.

(C) Manure losses are 65% of intake N; 35% is reapplied to crops.

(D) Assuming live animal to carcass conversion of 0.75. In practice, this would end up in dog food or other animal feed.

(E) Most carcass trimmings are recovered for animal feeds, including pet food, but were shown as fed back in this case.

(F) 16% of red meat in the United States is spoiled or wasted at the retail, food service, and consumer levels (Kantor et al., 1997).

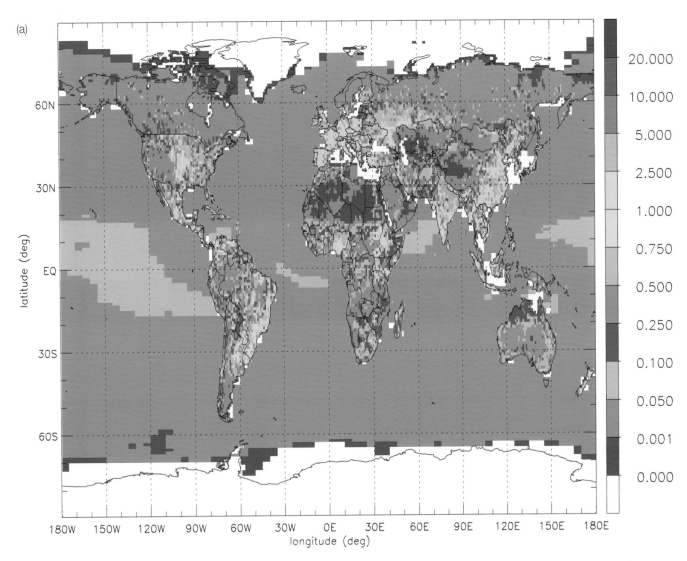

(a)

Figure 6.5. (a–c) Global 1° × 1° distribution (g N m⁻² y⁻¹) of (a) annual total NH₃ emissions, totaling 57.5 Tg N y⁻¹; (b) annual IAPS emissions, totaling 31.0 Tg N y⁻¹ (IAPS emissions are updated from Bouwman et al., 2002a; other emissions are taken mainly from Bouwman et al., 1997) and (c) global 1° × 1° distribution of the fraction of emissions from IAPS relative to total NH₃ emissions. Regions indicated with white colors have emissions lower than 0.00001 g m⁻² y⁻¹.

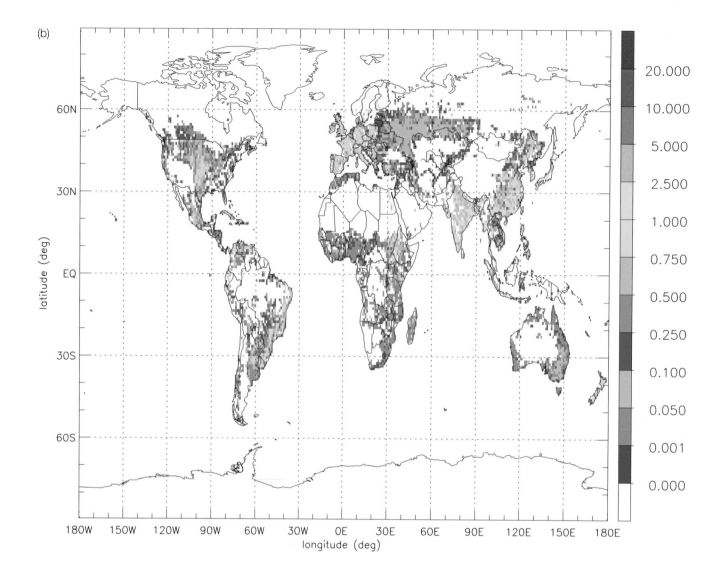

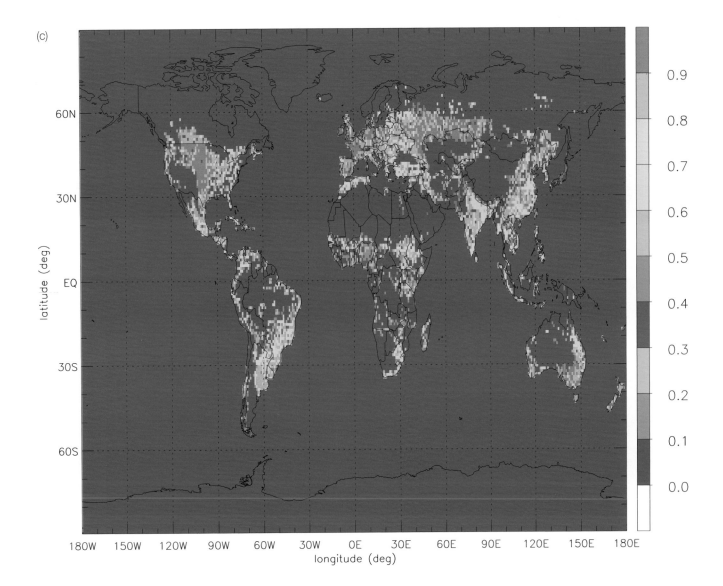

(a)

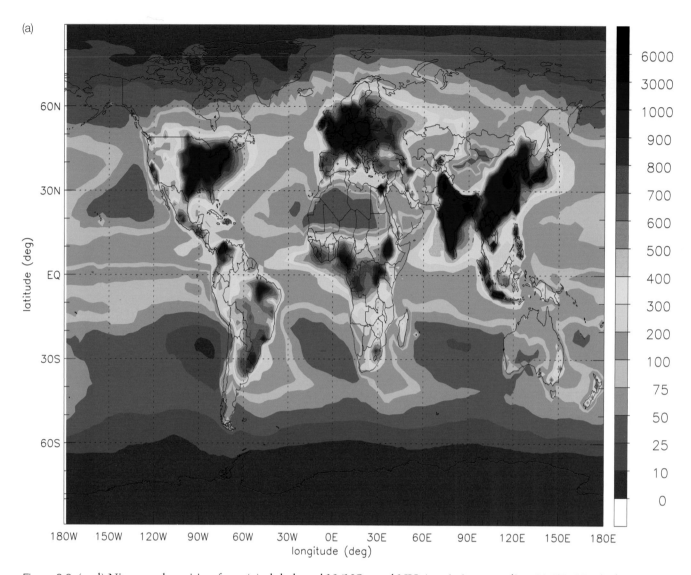

Figure 6.6. (a–d) Nitrogen deposition from (a) global total N (NOy and NHx) emissions, totaling 105 Tg N y^{-1}; (b) NHx deposition from IAPS, totaling 31 Tg N y^{-1}; (c) fraction of NHx deposition from IAPS, to total NHx deposition; and (d) fraction of NHx deposition from IAPS to total reactive N (NOy + NHx) deposition. Model calculations were performed using the TM5 model on a global 3° × 2° resolution using meteorology for the year 2000.

(b)

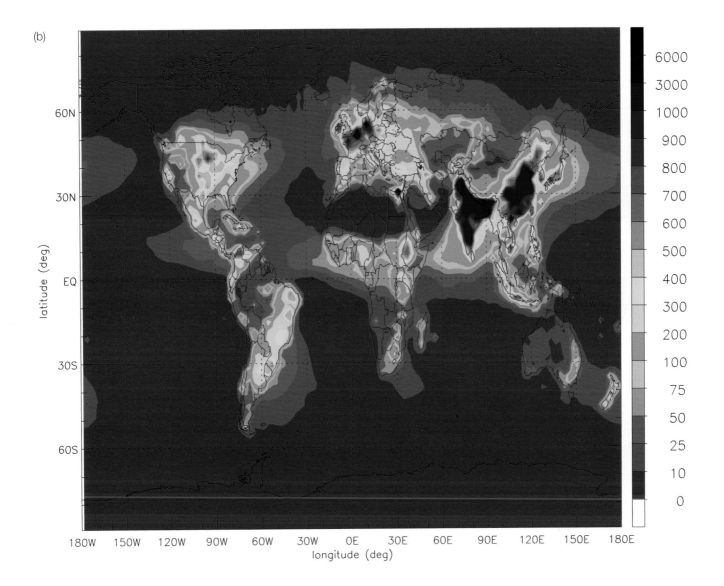

(c)

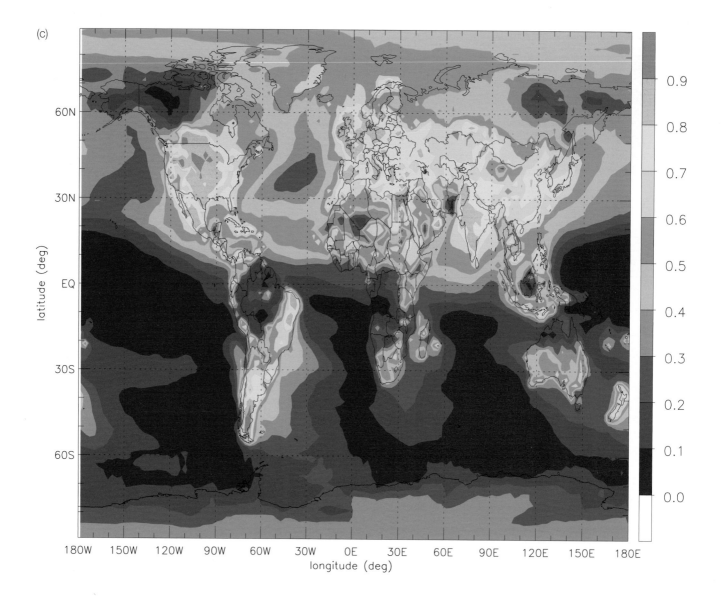

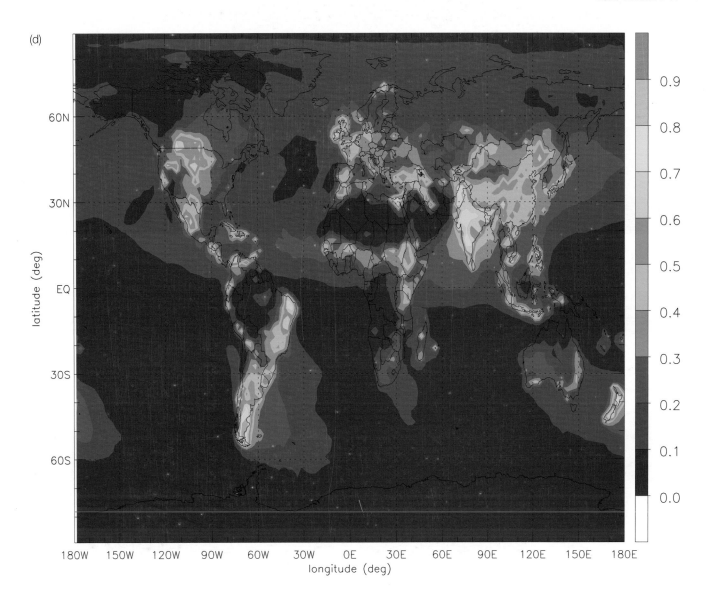

(a)

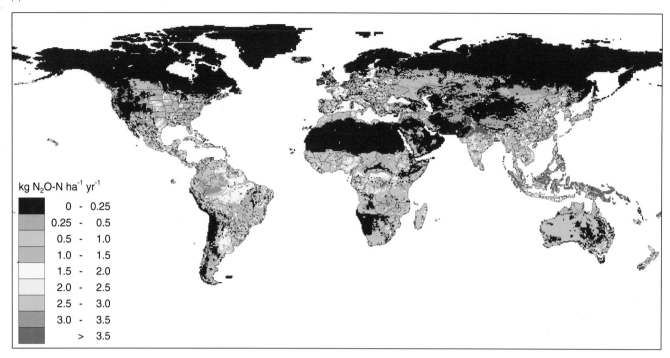

Figure 6.7. (a–c) Global 0.5° × 0.5° distribution (kg N ha⁻¹ yr⁻¹) of (a) annual total N₂O emissions from soils under natural vegetation (4.6 Tg N y⁻¹) and agricultural fields (6.0) (including arable land and grassland), totaling 10.6 Tg N y⁻¹; (b) annual emissions from IAPS, totaling 4.1 Tg N y⁻¹; and (c) percentage of N₂O emissions from IAPS relative to total N₂O emissions. (IAPS emissions are updated from Bouwman et al., 2002a. Other emissions are taken mainly from Bouwman et al., 1993.) The emissions from the IAPS include 2.9 Tg N₂O-N per year from fertilized cropland, 0.2 Tg N₂O-N per year from fertilized grassland, 0.5 Tg N₂O-N per year from housing and storage systems, and 0.5 Tg N₂O-N per year from grazing. This reflects the intensive use of nitrogen fertilizers and the cycling of nitrogen in the intensive livestock production system.

(b)

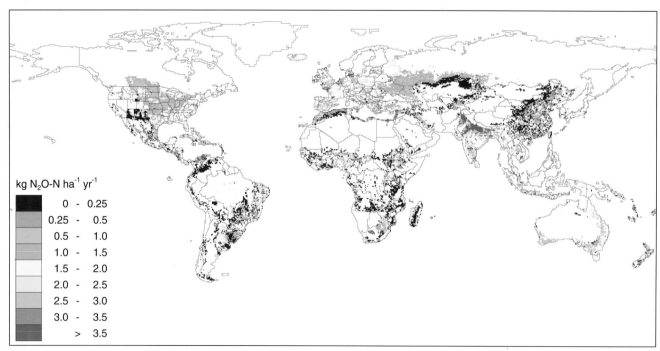

(c)

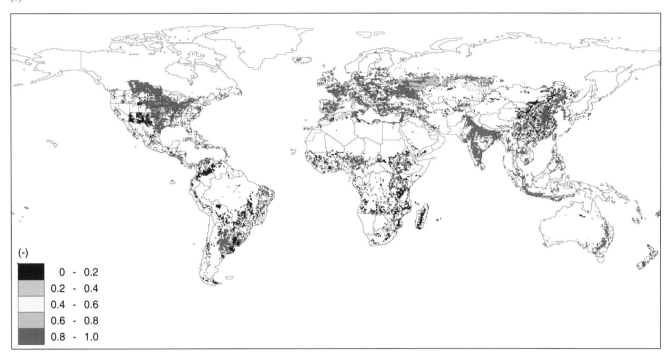

(-)

	0 - 0.2
	0.2 - 0.4
	0.4 - 0.6
	0.6 - 0.8
	0.8 - 1.0

(a)

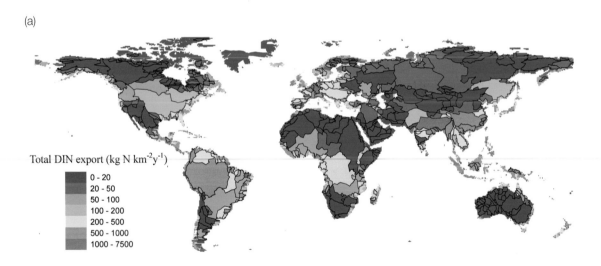

Total DIN export (kg N km^{-2}y^{-1})

	0 - 20
	20 - 50
	50 - 100
	100 - 200
	200 - 500
	500 - 1000
	1000 - 7500

Figure 6.8. Amount of DIN discharged (kg N km^{-2} y^{-1}) at the mouth of rivers from: (a) all watershed sources, totaling 24.8 Tg N yr^{-1} (Figure is modified from Dumont et al., 2005.); (b) animal production, totaling 3.0 Tg N yr^{-1}, excluding animals that are only pasture fed; and (c) percent of DIN transport to the coast from IAPS. (The figure was developed from model output of Dumont et al., 2005.) Note that both maps are based on 1995 input data (see Dumont et al., 2005 for detailed model and input database descriptions). For a discussion of uncertainties in the model results, see Dumont et al., 2005 and Bouwman et al., 2005.

(b)

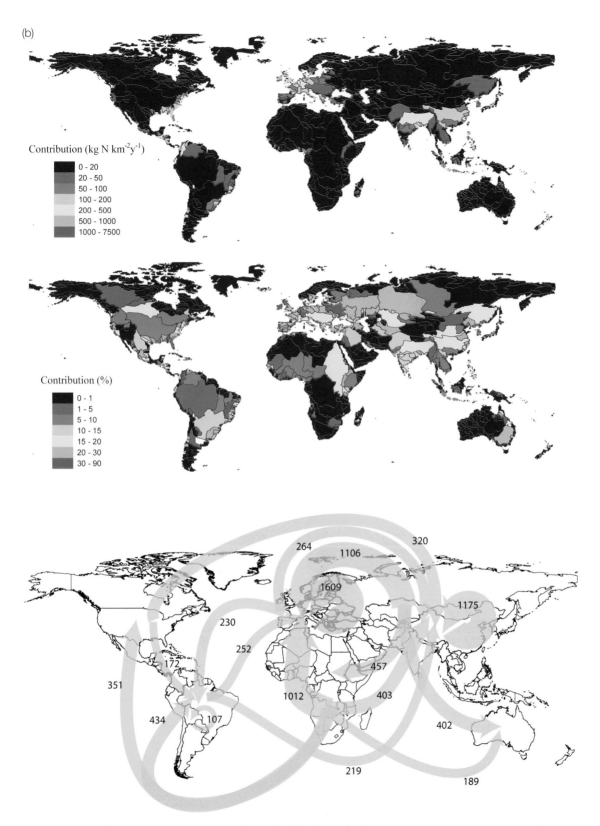

Box Figure 6.1. Phosphorus in internationally traded fertilizer, feed, and meat by continent. Data (2004) shown in thousands of tons of P; minimum requirement for drawing a line is 100,000 tons P. Total international P trade is composed of the following: fertilizer (7.9 Tg P); feed (2.1 Tg P); meat (0.1 Tg P). (Where Tg = 1 million tons.)

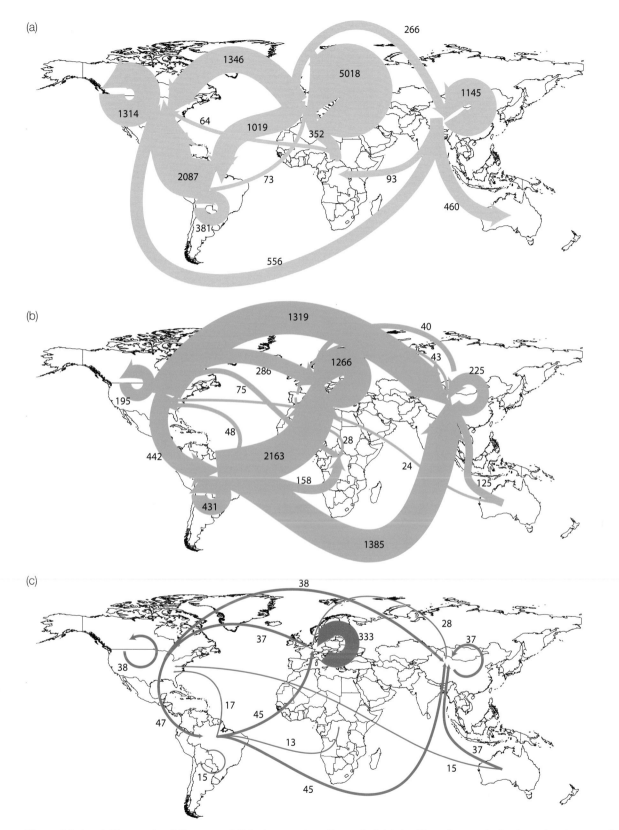

Figure 6.9. (a–c) N contained (k tonnes N) in internationally traded (a) fertilizer for use in feed production (total = 14.4 Tg N yr⁻¹); (b) feed for use in meat production (total = 8.4 Tg N yr⁻¹); and (c) meat (total = 0.8 Tg N yr⁻¹). The data for this figure are compiled assuming all traded fertilizer and 85% of traded ammonia are for agricultural purposes. It is also assumed that feed and nonfeed crops receive on average equal amounts of fertilizer, and thus that the percentage of fertilizer going to feed is the same as the percentage of crops being used as feed.

The Nitrogen Cascade

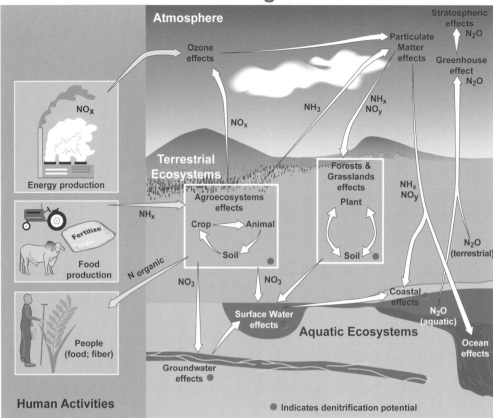

Figure 6.10. Illustration of the nitrogen cascade, showing the sequential effects that a single atom of N can have in various reservoirs after it has been converted from a nonreactive to a reactive form. Abbreviations: GH = greenhouse effect; PM = particulate matter.
Source: Galloway et al., 2003.

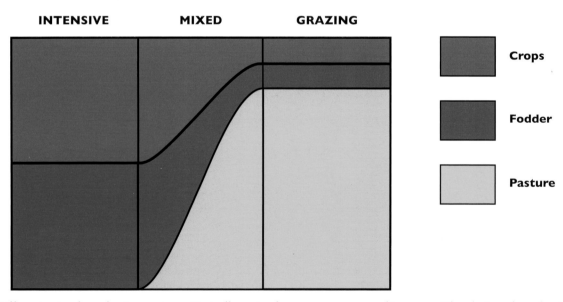

Figure 7.1. Different animal production systems. Basically, animals consume crops and/or graze. The choice of production system will determine the combination of crop and grazing lands used. Illustration: T. Gordon/tovedesign.se.

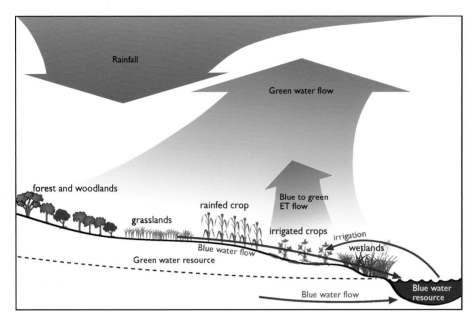

Figure 7.2. Rainwater partitioning of terrestrial water flows and resources. Precipitation is partitioned into liquid water flows (blue water) and water vapor (green water) through evapotranspiration. Illustration: R. Kautsky/azote.se.

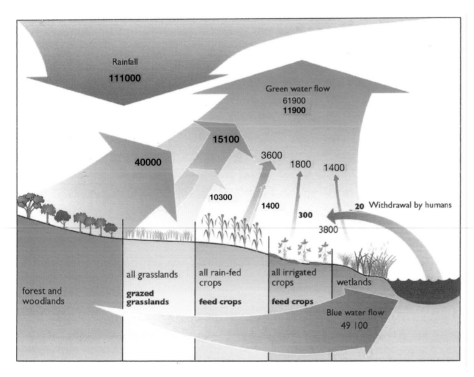

Figure 7.3. Global hydrological flows (km³) and livestock production's share (in bold text). Estimates are unavailable for forests and wetlands. (Falkenmark and Rockström 2006, Gordon et al., 2005, Oki and Kanae 2006, Postel et al., 1996, Rockström et al., 1999, Steinfeld et al., 2006). Illustration: R. Kautsky/azote.se.

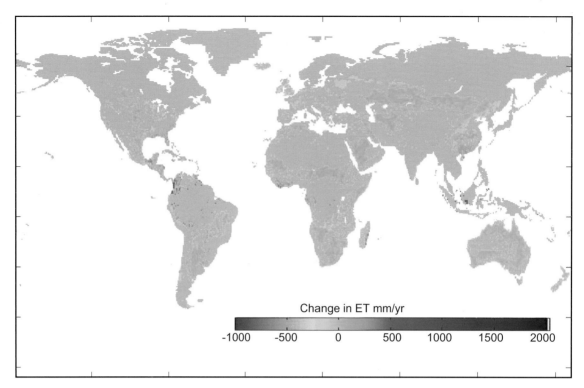

Figure 7.4. Reduction of evapotranspiration flows attributed to conversion of forests to grazing lands for livestock production. Green areas show grazing lands with no net change of green water flows, orange to red areas show reductions of green water flows, and gray areas are other vegetation classes.
Source: Adapted from Gordon et al., 2005.

Figure 7.5. Ecohydrological framework for connecting livestock production to consequences for ecosystem services. Livestock production decisions (A) can result in agricultural production system (agrosystem) changes (B). These agrosystem changes affect water partitioning processes (C) resulting in alterations of ecosystem water determinants (D), which then have impacts on ecosystem services (E). Illustration: T. Gordon/tovedesign.se.

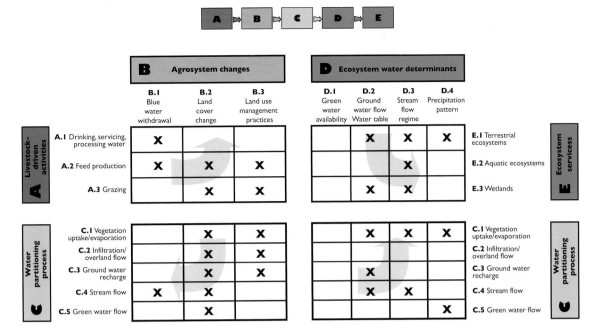

Figure 7.6. Composite matrix for analyzing ecosystem effects of hydrological alternations due to livestock production changes based on a five-step ecohydrological framework. Livestock production decisions (A) can result in agricultural production system (agrosystem) changes (B). These agrosystem changes affect water partitioning processes (C) resulting in alterations of ecosystem water determinants (D), which then have impacts on ecosystem functioning and thus the ecosystem services provided (E). The matrix is marked only for the examples described in Chapter 7. Illustration: T. Gordon/tovedesign.se.

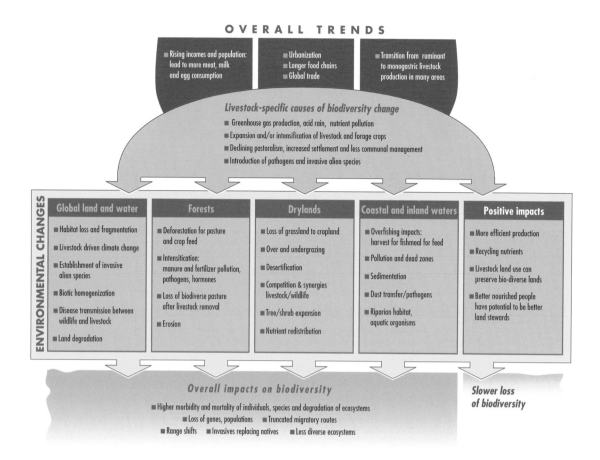

Figure 8.1. Conceptual model of livestock impacts on biodiversity at different spatial and temporal scales.

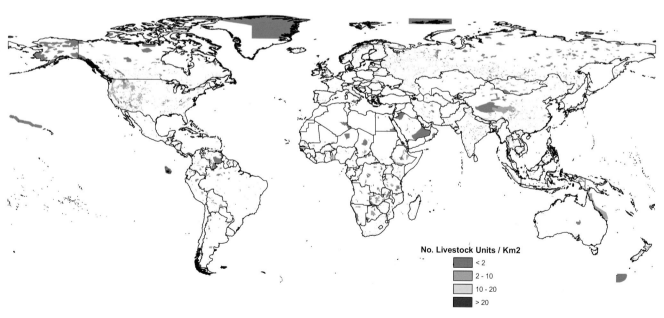

Figure 8.2. Density of livestock units (LU/km²) of cattle, sheep, goats, and buffalo within 30 km of protected areas. The actual protected areas are shown, colored by the density of livestock outside and adjacent to the protected area.

System changes 2000 to 2050
(MA Global Orchestration Scenario)

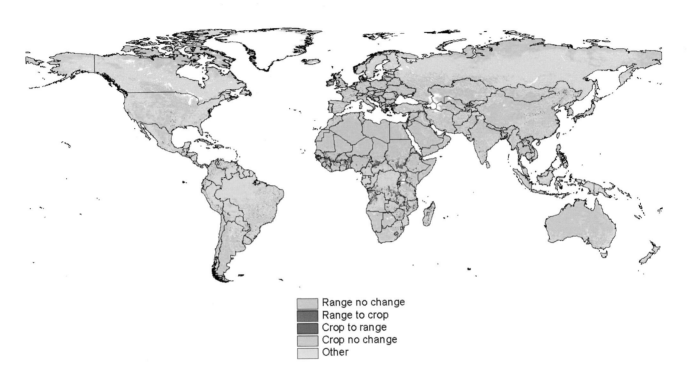

Range no change
Range to crop
Crop to range
Crop no change
Other

Figure 8.3. Global extent of rangelands and crop–livestock and landless systems, 2000 to 2050.

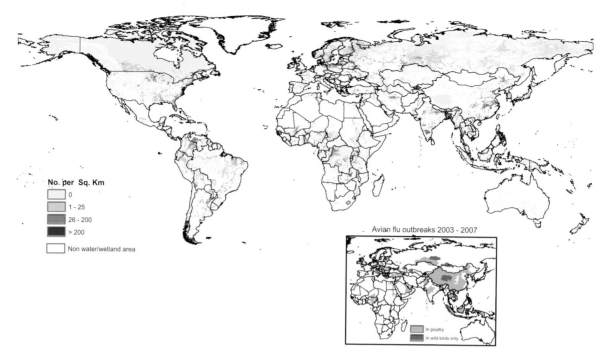

Box Figure 8.1. Conceptual model of possible avian flu transmission between poultry, wild birds, and other animal species.

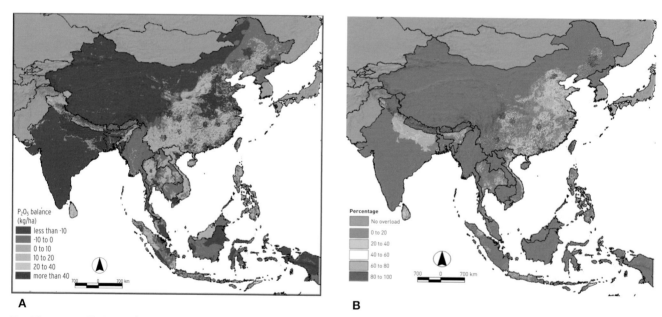

Box Figure 9.1. Estimated P_2O_5 mass balance (a) and contribution of livestock to total P_2O_5 supply on agricultural land, in area presenting a P_2O_5 mass balance of more than 10 kg per hectare (b) in selected Asian countries, 1998–2000.
Source: Gerber et al., 2005.

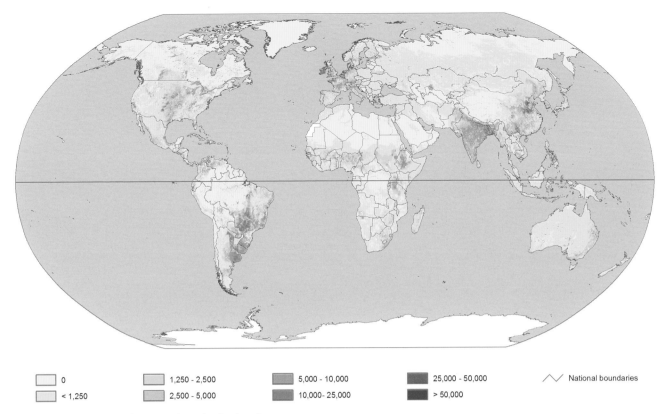

	0		1,250 - 2,500		5,000 - 10,000		25,000 - 50,000		National boundaries
	< 1,250		2,500 - 5,000		10,000- 25,000		> 50,000		

Figure 9.1. Liveweight density of cattle (kg km^{-2}).

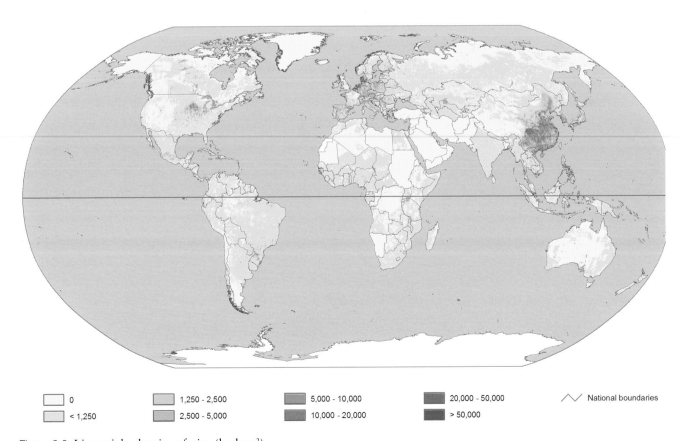

	0		1,250 - 2,500		5,000 - 10,000		20,000 - 50,000		National boundaries
	< 1,250		2,500 - 5,000		10,000 - 20,000		> 50,000		

Figure 9.2. Liveweight density of pigs (kg km^{-2}).

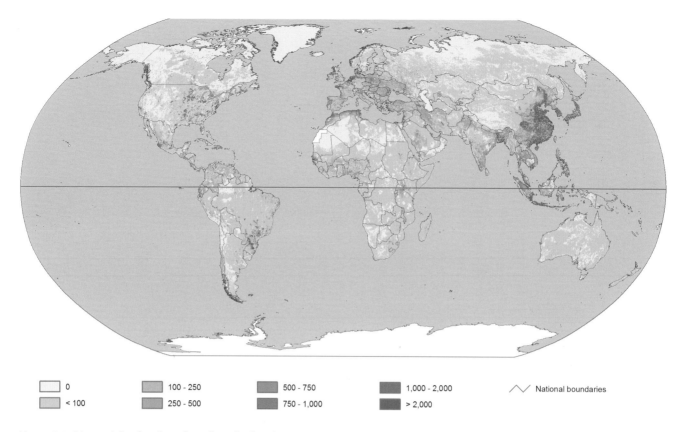

	0		100 - 250		500 - 750		1,000 - 2,000		National boundaries
	< 100		250 - 500		750 - 1,000		> 2,000		

Figure 9.3. Liveweight density of poultry (kg km^{-2}).

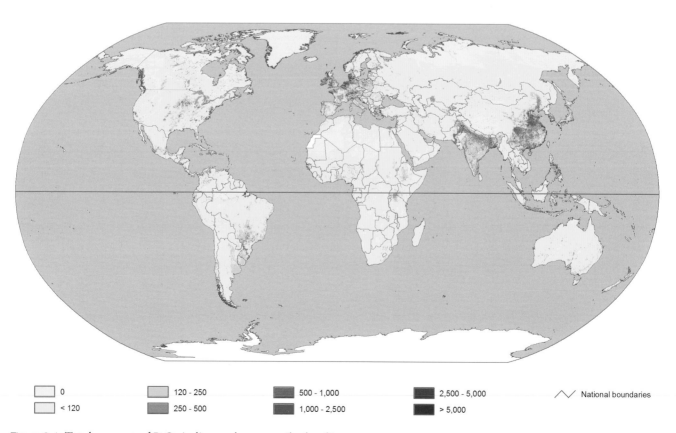

	0		120 - 250		500 - 1,000		2,500 - 5,000		National boundaries
	< 120		250 - 500		1,000 - 2,500		> 5,000		

Figure 9.4. Total amount of P_2O_5 in livestock excreta (kg km^{-2}).

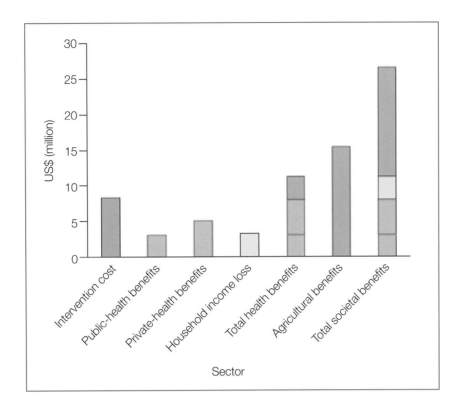

Figure 11.3. Distribution of benefits of livestock brucellosis vaccination in Mongolia
Source: Zinsstag et al., 2005a, with kind permission from Lancet for reproduction of this graph.

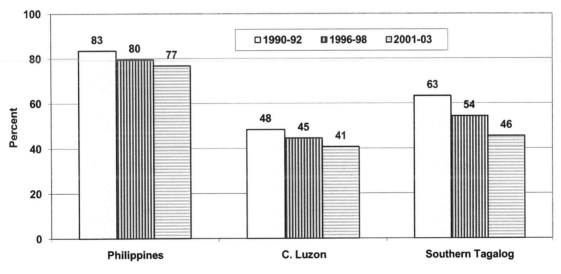

Figure 14.1. Trends in share of backyard hog inventories in the Philippines, 1990–2003
(shares computed over three-year averages).
Source: Costales et al., 2007, citing the Bureau of Agricultural Statistics 2005.

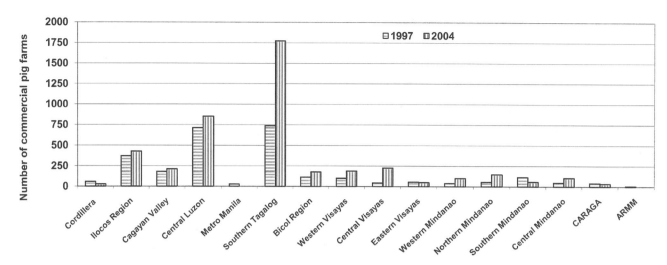

Figure 14.2. Regional distribution of the number of registered commercial pig farms in the Philippines, 1997 and 2004.
Source: Costales et al., 2007, citing Bureau of Agricultural Statistics 2005.

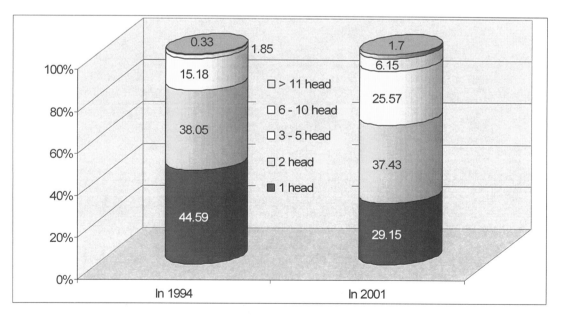

Figure 14.3. Change in farm size of pig household producers in Viet Nam, 1994–2001.
Source: ILRI-HAU-IFPRI-PPLPI/FAO 2007 citing the GSO, Agriculture Census 1994, 1995, 2001, and 2002.

developing as a result of the much broader use of avoparcin. In 1994, a year before the European Union ban on avoparcin, 12% of healthy Germans were found to carry glycopeptide-resistant strains of *Enterococcus faecium*, while in the United States, where avoparcin was never approved for growth promotion, no VRE could be detected in the guts of healthy persons, nor could VRE be found in animals or meat products (Klare 2003).

Physically Linked Resistance

Often overlooked fundamentals of microbiology are essential for fully understanding the public health ramifications of antimicrobial use in animal agriculture (Walsh 2000, Summers 2002). The microbiology of resistance is complex, often involving promiscuous resistance genes widely shared even among dissimilar bacteria (OIE/WHO/FAO 2006). Adding to the complexity is that mechanisms have evolved whereby bacteria transfer to one another multiple, physically linked resistance genes at the same time:

> A principal mechanism for the rapid spread of antibiotic resistance genes through bacterial populations is that such genes get collected on plasmids that are independently replicated within and passed between bacterial cells and species. Furthermore, some of these genes that reside on plasmids . . . may be further segregated within transposons that can actively cut themselves out of one DNA locale and hop into other locales, promiscuously moving their antibiotic resistance-conferring genetic cargo. (Walsh 2000, p. 777)

Some plasmids have collected a dozen or more resistance genes. A pathogen on the receiving end of the promiscuous transfer of a plasmid thereby can become resistant to treatment with multiple antibiotics with great efficiency. Physically linked resistance also means that a bacterium's exposure to just one antibiotic can increase the selection pressure for resistance to the other antibiotics to which that resistance is linked—a phenomenon termed coselection.

In the United States, where the FDA has approved at least nine different antibiotics and three arsenical feed additives, it is common for poultry feed to contain several different antimicrobials concurrently (FDA Green Book 2004, NAS/NRC 1999). Arsenical compounds and other metals, including zinc, selenium, cobalt, arsenic, iron, manganese, and copper, are routinely added to animal feeds as antimicrobials as well as for growth promotion (Steinfeld et al., 2006, Sapkota et al., 2007). Organic arsenicals are widely used in poultry production, though not in Europe, to spur growth and control a parasitic problem, coccidiosis. After the phaseout of antibiotic growth promoters in Sweden, zinc oxide was commonly added to feed to help prevent diarrhea in weaner pigs (Wierup 2001). Copper is a widely used growth promoter in both intensive and organic pig production; the concentrations used are 20- to 30-fold higher than the pigs' minimal nutritional requirements for copper. In Europe, the concentrations of copper used in particular pig age groups are regulated; copper feed supplements are unregulated in the United States (Hasman et al., 2006).

Copper resistance genes are widespread across multiple countries and within potential pathogens such as *E. faecium* and *E. faecalis* (Aarestrup et al., 2002, Hasman and Aarestrup 2005). It has been demonstrated that the high copper levels added to feed in pig production do select for copper-resistant *E. faecium*. Resistance to metals and to antibiotics can be physically linked. Therefore, exposing bacteria to metals in animal feed could co-select for resistance to antibiotics as well (Hasman et al., 2006, Sapkota et al., 2006).

Public Health Implications

It is generally acknowledged that any use of antibiotics can contribute to the emergence of resistance and promote dissemination of that resistance into the broader microbial ecosystem (O'Brien 2002, OIE/WHO/FAO 2004a). Food animal production is no exception. Through its intensive use of antibiotics and metals and the global trade in both animals and food derived from them, current food animal production facilitates the emergence and spread of resistance.

Resistance is transmitted between food animals and humans mainly via contaminated foods but also via other modes of transmission, such as farmers' or farmworkers' direct contact with animals (Levy 1976, Price et al., 2007), or through an environment where there are many reservoirs of resistant bacteria and genes (Aarestrup et al., 2008). Retail meat is commonly contaminated with antibiotic-resistant (often multidrug-resistant) bacteria, including foodborne pathogens such as *Salmonella* and *Campylobacter* in addition to resistant strains of enterococci and *E. coli* that may colonize the human gut (White et al. 2001; Zhao et al. 2006, 2008). But resistant bacteria are also detected in animal bedding, animal feed, air both inside and downwind of intensive operations, animal manure waste, and groundwater underneath and near such operations (Chee-Sanford et al., 2001; Chapin et al., 2005; Sapkota et al., 2006, 2007; Price et al., 2007).

Quantification of the transmission of resistance between food animals and humans is made more difficult not only by these multiple routes of transmission but also by the likelihood that resistance spreads via both resistant pathogens as well as via transferable genes in different commensal bacteria (Smith et al., 2002).

Controlling the public health risks of exposing humans to resistance from food animals therefore can occur either by limiting the selective pressure from antimicrobial usage or by limiting the spread of resistant bacteria or resistance genes. Options for the latter can include, for example, establishment of global criteria for certain resistant pathogens for use in controlling the trade of food animals, food products from those animals, and animal feed. Options for limiting the selective pressure from antimicrobial usage can range from educational efforts to influence farmer or veterinarian use of antimicrobials to taxes that change the relative cost of antimicrobials for use in animal feed and to regulatory or legislative bans or phaseouts of certain antimicrobials or their particular uses, such as for growth promotion (Aarestrup et al., 2008).

Sweden and Denmark have demonstrated that legislative bans or producer-initiated phaseouts of antimicrobial growth promotion can be quite effective at reducing antimicrobial use. Sweden's 1986 ban on the use of antimicrobial growth promotion led to a decrease of 55% in total antimicrobial use in food animals (Anderson et al., 2003). In the mid-1990s, chicken and then pork producers in Denmark voluntarily committed themselves to sharply reducing the use of antibiotics in their largely extensive animal production systems by phasing out antibiotic growth promotion (excepting the nonhuman antimicrobials, ionophores). Doing so reduced total antimicrobial use in food animals by nearly 54% (Wegener 2003).

The public health and other impacts of phasing out growth promotion have been most closely examined in Denmark, the world's third-largest pork producer and largest pork exporter. WHO undertook an expert review of the Danish experience and concluded the following:

> Denmark's program to discontinue use of antimicrobial growth promoters has been very beneficial in reducing the total quantity of antimicrobials administered to food animals. This reduction corresponds to a substantial decrease in the overall proportion of individual animals given antimicrobials, and in the duration of exposure among animals given antimicrobials. This represents a general change in Denmark from continuous use of antimicrobials for growth promotion to exclusive use of targeted treatment of specific animals for therapy under veterinary prescription. The program has also been very beneficial in reducing antimicrobial resistance in important food animal reservoirs. This reduces the threat of resistance to public health. (WHO 2003)

In terms of public health benefits, Dr. Henrik Wegener of the Danish Veterinary Institute observes that "ending the use of antimicrobial growth promoters has led to reductions in the prevalence of resistant bacteria in food and food animals, as well as in humans" in all the countries where this phaseout has occurred. Reduced resistance in enterococci to vancomycin, quinupristin-dalfopristin (a streptogramin), and erythromycin (a macrolide) antibiotics has been well documented following reductions in the growth promotion use of avoparcin, virginiamycin, and tylosin in pigs and/or broiler chickens. These reductions occurred with insignificant or no impacts on animal health or productivity in broiler chickens and finisher pigs; in weaner pigs, there have been manageable increases in diarrhea (scours). These minor impacts to the animal industry contrast with the significant public health benefits gained from reducing the selection pressure provided by antibiotic use and the subsequent decreases in reservoirs of resistance to clinically important antibiotics. Experience across Europe, Wegener concludes, "shows that the use of antimicrobials for growth promotion provides insignificant benefits to agriculture and that it can be terminated" (Wegener 2003, p. 433). Because Danish livestock production has been quite intensive and export-driven, its experience also suggests that antibiotic growth promotion could very well be terminated across all industrialized, intensive food animal production systems.

Models of the transmission of resistance from the farm environment to humans indicate that the greatest impact of animal antibiotic use occurs very early in the emergence of resistance, when antibiotic-resistant bacteria are still rare, possibly too low even for current surveillance methods to detect them (Smith et al., 2002). These observations support a more proactive public health approach to antibiotic use in food animals. One suggestion has been that regulation of animal antibiotic use might best occur therefore before resistance becomes a problem in human medicine, with prudent animal use being allowed only subsequent to the development of clinically significant resistance. It should be noted that this is precisely not the approach regulators have taken in the United States and elsewhere, where animal use has basically been allowed until human resistance has exceeded a threshold.

Summary and Policy Implications

Antibiotics are widely used in food animal production, often from classes of antibiotics important to human medicine. This practice selects for antibiotic-resistant populations of bacteria, thereby increasing environmental reservoirs of resistance that may then disseminate elsewhere in the bacterial ecosystem, including to humans. Antibiotic use in humans also contributes to the emergence of antibiotic resistance.

Adverse impacts on the human population stem not only from growth promotion but from the heavy therapeutic use of critically important human medicines in food animals as well. In other words, selection for antibiotic resistance will occur regardless of why that antibiotic is used. Darwin would not be surprised.

Bacteria efficiently evolve and spread antibiotic resistance via multiple mechanisms. Microbiological fundamentals help explain why modern intensive food animal production facilitates the emergence and spread of resistance, through its extensive use of antibiotics—as well as other feed additives to which antibiotic resistance is linked—and through the international trade of both animals and food products.

Because environmental reservoirs of multidrug resistance already exist in both human and nonhuman settings, a successful response to the global epidemic of antibiotic resistance likely will have to be comprehensive, addressing antimicrobial use across the various animal and human settings in the bacterial ecosystem. Experience in Denmark and elsewhere suggests policy-led efforts to reduce antimicrobial use in food animal production can be successful, with substantial public health benefits and minimal impact on the profits or productivity of producers. Environmental reservoirs of antibiotic resistance will remain, however. Public health prudence would suggest the need for greater public and private investment in food animal production that emphasizes, through better animal health, husbandry, and hygiene, avoiding the use of animal antibiotics in the first place.

References

Aarestrup, F. M. 2000. Characterization of glycopeptide-resistant *Enterococcus faecium* (GRE) from broilers and pigs in Denmark: genetic evidence that persistence of GRE in pig herds is associated with coselection by resistance to macrolides. *J Clin Microbiol* 38:2774–2777.

Aarestrup, F. M., A. M. Seyfarth, H. D. Emborg, K. Pedersen, R. S. Hendriksen, and F. Bager. 2001. Effect of abolishment of the use of antimicrobial agents for growth promotion on occurrence of antimicrobial resistance in fecal enterococci from food animals in Denmark. *Antimicrob Agents Chemother* 45 (7) : 2054–2059.

Aarestrup, F. M., H. Hasman, L. B. Jensen, M. Moreno, I. A. Herrero, L. Dominguez, M. Finn, and A. Franklin. 2002. Antimicrobial resistance among enterococci from pigs in three European countries. *Appl Environ Microbiol* 68:4127–4129.

Aarestrup F. M., H. C. Wegener, and P. Collignon. 2008. Resistance in bacteria of the food chain: epidemiology and control strategies. *Expert Rev Anti Infect Ther* 6 (5) : 733–750. Soil Association 2006. Information sheet: Animal welfare on organic farms. 10/04/2006. http://www.soilassociation.org/web/sa/saweb.nsf/librarytitles/20692.HTMl.

Aengst, C. 1984. *Zur Zusammensetzung des Staubes in einem Schweinemaststall*. (The composition of dusts in a fattening pig house) Thesis, University of Veterinary Medicine, Hannover, Germany.

Agricultural Occupational Health Board in Lower Saxony and Bremen. 2003. *Präventionsbericht 2003*. Hannover: Landwirtschaftliche Berufsgenossenschaft Niedersachsen-Bremen.

Agricultural Occupational Health Board in Lower Saxony. 1996. *Präventionsbericht 1996*. Hannover: Landwirtschaftliche Berufsgenossenschaft Niedersachsen.

Agriculture Business Week. Novartis supports judicious use of antibiotics. January 9, 2009. http://www.agribusinessweek.com/novartis-supports-judicious-use-of-antibiotics/.

Anderson, A.D., J. M. Nelson, S. Rossiter, and F. J. Angulo. 2003. Public health consequences of use of antimicrobial agents in food animals in the United States. *Microbial Drug Resistance* 9 (4) : 373–377.

Animal Health Institute. 2002. Press Release dated September 30, 2002. http://www.ahi.org.

Animal Health Institute. 2008. Sales of Disease fighting animal medicines rise. Press release dated November 18, 2008. http://www.ahi.org/files/Media%20Center/Antibiotic%20Use%202007.pdf.

Anonymous. 2005. *Luftgetragene biologische Belastungen und Infektionen am Arbeitsplatz Stall*. KTBL-Schrift 436. Darmstadt: Kuratorium für Technik und Bauwesen in der Landwirtschaft, 1–201.

Archer, J. 1979. *Animals under stress*. London: Edward Arnold.

Banhazi, T., C. Cargill, H. Payne, and G. Marr. 2000. *Report on national air quality survey to Pig research and Development Corporation*. Kingston: ACT.

Banhazi, T., M. O'Grady, C. Cargill, J. Wegiel, and N. Masterman. 1999. The effects of oil spraying on air quality in straw based shelters. In: P. D. Cranwell, ed. *Manipulating pig production VII*, Werribee, Australia.

Bergmann, V., and J. Scheer. 1979. konomisch bedeutungsvolle Verlustursachen beim Schlachtgeflügel. (Economically important losses in broiler production). *Mh Vet-Med* 34:543–547.

Beusker, N. 2007. *Welfare of dairy cows: lameness in cattle: a literature review*. Thesis, University of Veterinary Medicine, Hannover.

Blaha, T., and M.-L. Blaha. 1995. *Qualitätssicherung in der Schweinefleischerzeugung*. Stuttgart: G. Fischer Verlag, Jena.

Blancou, J., B. B. Chomel, A. Belotto, F. X. Meslin. 2005. Emerging or re-emerging bacterial zoonoses: Factors of emergence, surveillance, and control. Vet. Res. 36 (3):507–522.

Blom, J. Y. 1992. Environment-dependent disease. In: C. Phillips and D. Piggins, eds. *Farm animals and the environment*. Wallingford, UK: CAB International, 263–287.

Boerlin P., A. Wissing, F. M. Aarestrup, J. Frey, and J. Nicolet. 2001. Antimicrobial growth promoter ban and resistance to macrolides and vancomycin in enterococci from pigs, *Journal of Clinical Microbiology* 39 (11) : 4193–4195.

Bonfoh, B., P. Ankers, A. Sall, M. Diabaté, S. Tembely, Z. Farah, I. O. Alfaroukh, and J. Zinsstag. 2006. Schéma fonctionnel de services aux petits producteurs laitiers périurbains de Bamako (Mali). Operational plan for services to small-scale milk producers in peri-urban Bamako (Mali). *Etudes et recherches sahéliennes INSAH*, No. 12, 7–25.

Brouwer, R., K. Biersteker, P. Bongers, R. Remin, and D. Houthuijs. 1986. Respiratory symptoms, lung function and IgG4 levels against pig antigens in a sample of Dutch pig farmers. *American J Industrial Med* 10:283–285.

Budke, C. M., P. Deplazes, and P. R. Torgerson. 2006. Global socioeconomic impact of cystic echinococcosis. *Emerging Infectious Diseases* 12 (2):296–303.

Carabin, H. C., C. Budke, L. D. Cowan, A. L. Willingham, and P. R. Torgerson. 2005. Methods for assessing the burden of parasitic zoonoses: cysticercosis and echinococcosis. *Trends in Parasitology* 21:327–333.

Cargill, C., and T. Banhazi. 1998. *The importance of cleaning in all-in/all-out management systems*. Proceedings 15th IPVS Congress, Birmingham, England., 3, 15.

Cargill, C., and J. Hartung. 2001. Air quality: from an OH&S perspective. In: *Proceedings Australian Association of Pig*

Veterinarians, Pan Pacific Conference on Consistent Pork, Melbourne, 93–101.

Centers for Disease Control and Prevention. 2008. Letter from Lonnie King, DVM, Director, CDC's National Center for Zoonotic, Vector-Borne and Enteric Diseases to Bernadette M. Dunham, DVM, Director, Food and Drug Administration's Center for Veterinary Medicine, dated 7 November 2008. www.KeepAntibioticsWorking.org.

Chapin, A., A. Rule, K. Gibson, T. Buckley, and K. Schwab. 2005. Airborne multidrug-resistant bacteria isolated from a concentrated swine feeding operation. *Environmental Health Perspectives* 113:137–142.

Chapman, H. D., and Z., B. Johnson. 2002. Use of antibiotics and roxarsone in broiler chickens in the USA: analysis for the years 1995 to 2000. *Poultry Sci* 81 (3) :356–364.

Chee-Sanford, J. C., R. I. Aminov, I. J. Krapac, N. Garrigues-Jeanjean, and R. I. Mackie. April 2001. Occurrence and diversity of tetracycline resistance genes in lagoons and groundwater underlying two swine production facilities. *Applied and Env Microbiology* 67 (4) : 1494–1502.

Christiaens, J. P. A. 1987. Gas concentrations and thermal features of the animal environment with respect to respiratory diseases in pig and poultry. In: J. M. Buce and M. Sommer, eds. *Agriculture: environmental aspects of respiratory disease in intensive pig and poultry houses, including the implications for human health*. Proceedings of a meeting in Aberdeen 29–30 October 1986, Commission of the European Communities EUR 10820 EN, 29-43.

Clark, S., R. Rylander, and L. Larsson. 1983. Airborne bacteria, endotoxin and fungi in dust in poultry and swine confinement buildings. *Am-Ind Hyg Assoc J* 44:537–541.

Clarkson, M. J., D. Y. Downham, W. B. Faull, J. W. Hughes, F. J. Manson, J. B. Merritt, R. D. Murray, W. B. Russell, J. E. Sutherst, and W. R. Ward. 1996. Incidence and prevalence of lameness in dairy cattle. *Veterinary Record* 138:563–567.

Coleman, P. G., E. M. Fèvre, S. Cleaveland. 2004. Estimating the public health impact of rabies. Emerg Infect Dis [serial online]. http://www.cdc.gov/ncidod/EID/vol10no1/02-0744.htm.

Collaborating Centre for Antimicrobial Resistance in Foodborne Pathogens. http://www.who.int/salmsurv/links/gssamrgrowth reportstory/en/.

Collignon, P., and F. M. Aarestrup. 2007. Extended-spectrum b-lactamases, food, and cephalosporin use in food animals. CID 44:1391–1392.

Collignon, P., H. C. Wegener, P. Braam, C. D. Butler. 2005. The routine use of antibiotics to promote animal growth does little to benefit protein undernutrition in the developing world. *Clinical Infectious Diseases* 41:1007–1113.

Cormier, Y., G. Tremblay, A. Meriaux, G. Brochu, and J. Lavoie. 1990. Airborne microbial contents in two types of swine confinement buildings in Quebec. *Am. Ind. Hyg. Assoc. J.* 51:304–309.

Dantzer, R., and P. Mormede. 1983. Stress in farm animals: a need for reevaluation. *J Anim Sci* 57:6–18.

De Roth, L. L. Vermette, A. Blouin, and N. Lariviere. 1989. Blood catecholamines in response to handling in normal and stress-susceptible swine. *Appl Anim Behav Sci* 22:11–16.

Digoutte J. P. 1999. Present status of an arbovirus infection: Yellow fever, its natural history of hemorrhagic fever, Rift Valley fever [in French]. *Bull. Soc. Pathol. Exot.* 92:343–348.

Digoutte, J. P., and C. J. Peters. 1989. General aspects of the 1987 Rift Valley fever epidemic in Mauritania. *Res Virol* 140:27–30.

Dixon. B. May 2000. Antibiotics as growth promoters: risks and alternatives. *ASM News*. American Society for Microbiology. http://newsarchive.asm.org/may00/animalcule.asp.

Djamen, P., J. Lossouarn, and M. Olliver. 2005. Développement des filières et dynamiques du changement, quelles perspectives pour les élevages bovines de la vina (Cameroun). Symposium international sur le développement des filières agropastorales en Afrique. 1ere édition, du 21 au 27 février, Niamey, Niger.

Donham, K. J. 1987. Human health and safety for workers in livestock housing. In: *Latest developments in livestock housing*. Proceedings of CIGR, Illinois, USA, 86–95.

Donham, K. J. 1989. Relationship of air quality and productivity in intensive swine housing. *Agri-Practice* 10:15–26.

Donham, K. J. 1990. Health effects from work in swine confinement buildings. *Am J Ind Med* 17:17–25.

Donham, K. J. 1995. A review: the effects of environmental conditions inside swine housing on worker and pig health. In: V. Jennessy, and D. P. Cranwell, eds. *Manipulating pig production V*, 203–221.

Donham, K. J., M. J. Rubino, T. D. Thedell, and J. Kammermeyer. 1977. Potential health hazards of workers in swine confinement buildings. *J Occupational Medicine* 19:383–387.

Dosman, J. A., B. L. Grahm, D. Hall, P. Pahwa, H. McDuffice, and M. Lucewicz. 1988. Respiratory symptoms and alterations in pulmonary function tests in swine producers in Saskatchewan: results of a farm survey. *J Occupational Medicine* 30:715–720.

Doyle, M. E. Alternatives to Antibiotic use for growth promotion in animal husbandry. Food Research Institute (FRI) Briefings, University of Wisconsin–Madison, http://www.wisc.edu/fri/briefs/antibiot.pdf.

Dührsen, H. H. 1982. *Vergleichende Prüfung mehrerer Stalluft- und Stallbaumerkmale von 20 Schweinemastställen im Landkreis Dithmarschen in je drei Jahreszeiten unter Berücksichtigung der Hustenhäufigkeit*. (Comparison of some air quality parameters in 20 pig fattening units in the district Dithmarschen at three seasons in relation to the frequency of coughing). Thesis, University of Veterinary Medicine, Hannover.

Dutkiewicz, J., Z. J. H. Pomorski, J. Sitkowsaka, T. E. Krysinska, and H. Wojtowicz. 1994. Airborne microorganisms and endotoxin in animal houses. *Grana* 33:85–90.

Ekesbo, I. 1970. *Traditional and modern barn environments related to animal health and welfare*. Proceedings 11th Nordiske Veterinarkongres, Bergen, 56–63.

Elbers, A. R.W. 1991. *The use of slaughterhouse information in monitoring systems for herd health control in pigs*. Thesis, University of Utrecht, Netherlands.

Erwerth, W., G. Mehlhorn, and K. Beer. 1983. Die mikrobielle Kontamination der Luft in den Kälberställen einer Rindermastanlage. *Mh Vet-Med* 38:300–307.

European Union. 2003. EC regulation no. 1831/2003. On additives for use in animal nutrition. Official Journal of the European Union.

Ewbank, R. 1992. Stress: a general overview. In: C. Phillips and D. Piggins, eds. *Farm animals and the environment*. Wallingford, UK: CAB International, 255–262.

FAO. 2005. Responding to the "livestock revolution"—the case of livestock policies. Livestock Policy Brief 01, 8.

FAO. 2006. Livestock Report 2006. ftp://ftp.fao.org/docrep/fao/009/a0255e/a0255e.pdf.

Guarda, F., G. Tezzo, and C. Bianchi. 1980. Sulla frequenza e natura delle lesioni osservate in broilers al mace11o. (On frequency and kind of lesions observed in broilers at slaughter). *Clinica Veterinaria* 103:437–439.

Haglind, P., and R. Rylander, R. 1987. Occupational exposure and lung function measurements among workers in swine confinement buildings. *J Occupational Medicine* 29:904–907.

Hamann, J.1989. Faktoren der Genese boviner subklinischer Mastitiden (Factors of the genesis of bovine sub-clinical mastitis). In: *Pathogenesis and control of factorial diseases*. Proceedings 18th Congress, Deutsche Veterinärmedizinische Gesellschaft, Giessen, Bad Nauheim, 45–53.

Hammerum, A. M., O. E. Heuer, H. Emborg, L. Bagger-Skjøt, V. F. Jensen, A. Rogues, Y. Skov RL, Agersø, C. T. Brandt, A. M. Seyfarth, A. Muller, K. Hovgaard, J. Ajufo, F. Bager, F. M. Aarestrup, N. Frimodt-Møller, H. C. Wegener, D. L. Monnet. 2007. Danish Integrated Antimicrobial Resistance Monitoring and Research Program. *EID* 13 (11) : 1632–1639. www.cdc.gov/eid.

Hamscher, G., H. T. Pawelzick, S. Sczesny, H. Nau, and J. Hartung. 2003. Antibiotics in dust originating from a pig-fattening farm: a new source of health hazard for farmers? *Environmental Health Perspectives* 111:1590–1594.

Harrus, S., and G. Baneth. 2005. Drivers for the emergence and re-emergence of vector-borne protozoal and bacterial diseases. *International Journal for Parasitology* 35 (11–12): 1309–1318.

Hartung, J. 1994. The effect of airborne particulates on livestock health and production. In: I. Dewi, ed. *Pollution in livestock production systems*. Wallingford, UK: CAB International, 55–69.

Hartung, J. 2005. Luftverunreinigungen in der Nutztierhaltung. In: *Luftgetragene biologische Belastungen und Infektionen am Arbeitsplatz*. KTBL Schrift 436. Darmstadt, Kuratorium für Technik und Bauwesen in der Landwirtschaft, 7–19.

Hasman, H. and F. M. Aarestrup. 2005. Relationship between copper, glycopeptide, and macrolide resistance among *Enterococcus faecium* strains isolated from pigs in Denmark between 1997 and 2003. *Antimicrobial Agents and Chemotherapy* 49 (1) : 454–456.

Hasman, H., and F. M. Aarestrup. 2002. *tcrB*, a gene conferring transferable copper resistance in *Enterococcus faecium*: occurrence, transferability, and linkage to macrolide and glycopeptide resistance. *Antimicrobial Agents and Chemotherapy* 46:1410–1416.

Hasman, H., I. Kempf, B. Chidaine, R. Cariolet, A. Kjaer Esbøll, H. Howe, H. C. Bruun Hansen, and F. M. Aarestrup. 2006. Copper resistance in *Enterococcus faecium*, mediated by the *tcrB* gene, is selected by supplementation of pig feed with copper sulfate. *Applied and Environmental Microbiology* 72 (9) : 5784–5789.

Hatanaka, M., C. Bain C., and L. Busch. 2005. Third-party certification in the global agrifood system. *Food Policy* 30:354–369.

Health Protection Agency. 24 September 2007. Infections caused by ESBL-producing *E. coli*. http://www.hpa.org.uk/webw/HPAwebandHPAwebStandard/HPAweb_C/1195733708724?p=1171991026241.

Helfand, M. S., R. A. Bonomo. 2006. Extended-spectrum b-lactamases in multidrug-resistant *Escherichia coli*: changing the therapy for hospital-acquired and community-acquired infections. *Clin Infect Dis* 43:1415–1416.

Heuer, H., A. Focks, M. Lamshöft, K. Smalla, M. Matthies, and M. Spiteller. 2008. Fate of sulfadiazine administered to pigs and its quantitative effect on the dynamics of bacterial resistance genes in manure and manured soil. *Soil Boil Biochem* 40:1892–1900.

Heuer, H, C. Kopmann, C. T. T. Bihh, E. M. Top, K. Smalla. 28 November 2008. Spreading antibiotic resistance through spread manure: characteristics of a novel plasmid type with low %G+C content. *Environmental Microbiology* http://www.ncbi.nlm.nih.gov/pubmed/19055690?ordinalpos=2anditool=EntrezSystem2.PEntrez.Pubmed.Pubmed_ResultsPanel.Pubmed_DefaultReportPanel.Pubmed_RVDocSum.

Hilliger, H. G. 1990. *Stallgebäude, Stalluft und L ftung*. (Housing, air and ventilation). Stuttgart: Ferdinand Enke Verlag.

Hirst, J. M. 1995. Bioaerosols: introduction, retrospect and prospect. In: C. S. Cox and C. M. Wathes, eds. *Bioaerosols handbook*. Boca Raton: CRC Press.

Institute of Medicine. 1998. Antimicrobial resistance: issues and options. In P. F. Harrison and J. Lederberg, eds. Division of Health Sciences Policy, Forum on Emerging Infections. Washington, DC: National Academy Press. www.nap.edu.

Institute of Medicine, Board on Global Health. 2003. Microbial threats to health: emergence, detection and response. Washington, DC: National Academies Press, 207. http://books.nap.edu/books/030908864X/html/207.html#pagetop.

Iversen, M. 1999. Humans effects of dust exposure in animal confinement buildings. *Proceedings of the Dust Control in Animal Production Facilities International Symposium*, Jutland 31 May to 2 June 1999, 131–139.

Jackson, A., and M. Mahon. 1995. *Occupational health and safety for the Australian pig industry*. PRDC and Queensland Farmers' Federation.

Jasmin, G., and M. Cantim. 1991. Stress revisited: neuroendocrinology of stress. In: G. Jasmin, ed. *Methods and achievements in experimental pathology*, vol. 14. Basel: Verlag Karger.

JETACAR. 1999. The use of antibiotics in food-producing animals: antibiotic-resistant bacteria in animals and humans. Joint Expert Advisory Committee on Antibiotic resistance (JETACAR). Canberra: Commonwealth Department of Health and Aged Care and the Commonwealth Department of Agriculture, Fisheries and Forestry, September 1999. http://www.health.gov.au/pubs/jetacar.htm.

Jouan, A. F., O. Adam, B. Riou, N. O. Philippe, Merzoug, T. Ksiazek, B. Leguenno, and J. P. Digoutte. 1990. Evaluation des indicateurs de sante dans la region du Trarza lors de l'épidémie de fièvre de la Vallée du Rift en 1987. *Bull Soc Pathol Exot Filiales* 83:621–627.

Kahn, L. H. 2006. Confronting zoonoses, linking human and veterinary medicine. *Emerg Infect Dis* 12: 556–1561.

Karesh, W. B., R. A. Cook, E. L. Bennet, and J. Newcomb. 2005. Wildlife trade and global disease emergence. *Emerg. Infect. Dis.* 11 (7):1000–1002.

Kieke, A. L., M. A. Borchardt, S. K. Spencer, M. F. Vandermause, K. E. Smith, S. L. Jawahir, E. A. Belongia. 1 November 2006. Use of streptogramin growth promoters in poultry and isolation of streptogramin-resistant *Enterococcus faecium* from humans. *J Infect Dis* 194 (9) : 1200–1208.

Klare, I., C. Konstabel, D. Badstübner, G. Werner, and W. Witte. 2003. Occurrence and spread of antibiotic resistances in Enterococcus faecium. *Int. J. Food Microbiol.* 88 (2–3):269–290.

Knobel, D. L., S. Cleaveland, P. G. Coleman, E. M. Fevre, M. I. Meltzer, and M. E. Miranda. 2005. Re-evaluating the burden of rabies in Africa and Asia. *Bull. World. Health Organ.* 83 (5): 360–368.

Knobler, S., A. Mahmoud, S. Lemon, A. Mack, L. Sivitz, and K. Oberholtzer, eds. 2004. Learning from SARS: preparing for the next disease outbreak. Workshop summary. Washington, DC: National Academies Press.

Konermann, H. 1992. Eutererkrankungen des Rindes: Ursachen und Bekämpfungsmöglichkeiten. (Mastitis in cattle: aetiology and control). In: *Milchviehhaltung. (Dairy production).* Landwirtschaftsverlag Murister. Baubriefe Landwirtschaft 33, 12–20.

Krönke, F. 2001. Perception of ill-health in a FulBe pastoralist community and its implications on health interventions in Chad. Ph.D thesis, University of Basel, Switzerland.

Ladewig, J., and E. von Borell. 1988. Ethological methods alone are not sufficient to measure the impact of environment on animal health and animal wellbeing. In: J. Unshelm, G. van Putten, K. Zeeb, and I. Ekesbo, eds. *Proceedings of the International Congress on Applied Ethology in Farm Animals.* Skara Darmstadt: KTBL, 95–102.

Lapedes, D. N., ed. 1978. *Dictionary of scientific and technical terms.* New York: McGraw-Hill.

Leach, K. A., D. N. Logue, S. A. Kempson, J. E. Offer, H. E. Ternent, and J. M. Randall. 1998. Claw lesions in dairy cattle: development of sole and white line lesions during the first lactation. *Vet J* 155:215–255.

Levy, S. 1998. The challenge of antibiotic resistance. *Scientific American* (3):46–48.

Levy, S. B., G. B. FitzGerald, and A. B. Macone. 1976. Changes in intestinal flora of farm personnel after introduction of a tetracycline-supplemented feed on a farm. *N. Engl. J. Med.* 295 (11):583–588.

Lotthammer, K. H. 1989. Tierärztliche Aspekte der Kälberaufzucht und Rindermast. (Veterinary aspects of calf rearing and beef production). In: *Kälberaufzucht, Jungviehhaltung, Rindermast. (Calf and heifer rearing, beef production).* Landwirtschaftsverlag Münster. Baubriefe Landwirtschaft, 31, 12–17.

Lundervold, M., E. J. Milner-Gulland, C. J. O'Callaghan, C. Hamblin, A. Corteyn, and A. P. Macmillan. 2004. A serological survey of ruminant livestock in Kazakhstan during post-Soviet transitions in farming and disease control. *Acta Vet Scand* 45 (3–4) : 211–224.

Madsen, E. B., and K. Nielsen. 1985. A study of tail tip necrosis in young fattening bulls on slatted floors. *Nord Vet-Med* 37:349–357.

Majok, A. A., and C. W. Schwabe. 1996. Development among Africa's migratory pastoralists. Westport: Greenwood Publishing.

Malberg, P., and K. Larsson. 1992. *Acute exposure to swine dust causes bronchial hyperresponsiveness.* Stokloster Workshop 3, Stoklostwer, Sweden April 6–9.

Mayr, A. 1984. Allgemeine Infektions- und Seuchenlehre. (Principles of infection and epidemics). In: A. Mayr, ed. *Medizinische Mikrobiologie, Infektions- und Seuchenlehre.* (Medical microbiology, infection and epidemics). Stuttgart: Ferdinand Enke Verlag, 1–142.

McDermott, J. J., and S. M. Arimi. 2002. Brucellosis in sub-Saharan Africa: Epidemiology, control, and impact. *Vet. Microbiol.* 90 (1–4):111–134.

McEwen, S. A., and P. Fedorka-Cray. 2002. Antimicrobial use and resistance in animals. *Clinical Infectious Diseases* 34 (Suppl 3) : S93–S106.

Mellon, M., C. Benrook, and K. L. Benbrook. 2001. *Hogging it: Estimates of antimicrobial abuse in livestock.* Boston: Union of Concerned Scientists. http://www.ucsusa.org/assets/documents/food_and_agriculture/hog_front.pdf.

Moberg, G. P. 1987. Problems in defining stress and distress in animals. *J Am Vet Med Ass* 191:1207–1211.

Monreal, G. 1989. Infekt se Faktorenkrankheiten beim Geflügel. (Infectious factorial diseases in poultry). In: *Pathogenesis and control of factorial diseases.* Proceedings 18th Congress, Deutsche Veterinärmedizinische Gesellschaft, Giessen, Bad Nauheim, 180–192.

Mount, L. E., and I. B. Start. 1980. A note on the effect of forced air movement and environmental temperature on weight gain in the pig after weaning. *Anim Prod* 30:295–298.

Mülling, C. K. W., and C. J. Lischer, C.J. 2002. New aspects on etiology and pathogenesis of laminitis in cattle. In: *Recent developments and perspectives in bovine medicine.* Hannover, 236–247.

Mundy, A., and J. Favole. 10 December 2008. FDA calls off ban on animal antibiotics. *Wall Street Journal* http://online.wsj.com/article/SB122887467038993653.html?mod=google news_wsj.

National Academy of Sciences/National Research Council. 1999. *The use of drugs in food animals: benefits and risks.* Washington, DC: National Academy Press. www.nap.edu/books/0309054346/html/30.html.

Navruzschoeva, G. 2005. Epidemiological monitoring and upgrading of specific prophylaxis of Anthrax. Autoreferat of the dissertation work, Moscow, 3–20.

Neumann, R. 1988 *Der Schwimmtest für Saugferkel als Grundbelastungsmodell zur Untersuchung der Wirkung der Umwelt auf die Infektionsabwehr unter besonderer Ber cksichtigung der experimentellen Infektion des Atmungstraktes.* (The swim test for suckling pigs as model to investigate the effect of the environment on infection defense with particular respect to the artificial infection of the respiratory tract.) Thesis, University Leipzig.

Nilsson, C. 1984. Experiences with different methods of dust reduction in pig houses. In: G. Hilliger, ed. *Dust in animal houses.* Proceedings of German Veterinary Association, 13–14 March, Hannover, 90–91.

Nilsson, C. 1992. Walking and lying surfaces in livestock houses. In: C. Phillips and D. Piggins, eds. *Farm animals and the environment.* Wallingford, UK: CAB International, 93–110.

Nowak, D. 1998. Die Wirkung von Stalluftbestandteilen, insbesondere in Schweineställen, aus arbeitsmedizinischer Sicht. *Dtsch tierärztl Wschr* 105:225–234.

Nowak, D., G. Denk, R. Jörres, D. Kirsten, F. Koops, D. Szadkowski, B. Wiegand, J. Hartung, and H. Magnussen. 1994. Endotoxin-related inflammatory response in nasal lavage fluid after nasal provocation with swine confinement dusts. *Am J Resp Crit Care Med* 149:A401.

Nunn, M., and P. Black. 2006. Intensive animal production systems: how intensive is intensive enough? 11th Symposium of the International Society for Veterinary Epidemiology and Economics. http://www.sciquest.org.nz/crusher_download.asp?article=10002927.

O'Brien, T. F. 2002. Emergence, spread, and environmental effect of antimicrobial resistance: How use of an antimicrobial anywhere can increase resistance to any antimicrobial anywhere else. *Clinical Infectious Diseases* 34 (Suppl 3): S85–S92.

Offer, J. E., D. McNulty, and D. N. Logue. 2000. Observations of lameness, hoof conformation and development of lesions in dairy cattle over four lactations. *Vet Rec* 147:105–109.

Office International des Epizooties (World Organisation for Animal Heath), World Health Organization, and Food and Agriculture

Organization of the United Nations. 2004a. Joint FAO/OIE/ WHO 1st expert workshop on non-human antimicrobial usage and antimicrobial resistance: scientific assessment. Geneva, Switzerland, 1–5 December 2003. http://www.who.int/food-safety/publications/micro/en/report.pdf.

Office International des Epizooties (World Organization for Animal Heath), World Health Organization, and Food and Agriculture Organization of the United Nations. 2004b. Joint FAO/ OIE/WHO 2nd workshop on non-human antimicrobial usage and antimicrobial resistance: management options. Oslo, Norway, 15–18 March 2004. http://www.who.int/foodsafety/ publications/micro/en/exec.pdf.

Office International des Epizooties (World Organisation for Animal Heath), World Health Organization, and Food and Agriculture Organization of the United Nations. 2006. Joint FAO/ OIE/WHO Expert Consultation on Antimicrobial Use in Aquaculture and Antimicrobial Resistance, Seoul, Republic of Korea, 13–16 June 2006. ftp://ftp.fao.org/ag/agn/food/aquaculture_rep_13_16june2006.pdf.

Omore A., S. J. Staal, L. Kurwijila, E. Osafo, K. G. Aning, N. Mdoe, and G. Nurah. 2001. Indigenous markets for dairy products in Africa: trade-offs between food safety and economics. Proceedings of Symposium on Dairy Development in the Tropics, 2 November 2001, Utrecht University, Utrecht (Netherlands), 19–24.

Papasolomontos, P. A., E. C. Appleby, and O. Y. Mayor. 1969. Pathological findings in condemned chickens: a survey of 1.000 carcasses. Vet Rec 85:459-464.

Pedersen, S. 1998. Staubreduzierung in Schweineställen. Dtsch tierärztl Wschr 105:247–250.

Perry, B., T. F. Randolph, and J. McDermott. 2002. Investing in animal health research to alleviate poverty. ILRI (International Livestock Research Institute), Nairobi, Kenya.

Price, L. B., J. P. Graham, L. G. Lackey, A. Roess, R. Vailes, and E. Silbergeld. 2007. Elevated risk of carrying gentamicin-resistant Escherichia coli among U.S. poultry workers. Environ Health Perspect 115 (2) : 1738-1742.

Radon, K., E. Monso, C. Weber, B. Danuser, M. Iversen, U. Opravil, K. Donham, J. Hartung, S. Pedersen, S. Garz, D. Blainey, U. Rabe, and D. Nowak. 2002. Prevalence and risk factors for airway diseases in farmers: summary of results of the European farmers' project. Ann Agric Environ Med 9:207–213.

Raymond, S., E. Curtis, L. Winfield, and A. Clarke. 1997. A comparison of respirable particles associated with various forage products for horses. Equine Practice 23–26.

Rodriguez-Bano, J., M. D. Navarro, L. Romero, et al. 2006. Bacteremia due to extended-spectrum blactamase–producing Escherichia coli in the CTX-M era: a new clinical challenge. Clin Infect Dis 43:1407–1414.

Roth, F., J. Zinsstag, D. Orkhon, G. Chimed-Ochir, G. Hutton, and O. Cosivi. 2003. Human health benefits from livestock vaccination for brucellosis: case study. Bull World Health Organ 81 (12):867–876.

Sapkota, A. R., F. C. Curriero, K. E. Gibson, and K. J. Schwab. 2007b. Antibiotic-resistant enterococci and fecal indicators in surface water and groundwater impacted by a concentrated Swine feeding operation. Environ Health Perspect 115 (7) : 1040–1045.

Sapkota, A. R., L. Y. Lefferts, S. McKenzie, and P. Walker. 2007. What do we feed to food-production animals? A review of animal feed ingredients and their potential impacts on human health. Environ Health Perspect 115:663–670.

Sapkota, A. R., L. B. Price, E. K. Silbergeld, and K. J. Schwab. April 2006. Arsenic resistance in Campylobacter spp. isolated from retail poultry products. Appl Environ Microbiol 72 (4) : 3069-3071.

Scheepens, C. J. M. 1991. Effects of draught as climatic stressor on the health status of weaned pigs. Thesis, University of Utrecht, Netherlands.

Schelling, E., S. Daoud, D. M. Daugla, P. Diallo, M. Tanner, and J. Zinsstag. 2005. Morbidity and nutrition patterns of three nomadic pastoralist communities of Chad. Acta. Trop. 95:16–25.

Schelling, E., K. Wyss, M. Bechir, D. D. Moto, and J. Zinsstag. 2005. Synergy between public health and veterinary services to deliver human and animal health interventions in rural low-income settings. BMJ 331:1264–1267.

Schmidt, M., M. Jørgensen, A. A. Møller-Madsen, H. Jensen, Z. Horvath, P. Keller, and S. P. Konggaard. 1985. Straw bedding for dairy cows. Report 593 from the National Institute of Animal Science, Copenhagen.

Scholz, F. 1991. Nomaden - Mobile Tierhaltung: zur gegenwärtigen Lage von Nomaden und zu den Problemen und Chancen mobiler Tierhaltung. Berlin: Das Arabische Buch.

Schulz, J. 2008. Zur Charakterisierung der Ausbreitungsentfernung von Bioaerosolen aus Masthühnerställen. (Estimation of airborne transmission distances for bio-aerosols). Thesis rer. nat., University of Bielefeld, Germany.

Schwabenbauer, K., and J. Rushton. 2007. Veterinary services for poultry production. In: FAO. 2008. Poultry in the 21st century: avian influenza and beyond. ftp://ftp.fao.org/docrep/fao/011/ i0323e/i0323e.pdf.

Seabrook, M.F., and N. C. Bartle. 1992. Human factors. In: C. Phillips and D. Piggins, eds. Farm animals and the environment. Wallingford, UK: CAB International, 111–125.

Seedorf, J., and J. Hartung. 2002. Stäube und Mikroorganismen in der Tierhaltung. KTBL-Schrift 393. Münster: Landwirtschaftsverlag GmbH.

Seedorf, J., J. Hartung, M. Schröder, K. H. Linkert, V. R. Phillips, M. R. Holden, R. W. Sneath, J. L. Short R. P. White, S. Pedersen, T. Takai, J. O. Johnsen, J. H. M. Metz, P. W. G. Groot Koerkamp, G. H. Uenk, and C. M. Wathes. 1998. Concentrations and emissions of airborne endotoxins and microorganisms in livestock buildings in northern Europe. Journal of Agricultural Engineering Research 70:97–109.

Selye, H. 1974. Stress without distress. Philadelphia: Lippincott.

Shaw, F. D., and R. K. Tume. 1992. The assessment of pre-slaughter and slaughter treatments of livestock by measurement of plasma constituents: a review of recent work. Meat Science 32:311–329.

Shearer, J. K., and S. R. van Amstel. 2002. Claw health management and therapy of infectious claw diseases. In: Recent developments and perspectives in bovine medicine, 258–267.

Silbergeld, E. K., J. Graham, and L. B. Price. 2008. Industrial food animal production, antimicrobial resistance, and human health. Annual Review of Public Health 29:151–169.

Sjölund, M., J. Yam, J. Schwenk, K. Joyce, F. Medalla, E. Barzilay, et al. December 2008. Human Salmonella infection yielding CTX-M β-lactamase, United States (letter). Emerg Infect Dis http://www.cdc.gov/EID/content/14/12/1957.htm.

Slingenbergh, J., and M. Gilbert. 2007. Do old and new forms of poultry go together? In: FAO. 2008. Poultry in the 21st century: avian influenza and beyond. ftp://ftp.fao.org/docrep/fao/011/ i0323e/i0323e.pdf.

Smith, D. L., A. D. Harris, J. A. Johnson, E. K. Silbergeld, and J. G.

Morris. 2002. Animal antibiotic use has an early but important impact on the emergence of antibiotic resistance in human commensal bacteria. *PNAS* 99 (9) : 6434–6439.

Steinfeld, H., P. Gerber, T. Wassenaar, V. Castel, M. Rosales, and C. de Haan. 2006. Livestock's long shadow: environmental issues and options. United Nations Food and Agriculture Organization. http://www.fao.org/docrep/010/a0701e/a0701e00.htm.

Steinfeld, H., and J. Mäki-Hokkonen. 1995. A classification of livestock production systems. *World Animal Revue,* No. 84/85.

Stöber, M. 1989. Zur Pathogenese multifaktoriell bedingter Krankheiten und buiatrisch-klinischer Sicht. (Pathogenesis of multifactorial diseases from the point of view of bovine medicine). In: *Pathogenesis and control of factorial diseases.* Proceedings 18th Congress, Deutsche Veterinärmedizinische Gesellschaft, Giessen, Bad Nauheim, 19–35.

Strauch, D. 1987. Hygiene of animal waste management. In: D. Strauch, ed. *Animal production and animal health.* Amsterdam: Elsevier Science, 155–202.

Summers, A. O. 2002. Generally overlooked fundamentals of bacterial genetics and ecology. *Clinical Infectious Diseases* 34:s3, S85–S92.

Takai, H., S. Pedersen, J. O. Johnsen, J. H. M. Metz, P. W. G. Groot Koerkamp, G. H. Uenk, V. R. Phillips, M. R. Holden, R. W. Sneath, J. L. Short, R. P. White, J. Hartung, J. Seedorf, M. Schröder, K. H. Linkert, and C. M. Wathes. 1998. Concentrations and Emissions of Airborne Dust in Livestock Buildings in Northern Europe. *Journal of Agricultural Engineering Research*, 70:59–77.

Thébaud, B., and S. Batterbury. 2001. Sahel pastoralists: opportunism, struggle, conflict and negotiation: a case study from eastern Niger. *Global Environmental Change* 11 (1):69–78.

Tielen, M. J. M. 1977. *Stallklima und Tiergesundheit in Schweinemastbetrieben. (Indoor climate and animal health in fattening piggeries).* Publication of the Animal Health Office in Nord-Brabant, Boxtel, Netherlands.

Unicomb, L. E., J. Ferguson, R. J. Stafford, R. Ashbolt, M. D. Kirk, N. G. Becker, M. S. Patel, G. L. Gilbert, M. Valcanis, and L. Mickan. 2006. Low-level fluoroquinolone resistance among *Campylobacter jejuni* isolates in Australia. *Clinical Infectious Diseases* 42:1368–1374.

U.S. Food and Drug Administration. 1997a. Notice of opportunity for hearing. Penicillin-containing premixes: opportunity for hearing. *Federal Register* 42:43772–43793.

U.S. Food and Drug Administration. 1997b. Notice of opportunity for hearing. Tetracycline (chlortetracycline and oxytetracycline)-containing premixes: opportunity for hearing. *Federal Register* 42:56264–56289.

U.S. Food and Drug Administration. 2000. Notice of opportunity for hearing. Enrofloxacin for poultry: opportunity for hearing. *Federal Register* 65:64954–64965.

U.S. Food and Drug Administration. 2001a. Abbott Laboratories' sarafloxacin for poultry; withdrawal of approval of NADAs. *Federal Register* 66:21400–21401.

U.S. Food and Drug Administration. 2001b. Enrofloxacin in poultry: opportunity for hearing: Correction. *Federal Register* 66:6623–6624.

U.S. Food and Drug Administration. 15 April 2002. Summary of data from hatchery inspections conducted September–October 2001. FDA Internal document obtained by Freedom of Information Request.

U.S. Food and Drug Administration. 2003. Evaluating the safety of antimicrobial new animal drugs with regard to their microbiological effects on bacteria of human health concern. Guidance for Industry #152. http://www.fda.gov/cvm/Guidance/fguide152.pdf.

U.S. Food and Drug Administration. 2004a. Center for Veterinary Medicine, 2004 Online Green Book. http://www.fda.gov/cvm/greenbook/elecgbook.html.

U.S. Food and Drug Administration. 2004b. Memo: summary of data from hatchery inspections conducted. Response to FOIA request F03-14816. 23 June 2004. U.S. Department of Health and Human Services, National Antimicrobial Resistance Monitoring System/Enteric Bacteria (NARMS/EB) Salmonella Annual Veterinary Isolates Data, U.S. Department of Agriculture, http://www.ars.usda.gov/Main/docs.htm?docid=6750andpage=4, 2006.

U.S. Food and Drug Administration. 28 July 2005. Final decision of the commissioner: withdrawal of approval of the new animal drug application for enrofloxacin in poultry. Docket no. 2000N-1571. http://www.fda.gov/oc/antimicrobial/baytril.pdf.

U.S. Food and Drug Administration. 2006. NARMS retail meat annual report, 2006. National Antimicrobial Resistance Monitoring System, Center for Veterinary Medicine. http://www.fda.gov/cvm/2006NARMSAnnualRpt.htm.

U.S. Food and Drug Administration, Department of Health and Human Services (FDA). 3 July 2008a. An order prohibiting the extralabel use of cephalosporin antimicrobial drugs in food-producing animals. *Federal Register* 73 (129). http://edocket.access.gpo.gov/2008/E8-15052.htm.

U.S. Food and Drug Administration, Department of Health and Human Services (FDA). 26 November 2008b. Revocation of order of prohibition. *Federal Register* 73 (229). http://edocket.access.gpo.gov/2008/E8-28123.htm.

U.S. General Accounting Office (GAO). 2004. Federal agencies need to better focus efforts to address risk to humans from antibiotic use in animals. http://www.gao.gov/cgi-bin/getrpt?GAO-04-490.

Valentin, A., V. Bergmann, J. Scheer, I. Tschirch, and H. Leps. 1988. Tierverluste und Qualitätsminderungen durch Hauterkrankungen bei Schlachtgeflügel. (Dead losses and losses of meat quality by skin disease in broilers). *Mh Vet -Med* 43:686–690.

Verhagen, J. M. F. 1987. *Acclimation of growing pigs to climatic environment.* Thesis, Agricultural University of Wageningen, Netherlands.

Vermunt, J. J., and M. E. Smart. 1994. The impact of a lameness management programme. In: VIIIth *Symposium on Disorders of the Ruminant Digit and International Conference on Bovine Lameness,* 275–298.

Walsh, C. 2000. Molecular mechanisms that confer antibacterial drug resistance. *Nature* 406 (6797):775–781.

Wathes, C. M. 1994. Air and surface hygiene. In: C. M. Wathes and D. R. Charles, eds. *Livestock housing.* Wallingford, UK: CAB International, 123–148.

Webster, A. J. F. 1982. Improvements of environment, husbandry and feeding. In: H. Smith and J. M. Payne, eds. *The control of infectious diseases in farm animals.* London: British Veterinary Association Trust, 28–36

Webster, A. J. F. 1985. Animal health and the housing environment. In: *Animal health and productivity.* Royal Agricultural Society of England, 227–242.

Webster, A. J. F. 2001. Effects of housing and two forage diets on the development of claw horn lesions in dairy cows at first calving and in first lactation. *Veterinary Journal* 162:56–65.

Webster, J. 2002. Effect of environment and management on the development of claw and leg diseases. In: *Recent developments and perspectives in bovine medicine*, 18–23th August 2002, Hannover, Germany, 248–256.

Wegener, H. C. 2003. Antibiotics in animal feed and their role in resistance development. *Current Opinions in Microbiology* 6:439–445.

Weiss, R. 4 March 2007. FDA rules override warnings about drug. *Washington Post*, http://www.washingtonpost.com/wp-dyn/content/article/2007/03/03/AR2007030301311.html.

Wells, A. 2003. *Reducing antimicrobial use: innovative approaches for meeting consumer demands while maintaining healthy animals and safe foods –case studies in animal production industries.* Interpretative Summaries from the Seventh DISCOVER Conference on Food Animal Agriculture Is there a Future for Antibiotics in Animal Agriculture? September 21–24, 2003, Nashville, Indiana. http://www.adsa.org/discover/past.html.

Wheelock, J. V., and Foster, C. September 2002. *Food safety and pig production in Denmark.* Report Commissioned by the Danish con and Meat Council: Verner Wheelock Associates, Shipton, United Kingdom. http://www.thepigsite.com/articles/715/food-safety-and-pig-production-in-denmark.

White, D. G., S. Zhao, R. Sudler, S. Ayers, S. Friedman, S. Chen, et al. 2001. The isolation of antibiotic-resistant *Salmonella* from retail ground meats. *N Engl J Med* 345:1147–1154.

Whyte, R. T. P. A. M. Williamson, and J. Lacey. 1993. Air pollutant burdens and respiratory impairment of poultry house stockmen. Livestock Environment, 4th International Symposium, University of Warwick, Coventry, ASAE St. Josepf, MI, UK, 6–9 July 1993, 709–717.

Wierup, M. 2001. The Swedish experience of the 1986 year ban of antimicrobial growth promoters, with special reference to animal health, disease prevention, productivity, and usage of antimicrobials. *Microbial Drug Resistance* 7 (2) : 183–190.

Windhorst, H. W. 2005. Ein neues Bild der Landwirtschaft. (A new view of agriculture). In: H. W. Windhorst, ed. *Der Agrarwirtschaftsraum Oldenburg im Wandel.* (Agricultural economy of Oldenburg in changing times). Die Violette Reihe, No. 3. Vechta: Vechtaer Druckerei und Verlag GmBH, 9–18.

World Bank. 2006. Economic impact of avian flu: Global program for avian influenza and human pandemic. http://web.worldbank.org/WBSITE/EXTERNAL/COUNTRIES/EASTASIAPACIFICE

World Health Organization. 2000. *WHO global principles for the containment of antimicrobial resistance in animals intended for food.* Report of a WHO Consultation with the participation of the Food and Agriculture Organization of the United Nations and the Office International des Epizooties Geneva, Switzerland, 5–9 June 2000. Full report available at http://www.who.int/emc/diseases/zoo/who_global_principles/index.htm.

World Health Organization. 2003. Impacts of antimicrobial growth promoter termination in Denmark: The WHO international review panel's evaluation of the termination of the use of antimicrobial growth promoters in Denmark. http://www.who.int/salmsurv/links/gssamrgrowthreportstory/en/

World Health Organization. 2006. The control of neglected zoonotic diseases: A route to poverty alleviation. Department for International Development, Animal Health Programme. WHO/SDE/FOS. http://www.who.int/zoonoses

World Health Organization. 2007. Critically important antimicrobials for human medicine: categorization for the development of risk management strategies to contain antimicrobial resistance due to non-human antimicrobial use. Report of the Second WHO Expert Meeting Copenhagen, 29–31 May 2007. http://www.who.int/foodborne_disease/resistance/antimicrobials_human.pdf.

Youssef, M. K. 1988. Animal stress and strain: definition and measurements. *Appl Anim Behav Sci* 20:119–126.

Zeitler, M. 1988. Hygienische Bedeutung des Staubes- und Keimgehaltes der Stallluft. (Hygienic significance of dust and bacteria in animal house air). *Bayerisches Landwirtschaftliches Jahrbuch* 65:151–165.

Zhao, S., P. F. McDermott, S. Friedman, J. Abbott, S. Ayers, A. Glenn, E. Hall-Robinson, S. K. Hubert, H. Harbottle, R. D. Walker, T. M. Chiller, and D. G. White. 2006. Antimicrobial resistance and genetic relatedness among *Salmonella* from retail foods of animal origin: NARMS retail meat surveillance. *Foodborne Pathog Dis* 3 (1) : 106–117.

Zhao, S., D. G. White, S. L. Friedman, A. Glenn, K. Blickenstaff, S. L. Ayers, J. W. Abbott, E. Hall-Robinson, and P. F. McDermott. 2008. Antimicrobial resistance in *Salmonella enterica* serovar Heidelberg isolates from retail meats, including poultry, from 2002 to 2006. *Appl Environ Microbiol* 74 (21) : 6656–6662.

Zinsstag, J., M. Ould Taleb, and P. S. Craig. 2006. Health of nomadic pastoralists: New approaches towards equity effectiveness. *Tropical Medicine and International Health* 11 (5):565–568.

Zinsstag, J., F. Roth, D. Orkhon, G. Chimed-Ochir, M. Nansalmaa, and J. Kolar, et al. 2005b. A model of animal–human brucellosis transmission in Mongolia. *Preventive Veterinary Medicine* 69:77–95.

Zinsstag J., F. Schelling, F. Roth, B. Bonfoh, D. de Savigny, and M. Tanner. 2007. Human benefits of animal interventions for zoonosis control. *Journal of Emerging Infectious Disease* 13 (4).

Zinsstag, J., E. Schelling, K. Wyss, and M. Bechir. 2005a. Potential of cooperation between human and animal health to strengthen health systems. *Lancet* 2142–2145.

Zuskin, E., B. Kanceljak, E. Schlachter, J. Mustajbegovic, S. Giswami, S. Maayani, Z. Marom, and N. Rienzi. 1991. Immunological and respiratory findings in swine farmers. *Environmental Research* 56:120–130.

12

The Livestock Revolution and Animal Source Food Consumption

Benefits, Risks, and Challenges in Urban and Rural Settings of Developing Countries

Charlotte G. Neumann, Montague W. Demment, Audrey Maretzki, Natalie Drorbaugh, and Kathleen A. Galvin

Main Messages

- **Animal source foods (ASFs) are important for optimal protein, energy, and micronutrient nutrition.** Animal source foods of a wide variety in the human diet provide high-quality, complete, readily digestible protein, energy, and vital micronutrients in bioavailable form. Meat is an excellent source of complete protein, vitamin B-12, riboflavin, heme iron, and zinc, while milk is an excellent source of complete protein, calcium, vitamin A, riboflavin, and vitamin B-12. Hence, ASFs, particularly meats of a wide variety, have the potential to address multiple macro- and micronutrient deficiencies and to enhance diet quality.

- **The inclusion of ASFs, particularly meat, in the diet promotes growth, cognitive function, physical activity, and health.** Consumption of ASFs is particularly important for women of reproductive age to improve optimal maternal nutritional and health status, fetal growth, and pregnancy outcome and for young children to ensure normal growth and development. Further, causal evidence from a recent Kenyan intervention study shows that even a modest amount of meat in the diet of schoolchildren improves cognitive function, school performance, and physical activity and increases muscle mass and micronutrient levels.

- **Nutritional risks and decreased function are associated with lack of ASF consumption.** Diets devoid of ASFs are often low in energy, quality protein, and multiple micronutrients. Such diets are associated with reduced physical activity and work capacity, poor growth, increased infection, reduced cognitive capacity, and nutritional anemia.

- **The livestock revolution has affected urban and periurban populations to a much greater extent than rural subsistence populations in developing countries, particularly in sub-Saharan Africa.** The livestock revolution is mainly driven by Asia and Latin America. Hence, while levels of production of meat may have increased, undernutrition remains a large problem for those without access to ASFs and with food insecurity, particularly among the urban and rural poor.

- **Urban populations in developing countries are undergoing a nutrition transition.** These populations are showing increases in chronic diseases, and traditional diets are giving way to greater consumption of energy, sugars, fats, and ASFs and to decreased intake of fruits, vegetables, and grains. This is occurring in combination with an altered lifestyle, decreased physical activity, with diet changes to street food and fast food, and with a decreased ability to grow fruits and vegetables in urban settings.

- **The livestock revolution and increased meat consumption are not equated with universal dietary improvement and optimal health.** Chronic cardiovascular disease, hypercholesterolemia, hypertension, stroke, obesity, and type 2 diabetes are increasing in industrial and developing countries, particularly in urban areas. Intake of high-fat meats and sugar is associated with increased risk of developing the above conditions.

- **Health risks are associated with overconsumption of ASFs.** Overconsumption of ASFs high in saturated fat has been associated with a number of chronic conditions and cancer. The risk for these is affected by individual differences in genetics, metabolism, lifestyle and environmental factors, and the general type of main diet. Therefore, it is difficult to precisely define an optimal range of meat consumption. Accordingly, the World Cancer Research Fund and American Institute for Cancer Research recommended in 2007 that red meat consumption be less than 500 g (18 oz.) per week, with little if any processed meat.

- **Factors at multiple levels—community, household, and individual—determine ASF consumption in rural, subsistence areas.** At the community level, lack of land for growing forage and for grazing, lack of or poor extension and animal health services, and poorly developed markets inhibit animal husbandry and ASF consumption. At the household and individual levels, poverty, cultural, and religious factors governing intrahousehold food distribution and an inability to preserve meat influence ASF consumption. Moreover, cultural beliefs and people's lack of knowledge about the value of ASFs in the diets of women and children influence their consumption.

- **Nonlivestock and small animals are important sources of ASF intake and nutritional improvement.** A wide variety of small animals, including rabbits, rodents, birds (wild range chickens), wild game, fish, insects, slugs, and mollusks, are important sources of ASFs globally. Large-scale livestock production is not necessary to improve nutritional status, particularly on a household level in rural populations. Many small animals and other ASF sources that can be raised by the household have the potential to improve diet quality.

- **Traditional pastoralism promotes intake of ASFs.** Traditional pastoralists have a substantial intake of ASFs, particularly of milk and dairy products, due to better access to these products. The transition of pastoralists in recent years to living in settled areas, with the cultivation of crops, has separated them from herds, with decreased access to ASFs and poorer diet quality. Availability of grazing lands has been steadily declining, particularly in Africa.

- **There is a need for increased and improved agricultural extension services by women for women in order to promote small-animal husbandry for household animal production for consumption.** Agricultural extension is vital to promoting knowledge about and improvements in household small-animal husbandry at a household level. Women along with school-aged children perform most small-animal rearing activities in many areas of the world. Knowledge about how to preserve and prepare ASFs and nutrition education about the value of ASFs in the diet is widely needed.

- **Adequate nutrition is essential for economic development and optimal human functioning within society.** Nutrition improvement is vital and should be an integral part of health, education, and development efforts. Food-based approaches in rural areas are most likely to be sustainable in improving diet quality with ASFs in contrast to "pill-based" approaches. The latter can be costly and nonsustainable, with serious problems of population coverage, targeting of recipients, and distribution.

Introduction

Animal source foods (ASFs) as part of a diet—particularly meat, fish, fowl of a wide variety, milk, and eggs—are important for protein and micronutrient nutrition globally. They play an important role in ensuring optimal health and function, and their consumption is particularly important for women of reproductive age, fetuses, and young children.

Increases in demand for and consumption of animal source foods have taken place since 1960, with an increased percent of dietary energy from ASFs (Speedy 2003; see also Chapters 1 and 2). (See Table 12.1.) In developing countries, there have been large increases in demand for meat, principally due to demand in urban and periurban Asia and Latin America created by burgeoning population growth and increased migration to cities (Gill 1999). This trend is seen to a much lesser extent in urban and periurban areas in sub-Saharan Africa, and growth of a cash economy livestock revolution is largely driven by a few countries such as China and Brazil (Speedy 2003; see also Chapter 2). Increased consumption of meat is raising concerns about the relationship between ASF consumption and the rise in chronic diseases everywhere, particularly in urban populations. At the same time, in many rural areas of the world ASFs are not readily available or affordable.

There are several important issues in relation to ASF consumption and human nutrition when considering the livestock revolution. First, much of the data regarding the livestock revolution and increases in ASF consumption by humans are likely overestimates, as they do not account for nonhuman use of food and the ways in which food may be lost within households and in storage. Second, data from developing countries may be unreliable.

Finally, distinct populations and contexts must be considered when discussing increases in demand for and consumption of meat, fish, and poultry (MFP) and other ASFs, as population-level data do not give sufficient information on the distribution of actual food consumed. One context includes urban and periurban populations in poor developing countries and emerging industrial countries (industrializing countries with large pockets of rural and urban poverty and vast populations of malnourished individuals and high morbidity and mortality rates), which are experiencing a "nutrition transition" with diet and lifestyle changes and with reliance on cash purchases of food in place of home production for food. A second context refers to smallholder and subsistence farmers and the urban poor in developing countries with relatively widespread chronic protein-energy malnutrition (PEM), superimposed micronutrient deficiencies (MNDs), widespread food insecurity, and limited household access to MFP, milk, and dairy products. Statistics that do not differentiate between urban and rural areas of countries do not capture these differences in access to and consumption of ASFs, particularly meat, fish, and poultry. A limited number of food intake surveys from various countries have documented rural–urban differences in ASF consumption.

Table 12.1 Percent of calories and protein from animal products, 1983 and 2003.

Region	Calories from animal product (%)		Protein from animal products (%)	
	1983	2003	1983	2003
Latin America and Caribbean	17	19	41	46
South Asia	7	9	16	20
East and Southeast Asia	7	10	24	30
Sub-Saharan Africa	8	7	24	21
Near East	12	11	26	26
North Africa	9	9	19	21
Developing	**9**	**14**	**21**	**31**
Developed	**28**	**26**	**57**	**57**
World	**15**	**17**	**34**	**38**

Source: FAOStat, 2009

Benefits of ASF Consumption for Human Health and Nutrition

This section reviews the constituents of ASFs and their contribution to diet quality and functional outcomes. The bulk of the evidence comes from observational studies that statistically control for a number of confounding and intervening variables. Because of their expense and complexity, randomized, controlled feeding intervention trials from which causation can be inferred are extremely rare. The majority of such trials have intervened with single or multiple micronutrients (tablets, capsules, etc.) rather than food-based approaches such as the use of actual ASFs, since use of micronutrient supplements is far less complicated than preparing, delivering, and serving test feedings to large groups of individuals.

Nutrient Content of ASFs

ASFs, particularly a broad assortment of meat and animal products, supply complete protein and readily bioavailable (absorbable) micronutrients (Neumann et al., 2002). The relatively high fat and protein content of meat increases energy density, which is particularly relevant for young children, given their relatively small gastric volume. Milk, other dairy products, eggs, and MFP provide high-quality, readily digestible, complete protein containing all essential amino acids (Williamson et al., 2005). Meat, fish, and poultry contain heme iron, which enhances nonheme iron and zinc absorption from cereals and legumes when mixed with those foods (Lynch et al., 1989, Swain et al., 2002, Yeudall et al., 2002). (It is hypothesized that amino acids such as asparagine, glycine, serine, cysteine, and possibly histidine facilitate the passage of iron through the mucosal layer of the small

intestine.) Although cereals and legumes may contain considerable amounts of iron, zinc, and calcium, these plant foods have high phytate and fiber content, which form insoluble compounds and reduce absorption of iron, zinc, and calcium (Reddy and Cook 1997).

ASFs are inherently richer and more absorbable sources than plant foods of specific micronutrients, including iron, zinc, riboflavin, vitamin B-12, and particularly calcium and vitamin A (Calloway et al., 1994). Animal foods can fill multiple micronutrient gaps at greater concentration and a lower volume and greater concentration of intake (Murphy and Allen 2003). For example, 100 grams of beef has a zinc content that is more than twice that of maize and beans, and it is up to 10 times as absorbable (Neumann and Harris 1999). Not all ASFs are of equal nutritional benefit, however. (See Table 12.2.) Meat and milk are not nutritionally equal; milk cannot be a substitute for meat, although it has similar vitamin B-12 and protein content. This is particularly important in populations where milk consumption is relatively high and meat consumption is particularly low. Red meat (beef, lamb, and pork) and some small fish have consistently higher zinc and iron content than other meats such as poultry and fish. Milk, eggs, and fish are important sources of preformed vitamin A in addition to iron and zinc, and fish and milk provide calcium and phosphorus (Hansen et al., 1998). Vitamin B-12 is provided almost exclusively by MFP, and milk and is not found in plant foods.

Other significant nutrients supplied by MFP include copper, riboflavin, magnesium, phosphorus, chromium, lysine, and selenium (Williamson et al., 2005). Lysine, an essential amino acid, is particularly low in cereal foods

(Rutherfurd and Moughan 2005, Williamson et al., 2005). Fish are a rich source of high-quality protein and micronutrients such as selenium, vitamin C, and vitamin D (Roos et al., 2003a, 2003b; Speedy 2003; Akpaniteaku et al., 2005). Sea fish and other sea animals are rich in iodine (Julshamn et al., 2001, Andersen et al., 2002) and zinc, and small fish when consumed whole are an excellent source of calcium, vitamin A, and iron (Roos et al., 2003b). Moreover, both shellfish and finfish are important sources of omega-3 fatty acids (Akpaniteaku et al., 2005, Torpy et al., 2006). These are especially present in oily fish such as herring, salmon, and sardines, but substantial amounts are also present in red meat (beef, pork, lamb, goat, etc.). These fatty acids can help lower heart rate and blood pressure and reduce other cardiovascular risk factors (Geleijnse et al., 2002, Dallongeville et al., 2003, Torpy et al., 2006).

Types of animal feed, forage, and the nutrient content of the soil and water in which they are grown affect the nutrient content of animal products, particularly with regard to content of vitamins, trace minerals, and fat composition. Forage grown in micronutrient-deficient soil and water deficient in iron, zinc, selenium, and iodine cause these deficiencies in animal products as well (He et al., 2005). The composition of polyunsaturated fatty acids (PUFAs), known to lower blood cholesterol levels in humans, in meat is influenced by animal feeding regimes (Wood et al., 1999, Williamson et al., 2005). The fatty acid profile of meat from nonruminant animals largely reflects their diet (Givens 2005). In monogastric animals such as swine, the inclusion of fish and vegetable oils in feed has been shown to result in increases of PUFAs in the meat (Higgs 2000, Givens and Shingfield 2004).

Ruminant animals, such as sheep, goats, cows, and camels, metabolize some fatty acids, so the fatty acid composition of their meat products reflects their diet to a lesser extent than that of nonruminant animals (Givens and Shingfield 2004, Givens 2005, Williamson et al., 2005). Trans fatty acids, which are associated with negative health effects such as increased risk for ischemic heart disease, are produced in small amounts through natural biological processes in meat of ruminant animals (Williamson et al., 2005). The animal's age and physical state at the time of slaughter also influence the iron content of meat (Williamson et al., 2005).

Table 12.2. Major micronutrients (per 100 g) contained in animal source foods.

Animal source foods	Iron (mg)	Zinc (mg)	Vitamin B-12 (µg)	Vitamin A[a] (µg RAE[b])	Calcium (mg)
Meat					
Beef, medium fat, cooked	0.32 (available)	2.05 (available)	1.87	15	8
Liver, beef	10	4.9	52.7	1500	8
Pork	1.8	4.4	5.5	2	11
Mutton	2	2.9	2.2	10	10
Goat (moderately fat)	2.3	4.0	1.13	0	11
Poultry	1.1	4.0	0.10	85	10
Milk					
Whole, unfortified	0.01	0.18	0.39	55	119
Fish					
Freshwater fish, raw	1.8	0.09	2.2	43 IU[c]	175
Small fish (<25 cm)	5.7	0.33	2.1	100	776
Other ASFs					
Rabbit	3.2	0.18	6.5	0	12
Termite (fresh)	1.0	NA	NA	0?	12
Caterpillars	2.3	At least 10	NA	NA	185
Hen egg, cooked	3.2	0.9 (raw)	2.0 (raw)	500	61

Nutrient content values are approximate and based on multiple sources.

NA = Not available.

[a]Vitamin A content varies with cooking method.

[b]Retinol activity equivalents.

[c]International units.

Sources: Leung et al., 1972, West et al., 1987, DeFoliart 1992, Nettleton and Exler 1992, American Academy of Pediatrics 1997, Pennington 1998, Grantham-McGregor and Ani 1999, Shils et al., 1999, Neumann et al., 2002, Murphy and Allen 2003, Roos et al., 2003b, USDA 2006.

Benefits for Human Health and Function

Many of the beneficial effects of ASFs on human health and function can be ascribed to their high content of complete protein, lysine content, relatively high energy density, and multiple micronutrient content. The micronutrients of particular interest are iron, zinc, calcium, selenium, vitamin B-12, and vitamin A, which have important biological roles. Intake of ASFs, particularly meat, has been linked to important functional outcomes, such as pregnancy outcome, growth, cognitive function, and physical activity, and to immune function, particularly resistance to infection. These nutrients are of vital importance during pregnancy and lactation to support maternal nutritional needs and for fetal and postnatal growth and development (Allen 2005).

Animal source foods are excellent complementary foods to support health and optimal growth and development in preschool and school-aged children (Brown et al., 1995, Marquis et al., 1997, Grantham-McGregor and Ani 1999, Black 2003). Schoolchildren whose diets contain MFP are more physically active, perform better on cognitive tests, and exhibit behaviors that are more conducive to learning than those who have no ASFs in their diet (Sigman 1995, Grantham-McGregor and Ani 1999, Black 2003). Several functional implications of various micronutrients found in ASFs are listed in Table 12.3.

The beneficial role of ASFs in the diets of young children and pregnant women is highlighted by findings from a three-country longitudinal observational study, the Human Nutrition Collaborative Research Support Program (NCRSP), which examined the functional outcomes of mild to moderate malnutrition in rural and semirural Kenya, Mexico, and Egypt from 1983 to 1987. Animal source food intake, especially MFP, was found to be generally very low, and diets were generally deficient in fat, zinc, iron, calcium, and vitamins B-12 and A (Calloway et al., 1992). There were positive and statistically significant relationships between ASF intake and birthweight, physical growth, and cognitive development in all three countries and between ASF intake and physical activity in the Kenyan sample (Neumann et al., 1992a, Calloway et al., 1992). Vitamin B-12 during pregnancy is essential for fetal brain development. Breast milk from women who ingest little ASF has lower concentrations of vitamin B-12 than that of women who consume ASFs (Specker et al., 1990, Neumann et al., 1992b), thus exposing the infant to vitamin B-12 deficiency and affecting brain formation and development.

Typical weaning diets have low energy density and protein content, are commonly micronutrient-deficient, and contain little or no ASF except for varying but usually small amounts of nonhuman milk. Because of small gastric volume, a child could not consume enough in order to obtain adequate nutrients. Even modest amounts of MFP incorporated into weaning diets would increase energy density, supply vital micronutrients, and improve protein quality.

In all three countries in the NCRSP study and in studies in Latin America, meat and milk intake and the intake of available iron, zinc, and iodized salt (in Kenya) were significant predictors of growth (Harrison et al., 1987, Allen et al., 1992, Neumann and Harrison 1994, Marquis et al., 1997). Studies with varying designs in disparate locations all show that cow or goat milk consumption by infants and young children is positively associated with physical growth, particularly in length or height (Walker et al., 1990, Guldan et al., 1993, Shapiro et al., 1998). Ownership of cows and water buffalo has been positively associated with children's nutritional status if milk is actually consumed by the children and not all sold commercially (Hitchings 1982, Shapiro et al., 1998). Milk and dairy products contain multiple micronutrients relevant to linear growth (calcium, phosphorus, and vitamin B-12) as well as protein of high biological value. However, it is an important caveat that consumption of animal milk by children younger than 18 months should not displace breastfeeding or interfere with exclusive breastfeeding in infants 4–6 months of age.

Both observational and interventional studies concerning the role of MFP in health, cognitive development, and school achievement were addressed by the NCRSP

Table 12.3. Description of functional areas affected by various micronutrient deficiencies.

Functional areas affected	Iron	Zinc	Vitamin B-12	Vitamin A	Calcium
Anemia	+ + + +	NE	+ + + +	+	NE
Immunodeficiency/infection	+ +	+ + + +	+	+ + + +	NE
Intrauterine malnutrition	+	+ +	+	NE	NE
Cognition	+ + +	+	+ + +	NE	NE
School performance	+ + + +	+	+	NE	NE
Physical activity	+ + +	+ +	NE	NE	NE
Work capacity	+ + + +	+ +	NE	NE	+ +

NE: not established.

Source: Modified from Neumann et al., 2002 and Neumann and Harris 1999.

and others as well as by the more recent Child Nutrition Project. Iron, zinc, and vitamin B-12 are important constituents of MFP. Iron deficiency, with or without anemia, plays a prominent role in impairing cognitive development and is operative in infancy, preschool, and school-aged children. Diets high in milk and devoid of meat also cause iron deficiency because of poor iron absorption in the presence of milk (Murphy and Allen 2003). The impairment, even with treatment for iron deficiency, is not reversible based on very long-term follow-up studies into adulthood (Lozoff et al., 2000). Iron deficiency is often associated with decreased physical activity, fatigue, decreased work capacity, and apathy, with adverse consequences for school success and future productivity. Iron is also a major component of myoglobin (muscle), important for physical activity. Zinc, prominently found in ASFs in a bioavailable form, also plays a major role in growth, physical activity, and cognitive function. A few intervention studies with zinc supplements have shown increased activity and improved performance on neuropsychological testing compared with controls (Black 1998).

Deficiency of vitamin B-12 (which is found solely in ASFs, particularly in meat and less so in milk) is also implicated in poor brain growth, loss of acquired motor milestones, impaired brain development and cognitive function in childhood, and a range of neurological impairments in adulthood (Dagnelie et al., 1989, Dagnelie and van Staveren 1994). Young infants being breastfed by mothers deficient in B-12 are at risk for impaired neurobehavioral development. Children raised on strict vegetarian diets with chronic vitamin B-12 deficiency develop decreased ability to concentrate, depression, problems with abstract thought, and memory impairment (Kapadia 1995). Cognitive performance has been found to be impaired in adolescents who consumed vegan-type diets as children (Louwman et al., 2000), with reversibility of impairment dependent upon the duration and type of neurological abnormality (Healton et al., 1991). Iodine deficiency, which has the most drastic and dramatic impact on cognitive development and growth, is seen in areas with no access to ocean products or iodized salt (Semba 2001).

ASFs, particularly meat, contain nutrients necessary for red blood cell formation, particularly heme iron and vitamin B-12. Nutritional anemia is common throughout the world, and where ASFs are consumed in negligible amounts, iron deficiency anemia and macrocytic anemia due to vitamin B-12 deficiency, which tends to be severe, are well documented (Neumann et al., 1992a, Allen et al., 1995, Casterline et al., 1997). Iron deficiency anemia is associated with extreme fatigue, anorexia, decrease in cognitive abilities, and decreased work capacity and physical activity in adults (Haas and Brownlie 2001).

Physical barriers to invasion by microorganisms are disrupted in both PEM and micronutrient malnutrition, contributing to increased morbidity and mortality (Neumann and Stephenson 1991, Bhaskaram 2002). Zinc, vitamins A and C, and protein are critical to the maintenance and integrity of skin, mucous membranes, and epithelial coverings that form a first line of defense against invasion by microorganisms (Neumann 1981, Neumann and Stephenson 1991, West and Darnton-Hill 2001, Kanazawa et al., 2002, Gong et al., 2004).

Key nutrients essential for the maintenance and integrity of the body's defenses against infection, particularly the immune system, are also found in ASFs, especially in meat. Iron, zinc, and vitamin B-12 deficiencies, superimposed on PEM, have been linked to immunodeficiency (cell-mediated response), diminished granulocyte function, and increased morbidity (Neumann et al., 1975, Keusch and Farthing 1986, Neumann and Stephenson 1991). Cell-mediated immunity is the first line of defense against viruses and certain bacterial and fungal infections (Keusch 1990, Erickson et al., 2000, Bhaskaram 2002).

Experimental Evidence of Positive Impact of Meat on Child Growth and Function

The positive associations observed in the NCRSP studies between meat intake and physical growth, cognitive function and school performance, and physical activity stimulated further study utilizing a feeding intervention study in Kenya from 1998 to 2001. Findings from this study show important causal benefits of meat and, to a lesser extent, milk on growth, cognition, behaviors, and activity in rural Kenyan school-aged children.

The study was designed as a randomized, controlled isocaloric feeding intervention study with three different additions to the local traditional plant-based dish *githeri*: the addition of either ground beef (meat group), whole milk (milk group), or fat to render the plain *githeri* group isocaloric. A control group received no intervention school feeding, but the families were compensated by gifts of milk goats at the end of the study, one per family. This design allowed for examination of cause and effect relationships between meat intake and functional outcomes and allowed for the control of intervening confounding variables.

Twelve schools in Embu district of Kenya were randomized to one of the three types of feeding or the control group, with three schools assigned to each condition. A midmorning snack was given every day that children attended school. Data were collected at baseline and longitudinally over seven three-month school terms (2.25 years). There were two cohorts with ~500 children and ~375 children, respectively, enrolled exactly one year apart.

At baseline, total protein intake was normal, but little or no animal source protein was consumed. Stunting and underweight were present in ~30% of children. Biochemical and food intake analyses confirmed multiple

MNDs of iron, zinc, vitamin B-12, vitamin A, riboflavin, and calcium at baseline.

The meat group exhibited the greatest statistically significant rate of increase in scores over time on the Raven's Progressive Matrices test, which measures abstract reasoning, problem solving, reasoning by analogy, and perceptual awareness of sequences compared with all other groups. (See Figure 12.1.) This test was used extensively in this area in a previous study (Neumann et al., 1992a). The plain *githeri* plus fat and meat groups performed significantly better over time on arithmetic tests than the milk and control groups. Improvements on zone-wide end-term school examinations showed that the meat group had statistically significant greater gains on total test scores and arithmetic scores compared with all the other groups. (See Figure 12.2.)

The greatest increase in the amount of time students spent during free play in high levels of physical activity and the greatest decrease in time spent in low levels of physical activity were seen in the meat group. Moreover, the meat group showed the greatest increase in initiative and in leadership behavior among peers during recess free play. These increases were all statistically significant.

The meat group and, to a lesser extent, the milk group showed the steepest and highly significant gain in arm muscle area (lean body mass) compared with all other groups (Grillenberger et al., 2003, 2006). (See Figure 12.3.) Increases in weight were observed in all children supplemented with any type of intervention feeding compared with the control group. Although no significant group differences in height were seen, in

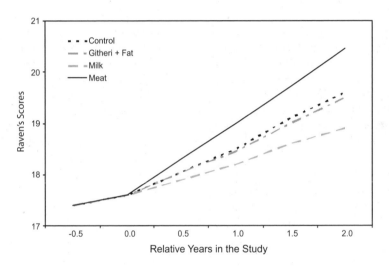

Figure 12.1. Changes in Raven's Progressive Matrices test scores by relative year in study (over a 2-year period) of Kenyan schoolchildren, 1998–2001.
Source: Neumann et al., 2007.

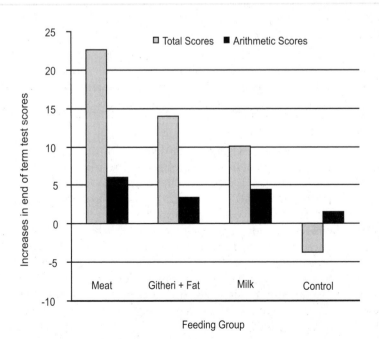

Figure 12.2. Increases in end of term test scores over time by feeding groups (Cohort II), Kenyan schoolchildren, 1998–2001.
Source: Neumann et al., 2007.

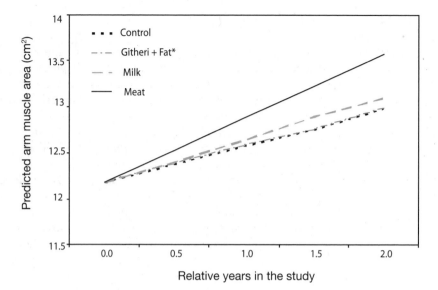

Figure 12.3. Increases in lean body mass (arm muscle area) by feeding group, Kenyan schoolchildren, 1998–2001. Noting *Githeri* + Fat and Control groups have nearly identical slopes. (All predictions correspond to a baseline age of 7.5 years.)

6- to 7-year-olds and in stunted children the milk group showed slightly improved linear growth rates.

A statistically significant improvement in vitamin B-12 status was observed in the meat and milk groups, with the greatest increase in vitamin B-12 blood concentrations and elimination of deficiency in the meat group (Siekmann et al., 2003). The milk group showed the second best significant improvement in vitamin B-12 status.

This was the first randomized, controlled intervention study to show positive causal effects of meat intake on cognitive function and school test scores, physical activity levels, behaviors, arm muscle area, and nutritional status. These findings demonstrate that inclusion of meat in the diet has multiple positive functional impacts. Schoolchildren need diets of adequate quantity and quality for optimal learning.

One other randomized trial has been published as an abstract from a recently completed study in Guatemalan children, which showed improvements in B-12 status and development following supplementation with beef or vitamin B-12 (Allen et al., 2007). Other evidence for the positive effects of ASFs on a variety of functional outcomes were reported in a quasi-experimental community-based dietary intervention in Malawi involving dietary diversification with an increase in the consumption of fish. Significant improvement in lean body mass of stunted children was found after 12 months (Yeudall et al., 2002). Nonrandomized trials in elderly individuals have provided some further evidence.

Health Risks Associated with Overconsumption of ASFs

While consumption of ASFs leads to improved nutrition, health status, and function, excessive consumption of meat and other ASFs has been associated with increased risk of certain chronic conditions in a number of epidemiological studies, largely conducted with populations in industrial countries. These conditions include atherosclerosis/coronary heart disease, hypercholesterolemia, hypertension, stroke, type 2 diabetes, obesity, and various types of cancer (Barnard et al., 1995, Willett 1995, Baghurst et al., 1997, Evans et al., 2002, van Dam et al., 2002, Holmes et al., 2003, Schulze et al., 2003, Fung et al., 2004, Steffen et al., 2005).

It is important to consider the levels of evidence and limitations of epidemiological evidence available to support associations between ASF consumption and chronic disease risk. Randomized, controlled trials provide the strongest evidence for causality, followed by prospective cohort studies, case-control studies, and associations from multivariate analyses. The weakest statistical evidence is provided by ecological studies that use aggregate data from large groups of people and examine exposure and outcomes on a group level. This design is weak because conclusions drawn from group data cannot necessarily be extended to individuals, as there is no way of verifying individual exposures or outcomes. "No single study or study design can provide absolute proof. The strongest evidence comes from a combination of different types of epidemiological investigations, with support from animal studies, metabolic, and mechanistic data" (Hu and Willett 1998).

Abundant evidence regarding health outcomes and consumption of ASFs comes from case-control studies and ecological associations; hence, the associations provided cannot necessarily demonstrate causation. Also, most studies have been performed in industrial countries, where nutrition is generally adequate. While there have been an increasing number of studies of the increasing prevalence of cardiovascular disease, obesity, and diabetes in urbanized populations in developing countries (Reddy and Yusuf 1998, Albala et al., 2001, Popkin et al., 2001, Galal 2002), most evidence for causal links comes from the United States and western Europe.

The increased risk of coronary heart disease has frequently been attributed to the consumption of red meat and high-fat dairy products (Hu and Willett 1998). Evidence supporting a relationship between meat consumption and the increased risk of developing hypertension, stroke, or obesity is generally lacking. Other coexisting factors no doubt play a role, such as little or no consumption of lean meat and dairy products and fruits and vegetables, along with decreased fiber consumption and/or decreased energy expenditure in urban settings (Willett 1995, Baghurst et al., 1997, Evans et al., 2002). Genetic predisposition is a determinant of cholesterol metabolism, hypertension, obesity, and diabetes (Pajukanta et al., 2004, Gable et al., 2006, Munroe et al., 2006, Voruganti et al., 2006). Fish intake has been associated with reduced risk for coronary heart disease, acute coronary syndromes, and stroke (Mozaffarian and Rimm 2006). "Modest consumption of fish (e.g., 1–2 servings/wk)," especially of species with high content of the fatty acids eicosapentaenoic acid and docosahexaenoic acid, "reduces risk of coronary death by 36%" (Mozaffarian and Rimm 2006).

Several large prospective studies have suggested an increase in risk for development of diabetes in relation to consumption of red meat and processed meats such as bacon, hot dogs, and sausage (van Dam et al., 2002, Schulze et al., 2003, Fung et al., 2004). This increase is explained by several mechanisms, including the presence of chemicals such as nitrites and the free radicals produced by advanced glycation (the bonding of sugars to a protein or lipid molecule with production of free radicals), rather than by the higher fat and cholesterol content of these products (Schulze et al., 2003). Insulin sensitivity may be impaired by the higher iron stores resulting from frequent meat intake (Tuomainen et al., 1997, Schulze et al., 2003). Moreover, physical activity, lifestyle, increase in body mass index, and other dietary factors such as low intake of fiber, fruits, and vegetables exert a large influence on the association between intake of processed meats or red meat and diabetes (Schulze et al., 2003).

The risk for development of cancer varies by type of ASF. While red meat has been associated with an increased risk for colon and prostate cancer development, it may be protective against rectal cancer (Hu and Willett 1998, World Cancer Research Fund/American Institute for Cancer Research 2007). While some studies show no link between red meat consumption and breast cancer (Hu and Willett 1998, Holmes et al., 2003), recent evidence suggests that red meat consumption by women may increase the risk for breast cancer (Cho et al., 2006, Taylor et al., 2007). Dairy products are implicated as increasing the risk of prostate cancer (Hu and Willett 1998). However, low-fat dairy products do not appear to increase risk for colon cancer; in fact, low fat milk may be protective, and poultry consumption possibly decreases the risk for colon cancer (Hu and Willett 1998) and is unrelated to the development of breast or prostate cancer (Hu and Willett 1998, Holmes et al., 2003). Eggs are probably unrelated to the development of most cancers and may actually be protective against rectal cancer (Hu and Willett 1998). Finally, fish may be protective against colon cancer (Hu and Willett 1998) and appears to be unrelated to breast or prostate cancer development (Hu and Willett 1998, Holmes et al., 2003). (See Table 12.4.)

Although evidence regarding the relationship of red meat intake and chronic diseases is inconsistent, there exists a fear of eating red meat in certain segments of the population, particularly in the United States and Europe. The high serum lipid levels and negative health effects associated with meat are probably more related to saturated fat and cholesterol content than to the type of meat. Thus the distinction between lean and nonlean meat may be more important than the distinction between red and white meat. In fact, lean red meat may actually have positive impacts on blood cholesterol levels (Davidson et al., 1999).

Preparation methods for meat may be a factor influencing the risk for chronic disease, particularly cancer. Certain cooking methods may produce mutagens, such as polycyclic aromatic hydrocarbons. These are produced particularly by use of high-temperature, rapid cooking methods (Baghurst et al., 1997). Close contact of the meat to fuel sources may result in exposure to toxic forms of hydrocarbons (Baghurst et al., 1997). Thus preparation methods may be more important than fat or cholesterol content in meat.

Urban populations in developing countries are undergoing a so-called nutrition transition, in which diets shift from traditional grains and legumes to more energy-dense high fat diets, which include more sugar, and ASFs containing more saturated fat, along with heavy alcohol consumption (WHO 1999, Popkin 2002, WHO/FAO 2003). Because of the limitations for growing fruit, vegetables, and legumes in urban settings, there is decreased consumption of fiber and antioxidants. There is an increased necessity in cities to purchase food in the cash economy and greater reliance on fried street foods and fast foods, which often include fatty meats (Smith 1998, Colecraft et al., 2004). These changes in diet, lifestyle, and diminished physical activity have been accompanied by the emergence of obesity, diabetes, cardiovascular diseases, and hypertension. Yet the situation is complex. Undernutrition, while declining in some regions, remains an important health problem, particularly among the rural poor and in vast rural areas (de Onis et al., 2000), and infectious diseases continue to cause a burden, along with malnutrition (Gopinath 1997, Gwatkin et al., 1999).

Vegetarianism is practiced to counter some of the problems associated with overconsumption of meat. However, vegetarianism can be nutritionally risky if complete protein quality and micronutrient supplementation are not available. Soy and quinoa do not provide

Table 12.4. A qualitative assessment of associations of animal product consumption with risk of chronic diseases.

Disease	Red meat (beef, pork, lamb)	Poultry	Fish	Egg[a]	Dairy products
Coronary heart diseases	Probably ↑	Possibly ↓	Probably ↓	Probably no increase in risk	Probably small ↑[b]
Stroke[c]	Possibly ↓	Possibly ↓	Uncertain[d]	Probably no increase in risk	Possibly ↓
Breast cancer	Possibly ↑	Probably no relation	Probably no relation	Probably no relation	Probably no relation
Colon cancer	Probably ↑	Possibly ↓	Possibly ↓	Probably no increase in risk	Probably no increase in risk
Prostate cancer	Probably ↑	Probably no relation	Probably no relation	Probably no relation	Probably ↑[e]

↓ = increase in risk.

↑ = decrease in risk.

[a]Up to one egg per day.

[b]It is possible that substitution of low-fat dairy (e.g., skim milk) for high-fat dairy products (e.g., whole milk) decreases risk of cardiovascular disease for individuals, but this is largely irrelevant for population disease rates because the dairy fat produced is almost inevitably consumed.

[c]The associations between animal products and risk of ischemic stroke, which is more common in the United States, are probably similar to those for coronary heart disease.

[d]Data are limited. It is possible that higher intake increases risk.

[e]It is unclear whether this is due to fat or calcium.

Source: Modified from Hu and Willett 1998.

complete proteins as they are lacking in methionine and must be combined with other grains to add the methionine needed to make a complete protein (Gibson 1994, Hunt 2003).

The Economic Implications of Overnutrition

Overnutrition, particularly obesity, has economic consequences through the resultant increased burden of chronic diseases and through increased costs to health systems for treatment. While these economic consequences have not been well quantified in developing countries, evidence from industrial countries shows an overwhelming increase in health care costs, morbidity, and mortality resulting from conditions to which overnutrition contributes. Substantial increases in health care costs have been seen in the United States, with chronic conditions as the main contributors to these costs (Thorpe et al., 2004, Thorpe 2006). Obesity has been associated with increasing prevalence of diabetes, cardiovascular disease, osteoarthritis, hypertension, high blood cholesterol, asthma, and sleep apnea (Turley et al., 2006). Obesity reduces life expectancy significantly, especially among younger adults, relative to individuals with normal weight (Fontaine et al., 2003, Flegal et al., 2005, Adams et al., 2006). Associations between obesity and mental health problems and reduced employment have

been found in U.S. and Canadian studies (Klarenbach et al., 2006, Tunceli et al., 2006, Kasen et al., 2008).

Data are now becoming available from developing countries regarding the increasing burden imposed by chronic conditions, such as obesity, diabetes, and hypertension, that are the result of increased consumption of fatty meat, dairy products, and refined sugar in combination with changes in lifestyle with reduction in physical activity. In African countries, substantial losses in national income result from deaths related to chronic diseases (Abegunde and Stanciole 2006). "The cost of treating diabetes accounts for approximately 10% of the national income in most countries in sub-Saharan Africa," notes one study (Thiam et al., 2006). Likewise, increased health care costs due to diabetes are being reported in India and in Latin American and Caribbean countries (Arredondo and Zuniga 2004, Suarez-Berenguela et al., 2006, Kapur 2007). Premature mortality and disability due to hypertension and diabetes in these areas are estimated to result in annual economic losses of $9.3 billion, 0.28% of their gross domestic product (GDP) (Suarez-Berenguela et al., 2006).

Hence, increased saturated fat consumption and changes in lifestyle have potentially devastating effects on the social and economic well-being of individuals and nations through increased prevalence of chronic conditions, with loss in productivity, increased burden on health systems, and loss of years of healthy life due to mortality.

Factors Contributing to Low ASF Consumption in Rural Areas

As noted earlier, the livestock revolution has mainly affected urban and periurban areas, particularly in Asia and Latin America and to a much lesser extent in sub-Saharan Africa. This section focuses on the vast number of poor rural, smallholder subsistence farmers and poor people in urban areas who are still a large part of the population in many countries in sub-Saharan Africa, Latin America, South Asia, and Southeast Asia. In many of these areas, the livestock revolution—with its dramatic increases in availability, demand for, and consumption of livestock—has scarcely taken place. In rural areas, families live on small, scattered subsistence farms and may be landless. They are largely outside the cash economy, and markets are underdeveloped and often scattered and inaccessible. Recurring droughts, political strife, and the refugee status of people forced from their homes and farms affect large groups. Poverty, lack of property rights and land access, poor health, low employment, and a lack of distribution networks, market access, storage facilities, and infrastructure negatively affect food security (Misselhorn 2005; see also Chapter 3). Malnutrition, with over 50% of children suffering from stunting and anemia in addition to food insecurity and high disease burden, is the most serious global health problem and the single largest contributor to child mortality (Black et al., 2003, World Bank 2006). Coexisting multiple micronutrient deficiencies are present in over 30% of people in the developing world (World Bank 2006).

Food shortages, poor diet quality, multiple nutrient inadequacies, and heavy infection burdens make it difficult to meet the dietary needs of the population (WHO/FAO 2003). Decimation of the agricultural labor force by HIV/AIDS has worsened the situation, particularly in areas of sub-Saharan Africa. The diet, particularly of rural populations, is mainly plant based and includes little or no meat, fish, or poultry, with serious nutrition implications, as noted earlier. High infection and parasite burdens are catabolic and contribute to both macro- and micronutrient malnutrition and malabsorption. Low calcium and phosphorus intake due to little or no milk or other dairy products results in rickets and osteoporosis/osteomalacia, which affects the health and physical activity in children and adults, respectively (Nyakundi et al., 1994, Dunnigan et al., 2005).

In Africa, India, and rural areas of Southeast Asia, little to no meat is included in the rural diet, in contrast to other countries such as Mongolia and certain ones in Latin America, where meat accounts for as much as 30% to 40% of the diet (Gill 1999). Worldwide, the least amount of meat is consumed in Sierra Leone, the Democratic Republic of Congo, Mozambique, Rwanda,

Malawi, Guinea, Burundi, India, Bangladesh, and Sri Lanka (Speedy 2003). In these countries, meat consumption has not increased with the livestock revolution (Gopalan 1996). Levels of meat consumption throughout Africa from 1985 to 1995 remained stable or decreased (Gill 1999), and while increases have occurred in countries such as India and Indonesia since 1992, they are much smaller than those seen in China (see Chapter 2).

While increased consumption of livestock products has the potential to benefit the poor nutritionally and economically (Delgado et al., 1999), there is little evidence of increased ASF consumption in rural areas as a result of the livestock revolution. Very limited information is available on the demand for and consumption of MFP in such rural areas. Besides aggregate statistical information from the United Nations Food and Agriculture Organization (FAO) and similar groups, most information comes from modest studies conducted in small geographic areas. Evidence from Demographic and Health Surveys shows that in rural areas a smaller percentage of women consistently consume MFP than do women in urban areas of Ethiopia, Zimbabwe, Uzbekistan, Cambodia, Nepal, India, and Honduras (IIPS and ORC Macro 2000, Analytical and Information Center Ministry of Health of the Republic of Uzbekistan et al., 2004, Central Statistical Agency [Ethiopia] and ORC Macro 2006, National Institute of Public Health National Institute of Statistics [Cambodia] and ORC Macro 2006, Secretaría de Salud [Honduras] et al., 2006, CSO [Zimbabwe] and Macro International Inc. 2007, MOHP [Nepal] et al., 2007). (See Table 12.5.)

In many areas with low MFP consumption, dairy products in various forms (yogurt, milk, etc.) and eggs may be consumed. Religious sanctions influence meat consumption in India, where 72.5% of respondents in a rural area did not consume meat daily (Thammi Raju and Suryanarayana 2005). In India and Sri Lanka, milk consumption is ~47.5 kg/capita/year and ~35.9 kg/capita/year, respectively, compared with 118 kg/capita/year in the United States (Speedy 2003). Pastoralist populations worldwide traditionally consume relatively high levels of milk and other milk products (FAO 2001). Milk and dairy products may partially compensate for MFP for protein quality and vitamin B-12, but not for iron and zinc.

In addition to published data on aggregate meat demand and consumption, there have been isolated studies of food consumption among various tribal or regional groups in rural Africa. Among Lese women farmers in rural Democratic Republic of Congo, meat and fish availability is seasonal and may constitute as little as 5% of total energy intake during certain seasons (Bentley et al., 1998). Evidence from two longitudinal studies in the same area and population in rural Embu district of Kenya shows a modest increase in consumption of ASFs and energy over a 15-year period (1984–1986 and 1998–2000). (See Table 12.6.) As stated earlier, the Embu diet

Table 12.5. MFP consumption data from DHS surveys

Country	Mothers of children under age 3 who consumed meat, fish, shellfish, poultry, or eggs in the preceding 24 hours (percent)	
	Urban	Rural
Ethiopia	31.5	12.8
Zimbabwe	71.8	32.3
Cambodia	96.1	93.4
Nepal	42.2	28.3
India	41.7[a]	28.5[a]
Honduras	81.7	75.5
	Median number of days in past week in which women consumed red meat	
Uzbekistan	7	5.9

[a]In the Indian DHS, the percentage values refer to the share of mothers who consumed these foods in the last week.

Sources: IIPS and ORC Macro 2000, Analytical and Information Center Ministry of Health of the Republic of Uzbekistan et al., 2004, Central Statistical Agency [Ethiopia] and ORC Macro 2006, National Institute of Public Health National Institute of Statistics [Cambodia] and ORC Macro 2006, Secretaría de Salud [Honduras] et al., 2006, CSO [Zimbabwe] and Macro International Inc. 2007, MOHP [Nepal] et al., 2007.

Table 12.6. Comparison of usual intake of energy and ASF in Embu District, Kenya, in 1984–1986 and 1998–2000 in two study populations (mean ± SD)

Energy and ASF intake	Men 1984–1986, n = 146 1998–2000, n = 155	Women 1984–1986, n = 186 1998–2000, n = 190	Children 1984–1986, n = 119 1998–2000, n = 167
Energy (kcal/d)			
1984–1986	1924 ± 409	1762 ± 375	1427 ± 224
1998–2000	2450 ± 867	2168 ± 732	1534 ± 529
Total protein (g/d)			
1984–1986	60.8 ± 34.2	51 ± 28	43 ± 21
1998–2000	78.8 ± 31.2	68.6 ± 26.7	48.0 ± 18.9
Animal protein (g/d)			
1984–1986	5.0 ± 11.4	3.9 ± 8.1	2.6 ± 5.2
1998–2000	7.2 ± 8.5	6.3 ± 5.6	3.7 ± 3.5
Percent fat			
1984–1986	12.3	13.2	13.2
1998–2000	—	—	—

Sources: Neumann et al. 1992a, 2003a, 2003b.

is generally low in ASFs, particularly MFP, with multiple MNDs documented (Neumann et al., 1992a, 2003a, 2003b), like the diets in many other rural areas of developing countries.

Even within urban areas there is unequal distribution of MFP consumption, with malnutrition and a relative lack of ASFs in the diet (Abidoye and Soroh 1999, Haddad et al., 1999). In urban Kenya, 99% of households consumed meat within a given month, and adults consumed an average of ~4 oz./day (~131 g/day) of meat. However, MFP remain luxury foods, with consumption increasing with rising income and with high- and middle-income households consuming significantly more chicken, beef, and eggs within the home than low-income households do (Gamba 2005). High-income families consumed three times as much beef, for example, as low-income households did. The data are not broken down by gender or age.

Undernutrition is prevalent in households that relocate from rural villages to urban slums or rapidly growing periurban commercial centers in search of employment (United Nations 1988). Urban poverty and undernourishment have grown with increasing urbanization (Haddad et al., 1999). Evidence from urban and rural areas of Angola, Central African Republic, and Senegal reveals that poor children in urban areas were just as likely to be stunted or underweight as poor children in rural areas (Kennedy et al., 2006). In many countries the percentage of impoverished people in cities is higher than in the rural areas (Habitat 1996). Newcomers to urban areas are forced to enter the cash economy to purchase food, pay rent, and take transportation to work, yet poverty and loss of the traditional means of food production in cities result in increased nutritional problems (Smith 1998).

Socioeconomic, Cultural, and Gender Factors Influencing ASF Consumption

Gender and age influence the quantity and quality of food received by various family members (Gittelsohn and Vastine 2003). Adult men receive priority in the household distribution of meat products in many areas. In rural Bangladesh, men are favored over women in distribution within the household of the most nourishing ASFs (Roos et al., 2003a). While women and children play major roles in many of the small-animal raising activities, such as small-animal husbandry, they consume relatively less MFP than men (Kaasschieter et al., 1992, Gueye 2000, Nielsen et al., 2003). Moreover, agricultural extension services tend to favor male farmers and are conducted mainly by men, often bypassing women.

Intake of MFP is frequently low or absent from the diet of infants and young children, who are weaned onto dilute cereal porridges with low and incomplete protein and micronutrient content (Neumann et al., 2004). Moreover, the cost of purchasing milk and MFP may be prohibitive. Nonhuman milk is consumed only in small quantities postweaning, if at all. After weaning, children tend to eat from the parent's plate or from a family pot, which may contain only small amounts of MFP. The young child's share is likely to be minimal, as MFP is more likely to be consumed by older siblings and adults, particularly men.

Food taboos play a role in avoidance of certain ASFs. Prohibited items are mainly flesh or other protein-rich ASFs, which are particularly taboo for pregnant women and young children, further exacerbating the lack of adequate animal protein in the diet (Simoons 1961, Ogbeide 1974, Bentley et al., 1999, Onuorah and Ayo 2003). In some cultures, it is believed that young children cannot digest these foods; other cultures have specific beliefs regarding the possible consequences of eating certain ASFs. A prevailing belief in several cultures is that consumption of eggs by children will make them become thieves (Onuorah and Ayo 2003, Pachón et al., 2007). Also,

during illnesses such as measles, all ASFs—particularly milk as well as MFP—may be eliminated from the diet, particularly in children, withdrawing important sources of nutrients (Maina-Ahlberg 1984).

Cattle, sheep, and goats are often considered a household's financial security or reserve and are not regularly slaughtered for domestic consumption because of their vital nonnutritional functions. They supply draught animal power, dung fuel, and the recycling of household wastes, as well as income, employment, and socioeconomic status (Sansoucy 1995). These animals are sold only reluctantly to cover expenses such as school fees, funerals, and emergency medical care, but they may be slaughtered or given away to fulfill social obligations.

Meat Preservation

The rapid deterioration in safety and palatability of meat and spoilage that occurs within a few hours after animals are slaughtered is an important factor limiting MFP consumption in rural diets. Lack of affordable and feasible household or community meat preservation methods to ensure an adequate, ongoing MFP supply as well as eliminating waste are serious and critical problems where refrigeration of any type is not widely available (Brown 2003). When the meat supply is seasonal or sporadic and cannot be consumed by a household or the local population within a short period of time, indigenous techniques for preserving ASF and plant foods are used. In arid and semiarid areas these include solar drying, with or without the addition of salt, or salting alone. In more temperate and humid areas, drying may be aided by use of a wood fire for smoking.

Box 12.1 provides an example of a successful Nutri-Business enterprise by rural Kenyan women to preserve plant and meat foods. Village women's groups are involved in using cooked maize, beans, greens, and other vegetables to dry and prepare a weaning mix for local use and for sale to promote income generation. Meat products are now being added to weaning mixes as a strategy to improve MFP intake in children.

Other Livestock Sources in Rural Areas

Among poor rural populations in developing countries, animal protein may be obtained from noncommercial meat sources or farm production. The few existing studies suggest that levels of consumption of bushmeat, fish, and other animal protein sources may compensate to varying degrees for the lack of livestock meat in the domestic diet.

In many rural communities with marginal food security, freshwater fish from lakes, ponds, rivers, and streams provide an essential source of affordable ASF with complete protein, fat, omega-3 fatty acids, and an abundance of other necessary nutrients such as calcium, iron, and vitamin A (Akpaniteaku et al., 2005, Roos et al., 2007). Fish account for 20% of animal-derived protein in low-income, food-deficit countries, compared with 13% in

Box 12.1. ASF Preservation in Rural Africa: The NutriBusiness Case Study

The NutriBusiness approach is a strategy involving a community cooperative that could be used to address both ASF preservation and income generation in rural areas. Maretzki and Mills (2003) have proposed a model in which enough rabbits are produced by the sharholders so that half can be used directly by each household and the remainder can be processed into a novel food product for sale by the cooperative. They outline a plan for processing and marketing rabbit and sweet potato solar-dried chips by a women's community cooperative in an approach likely to lead to increased ASF consumption by children. The NutriBusiness concept has been successfully used by women's cooperatives in Kenya for village-level production of dry porridge mixes suitable as a complement to breast milk for children over 4 months old (Maretzki 2007).

The solar drier used by the Kenyan cooperatives is also capable of drying a cooked meat and starchy grain or tuber mixture to produce a crispy ASF product (chiparoo) that can be crumbled into an infant's gruel or eaten as a snack food by older children (Kieras 1999, Kieras et al. 2003, Harper 2006, Mills et al. 2007). This concept is being adopted in other African countries, including Namibia. Chiparoos can be made from any available animal muscle tissue combined with cooked grain or tuber and a small amount of lime or lemon juice to limit growth of harmful bacteria during drying.

A number of cultural, economic, and technical issues must be addressed in a NutriBusiness ASF operation:

- Meat should be obtained from small animals, fish, or birds whose husbandry is in the cultural domain of women or children.
- Product marketing should target individuals such as children and pregnant or lactating women who have heightened nutritional needs.
- When funds for purchasing the dryer are acquired, technical support and maintenance must also be considered and included.
- Individuals operating equipment must be adequately trained and supervised.
- A NutriBusiness cooperative should use transparent decision-making and financial management systems to enable shareholders to develop and implement a business plan, obtain credit, assure cash-flow, and return a profit to the group.
- Instruction on how to market and sell a novel ASF food is needed, along with training to develop simple business management, bookkeeping, and inventory skills.
- Nutrition education should be provided to families and communities on the value and need for ASF to promote optimal health.

industrial countries (as categorized by FAO) (Delgado et al., 2003). Farming of tilapia is increasing in many African rural areas. Dried fish is often used to flavor sauces and stews and furnishes an abundance of macro- and micronutrients.

There has been a relatively recent increase in demand for fish to create fishmeal and fish oil for utilization as feed for farmed carnivorous fish and livestock (see Chapter 3). It is estimated that approximately one-third of wild-caught fish globally are used for this purpose (Delgado 2003). Given the nutritional value of fish for human nutrition, the large-scale diversion of fish for livestock feed from direct human consumption needs to be balanced with fish used as a nutritionally beneficial human food. Furthermore, large-scale fish production may cause environmental problems and challenges.

Bushmeat continues to provide an important protein source for household meat and fowl consumption, especially in rural areas of Latin America, Asia, and Africa. In Zimbabwe, Kenya, and Botswana, bushmeat is more affordable than other domestic meat sources (TRAFFIC 2000). While actual levels of consumption are not well documented worldwide, there have been reports of high consumption in central Africa (TRAFFIC 2000), with meat consumption levels approaching those in industrial countries (U.S. House of Representatives 2002). The type of meat hunted and consumed varies geographically, with a diversity of meat—from buffalo, primates, elephants, and birds to rodents and other small animals— used (Stein and BCTF 2001, East et al., 2005).

Bushmeat is considerably cheaper than "domestic meat," ranging from 30% to 80% less expensive in certain countries, with households consuming between 14.1 kg and 18.2 kg of bushmeat per month in a number of countries in East and Central Africa (TRAFFIC 2000, Bennett and Rao 2002). Bushmeat may be the only affordable and accessible meat source in some rural areas (TRAFFIC 2000, de Merode et al., 2004). Thirty years of data from Ghana indicate bushmeat's role as a dietary staple, with increased hunting corresponding to years in which fish supplies were poor (Brashares et al., 2004). However, in urban areas bushmeat may cost considerably more than domestic meat, as it may be considered a luxury item for the wealthy and the restaurant trade (TRAFFIC 2000).

While game meat may be legally obtained through such activities as game ranching, destruction of problem animals, ecological culling programs, or licensed hunting, poaching or trade in bushmeat is illegal in some countries (TRAFFIC 2000). The endangerment of various animal species is of increasing concern (Wilkie and Carpenter 1999, Bennett and Robinson 2000, TRAFFIC 2000, U.S. House of Representatives 2002). Alarmingly, wild meat hunting and trade have been linked to health hazards with the emergence or reemergence of certain serious and often fatal zoonoses—including HIV/AIDS,

SARS, hepatitis E, Ebola hemorrhagic fever, and monkey pox—when hunters and villagers come in close contact with infected primates (EESI 2002, Bell et al., 2004, Bengis et al., 2004, Leroy et al., 2004, Wolfe et al., 2004, Li et al., 2005, Nalca et al., 2005). Antibody studies of people in close contact with animals, particularly primates, show evidence of infection through elevated antibody titers for the infections just mentioned (Li et al., 2005; Nalca et al., 2005).

Other sources of ASFs not often documented include dried blood, worms, insects, grubs, wild birds, rabbits, rats, and other small animals, particularly rodents (Ekvall 1963, National Research Council 1991, Obi et al., 2005). In various parts of the world, although this is seasonal, insects as well as mollusks, caterpillars, water bugs, crickets, termites, ants, silkworms, and other small animals serve as important and excellent sources of animal protein and micronutrients (DeFoliart 1992, Illgner and Nel 2000). For example, insects have high levels of protein, with particularly high levels of the amino acids lysine and threonine (DeFoliart 1999). Moreover, many edible insects have high levels of iron, zinc, and fat, as indicated earlier in Table 12.3 (DeFoliart 1999).

The Contribution of Pastoralism to ASFs in the Diet

In the world's grasslands, in both hot and cold climates, herding domestic livestock is the major human land use, and livestock products are the main source of food. This is an efficient and reliable way to convert sunlight into human food in extreme and variable land environments. Rainfall is the most prevalent water source for grass productivity for most grazing lands. However, some grazing lands are found where persistent cold temperatures limit pasture growth more than rainfall. Although extensive grazing lands support only a small fraction (3%) of the world's people, they contain 35% of the world's sheep, 23% of the goats, and 16% of the cattle, including yaks. About half of the world's pastoralists are in Africa (55%), with 20% in Asia, 15% in the Americas, and 10% in Australia (Child et al., 1984). Globally, the main livestock species kept for food are cattle, donkeys, goats, and sheep (Casimir 1988, Fratkin et al., 1999). In the Middle East and East-central Asia, however, camels and horses are kept primarily for transport. Reindeer are herded in northern Eurasia, dromedary in Africa and West Asia, and yak in Asia (Blench and Sommer 1999, Williams 2002).

The livestock that are herded provide meat, milk, and other dairy products. China, for example, supports the world's largest population of sheep and goats (Williams 2002). In Inner Mongolia, which has a fourth of China's grassland area, ethnic Mongols herd livestock to feed China's growing population. China's domestic policy promotes dietary intake of protein and the expansion of exports of meat and leather products (Williams 2002). Table 12.7 describes the current global status of pastoralist peoples.

Table 12.7. Regional pastoral systems and present status.

Zone	Main species	Status
Sub-Saharan Africa	Cattle, camels, sheep, goats	Reducing because of advancing agriculture
Europe	Small ruminants	Declining everywhere because of enclosure and advancing agriculture
North Africa	Small ruminants	Reducing because of advancing agriculture
Near East and South-Central Asia	Small ruminants	Declining locally because of enclosure and advancing agriculture
India	Camels, cattle, buffaloes, sheep, goats, ducks	Declining because of advancing agriculture, but periurban livestock production is expanding
Central Asia	Yak, camels, horses, sheep, goats	Expanding following decollectivization
Circumpolar zone	Reindeer	Expanding following decollectivization in Siberia, but under pressure in Scandinavia
North America	Sheep, cattle	Declining because of increased enclosure of land and alternative economic opportunities
Central America	Sheep, cattle	Declining because of increased enclosure of land and alternative economic opportunities
Andes	Llamas, alpaca, sheep	Contracting llama production because of expansion of road systems and European-model livestock production, but increased alpaca wool production
South America	Cattle, sheep	Expanding where forests are converted to savanna, lowlands but probably static elsewhere

Source: FAO 2001.

In Asia and Africa, animals are kept primarily for food and secondarily for the market. Subsistence herders need to maximize production of edible animal products (milk, meat, ghee, and cheese) from more than one species of livestock to feed their families. However, there is a need for additional animals to produce food or other animal products (wool, cashmere, hides) to sell for cash to purchase other necessities (Behnke 1983, 1985). As the conversion of grass to milk is approximately six times more efficient than conversion of grass into meat, milk can support more people foodwise than meat alone can (Spedding 1971, Western 1982, Galvin et al., 2004).

Grasslands with adequate rainfall are essential for pastoralism, and a problem among African pastoralist populations is that they have been forced with their animals onto drylands where water availability limits supplies of grass and forage. Moreover, with privatization of formerly commonly pooled grazing lands and encroachment onto grazing lands because of population growth, food security is a constant problem (Millennium Ecosystem Assessment 2005).

In Mongolia, sheep, goats, cattle (mostly yak), horses, and camels are herded, with yaks being the main source of dairy products and with horses and camels used mainly for transportation. Dairy products and meat are the main foods consumed, with meat and fat being the most favored foods (Goldstein and Beall 1994). Mongolians consume the largest percent of their fat from ASFs in the world (88%). The nomad's standard of living declined following Mongolian independence in 1991, when many food products (wheat flour, sugar, cooking oil) were no longer available from the Soviet bloc and when the export of pastoralist livestock products was no longer possible; yet most pastoralist people were able to still feed themselves, in contrast to Mongolian city-dwellers, who were unable to obtain many foodstuffs (Harper 1994). The situation has improved in the last decade as Mongolia finds its way in a global market economy.

The subsistence patterns of nomads in Turkey, Iran, and Afghanistan include not only the herding of sheep and goats but also cereal production (Casimir 1988, Salzman 2004). Wheat-containing foods are staples of Turkey, Iran, and Afghanistan, along with a variety of dairy foods, largely produced and consumed at home. Animals and animal products such as wool are often sold for other foodstuffs (such as cereals, fruits, nuts, and tea). In Afghanistan, sheep and goats provide meat, wool, and milk products (Casimir 1988). When fresh buttermilk is scarce, milk products with a long storage life (dried casein and ghee) supplement the diet. Nutritional studies of Pashtun nomads in Afghanistan suggest that though the diet is predominantly wheat-based, buttermilk is available for about six months of the year and supplies the bulk of nutrients. Recent wars have produced many refugees and a decline in pastoralists (Stewart 2004).

With an ongoing trend toward fragmentation of pastoral lands due to such factors as the privatization of lands, which restricts movement of people and their herds, the social and economic prospects of pastoralists have changed significantly (Galvin et al., 2008). Thus access to forage and water is restricted, with resultant land use intensification, economic diversification, and a decline in well-being (Galvin et al., 2006). In several areas of the world, pastoralists are becoming less mobile, with many now living on smaller parcels of land and moving less with their herds. Other pastoralists have moved to population centers for wage labor opportunities or to be near schools and health services. They are separated from the livestock herds and can no longer depend on livestock as a daily food source. Thus pastoralists living in town have a different diet than those remaining with the herd, and children are more likely to be malnourished, since ASFs are less readily available and cereals replace dairy products in the diet. In some cases a mixed system exists of maintaining a homestead with agriculture along with family members still tending herds at a distance from home (see Chapter 15).

Pastoralists who have long traded or sold livestock products in order to purchase grain and other items, especially during droughts and in dry seasons (Sato 1997), are now increasingly entering the market (Fratkin et al., 1999) for two main reasons. They need to sell more animals to buy sufficient food grains as prices have increased (Fratkin 2004), and some groups are selling more livestock for additional cash and reinvestment in their herds in terms of veterinary services and different types of livestock. This development coincides with a global change toward increased livestock consumption in urban areas and more commoditization of livestock products (Fratkin et al., 1999, Fratkin 2004, Misselhorn 2005, ILRI 2006). There is now a focus on mixed crop–livestock systems in high-potential areas. In order for the poor to benefit from the livestock revolution (ILRI 2006), opportunities must be created without eroding the natural resource base and the livelihoods of mixed farmers and pastoralists. It is ironic that as pastoral livestock assets decline there is greater demand for their products. More detailed information focused on African pastoralists is presented in Box 12.2. Chapter 15 provides a more detailed description of changes in pastoralism globally, with an emphasis on various economic aspects and strategies to improve pastoralists' situation.

Interventions to Improve Nutritional Status

In Urban Populations Undergoing a Nutrition Transition in Developing Countries

As noted earlier, urban and periurban populations in developing countries are undergoing dietary and lifestyle changes that are accompanied by the emergence of obesity, diabetes, cancer, and cardiovascular disease. Increased intake of total energy, saturated fat, and refined

Box 12.2. Case Study: Pastoralism in Africa

Traditionally, African pastoralists have depended on herds for daily subsistence, with their diet consisting of livestock products—milk, meat, and blood—supplemented by grains and other grown or purchased foods. Animal products, milk, or hides were sold to purchase additional foods. Meat was traditionally reserved for special occasions or times of need or upon the death of an animal (Galvin 1992).

The mainstay of the pastoral diet was milk (Galvin 1992), providing 60–75% of daily calories (Galvin et al., 1994, Fratkin 2001). During the 1980s, Rendille camel pastoralists obtained up to 75% of their calories from milk (Field and Simpkin 1985, Galvin et al., 1994). During the wet season, when livestock give birth, milk was abundant, providing ~90% of dietary energy to the Turkana of northern Kenya and 80% to the Maasai (Nestel 1986, Galvin 1992). West African pastoralists relied more on grains, with milk and milk products providing less than half of dietary energy for various pastoral groups in Niger, Mali, and Senegal; yet milk was an important wet season food (INSEE-Coopération-SEDES 1966, Benefice et al., 1984, Wagenaar-Brouwer 1985, Bernus 1988, Galvin 1992). In the Maasai and similar cultures, herd animals represent a visible, male-controlled asset, giving social status and economic security to a household. Cattle are valued for family survival as they can be moved long distances to avoid droughts, often resulting in crop failure.

While African pastoralism may evoke images of tall, proud men herding cattle through the wildlife-rich savannas of Africa and living off their herds, in reality livestock are providing less of pastoralists' livelihoods, and they are becoming among the poorest people on earth. As pastoral populations increase, livestock numbers remain unchanged or plummet during drought. As land access becomes constrained, pastoralists are becoming less reliant on herds for their livelihoods and food. Hence, diets once rich in ASF from milk and meat, though often deficient in calories, are changing to grain-based diets.

Among settled pastoralists, maizemeal replaces milk as the staple. While diets of nomadic pastoralists exceed daily protein requirements, diets of settled pastoralists supply less protein (Fratkin 2004). During drought years, nomadic Ariaal and Rendille children were found to consume 3–10 times more milk than settled children (Nathan et al., 1996, Fratkin et al., 1999), who are three times more likely to be severely malnourished (Nathan et al., 1996). Similar conditions have been documented among settled Turkana who keep some livestock but derive most of their subsistence from cultivation (Little and Gray 1990).

sugar in the diet in combination with decreased physical activity are contributing factors. The importance of inclusion of lean MFP and low-fat dairy products needs to be emphasized through consumer and nutrition education to reduce the risk of chronic disease. Many urban consumers now obtain foods in processed forms in food shops or even supermarkets and may be unaware of the contents of the foods purchased in markets. Appropriate food labeling of ingredients, particularly fat content, with accurate health and nutrition information is very important whenever possible and feasible. Such labeling is gradually increasing and can encourage healthier diets in people who are literate and can afford these products (Shrimpton and Hawkes 2006). To be effective, labeling must be understandable, interpretable, compatible with the education level of the population, and culturally appropriate. For illiterate or low literate populations, pictorial messages and attractive street billboards and posters in stores are already being used to promote fortified products. In many countries, food labeling is currently voluntary but increasingly mandated by law.

Large-scale integrated education efforts, with participation of the media, can potentially be effective for addressing dietary and lifestyle changes. Billions of dollars are spent annually to market soft drinks and fast food, encouraging the transition from traditional diets (Shrimpton and Hawkes 2006). In the informal sector, street foods are available and affordable, albeit of substandard nutritional value. Public health efforts with meager financial backing need to rely on social marketing and community mobilization to combat chronic diseases associated with the nutrition transition. Opportunities and specific recommendations for physical activity within safe urban environments are needed, such as safe parks and schoolyards. Urban gardens using empty lots would help add affordable fresh fruits and vegetables to the diet. Raising small animals for household consumption has been implemented by a number of nongovernmental organizations (NGOs) and communities. The use of affordable fortified foods and/or affordable multiple micronutrient supplements, particularly for vegetarians, needs to be encouraged. The elementary, middle, and high schools should include physical, health, and nutrition education in the curriculum.

Urban dwellers, dependent on street foods and "eating out" in restaurants, need consumer and nutrition education. Given limitations of butcher shops as to the cost, types, and qualities of meat cuts available, individuals have to learn how to select lean cuts of MFP or trim off fat. Primary and secondary schools and community groups are possible venues for appropriate consumer and home economics education.

Although precise desirable amounts are not known, there is sufficient knowledge and experience to make general recommendations for intake of MFP in modest quantities several times a week and for the use of lean

MFP. The U.S. Dietary Guidelines recommend around 165 g from the meat group daily and 3 cups (24 ounces) from the milk group daily and the use of lean, low-fat, or fat-free ASF products (U.S. DHHS and USDA 2005). The most recent recommendations for optimal meat intake quoted widely are those of the World Cancer Research Fund, which recommends that red meat consumption be less than 500 g (18 oz.) per week, with little if any processed meat (World Cancer Research Fund/American Institute for Cancer Research 2007).

The American Academy of Pediatrics (AAP) Committee on Nutrition (AAP 2004) recommends 0.95 g/kg body weight daily intake of protein for children 3–8 years old and 1.10 g/kg/day for those 1–2 years old. No specific guidelines are available from international organizations, and AAP recommended levels may be too high for children in developing countries. As described in an earlier section, for Kenyan children ages 6–9 years, quantities of 60–85 g of meat added to a traditional plant dish given once per school day showed beneficial impacts on growth, development, physical activity, and vitamin B-12 (Neumann et al., 2003b, Grillenberger et al., 2006). If a source of ascorbic acid accompanies the meal, absorption of iron, calcium, and zinc from the accompanying meat will be increased, and 75–100 g of meat would be sufficient for children in primary school (6–9 years old) to meet recommended intakes of zinc, iron, and vitamin B-12 (Grillenberger 2006).

In Rural and Poor Urban Populations of Developing Countries with Deficient Diets

Various strategies have been undertaken to address the protein-energy malnutrition and multiple micronutrient deficiencies found among the urban poor and particularly among the rural poor globally. Supplement-based approaches using pills or sprinkles have been widely used because of their apparent cost-effective system of micronutrient delivery, but questions remain as to their sustainability and ability to reach widely scattered rural subsistence populations and target recipients (Nantel and Tontisirin 2002). Record-keeping, overdosages, and nutrient interactions (as in the case of zinc impeding iron absorption) can be problematic (Rossander-Hulten et al., 1991). Food fortification, plant breeding, biotechnology, and improved food processing are other approaches. Extension education, which includes home economic and animal husbandry education—preferably by women and schoolchildren extension workers working with women and children—is needed (Ranum 2001).

Food-based solutions have received less attention than the use of supplements and food fortification, yet as part of a national strategy these have more potential for impact than approaches that address single nutrients or single communities (Underwood and Smitasiri 1999, Welch and Graham 1999). While food-based solutions are more complex and interdisciplinary in nature and require long-term commitments, they are more likely to address malnutrition at its source, leading to long-term sustainable improvements. This latter approach is part of a development process that can also lead to long-term economic growth (Demment et al., 2003). The addition of modest amounts of MFP and other ASFs to the diet can greatly improve the health, micronutrient, and overall nutrient status and function of rural populations, particularly of women and children (Marquis et al., 1997; Neumann et al., 2002, 2003b; Allen 2005).

At both the household and the community level, various approaches for increasing access to and availability of ASFs are being used. Promotion of small-animal husbandry, particularly by women and children, primarily for household consumption and secondarily for income generation, is being promoted by a number of NGOs as a strategy for improving nutritional outcomes in populations without access to ASFs. This approach is being tried in multiple sites in Ghana through a project by the Global Livestock Collaborative Research Support Program, called Enhancing Child Nutrition through Animal Source Food Management (Global Livestock CRSP 2005). Household livestock ownership and production have been shown to positively affect production and consumption of ASFs, overall dietary intake, household income, and nutritional status (Leonard et al., 1994, Vella et al., 1995, Shapiro et al., 1998). At the household level, small animals can provide a variety of products, including meat, milk, butter, yogurt, and fat, to meet nutritional needs. After meeting those needs, ASFs can be sold for income generation. Microcredit can help promote nutrition improvement with income generation by providing small loans to start small businesses. Gender issues need to be addressed with regard to intrahousehold distribution of ASF, as the most vulnerable household members—young children, women of reproductive age, and HIV-positive individuals—are often denied ASFs.

More household and community initiatives for food preservation are needed to prevent spoilage and wastage and to ensure a steady supply of ASFs, especially for poor households who cannot afford to purchase MFP in the cash economy. Creative preservation techniques for ASFs have included blood biscuits as well as cereals fortified with dried blood, used in Latin America and in parts of Africa, resulting in improved iron status (Calvo et al., 1989, Olivares et al., 1990, Walter et al., 1993, Gewa 2003, Kikafunda and Sserumaga 2005). Smoking and solar drying are options to produce safe, shelf-stable products under controlled conditions. Small-scale community development approaches, such as NutriBusiness (as described in Box 12.1), for preservation of ASFs and promotion of their use for young child feeding have been successful (Muroki et al., 1997, Maretzki and Mills 2003). Such approaches address not only the problem of improving children's health, nutrition, and development, but also interrelated problems of rural poverty and gender inequity.

At a farm and community level, improvement of the nutrient content of soil in which forage is grown is needed to ensure the presence of adequate nutrients in the food chain. Affordable fertilizer and more sustainable agricultural practices through improved extension services are badly needed. Appropriate models for small livestock development utilize zero-grazing. Aquaculture to produce small fish for human domestic consumption is gaining in popularity and needs to be better balanced with fish production for animal feed. Schools can also be used to improve nutrition status. School gardens and small animal husbandry projects can increase children's knowledge as well as access to ASFs.

Agricultural extension services (government and NGOs) need to be extended to women and schoolchildren who perform much of the raising of small animals at a household level. Several NGOs such as Heifer Project International, Farm Africa, and World Vision already have successful programs to promote raising small livestock using appropriate technologies and education on animal husbandry for individuals and communities. In the Paravet program in Kenya, women veterinarians train local women to take care of animals (Mugunieri et al., 2004, 2005). Appropriate nutrition education emphasizing the preparation and value of different foods for dietary improvement would greatly enhance these programs (Neumann et al., 2002).

Linking Improved Diet Quality to Development

One major constraint on the development of human capital and capacities is the loss of human potential, both physical and mental, due to poor nutrition. The links among poverty, inadequate diet quantity and quality, little or no intake of ASFs (particularly meat, fish, and poultry), and children's cognitive and physical development are clear (Demment et al., 2003). These links extend from individuals to households, communities, and nations, with poverty, malnutrition, and disease all affecting outcomes. In the literature, causality is argued both ways: that poverty creates malnutrition and malnutrition causes poverty. Poverty, macro-, and micronutrient deficiencies exercise synergistic effects in a negative feedback cycle in which the forces of one condition reinforce the other. This relationship has been described as the poverty micronutrient malnutrition trap (Demment et al., 2003).

Although widespread micronutrient malnutrition is well documented in all age groups, nutrition in utero, early in life, and during childhood largely determines the cognitive potential through which education builds functional capacity. Not only do poor health and inadequate nutrition diminish children's cognitive development by reducing their ability to participate and benefit in learning experiences, but malnourished children have reduced physical capacity, less exploratory behavior, and physiological changes that compromise their ability to learn

(Del Rosso and Marek 1996, MacDonald et al., 2002). Malnutrition affects national development in two ways: through individual productivity losses, leading to reductions in national productivity, and through increased demands on social and health services and public revenues that indirectly counter economic productivity gains (Khan 1984).

Low birthweight term infants who suffered intrauterine malnutrition are hypothesized to be at greater risk for certain chronic diseases in adulthood, with subsequent decreased economic productivity and higher costs due to treating disease (Barker 1996, 1998). Malnutrition incurs great societal and economic costs for individuals and national economies through its effects on growth, development, and school and work performance, which affect productivity and adult health over the long term (Darnton-Hill et al., 2005, Hunt 2005).

At a macroeconomic level, estimates of economic losses from malnutrition on human productivity are in the range of 10–15%; for GDP, 5–10%; and for children's disability-adjusted life years, 20–25% (WHO 2000). It has been estimated that as much as 5% of GDP may be lost due to multiple micronutrient deficiencies (McGuire et al., 1994). Addressing these comprehensively would cost less than one-third of 1% of GDP (Bouis et al., 1999). More recent estimates for 10 countries of the average annual productivity losses of cognitive and physical productivity due to iron deficiency are a median of 4.05% of the GDP (Horton and Ross 2003). Thus investments in health and nutrition enable formation of human capital, commencing in the early years of life with ensuing increased efficiency of education; raised skilled and unskilled labor efficiency; and promotion of longevity (Galor and Mayer 2002). Hence programs addressing PEM and MNDs and anemia could greatly increase economic productivity.

Development goals are too often narrowly defined in economic terms under the assumption that nutritional status is directly linked to income and that benefits of macroeconomic growth will trickle down, having a positive nutritional effect at the household and individual levels. Martorell (1996) instead suggests that it is nutritional status that largely enables economic development. He presents a conceptual looped model of development in which improved nutrition leads to improvements in cognitive and physical development, which together generate increases in human capital and in turn fuel economic growth, social sector investments in nutrition, and continued improvement in nutritional status. (See Figure 12.4.)

For most developing countries, agriculture, including animal production, is the largest sector of the economy and includes the majority of the workforce. For that reason, making agriculture more efficient and integrating human nutrition improvement and development are critical to long-term economic growth (Mellor 1990).

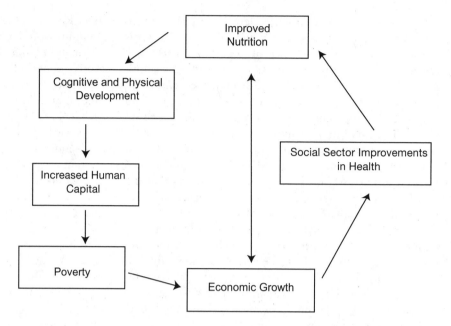

Figure 12.4. Martorell's model of development.

For those poor who spend a large proportion of their income on food or of their time and effort in producing food, increasing production efficiency through improved agricultural and livestock sector methods lowers food prices and increases food availability and security, which can have strong positive effects on human welfare. If cash-poor or subsistence farming households recognize the importance of meat, fish, and poultry as a source of micronutrients and fresh quality protein, and if MFP are accessible and available, they would be able to use increased disposable income to improve dietary diversity and micronutrient intake and also to invest in farm inputs to raise improved animals and livestock for their own consumption and income generation.

Several caveats exist. First, the profile of poverty for the urban and rural poor is highly heterogeneous, and the impact of reductions in food prices can be different for different populations of the poor. Some of the rural poor may obtain the most income, if any, from the sale of agricultural produce; hence lower prices may result in lower income. If these price reductions are accompanied by macropolicy adjustments at the national level, such as real exchange rate depreciations, then rural incomes can increase (Demery and Squire 1996). However, impacts of increased efficiencies of agricultural production are generally positive for most poor people (De Janvry and Sadoulet 2000). This leads directly to the second caveat. One of the major macroeconomic constraints on developing countries' economic growth are the trade tariffs and subsidies that protect producers in industrial countries, reduce overall economic growth in developing countries (Dollar and Kraay 2004), and block the expansion of developing countries' agricultural systems (Binswanger and Lutz 2000, IMF 2001). These subsidies

are greater than the combined GDP of all sub-Saharan countries.

The way forward has many potential routes. These range from broad approaches directed at economic growth that reduce poverty to targeted programs that supply micronutrients as supplements. Economic growth is clearly of great importance and depends on the contribution of human capital, which is determined in large part by nutrition and health. Reducing trade barriers and subsidies is certainly a widely accepted way to help poor nations' economies. Animal source foods, particularly MFP, offer a sustainable approach to addressing multiple MNDs and PEM and should be integrated into food-based approaches to improve nutrient status. Their production can grow economies (Delgado et al., 1999) with the development of markets, and, what is as important, they have potential to have a positive impact on economic productivity by alleviating nutritional deficiencies.

Two elements are lacking. First, while the past approaches of narrow nutrient-specific goals or targeted local interventions may have had effects on their targets, they have been neither sustainable nor generalizable (Underwood and Smitasiri 1999). A complex of local and national factors—many of which are interdependent and interactive—determines the process by which food reaches the poor. For governments and donors to address the problem efficiently, those working in nutrition, livestock, fishing, agronomy, and agricultural sciences must collaborate to develop an integrated conceptual model. This model would provide a systematic mechanism to analyze food systems and determine the constraints that prevent the delivery of sufficient energy and micronutrients to the individual and would provide a holistic platform for donors (Underwood and Smitasiri 1999). Of fundamental importance throughout all of nutrition improvement is reduction of the infection burden that is detrimental to nutritional status.

Second, and most important, there is not sufficient awareness at the level of policy makers in industrial as well as developing countries of the true economic costs of malnutrition. The United States has made development and poverty alleviation a high priority for promotion of peaceful foreign relations, signaling perhaps for the first time that the clear link between international poverty and domestic security justifies a major increase in development assistance. It is time to reevaluate our approach in light of the emerging importance of the interactions between poverty and malnutrition and to commit new funds to support innovative and integrated approaches. Clearly the evidence, outlined here and elsewhere (Martorell 1996, Arcand 2001, Galor and Mayer 2002) indicates that if the goal is to eliminate poverty, development must address the importance of nutrition in building human capital as the first and fundamental step to reducing poverty and promoting social and economic development.

References

AAP (American Academy of Pediatrics). 1997. *Pediatric nutrition handbook.* 4th ed. Grove Village, IL: American Academy of Pediatrics.

AAP (American Academy of Pediatrics). 2004. *Pediatric nutrition handbook.* ed. R. E. Kleinman. 5th ed. Grove Village, IL: American Academy of Pediatrics.

Abegunde, D., and A. Stanciole. 2006. An estimation of the economic impact of chronic noncommunicable diseases in selected countries. Working paper. Geneva: World Health Organization.

Abidoye, R. O., and K. W. Soroh. 1999. A study on the effects of urbanization on the nutritional status of primary school children in Lagos, Nigeria. *Nutrition and Health* 13 (3): 141–151.

Adams, K. F., A. Schatzkin, T. B. Harris, V. Kipnis, T. Mouw, R. Ballard-Barbash, A. Hollenbeck, and M. F. Leitzmann. 2006. Overweight, obesity, and mortality in a large prospective cohort of persons 50 to 71 years old. *New England Journal of Medicine* 355:763–778.

Akpaniteaku, R. C., M. Weimin, and Y. Xinhua. 2005. Evaluation of the contribution of fisheries and aquaculture to food security in developing countries. *NAGA: The WorldFish Center Quarterly* 28:28–32.

Albala, C., F. Vio, J. Kain, and R. Uauy. 2001. Nutrition transition in Latin America: The case of Chile. *Nutrition Reviews* 59:170–176.

Allen, L. H. 2005. Multiple micronutrients in pregnancy and lactation: An overview. *American Journal of Clinical Nutrition* 81:1206S–1212S.

Allen, L. H., J. R. Backstrand, A. Chávez, and G. H. Pelto. 1992. People cannot live by tortillas alone: The results of the Mexico Nutrition CRSP. U.S. Agency for International Development, University of Connecticut, Instituto Nacional de la Nutrición Salvador Zubirán.

Allen, L. H., J. L. Rosado, J. E. Casterline, H. Martinez, P. Lopez, E. Munoz, and A. K. Black. 1995. Vitamin B-12 deficiency and malabsorption are highly prevalent in rural Mexican communities. *American Journal of Clinical Nutrition* 62 (5): 1013–1019.

Allen, L. H., M. Ramirez-Zea, C. Zuleta, R. M. Mejia, K. M. Jones, M. W. Demment, and M. Black. 2007. Vitamin B-12 status and development of young Guatemalan children: Effects of beef and B-12 supplements. In *Experimental Biology*. Washington, DC.

Analytical and Information Center Ministry of Health of the Republic of Uzbekistan, State Department of Statistics Ministry of Macroeconomics and Statistics [Uzbekistan], and ORC Macro. 2004. Uzbekistan health examination survey 2002. Calverton, MD: Analytical and Information Center, State Department of Statistics, and ORC Macro.

Andersen, S., B. Hvingel, and P. Laurberg. 2002. Iodine content of traditional Greenlandic food items and tap water in East and West Greenland. *International Journal of Circumpolar Health* 61:332–340.

Arcand, J. 2001. Undernourishment and economic growth. In *FAO Economic and Social Development Papers No. 147*. Rome: Food and Agriculture Organization.

Arredondo, A., and A. Zuniga. 2004. Economic consequences of epidemiological changes in diabetes in middle-income countries: The Mexican case. *Diabetes Care* 27:104–109.

Baghurst, P., S. Record, and J. Syrette. 1997. Does red meat cause cancer? *Australian Journal of Nutrition and Dietetics* 54:S1–44.

Barker, D. J. 1996. Growth in utero and coronary heart disease. *Nutrition Reviews* 54:S1–7.

Barker, D. J. 1998. *Mothers, babies and diseases in later life.* London: Churchill-Livingstone.

Barnard, N. D., A. Nicholson, and J. L. Howard. 1995. The medical costs attributable to meat consumption. *Preventive Medicine* 24:646–655.

Behnke, R. H. 1983. Production rationales: The commercialization of subsistence pastoralism. *Nomadic Peoples* 14:3–27.

Behnke, R. H. 1985. Measuring the benefits of subsistence versus commercial livestock production in Africa. *Agricultural Systems* 16:109–135.

Bell, D., S. Roberton, and P. R. Hunter. 2004. Animal origins of SARS Coronavirus: Possible links with the international trade in small carnivores. *Philosophical Transactions of the Royal Society of London, Series B. Biological Sciences* 359:1107–1114.

Benefice, E., S. Chevassus-Agnes, and H. Barral. 1984. Nutritional situation and seasonal variations for pastoralist populations of the Sahel (Senegalese Ferlo). *Ecology of Food and Nutrition* 14:229–247.

Bengis, R. G., F. A. Leighton, J. R. Fischer, M. Artois, T. Morner, and C. M. Tate. 2004. The role of wildlife in emerging and re-emerging zoonoses. *Revue Scientifique et Technique* 23:497–511.

Bennett, E. L., and M. Rao. 2002. Wild meat consumption in Asian tropical forest countries: Is this a glimpse of the future for Africa? In *Links between biodiversity, conservation, livelihoods and food security: The sustainable use of wild species for meat,* ed. S. Mainka and M. Trivedi, 39–44. Gland, Switzerland, and Cambridge, U.K.: World Conservation Union–IUCN.

Bennett, E. L., and J. G. Robinson. 2000. Hunting of wildlife in tropical forests: Implications for biodiversity and forest peoples. In *Toward environmentally and socially sustainable development.* Environment department papers, No. 76. Washington, DC: World Bank.

Bentley, G. R., A. M. Harrigan, and P. T. Ellison. 1998. Dietary composition and ovarian function among Lese horticulturalist women of the Ituri Forest, Democratic Republic of Congo. *European Journal of Clinical Nutrition* 52:261–270.

Bentley, G. R., R. Aunger, A. M. Harrigan, M. Jenike, R. C. Bailey, and P. T. Ellison. 1999. Women's strategies to alleviate nutritional stress in a rural African society. *Social Science & Medicine* 48:149–162.

Bernus, E. 1988. Seasonality, climatic fluctuations, and food supplies. In *Coping with uncertainty in food supply*, ed. I. de Garine and G. A. Harrison, 318–336. Oxford: Clarendon Press.

Bhaskaram, P. 2002. Micronutrient malnutrition, infection, and immunity: An overview. *Nutrition Reviews* 60:S40–45.

Binswanger, H., and E. Lutz. 2000. Agricultural trade barriers, trade negotiations and interests of developing countries. Tomorrow's agriculture: Incentives, institutions, infrastructure, and innovations. In *International Association of Agricultural Economists Meeting in Berlin, August 2000*. Berlin, Germany.

Black, M. M. 1998. Zinc deficiency and child development. *American Journal of Clinical Nutrition* 68:464S–469S.

Black, M. M. 2003. Micronutrient deficiencies and cognitive functioning. *Journal of Nutrition* 133:3927S–3931S.

Black, R. E., S. S. Morris, and J. Bryce. 2003. Where and why are 10 million children dying every year? *Lancet* 36:2226–2234.

Blench, R., and F. Sommer. 1999. Understanding rangeland biodiversity. Working paper no. 121. London: Overseas Development Institute.

Bouis, H. E., R. D. Graham, and R. M. Welch. 1999. The CGIAR Micronutrients Project: Justification, history, objectives, and summary findings. Paper at IRRI-hosted IFPRI Workshop: Improving Human Nutrition through Agriculture: The role of international agricultural research, at Los Banos, Philippines.

Brashares, J. S., P. Arcese, M. K. Sam, P. B. Coppolillo, A. R. E. Sinclair, and A. Balmford. 2004. Bushmeat hunting, wildlife declines, and fish supply in West Africa. *Science* 306:1180–1183.

Brown, D. L. 2003. Solutions exist for constraints to household production and retention of animal food products. *Journal of Nutrition* 133: 4042S–4047S.

Brown, K. H., H. Creed-Kanashiro, and K. G. Dewey. 1995. Optimal complementary feeding practices to prevent childhood malnutrition in developing countries. *Food and Nutrition Bulletin* 16:320–339.

Calloway D. H., S. Murphy, J. Balderston, O. Receveur, D. Lein, and M. Hudes. 1992. Village nutrition in Egypt, Kenya and Mexico: Looking across the CRSP projects. Final Report to the U.S. Agency for International Development. Berkeley, CA: University of California.

Calloway, D. H., S. P. Murphy, S. Bunch, and J. Woerner. 1994. *WorldFood dietary assessment system user's guides.* Berkeley, CA: University of California.

Calvo, E., E. Hertrampf, S. de Pablo, M. Amar, and A. Stekel. 1989. Haemoglobin-fortified cereal: An alternative weaning food with high iron bioavailability. *European Journal of Clinical Nutrition* 43:237–243.

Casimir, M. J. 1988. Nutrition and socio-economic strategies in mobile pastoral societies in the Middle East with special reference to West Afghan Pashtuns. In *Coping with uncertainty in food supply*, ed. I. de Garine and G. A. Harrison. Oxford: Clarendon Press.

Casterline, J. E., L. H. Allen, and M. T. Ruel. 1997. Vitamin B-12 deficiency is very prevalent in lactating Guatemalan women and their infants at three months postpartum. *Journal of Nutrition* 127:1966–1972.

Central Statistical Agency [Ethiopia] and ORC Macro. 2006. *Ethiopia demographic and health survey 2005*. Addis Ababa and Calverton, MD: Central Statistical Agency and ORC Macro.

Child, R. D., H. F. Heady, W. C. Hickey, R. A. Peterson, and R. D. Piper. 1984. *Arid and semi-arid lands: Sustainable use and management in developing countries*. Morriton, AR: Winrock International.

Cho, E., W. Y. Chen, D. J. Hunter, M. J. Stampfer, G. A. Colditz, S. E. Hankinson, and W. C. Willett. 2006. Red meat intake and risk of breast cancer among pre-menopausal women. *Archives of Internal Medicine* 166:2253–2259.

Colecraft, E. K., G. S. Marquis, A. A. Bartolucci, L. Pulley, W. B. Owusu, and H. M. Maetz. 2004. A longitudinal assessment of the diet and growth of malnourished children participating in nutrition rehabilitation centres in Accra, Ghana. *Public Health Nutrition* 7:487–494.

CSO (Central Statistical Office) [Zimbabwe] and Macro International Inc. 2007. *Zimbabwe demographic and health survey 2005–2006*. Calverton, MD: CSO and Macro International Inc.

Dagnelie, P. C., and W. A. van Staveren. 1994. Macrobiotic nutrition and child health: Results of a population-based, mixed-longitudinal cohort study in the Netherlands. *American Journal of Clinical Nutrition* 59:1187S–1196S.

Dagnelie, P. C., W. A. van Staveren, F. J. Vergote, P. G. Dingjan, H. van den Berg, and J. G. Hautvast. 1989. Increased risk of vitamin B-12 and iron deficiency in infants on macrobiotic diets. *American Journal of Clinical Nutrition* 50:818–824.

Dallongeville, J., J. Yarnell, P. Ducimetiere, D. Arveiler, J. Ferrieres, M. Montaye, G. Luc, et al. 2003. Fish consumption is associated with lower heart rates. *Circulation* 108:820–825.

Darnton-Hill, I., P. Webb, P. W. Harvey, J. M. Hunt, N. Dalmiya, M. Chopra, M. J. Ball, M. W. Bloem, and B. de Benoist. 2005. Micronutrient deficiencies and gender: Social and economic costs. *American Journal of Clinical Nutrition* 81:1198S–1205S.

Davidson, M. H., D. Hunninghake, K. C. Maki, P. O. Kwiterovich, Jr., and S. Kafonek. 1999. Comparison of the effects of lean red meat vs lean white meat on serum lipid levels among free-living persons with hypercholesterolemia: A long-term, randomized clinical trial. *Archives of Internal Medicine* 159:1331–1338.

De Janvry, A., and E. Sadoulet. 2000. Rural poverty in Latin America: Determinants and exit paths. *Food Policy* 25:389–409.

de Merode, E., K. Homewood, and G. Cowlishaw. 2004. The value of bushmeat and other wild foods to rural households living in extreme poverty in Democratic Republic of Congo. *Biological Conservation* 118:573–581.

de Onis, M., E. A. Frongillo, and M. Blossner. 2000. Is malnutrition declining? An analysis of changes in levels of child malnutrition since 1980. *Bulletin of the World Health Organization* 78:1222–1233.

DeFoliart, G. R. 1992. Insects as human food. *Crop Protection* 11:395–399.

DeFoliart, G. R. 1999. Insects as food: Why the Western attitude is important. *Annual Review of Entomology* 44:21–50.

Del Rosso, J. M., and T. Marek. 1996. *Class action: Improving school performance in the developing world through better health and nutrition*. Washington, DC: World Bank.

Delgado, C. L. 2003. Rising consumption of meat and milk in developing countries has created a new food revolution. *Journal of Nutrition* 133:3907S–3910S.

Delgado, C. L., M. W. Rosegrant, H. Steinfeld, S. Ehui, and C. Courbois. 1999. Livestock to 2020: The next food revolution. In *A 2020 Vision for Food, Agriculture, and the Environment*. Washington, DC: International Food Policy Research Institute.

Delgado, C. L., N. Wada, M. W. Rosegrant, S. Meijer, and M. Ahmed. 2003. Outlook for fish to 2020: Meeting global demand. In *A 2020 Vision for Food, Agriculture, and the Environment Initiative*. Washington, DC, and Penang, Malaysia: International Food Policy Research Institute and WorldFish Center.

Demery, L., and L. Squire. 1996. Macroeconomic adjustment and poverty in Africa: An emerging picture. *World Bank Research Observer* 11:39–59.

Demment, M. W., M. M. Young, and R. L. Sensenig. 2003. Providing micronutrients through food-based solutions: A key to human and national development. *Journal of Nutrition* 133:3879S–3885S.

Dollar, D., and A. Kraay. 2004. Trade, growth and poverty. *Economic Journal* 114:F22–49.

Dunnigan, M. G., J. B. Henderson, D. J. Hole, E. Barbara Mawer, and J. L. Berry. 2005. Meat consumption reduces the risk of nutritional rickets and osteomalacia. *British Journal of Nutrition* 94:983–991.

East, T., N. F. Kumpel, E. J. Milner-Gulland, and J. M. Rowcliffe. 2005. Determinants of urban bushmeat consumption in Rio Muni, Equatorial Guinea. *Biological Conservation* 126:206–215.

EESI (Environmental and Energy Study Institute). 2002. *"Bushmeat" and the origin of HIV/AIDS: A case study of biodiversity, population pressures and human health*. Washington, DC: Environmental and Energy Study Institute.

Ekvall, R. B. 1963. 175. A note on 'live blood' as food among the Tibetans. *Man* 63:145–146.

Erickson, K. L., E. A. Medina, and N. E. Hubbard. 2000. Micronutrients and innate immunity. *Journal of Infectious Diseases* 182:S5–10.

Evans, R. C., S. Fear, D. Ashby, A. Hackett, E. Williams, M. Van Der Vliet, F. D. Dunstan, and J. M. Rhodes. 2002. Diet and colorectal cancer: An investigation of the lectin/galactose hypothesis. *Gastroenterology* 122:1784–1792.

FAO (Food and Agriculture Organization). 2001. Pastoralism in the new millennium. In *FAO Animal Production and Health Paper 150*. Rome: FAO.

Field, C., and S. Simpkin. 1985. *The importance of camels to subsistence pastoralists in Kenya. Integrated Project in Arid Lands (IPAL)*. Technical Report E-7. Nairobi, Kenya: UNESCO.

Flegal, K. M., B. I. Graubard, D. F. Williamson, and M. H. Gail. 2005. Excess deaths associated with underweight, overweight, and obesity. *Journal of the American Medical Association* 293:1861–1867.

Fontaine, K. R., D. T. Redden, C. Wang, A. O. Westfall, and D. B. Allison. 2003. Years of life lost due to obesity. *Journal of the American Medical Association* 289:187–193.

Fratkin, E. 2001. East African pastoralism in transition: Maasai, Boran, and Rendille cases. *African Studies Review* 44:1–25.

Fratkin, E. 2004. *Ariaal pastoralists of Kenya: Studying pastoralism, drought, and development in Africa's arid lands*. Boston: Pearson Education.

Fratkin, E., E. A. Roth, and M. A. Nathan. 1999. When nomads settle: The effects of commoditization, nutritional change, and formal education on Ariaal and Rendille pastoralists. *Current Anthropology* 40:729–735.

Fung, T. T., M. Schulze, J. E. Manson, W. C. Willett, and F. B. Hu. 2004. Dietary patterns, meat intake, and the risk of type 2 diabetes in women. *Archives of Internal Medicine* 164:2235–2240.

Gable, D. R., J. W. Stephens, J. A. Cooper, G. J. Miller, and S. E. Humphries. 2006. Variation in the UCP2-UCP3 gene cluster predicts the development of type 2 diabetes in healthy middle-aged men. *Diabetes* 55:1504–1511.

Galal, O. M. 2002. The nutrition transition in Egypt: Obesity, undernutrition and the food consumption context. *Public Health Nutrition* 5:141–148.

Galor, O., and D. Mayer. 2002. *Food for thought: Basic needs and persistent educational inequality*. Washington, DC: Pan American Health Organization.

Galvin, K. A. 1992. Nutritional ecology of pastoralists in dry tropical Africa. *American Journal of Human Biology* 4:209–221.

Galvin, K. A., D. L. Coppock, and P. W. Leslie. 1994. Diet, nutrition, and the pastoral strategy. In *African pastoralist systems: An integrated approach*, ed. E. Fratkin, K. A. Galvin, and E. A. Roth, 113–131. Boulder, CO: Lynne Rienner.

Galvin, K. A., P. K. Thornton, R. B. Boone, and J. Sunderland. 2004. Climate variability and impacts on East African livestock herders: The Maasai of Ngorongoro Conservation Area, Tanzania. *African Journal of Range and Forage Science* 21: 183–189.

Galvin, K. A., P. K. Thornton, J. Roque dePinho, J. Sunderland, and R. Boone. 2006. Integrated modeling and assessment for resolving conflicts between wildlife and people in the rangelands of East Africa. *Human Ecology* 34:1–29.

Galvin, K. A., R. Reid, R. Behnke, and N. T. Hobbs. 2008. *Fragmentation in semi-arid and arid landscapes: Consequences for human and natural systems*. Dordrecht, Netherlands: Springer.

Gamba, P. 2005. Urban consumption patterns for meat: Trends and policy implications. In *Tegemeo Working Paper 17/2005*. Nairobi: Tegemeo Institute of Agricultural Policy and Development, Egerton University.

Geleijnse, J. M., E. J. Giltay, D. E. Grobbee, A. R. Donders, and F. J. Kok. 2002. Blood pressure response to fish oil supplementation: Metaregression analysis of randomized trials. *Journal of Hypertension* 20:1493–1499.

Gewa, C. M., MPH, doctoral candidate. 2003. Personal communication. Los Angeles.

Gibson, R. S. 1994. Content and bioavailability of trace elements in vegetarian diets. *American Journal of Clinical Nutrition* 59:1223S–1232S.

Gill, M. 1999. Meat production in developing countries. *Proceedings of the Nutrition Society* 58:371–376.

Gittelsohn, J., and A. E. Vastine. 2003. Sociocultural and household factors impacting on the selection, allocation and consumption of animal source foods: Current knowledge and application. *Journal of Nutrition* 133:4036S–4041S.

Givens, D. I. 2005. The role of animal nutrition in improving the nutritive value of animal-derived foods in relation to chronic disease. *Proceedings of the Nutrition Society* 64:395–402.

Givens, D. I., and K. J. Shingfield. 2004. Foods derived from animals: The impact of animal nutrition on their nutritive value and ability to sustain long-term health. *Nutrition Bulletin* 29:325–332.

Global Livestock CRSP (Collaborative Research Support Program). 2005. *Annual report 2005*. Davis: University of California.

Goldstein, M. C., and C. M. Beall. 1994. *The changing world of Mongolia's nomads*. Berkeley: University of California Press.

Gong, H., Y. Takami, T. Amemiya, M. Tozu, and Y. Ohashi. 2004. Ocular surface in Zn-deficient rats. *Ophthalmic Research* 36:129–138.

Gopalan, C. 1996. Current food and nutrition situation in South Asian and South-East Asian countries. *Biomedical and Environmental Sciences* 9:102–116.

Gopinath, N. 1997. Nutrition and chronic diseases—Indian experience. *Southeast Asian Journal of Tropical Medicine and Public Health* 28:113–117.

Grantham-McGregor, S. M., and C. C. Ani. 1999. The role of micronutrients in psychomotor and cognitive development. *British Medical Bulletin* 55 (3): 511–527.

Grillenberger, M. 2006. *Impact of animal source foods on growth, morbidity and iron bioavailability in Kenyan school children.* Wageningen, Netherlands: Wageningen University.

Grillenberger, M., C. G. Neumann, S. P. Murphy, N. O. Bwibo, P. van't Veer, J. G. Hautvast, and C. E. West. 2003. Food supplements have a positive impact on weight gain and the addition of animal source foods increases lean body mass of Kenyan schoolchildren. *Journal of Nutrition* 133:3957S–3964S.

Grillenberger, M., C. G. Neumann, S. P. Murphy, N. O. Bwibo, R. E. Weiss, L. Jiang, J. G. Hautvast, and C. E. West. 2006. Intake of micronutrients high in animal-source foods is associated with better growth in rural Kenyan school children. *British Journal of Nutrition* 95:379–390.

Gueye, E. H. F. 2000. Women and family poultry production in rural Africa. *Development in Practice* 10:98–102.

Guldan, G. S., M. Y. Zhang, Y. P. Zhang, J. R. Hong, H. X. Zhang, S. Y. Fu, and N. S. Fu. 1993. Weaning practices and growth in rural Sichuan infants: A positive deviance study. *Journal of Tropical Pediatrics* 39:168–175.

Gwatkin, D. R., M. Guillot, and P. Heuveline. 1999. The burden of disease among the global poor. *The Lancet* 354:586–589.

Haas, J. D., and T. Brownlie. 2001. Iron deficiency and reduced work capacity: A critical review of the research to determine a causal relationship. *Journal of Nutrition* 131:676S–688S.

Habitat (United Nations Centre for Human Settlements). 1996. *An urbanizing world: Global report on human settlements, 1996.* Oxford: Oxford University Press, for the United Nations Centre for Human Settlements.

Haddad, L., M. T. Ruel, and J. L. Garrett. 1999. Are urban poverty and undernutrition growing? Some newly assembled evidence. *World Development* 27:1891–1904.

Hansen, M., S. H. Thilsted, B. Sandstrom, K. Kongsbak, T. Larsen, M. Jensen, and S. S. Sorensen. 1998. Calcium absorption from small soft-boned fish. *Journal of Trace Elements in Medicine and Biology: Organ of the Society for Minerals and Trace Elements* 12:148–154.

Harper, C. 1994. An assessment of vulnerable groups in Mongolia: Strategies for social planning. World Bank Discussion Paper 229. China and Mongolia Department. Washington, DC: World Bank.

Harper, M. T. 2006. *Improving texture and oxidative stability in Chiparoos.* College of Agricultural Sciences. University Park: Pennsylvania State University.

Harrison, G., O. Galal, A. Kirksey, and N. Jerome. 1987. *Egypt project: The Collaborative Research Support Program on food intake and human function: Final report.* Washington, DC: U.S. Agency for International Development.

He, Z. L., X. E. Yang, and P. J. Stoffella. 2005. Trace elements in agroecosystems and impacts on the environment. *Journal of Trace Elements in Medicine and Biology: Organ of the Society for Minerals and Trace Elements* 19:125–140.

Healton, E. B., D. G. Savage, J. C. Brust, T. J. Garrett, and J. Lindenbaum. 1991. Neurologic aspects of cobalamin deficiency. *Medicine (Baltimore)* 70:229–245.

Higgs, J. D. 2000. The changing nature of red meat: 20 years of improving nutritional quality. *Trends in Food Science & Technology* 11:85–95.

Hitchings, J. 1982. *Agricultural determinants of nutritional status among Kenyan children with model of anthropometric and growth indicators.* Palo Alto, CA: Stanford University.

Holmes, M. D., G. A. Colditz, D. J. Hunter, S. E. Hankinson, B. Rosner, F. E. Speizer, and W. C. Willett. 2003. Meat, fish and egg intake and risk of breast cancer. *International Journal of Cancer* 104:221–227.

Horton, S., and J. Ross. 2003. The economics of iron deficiency. *Food Policy* 28:51–75.

Hu, F. B., and W. C. Willett. 1998. *The relationship between consumption of animal products (beef, pork, poultry, eggs, fish and dairy products) and risk of chronic diseases: A critical review.* Unpublished report for the World Bank. Boston, MA: Harvard School of Public Health.

Hunt, J. R. 2003. Bioavailability of iron, zinc, and other trace minerals from vegetarian diets. *American Journal of Clinical Nutrition* 78:633S–639S.

Hunt, J. M. 2005. The potential impact of reducing global malnutrition on poverty reduction and economic development. *Asia Pacific Journal of Clinical Nutrition* 14:10–38.

IIPS (International Institute for Population Sciences) and ORC Macro. 2000. *National Family Health Survey (NFHS-2), 1998–99: India.* Mumbai: International Institute for Population Sciences.

Illgner, P., and E. Nel. 2000. The geography of edible insects in sub-Saharan Africa: A study of the mopane caterpillar. *Geographical Journal* 166:336–351.

ILRI (International Livestock Research Institute). 2006. *Changing livestock landscapes: Drivers of change are creating a "new livestock economy" that could spur pro-poor growth.* Paper at International Workshop on Smallholder Livestock Production in India: Opportunities and Challenges, 31 January–1 February, New Delhi, India.

IMF (International Monetary Fund). 2001. *Global trade liberization and the developing countries.* Washington, DC: International Monetary Fund.

INSEE-Coopération-SEDES. 1966. Etude démographique et économique en milieu nomade. Paris: Républic du Niger.

Julshamn, K., L. Dahl, and K. Eckhoff. 2001. Determination of iodine in seafood by inductively coupled plasma/mass spectrometry. *Journal of AOAC International* 84:1976–1983.

Kaasschieter, G. A., R. de Jong, J. B. Schiere, and D. Zwart. 1992. Towards a sustainable livestock production in developing countries and the importance of animal health strategy therein. *Veterinary Quarterly* 14:66–75.

Kanazawa, S., T. Kitaoka, Y. Ueda, H. Gong, and T. Amemiya. 2002. Interaction of zinc and vitamin A on the ocular surface. *Graefes Archive for Clinical and Experimental Ophthalmology* 240:1011–1021.

Kapadia, C. R. 1995. Vitamin B12 in health and disease, I: Inherited disorders of function, absorption, and transport. *Gastroenterologist* 3:329–344.

Kapur, A. 2007. Economic analysis of diabetes care. *Indian Journal of Medical Research* 125:473–482.

Kasen, S., P. Cohen, H. Chen, and A. Must. 2008. Obesity and psychopathology in women: A three decade prospective study. *International Journal of Obesity (London)* 32:558–566.

Kennedy, G., G. Nantel, I. D. Brouwer, and F. J. Kok. 2006. Does living in an urban environment confer advantages for childhood nutritional status? Analysis of disparities in nutritional status by wealth and residence in Angola, Central African Republic and Senegal. *Public Health Nutrition* 9:187–193.

Keusch, G. T. 1990. Micronutrients and susceptibility to infection. *Annals of the New York Academy of Sciences* 587:181–188.

Keusch, G. T., and M. J. Farthing. 1986. Nutrition and infection. *Annual Review of Nutrition* 6:131–154.

Khan, Q. M. 1984. The impact of household endowment constraints on nutrition and health–A simultaneous equation test of human-capital divestment. *Journal of Development Economics* 15:313–328.

Kieras, S. J. 1999. *Safety assessment and sensory analysis of a meat-based dried snack food for children in sub-Saharan Africa.* University Park, PA: Pennsylvania State University.

Kieras, S. J., E. W. Mills, S. J. Knabel, and A. N. Maretzki. 2003. Validation of pathogen destruction during manufacture of a meat-based potato snack (Chiparoo). *Journal of Food Processing and Preservation* 26:385–399.

Kikafunda, J. K., and P. Sserumaga. 2005. Production and use of a shelf-stable bovine blood powder for food fortification as a food-based strategy to combat iron deficiency anaemia in Subsaharan Africa. *African Journal of Food Agriculture Nutrition and Development* 5:1–17.

Klarenbach, S., R. Padwal, A. Chuck, and P. Jacobs. 2006. Population-based analysis of obesity and workforce participation. *Obesity* 14:920–927.

Leonard, W. M., K. M. DeWalt, J. E. Uquillas, and B. R. DeWalt. 1994. Diet and nutritional status among cassava producing agriculturalists of Coastal Ecuador. *Ecology of Food and Nutrition* 32:113–127.

Leroy, E. M., P. Rouquet, P. Formenty, S. Souquière, A. Kilbourne, J. M. Froment, M. Bermejo et al. 2004. Multiple Ebola virus transmission events and rapid decline of Central African wildlife. *Science* 303:387–390.

Leung, W.-T. W., R. R. Butrum, F. H. Chang, M. N. Rao, and W. Polacchi. 1972. *Food composition table for use in East Asia.* Rome and Washington, DC: Food and Agriculture Organization and U.S. Department of Health, Education, and Welfare.

Li, T. C., K. Chijiwa, N. Sera, T. Ishibashi, Y. Etoh, Y. Shinohara, Y. Kurata, et al. 2005. Hepatitis E virus transmission from wild boar meat. *Emerging Infectious Diseases* 11:1958–1960.

Little, M. A., and S. J. Gray. 1990. Growth of young nomadic and settled Turkana children. *Medical Anthropology Quarterly* 4:296–314.

Louwman, M. W., M. van Dusseldorp, F. J. van de Vijver, C. M. Thomas, J. Schneede, P. M. Ueland, H. Refsum, and W. A. van Staveren. 2000. Signs of impaired cognitive function in adolescents with marginal cobalamin status. *American Journal of Clinical Nutrition* 72:762–769.

Lozoff, B., E. Jimenez, J. Hagen, E. Mollen, and A. W. Wolf. 2000. Poorer behavioral and developmental outcome more than 10 years after treatment for iron deficiency in infancy. *Pediatrics* 105:E51.

Lynch, S. R., R. F. Hurrell, S. A. Dassenko, and J. D. Cook. 1989. The effect of dietary proteins on iron bioavailability in man. *Advances in Experimental Medicine and Biology* 249:117–132.

MacDonald, B., L. Haddad, R. Gross, and M. McLachlan. 2002. Nutrition: Making the case. In *Nutrition: A Foundation for Development*, 1–4. Geneva: United Nations Administrative Committee on Coordination/Sub-Committee on Nutrition.

Maina-Ahlberg, B. 1984. Beliefs and practices related to measles and acute diarrhoea. In *Maternal and child health in rural Kenya: An epidemiological study*, ed. J. K. Van Ginneken and A. S. Muller, 323–331. London: Croom Helm.

Maretzki, A. N. 2007. Women's NutriBusiness cooperatives in Kenya: An integrated strategy for sustaining rural livelihoods. *Journal of Nutrition Education and Behavior* 39:327–334.

Maretzki, A. N., and E. W. Mills. 2003. Applying a NutriBusiness approach to increase animal source food consumption in local communities. *Journal of Nutrition* 133:4031S–4035S.

Marquis, G. S., J. P. Habicht, C. F. Lanata, R. E. Black, and K. M. Rasmussen. 1997. Breast milk or animal-product foods improve linear growth of Peruvian toddlers consuming marginal diets. *American Journal of Clinical Nutrition* 66:1102–1109.

Martorell, R. 1996. The role of nutrition in economic development. *Nutrition Reviews* 54:S66–71.

McGuire, J., R. Galloway, and World Bank. 1994. Enriching lives: Overcoming vitamin and mineral malnutrition in developing countries. In *Development in Practice Series*. Washington, DC: World Bank.

Mellor, J. W. 1990. Agriculture on the road to industrialization. In *Agriculture in the Third World*, ed. C. K. Eicher and J. M. Staaz, 70–88. Baltimore, MD: Johns Hopkins University Press.

Millennium Ecosystem Assessment. 2005. Ecosystems and human well-being: Desertification synthesis. Washington, DC: World Resources Institute.

Mills, E. W., K. Seetharaman, and A. N. Maretzki. 2007. A NutriBusiness strategy for processing and marketing of animal-source foods for children. *Journal of Nutrition* 137:1115–1118.

Misselhorn, A. A. 2005. What drives food security in southern Africa? A meta-analysis of household economy studies. *Global Environmental Change* 13:33–43.

MOHP (Ministry of Health and Population) [Nepal], New ERA, and Macro International Inc. 2007. Nepal demographic and health survey 2006. Kathmandu, Nepal: MOHP, New ERA, and Macro International Inc.

Mozaffarian, D., and E. B. Rimm. 2006. Fish intake, contaminants, and human health: Evaluating the risks and the benefits. *Journal of the American Medical Association* 296:1885–1899.

Mugunieri, G. L., P. Irungu, and J. M. Omiti. 2004. Performance of community-based animal health workers in the delivery of livestock health services. *Tropical Animal Health and Production* 36:523–35.

Mugunieri, L. G., J. M. Omiti, and P. Irungu. 2005. Animal health service delivery systems in Kenya's marginal districts. In *The future of smallholder agriculture in Eastern Africa: The roles of states, markets, and civil society*, ed. S. W. Omamo, S. Babu, and A. Temu, 401–453. Kampala, Uganda: IFPRI Eastern Africa Food Policy Network.

Munroe, P. B., C. Wallace, M. Z. Xue, A. C. Marcano, R. J. Dobson, A. K. Onipinla, B. Burke et al. 2006. Increased support for linkage of a novel locus on chromosome 5q13 for essential hypertension in the British Genetics of Hypertension Study. *Hypertension* 48:105–111.

Muroki, N. M., G. K. Maritim, E. G. Karuri, H. K. Tolong, J. K. Imungi, W. Kogi-Makau, S. Maman, E. Carter, and A. N. Maretzki. 1997. Involving rural Kenyan women in the development of nutritionally improved weaning foods: Nutribusiness strategy. *Journal of Nutrition Education* 29:335–342.

Murphy, S. P., and L. H. Allen. 2003. Nutritional importance of animal source foods. *Journal of Nutrition* 133:3932S–3935S.

Nalca, A., A. W. Rimoin, S. Bavari, and C. A. Whitehouse. 2005. Re-emergence of monkeypox: Prevalence, diagnostics, and countermeasures. *Clinical Infectious Diseases: An Official Publication of the Infectious Diseases Society of America* 41:1765–1771.

Nantel, G., and K. Tontisirin. 2002. Policy and sustainability issues in forging effective strategies to combat iron deficiency. *Journal of Nutrition* 132:839S–844S.

Nathan, M. A., E. M. Fratkin, and E. A. Roth. 1996. Sedentism and child health among Rendille pastoralists of northern Kenya. *Social Science and Medicine* 43:503–515.

National Institute of Public Health National Institute of Statistics [Cambodia] and ORC Macro. 2006. *Cambodia demographic and health survey 2005.* Phnom Penh, Cambodia, and Calverton, MD: National Institute of Public Health National Institute of Statistics and ORC Macro.

National Research Council. 1991. *Microlivestock: Little-known small animals with a promising economic future.* Washington, DC: National Academy Press.

Nestel, P. S. 1986. A society in transition: Developmental and seasonal influences on the nutrition of Masai women and children. *Food and Nutrition Bulletin* 8:2–18.

Nettleton, J. A., and J. Exler. 1992. Nutrients in wild and farmed fish and shellfish. *Journal of Food Science* 57:257–260.

Neumann, C. G. 1981. Malnutrition and infection. In *Infection: The physiologic and metabolic responses of the host,* ed. M. C. Powanda and P. G. Canonico, 319–357. Amsterdam: Elsevier/North-Holland Biomedical Press.

Neumann, C. G., and D. M. Harris. 1999. *Contribution of animal source foods in improving diet quality for children in the developing world.* Washington, DC: Prepared for The World Bank.

Neumann, C. G., and G. G. Harrison. 1994. Onset and evolution of stunting in infants and children. Examples from the Human Nutrition Collaborative Research Support Program. Kenya and Egypt studies. *European Journal of Clinical Nutrition* 48:S90–102.

Neumann, C. G., and L. S. Stephenson. 1991. Interaction of nutrition and infection. In *Hunter's tropical medicine,* ed. G. T. Strickland, 947–950. Philadelphia: W. B. Saunders Company.

Neumann, C. G., G. J. Lawlor, Jr., E. R. Stiehm, M. E. Swenseid, C. Newton, J. Herbert, A. J. Ammann, and M. Jacob. 1975. Immunologic responses in malnourished children. *American Journal of Clinical Nutrition* 28:89–104.

Neumann, C. G., N. O. Bwibo, and M. Sigman. 1992a. *Final report phase II: Functional implications of malnutrition, Kenya Project.* Human Nutrition Collaborative Research Support Program. Los Angeles: University of California.

Neumann, C. G., S. Oace, S. Murphy, N. O. Bwibo, and C. Calloway. 1992b. *Low B-12 content in breast milk of rural Kenyan women on predominantly maize diets.* Stockholm, Sweden.

Neumann, C., D. M. Harris, and L. M. Rogers. 2002. Contribution of animal source foods in improving diet quality and function in children in the developing world. *Nutrition Research* 22:193–220.

Neumann, C. G., N. O. Bwibo, and C. Gewa. 2003a. *GL-CRSP final report: Role of animal source foods to improve diet quality and growth and development in Kenyan schoolers.* Los Angeles: UCLA School of Public Health.

Neumann, C. G., N. O. Bwibo, S. P. Murphy, M. Sigman, S. Whaley, L. H. Allen, D. Guthrie, et al. 2003b. Animal source foods improve dietary quality, micronutrient status, growth and cognitive function in Kenyan school children: Background, study design and baseline findings. *Journal of Nutrition* 133:3941S–3949S.

Neumann, C. G., S. P. Murphy, C. Gewa, M. Grillenberger, and N. O. Bwibo. 2007. Meat supplementation improves growth, cognitive, and behavioral outcomes in Kenyan children. *Journal of Nutrition* 137: 1119–1123.

Neumann, C. G., C. Gewa, and N. O. Bwibo. 2004. Child nutrition in developing countries. *Pediatric Annals* 33:658–674.

Nielsen, H., N. Roos, and S. H. Thilsted. 2003. The impact of semi-scavenging poultry production on the consumption of animal source foods by women and girls in Bangladesh. *Journal of Nutrition* 133:4027S–4030S.

Nyakundi, P. M., D. W. Kinuthia, and D. A. Orinda. 1994. Clinical aspects and causes of rickets in a Kenyan population. *East African Medical Journal* 71:536–542.

Obi, A., J. Bashi, M. Tshilamatanda, and H. van Schalkwyk. 2005. Food flows at the community level: Examining the market potential of rabbits and cane rats and implications for poverty alleviation and food security. In *International Food & Agribusiness Management Association World Food & Agribusiness Symposium.* Chicago.

Ogbeide, O. 1974. Nutritional hazards of food taboos and preferences in Mid-West Nigeria. *American Journal of Clinical Nutrition* 27:213–216.

Olivares, M., E. Hertrampf, F. Pizzarro, T. Walter, M. Cayazzo, S. Llaguno, P. Chadud, et al. 1990. Hemoglobin-fortified biscuits: Bioavailability and its effect on iron nutriture in school children. *Archivos Latinoamericanos de Nutrición* 40:209–220.

Onuorah, C. E., and J. A. Ayo. 2003. Food taboos and their nutritional implications on developing nations like Nigeria—a review. *Nutrition and Food Science* 33:235–240.

Pachón, H., K. B. Simondon, S. T. Fall, P. Menon, M. T. Ruel, C. Hotz, H. Creed-Kanashiro, et al. 2007. Constraints on the delivery of animal-source foods to infants and young children: Case studies from five countries. *Food and Nutrition Bulletin* 28:215–229.

Pajukanta, P., H. E. Lilja, J. S. Sinsheimer, R. M. Cantor, A. J. Lusis, M. Gentile, X. J. Duan, et al. 2004. Familial combined hyperlipidemia is associated with upstream transcription factor 1 (USF1). *Nature Genetics* 36:371–376.

Pennington, J. A. T. 1998. *Bowe's and Church's food values of portions commonly used.* 17th ed. Philadelphia: Lippincott-Raven Publishers.

Popkin, B. M. 2002. The dynamics of the dietary transition in the developing world. In *The nutrition transition: Diet and disease in the developing world,* ed. B. Caballero and B. M. Popkin, 111–128. London: Academic Press.

Popkin, B. M., S. Horton, S. Kim, A. Mahal, and J. Shuigao. 2001. Trends in diet, nutritional status, and diet-related noncommunicable diseases in China and India: The economic costs of the nutrition transition. *Nutrition Reviews* 59:379–390.

Ranum, P. 2001. Solving micronutrient deficiency problems. *Cereal Foods World* 46:441–443.

Reddy, K. S., and S. Yusuf. 1998. Emerging epidemic of cardiovascular disease in developing countries. *Circulation* 97:596–601.

Reddy, M. B., and J. D. Cook. 1997. Effect of calcium intake on nonheme-iron absorption from a complete diet. *American Journal of Clinical Nutrition* 65:1820–1825.

Roos, N., M. Islam, and S. H. Thilsted. 2003a. Small fish is an important dietary source of vitamin A and calcium in rural Bangladesh. *International Journal of Food Sciences and Nutrition* 54:329–339.

Roos, N., M. M. Islam, and S. H. Thilsted. 2003b. Small indigenous fish species in Bangladesh: Contribution to vitamin A, calcium and iron intakes. *Journal of Nutrition* 133:4021S–4026S.

Roos, N., M. A. Wahab, C. Chamnan, and S. H. Thilsted. 2007. The role of fish in food-based strategies to combat vitamin A and mineral deficiencies in developing countries. *Journal of Nutrition* 137:1106–1109.

Rossander-Hulten, L., M. Brune, B. Sandstrom, B. Lonnerdal, and L. Hallberg. 1991. Competitive inhibition of iron absorption by manganese and zinc in humans. *American Journal of Clinical Nutrition* 54:152–156.

Rutherfurd, S. M., and P. J. Moughan. 2005. Digestible reactive lysine in selected milk-based products. *Journal of Dairy Science* 88:40–48.

Salzman, P. C. 2004. *Pastoralists. Equality, hierarchy, and the state.* Boulder, CO: Westview Press.

Sansoucy, R. 1995. *Livestock—A driving force for food security and sustainable development.* Rome: Food and Agriculture Organization.

Sato, S. 1997. How the East African pastoral nomads, especially the Rendille, respond to the encroaching market economy. *African Study Monographs* 18:121–135.

Scholl, T. O., M. L. Hediger, R. L. Fischer, and J. W. Shearer. 1992. Anemia vs iron deficiency: Increased risk of preterm delivery in a prospective study. *American Journal of Clinical Nutrition* 55:985–988.

Schulze, M. B., J. E. Manson, W. C. Willett, and F. B. Hu. 2003. Processed meat intake and incidence of type 2 diabetes in younger and middle-aged women. *Diabetologia* 46:1465–1473.

Secretaría de Salud [Honduras], Instituto Nacional de Estadística, and Macro International. 2006. Encuesta Nacional de Salud y Demografía 2005–2006. Tegucigalpa, Honduras: Secretaría de Salud [Honduras], Instituto Nacional de Estadística, and Macro International.

Semba, R. D. 2001. Iodine deficiency disorders. In *Nutrition and health in developing countries*, ed. R. D. Semba and M. W. Bloem, 343–363. Totowa, NJ: Humana Press.

Shapiro, B. I., J. Haider, A. G/Wold, and A. Misgina. 1998. The intrahousehold economic and nutritional impacts of market-oriented dairy production: Evidence from the Ethiopian highlands. Paper at Agro-ecosystems, natural resources management and human health related research in East Africa, May 11–15, at International Livestock Research Institute, Addis Ababa, Ethiopia.

Shils, M. E., J. A. Olson, M. Shike, A. C. Ross, and eds. 1999. *Modern nutrition in health and disease.* 9th ed. Baltimore: Williams & Wilkins.

Shrimpton, R., and C. Hawkes. 2006. Diet related chronic diseases and the processing, labelling and marketing of food. *SCN News* 33:33–38.

Siekmann, J. H., L. H. Allen, N. O. Bwibo, M. W. Demment, S. P. Murphy, and C. G. Neumann. 2003. Kenyan school children have multiple micronutrient deficiencies, but increased plasma vitamin B-12 is the only detectable micronutrient response to meat or milk supplementation. *Journal of Nutrition* 133:3972S–3980S.

Sigman, M. 1995. Nutrition and child development: More food for thought. *Current Directions in Psychological Science* 4:52–55.

Simoons, F. J. 1961. *Eat not this flesh: Food avoidances in the Old World.* Madison: University of Wisconsin Press.

Smith, D. W. 1998. Urban food systems and the poor in developing countries. *Transactions of the Institute of British Geographers* 23:207–219.

Specker, B. L., A. Black, L. Allen, and F. Morrow. 1990. Vitamin B-12: Low milk concentrations are related to low serum concentrations in vegetarian women and to methylmalonic aciduria in their infants. *American Journal of Clinical Nutrition* 52:1073–1076.

Spedding, C. R. W. 1971. *Grassland ecology.* Oxford, U.K.: Oxford University Press.

Speedy, A. W. 2003. Global production and consumption of animal source foods. *Journal of Nutrition* 133:4048S–4053S.

Steffen, L. M., C. H. Kroenke, X. Yu, M. A. Pereira, M. L. Slattery, L. Van Horn, M. D. Gross, and D. R. Jacobs, Jr. 2005. Associations of plant food, dairy product, and meat intakes with 15-y incidence of elevated blood pressure in young black and white adults: The Coronary Artery Risk Development in Young Adults (CARDIA) Study. *American Journal of Clinical Nutrition* 82:1169–1177.

Stein, J. T., and BCTF (Bushmeat Crisis Task Force). 2001. *Species affected by the bushmeat trade.* Bushmeat Crisis Task Force 2001.

Stewart, R. 2004. *The places in between.* London: Picador.

Suarez-Berenguela, R. M., A. Gordillo, and P. Vane. 2006. *Economic impact of obesity in Latin America and the Caribbean.* Washington, DC: Pan American Health Organization/World Health Organization.

Swain, J. H., L. B. Tabatabai, and M. B. Reddy. 2002. Histidine content of low-molecular-weight beef proteins influences nonheme iron bioavailability in caco-2 cells. *Journal of Nutrition* 132:245–251.

Taylor, E. F., V. J. Burley, D. C. Greenwood, and J. E. Cade. 2007. Meat consumption and risk of breast cancer in the UK Women's Cohort Study. *British Journal of Cancer* 96:1139–1146.

Thammi Raju, D., and M. V. A. N. Suryanarayana. 2005. Meat consumption in Prakasam district of Andhra Pradesh: An analysis. *Livestock Research for Rural Development* 17 (article 130).

Thiam, I., K. Samba, and D. Lwanga. 2006. Diet related chronic disease in the West Africa Region. *SCN News* 33:6–10.

Thorpe, K. E. 2006. Factors accounting for the rise in health-care spending in the United States: The role of rising disease prevalence and treatment intensity. *Public Health* 120:1002–1007.

Thorpe, K. E., C. S. Florence, and P. Joski. 2004. Which medical conditions account for the rise in health care spending? *Health Affairs* W4:437–445.

Torpy, J. M., C. Lynm, and R. M. Glass. 2006. JAMA patient page. Eating fish: health benefits and risks. *Journal of the American Medical Association* 296:1926.

TRAFFIC. 2000. *Food for thought: The utilization of wild meat in Eastern and Southern Africa.* Harare, Zimbabwe: TRAFFIC East and Southern Africa.

Tunceli, K., K. Li, and L. K. Williams. 2006. Long-term effects of obesity on employment and work limitations among U.S. Adults, 1986 to 1999. *Obesity* 14:1637–1646.

Tuomainen, T. P., K. Nyyssönen, R. Salonen, A. Tervahauta, H. Korpela, T. Lakka, G. A. Kaplan, and J. T. Salonen. 1997. Body iron stores are associated with serum insulin and blood glucose concentrations. Population study in 1,013 eastern Finnish men. *Diabetes Care* 20:426–428.

Turley, M., M. Tobias, and S. Paul. 2006. Non-fatal disease burden associated with excess body mass index and waist circumference in New Zealand adults. *Australian and New Zealand Journal of Public Health* 30:231–237.

Underwood, B. A., and S. Smitasiri. 1999. Micronutrient malnutrition: Policies and programs for control and their implications. *Annual Review of Nutrition* 19:303–324.

United Nations. 1988. Rapid urbanization poses challenge to health, nutrition. *SCN News*.

USDA (U.S. Department of Agriculture). 2006. *National Nutrient Database for Standard Reference, Release 19*. Available at http://www.ars.usda.gov/Services/docs.htm?docid=15973

U.S. DHHS (Department of Health and Human Services) and USDA (U.S. Department of Agriculture). 2005. Dietary guidelines for Americans 2005. Home and Garden Bulletin No. 232. Washington, DC: U.S. Department of Health and Human Services and U.S. Department of Agriculture.

U.S. House of Representatives. 2002. Subcommittee on Fisheries Conservation, Wildlife and Oceans of the Committee on Resources. *The developing crisis facing wildlife species due to bushmeat consumption*. 2nd session. July 11.

van Dam, R. M., W. C. Willett, E. B. Rimm, M. J. Stampfer, and F. B. Hu. 2002. Dietary fat and meat intake in relation to risk of type 2 diabetes in men. *Diabetes Care* 25:417–424.

Vella, V., A. Tomkins, J. Nviku, and T. Marshall. 1995. Determinants of nutritional status in southwest Uganda. *Journal of Tropical Pediatrics* 41:89–98.

Voruganti, V. S., G. Cai, S. A. Cole, J. H. Freeland-Graves, S. Laston, C. R. Wenger, J. W. Maccluer, et al. 2006. Common set of genes regulates low-density lipoprotein size and obesity-related factors in Alaskan Eskimos: Results from the GOCADAN Study. *American Journal of Human Biology* 18:525–531.

Wagenaar-Brouwer, M. 1985. Preliminary findings on the diet and nutritional status of some Tamasheq and Fulani groups in the Niger Delta of central Mali. In *Population, health and nutrition in the Sahel*, ed. A. G. Hill, 226–253. London: Routledge and Kegan Paul.

Walker, S. P., C. A. Powell, and S. M. Grantham-McGregor. 1990. Dietary intakes and activity levels of stunted and non-stunted children in Kingston, Jamaica, I: Dietary intakes. *European Journal of Clinical Nutrition* 44:527–534.

Walter, T., E. Hertrampf, F. Pizarro, M. Olivares, S. Llaguno, A. Letelier, V. Vega, and A. Stekel. 1993. Effect of bovine-hemoglobin-fortified cookies on iron status of schoolchildren: A nationwide program in Chile. *American Journal of Clinical Nutrition* 57:190–194.

Welch, R. M., and R. D. Graham. 1999. A new paradigm for world agriculture: Meeting human needs productive, sustainable, and nutritious. *Field Crops Research* 60:1–10.

West, K. P., Jr., and I. Darnton-Hill. 2001. Vitamin A deficiency. In *Nutrition and Health in Developing Countries*, ed. R. D. Semba and M. W. Bloem, 267–306. Totowa, NJ: Humana Press.

West, C. E., F. Pepping, I. Scholte, W. Jansen, and H. F. F. Albers. 1987. *Food composition table for energy and eight important nutrients in foods commonly eaten in East Africa*. Wageningen, Netherlands, and Dar es Salaam, Tanzania: Technical Centre for Agricultural and Rural Cooperation of ACP/ECP Convention of Lomé and Food and Nutrition Cooperation.

Western, D. 1982. The environment and ecology of pastoralists in arid savannas. *Development and Change* 13:183–211.

WHO (World Health Organization). 1999. *The world health report 1999: Making a difference*. Geneva: World Health Organization.

WHO (World Health Organization). 2000. *ACC/SCN report: Attacking the double burden of malnutrition in Asia and the Pacific*. Geneva: ACC/SCN.

WHO (World Health Organization)/FAO (Food and Agriculture Organization). 2003. Diet, nutrition and the prevention of chronic diseases. In *Report of a Joint WHO/FAO Expert Consultation*. WHO Technical Report Series 916. Geneva: World Health Organization.

Wilkie, D. S., and J. F. Carpenter. 1999. Bushmeat hunting in the Congo Basin: An assessment of impacts and options for mitigation. *Biodiversity and Conservation* 8:927–955.

Willett, W. C. 1995. Diet, nutrition, and avoidable cancer. *Environmental Health Perspectives* 103 Suppl 8:165–170.

Williams, D. M. 2002. *Beyond great walls. Environment, identity, and development on the Chinese grasslands of Inner Mongolia*. Stanford, CA: Stanford University Press.

Williamson, C. S., R. K. Foster, S. A. Stanner, and J. L. Buttriss. 2005. Red meat in the diet. *Nutrition Bulletin* 30:323–355.

Wolfe, N. D., W. M. Switzer, J. K. Carr, V. B. Bhullar, V. Shanmugam, U. Tamoufe, A. T. Prosser, et al. 2004. Naturally acquired simian retrovirus infections in Central African hunters. *Lancet* 363:932–937.

Wood, J. D., M. Enser, A. V. Fisher, G. R. Nute, R. I. Richardson, and P. R. Sheard. 1999. Manipulating meat quality and composition. *Proceedings of the Nutrition Society* 58:363–370.

World Bank. 2006. Repositioning nutrition as central to development: A strategy for large-scale action. In *Directions in Development*. Washington, DC: World Bank.

World Cancer Research Fund/American Institute for Cancer Research. 2007. *Food, nutrition, physical activity, and the prevention of cancer: A global perspective*. Washington, DC: American Institute for Cancer Research.

Yeudall, F., R. S. Gibson, C. Kayira, and E. Umar. 2002. Efficacy of a multi-micronutrient dietary intervention based on haemoglobin, hair zinc concentrations, and selected functional outcomes in rural Malawian children. *European Journal of Clinical Nutrition* 56:1176–1185.

13

Social Consequences for Mixed Crop–Livestock Production Systems in Developing Countries

Achilles C. Costales, Ugo Pica-Ciamarra, and Joachim Otte

Main Messages

- **Smallholders predominate in mixed crop–livestock production systems.** Economic activity in livestock in mixed production systems in the developing world is predominantly undertaken by smallholders, making a living on farms of less than 2 hectares. These small farms account for a significantly larger share of meat and milk outputs in developing countries, and they contribute significantly to rural self-employment, given the labor intensity of such production systems. Projections indicate that small farms will continue to be a prominent feature in rural areas in the next decades.
- **Informal market chains are the main link between rural smallholders of livestock and the growing demand for meat and dairy products in the major urban centers and smaller rural towns.** They handle a far larger share of market output than the formal market chains that are linked with supermarkets and other outlets of higher-end meat and milk products. They generate more employment per unit of output and in the aggregate along the processing and distribution chain than the more capital-intensive formal market chain does. Formal and informal market chains are, however, treated unequally, with the perception of "superiority" being attached to the formal mode of trading. Although the stigma of illegality or inferiority of products is connected to informal transactions, the negative attitude toward informal market chains is an unfavorable starting point in public policy and for investments in integrating the participants into the whole scheme of market development for rural smallholder livestock keepers.
- **The impact of the drivers that are changing the livestock landscape—including demand and supply factors as well as institutional and policy changes—vary, depending on a country's resource endowments.** This is connected to the development of

the livestock sector through the role of agriculture and the sector in the economy coupled with the existing economic and institutional framework. Positive livestock sector development occurs when its growth and transformation are supported by a strong connection between the demand for and supply of livestock products, benefiting the majority of rural producers and leading to smallholder-based rural development.

- **Livestock sector stagnation or involution could occur when demand growth is confronted by a sluggish and even contracting livestock sector, where both the demand and supply sides are unable to respond to and lead any significant social transformation.** This can occur in countries where governments have not designed policies that allow farmers to compete in the downstream markets for meat and dairy products, so that both small-scale and large-scale producers have few or no incentives to respond to market signals in rural areas.
- **Positive but inequitable livestock sector growth could happen when the more "traditional" and more "modern" segments of the sector are not given relatively equal opportunities for growth in their respective markets.** In this case, growth would be characterized by a dualistic pattern of development, where few and large market-integrated producers appropriate the benefits arising from the drivers of change in the livestock sector, while the majority of rural smallholder producers, processors, and distributors are excluded from the market transactions that would let them share in the benefits of the changing livestock landscape. At times, they are even forced out of the market without similarly remunerative livelihood or employment opportunities outside the sector.
- **Policy directions for more equitable livestock sector transformation can turn the changing livestock landscape into an opportunity rather than a threat**

for smallholder livestock producers and for the stakeholders along the market chains they use. To accomplish this, two key elements need to be emphasized. First, livestock production activities can be organized efficiently in a decentralized manner by exploiting the capital, physical labor, and entrepreneurship of small rural producers. This smallholder-based rural development strategy, if scaled up to the level of countries, would certainly contribute to broad-based agricultural growth, rural employment, and poverty alleviation. Second, policy makers, international organizations, and development practitioners should focus not only on technical issues but also on the economic and institutional context in which the technologies are to be used. Building this supporting context would significantly assist the less well endowed livestock holders to benefit from the changing livestock landscape and to contribute to agricultural and rural growth.

Introduction

Livestock systems in many developing countries are rapidly evolving in response to changes in the economic and institutional environment in which livestock production, processing, and distribution occur (see Part I). Significant changes are also taking place in the organization of production, procurement of supplies, processing, and the distribution of products. The way in which the livestock sector of individual countries will be affected by and respond to these trends will to a large extent be determined by the country's "initial conditions"—the role of the agricultural and livestock sector in the economy, the country's basic resource endowments (land, labor, capital), and the economic and institutional framework.

Within countries, the consequences of the changing landscape will differ for the various livestock production systems (extensive, intensive, and mixed production systems) and for the different actors within these systems. In particular, expanding markets provide opportunities for livestock producers and associated market agents in developing countries to increase their incomes. However, the increasing importance attached to product quality and food safety by more affluent consumers, particularly those in urban areas, could mean that some production systems are relegated to markets for low-value products or squeezed out of markets altogether. The individuals excluded from these markets will not necessarily find alternative jobs in other productive sectors.

Mixed Crop–Livestock Production Systems

Human populations in developing countries are, in general, predominantly rural, and a significant proportion of the economically active population is engaged in agriculture. The United Nations Food and Agriculture Organization's estimates from 1990, with projections to 2010, for most regions in the developing world show that the agricultural population continues to increase in absolute terms, although its proportion in the total economically active population is on a decline. By 2010, some 209 million people will be active in agriculture in sub-Saharan Africa, 383 million in South Asia, and 151 million in East and Southeast Asia. In these regions, agriculture will be providing about 60% of total employment in sub-Saharan Africa, 53% in South Asia, and 43% in East and Southeast Asia (FAO 2006).

Within agriculture, the great majority of self-employment is generated by small-sized farms, either irrigated or rainfed, engaging in the production of crops and livestock. Of the more than 450 million farms in the world, about 85% are smaller than 2 hectares. (See Table 13.1.)

The "average" smallholder, however, differs among developing regions. In Asia and sub-Saharan Africa, the average farm size is 1.6 hectares. Within Asia, the average farm size is as low as half a hectare in Bangladesh and China (von Braun 2005). In regions with relatively abundant land or that are sparsely populated, such as Latin America and the Caribbean, the average farm size is 67 hectares. This figure hides the extent of smallholder farms, however, as this region has the highest inequality in landholding size (von Braun 2005). The differences in average farm sizes across regions are also a function of land quality and water availability, with viable farm sizes in the more arid regions tending to be relatively larger.

Table 13.1. Estimates of world farm size distribution, late 1990s.

Farm size class (ha)	Number of farms within size class (million)	Proportion of farms within size class (%)
<1	334.0	73.2
1–2	53.3	11.7
2–5	40.3	8.8
5–10	13.8	3.0
>10	14.7	3.3
Total	456.1	100.0

Source: von Braun 2005.

This situation is not going to change significantly in the next decade, and small farms will continue to predominate well into 2015, particularly in sub-Saharan Africa, South Asia, and Southeast Asia, despite the decline in numbers of farmers in the latter. With the continuing fragmentation of land, most farmers will have to make a living from smaller and smaller farms. (See Table 13.2.)

The general trend of agricultural land availability noted previously is borne out by country-specific statistics on average farm size. From the 1970s to the 1990s, average farm sizes continued to fall in Ethiopia, the Democratic Republic of Congo, India, Pakistan, China, Indonesia, and the Philippines. (See Figure 13.1.) In these countries, except for Pakistan, average farm sizes have been reduced to 2 hectares or less.

Livestock are an important source of income for small mixed farms. Despite comparable data not being available on a global scale, household surveys suggest that livestock contribute 5–20% of the household total income in mixed rainfed production systems and approximately 25–35% in irrigated zones. In pastoral areas, household income from livestock can rise as high as 70–80% (Davis et al., 2007, Maltsoglou and Taniguchi 2004, Roxas et al., 1997).

On a global scale, mixed farming systems account for close to 90% of milk and 70% of ruminant meat output. (See Table 13.3) Although globally intensive landless systems are dominant in the production of pig and poultry meat and eggs, mixed farming systems still account for more than one-third of world output in these commodities. Within the developing world, the contribution of mixed systems to the production of pig and poultry meat and of eggs is estimated at 45% and 39%, respectively.

There is a very close connection between mixed crop–livestock production units and smallholder farms in the developing world. Groenewold (2004) has shown that in the livestock production systems of developing-country regions of sub-Saharan Africa, Near East and North Africa, South Asia, Southeast Asia, and Latin America and the Caribbean, 84% of the agricultural populations are in mixed systems. The average size of the agricultural land resource base is small, computed at 0.33 ha/agricultural inhabitant, ranging from just 0.15 ha per inhabitant

Table 13.2. Hectares of agricultural land available per economically active person in agriculture, 1985 and 2000, with projections for 2015.

Region	1985	2000	2015
Sub-Saharan Africa	7.0	5.2	4.1
Near East/North Africa	12.2	10.4	9.3
South Asia	0.8	0.7	0.6
East and Southeast Asia	1.4	1.2	1.3
Latin America and Caribbean	15.9	16.5	18.1

Source: Elaborated with data from FAO 2006 and FAO projections.

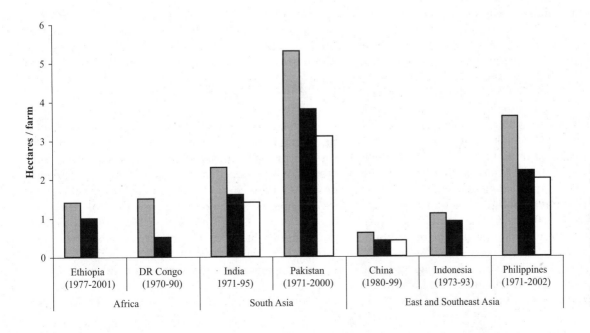

Figure 13.1. Trends in average farm size for selected developing countries in sub-Saharan Africa and Asia. *Source*: Nagayets 2005.

Table 13.3. Percent of major production systems in volume of meat, milk, and egg output by geographic regions in 2004.

Commodity/production system	Sub-Saharan Africa	Near East/ North Africa	South and Southeast Asia	Latin America and Caribbean	OECD Nations	Global Total
Ruminant meat						
Land-based extensive	50	6	15	49	21	25
Mixed crop–livestock	50	91	85	50	68	69
Landless intensive	0	3	0	1	11	5
Total	100	100	100	100	100	100
Pig and poultry meat						
Land-based extensive	19	1	0	5	0	1
Mixed crop–livestock	56	26	52	23	21	36
Landless intensive	25	73	48	72	79	62
Total	100	100	100	100	100	100
Milk						
Land-based extensive	62	6	3	45	10	12
Mixed crop–livestock	38	94	97	55	90	88
Landless intensive	0	0	0	0	0	0
Total	100	100	100	100	100	100
Eggs						
Land-based extensive	14	1	0	4	0	1
Mixed crop–livestock	60	28	51	15	22	38
Landless intensive	26	71	49	81	78	61
Total	100	100	100	100	100	100

Source: Groenewold 2004.

in South Asia and Southeast Asia to 1.15 ha per inhabitant in Latin America and the Caribbean. A large proportion of the arable lands in these mixed systems is rainfed (71%), with the remainder being irrigated. In terms of the physical volume of output, measured in metric tons of milk, meat, and eggs, differences in physical product form are to be noted. Milk accounted for the largest proportion (69%), followed by meat (24%). In the meat category, the bulk of the output consists of monogastric meat (61%), with ruminant meat accounting for the remaining output.

Whereas Table 13.3 is likely to overestimate the contribution of smallholder crop–livestock producers to national production, it is important to note that in countries with particularly skewed land distribution, smallholders largely contribute to agricultural production. For instance, in Kenya, smallholder dairy farmers produce approximately 85% of milk output and account for about 80% of marketed milk. In Ethiopia, rural small-scale mixed farms in the highlands account for around 63% of total milk output, compared with 22% from the extensive pastoral/agropastoral producers in the lowlands (Staal et al., 2006). In India, smallholders provide around 80% of total milk output. The corresponding figure for Pakistan ranges between 65% and 70% (Staal et al., 2006). In Viet Nam, where pigs generate the highest livestock income, smallholder households are estimated to produce 80% of total output (Tung et al.,

2005). In the Philippines, backyard producers still hold around 77% of the national pig inventory (Costales et al., 2007). Thus, even in the rapidly growing economies of Southeast Asia, such as Thailand, the Philippines, and Viet Nam, small farms continue to constitute the majority of farms and hold a large share of the national livestock inventory, despite the "scaling-up" of the sector and the gradual exit of small-scale producers over time. (See Table 13.4.)

Small farms not only contribute significantly to total output, as they are labor-intensive, they also contribute to overall employment in rural areas. On the one hand, because of the minimal or nonexistent alternative farm or nonfarm employment opportunities, the market wage does not reflect the opportunity cost of labor (which would likely be approaching zero), and thus family members work beyond levels that wage laborers would be willing to do. On the other hand, farmers, with little investment capital at their disposal, typically make use of labor-intensive rather than capital-intensive technologies and thus largely contribute to employment in rural areas.

Comparable examples are available for dairy production in India, Pakistan, Kenya, and Ethiopia. Considering the livelihood and employment dimension, small farms (one to three head of dairy cattle) in these countries generate significantly more direct (and indirect) full-time equivalent employment than large farms do per 1,000 liters of milk produced. (See Table 13.5.) Projecting the

Table 13.4. Changes in relative shares of small-sized pig farms in Southeast Asia.

Thailand* (<20 sows)		Philippines** (<20 pigs)		Viet Nam* (<6 pigs)	
Year	Share (%)	Year	Share (%)	Year	Share (%)
1988	97	1990	83	1994	98
2003	88	2003	77	2001	92

*Proportion of total number of pig farms

**Proportion of total pig inventory

Source: Poapongsakorn et al., 2003, Costales et al., 2007, Tung et al., 2005.

relative shares of farm employment in dairy for Pakistan and Kenya to sector level on the basis of employment generation capacity per 1,000 liter of milk output indicates that large farms account for only 12–13% of dairy farm employment, with the rest of the jobs being on small and medium-sized farms (Staal et al., 2006).

Small crop–livestock producers are also invariably linked to informal markets in which the supply chains are in general more labor-intensive per unit of output (throughput) than more capital-intensive formal chains. Comparative estimates of dairy chain employment in India, Pakistan, Kenya, and Ethiopia for formal and informal market chains show that on a 1,000-liter throughput basis, various formal processing and distribution chains on average generate employment for 1 to 17 regular workers, while different types of informal chains on average employ anywhere from 3 to 26 workers on a full-time equivalent basis. While remuneration in the formal sector is much higher, the returns to labor in the informal chains are still above rural nonfarm wages (Staal et al., 2006).

Table 13.6 presents estimates of the employment-generating capacity of the informal chains as a ratio to that of the formal enterprises. For India, producers of traditional sweet dairy delicacies (such as *halwai*) are taken to represent the informal market processors, while creameries are taken as example of processors in the formal sector. For Pakistan, Kenya, and Ethiopia, the informal market chains involve the selling of raw milk and the

processing and distribution of traditional milk products. In all countries, the final output of the formal market chains are highly processed products such as pasteurized milk, ultra-high temperature (UHT) processed milk (which has a longer shelf life), noncream milk, and other products. In general, the informal chains are more labor-intensive than the formal chains. Though the differences are not extreme, they can range from 1.2:1 for informal-to-formal chain employment ratio per unit of output in Kenya to a relatively high 2.5:1 ratio per unit output in Pakistan. Thus, on per volume throughput basis, employment generation at the level of the supply chains in the dairy sector is stronger along the informal chains.

While on a per volume output basis the differences in employment generation may not be large between these chains, total employment generation depends on the relative sizes of the informal and formal markets in terms of market shares of output. Table 13.7 shows that in important developing countries of sub-Saharan Africa, South Asia, and Southeast Asia, at least for dairy products, informal markets dominate, both in terms of production as well as in market chain employment. Given the dominance of informal chains in developing countries, they are expected to generate more employment opportunities for communities than formal chains.

In conclusion, traditional mixed crop–livestock production farms, predominantly consisting of small farms, significantly support the rural economy in developing countries. This is due also to their contribution

Table 13.5. Employment generated in dairy production by scale of operation in India, Pakistan, Kenya, and Ethiopia (in full-time equivalent employed workers/1,000 liters of milk offtake).

Country	Small farm (1 to 3 head)	Large farm (>3 head)	Small-to-large farm employment ratio
India	230	25	9.2:1
Pakistan	242	73	3.3:1
Ethiopia	224	n.a.	n.a.
Kenya	105	49	2.1:1

n.a. = not available

Source: Staal et al., 2006.

Table 13.6. Ratio of informal-to-formal market chain employment generation per 1,000-liter output in India, Pakistan, Kenya, and Ethiopia.

Country	Informal-to-formal chain employment ratio
India	1.5:1
Pakistan	2.5:1
Kenya	1.2:1
Ethiopia	1.5:1

Source: Staal et al., 2006.

to livestock production and to livestock-related downstream employment. This is in addition to their value to the nonlivestock dimension of their economic activities mediated through consumption linkages. The changing environment for smallholder-based mixed crop–livestock production systems can therefore have profound impacts on societies in which agriculture still plays an important role and where mixed systems contribute a significant proportion of livestock output. The consequences will depend on the impact of the changes on livestock supply chains that traditional/mixed livestock producers mainly use. This applies to traditional informal supply chains and how they respond to the growth and reconfiguration of more modern formal supply chains within domestic markets and to the competition that these formal chains bring.

Stylized Livestock Sector Development Pathways

Countries worldwide present a variety of different agricultural and livestock development paths. For example, the livestock sector is growing fast in Brazil and India but slowly in Zambia. In Thailand, growth in poultry production is driven by large-scale producers. In India, small dairy farmers have been able to adapt to the changing livestock landscape and contribute to meeting the growing demand for livestock products of urban consumers.

Although each country follows its own particular trajectory, it is possible to distinguish three potential development paths a traditional crop–livestock production system might take: a positive and equitable development path, livestock sector stagnation/involution, and a positive but inequitable development path.

Positive Livestock Sector Development Path

The forces shaping livestock production can constitute a stimulus for growth of the rural livestock sector and thereby improve rural incomes and reduce poverty if the conditions are such that the increase in demand for livestock products translates into expanded markets and incentives for a majority of rural producers. The following investments in rural production and rural-to-urban distribution systems should result in increased productivity of rural labor, leading to increasing returns to own resources (labor and capital) and broad-based increases in household and per capita incomes. This in turn can generate demand for nonfarm goods and services (consumption linkages) and for production inputs and agricultural/livestock services (backward production linkages), and spur economic activities along the market chains for livestock products (forward production linkages in transport and commerce services).

This development path would only gradually change the structure of production. It is based on the assumption that industrialization and urbanization are not necessarily inseparable, as assumed in the traditional paradigm of economic development, but that there could be a movement of the modern production base into the rural sector. In this development pathway, livestock production activities could be organized in a "deconcentrated" manner by exploiting both the physical labor and the entrepreneurial ability of rural people, thus supporting a smallholder-based rural industrialization (Hayami 1998). A number of studies have documented this development path in East Asia and some industrial countries (Hayami 1998), which, if found to be feasible

Table 13.7. Share of marketed milk sold through informal and formal markets and share of employment in informal and formal dairy market chains, selected developing countries.

Country	GDP/person (2003 PPP USD)	Share of marketed milk (%)		Share of employment (%)	
		Informal	Formal	Informal	Formal
Pakistan	2,097	93	7	94	6
India	2,892	85	15	86	14
Bangladesh	1,770	97	3	n.a.	n.a.
Kenya	1,037	80	20	87	13
Tanzania	621	88	12	n.a.	n.a.
Ethiopia	711	89	11	98	2

n.a. = not available

Source: Knips 2006, Staal et al., 2006, World Bank 2006.

on a broad scale, would alleviate the major difficulty in the trade-offs between growth and equity that confronts developing countries.

Livestock Sector Stagnation/Involution

In some developing countries, responses to the changing environment may almost entirely bypass the rural livestock sector and thereby induce its "involution." This pathway is likely to be followed when rural livestock production systems are largely disconnected from the growing urban markets for products. Potential consumption and production growth linkages will not materialize, and the continued growth of the agricultural population and rural labor force, coupled with the stagnation in agricultural and livestock production activities, will lead to worsening agricultural land-to-person ratios. This leads to deterioration of the productivity of rural labor, a fall in real wages, and ultimately a decline in rural per capita incomes. The pressure to move to other low-income nonfarm occupations and migration to town centers will reflect the lack of opportunities in livestock and agriculture rather than growth in other sectors of the economy.

Essentially, this "development pathway" has been associated with indiscriminate liberalization and privatization policies that have not been accompanied by adequate transition policies allowing smallholders to be competitive in nearby markets. This is not to say that liberalization and privatization generally lead to rural stagnation and involution, as there are a number of countries where these have successfully led to rural poverty reduction (Araujo et al., 2005, Besley and Cord 2007). Where accompanying support policies were absent, the closure of agricultural state enterprises that provided a secure output market for agricultural producers and the privatization of animal health services that provided cheap if not free services to smallholders, have led to the loss of markets and to a shortage of services in rural areas that have rarely been filled by the private sector. Smallholders were thus largely unable to satisfy the growing urban demand for livestock products, and in several cases imports of both dairy and meat products have substantially increased. It is only in recent years that policy makers in developing countries are experimenting with new market-friendly policies that aim to trigger the functioning of rural markets in low-income settings (De Janvry et al., 1997). Examples are the institutionalization of community animal health workers, market-smart subsidies to private animal health service providers, and legal reforms in the financial sector that allow banks to accept movable properties, including livestock, as collateral on loans.

Positive yet Inequitable Development Paths

There may be variants to the two development pathways just described. The livestock subsector could be growing in aggregate but this could be due to rapid growth of only the segment of the sector connected to the major urban consumption centers (as well as urban/periurban processors). Where that connection does not hold in the more remote locations, weak production and consumption linkages become apparent within the rural economy. Or there could be an overall decline in the livestock sector, but producers with good access to markets and services might in fact be experiencing growth in their own particular products, while for other producers there could be an even more rapid decline.

These scenarios depict situations in which changes in the landscape surrounding the livestock sector create opportunities that benefit a few specialized and strategically located producers in the modern segment but exclude the majority of small livestock keepers in the rural areas who have poor access to input and output markets, credit and extension services, and other information. This development path would create a dual livestock economy structure: a modern segment of a few capital-intensive large firms in livestock production and processing, generating limited employment per unit of output, and a traditional segment consisting of many small livestock keepers in the rural areas. In the medium to long term, industry growth will be accompanied by the exit of the small rural producers who face relatively higher transport and marketing costs per unit of output in the competition to supply the larger mainstream market, apart from the cost of meeting standards of product quality.

The likelihood of a country following one pathway rather than another is influenced by the strength of linkages within the agricultural subsectors and between agriculture and the rest of the economy, whereby growth in one subsector is able to induce a corresponding growth in the others. Table 13.8 provides an overview of empirically derived relative strength of linkages for different regions and continents; the ratios indicate the magnitude of additional income growth created elsewhere for any unit percentage growth in agricultural income. The sources of linkages are consumption and production economic activities, with their relative shares adding up to the total (100%). As can be seen, agricultural growth linkages are strongest in Asia (0.64), and the relative share of consumption linkages outweighs those in production. The results for Asia indicate that the farm and nonfarm sectors are closely interrelated. By implication, growth (or decline) in one sector is likely to have repercussions on other sectors. By contrast, in Latin America agricultural growth linkages are weakest, at 0.26. This means that growth in agricultural income is mainly mediated within the sector, resulting in a relative "insulation" of the sector (or subsector) and thus higher potential for "inequitable" growth.

Two caveats are necessary in applying these stylized livestock sector development paths to sample countries. The first caveat is that the three development pathways are not necessarily exclusive routes, as it is possible that

Table 13.8. Agricultural growth linkages in Asia, Africa, and Latin America.

Region	Initial agricultural income increment	Magnitudes of additional income growth			Source of linkages and relative share (%)	
		Other agriculture	Rural nonfarm	Total	Consumption	Production
Asia	1.00	0.06	0.58	0.64	81	19
Africa	1.00	0.17	0.30	0.47	87	13
Latin America	1.00	0.05	0.21	0.26	42	58

Source: Haggblade et al., 2005.

within the same country, for example, smallholder dairy producers will be "crowded out" as the dairy industry expands, whereas small pig farmers will be able to benefit from the opportunities offered by the growth in demand for pork, both of which may constitute aspects of livestock sector developments. It could also happen that some regions within a country will be able to benefit from the changing landscape—for example, those better endowed with market infrastructure or governed by more efficient local governments—whereas other regions will not.

The second caveat relates to the often presumed dichotomy between formal and informal supply chains, and the presumed superiority of the former. First, it is important to note that formal and informal supply chains are not mutually exclusive and often exist side by side, catering to various types of consumers.

Second, the traditional informal supply chains are relatively well established and can maintain their market position in terms of serving consumer demand in rural communities and small towns, where average household incomes are relatively low and consumer preferences lean toward fresh produce or traditional processed products (raw milk, live poultry, warm meat, traditional sausages). In these situations, further home processing can easily be undertaken (boiling milk, home slaughtering of poultry, preserving meat). In supplying this market, the presumed superiority of formal market chains does not come into play. The size of this market in terms of market share of livestock products needs to be more firmly determined, and its growth path predicted. The size can increase (in absolute terms) with the increase in rural populations. It can also decline as competing products of more modern formal supply chains make inroads with consumers in small towns. In the competition for market share in the more distant larger urban centers, however, where the more rapid growth in demand for livestock products is occurring and where consumer demand for quality and food safety are becoming more significant, the traditional informal market chains are at a disadvantage vis-à-vis the more modern organized formal supply chains. In this context, the "superior" position of the formal supply chains applies, as the market chains

distinguish between products coming from formal and informal supply sources.

Third, a movement from an informal toward a formal supply chain does not necessarily imply a significant reduction of employment. On a per unit throughput basis, the differences in employment generation capacity between formal and informal chains are in fact not necessarily very large and may be more dependent on scale than degree of formality. It is rather the relatively large share of the informal market chains in total marketed volume that makes the employment shares disproportionate. As the formal supply chains grow and gain in market share, the employment levels will also grow in absolute terms in various value-adding activities along the chain.

Fourth, informal is not synonymous with illegal. Often, informal small traders or processors would be willing to formalize their transactions, for example, to register as enterprises for tax purposes and thereby be able to obtain loans from formal credit institutions. Even when market opportunities exist, however, the presence of institutional bottlenecks can make formalization unprofitable. For instance, becoming a formal business entity might not only mean the obligation to pay proper license fees and taxes but also imply a greater vulnerability to having to pay bribes to rent seekers who hold discretionary powers in making business transactions happen or not.

Case Studies: Dairy Sector Development in India, Zambia, and Brazil

Positive Development: The Dairy Sector in India

India is a low-income economy with a gross domestic product (GDP) per capita of $3,460 in 2005 (at purchasing power parity, PPP). Agriculture is a dominant sector of the Indian economy and accounts for over 18% of GDP (World Bank 2006) and, according to the most recent figure (1995), for about 67% of employment. During the past five years the agricultural sector has witnessed advances in the production and productivity of food grains, oilseeds, commercial crops, fruits, vegetables, poultry, and dairy. Today, India is the largest producer of milk in the world, at 91,940 million tons recorded in 2005 (FAO 2006).

The Indian agricultural sector is dominated by small to very small farms, with over 80% of farmers, or about 93 million farms, cropping less than 2 hectares. Average farm size has been constantly falling, from 1.6 ha in 1991 to 1.4 ha in 1995, while over the same period the number of farms of less than 2 ha increased by almost 10% (Nagayets 2005). With respect to the dairy subsector, the number of farmers engaged in dairy production increased by around 40% between 1991 and 2000 (see Figure 13.2), while average herd size declined. Small and marginal farmers raising one to three head of cattle or buffalo continue to account for close to 60% of all dairy farms (GOI, Basic Animal Industry Statistics, 2002, in Sharma et al., 2003). Figure 13.3 shows, however, that qualitative changes are taking place in the composition of the dairy herd in Indian farms, with the number of buffalo and crossbred cattle increasing rapidly while numbers of animals of local cattle breeds are in absolute decline. (Crossbreeds refer to animals with varying degrees of high potential dairy genes [*Bos taurus*, usually Holstein and brown Swiss] and one of the many Indian breeds. Local breeds refers to the original Indian cattle, mostly *Bos indicus*, which have a relatively low milk yield potential but are well adapted to local conditions.)

On the marketing side of the dairy industry, Figure 13.4 shows that while the bulk of marketed dairy output comes from rural producers, the greater part of marketed output finds its way to urban market destinations (86% of marketed milk). Milk sold to urban markets flows largely through informal chains (82%). However, formal chains for the supply of milk are emerging through processing cooperatives (11%) and private companies (7%).

A close look at rural dairy market chains in India shows that segments in the informal sector connect with formal market chains where controls are in place for smallholder milk producers to conform to the standards of private processors. Figure 13.5 shows that in the state of Rajasthan, independent smallholders selling their output through the dominant traditional informal market chain exist side by side with smallholders whose milk flows through the formal channel, with informal agreements made with collecting agents who have formal supply contracts with milk companies or private processors. These agreements involve commitments in the supply of output by smallholders and the direct provision of inputs and livestock services by the formal dairy processors or their indirect provision through the contracted village agents.

In conclusion, rapid growth in the livestock industry could coincide with the continued proliferation of small farms with small livestock holdings. It is imperative, however, that there be a technological transformation in these small farms themselves (e.g., improved breeds, higher quality feeds) to improve productivity significantly and increase income per unit of land, per head of livestock, and per unit of labor input. Furthermore, the growth and development of the livestock industry could well coincide with a still large segment of the urban market being served by informal supply chains supplying differentiated products particularly demanded by the greater bulk of the consuming public. As consumers

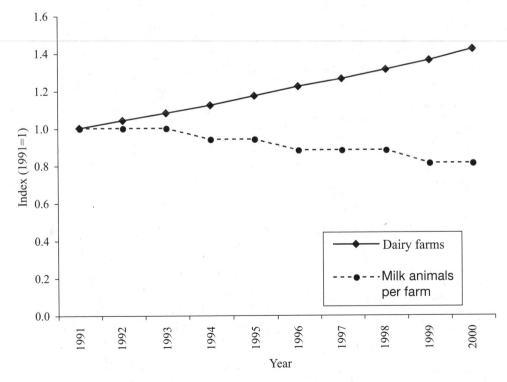

Figure 13.2. Growth in dairy farm numbers and average herd size in India (1991 = 1).
Source: Hemme et al., 2003.

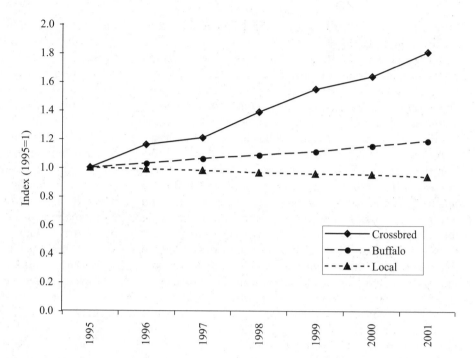

Figure 13.3. Differential growth of different types of dairy animals in India (1995 = 1).
Source: Hemme et al., 2003.

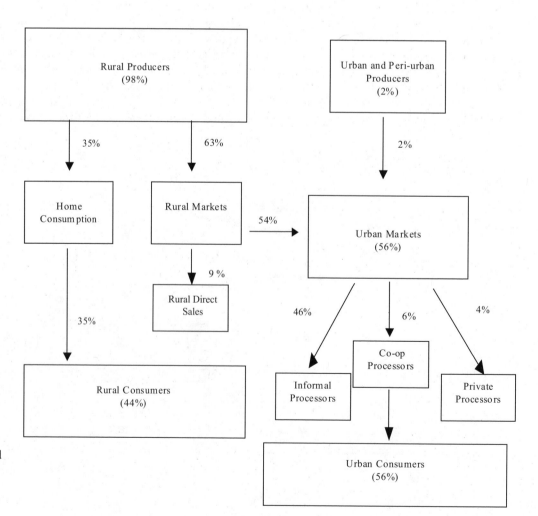

Figure 13.4. Rural and urban/periurban formal and informal market flows for dairy in India.
Source: Staal et al., 2006.

Independent / Informal Chain

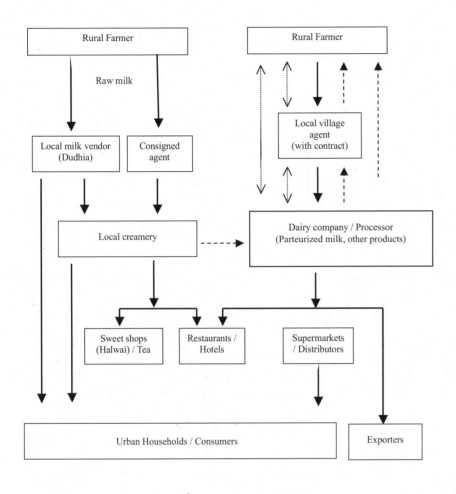

Figure 13.5. Independent and contractual market linkages for rural smallholder dairy producers in Rajastan, 2005.
Source: Hemme et al., 2003, Birthal et al., 2006.

become more discriminating and as formal organizations become more convenient and credible suppliers of high-quality and food-safe livestock products, the vertical coordination along formal lines in supplying processing firms will obtain increasing importance. This results in organizing groups of smallholders at the farm level (through formal contracts or informal agreements), with their main market links more likely becoming a necessity for smallholders to link up with formal market chains.

Stagnation and Involution Pathway: The Dairy Sector in Zambia

Zambia is a low-income rural economy, with a GDP per capita of $950 in 2005 (at PPP). The country has gone through three decades of declining living standards (per capita income in 2000 was just 60% of the level in the late 1960s), and today over 83% of the rural population and 56% of urbanites live below the national poverty line (World Bank 2006). Agriculture contributes about 22% to national GDP and about 75% to overall employment (World Bank 2006). In value terms, livestock accounts for about 39% of national agricultural output (FAO 2006). The agricultural sector comprises about 450,000 small rainfed mixed crop–livestock farms with less than 10 ha of land, using basic production technologies and largely relying on family labor. Some 145,000 emergent, medium-scale farmers crop between 10 and 60 hectares, making use of draft power and some purchased inputs; fewer than 750 large-scale farmers have holdings of 60 ha or more and use mechanized farming techniques; there are only a dozen large corporate farms operating in Zambia (Government of Zambia 2002, FAO and WFP 2005).

The radical liberalization and privatization programs carried out in the early 1990s transformed Zambia into one of the most liberalized and deregulated economies in Africa. These reforms had negative impacts on agriculture. In the livestock sector, following the government's

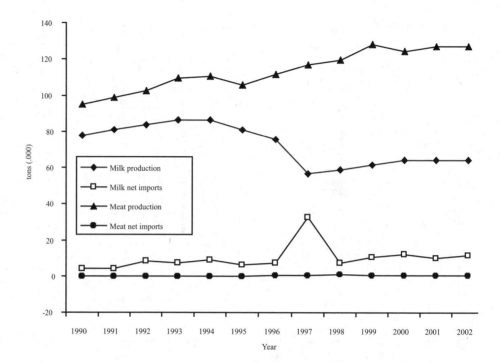

Figure 13.6. Milk production and net imports in Zambia, 1990–2002. *Source*: FAO 2006.

withdrawal from the provision of livestock services, animal diseases spread and livestock mortality increased significantly (IFAD 2006). The resulting loss of livestock negatively affected overall agricultural production, as animals had been used for land preparation, and manure was used for soil fertilization. The adverse impacts were more strongly felt in the dairy sector, which is dominated by smallholders, and less in cattle production, where almost 40% of producers are commercial and semicommercial and therefore able to adapt rapidly to changing market dynamics (McLeod and Chilonda 2004).

Figure 13.6 illustrates the trends in the aggregate production and net imports of milk and meat, while Figure 13.7 shows the trend in per capita production for the 1990 to 2002 period.

During the 1990–2002 period, per capita milk production in Zambia declined from 9.5 to 6.0 liters per year while net imports of milk steadily increased (FAO 2006). At present about 25% of the milk consumed in the country is imported as milk powder from New Zealand and as ultra-high temperature milk from South Africa and other countries (Emongor et al., 2004).

Dairy production has been and remains largely dominated by traditional smallholder producers, which represent 99% of all milk producers. These smallholders own over 90% of Zambia's dairy cattle and account for about 65% of all milk produced in the country. Large commercial producers account for 22%, and small to medium-scale periurban producers account for 13% of total milk production (Emongor et al., 2004, Mukumbuta and Sherchand 2006).

Some 70–75% of national milk production in Zambia flows through informal channels. Smallholders either produce for home consumption or sell milk informally to neighbors, traditional small traders, and hawkers. Small periurban producers, who are more market oriented than rural livestock keepers, tend to sell milk directly to consumers and, in some cases, to small informal vendors. Those few that are members of cooperatives sell part of their surplus milk at collection points, but only about 3% of production is marketed through this channel. About 70 large-scale commercial farms provide milk to the 19 milk processors in the country, which are mainly of small to medium size. Some of the largest processing firms, including Parmalat, process milk into yogurt, cheese, and butter (Emongor et al. 2004, Mukumbuta and Sherchand 2006).

Given an inadequate supply of raw liquid milk, processing companies make up the deficit by recombining imported powder milk. They market milk through supermarkets (50%), cash-and-carry stores (23%), and small formal retailers (22%). Overall, this system handles less than 25% of all sales of domestic milk output (Emongor et al. 2004). Figure 13.8 displays the dairy production and marketing chain in Zambia, which is dominated by small mixed crop–livestock farmers producing milk for home consumption and marketing surplus milk through informal market transactions.

Thus, in the face of liberalization and privatization, the livestock sector in Zambia has not undergone any significant structural change over the last 15 years; the dairy sector has even shown signs of involution, with decreased aggregate and per capita production leading to increased net imports. The decline of milk production in Zambia coincides with the deterioration in the overall economy, where real per capita incomes underwent a downward spiral until 1995. Nevertheless, the adverse infrastructural and market environment of the

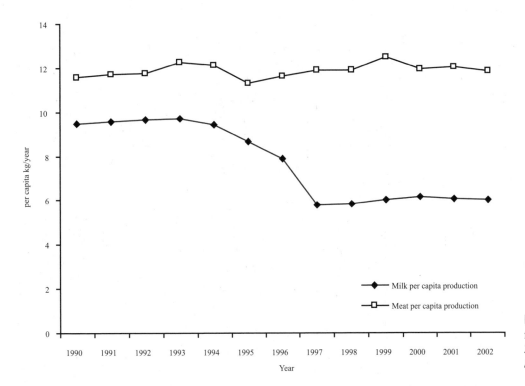

Figure 13.7. Per capita milk and meat production in Zambia, 1990–2002. *Source*: FAO 2006.

rural smallholder dairy sector presented constraints to more rapid recovery as per capita incomes improved. Whereas part of the involution of the dairy sector might be explained by the policy reforms carried out in the 1990s, the negative growth rates of milk production and consumption are particularly glaring, as on the one hand there is a demand for milk that is not met by domestic production. On the other hand, Zambia's resources for agricultural production are largely idle, with only about 14% of the cultivable land currently being exploited and only 9% of the irrigation potential being used (McLeod and Chilonda 2004).

The World Bank (2007) reports that one of the main challenges facing rural farming households in Zambia, of which 71% own livestock, relate to low and declining levels of smallholder production systems. Low productivity and low returns to labor and other resources bring about chronically low levels of farm income. Among the main constraints relating to low productivity were poor public rural market infrastructure (e.g., roads) and support mechanisms, which impede the private provision of goods and services to rural farmers, and incomplete implementation of policies, which leads to underprovision of public goods (e.g., rural irrigation) and services (e.g., extension and research). While public funds had been allocated for agricultural development and commercialization programs, the government has traditionally chosen to make interventions in the supply of inputs and in crop marketing arrangements, activities that are primarily driven by short-term political interests (World Bank 2007).

Inequitable Development: The Brazilian Dairy Sector

Brazil is a middle-income economy with an annual per capita GDP of about $8,200 in 2005 (at PPP). The services and industry sector account for about 90% of national GDP and for 80% of total employment, with agriculture contributing 10% to national value added and 20% to overall employment (World Bank 2006).

Since the structural reforms of the early 1990s, the Brazilian economy has shown unstable growth, averaging about 2% per annum. Agriculture has been the fastest-growing sector in the economy, growing at about 3% per year, with industry and services growing at 1.2 and 2.2% per annum, respectively (World Bank 2006).

Within agriculture, the livestock sector has been performing particularly well. According to FAO (2006), livestock accounted for about 39% of agricultural value added in 1990 and for 45% in 2002. While per capita cereal availability slightly decreased between 1990 and 2002, meat and milk per capita availability increased by 71% and 27%, respectively. (See Figures 13.9 and 13.10.)

Currently Brazil is the eighth largest milk producer in the world (23.5 million tons in 2005) and is expected to become the world's second largest producer within five years. Although Brazil imported dairy products in the second half of the 1990s, the country has moved close to self-sufficiency, and export sales are steadily increasing. Today milk is the sixth most important agricultural product in Brazil (FAO 2006, Phillips 2006).

These changes in the dairy sector have been driven by the growing demand for dairy products in the country

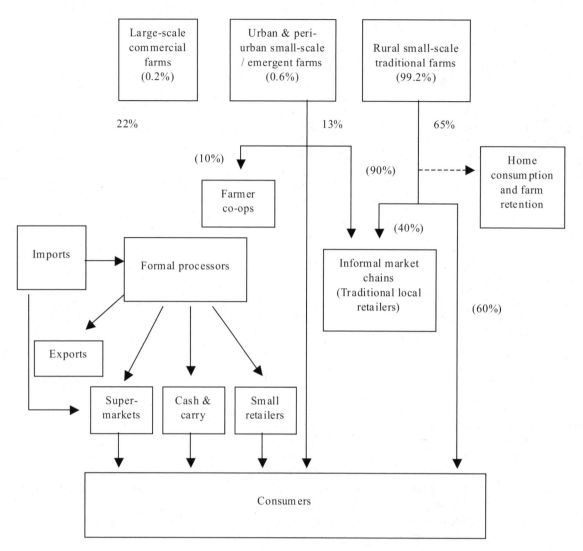

Figure 13.8. Informal and formal market chains for milk in Zambia.
Source: Elaborated from Emongor et al., 2004, Mukumbuta and Scherchand 2006.

and by the liberalization program carried out by the government since the late 1980s. Between 1989 and 1995, after four decades of market regulation, Brazil liberalized the domestic dairy sector, including retail and farm-gate prices, and opened it to international competition when the country joined MERCOSUR in 1991 and the World Trade Organization in 1995 (Farina 2002, WTO 1994). The local milk supply chains were thus forced to compete with industrial firms and multinationals entering the Brazilian markets.

The new policy regime severely affected small farmers; the last 10–20 years have seen a general decline in the number of small dairy farms in the order of 3–4% per year. In 2003 there were about 1.4 million dairy farms, down from 1.8 million in 1996, and average herd size had grown from 9 to 14 dairy cows (Hemme et al., 2005). Between 1997 and 2001 more than 75,000 Brazilian dairy farmers (around 5%) were "delisted" by the 12 largest milk processors. Most of these farmers

presumably went out of business (FAO 2003). Milk supply data for the 12 major dairy industries indicate that "between 1996 and 1998, the number of suppliers fell by 28% and supply by farm increased by 37%" (FAO 2003). Barros et al. (2004) report that suppliers to the 12 major processing firms—which account for about 50% of inspected milk processed—declined from 152,500 in 1998 to 89,300 in 2002, a net reduction of nearly 71%, with the average supplier providing 117 liters/day in 1998 and 207.2 liters/day in 2002.

Despite these numbers, 80% of dairy farmers maintain 10 or fewer dairy cows. But overall these farmers manage 50% of the national dairy herd. Only about 10% of dairy farmers keep more than 30 dairy cows; but these farmers manage 30–35% of the national dairy herd and contribute about 30–40% of national milk production. Matthey et al. (2004) state that more than 30% of Brazil's milk is produced by small farmers and marketed through local markets and stores; Phillips

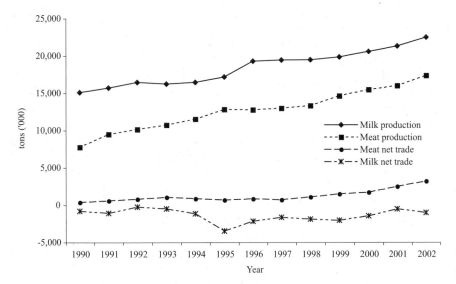

Figure 13.9. Net trade and production of meat and milk in Brazil, 1990–2003. *Source*: Elaborated from FAO 2006.

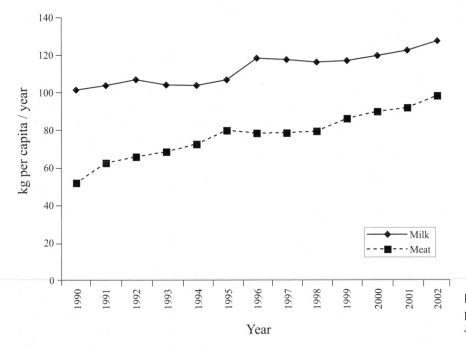

Figure 13.10. Per capita meat and milk production in Brazil, 1990–2002. *Source*: Elaborated from FAO 2006.

(2006) contends that around 40% of milk is produced on small, nonspecialized farms; Costa et al. (2004) state that about 41% of total milk production is informally sold to consumers; and *Leite Brasil* estimates that in 1997–2000, 43% of production was channeled through informal markets (Ostrowski and Deblitz 2001).

Liberalization and the ensuing competition in milk production are certainly among the root causes of the reduction in the number of small dairy farms in Brazil, which at the same time has led to structural changes in the processing industry. Since the market has been liberalized, the milk producer price has decreased by 44%, forcing the more inefficient (typically small-scale) producers out of the formal market. Many of the latter diverted sales into the informal market; between 1990 and 1998, informally marketed milk increased by about 4

million liters (a gain of nearly 10 percentage points of market share) while the amount of formally marketed milk increased by less than 2 million liters. (See Table 13.9.) Barros et al. (2003) argue that since the regional markets are the outlets of the informal sector, milk marketed through these chains could pass without inspection by the federal agency. Smallholders find fewer restrictions in selling their milk directly rather than through the large companies of the formal sector. The existence of the informal market also reduced the power that large companies exercised in setting the price of milk.

Competition among processors has increased since the introduction of UHT technology in the late 1980s. UHT milk can be transported over long distances, and inefficient cooperatives, previously protected from competition because pasteurized milk required a cold chain, left the market or were acquired by large private firms. In the late 1980s almost all formally marketed milk was collected and processed by strong national cooperatives. In the second half of the 1990s this share had dropped to 60%, and today cooperatives collect about 40% of all milk processed in the country (Costa et al., 2004, Farina 2002). (The current processing industry differs from those prevalent in the United States, Europe, and Australia, where large cooperatives dominate milk processing and marketing [FAO 2003].)

Increasing competition among processors has led many to develop new, more efficient chain management involving refrigeration tanks at farm level. Since the smallest tank in Brazil holds 200 liters while average farm production is 50 liters, most small farmers cannot comply with the new system and are forced out of the formal market (Farina 2002).

Finally, in order to increase milk quality, the government requires that all producers acquire on-farm cooling and storage to meet international standards, with the aim of eliminating type C milk from the market. Type C milk is marketed through informal channels and is sold in plastic bags and requires boiling before drinking. This policy is expected to force more small dairy producers out of the formal market (Matthey et al., 2004).

Whereas the economic reforms of the 1990s have led to significant changes in both dairy production and processing, only minor changes have occurred in dairy distribution. The three main actors in the Brazilian food retailing industry are supermarkets or hypermarkets (with five cash registers or more); self-services shops, including independent supermarkets; and traditional retailers. Supermarket chains and independent supermarkets are by far the dominant food retailers as they account, respectively, for about 42% and 44% of sales through about 16% and 1.5% of all stores. Traditional retailers account for 13% of all sales and 82% of stores (Farina 2002).

The last 10 years have witnessed a marginal increase in the number of traditional stores and independent supermarkets and a decrease in the number of supermarkets and hypermarkets, suggesting that traditional retailers can exist alongside larger stores. Farina and Nunes (2002) argue that increases in the number of traditional retailers and independent supermarkets originate from the management strategies of large food processors that,

in order not to have to deal with large monopolistic retailers, are willing to supply independent supermarkets and small retailers with dairy products.

A significant change in the milk marketing system has been the rapid concentration within the supermarket industry. In 1994 the 10 largest supermarket chains earned about 24% of total supermarket revenue, whereas in 2000 they earned 47%. Furthermore, the reconfiguration of the procurement, processing, and distribution systems in Brazil led to an increasing segmentation and dualistic development of the market into the formal and informal supply chains. The Brazilian milk supply chain and its components are represented in Figure 13.11.

Trends in the dairy sector, with production increasingly in the hands of fewer farms and an increasing concentration of processing, suggest that the development of the sector is following an unequal development path, excluding smallholders from the direct benefits of the growing demand for dairy products, particularly in mainstream urban markets.

The modernization process of the Brazilian dairy industry certainly resulted in casualties, directly affecting those who could not participate in formal dairy chains; the decline in the aggregate number of dairy farms over the last two decades bears testimony to their exit from the dairy sector. Without further information on their shift in livelihood to alternative sources of employment and incomes (e.g., from mixed farms to pure crop production or to the nonfarm sector), however, it is difficult to judge the poverty impacts of the fall in dairy farms. The national reduction in the $1-a-day poverty index from 14% to 8.2% between 1990 and 2001 does not necessarily indicate that the potential of dairy farmers in the business has significantly contributed to changes in the national poverty headcount.

Conclusions

Smallholders engaged in mixed crop–livestock farming currently constitute the majority of rural people in developing countries, and projections clearly indicate that they will remain a prominent feature of rural areas well into the next decades.

In developing countries, smallholder livestock producers account for a significant share of meat, particularly ruminant meat, and milk output. They also contribute significantly more to rural self-employment than large-

Table 13.9. Market shares of formal and informal dairy chains in Brazil, 1990–98.

Market segment	Volume (million liters equivalent)			Market share (%)	
	1990	1998	Change (%)	1990	1998
Formal	9,609	11,345	18.1	66.3	56.5
Informal	4,875	8,732	79.1	33.7	43.5
Total	14,484	20,077	38.6	100.0	100.0

Source: Barros et al., 2003.

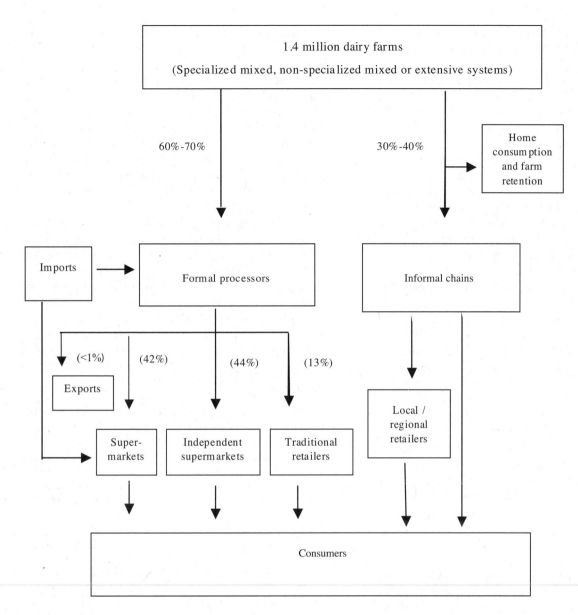

Figure 13.11. Formal and informal milk supply chains in Brazil. *Source*: Elaborated from Costa et al., 2004; Farina 2002; Phillips 2006.

scale industrial producers do. At the market chain level, in most developing countries small-scale processors and distributors also handle a larger share of market output than their large-scale counterparts do.

Although in most cases small-scale producers and processors operate informally, this is not necessarily synonymous with illegally. These producers are often willing to formalize their transactions but are unable to do so because of an "unfriendly" institutional and regulatory infrastructure that makes formalization difficult and unprofitable.

Drivers of change in the livestock sector that include changes in demand, production/supply technology, national policies, and international setting will certainly affect smallholder mixed crop–livestock farmers via changes in market shares of the associated processing and marketing businesses.

The impact of these drivers will vary, depending on a country's resource endowment, the role of agriculture and the livestock sector in the economy, and the economic and institutional frameworks. These drivers may also either encourage or inhibit the development of mixed crop–livestock production systems.

Three dominant patterns of livestock sector development affecting mixed livestock production systems can be identified:

• *Positive livestock sector development*, supported by a strong connection between the demand for and supply of livestock products, benefiting the majority of rural producers and leading to smallholder-based rural development. An example would be the development of the dairy sector in India, characterized by an increasing number of small dairy farms using improved production technology and possessing the ability to compete with large producers.

- *Livestock sector stagnation/involution*, characterized by a sluggish and even contracting livestock sector, both unable to respond to and lead any significant social transformation. This situation may occur in countries where governments have not designed policies that allow farmers to compete in the downstream markets for meat and dairy products, so that both small-scale and large-scale producers have few if any incentives to respond to market signals in rural areas. Zambia represents this development path, as demonstrated in the last 15 years or so in the livestock sector and dairy production in particular. The country has been unable to keep pace with population growth, let alone respond to increasing demand for quality products.
- *Positive but inequitable livestock sector growth*, characterized by a dualistic pattern of development where few and large market-integrated producers benefit from the drivers of change in the livestock sector while the majority of smallholder producers, processors, and retailers are not only unable to benefit from the changing livestock landscape but at times are forced out of the market. A representative case is the Brazilian dairy sector, which on aggregate has been performing particularly well since the structural reforms of the 1990s. But at the same time the number of dairy farms has declined by 3–4% per year while informally marketed milk increased in market share by around 10%.

There is thus both good news and bad news. The good news is that the changing landscape can benefit smallholders who, given an appropriate institutional and infrastructural setting, are able to compete with large producers and contribute to rural employment and agricultural growth. The bad news is that in some cases the changing landscape is either not benefiting smallholder livestock producers at all or is mainly benefiting the already well-off and educated producers. Many smallholder producers might be forced out of the sector without being able to find productive employment in other sectors and may induce migration to urban areas as casual laborers.

Furthermore, modern supply chains tend to cater to higher-income groups and more advanced producers while low-income consumers represent a huge market ($13 billion at PPP) whose needs often go unmet (versus $10 billion in mature markets). Therefore, bowing to pressures exerted by high-income consumers will harm both poorer producers and consumers.

There is certainly room for policies to turn the changing livestock landscape into an opportunity rather than a threat for smallholder livestock producers and

their direct and indirect employees. Considering this perspective, two key encouraging elements deserve to be emphasized.

The first is that it has been shown that livestock production activities can be efficiently organized in a decentralized manner by exploiting the capital, physical labor, and entrepreneurship of small rural producers. This development strategy, if scaled up to the level of countries, would certainly contribute to broad-based agricultural growth, rural employment, and poverty alleviation, ultimately mitigating the presumed trade-offs between growth and equity.

The second is that policy makers, international organizations, and development practitioners are increasingly appreciating the poverty-reduction potential of the livestock sector and are becoming aware that technical solutions alone are rarely able to deliver efficient and equitable outcomes. Policies are therefore beginning to focus not only on technical issues, but also on the economic and institutional context in which these technologies are supposed to be used.

References

Araujo, C., C. A. Bonjean, J. L. Combes, and P. C. Motel. 2005. Devaluation and cattle market integration in Burkina Faso. *Journal of African Economics* 14 (3): 359–384.

Barros, G. S. C., S. G. De Zen, M. R. P. Bacchi, S. H. G. de Miranda, C. Narrod, and M. Tiongco. 2003. *Policy, technical and environmental determinants and implications of the scaling-up of swine, broiler, layer and milk production in Brazil.* Livestock Industrialization Project, AGAL-LEAD. Rome: Food and Agriculture Organization.

Barros, G. S. C., S. G. Fisher, S. H. F. Silva, and L. A. Ponchio. 2004. *Mudanças estructurais na cadia do leite: Reflexos sobre os preços.* São Paulo: Centro de Estudo Avançados em Economia Aplicada.

Besley, T., and L. J. Cord. 2007. *Delivering on pro-poor growth: Insights and lessons from country experiences.* Washington, DC: World Bank and Palgrave Macmillan.

Birthal, P. S., A. K. Jha, M. Tiongco, C. Delgado, C. Narrod, and P. K. Joshi. 2006. *Equitable intensification of market-oriented smallholder dairy production in India through contract farming.* IFPRI Project on Contract Farming of Milk and Poultry in India. Rome: Food and Agriculture Organization–AGAL.

Costa, D., D. J. Reinemann, N. Cook, and P. Ruegg. 2004. *The changing face of milk production, milk quality and milking technology in Brazil.* Babcock Institute Discussion Paper 2004–02. Madison: University of Wisconsin.

Costales, A., C. Delgado, M. A. Catelo, M. L. Lapar, M. Tiongco, S. Ehui, and A. Z. Bautista. 2007. *Scale and access issues affecting smallholder hog producers in an expanding peri-urban market: Southern Luzon, Philippines.* IFPRI Research Report No. 151. Washington, DC: International Food Policy Research Institute.

Davis, B., P. Winters, C. Gero, K. Covarrubias, E. Quinones, A. Zezza, K. Stamoulis, G. Bonomi, and S. Di Giuseppe. 2007. *Rural income generating activities: A cross-country comparison.* ESA Working Paper No. 07-16. Rome: Food and Agriculture Organization.

De Janvry, A., N. Key, and E. Sadoulet. 1997. *Agricultural and rural development policy in Latin America: New directions and new challenges.* FAO Agricultural Policy and Economic Development Series 2. Rome: Food and Agriculture Organization.

Emongor, R. A., A. Louw, J. F. Kirsten, and H. Madevu. 2004. *Re-governing markets: Securing small producer participation in restructured national and regional agri-food systems. Zambia country report.* London: International Institute for Environment and Development.

FAO (Food and Agriculture Organization). 2003. *Project on livestock industrialization, trade and social-health-environment impacts in developing countries: Final research report of phase II,* ed. C. S. Delgado, C. A. Narrod, and M. M. Tiongco. Rome: FAO.

FAO (Food and Agriculture Organization). 2006. FAO statistical databases. http://faostat.external.fao.org.

FAO and WFP (Food and Agriculture Organization and World Food Programme). 2005. *Special report: FAO/WFP crop and food supply assessment mission to Zambia.* Rome: FAO and WFP.

Farina, E. M. M. Q. 2002. Consolidation, multi-nationalisation, and competition in Brazil: Impacts on horticultural and dairy products systems. *Development Policy Review* 20:441–457.

Farina, E. M. M. Q., and R. Nunes. 2002. A evolução do sistema agroalimentar no Brasil e a redução de preços para o consumidor: Os effectos da atuação dos grandes compradores. Mimeo. Santiago, Chile, and São Paulo, Brazil: U.N. Economic Commission for Latin America and the Caribbean and University of São Paulo.

Government of Zambia. 2002. *Poverty reduction strategy paper.* Lusaka: Government of Zambia.

Groenewold, J. P. 2004. *Classification and characterization of world livestock production systems: Update of the 1994 livestock production systems with recent data.* Rome: Food and Agriculture Organization–AGAL.

Haggblade, S., P. Hazell, and T. Reardon. 2005. *The rural nonfarm economy: Pathway out of poverty or pathway in?* International Food Policy Research Institute, Overseas Development Institute, and Imperial College London. Proceedings from the research workshop on the future of small farms, June 26–29, Withersdane Conference Centre, Wye, U.K.

Hayami, Y. 1998. *Toward the rural-based development of commerce and industry: Selected experiences from East Asia.* Washington, DC: World Bank.

Hemme, T., O. Garcia, and A. Saha. 2003. *A review of milk production in India with particular emphasis on small-scale producers.* PPLPI Working Paper No. 2. Kiel, Germany: International Farm Comparison Network–Dairy Network.

Hemme, T., E. Deeken, et al. 2005. IFCN Dairy Report 2005. International Farm Comparison Network. Global Farm GbR, 2005, Braunschweig. Germany.

IFAD (International Fund for Agricultural Development). 2006. *Geography, agriculture and the economy in Zambia.* Rome: IFAD.

Knips, V. 2006. *Developing countries and the global dairy sector, part II: Country case studies.* PPLPI Working Paper No. 31. Rome: Food and Agriculture Organization.

Maltsoglou, I., and K. Taniguchi. 2004. *Poverty, livestock and household typologies in Nepal.* PPLPI Working Paper No.13. Rome: Food and Agriculture Organization.

Matthey, H., J. F. Fabiosa, and F. H. Fuller. 2004. *Brazil: The future of modern agriculture?* MATRIC Briefing Paper 04-MBP 6. Ames: Iowa State University.

McLeod, A., and P. Chilonda. 2004. *Mission report of an assessment of the livestock sub-sector in Zambia, with special reference to policies.* Rome: Food and Agriculture Organization.

Mukumbuta, L., and B. Sherchand. 2006. *Enabling smallholder prosperity: Zambia's smallholder milk collection centers.* Report submitted to USAID, Zambia Agribusiness Technical Assistance Center, Lusaka, Zambia.

Nagayets, O. 2005. *Small farms: Current status and key trends.* International Food Policy Research Institute, Overseas Development Institute, and Imperial College London. Proceedings from the research workshop on the future of small farms, 26–29 June. Withersdane Conference Centre, Wye, U.K.

Ostrowski, B., and C. Deblitz. 2001. *La competitividad en producción lechera de los países de Chile, Argentina, Uruguay y Brasil.* Livestock Policy Discussion Paper No. 4, Livestock Information and Policy Branch, AGAL. Rome: Food and Agriculture Organization.

Phillips, C. 2006. Dairy challenges in an evolving export market: Competition from South America. Paper presented at the 2006 Australian Bureau of Agricultural and Resource Economics Outlook Conference, Canberra, Australia.

Poapongsakorn, N., V. Na Ranong, C. Delgado, C. Narrod, P. Siriprapanukul, N. Srianant, P. Goolchai, et al. 2003. *Policy, technical, and environmental determinants and implications of the scaling-up of swine, broiler, layer and milk production in Thailand.* Livestock Industrialization Project, AGAL-LEAD. Rome: Food and Agriculture Organization.

Roxas, D. B., M. Wanapat, and M. Winugroho. 1997. Dynamics of feed resources in mixed farming systems in Southeast Asia. In *Crop residues in sustainable mixed crop/livestock farming systems,* ed. C. Renard. Oxon, U.K.: CAB International.

Sharma, V. P., C. Delgado, S. Staal, and V. S. Singh. 2003. *Policy, technical, and environmental determinants and implications of the scaling-up of milk production in India.* Livestock Industrialization Project, AGAL-LEAD. Rome: Food and Agriculture Organization.

Staal, S. J., A. N. Pratt, and M. Jabbar. 2006. *A comparison of dairy policies and development in South Asia and East Africa. Country case studies from South Asia and East Africa: Kenya, Ethiopia, Pakistan and India.* PPLPI Working Paper. Rome: Food and Agriculture Organization.

Tung, D. X., N. T. Thuy, and T. C. Thang. 2005. *Pork Development in Vietnam.* Hanoi, Viet Nam: Ministry of Agriculture and Rural Development.

von Braun, J. 2005. Small-scale farmers in a liberalized trade environment. In *Small-scale farmers in a liberalized trade environment,* ed. T. Huvio, J. Kola, and T. Lundstrom, pp. 21–52. Seminar Proceedings, Haikko, Finland, 18–19 October 2004. Department of Economics and Management Publications No. 38. Agricultural Policy. University of Helsinki, Finland.

World Bank. 2006. *World development indicators.* Washington, DC: World Bank.

World Bank. 2007. *Zambia smallholder agricultural commercialization strategy.* Report No. 36573-ZM. Sustainable Development, AFTS1. Country Department 2, African Region. Zambia.

WTO (World Trade Organization). 1994. *Brazil: Imposition of provision and definitive countervailing duties on milk powder and certain types of milk from the European Economic Community.* Report of the Panel adopted by the Committee on Subsidies and Countervailing Measures on 28 April 1994. Geneva: WTO.

14

Socioeconomic Implications of the Livestock Industrialization Process

How Will Smallholders Fare?

Clare Narrod, Marites Tiongco, and Christopher Delgado

Main Messages

- **The recent exceptional growth in demand for animal products in the developing world has been accompanied by a movement of many small-scale producers out of the livestock sector, at least in the industrial world.** Small-scale producers cannot compete with the larger operations that benefit from economies of scale because of the dominant view that smallholders are "inefficient." Thus many smallholders lose their opportunities for growth and poverty alleviation unless there are mechanisms or institutions to link them in some way to the larger supply chain and ensure they meet all the changing requirements of the marketplace. Case studies show that smallholders have a chance to compete if they can cost family labor at less than the full opportunity cost of hired labor doing the same tasks on larger farms. In addition, the relative competitiveness of smallholders is largely determined by farm-specific abilities to overcome barriers to information and assets, such as credit and market information asymmetries.

- **Smallholder livestock farming in developing countries will be driven by collective action.** The future of this type of farming will be largely driven over the medium to long run by the issue of whether collective action such as cooperatives or contract farming schemes can sufficiently reduce the transaction costs that smallholders face so that they can continue to compete with larger farms.

- **Linking smallholders to the supply chain leads to greater sustainability.** Given that a more complex, demand-responsive supply chain may play a larger role in the long-term sustainability of smallholders being profitable, it is important to look at potential ways to link smallholders to high-value supply chains in their countries. It has been suggested that the requirements of the emerging food system, with its high demands for food safety, traceability,

and compliance, often work against smallholders owing to high coordination costs.

- **Several market failures may hinder smallholders from being integrated to the larger supply chain: information asymmetry and high transaction costs, lack of coordination, and regulatory failures.** To ensure smallholders' access to markets, producers need support through extension services or technical assistance; good infrastructures; good sources of information; certification, grades, and standards; and good mechanisms for coordination of getting their supplies to markets.

The Scaling Up of the Livestock Sector[1]

As described in earlier chapters, growing population and rising per capita income have led to an increasing worldwide demand for animal products (Delgado et al., 1999, 2003; Narrod and Pray 1995, 2001; Hayami and Otsuka 1994). The rapid rise in per capita income in middle-income countries has enabled consumption to move away from low-income elastic food consumption such as cereals to high income-elastic foods such as fruits, pulses, and animal products (Hayami and Otsuka 1994). Most countries have attempted to meet this demand by raising domestic production rather than relying on trade.

Over the past three decades, technological change driven largely by the recognition of increased demand occurred rapidly for the poultry and swine industries. The move to confined operations dramatically increased what one farmer could manage, which led to a significant increase in labor productivity (Fuglie et al., 2000, Narrod and Fuglie 2000). The ability to control for the spread of livestock diseases through improved vaccines and pharmaceuticals also helped to expand the number

---

1. Much of this section is based on Delgado et al., 2003 and 2008 and from Tiongco et al., 2006.

of large-scale operations where farmers were able to achieve significant economies of scale and unit-cost reductions. Thus technology change had an impact on the structure of the industry, resulting in a rapid change in the supply chains associated with the production and marketing of animal products.

It has been suggested that this rapid growth in livestock production and the scaling up of this sector has forced many smallholders out of production because they are no longer competitive with larger operations that benefit from technical and allocative economies of scale. The profitable adoption of improved genetics, compound feeds, or organization, especially for poultry and pigs, simply requires larger farm sizes (Martinez 2002, Morrison Paul et al., 2004).

Delgado et al. (2003, 2008) suggest that, although all this may be true, in most cases the arguments focus on anecdotal experiences in the industrial world; there has been limited effort to detail the causes of this situation and whether smallholders can be competitive under certain institutional arrangements. To understand these cost advantages better, Delgado et al. (2003) and Tiongco et al. (2006) looked closely at the role of various elements in the livestock sector, such as economies of scale in input supply, increasing transaction costs, smallholder competitiveness, and the way contract farming can help reduce these costs and their effect on profitability.

The importance of understanding cost advantages in input procurement is heightened by the increased growth of monogastrics (animals with a single stomach compartment, such as poultry and pigs) and the fact that up to 70% of the cost of production of monogastric livestock is feed costs (Delgado et al., 1999). If larger-scale farms can regularly secure access to feed of a given quality at a lower price per unit, they gain a great cost advantage over small-scale producers largely because it costs less per kilogram to deliver a full truckload versus one bag of feed.

For instance, if large vertically integrated operations pay no sales taxes on feed but small independent producers do, the latter are somewhat at a disadvantage because of a policy distortion that is not a true economy of scale. Similarly, most monogastric producers use a concentrated feed mixture (grain plus various sources of micronutrients and pharmaceuticals) to ensure optimal use of feed (low feed conversion rate). Small farmers often have to rely on buying this product, though they often cannot be assured about what is in it. Where public enforcement of the truthfulness of ingredient labels is lax or branding is unreliable, as is often the case in developing countries, producers who are large enough to mix their own feed are more certain to get good-quality feeds. This results in smallholders facing a higher "transaction cost" in terms of having less information about what their feeds contain.

On the output market side, if buyers and sellers can easily ascertain the quality of the item being sold and the prices in alternative markets, competitive forces would eventually equate market prices across different categories of farmers. But if buyers cannot be sure of the true quality of the good they are purchasing, they will presumably be less willing to pay a premium for it based on claimed quality. Smallholders in developing countries often have limited market access due to a number of hidden transaction costs associated with the extra costs of searching, bargaining, monitoring, and enforcing exchange, which often act as barriers to smallholder participation in many markets.

Smallholders often have trouble selling milk outside the local market, for example, because purchasers in anonymous markets cannot be sure without a bacteriological test that the unbranded milk is safe. Large-scale producers and cooperatives of small-scale ones may be able to establish trust and reputation in markets, since they will be able to depend on repeat sales of quasi-branded product to the same clients who can identify the source of the milk (Staal et al., 1997), while independent producers may not be able to. The clients can judge the quality of the next purchase based on their history of purchases. Thus larger producers, through regular large sales, may also be able to get higher prices per unit of output by developing a steady clientele who gain confidence in the product's quality (Staal et al., 1997). Generally, rising demand for food safety and quality from the domestic and export markets is likely, all things being equal, to exacerbate transaction costs in animal product exchange arising from asymmetries in access to information.

Vertical coordination helps reduce many of these transaction costs. Cost savings often occur from purchasing inputs or selling outputs in bulk (economies of scale), from avoiding taxation of inputs (policy subsidy), or from reduction of the transaction costs between actors along the supply chain. Many types of livestock smallholders are linked to vertically integrated operations through contracts. Martinez has suggested that vertical coordination in the pork and poultry sectors in the United States has in part occurred because "contracting and vertical integration produced a means for reducing transaction costs associated with relationship-specific transactions, especially in regions of expanding production. Contracts would provide some safeguards against opportunistic behavior, and vertical integration eliminated the exchange relationship. For attributes that are difficult to measure, gaining additional control over related production inputs may reduce measuring costs by reducing the need to measure quality" (Martinez 2002, p. iii).

For example, if the purchaser of hogs in Asia is concerned about off-flavors and excess back-fat of pork, he has to at least know whether quality feed was used, because these problems can only be assessed after slaughter. A smallholder independent producer faces a higher "transaction cost" in terms of lack of information about the true quality of feed being purchased, or unreliable information on the percentage of back-fat of a hog being sold, resulting in higher prices for inputs and lower prices received for outputs. A contractual relationship can build trust on the basis of reputation, thus allowing farmers to secure higher prices for their output than they would if they operate independently.

The key is that vertical coordination potentially allows a net reduction to all parties of the costs of doing business through the reduction of transaction costs of exchange. These high transaction costs can inhibit smallholders' participation or entry into competitive markets. Although transaction costs are very difficult to measure, one approach is to take the difference between the prices received by large and small producers, while controlling for differences in transport costs and observable product quality.

Impact of Structural Changes in the Livestock Sector on Smallholder Profitability[2]

Delgado et al. (2003, 2008) and Tiongco et al. (2006) have tried to separate some of the issues associated with structural changes and transaction costs and their impact on profitability for smallholders through a series of case studies between 2002 and 2006. The studies involved household surveys from Brazil, India, Philippines, and Thailand to capture various factors affecting profitability, including transaction costs and the internalization of efforts to mitigate environmental externalities by producers. These nations were chosen because they are all fast-changing developing countries where cities, population, urban incomes, and the consumption of livestock products have been growing rapidly since the early 1980s. They also have all seen the rise of large periurban livestock operations of one form or the other, typically not far from major cities, at least in the initial stage of livestock industrialization. In addition, they typically also have vibrant smallholder livestock sectors producing similar livestock products nearby.

This section uses one example of each commodity in a select country to illustrate the general types of changes that have occurred. More specific details of these studies and specific commodities within these countries can be found in Delgado et al. (2003) and Tiongco et al. (2006) as well as in the numerous country studies that contributed to this work.

The poultry industry in Thailand has undergone significant structural changes since the start of the livestock revolution in the 1970s (Poapongsakorn et al., 2003). It started when the Charoen Pokphand Company established its chicken breeding business through a joint

venture with U.S.-based Arbor Acres, bringing improved parent stock into Thailand. As regional demand grew, other production technologies and modern contractual arrangements were introduced into the eastern provinces. Since then, contracting between producers and private industry, particularly feed companies, has become common in Thailand. These coordinating mechanisms in the poultry industry in Thailand allowed integrators to achieve economies of scale, quality control, and timeliness of inputs in various stages as well as to provide incentives for contract growers to increase productivity and adopt improved technologies. The Thai broiler market is now dominated by a few integrators who control the complete supply chain—from grandparent stock farms to hatcheries, feed mills, growing farms, the food retail business, and frozen chicken exports.

In the Philippines, expansion of the pig sector had concentrated near the urban and periurban areas (Costales et al., 2003). Despite this scenario, smallholder pig production has grown nationwide, but even faster growth was experienced in large-scale livestock production. Vertical integration and contract growing have not yet become the norm throughout the Philippines, as in the case of the poultry industry. However, the truly larger-scale commercial firms in the hog sector employ vertical integration in operations: from breeding and contract production to slaughter and processing of branded meat products. These large firms have access to a different and higher-value market than medium-sized commercial and smallholder farms do. They increasingly service the growing institutional food sector in large cities: supermarkets, up-scale restaurants, and hotels. They can meet the food safety certification and consistent quality standards required by the institutional food sector. However, no integrator in the Philippines currently has the dominant market share.

India's dairy sector has undergone significant changes over the last three and a half decades (Sharma et al., 2003, Birthal et al., 2006). Milk production has increased from 20 million tons in the 1950s and 1960s to over 90 million tons in 2004 (Birthal et al., 2006). Growth in milk production resulted from increases in animal numbers and in productivity, along with the implementation of government support programs, particularly the development of rural milk sheds through milk producers' cooperatives and the marketing of milk from rural to urban demand centers. Birthal and Taneja (2006) estimated that 56% of the output increase between 1992/93 and 2003/04 stemmed from an increase in numbers and 37% from productivity improvement. Crossbreeding technology in cows was an important driver of productivity growth in India. The share of crossbreds in total cattle population has increased continuously, from 4.6% in 1982 to 7.4% in 1992 and to 13.3% in 2003 (Birthal et al., 2006).

Producing milk in rural areas through smallholder producer cooperatives and moving industrially processed

2. Much of this section comes from Delgado et al., 2003 and 2008 and from Tiongco et al., 2006.

milk from these smallholders to urban demand centers became the cornerstone of the government's dairy development policy. This initiative gave a boost to dairy development and initiated the process of establishing the much-needed linkages between rural producers and urban consumers (Sharma et al., 2003). Until 1991, the Indian dairy industry was highly regulated and protected. Milk processing and product manufacturing were mainly restricted to small firms and cooperatives. High import duties, nontariff barriers, restrictions on imports and exports, and stringent licensing provisions provided incentives to Indian-owned small enterprises and cooperatives to expand production in a protected market.

All of the studies conducted in 2002–06 indicated that indeed the share of small farms has declined relative to the total number of farms. It is possible that large-scale producers have displaced smallholders because of the scaling up of livestock production. In Thailand, for example, the number of small pig farms fell from 1.3 million in 1978 to 420,000 in 1998, which could be interpreted as a significant restructuring (World Bank 2005) of the industry. There has been a reduction in the number of animals on small farms (which includes backyard ones) and medium-sized farms in Thailand of more than 4% per annum during the last 10 years. (See Table 14.1.) Although there are no official statistics on backyard farmers there, anecdotal evidence suggests that their contribution in the livestock sector has also significantly decreased (Poapongsakorn et al., 2003). However, this does not imply that small farms have exited from the sector. It could mean that these farmers have diversified into something better—either high-value agriculture or nonfarm work—or that they have expanded their operations to a larger scale.

In the Philippines, large-scale farms have overtaken backyard farms in Central and Southern Luzon as shown in Figure 14.1 in the color well by the declining share of backyard hog inventories over time. Moreover, while backyard farms' share at the national level remains high at 77%, backyard producers in these two regions now account for less than half of production (Costales et al., 2007). In contrast, the number of commercial pig farms has increased by 60% over a seven-year period, particularly in Southern Tagalog, where the number of commercial farms more than doubled. (See Figure 14.2 in the color well.)

In Viet Nam, the number of pig farms having just one pig gradually decreased from 1994 to 2001. Farms with more than 6–10 pigs accounted for only 2% of the total pig farms in 1994 but 6% in 2001. (See Figure 14.3 in the color well.) Similarly, 548 commercial farms had more than 10 pigs in 2001, which was about 2% of total. But small-scale pig farms there had lower profitability compared with medium-scale farms in both areas in the Red River Delta in terms of gross margin. This is due to the fact that the medium-scale farms kept

improved breeds fed with high-quality feeds, leading to better performance than in small-scale farms (Nho et al., 2001).

As for broilers, independent small-scale producers in India made more profit per unit of output than large-scale independents. (See Table 14.2.) Small-scale contractors did not perform well in terms of per unit profit compared with large-scale contractors. These findings were supported in a recent study in Karnataka, India, that looked at the effect on profitability of contract farming of broiler production (Fairoze et al., 2006). Similarly, independent smallholders in the Philippines also had significantly higher profits per kilogram than large-scale independents. Moreover, small contract broiler farms there had almost the same per-unit profits as large contract farms.

In Thailand, large independent broiler farms made higher profits than medium-sized independent farms. Fee contract farmers in the Thai broiler sample had similar per-unit profits for large scale and small scale. In Brazil, however, small and large broiler farms have surprisingly similar average profits per kilogram. This may reflect the fact that almost all broiler production in Brazil is vertically integrated, with the integrators supplying the main inputs for production, giving producers access to better inputs and modern technology, as well as the fact that the integrators passed on some of the cost savings of dealing with larger producers to them (Camargo Barros et al., 2003). This implies that smallholders might be excluded in the process of contractual arrangements, as integrators would prefer to contract out with large-scale farmers so as to minimize production and transaction costs.

Tiongco et al. (2006) observed that an integrator's transaction costs are incurred on a per-grower basis and do not depend on the size of the farm. Moreover, small farms usually require more technical assistance from the integrator per unit of output. For example, an on-farm visit may require the same amount of time regardless of the scale of production. It was also observed that there was no significant difference between small and large farms in the growing fees paid by integrators per unit of output. Holding this fee constant, integrators would rather contract with larger producers to lower their cost of procurement or default.

Table 14.3 summarizes profit per liter of milk across farm sizes in three countries. In terms of milk production, India's small-scale dairy farms had higher average profits per liter than large-scale farms. In Thailand, medium-scale dairy farms made about 20% more profit, at $0.15 per liter, than either small or large-scale dairy farms, which have the same average profits. For Brazil, however, the profit per liter of small dairy farms was lower than that of large farms. These observations are mixed, but smallholders will at least have a chance to compete with larger-scale producers since they have the ability to produce at a lower per unit cost of production

Table 14.1. Farm size in Thailand by type of animal.

Number of swine in holding	Percentage share in number of pigs				Percentage share in number of pigs in holding			
	1978	1988	1993	1998	1978	1988	1993	1998
1–4	n.a.	n.a.	25.26	10.07	79.35	80.15	21.16	60.66
5–19	n.a.	n.a.	29.89	22.34	17.83	16.46	39.52	30.10
20–49	n.a.	n.a.	20.80	16.72	2.29	2.51	25.50	7.29
50–99	n.a.	n.a.	8.28	4.24	0.35	0.54	8.60	0.82
100 and over	n.a.	n.a.	15.82	46.63	0.18	0.34	5.22	1.13
Total (percent)	n.a.	n.a.	100.00	100.00	100.00	100.00	100.00	100.00
Total (number of pigs)	5,314,000	4,684,926	6,182,953	5,731,360	1,263,000	778,113	590,616	465,509

Number of dairy cows in holding	Percentage share in number of dairy cows				Percentage share in number of dairy cows in holding			
	1978	1988	1993	1998	1978	1988	1993	1998
1–4	n.a.	n.a.	9.21	5.76	n.a.	54.73	11.88	34.67
5–19	n.a.	n.a.	44.04	48.90	n.a.	39.85	39.15	49.56
20–49	n.a.	n.a.	26.66	34.90	n.a.	4.53	31.42	15.07
50 and over	n.a.	n.a.	20.09	10.44	n.a.	0.89	17.54	0.69
Total (percent)	n.a.	n.a.	100.00	100.00	n.a.	100.00	100.00	100.00
Total (number of dairy cows)	n.a.	93,654	257,895	240,206	n.a.	15,966	11,616	21,540

Number of chickens in holding	Percentage share in number of chickens				Percentage share in number of chickens in holdings			
	1978	1988	1993	1998	1978	1988	1993	1998
1–19	n.a.	n.a.	9.15	10.15	69.30	69.77	64.24	61.35
20–99	n.a.	n.a.	16.91	20.15	30.00	29.11	33.00	36.09
100–499	n.a.	n.a.	4.80	4.96	0.53	0.84	2.03	1.99
500–999	n.a.	n.a.	1.54	1.00	0.05	n.a.	0.15	0.08
1000–9999	n.a.	n.a.	29.11	31.16	0.09	0.28	0.50	0.42
10,000 and over	n.a.	n.a.	38.49	32.58	0.03	n.a.	0.09	0.06
Total (percent)	n.a.	n.a.	100.00	100.00	100.00	100.00	100.00	100.00
Total (number of chickens)	54,157,000	86,679,292	154,921,930	169,102,499	2,638,000	3,249,177	2,617,412	3,174,410

Note: Data for 1993 and 1998 are not comparable since they came from different sources—*Agricultural survey* and *Inter-censal survey*—conducted by the National Statistics Office.

n.a. = not available

Sources: Poapongsakorn et al. 2003, citing 1988 Inter-censal survey of agriculture by the National Statistical Office; *1993 Agricultural survey* by the National Statistical Office; *1998 Inter-censal survey of agriculture* by the National Statistical Office.

(by using family labor) or at least achieve similar profits per unit of output as large-scale farmers.

Table 14.4 compares the profit per kilogram live weight of swine output across farm sizes in three countries. In the case of swine production in Brazil, the profit per kilogram of output was higher for large-scale (independent and integrated farm) producers than for small-scale producers (Camargo Barros et al., 2003). Further, the integrated farms actually made positive profits per unit of output, while farms under cooperatives and independent farms incurred losses. In the Philippines, however, independent smallholder swine farmers had higher profits per unit of output than large-scale independents.

In Thailand, small-scale independent swine farms had lower average profits per unit of output than medium-sized farms and large or commercial farms. On the other hand, large contract swine producers generated the lowest farm profit per unit of output.

Contrary to dairy, the relationship between profitability and size for swine producers is not so clear. Since contract farmers were given a fixed growing fee, and since their fee was not adjusted relative to increases in costs of other inputs, their net revenue was quite low relative to independent farms. However, these contract farmers have benefited from contract farming in terms of improved technology, assured quality of inputs and

Table 14.2. Average profit per unit of output of liveweight broiler across farm sizes, by country and production arrangement, 2002.

		Farm size			
		Smallholder <10,000 birds		Large/commercial ≥10,000 birds	
Country		Independent	Contract	Independent	Contract
India					
Average profit without family labor cost	Rupees/bird	13.13	1.03	10.93	3.16
	$/kg*	(0.11)	(0.01)	(0.09)	(0.03)
	Rupees/bird	11.36		9.98	
	$/kg*	(0.10)		(0.09)	
Average profit with family labor cost	Rupees/bird	12.40	0.04	10.80	3.01
	$/kg	(0.11)	(0.003)	(0.09)	(0.03)
	Rupees/bird	10.59		9.85	
	$/kg	(0.09)		(0.08)	
Philippines					
Average profit without family labor cost	Pesos/kg $/kg	1.59 (0.03)	4.05 (0.08)	1.07 (0.02)	3.96 (0.08)
Average profit with family labor cost	Pesos/kg $/kg	1.34 (0.03)	3.98 (0.08)	1.06 (0.02)	3.95 (0.08)
Thailand		Forward contract and independent	Per-bird wage contract	Forward contract and independent	Per-bird wage contract
Average profit	Baht/kg live-weight	0.71	1.35	2.48	1.51
	$/kg live-weight	(0.02)	(0.03)	(0.06)	(0.04)
Brazil					
Average profit	Real/kg live-weight	0.05		0.06	
	$/kg live-weight	(0.02)		(0.02)	

*Assuming one bird weighs 2.4 kg liveweight.

Numbers in parentheses are average profit in dollars per unit of output. The currency conversion rates used are based on 2002 foreign exchange rates: for Thailand, $1 = 42.96 baht; for India, $1 = 48.61 rupees; for Brazil, $1 = 2.92 reals; for the Philippines, $1 = 51.60 pesos.

Source: Delgado et al., 2003, 2008.

outputs, technical assistance, veterinary services, and, most of all, an income guaranteed to be stable.

The overall implication from comparing relative profit efficiency across countries for these different commodities and degree of vertical integration is that small farms are no less efficient in all cases at securing profits per unit of output when family labor and environmental externalities are controlled for.

For smallholders to survive livestock industrialization, the key then is for them to have the ability to use their farm resources more efficiently than large-scale producers do. If large-scale producers are more efficient on average, then they will be able to drive their costs down

and survive on smaller unit profits but bigger volume of sales. If this is the case, it is possible that smallholders will be driven out of the market because of their small volume of production.

It remains unclear if smallholders have been driven out of livestock markets because of the scaling up of the sector, as no one has looked at where they have gone. In the case of Thailand, available information from secondary sources indicates that the percentage of poor livestock households has declined proportionately more than the percentage of poor agricultural households overall (Poapongsakorn et al., 2003). (See Table 14.5.) Moreover, the Gini coefficient for livestock farmers is

Table 14.3. Average profit per liter of milk across farm sizes by country, 2002.

Country			Farm size						
			Small	Medium		Large/commercial			
			Liters per day						
India	Region	Unit	<10	10–20	20–40	40–80	80–150	>150	All
Average profit without family labor cost	North	Rs/liter	2.21	1.53	1.10	0.88	0.63	0.38	1.47
	West	Rs/liter	3.09	1.72	1.09	0.71	0.48	0.33	1.27
	Pooled	Rs/liter	2.45	1.62	1.09	0.71	0.48	0.38	1.37
	Pooled	$/liter	(0.05)	(0.03)	(0.02)	(0.01)	(0.01)	(0.01)	(0.03)
Average profit with family labor costed	North	Rs/liter	0.46	0.50	0.49	0.39	0.13	0.25	0.44
	West	Rs/liter	1.45	0.37	0.69	0.62	0.47	0.38	0.66
	Pooled	Rs/liter	0.52	0.42	0.53	0.49	0.40	0.29	0.43
	Pooled	$/liter	0.011	0.009	0.011	0.010	0.008	0.006	0.009
Thailand			1–20 cows	21–60 cows		61–100 cows			All
Average profit		Baht/liter	5.10	6.25		5.35			5.63
		$/liter	(0.12)	(0.15)		(0.12)			(0.13)
Brazil			<50 head	50–70 head		>70 head			
Average profit		Real/liter	0.04	0.04		0.05			
		$/liter	(0.01)	(0.02)		(0.02)			

Numbers in parentheses are average profit in dollars per liter. The currency conversion rates used are based on 2002 foreign exchange rates: for Thailand, $1 = 42.96 baht; for India, $1 = 48.61 rupees; and for Brazil, $1 = 2.92 reais.

Source: Delgado et al., 2003, 2008.

shown to be much lower than the Gini for agriculture as a whole (Poapongsakorn et al., 2003). (A Gini coefficient is tantamount to a cumulative average of income inequality by class, where a perfectly equal distribution across classes would yield a Gini of 0 and a perfectly unequal distribution would score 1.) This suggests that the present income distribution among the livestock-keeping households has become relatively more equitable over time. This observation needs further investigation by empirically testing the effect of livestock industrialization on the livelihoods and employment of smallholders and on rural poverty. Equally important, though not looked at in this empirical work, is how smallholders are linked to the rest of the supply chain, as discussed in the next section.

Though this discussion has focused on producers, consumers as a whole have benefited from the livestock industrialization process through the reduction in meat prices, which are projected to continue to decline at a rate of about 3% per year (see Table 14.6.). For the poor, this has had a tremendous effect on their health and nutrition, as livestock products contain many types of micronutrients found only in livestock products (Neumann et al., 2002; see also Chapter 12).

Linking Livestock Production to the Supply Chain in an Era of Increased Food Safety and Quality Concerns

As the preceding section makes clear, in some cases and for some countries smallholders have been profitable under certain conditions. Profitability, however, is not the only key to market access. Today many smallholders are finding it difficult to link themselves to the new and emerging food systems that are trying to meet the high demands of food safety, traceability, and compliance. The food safety requirements and food systems often favor larger operations over smallholders in supplying the supply chain, due to high coordination costs.

The problem is exacerbated by geographic dispersion, low education, and poor access to capital and information (Humphrey 2005, Rich and Narrod 2005). Because of the high transaction and marketing costs of using smallholders, large-scale producers are increasingly either sourcing from medium-sized and large growers

Table 14.4. Profit per kilogram live weight of output of swine across farm sizes by country and production arrangement, 2002.

Country			Farm size			
		Small <100 head		Large/commercial ≥100 head		
Philippines		Independent	Contract	Medium Independent 100–500 head	Large Independent	Contract
Average profit without family labor cost	Pesos/kg	26.60	2.08	19.61	19.83	2.33
	$/kg	(0.52)	(0.04)	(0.38)	(0.38)	(0.05)
Average profit with family labor cost	Pesos/kg	26.45	2.05	19.58	19.82	2.33
	$/kg	(0.51)	(0.04)	(0.38)	(0.38)	(0.05)
Thailand		Small ≤500 head		Large/commercial >1000 head		
		Independent	Contract	Independent		Contract
Average profit	Baht/kg	11.9	11.5	15.4		1.7
	$/kg	(0.28)	(0.27)	(0.36)		(0.04)
Brazil		Small 100–1000 head		Large >1000 head		
Average profit	Real/kg	−0.25		−0.15		
	$/kg	(−0.09)		(−0.05)		

Numbers in parentheses are average profit in dollars per unit of output. The currency conversion rates used are based on 2002 foreign exchange rates: for Thailand, $1 = 42.96 baht; for Brazil, $1 = 2.92 reais; and for the Philippines, $1 = 51.60 pesos.

Source: Delgado et al., 2008.

trained and trusted to deliver both traceability and food safety or just producing the livestock product themselves (Narrod et al., 2006).

Currently in most developing countries, producers supply three different markets: domestic traditional markets, modern urban markets, and export markets. (See Table 14.7.) These markets differ in several organizational respects, particularly in their demand for food safety (World Bank 2006). In traditional local markets, there is little consumer concern for food safety and thus little private effort or government control over safety. Incentives for farmers and enterprises to improve food safety are weak. There is no attempt to standardize or coordinate the delivery of products, and there is no technical assistance to aid farmers in improving the situation. Trust between buyers and sellers is not very important, and price largely determines competitiveness.

In the emerging modern urban markets in the developing countries, there is growing consumer awareness for food safety and, as a result, retailers setting up methods to control and sell products that meet a certain set of safety attributes. The buyers are willing to pay a bit more for quality and safety, but for the most part the returns to such investments are low. Trust between buyers and sellers is, however, becoming increasingly important.

In export markets, there is a high level of consumer concern for food safety, and thus retailers impose strict requirements through their supply chain and require traceability. Also, products need to adhere to a strict supply schedule. The supply chain itself is demand-driven, and in order to ensure consistency and timely delivery, contracts are often used. Trust between the buyers and sellers is crucial for building long-term relationships. Competitiveness depends on deliveries of large

Table 14.5. Percentage of livestock households, agricultural households, and their farm income, Thailand, 1986–2000.

				% of household income						
	% of households			Agriculture households	Livestock households	Poor agriculture households	Poor livestock households	Gini coefficient ratio		
	Livestock	Poor livestock	Poor	Farm income	Farm income	Farm income	Farm income	All	Agriculture	Livestock
Year	Agriculture	Livestock	Agriculture	Farm income / Total income	Farm income / Total income	Farm income / Total income	Farm income / Total income	House-holds	House-holds	House-holds
1986	0.4	30.0	46.3	21.8	9.4	19.6	12.3	0.50	0.32	0.19
1988	0.6	33.8	41.2	31.6	26.0	23.6	23.7	0.49	0.34	0.57
1990	0.6	19.8	34.1	36.2	29.7	22.6	19.9	0.51	0.34	0.31
1992	1.0	18.4	31.6	26.4	12.4	17.8	11.1	0.54	0.33	0.21
1994	0.4	2.0	25.1	20.9	5.6	12.1	0.0	0.52	0.34	0.13
1996	0.5	13.9	17.4	23.2	8.5	11.4	0.4	0.52	0.35	0.25
1998	0.5	5.7	17.3	26.1	16.2	9.2	0.0	0.51	0.36	0.22
1999	0.6	–	23.1	26.9	19.6	17.7	–	0.53	0.36	0.18
2000	1.4	1.4	22.7	50.2	58.6	12.5	8.3	0.53	0.37	0.52

1986–99 definition of livestock occupation is livestock workers, dairy farm workers, and poultry farm workers; 2000 definition of livestock occupation is market-oriented animal producers and related workers.

Source: Poapongsakorn et al., 2003, calculated from National Statistical Office, Socioeconomic Survey data tape.

Table 14.6. Real price changes, 1997–2020, of selected commodities under baseline scenario.

Commodity	Price change 1997–2020 (%)
Beef	–3
Pigmeat	–3
Poultry meat	–2
Eggs	–3
Milk	–8

Projections to 2020 are from the July 2002 version of IFPRI's IMPACT model.

Source: Adapted from Rosegrant et al., 2002.

quantities that are flexible to the changing demands of the markets.

Two specific examples from India and Viet Nam illustrate how smallholders are linked to the rest of the supply chain. The supply chain for milk in Rajasthan, India, includes both organized and unorganized sectors for the marketing of milk and dairy products (see Birthal et al., 2006 for more details). (See Figure 14.4.) The organized sector is an integrated structure that includes the dairy cooperatives and private dairy processors that procure, process, and market produce from dairy farmers through intermediaries called agents. The informal or traditional sector involves independent milk producers who sell fresh milk directly to rural and urban consumers, vendors, or processors through their agents. The local milk vendor who buys milk from producers sells milk directly to urban consumers and/or to processors, sweet shops, restaurants, or hotels. The sale to rural consumers is limited and mostly done by smallholders. Large producers in general sell milk to sweet shops, restaurants, and hotels and in urban milk markets if they exist. The advantage of contracting is that producers are able to sell their output to a single buyer and avoid both high marketing costs in terms of searching for buyers and the need for timely delivery of output if they sell their milk to a number of small buyers (Tiongco et al., 2006).

The organized sector uses either market specification or resource-providing contracts to source milk from producers. Dairy cooperatives, however, are more dominant and have evolved over time as an important institution for dairy development. Most private dairy processors source milk through direct or indirect contracts with producers. Both cooperatives and private processors provide inputs (such as feed, medicine, vaccines, and mineral mixture) and services (veterinary and technical) to the willing producers to improve production and quality of milk, as well as to mitigate production risks.

Since dairying in India is largely concentrated among smallholders, contracting with a large number of small dairy producers would be costly for processors. The processors typically contract with a local villager who acts as an intermediary (agent) between the firm and the small producers. The agent motivates producers to supply milk to the firm; procures milk; helps the firm in

Table 14.7. Three types of markets and their characteristics.

Market characteristics	Traditional local markets	Emerging modern urban domestic markets (supermarkets, tourist hotels/restaurants, educated affluent consumers)	Export markets in industrial countries (retail markets, modern food services)
		Type of market	
Food safety control	Little consumer awareness, concern. Little private effort. Government control.	Emerging consumer awareness, concern. Retailers try to control and sell "safety."	High consumer concern. High retailer requirements imposed on suppliers.
Standardization, grading, supply	Virtually absent. Irregular supply.	Emerging importance of grading, stable supply.	High requirements of grading, consistency, supply schedule.
Supply-chain organization	Supply-driven. Transaction-based. Little or no net benefit from coordination. Little durability in relation between private actors. No technical cooperation.	Efforts by retailers to control quality, safety, and reliability of supply. Net financial benefits from coordination still fragile. Emerging coordination, occasional technical support.	Strongly demand-driven. Durable relations within supply chain, often on contractual basis. Cooperation between buyers, exporters, growers on technology, information, sometimes on finance.
Price level for grower and consumer	Relatively low. Limited willingness to pay for quality and safety.	Moderate. Moderate willingness to pay for quality and safety.	Relatively high. High willingness to pay for quality and safety.
Value added	Very low.	Low/moderate.	Moderate/high.
Trust between buyers and sellers	Not very important.	Of emerging importance.	Crucial factor for long-term successful relations.
Main determinants of competitiveness	Supply at low cost.	Sufficient quantity of improved quality.	Large quantity required. Efficient, effective coordinated supply chains. Flexible response to changing demand. Market and product innovation.
Participation of small-scale producers	No constraints.	Emerging constraints in meeting requirements of quality, safety, consistency of product, regular supply.	Only if well-organized in out-grower schemes and able to guarantee safety and uniform quality.

Source: Adapted from World Bank 2006.

the dissemination of market information, the distribution of inputs, and the provision of services; and pays the producers.

The agreement between the processor and the agent is informal but includes specific terms and conditions in terms of quantity, quality, and price of milk; contract duration; payment procedures; the sharing of costs and risks; dispute settlement; and so on. The contract is generally for three years and can be renewed or reneged with prior notice. The agent (who is also a milk producer) provides space for milk collection, and the firm provides necessary equipment such as weighing scales, milk analyzers, milk coolers, and water geysers at no cost to the agent. Operational expenses, except electricity and

water usage costs, are borne by the agent. The costs of transporting milk from the collection center to the firms' processing plant are borne by the firm. The contract agreement also provides for risk-sharing mechanisms. Unintentional risks to firm-installed assets (machine failure and wear and tear) are borne by the firm. Risks arising from non-compliance of specified quality standards are the responsibility of the agent.

The dairy farmers who have informal contracts with the agent can sell any amount of milk, provided the supply is on a regular basis and of good quality (based on fat and solids nonfat content) of milk. The price is determined on the basis of fat and solids nonfat contents (3–4% in cow milk and 6–8% in buffalo milk). In

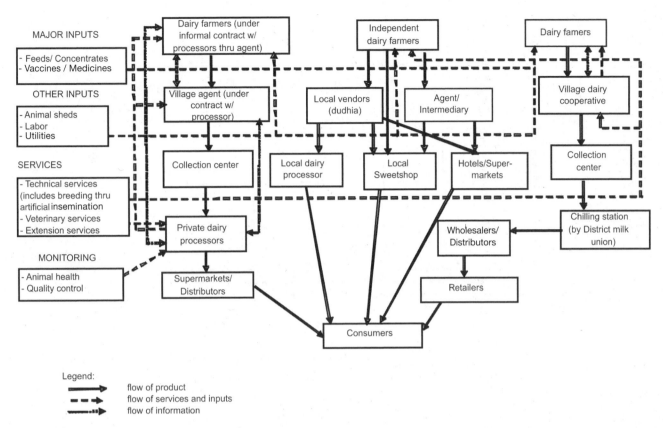

Figure 14.4. Milk supply chain in Rajasthan, India.
Source: Birthal et al., 2006.

determining the milk price, the firms consider the prices paid by competitors in the neighborhood. Payments to the producers are generally made at an interval of 10–15 days. Most producers receive payments in full and on time (Birthal et al., 2006).

Swine production in northern Viet Nam is still dominated by smallholders, and the Ministry of Agriculture and Rural Development is promoting commercial farms (see ILRI-HAU-IFPRI-PPLPI/FAO 2007 for more details). The majority of these pig raisers are independent producers. They buy inputs such as piglets, feeds, and medicines from the input suppliers, who offer them attractive perks such as discounts for bulk purchasing of inputs, credit at zero interest rates, good-quality inputs, or the marketing of output. (See Figure 14.5.) Normally, the input traders are well connected with output traders so they can certify the quality of the output. Sometimes output traders provide payment through input suppliers, which is subtracted from the amount borrowed by the producer. The independent pig raisers normally get veterinary services from a paraveterinarian, and medicines are procured from medicine shops. Two banks usually provide formal credit to pig producers: the Vietnam Bank for Agriculture and Rural Development and the Bank for Investment and Development of Vietnam (which provides credit only to commercial farms).

Depending on the type of output, smallholders can sell to different buyers. Farmers normally sell piglets either to their neighbors or to traders, while slaughter hogs are typically sold to local butchers or traders. Traders then sell slaughter hogs to wholesalers for marketing to urban areas or other provinces.

There are also formal contractual arrangements between feed companies (such as Jappa Comfeed of Indonesia and the Charoen Pokphand Company of Thailand) and large-scale producers. These feed companies provide contract growers' inputs such as breeding stock (sows and boars), weanlings, feeds, medicines, and technical assistance. The contract growers receive a fixed fee for raising pigs, based on performance standards (for example, feed conversion ratio or mortality rate) that are agreed when signing the contract. The contract growers provide land, pigpens, tools and equipment, and management skills to the feed companies they have contracted with.

There are also informal contracts between cooperatives and their members. The cooperative provides producers' inputs such as feeds and medicines at a cost and gets paid either in cash or with credit by producers. It also provides technical assistance for a minimal fee and helps producers to find buyers.

In both these cases—milk supply in Rajasthan and swine in Viet Nam—smallholders were linked to the

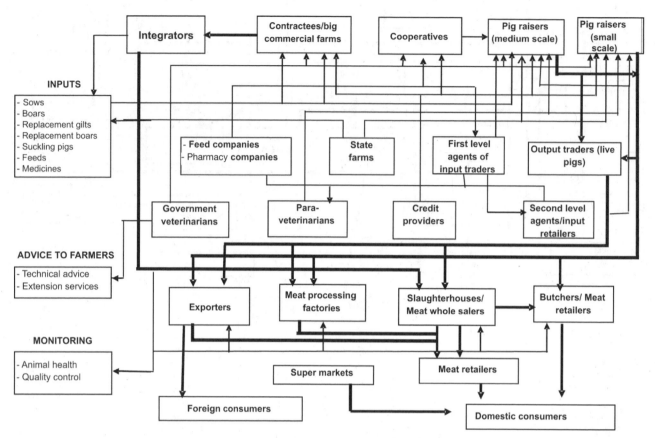

Figure 14.5. Swine supply chain in Viet Nam.
Source: ILRI-HAU-IFPRI-PPLPI/FAO 2007.

markets by either integrators or intermediaries. But their ability as a whole to have a market voice is rather small.

Smallholders' Difficulties in Getting Market Access to Supply Chains

Smallholders face problems both in meeting the product standards of growing urban and export markets and in delivering a regular supply. Consequently, they are often left out due to their low productive capacity, remoteness, and limited competitiveness with larger growers. Organizational challenges further impede private-sector inclusion of smallholders. Although the public sector has traditionally provided services such as extension, research, infrastructure, and marketing outlets, the movement toward demand-driven agriculture limits the government's ability to fully assist smallholders in the manner demanded by the marketplace. Rich and Narrod (2005) suggest that close coordination of the supply chain works against the smallholder for three reasons: information asymmetry and transaction costs, organizational constraints, and regulatory failure.

Basically there are five main elements essential for farmers to be linked to the supply chain to ensure market access. First, producers need access to extension services or technical assistance so that they stay abreast of changing knowledge of specialized techniques to ensure the safety of high-value products. Second, they need access to good infrastructures so they can manage flows between chain links quickly and efficiently and therefore meet the rigid deadlines imposed by buyers and reduce distribution costs. Third, they need access to good sources of information on changing market needs and the ability to integrate this information rapidly across the supply chain when it changes. Fourth, producers need to be able to ensure that their products meet certain standards through various types of certification, grade, and standards. These systems need to be not only consistent but also credible to meet buyer and customer demands. And fifth, producers need to have good mechanisms for coordination of their supplies to the markets so as to ensure the consistent delivery of high-quality products to the market to meet fluctuations in demand.

Table 14.8 identifies the major important supply chain processes that smallholders need to have access to in order to stay in the supply chain and the market failures along that supply chain that affect their ability to participate. Many smallholders currently experience market failures in the ability to achieve the processes just described. Their access to good extension services is often quite variable. Smallholders tend to face high transportation costs due to geographical locations and poor infrastructure linking them directly to markets. They often have imperfect information on the needs of buyers and customers in high-value markets. Further, their ability to meet public or private standards is limited, as there is often a large divergence between public and private standards, and their own public sectors often have a low capacity to enforce public standards. In terms of coordination mechanisms, smallholders often have limited ability to enforce contracts, and there is often a divergence in market power between supply chain actors.

Conclusions

It has often been argued that larger-scale production is a prerequisite to higher sustained profitability for livestock producers, all other things being equal. However, the evidence indicates that smallholders have succeeded in maintaining operations in some countries and for some commodities when family labor and environmental externalities are taken into account. One way they have kept a foot in livestock production is by moving into contract farming. There are mixed results about whether contract farming is more profitable than independent farming.

Clearly in the case of poultry production in India, the empirical evidence indicates that independent farmers are more profitable than contract farmers. Although this may be true, it also may reflect the fact that most of India's poultry is still slaughtered in live markets. Case studies have shown that contract farmers had a less-than-desired profit per unit of output than independents did, but this was outweighed by the performance

of contracts in terms of reduced transaction costs and stability of income. Broiler contract farmers, in particular, have benefited from the contractual arrangement in terms of increased knowledge (technical skills), stable incomes, and reduced transaction costs associated with the procurement of inputs and marketing of outputs.

In the case of India's dairy production, contract farming has resulted in higher profits per unit of output than independent farming. The study indicated that contract producers realized higher profit per liter of milk than independent producers who sell their produce to vendors and in the open market. The contract price, net of transaction costs, was higher than the open market price or the price paid by other buyers in the local market. The benefits of contract farming in dairy were skewed toward large producers mainly due to economies of scale in the disposal of milk. At a similar scale of production, however, smallholders derived significant benefits from contract farming through reduction in transaction costs. In the case of dairy cooperatives in India and Thailand, sustainability appears to be due in part to government support schemes. It is unclear how profitable they would be without such support.

For some commodities, such as broilers and layers, smallholders in all four countries studied have been rapidly losing their market share. Although smallholders have abandoned poultry activities in these countries, some producers who were formerly smallholders from the early 1990s are now large farmers, sometimes migrating to new areas such as central western Brazil or expanding where they are in southern India.

In countries where institutions for supporting smallholders are weak, where the supply chain has become

Table 14.8. Institutional roles in the supply chain management of high value agricultural products (HVA): support processes.

Supply chain support processes	Needed roles for supply chain management	Market failures affecting smallholders
Extension services	Knowledge of specialized techniques for high-value products	Variable smallholder access to public or private extension; limited public knowledge of new techniques; underfunding of services
Infrastructure development	Manage flows between chain links quickly and efficiently to meet rigid deadlines by buyers; reduce distribution costs to remain competitive with other supply chains	High transportation costs, low access to smallholder areas, poor infrastructure, erratic information flows, crowding out by public sector
Information services	Integrate information flows across supply chain actors	Imperfect information by smallholders on needs of buyers and customers in high value agricultural products
Certification, grades, and standards	Consistent, credible application of rigid standards on food safety and quality specifications to meet buyer and customer demands	Smallholders' ability to meet public or private standards is limited; divergence between public and private standards; low capacity to enforce public standards
Coordination mechanisms	Mechanisms must ensure consistent delivery of high-quality products	Limited enforcement of contracts; divergence in market power between chain actors

Source: Adapted from Rich and Narrod 2005.

more complex, and where there are efforts to control a disease such as avian influenza or foot and mouth disease (such as in Thailand and Brazil), smallholders are increasingly moving out of livestock production and using their productive resources for other activities. Currently, there exists limited evidence on where smallholders have moved or if they have exited market-oriented livestock-raising activities entirely. Although there have been tangible benefits to consumers from lower-cost livestock products, the evidence for gains to producers, particularly smallholders, is minimal.

Increasingly it appears that smallholders' ability to maintain competitiveness in the livestock market is dictated by the ability to be linked to the supply chain and obtain certain supply chain management necessities. If market failures are preventing smallholders' access to important elements such as extension and information services, good infrastructure, certification of produce, and good coordination mechanisms, it is very possible that they will lose much of their current market access unless appropriate institutional arrangements are put in place to ensure they can get the services they need.

References

Birthal, P. S., and V. K. Taneja. 2006. Livestock sector in India: Opportunities and challenges for smallholders. Smallholder livestock production in India: Opportunities and challenges. International Workshop, 31 January–1 February. Indian Council of Agricultural Research, New Delhi, and the International Livestock Research Institute, Nairobi. National Agricultural Science Complex, DPS Marg, Pusa, New Delhi.

Birthal, P. S., A. K. Jha, M. Tiongco, C. L. Delgado, C. Narrod, and P. K. Joshi. 2006. Equitable intensification of market-oriented smallholder dairy and poultry production in India through contract farming. *Final report of IFPRI-FAO contract farming of milk and poultry in India project*, Annex I.

Camargo Barros, G. S., S. D. Zen, M. R. Piedade Bacchi, S. H. Galvão de Miranda, C. Narrod, and M. Tiongco. 2003. Policy, technical, and environmental determinants and implications of the scaling-up of swine, broiler, layer and milk production in Brazil. *Final report of IFPRI-FAO/LEAD livestock industrialization project: Phase II*, Annex V.

Costales, A. C., C. Delgado, M. A. O. Catelo, M. Tiongco, A. Chatterjee, A. delos Reyes, and C. Narrod. 2003. Policy, technical, and environmental determinants and implications of the scaling-up of broiler and swine production in the Philippines. *Final report of IFPRI-FAO/LEAD livestock industrialization project: Phase II*, Annex I.

Costales, A. C., C. Delgado, M. A. O. Catelo, L. Lapar, M. Tiongco, S. Ehui, and A. Z. Bautista. 2007. *Scale and access issues affecting smallholder hog producers in an expanding peri-urban market: Southern Luzon, Philippines*. IFPRI Research Report No. 151. Washington, DC: International Food Policy Research Institute.

Delgado, C., M. Rosegrant, H. Steinfeld, S. Ehui, and C. Courbois. 1999. *Livestock to 2020: The next food revolution*. Food Agriculture and Environment Discussion Paper 28. Washington, DC: International Food Policy Research Institute.

Delgado, C., C. Narrod, and M. Tiongco. 2003. Policy, technical, and environmental determinants and implications of the scaling-up of livestock production in four fast-growing developing countries: A synthesis. *Final research report of IFPRI-FAO/LEAD livestock industrialization project: Phase II.*

Delgado, C., C. Narrod, and M. Tiongco. 2008. *Determinants and implications of the growing scale of livestock farms in four fast-growing developing countries*. IFPRI Research Report No. 157. Washington, DC: International Food Policy Research Institute.

Fairoze, M., L. Achoth, P. Rashmi, M. Tiongco, C. Delgado, C. Narrod, and P. Chengappa. 2006. Equitable intensification of market-oriented smallholder poultry production in India through contract farming. *Final report of IFPRI-FAO contract farming of milk and poultry in India project*, Annex II. Washington, DC: International Food Policy Research Institute.

Fuglie, K., C. Narrod, and C. Neumeyer. 2000. Public and private investment in animal research. In *Public–private collaboration in agricultural research: New institutional arrangements and economic implications*, ed. K. O. Fuglie and D. Schimmelpfenning, 117–151. Ames: Iowa State University Press.

Hayami, Y., and K. Otsuka. 1994. Beyond the green revolution: Agricultural development strategy into the new century. In *Agricultural technology: Policy issues for the international community*, ed. J. R. Anderson, 15–42. Wallingford, U.K.: CAB International.

Humphrey, J. 2005. *Shaping value chains for development: Global value chains in agribusiness*. Eschborn, Germany: Deutsche Gesellschaft fur Technische Zusammenarbeit.

ILRI-HAU-IFPRI-PPLPI/FAO. 2007. Contract farming for equitable market-oriented swine production in Northern Vietnam. *Report of the contract farming of swine production in Northern Vietnam project*. Nairobi: International Livestock Research Institute.

Martinez, S. W. 2002. *Vertical coordination of marketing systems: Lessons from the poultry, egg, and pork industries*. Agricultural Research Report No. 807. Washington, DC: Economic Research Service, U.S. Department of Agriculture.

Morrison Paul, C., R. Nehring, D. Banker, and A. Somwaru. 2004. Scale economies and efficiency in U.S. agriculture: Are traditional farms history? *Journal of Productivity Analysis* 22:185–205.

Narrod, C., and K. Fuglie. 2000. Private-sector investment in livestock breeding. *Agribusiness: An International Journal* 4:457–470.

Narrod, C., and C. Pray. 1995. Technology transfer in the poultry industry: Factors and externalities associated with increased production. In *Animal Wastes and the Land-Water Interface*, ed. Kenneth Steele. Boca Raton, FL: CRC Press.

Narrod, C., and C. Pray, 2001. Technology transfer, policies, and the global livestock revolution. Proceedings of the International Agricultural Trade Research Consortium Symposium on Trade in Livestock Products, Auckland, New Zealand.

Narrod, C., D. Roy, J. Okello, B. Avendaño, K. Rich, and A. Thorat. 2006. *The role of public–private partnerships and collective action in ensuring smallholder participation in high value fruit and vegetable supply chains*. Collective Action and Property Rights (CAPRI) Working Paper No. 70. Research Workshop on and Market Access for Smallholders, 2–5 October 2006. Cali, Colombia.

Neumann, C., D. M. Harris, and L. M. Rogers. 2002. Contribution of animal source foods in improving diet quality and function in children in the developing world. *Nutrition Research* 22:193–220.

Nho, L. T., D. X. Tung, D. H. Giang, and L. D. Cuong. 2001. Economic efficiency in fattening pig farms in Nam Sach-Hai Duong and Thai Thuy-Thai Binh. *Proceedings of national workshop of animal husbandry and health*. Hanoi, Viet Nam. Center for Agricultural Policy (CAP) and the Institute for Policy and Strategy for Agriculture and Rural Development (IPSARD).

Poapongsakorn, N., V. NaRanong, C. Delgado, C. Narrod, P. Siriprapanukul, N. Srianant, P. Goolchai, et al. 2003. Policy, technical, and environmental determinants and implications of the scaling-up of swine, broiler, layer and milk production. *Final report of IFPRI-FAO/LEAD livestock industrialization project: Phase II, Thailand*, Annex IV.

Rich, K. M., and C. A. Narrod. 2005. *Perspectives on supply chain management of high value agriculture: The role of public–private partnerships in promoting smallholder access*. Draft.

Rosegrant, M. W., X. Cai, and S. Cline. 2002. *World water and food to 2025: Dealing with scarcity*. Washington, DC: International Food Policy Research Institute.

Sharma, V. P., S. Staal, C. Delgado, and R.V. Singh. 2003. Policy, technical, and environmental determinants and implications of the scaling-up of milk production in India. *Research report of IFPRI-FAO/LEAD livestock industrialization project: Phase II*, Annex III.

Staal, S., C. Delgado, and C. Nicholson. 1997. Smallholder dairying under transactions costs in East Africa. *World Development* 25:779–794.

Tiongco, M., C. Narrod, and C. Delgado. 2006. Equitable intensification of market-oriented smallholder dairy and poultry production. In *India through contract farming: A synthesis*. Final Report, IFPRI-FAO Contract Farming of Milk and Poultry in India Project.

World Bank. 2005. *Managing the livestock revolution: Policy and technology to address the negative impacts of a fast-growing sector*. Agriculture and Rural Development Report No. 32725-GLB. Washington, DC: World Bank.

World Bank. 2006. China's compliance with food safety requirements for fruits and vegetables: Promoting food safety, competitiveness and poverty reduction. Beijing and Washington, DC: World Bank. http://siteresources.worldbank.org/INTRANET-TRADE/Resources/Topics/Standards/CHINA_Jan06part1.pdf.

15

Extensive Livestock Production in Transition

The Future of Sustainable Pastoralism

Jonathan Davies, Maryam Niamir-Fuller, Carol Kerven, and Kenneth Bauer

Main Messages

- **Pastoralism is economically viable, but current development policies do not necessarily promote effective and sustainable pastoralism.** The growth and viability of pastoralism are constrained in some countries by inadequately designed and inappropriately chosen policies that have sought to transform rather than enhance it. A clear example is the overemphasis on meat and livestock marketing rather than milk processing and marketing in Africa. Milk in pastoral economies of eastern Africa may be worth more than twice as much as meat, and in places such as Somalia or Botswana the income generated through milk sales is more important than income from livestock sales. Yet the trend in many countries is toward increased livestock sales, which is associated in some cases with decreasing livestock mobility.

- **Pastoralism is commercializing in many countries, but this process presents new risks to pastoral livelihoods and hidden environmental costs.** The practice of employing shepherds and having absentee herd owners is growing; in Jordan, for example, over half the livestock holdings employ herders. Commercialization is primarily driven by a combination of growing supply, as pastoralists increasingly seek goods from outside the pastoral system, and burgeoning global demand for livestock products. As pastoralism becomes more market oriented, a combination of policy incentives and market forces is driving a transformation of the production system. Pastoralism is becoming more meat oriented, which can diminish the gross productivity of the system and reduce the volume of milk in both the household diet and the local economy, where it plays a role in cementing social relationships and institutions.

- **Pastoralism is conducive to environmental stewardship, and where rangelands are degraded this** often reflects a breakdown of mobile pastoralism. Not only are pastoralists capable of environmental stewardship, but also many rangelands are grazing dependent. Rangeland degradation is often due to constraints on pastoral management, such as the loss of key resources or the restriction of mobility. Inappropriate policies and disincentives still contribute to poverty and environmental degradation in the rangelands. Yet this situation is reversed in countries that have created a more enabling policy environment for pastoral development and investment, as in Tanzania, Mongolia, and Sudan. Given the widespread perception that large-scale nature conservation efforts have not been entirely successful, pastoralists—with their extensive traditional knowledge systems and intricate institutional and legal arrangements—can provide important and valuable conservation solutions.

- **Sustainable management of the rangelands in many countries is set back by the political, social, and economic marginalization of pastoralists.** Marginalization manifests itself in diverse ways, from the weak availability of data on pastoral areas and concomitant weaknesses in understanding pastoralism to the low degree of consultation with pastoralists in development planning and investment. Rangeland areas often lag behind in terms of literacy rates or health conditions, and the inhabitants have been ill equipped to adapt to emerging trends, such as economic transformations, globalization, democratization, and climate change. But there is evidence that this is changing with education and the ongoing processes of decentralization and self-organization of pastoralists. Some advances have been made in providing appropriate services to pastoralists, including education, health, and financial services, and the infrastructure in many pastoral areas is gradually improving. Yet pastoralists often remain excluded from policy making or are represented by individuals with inadequate understanding of the pastoral reality.

- **Changing development paradigms have begun to favor pastoralist development by empowering pastoralists.** The most important shift may be that development is beginning to focus on pastoralists rather than pastoralism: seeking to raise capacities and empower individuals and communities rather than to provide technical solutions to problems as perceived by outsiders. Through such approaches pastoralists can identify their own problems and find appropriate solutions that are consistent with their own livelihood goals. These changes are supported by the greater awareness of and emphasis on rights-based development. They also coincide with greater understanding of rangeland ecosystems and growing respect for pastoral traditional knowledge and customary institutions. Good development practice now entails empowering pastoralists and enabling them to enhance their production system. It includes promoting pastoralist voices and ensuring that customary institutions are able to apply traditional knowledge for the effective management of rangeland environments. One important knowledge gap involves the empowerment of pastoralist women; sustainable pastoral development cannot be achieved without much more attention to their decision-making roles and their livelihood choices.

Pastoral Development: Modernization or Substitution?

From the remote hill farms of northern Europe to the hot lowlands of sub-Saharan Africa, extensive livestock production is central to the livelihoods of millions of rural people. Yak producers in western China, alpaca farmers in Peru, and camel herders in Somalia all live on marginal lands where productive potential is relatively low and the climate is adverse. This commonality is neither an accident nor an anomaly: extensive livestock production systems are uniquely adapted to such marginal environments and challenging climates, and they have been developed to fill just such challenging niches and to harvest resources that have been inaccessible to other production systems.

Land degradation is often observed in marginal rangeland environments, and this degradation is frequently blamed on livestock production. Yet livestock production has existed in these places for centuries and even millennia without apparently leading to catastrophic degradation. So what has changed that has led a once environmentally sustainable practice to degrade its environment? And are such changes avoidable, reversible, or perhaps even desirable?

Opinions over the role of livestock in land degradation are divided. Extensive livestock production, and in particular pastoralism, is considered by some to be a principal cause of land degradation but by others to be the solution to it. Such polarized points of view suggest either differences of perception as to what constitutes degradation or differences in interpretation of the facts. The different perspectives reflect deep misunderstandings of rangeland environments and pastoral production systems, as well as the competing interests of different resource users.

What Is Pastoralism?

Pastoralism is a global phenomenon, found from the Asian steppes to the Andean regions of South America and from the mountains of western Europe to the African savanna. It is practiced on 25% of the world's land area, provides 10% of global meat, and supports an estimated 200 million pastoral households and the herds of nearly a billion camelids, cattle, and smaller livestock, in addition to yaks, horses, reindeer, and other ungulates (FAO 2001a).

Pastoralism is an ancient form of human activity, and present-day pastoral peoples carry forward an array of diverse cultures, ecological adaptations, and management systems that have changed with modernity. Pastoralism is frequently associated with a particular group of producers—for example, the Masai or the Bedouin—so that it is often difficult to dissociate the system from the practitioners. Reflecting such semantic associations, Baxter (1994) asserts that the term *pastoralist* connotes ethnic groups rather than a production system, so that not every self-described pastoralist practices pastoralism.

Alternatively, according to the widely used definition of Swift (1988), pastoralist households are those in which at least 50% of household gross revenue comes from livestock or livestock-related activities. But such a definition is not sufficiently precise and could easily include intensive livestock systems, or indeed could define out many practitioners of pastoralism who have significant other sources of income, whether seasonal or permanent. So researchers often include other criteria in the definition of pastoral systems, such as the use of common property and private resources, the practice of mobility, the use of family labor, or husbandry of a mixture of indigenous and cross-bred livestock breeds.

Defining pastoralism is fraught with difficulties, given the huge variety of systems worldwide and the challenges of interpreting the various descriptive terms, such as nomadism or transhumance, between languages. Such terms often overlap in their meaning and cause confusion. So for the sake of clarity, in this chapter the term *pastoralism* refers to extensive production of herbivorous livestock using pasture (or browse). The term *mobile pastoralism* is used to describe systems in which livestock mobility is a central management strategy, typically associated with the use of common property (as well as private) resources. The term *pastoralist* refers simply to one who practices pastoralism.

The term *ranching* is used here to describe extensive systems in which landownership is more individualized and where organized livestock mobility (other than paddock grazing) is not a deliberate management strategy. It is clear that such definitions are not universally applied, however, and ranching systems in the United States and Australia may (increasingly) have elements of communal landownership or managed seasonal migration.

The term *nomadism* is only used where examples are borrowed from literature, although the precise definition of nomadism varies from country to country. Where the term is used in the literature, it is sometimes synonymous with mobile pastoralism (that is, it inappropriately replaces the term *transhumance*), while other times it indicates the absence of any permanent home. Where clarity over the nature of mobility is required, the term *transhumance* is used to imply a seasonal movement of livestock between distinct resource areas, such as wet and dry season pastures, or high and low altitude zones.

There are many types and degrees of pastoral mobility, which may vary according to environmental conditions or the given stage of a household's life cycle. Mobility can be seasonal, regular as a pendulum between two well-defined pasture areas following marked transhumant routes that have not changed for centuries, or near-random, following erratic rain clouds and rarely the same from one year to another. Movement is not necessarily undertaken only for ecological reasons; it can be for trade, because of conflict, or to seal new political alliances (Niamir-Fuller 1999).

These divergent patterns make it difficult to classify pastoralists' mobility. At one extreme, a pastoral Wodaabe household in Niger may move its camp every few days throughout the year to ensure that pasture is grazed vigorously and then rested. Yet the same household, after a catastrophic drought in which it loses all its animals, may settle and live from agriculture, food aid, or migrant labor while it builds up its herd again. Thus, for a time it becomes sedentary. But as soon as the herd is viable again, the household will resume its mobile patterns to find pasture for the animals. At the other extreme are agropastoral households in Kyrgyzstan, where sheep are led on a daily basis to graze upland pastures during summer months. Although mobility of the flock is essential for exploiting high altitude pastures in this Central Asian nation, the household itself is stationary.

The high degree of environmental uncertainty that characterizes drylands and the extensive nature of livestock production demand a high degree of social organization and control between pastoralists. Outsiders often perceive this as chaotic and, by failing to recognize the intricacies of customary institutions for resource management, assume that resources are grabbed in a headlong free-for-all. In reality, pastoralists manage rangelands through a complex of common and individual property rights, where rights holders abide by rules and enforce them. Where pastoralists' land has become degraded, it is usually a combination of a weakening of these customary management institutions and the loss of key resources that make up the pastoral system (McGahey et al., 2008, Niamir-Fuller 1999, Scoones 1995).

The way pastoralism is practiced around the world varies greatly, from highly technologically advanced systems in Australia or the United States to near-subsistence systems in parts of Africa. The degree of social and political support for pastoralism is equally diverse, with some African governments strongly opposed to it while some European countries are investing in mobile pastoralism to manage and conserve biological diversity.

The Economic Values of Pastoralism

In many developing countries pastoralists contribute significantly to their national economies (see Table 15.1), although production data are seldom disaggregated, making precise calculations of economic importance difficult.

In wealthier economies, the relative contribution of pastoralism to the national economy is lower, although its importance for rural inhabitants remains high. For instance, Botswana's rural economy has traditionally been dominated by the production of cattle, which was the leading export before independence in 1966 (Hubbard 1986). By the mid-1990s, as the mining industry grew, livestock accounted for only 3% of gross domestic product (GDP) and about 28% of the total agricultural gross product (Panin 2000). Yet raising animals remained a critical part of Botswana's economy: in the early 1990s livestock production accounted for almost half of rural household incomes (Panin and Mahabile 1996).

Likewise, there are strong economic rationales behind extensive livestock production on the vast rangelands of Central Asia, where pastoralists operate in an extreme environment and use low-cost by-products from intensive crop production. The Soviet administration understood this when it transformed livestock production into a major contributor to the rural and national economies of Kazakhstan, Kyrgyzstan, and Turkmenistan (Kerven et al., 1996). By the end of the Soviet era, Kazakhstan was producing 25% of the Soviet Union's lamb and a fifth of its wool, most of which came from pastoral areas (Diddy and Menegay 1997). Moreover, extensive grazing systems in the semiarid regions of Kazakhstan, Uzbekistan, and Turkmenistan were estimated to have 50% lower production costs than other, more intensive Soviet livestock systems (Kerven et al., 1996). Since independence in the early 1990s, however, extensive livestock production in these countries has declined significantly, with meat and wool outputs falling sharply as the governments have curbed investments in the once highly subsidized dryland pastoral sector (Kerven 2003).

Table 15.1. Economic significance of pastoralism in various national economies.

Country	Contribution to national economy
Pakistan	Livestock sector produces almost half of the country's agricultural GDP. About half of the country's meat market is supplied by small ruminants from pastoral or agropastoral systems (Nawaz and Khan 1995).
Mongolia	Pastoralism accounts for one-third of GDP and is the second largest source of export earnings (21%) (UNDP 2003).
Iran	Mobile pastoralists account for only 2% of the entire population yet contribute 14% of Iran's milk and 17% of its meat (FAO 2004).
Kazakhstan	The livestock sector, predominantly found in the drylands, provides 42% of agricultural GDP, down from 60% in the Soviet era when the agricultural sector as a whole made up 31% of net material production (a Soviet measure of output) (World Bank 2005).
Kenya	Pastoralists own over 60% of the national livestock herd, which produces 10% of GDP and 50% of agricultural GDP. The estimated value of livestock offtake from pastoral systems in Kenya is $67–107 million (Republic of Kenya 2000).
Ethiopia	Pastoralist-dominated livestock sector contributes more than 20% of GDP (Aklilu 2002).
Morocco	Contribution of rangelands to agricultural GDP estimated at 25% (Berkat 1995).
Namibia	Livestock production, most of which takes place in semiarid rangelands, contributes 70% of all agricultural exports. The total value of agricultural exports has greatly exceeded the cost of food imports since independence (Werner 2003).
Nigeria	Livestock sector, which is predominant in the drier northern regions, contributes 12.7% of agricultural GDP (Yahuza 2001).
Uganda	Pastoralist and smallholder livestock producers contribute 18% of agricultural GDP, or 8.5% of total GDP, providing the country's fourth biggest foreign exchange earner (Muhereza 2004).

Extensive livestock production is an effective means of exploiting rangelands because returns can be attained at relatively low cost, making the systems efficient and adaptable. In Algeria, for example, pastoralist livestock production accommodates unpredictable climatic and economic conditions by moving along a continuum from labor-intensive milk-based production to more market-oriented meat production (Homewood 1993). Still, the output does not satisfy the national demand for red meat, which has led some groups in Algeria to recommend against investing in the sector at all and to focus instead on intensification of fodder production in cropping zones (Boutonnet 1991). Such policies, however, fail to recognize the lower productivity costs of extensive livestock systems and the multiple and complementary outputs they yield.

Land use by pastoralists is shaped by relatively low inherent productivity due to aridity, altitude, or temperature and sometimes a combination of all three factors. Since the output per area of land is naturally constrained, the density of human population is accordingly limited, as there is little capacity to absorb more labor productively (Behnke 1995, Sikana and Kerven 1991). As pastoralism is a production system for marginal lands, efforts to compare extensive livestock production with the sort of figures that can be attained in less arid or less harsh conditions are irrelevant. To some extent, then, discussions of the relative economic significance of pastoralism can be misleading, since they encourage comparisons with agricultural systems that would not be viable in the same rangeland environments. Mobile pastoralism is an effective way of harnessing marginal resources, regardless of how it competes economically with production systems in nonrangeland ecological zones.

Most studies on pastoral economics use tools that focus overtly on productivity and commercial offtake, neglecting nonmonetized products or services. This overlooks the significant indirect values of pastoralism and hence the costs of pursuing strategies that encourage commercial ranching. For example, in South Africa Shackleton et al. (2000) have shown a net annual value per household for livestock goods and services of $765 for cattle-owning households and $79 for goat-owning households. But in this region meat and livestock sales contribute less than 25% of value, while milk, draft power, transport, and manure account for more than 75%. As a result, standard valuations of communal livestock systems miss three quarters of the direct use value and support the conclusion that livestock systems are unproductive and less efficient than commercial systems. The misperception that pastoralism does not produce significant economic values means that governments may fail to make the necessary public investments in market infrastructure, roads, security, education, and human and institutional capacity building (McPeak and Little 2006).

The significance of pastoralism lies in its direct contribution to national economies, in its complementarity with other forms of production, and in its indirect values, such as environmental services and inputs to tourism (Hesse and MacGregor 2006). Unfortunately, data on these values are extremely hard to come by: livestock data are rarely disaggregated to show the direct value of pastoralists and nonpastoralists, and indirect values are routinely overlooked (Hatfield and Davies 2006). Nevertheless, it is clear that pastoralism allows for complementarities between economic production and environmental protection, between common and private property regimes, between extensive and intensive land uses, and between market and subsistence economies (Jodha 2000).

Policy debates continue to be misinformed by incomplete and inappropriate economic valuations of pastoralist production. The effect is to compare pastoral and ranching systems in limited terms such as thresholds and stocking rates. For instance, the productivity of animals in arid or semiarid regions is frequently compared with that of animals raised in more-temperate regions, overlooking the fact that mobile production of indigenous, locally resilient breeds can yield financial returns in conditions where high-yielding exotic animals or crossbreeds would fail to thrive (Nimbkar 2006, UNCBD 2001).

Two sets of factors are responsible for these distinctive responses: the physiology of indigenous livestock and the broad mix of products derived from pastoral and agropastoral herds. Experimental research on output from indigenous African stock subjected to various levels of nutritional stress suggests that these animals respond very differently from European breeds of beef cattle to the nutritional deprivation associated with increasing stocking density (Behnke and Abel 1996). In areas where intensification delivers reliable support in the form of fodder and veterinary services, crossbreeds have been found to perform better than native breeds. In dryland areas where such intensive husbandry is impractical and expensive, however, exotics may perform poorly.

Modernization and the Transformation of Pastoralism

It is difficult to isolate the root causes of change in livestock production systems since most changes are part of a multiple series of links in an elaborate chain of cause and effect. Certain factors such as demographic shifts, climate change, or the globalization of trade arguably have disproportionate effects in precipitating changes, yet even these factors may be subject to feedbacks from the outcomes they generate. This section discusses some critical drivers of change, with the caveat that the distinction between cause and effect is not always clear.

Identifying the Value of Pastoral Systems

Many pastoral systems are oriented toward the production of multiple goods and services such as milk, hair, meat, blood, manure, transport, draft power, food storage, capital reserve, and a hedge against inflation, drought, and other risks (Niamir-Fuller 1999). The range of goods produced differs widely, with greater emphasis on fiber production in the pastoral systems of Central Asia or South America and an emphasis on dairy products in Southern Asia and sub-Saharan Africa.

Pastoralists are traditionally, and increasingly, market-oriented, and they take a pragmatic approach to stock management (Kerven 1992, McPeak and Little 2006, Jodha 2000). In many developing countries, notably those of North Africa and West Asia, there is a trend toward commercialized meat production, which is associated with decreased mobility (Bourbouze 2004). Pastoralism in Jordan has commercialized and diversified to the extent that professional paid shepherding has become common, and more than half the holdings on the Jordanian steppe employ shepherds, 98% of whom are paid in cash (Blench 1995).

Indeed, increased meat production and "livestock offtake" are major policy goals of a number of sub-Saharan African countries, despite the much greater significance of milk in the pastoral economy. Data from Kenya indicate that the total value of milk in the pastoral economy is around double the value of meat and livestock sales (Nyariki 2004); in the Afar region of Ethiopia, the difference may be as high as four times (Hatfield and Davies 2006). In the Ogaden region of Ethiopia—a traditionally food-insecure area—the sale of livestock milk products generates more than 60% of the income needed to meet basic needs among pastoral households (Hussein 1999). In the subsistence economy of Botswana, milk products account for 70% and 67%, respectively, of the total value of cattle and goat output, while sales of live animals only account for 23% (Panin 2000).

In the view of many policy makers, however, meat production systems are seen as more "modern," and the goal of converting pastoral systems into "ranching" operations has been used as a rationale for transforming mobile communities into sedentary ones (Lane and Moorehead 1995). Meat production is perceived as synonymous with commercial livestock production and has been promoted as a means of controlling livestock numbers and boosting livestock marketing. This commercialization represents a shift into a less productive system that reduces the capacity of the rangeland to support people. As production is commercialized, capital investment and purchased inputs replace labor in a process of substitution rather than intensification (Behnke and Abel 1996).

Extensive production techniques, which necessarily rely on higher degrees of mobility, can be more productive per hectare: studies estimate that pastoralists obtain more than 2.5 times more energy from meat and milk offtake than from meat offtake alone (Western 1982, Western and Finch 1986). This is supported by data in

Table 15.2, which illustrates the higher levels of productivity in some African pastoral systems compared with meat-oriented alternatives.

It is important to note that commercialization does not automatically imply transforming pastoral systems into meat-only production enterprises. There is significant potential in many pastoralist systems to increase the range of livestock produce that is marketed or to orient dairy and fiber production more toward the marketplace. The markets certainly exist: the global market for camel milk alone is estimated at $10 billion, with the vast majority of the production taking place in the drylands (FAO 2006b). Commercialization of dairy production has been constrained in some pastoralist areas by lack of access to milk collection centers and also by some cultural taboos on the sale of milk. However, in Africa's two pastoralist-dominated countries—Mauritania and Somalia—commercial milk collection from nomadic producers has become routine (Nori et al., 2006, Tiviski 2006).

Important global markets also exist for pastoral-produced fiber, and in particular for specialist and niche products. Examples include alpaca fiber from the Andean region, Astrakhan (karakul) lamb pelts that have traditionally been produced in desert pastoral systems of Central Asia, and cashmere, which is predominantly produced by pastoralists in China, Mongolia, and Afghanistan. China has become the world's leading cashmere producer, with around half of the world market share, and it imports large quantities of raw cashmere from other Central Asian countries and adds value through its highly efficient manufacturing sector (Kerven et al., 2005). Mongolian competitors on the global market have seen the capacity of their wool processing

Table 15.2. Comparisons between Ranching and Traditional Pastoral Production Systems in Africa.

Country	Comments
Zimbabwe	Studies show that the value of communal area (CA) cattle production far exceeds returns from ranching. If actual stocking rates are used, CA returns are 10 times higher per hectare.
Botswana	Communal area production (in cash, energy, and protein terms) per hectare exceeds returns from ranches by at least three times per hectare, even though technical production parameters are lower. The difference in soil erosion levels between the two production systems is negligible, despite differences in stocking rate.
Mozambique	Traditional systems have higher overall returns per hectare because of the multiple benefits of draft, transport, manure, milk, and meat compared with the single beef output from ranches.
South Africa	Cattle production systems in the Transkei show higher returns per hectare, but lower productivity indicators, compared with ranches in the commercial white farming sector.
Kenya	Gross output levels in individual ranches and undeveloped group ranches are comparable. Similarly high levels of productivity were found among livestock in Sukumaland.
Tanzania	The productivity of pastoral herds in the Ngorongoro Conservation Area were found to be comparable to commercial herds. The patterns of productivity were similar to those found in Kenyan Maasai herds. Maasai multiproduct outputs are higher than ranches on a per hectare basis.
Uganda	Recalculations of figures to include full range of costs and benefits show that dollar returns per hectare under pastoralism are two times higher than for ranching. Dollar returns per animal are one-third higher.
Ethiopia	The pastoral Borana system has higher returns of both energy and protein per hectare than industrialized ranching systems in Australia: Australian Northern Territory ranches only realize 16% as much energy and 30% as much protein per hectare as the Borana system.
Mali	Transhumant pastoral systems yield on average at least twice as much protein per hectare per year as both sedentary agropastoralists and ranchers in the United States and Australia.

Source: Scoones 1995, Wagenaar 1984.

industry fall in the face of Chinese competition and now also export raw materials to their neighbor (Halbach and Ahmad 2005).

Livestock product processing can be a good investment in countries with large pastoralist populations and a comparative advantage in the low cost of production in extensive systems. However, world trade in livestock products is competitive, and underinvestment in the sector leads to a loss of market share. For example, Afghanistan's mechanized wool-washing facilities have deteriorated through neglect to the extent that the country's traditional carpets are increasingly manufactured from Australian wool imported via Pakistan. Mismanagement and low marketing expertise have also seen Afghanistan's Astrakhan pelt industry lose market share, after having dominated international markets in the 1950s (FAO 2003a).

Ethiopia provides another salutary example of underinvestment in the pastoralist sector. Livestock contribute more than 20% of Ethiopia's total GDP, and probably much more if other intermediate values of livestock are properly assessed. Yet the government allocated less than 0.3% of its recurrent expenditures on livestock between 1993 and 1999 (Aklilu 2002). Partly as a consequence of its failure to recognize the value of this production, smuggling of live animals, hides, and skins from Ethiopia into neighboring countries is costing the treasury an estimated $100 million each year (BBC 2001).

Globalization and International Trade

International livestock trade has been important to pastoralists for a long time, despite long-term underinvestment in many cases. The Horn of Africa, for example, has traditionally been a globally important region for livestock exports, much of which originates from pastoralist production systems. Stock from Ethiopia has traditionally been traded to a large extent through the Somali ports of Berbera and Boosaso, destined for the Gulf States. Export figures for Somalia in 1998 show that 95% of all goat exports and 52% of sheep exports for eastern Africa were channeled through the country, and 50–60% of the small stock leaving these ports originated in Ethiopia (Zaal and Poderman 2000).

Trade in livestock from Somalia to the Gulf States was tremendously important to the East African economy until the late 1970s, when Somalia's market share began to be eroded by Australia (Reusse 1982). More recently, with the collapse of a centralized government and loss of West Asian market share, cross-border trade routes have developed, particularly with Kenya and Ethiopia (Little 2001). Although exports from the region were cut as a result of insecurity in the late 1980s and a Saudi ban on Somali livestock imports, trade in sheep and goats out of northern Somali ports has recovered in recent years (Little 2001, Nori et al., 2006).

Considering the demand for livestock produce from North African markets, opportunities to increase livestock trade from sub-Saharan Africa should be great. Further export growth to North African states is constrained by tight import regulations, however: by high import tariffs in Morocco (around 250%), by the influence exerted by large, wholesale butchering companies in Algeria, and by state institutions in Tunisia that have been set up to avoid price explosions during religious festivals. Moreover, in Morocco and Algeria the market for red meat is tightly controlled by urban herders owning large flocks (Alary and El Mourid 2005).

International livestock trade is often poorly monitored, particularly when it occurs across borders, as so often is the case in pastoralist areas. For example, large numbers of livestock from the pastoralist systems of southern Somalia and southeast Ethiopia are traded in Kenya, with the majority going through the Kenyan market of Garissa: a trade that is estimated to meet around a quarter of Kenya's demand for beef (Aklilu 2002), with the Garissa market generating more than $15 million in cattle sales alone in 1998. This compares very favorably with other districts that cultivate coffee and cash crops, yet the pastoral sector remains greatly undervalued in national policy deliberations (Little 2001).

Globalization of trade has also had negative impacts on pastoral producers in developing countries, particularly as pastoralists from countries with stronger economies take a greater market share and produce a greater diversity of pastoral products. Wool producers in India's Rajasthan region struggle to compete against imports from Australia and Canada, which both seek high import quotas to the country (Singh and Köhler-Rollefson 2005). As noted earlier, Australian wool has also replaced local supplies in Afghanistan's famous carpet manufacturing industry (FAO 2003a). Syria used to export half a million Awassi sheep annually to the Gulf States, but it has lost much of its market share to Australia, which now produces the same breed (FAO 2003b).

Developing countries face great challenges in getting access to international markets for livestock and livestock products, where standards are high and expensive to achieve and maintain. It is notable that livestock export growth in developing countries has largely been directed to other nonindustrialized countries, where food safety standards are relatively less stringent (see Chapter 2). The real, inflation-adjusted price for beef has remained stable in European markets for the past three or four decades, while the cost of meeting sanitary and phytosanitary standards has greatly increased as a consequence of European consumer concerns over food safety (Stevens and Kennan 2005, Perry et al., 2005). It is important to note that many of these food safety threats (e.g., mad cow disease) originated in very intensive livestock production systems.

Even where preferential trade arrangements are made, producers are often unable to fill their quotas as domestic demand increases and takes goods away from the export market, as in the case of Botswana (Stevens and Kennan 2005). Between 1995 and 2000, Cotonou beneficiaries in Southern Africa (Namibia, Swaziland, and Zimbabwe) have together fulfilled just 64% of their European export quota (Perry et al., 2005). Likewise, in Botswana the combination of escalating export costs, flat prices, and domestic competition for beef supplies meant the hugely subsidized livestock sector recorded financial losses in seven out of eight years up to 2005 (Stevens and Kennan 2005).

Many pastoral areas span international boundaries, and therefore cross-border trade is common. However, rather than facilitating this trade to stimulate local economies and investing in those economies, some governments impose tariffs and restrictions that effectively ensure the trade remains in the informal sector, relying on bribes at border posts where necessary. The cost to pastoralists of these policy failures is that "legal ambiguities" generate inefficiencies in the market, which creates opportunities for rent-seeking behavior that disrupts markets (McPeak and Little 2006). More open, integrated, and competitive markets are required, yet in government circles, transborder commerce is often portrayed as illegal smuggling and consequently still remains subject to disruptive border closures and animal confiscations (Little and Mahmoud 2005).

Cross-border trade is common in Central Asia too: yaks and cattle are driven from Tajikistan to the urban markets of southern Kyrgyzstan; sheep from western Kyrgyzstan supply the populous Ferghana valley of Uzbekistan; horses are trekked across the mountains of northern Kyrgyzstan for sale in the richer communities of south Kazakhstan; cashmere goat fiber is trucked over the borders of eastern Kazakhstan and Kyrgyzstan to China; meat from northern Kazakhstan goes to Russia; karakul lamb pelts are sold without state permission from Turkmenistan to Russia. Nevertheless, the true value of livestock exports to Central Asian countries and their populations has not been computed (Kerven 2006).

Pastoralists also serve critical roles in distributing foodstuffs across borders and ecological niches, as in the Himalayan regions bordering Nepal, India, Bhutan, and China, where pastoral movements have traditionally been integrated into regional systems of food distribution. The recent creation of community forest groups in Nepal has led to the imposition of restrictions in access to grazing lands in Humla District, which has disrupted the long-established patterns of transporting commodities (Bauer 2004). The outcome of these grazing restrictions is that "eighty percent of the traders who peddled their wares on the backs of mountain goats left the business" (Wagle and Pathak 1997). This breakdown

in pastoral mobility and its associated food trade was largely responsible for the famine that claimed hundreds of lives in 1999 (Bauer 2004).

Cross-border trade is also essential to food security in eastern Africa, where the export of livestock finances the import of essential foodstuffs such as rice, wheat flour, and cooking oil. When the export of animals is interrupted, food supplies dwindle and prices rise (Little 2001, Nori et al., 2006). Similarly, cross-border trade is an essential economic service provided by pastoralists in West Africa, in particular the north–south trade between the arid livestock-producing areas and the humid tropical countries such as Côte d'Ivoire, Ghana, Mali, and Benin, which was a factor in the introduction of "pastoral transhumance permits" in West Africa during the 1980s. The results of this initiative were mixed, however, and the benefits for pastoralists of cross-border trade are constrained by weak spatial integration, which results in a greater proportion of the proceeds of livestock sales accruing to nonpastoral agents (Little 2001).

Marketing and Commercialization of Pastoralism

The economic contribution of pastoralism, though significant in many countries, is almost invariably constrained by weak infrastructure, poor market access, and, increasingly, loss of grazing land. Policy makers in developing countries face dilemmas over whether to invest in livestock production for export or to satisfy domestic demand. Both may be desirable, but the policies and investments required for each may differ greatly (Scoones and Woolmer 2006).

Local and informal market exchanges frequently dominate the pastoralist economy, as in Morocco, where 95% of livestock transactions occur in weekly *souks* (rural markets), with extensive market chains and multiple actors (Ait-Baba 2003). The proliferation of informal markets in some countries reflects the imposition of policy and legal constraints to marketing, as in Sudan and Ethiopia, where bureaucracy, tariffs, and government barriers to cross-boundary livestock trade produce inefficiencies in formal livestock markets (Aklilu 2002).

Going the informal route can reduce certain costs, but it may also reduce profits, as in Kenya, where prices for milk in informal markets were found to be half those of formal markets, with prices falling even more precipitously in marketplaces further from urban centers (Staal and Jabbar 2000).

Marketing institutions in East African pastoral areas are generally poorly developed, particularly for procuring grain from livestock sales during a crisis or for restocking after one (McPeak and Barrett 2001). In West Asia and North Africa, where increasing affluence and population growth are increasing the demand for livestock products, poor marketing infrastructure in the most marginal areas—where the dependence on livestock production is greatest—leads to a loss in market share and a shortfall in national production (Aw-Hassan et al., 2005).

Despite the encouraging data on the contribution of pastoralism to national economies, pastoralists frequently do not receive commensurate benefits from trade in livestock products. Kenyan pastoralists, for example, receive on average around 40% of the terminal market value price of their livestock, when a level of around 70% would normally be expected (Kibue 2006). Livestock marketing in pastoral regions is complicated by transaction costs arising from the long distances that the pastoralist must travel and the poor infrastructure generally found in the marketplace (Scoones 1995). Transaction costs erode the returns and discount the labor that pastoralists invest in livestock production; if such costs are too high, they can deter producers from participating in markets altogether (Drabenstott 1995). Indeed, a strong correlation has been found in pastoral areas between transaction costs and participation in markets (McPeak 2002). The impact of transaction costs is therefore not simply a problem for the producer but also for the livestock industry at large and the national economy more generally.

In infrastructure-poor dryland areas, where the cost of procuring food can be prohibitive, even the opening of one or two roads can have a profound impact on livestock marketing. The opening of the Karakoram highway in Pakistan greatly affected animal husbandry in the area by facilitating importation to the region of grain from the plains, enabling livestock producers to use more land for producing winter fodder and to adopt improved fodder technologies (Ehlers and Kreutzman 2000). Similarly, the completion of a tarmac road from Mogadishu to Boosaso (Somalia) is considered a major trigger for the developing camel milk marketing in the region (Nori et al., 2006).

In particular, transaction costs associated with information and organization can create a power imbalance between trader and producer, lowering returns for pastoralist producers (Ait-Baba 2003). Still, in many pastoralist societies, access to information is not enough on its own; producers also need to develop the capacity to use that information (McPeak and Little 2006). It is therefore important to recognize the services provided by traders in facilitating livestock sales from pastoral areas, and their proliferation is testimony to their usefulness. In Tunisia, for example, traders account for 47% of all livestock sellers, whereas producers account for only a third. These middlemen relieve producers of the high costs and risks of traveling to markets, keeping up to date with market information, and maintaining trade links with buyers. Traders are often from pastoralist communities or at least of the same ethnic group, giving them greater access to producers. In India, for example, Raika middlemen and traders fill much of the market chain for wool and livestock sales, enabling greater market offtake for

households whose workforce is preoccupied with rearing the stock (Singh and Köhler-Rollefson 2005).

The multiple roles of livestock should also be kept in mind when considering the costs of marketing in pastoralist systems, since there may be a multitude of factors that influence the decision to sell livestock or products. Improving market access and infrastructure may improve the returns to livestock producers, but it may not automatically lead to high sales if livestock are being reared for other purposes, such as to trade internally for goods and services provided within the pastoral economy (Perrier 1995). Nevertheless, there is much evidence of increasing commoditization of pastoral livestock and livestock products, which demands attention to marketing constraints and costs, whether for raising profits, reducing rent seeking, or increasing total economic output of pastoral systems.

The Changing Face of Pastoralism

Commoditization

Growth in pastoral populations is fuelling the growing supply of livestock products in the marketplace as competition over rangeland resources raises the demand for calories in the form of grain (Desta and Coppock 2005, Dietz et al., 2001). This population pressure is exacerbated by decreases in the total available pastureland, which not only leads to greater market orientation but an increasing tendency to ignore environmental consequences (Jodha 2000).

This increase in supply is complemented, or perhaps augmented, by the unprecedented global growth in consumer demand for meat, exemplified by an increase in demand of 70 million metric tons between the 1970s and 1990s in developing countries alone, which has profound social, environmental, and economic implications (Delgado et al., 1999, FAO 2006b). The growing demand is increasingly met by livestock production in developing regions (see Chapter 2), which now produce more meat than the industrial world does. However, global animal production growth has so far been driven by increases in herd size and individual productivity, which is particularly promoted by greater use of purchased feedstuffs. Although developing-country growth has been more marked in the meat sector, the dairy sector is also beginning to catch up with production in industrial countries.

Globally, livestock is becoming agriculture's most important subsector in supplying commodities to growing markets (ILRI 2006), and pastoralists living on extensive rangelands can play a major role in satisfying this burgeoning demand, particularly for dairy products. However, much of the current global production growth is linked to intensification of production. (See Chapter 2.) Nevertheless, pastoralists are increasingly contributing to their national economies, in addition to satisfying their subsistence needs. Syria's pastoralists, for example, are almost self-sufficient in terms of daily food, yet they supply the country's urban areas with a large part of the livestock they need. Despite this, Syria still falls 25–30% short of its demand for animal products, and market opportunities remain to be exploited (Swaid 1997).

Syria's experience is similar to that of the North African states, where demand for meat far outstrips supply, with Algeria, Morocco, Tunisia, Libya, and Egypt now all net importers of mutton (FAO 2006a). Peru is also typical of many developing countries in that rapid urban growth is accelerating growth in demand for animal products, which is being satisfied through imports and is behind the policy goal of increasing offtake from the country's pastoral systems in the Andes (Leon-Velarde et al., 2000). In Turkmenistan and Kazakhstan, the rising demand for meat and dairy produce, particularly processed goods, has been driven by the rapidly growing oil and gas industries, which have greatly increased household incomes (World Bank 2005). Such demand offers important opportunities for the domestic livestock industries in these countries, provided they can compete with imports.

Making production more consumer-oriented implies major costs in terms of collection infrastructure and skills development, particularly where dairy goods are to be marketed. The dilemma for development planners is whether to invest in local processing or bulk collection of raw materials. The former is beset by issues of hygiene and technical constraints, except in cases where traditional products are already available, while the second approach is hampered by the cost of infrastructure (Heinz and Dugdill 2000). The preference in Afghanistan is to promote on-farm processing of milk into cheese, yogurt, curd, butter, and ghee, which allows the profits of value addition to be captured at the household level and also allows producers to be involved directly in marketing, reducing the number of links in the marketing chain (Khan and Iqbal 2000). A similar change is under way in Peru, where Andean pastoralists are being encouraged to expand partial or complete processing on site (Leon-Velarde et al., 2000).

Increased demand for livestock produce and increased pressure to generate cash income lead to attitudinal changes regarding the use of different livestock products. Many pastoralist societies have cultural constraints over the commercial use of dairy products, but milk is increasingly marketed, indicating important social changes. Such attitudes were prevalent in the Pashtun areas of Afghanistan and in northwest India, for example, where milk was produced for consumption or for distribution to poor households and the sale of fresh produce was considered shameful. However, some products, including butter, cheese, and qurut (dried whey), are more readily sold than others, and there is considerable potential to increase dairy production. Although challenging, there are profits to be made from investment in

collection, processing, and packaging of dairy products (Halbach and Ahmad 2005).

Commercialization and the shift in emphasis to meat production are influencing breed selection in some pastoral systems, as in Algeria, where the Ouled Djellal sheep is becoming the predominant breed. As a result of the demands for this breed, sheep rearing is now developing in traditionally agricultural zones (Madani et al., 2003), so the transformation is both a consequence of change and a driver of new changes. Meanwhile, Algeria's Barbarine sheep population decreased by 60% between 1990 and 2000, and the D'man population dropped by 50% in the same period (Laaziz 2005). In neighboring Morocco, flock composition is also changing, with bovine numbers increasing in the arid zones and herd structure shifting from mixed (i.e., ovine, caprine, and camelid) to one species (Abdelguerfi and Laouar 2000). The increasing demand for meat raises the slaughter value of livestock in some countries, as in Afghanistan where the Astrakhan pelt (lamb skin) industry has lost out to the partial shift in production toward rearing sheep for meat (FAO 2003a).

Changes in the production system also lead to changes in the lifestyle of pastoralists, as in the case of Syria's Jabbans—mobile herders who produce cheese and who are moving from the Khanasser valley toward the steppe, where more dairy sheep are raised (Abdelali-Martini et al., 2006). However, as Syria's demand for dairy products grows, so the pastoralist sector is shifting its focus from sheep toward cow products (FAO 2003b): in the four years leading up to 2003, yogurt supply increased by over 50% (Hilali et al., 2005).

Marketing of livestock is becoming more crucial to pastoralist welfare and production strategies. The Somali pastoralist economy, for example, has shifted from being centered on subsistence production to being export oriented (UNDP 1998). While the export of livestock is not a recent phenomenon (e.g., Somalis sent livestock to the British army's coaling station in Aden during the late nineteenth and twentieth centuries), the more recent economic boom in the oil-producing Gulf countries has expanded livestock exports. Ethiopia's Borana pastoralists have experienced a similar change, where the growing human population has led to an increase in commercial livestock trade, which enables pastoralists to raise the human carrying capacity of their production system (Desta and Coppock 2005). But simple risk-reducing mechanisms that are taken for granted in many regions, such as access to financial institutions for savings and credit, are widely absent from pastoralist areas. Thus even when markets can be made to work, there is often little incentive to sell animals and produce other than to satisfy immediate household needs.

Pastoralism is often an adaptation to uncertain environments, such as arid and semiarid lands where precipitation is highly variable in terms of both when and where it comes. In such environments communities need insurance against the vagaries of the climate, and particularly the risk of food insecurity. In areas where markets are weak or absent, this need has customarily been met through mechanisms for redistribution or social exchange and has led to the establishment of elaborate mechanisms or institutions for providing social support (Posner 1980). The "good" that such institutions provide has been referred to as bonding social capital (Putnam 2000).

Many pastoralist communities are experiencing change in the extent to which livestock are treated as commodities for sale, although the effect of increased commoditization of livestock on pastoralist social institutions is far from clear. If bonding social capital is a constraint to trade, then increasing trade may indicate a weakening of bonding social capital. Just as the market economics of pastoralism has been undervalued in the past, so too has the economic value of social capital–based insurance strategies, and it is possible that increased commoditization will lead to loss of resilience of pastoral communities as traditional mechanisms for risk spreading are eroded. Attempts to value systems of reciprocity have sometimes foundered on the misapprehension that they provide insurance against the loss of individual stock. In reality, such institutions that rely on social bonds and claims provide insurance against outright destitution of the household (Jahnke 1982) and as such are closer to a social welfare system.

In some communities, social bonds can inhibit commoditization of livestock through the transaction costs that they impose on livestock sales (Pryor 1977) or through cultural norms and taboos on the sale of certain livestock goods, such as camel milk. This is in contrast to the facilitating effect of bridging social capital (Putnam 2000), which refers to the richness of networks and knowledge and is enhanced through education. Bridging social capital enhances market access and reduces transaction costs, so individuals or communities that are rich in this capital face lower transaction costs and are better equipped to engage in commercial activities (Sadoulet and de Janvry 1995).

In certain cases, there are strong disincentives to increased commercialization. In Botswana, for example, smallholders rear cattle and goats primarily for milk that is consumed in the household rather than commercially traded, since trade yields only marginal gains. Increased commercial involvement has not meant that herds can be managed more profitably per head, or even that animals can be sold at a higher rate. It simply means that purchased inputs displace domestic labor in the production process and that specialized single-commodity production replaces the production of a diverse array of goods for home consumption—goods that then need to be purchased (Behnke 1987). Increasing commercialization of livestock would require greater allocation of

milk to calves, forcing households in turn to substitute milk in the household diet—a significant and inhibitory cost in nutritional and social terms. A similar constraint is widely observed in South Africa, where livestock are produced for the sake of consumption to offset the high costs of purchasing food (Shackleton et al., 2000).

Pastoralists routinely make trade-offs between livestock and livestock products and between the use value and commercial value of their assets. They trade off the value of selling livestock produce against the cost of purchasing replacement food. As commercialization increases, pastoralism is increasingly influenced by market forces, which inevitably means that pastoral production has to respond to market forces and to consumer preferences. To balance the demands of market orientation, economic efficiency, and mobility, however, pastoralists will need to develop new and appropriate technologies and production practices. Policy support is needed to enable pastoralists to engage more effectively in markets and to reduce the risks that come with increased commoditization. Market access for pastoralists at an international level also needs to be expanded, for example by achieving World Trade Organization agreement to curtail subsidies to livestock producers in richer countries.

Intensification, Extensification, and Adaptation

Pastoralism is typically practiced in environments that have low but highly variable precipitation and often in regions of cold temperatures and high altitude. Extensive livestock systems are adapted to this environmental variability and the short growing seasons that are common in rangelands, and mobile pastoralism is perhaps the epitome of adaptability and management of uncertainty. Mobility is a key component in the success of pastoral management strategies and has persisted despite attempts to sedentarize pastoralists and the failure to invest in their system. Indeed, mobility (whether it is of pastoralists or other mobile resource users, such as hunter-gatherers or fisherfolk) is not tolerated in many legal systems (Niamir-Fuller 1999).

Patterns of mobility in pastoral systems around the world are changing in response to an array of forces, with more mobility in some regions and less in others. An example of reduced mobility is found in Algeria, where animal movements have become limited to between 10 and 50 kilometers and where long-distance transhumance is now only carried out by 5% of the steppe population (Nedjraoui 2001). Increased or enhanced mobility, in contrast, is seen in Spain, where a 1995 Act of Parliament legitimizes the country's 120,000 kilometers of cañadas, or transhumance corridors, to ensure that pastoral flocks continue their transhumance and, in so doing, continue to preserve the country's biodiversity (Jefatura del Estado 1995).

Yet the extent of mobility has reduced in many countries, with herders switching from truly nomadic to a transhumant system. In the Syrian Al Badia steppe, for example, there are between 900,000 and 1.5 million transhumant herders, although the number of nomadic herders has decreased to about 10,000 (FAO 2003b). In the Jordan Badia, barely 2% of livestock herders are still nomadic, and many have become seminomadic and live in permanent houses, although their stock is still moved according to the availability of forage (Blench 1995). Indeed, some Jordanian Bedouin have taken up a variety of seasonal and permanent employment outside the pastoral sector; for them, sheep production has become more symbolic than actual (Abu-Rabia 1994).

In Iran, the fragmentation of traditional migratory routes has precipitated the now routine use of trucks to transport livestock (FAO 2004). Similarly, pastoralism in the Al-Badia desert of Jordan and Syria is increasingly mechanized, and 80% of producers own a truck, tractor, or car (Masri 2001). Traditionally, when water supplies dried up in late spring the flocks were moved to rainfed areas where water, crop residues, fallow, and mountain grazing were available. Nowadays, trucks are used to provide water and feed to the herds or to move whole herds to pastures, resulting in overexploitation of pastures where formerly the scarcity of water limited the grazing and allowed the vegetation to recover (FAO 2003b). Increased trucking of stock among Syria's Badia has thus come at an environmental cost, with trucks triggering breakup of the substrate, causing soil loss and the development of soft sand or mud that makes routes impassable, in some cases leading to erosion zones a kilometer wide (FAO 2003b). Likewise, in Algeria the mechanization of water transportation and reliance on supplemental feed has meant flocks spend too long in given areas, disturbing the natural balance and intensifying degradation (Sidahmed et al., 2000).

Reduced mobility and increased commercialization often go hand in hand, leading to further changes in the production system. As herders have become less mobile in South Tunisia, there has been a switch in the ovine race used from the drought-resistant, fat-tailed Barbarine to thin-tailed, selective grazing Bergui or Queue Fine. Such changes are also driven, as mentioned previously, by changing urban demand and consumer tastes (Alary and El Mourid 2005).

The process of intensification in many countries also entails fencing communal pastures, which can impose environmental costs through reduced livestock mobility and the shift from periodic intense grazing to sustained steady grazing. This is seen in the "ranching" systems that were imposed for meat production in Namibia, Botswana, Kenya, and more recently China (Niamir-Fuller 1999, Zhaoli et al., 2005, Bauer 2005).

As grazing pressures increase and livestock keeping becomes more intensive, the demand for feed supplements can increase. In Afghanistan, 18% of livestock income is used to buy supplementary feed in "normal

years," and some pastoralists spend up to $50 per animal on supplementary feed over the winter (FAO 2003a). Feed expenses have become critical investments in many pastoralist systems. With the availability of subsidized feeds in Jordan, feed purchases have become the single most important expenditure in almost every household that keeps livestock (Blench 1995). As a result of the increase in purchased feed supplements, pasture now only accounts for 5% of the feed input for small ruminants in Egypt and 30% in Morocco compared with 70–80% in the 1960s (Thomson 1997, MADR 2003).

The reverse of this trend toward intensification has been seen in recent years in Tajikistan's pastoralist sector. This sector declined during the 1990s as civil conflict restricted access to pastures and the state/collective sector was generally disrupted. The agricultural sector more generally declined, and land that was used for fodder production was diverted to other uses. Although efforts have been made to restock the country, breed improvement is no longer practiced, and the availability of or access to veterinary services is proving a challenge for the private sector (FAO 2001b).

In parts of Central Asia, after the dissolution of the Soviet Union and the end of central planning and subsidized irrigated agriculture, a significant number of people returned to the nomadic pastoral livelihood system their grandparents had been forced out of in the 1930s (Mearns 1993, Mearns and Swift 1996). In Mongolia, for example, many people turned from failed state enterprises to individualized (family) herding in the early 1990s, although many of these "new herders" were not skilled or committed enough to survive in herding and left again or were driven out by bad winters at the start of the millennium. The situation in the region is mixed, and despite the return to pastoralism in Mongolia, in many other Central Asian countries the opposite occurred (Mearns 1993, Potkanski 1993, Fernandez-Gimenez 2002).

Extensification of livestock systems is occurring in parts of the United States, notably in the drier rangelands states such as Texas. In such regions, monsoon rain patterns frequently leave ranches without water, which has encouraged the creation of pasture sharing or "grassbanking" between producers. This drought management tool resembles other mobile pastoralist systems in which herds are moved over great distances in response to the heterogeneous dispersion of water and pasture (Malpai Borderlands Group 2007). This initiative has contributed to development of a National Grassbank Network: "a partnership that leverages conservation practices across multiple land ownerships based on the exchange of forage for tangible conservation benefits." (See www.grassbank.net.)

In Romania, transhumance continued during communism largely because the remote mountain regions had limited agricultural potential and were deemed unsuitable for collectivization. The government did not allow international goods into Romania, so there was a high demand for fresh produce on the open, but controlled, markets. The great wealth that some transhumance shepherds accrued gave them leverage to negotiate with those in power. Since the revolution in 1989, however, Romanian transhumance has been affected by the opening of the markets to imports, the encouragement of large-scale cropping in the valleys, and development of forest enterprises (Huband et al., 2004).

In more industrialized countries, pastoralists may benefit from direct government support. For example, European Council Regulation 1698/2005 outlines support for protection of the environment and the countryside through appropriate land management and emphasizes the preservation and development of high-nature-value farming systems (which includes mountain pastoralism) as one of the priority areas of Rural Development (European Union 2005). Support for this is also becoming consumer driven, with the growth in the organic agriculture movement and civil society movements such as the Via Campesina (www.viacampesina.org) and the Slow Food Movement (www.slowfood.com). The Terra Madre experience shows that the appreciation of extensive livestock and pastoral animal products is on the rise (www.terramadre2006.org). Pastoralist communities are undergoing significant transformation, but they are also confronted with a growing range of opportunities for commercializing production. Understanding the drivers of such changes will enable more appropriate policy support for the development of extensive livestock-producing regions.

New Challenges and Opportunities for Extensive Livestock Systems

Securing Land Tenure to Promote Investment

Many pastoralist communities are experiencing significant change in land use as pressure over land increases and in response to market and other forces. The importance of mobility in many pastoral systems creates a challenge for securing land in many countries, since mobility often relies on a degree of communal ownership, regulated through customary institutions that are often not recognized in statutory law. Many countries lack statutory protection of common property rights and instead give priority to land privatization based on individual title, although statutory recognition of common property arrangements is now found in countries as diverse as Scotland and Uganda (Fuys et al., 2007).

In recent years, the adoption by most developing nations of principles such as land tenure reform, decentralization, devolution, and democracy has increased some mobile populations' security of access to land, resources, and services. In Bolivia, for example, where indigenous people constitute over half the rural population, a new land law in 1996 created the concept of community lands of origin and enabled the restitution of large territories in favor of the original inhabitants (Kay and Urioste 2005). The International Covenant on Economic, Social and Cultural Rights states that indigenous populations are particularly threatened by and vulnerable to loss of access to their ancestral lands (FAO 2005). Nevertheless, pastoralists and other "marginal" populations continue to be disadvantaged because of structural impediments and the requirement of many legal systems for strict definitions of boundaries, single systems of rights (versus nested or multiple rights), and strict codification.

Pastoralists continue to face the loss of key rangeland resources, which has major implications for pastoralist systems, as in Ethiopia, where the grazing land and water access has been lost in the Afar region, disrupting the pastoral economy (Motzfeldt 2005). Since land in Ethiopia is officially state owned, and since the government system does not recognize pastoralists' rights over unimproved, unsettled land, cultivators from neighboring regions are favored, which leads to the spread of farming and the loss of access to forests and pastures (Markakis 2004). Afar pastoralists have had little engagement with Ethiopian politics and they lack effective representatives to raise their interests. Despite Ethiopia's efforts to develop ethnic federalism, and despite the presence of a standing committee on Pastoralist Affairs in the Parliament, pastoralist politicians in the region have been unable to introduce legislation that would be of significant benefit to their constituents (Markakis 2004).

Uganda's Karimojong pastoralists face a similar curtailment of their rights, with "36% of Karamoja . . . gazetted as government land for forest and wildlife protection while the remaining 64% of the land is designated as a controlled hunting area. Karamoja is therefore made up of one national park and three game reserves comprising 7,349 square kilometers, nineteen forest resources taking up 2,307 square kilometers and three controlled hunting areas occupying 19,922 square kilometers. It is estimated that by 1996, 22,010 square kilometers had been licensed to 13 private companies who either engaged in mining of marble or gemstones, or hold exclusive or special prospective licenses" (Uganda Land Alliance 2000, p. 1).

The loss of key resource patches often strikes a critical blow to pastoralist systems, since these zones often provide buffer resources that are only occasionally tapped in times of duress; as such, the resources are declared vacant and title is given to other users. Laws that favor settled populations often lead to the loss of pastoral land to cultivators, the removal of seasonal resources, or blocks on transhumance routes (Toulmin 2006). In Somalia, drought contingency plans have been compromised by the loss of wetland ecosystems that have been put under irrigation (Adams et al., 2006). Kenya's Turkana have been similarly disadvantaged by the loss of key wetland pastures owing to the damming of the River Turkwel, leaving many of the community dependent on food aid (Adams 1992, Hawley 2003). Not only does the loss of resource patches compromise pastoral risk management strategies, but as a consequence pastoralists are squeezed into smaller areas of land, which raises their reliance on feed supplements (Dutilly-Diane et al., 2005).

Despite the environmental and economic rationale of pastoral mobility, the constraints on movement are manifold. Not only is there an absence of supporting legal frameworks for mobility, such as mechanisms for regulating transhumance, but also prevailing attitudes are stacked against mobility. Some policy makers maintain that pastoralists need to settle in order to benefit from services and that it is too difficult to deliver services to mobile pastoralists. In this way, administrative exigency can lead to the adoption of policies that curtail mobility and instead support the sedentarization of pastoralists (UNDP 2003).

In recent years, however, some positive developments have helped secure pastoral rights. The Senegalese government, for example, has for the first time officially transferred the ownership of pastoral lands to pastoral associations in the southern Ferlo region (Ly and Niamir-Fuller 2005). Revisions of China's national grasslands law in 2004 allowed communal control of pastures by village-level groups (Banks et al., 2003). Other progressive policies are found in the pastoral codes in West Africa (e.g., Mauritania's pastoral code, which is explicit in its support of pastoral mobility; Mali's Law 01-004 of 27 February 2001; and Guinea's Law 95/051, 1995), providing a legal mechanism for the recognition of pastoral land right. Unfortunately, pastoral codes in West Africa have been tainted by lack of consultation with and ownership by pastoralists. And implementation has proved to be problematic as a result of contradictory sectoral legislation (Hesse and Thébaud 2006).

Changing Land Use Regimes and Land Use Incentives

Recent changes in land use may be the result of new land users arriving with new production systems, as in the case of Tanzania's Barabaig, who have lost large areas of land to large-scale mechanized wheat production systems (Lane 1990). But they may also indicate pastoralists seizing opportunities, as in the case of some Masai in Tanzania who have resorted to farming as a means to protect their land from encroachment (Conroy 2001).

Increasing crop cultivation is witnessed on key dryland resource pockets (such as streambeds and valley bottoms) in Kenya and Ethiopia and is becoming a key driver of resource conflict in pastoralist areas. In areas where cultivation is increasing, traditional systems for resolving conflict are proving less effective and conflicts are becoming more frequent (Yirbecho et al., 2004).

Tunisia provides an apt illustration of the competing pressures and the consequences of privatization. Tunisia's collective arid rangelands cover more than 1 million hectares and receive between 50 and 150 millimeters of rainfall per year. Until recently, the region was still exploited by nomadic and seminomadic herders practicing seasonal transhumance, but rangeland management and livestock production systems have changed in the face of increased privatization of the best rangelands for arboricultural development (Nasr et al., 2000).

Other examples show that the interactions between irrigated agriculture and pastoralism in drylands are complex and, in some places, complementary. In Sudan, farmers practice "preventive clearing": when pastoralists are absent, farmers preventively clear land in order to secure property rights, given that both nomads and farmers will normally respect the security of usufruct property rights. Such encroachment by farmers is often backed by formal legislation (McGahey et al., 2008). In Kenya, cultivation of the Masai Mara has increased tremendously over the past 30 years, more than doubling every decade. Far from reducing pastoralism, however, this has been associated with a similarly dramatic increase in livestock marketing and little change in the total livestock population (Norton-Griffiths et al., 2006).

It is fairly incontrovertible that irrigated agriculture in drylands is more productive per hectare than nonirrigated land. The more relevant question in terms of comparing the productivity of irrigated agriculture and other uses of land in drylands is the scale of land area that needs to be irrigated and which parts of the land are being used to assess productivity. For example, in the Kilimanjaro region of Kenya, conversion of the scarce main wetland systems upon which the wider landscape depends might constitute 2–3% of the landscape but undermine productivity of the livestock system on most of the remaining 97% of land area (Hatfield and Davies 2006).

More holistic valuations will help prevent economically damaging policies that encourage replacing pastoralism with crop cultivation. This requires better understanding of consumption values, the current and the potential share of livestock products in foreign exchange earnings, and the value of the wide range of indirect goods and services that pastoralism provides. When multiple uses are considered, alternative rangeland livestock systems appear highly undesirable.

Lessons in Pastoral Development

Sedentarization has been at the heart of development of pastoral systems since the colonial times. Failure of such schemes has led to disillusionment on the part of decision makers, who in turn have blamed pastoralists for being resistant to change (Andersen 1999), even while the policy has led to overgrazing and land degradation. Livestock development projects have long sought to increase livestock productivity (usually for export) rather than to enhance livelihoods. Drawing on the classical ranching model from the United States, interventions from the 1960s to the 1980s encouraged sedentarization, destocking, and water development. These interventions did not necessarily improve livestock productivity, however, and some were very destructive, leading to increases in livestock populations in the immediate vicinity of water points and environmental degradation (Sandford 1983, Thébaud 1990).

The early 1980s saw the advent of integrated rural development projects, which were less coercive and more service oriented and had a nodding appreciation for local perspectives, yet they retained an implicit sedentarization agenda. The "blueprint" approach to pastoral development persisted, and land-use "guidelines" still tended to be discussed with land users only after their creation. Attempts were made to modify institutional structures for natural resource management, and legally registered pastoral associations emerged, with the responsibility for managing (but not owning) defined land areas. These new institutions had undefined relationships to customary ones, leading to ineffectiveness or further breakdown of customary institutions (Niamir-Fuller 1999).

In the 1980s and 1990s, government programs and development projects in Africa and parts of Asia and Latin America experimented with new forms of pastoral organization, such as pastoral associations, and group ranching. Many of these experiments failed for a variety of reasons:

- They induced a period of rapid loss of social capital among pastoral communities, with strong tendencies toward loss of the traditional power base, individualization, brigandry, and an increasing gap between the rich/powerful and the poor.
- The new forms of social organization were underpinned intuitively by relatively alien concepts, such as hierarchical or horizontal organizations, modern education being more valuable than traditional seniority, and planning/management concepts, all of which were never explicitly identified or overtly discussed and accepted.
- The diverse social and economic services provided by the institutions being replaced or supplanted were not acknowledged, and there was no effort to provide commensurate alternatives, such as insurance, finance, and other services (Swift et al., 2003).

Recognition of the importance of mobility has increased since the 1990s, and greater efforts have been made to develop, rather than replace, pastoralism. For example, efforts were made in Francophone West Africa to create official transhumance routes, with permits (e.g., ECOWAS Transhumance Certificates), supervised cross border movements, watering points, and quarantine stations (Thébaud 2001). These routes still exist, although infrastructure is inadequately maintained due to lack of funding.

Since the 1990s, development assistance projects have pursued natural resource management on a more localized scale and have been more strongly influenced by common property theory. Such projects have built local-level institutions for natural resource management, although they have been critiqued for overlooking informal local institutions and ignoring differences between the interests of leaders and nonleaders (Toulmin et al., 2002). The focus on the village (or groups of villages) has not fully suited the spatial arrangements of pastoralists, and the promotion of exclusionary mechanisms in land-tenure systems illustrates the continuing underappreciation of resource variability in dryland areas.

As mentioned earlier, customary institutions for decision making and resource allocation are integral to pastoralism—they are key, for example, for effective resource management, coordination of mobility, and the effective functioning of risk management mechanisms such as social safety networks. If these institutions are not supported, the flexibility of pastoral production systems will erode (Niamir-Fuller 1999).

As development planners have begun to acknowledge the role of traditional knowledge and customary institutions, and as they have accommodated more participatory and empowering approaches to development, they are opening the door for the self-organization of pastoralists. This is complemented by the steady increase in the total number of educated pastoralists and the improved access to democratic institutions associated, for example, with decentralization (Lister 2004). Such processes are crucial, because in many countries pastoralists are unaware of prevailing legal procedures and they lack understanding of their rights and responsibilities as citizens (Bonfiglioli 1992).

There has been a clear trend over the past 40 years from top-down planning toward grassroots consultation with pastoralists and from the assumption that pastoralism is archaic toward recognition that pastoralism can be compatible with development and progress. Changes in pastoral development programming are influenced by broader changes in the approach to development, including the shift toward rights-based programming and empowering approaches.

Advances in Range Science and the New Range Ecology

Beyond the social and economic issues just considered, the status of pastoralism also impinges upon and is affected by global concerns over biodiversity and natural resource management. In this context, advances in range science hold potential as future drivers of change. Research has shown that standard concepts of carrying capacity are inappropriate in nonequilibrium environments such as the semiarid to arid rangelands and that opportunistic pastoral systems, involving mobility and fluctuation in herd size, are more sustainable than constant stocking rates (Behnke and Abel 1996).

A large body of literature shows that, under proper management, livestock can be beneficial to rangeland production, biodiversity, and subsequent moisture availability. Indeed, this is a logical conclusion given the widely accepted concept of coevolution (in this case, grazers and grasses) backed by characteristics such as ground-level growth points in grasses that can withstand almost complete "harvesting" of the plant (Behnke and Abel 1996, Dijkman 1998, Breman and de Wit 1983, Savory 1999). Where grazing is sustained in one place, such as around wells and markets, degradation becomes apparent, but it is much less evident in open rangelands where mobility is unrestricted. High concentrations of livestock are beneficial for breaking up the soil surface and for gut scarification of grass seeds, but mobility must continue to ensure that pasture is rested and that seeds are dispersed (Savory 1999).

Grazing pressure can be timed to maximize plant productivity and overall biodiversity and can be used to reverse degradation. Appropriate use of enhanced livestock mobility for reclamation of degraded rangelands, which takes into account the time as well as the number of animals that need to be managed, has been shown to increase soil cover, to increase infiltration (decreased losses of water from runoff and evaporation), and subsequently to lead to more water being available to recharge streams, boreholes, and springs that can support livestock and wildlife (Savory 1999). This approach is being increasingly used by ranchers across the "developed drylands" of the United States, Australia, Canada, Mexico, and South Africa. Results from the United States have been impressive, with a 300% increase in the types of perennial species and an increase in beef productivity from 66 kg/ha to 171 kg/ha (Stinner at al., 1997). Such systems of planned management—of intensive grazing combined with mobility—have traditionally been a central part of mobile pastoralist systems (McGahey et al., 2008).

In most of Europe pastoralism takes place in areas of high nature value (HNV), and it is undergrazing rather than overgrazing that leads to degradation and the loss of pastoralism-related HNV. In the British landscape, grassland, heath, wood pasture, floodplain, and coastal marshes are all understood to require livestock grazing to conserve the associated wildlife habitats. Grazing plays a role in the maintenance of the species habitats through preventing scrub encroachment and removing plant material without cutting and burning, thus providing area for species mobility (English Nature 2005). European pastoral systems—ranging from reindeer husbandry in Norway to sheep grazing in Spain and cattle grazing in the Swiss Alps—illustrate how large-scale grazing systems can lend themselves to biodiversity improvements and habitat creation. There is a widespread perception that large-scale nature conservation efforts have not been successful and that pastoralists, through their intricate institutional and legal arrangements, can provide a solution to conservation (Gueydon and Roder 2003).

Transhumance is in decline in some European mountain regions (Baldock et al., 1996) but in central and southern Europe many vigorous systems remain (Mangas Navas 1992, Gómez Sal and Rodríquez-Pascual 1992). Mountain ecosystems are among the most highly valued in Europe, not only because of their high biodiversity but also because of their aesthetic appeal and high tourist value. In many parts of Europe there is potential for farmers to enhance their extensive livestock systems to optimize the HNV they protect, for example through reopening of traditional transhumance corridors. These farmers can target organic niche markets (meat and milk) as well as participate in the tourist industry. There are hopeful examples from the European Union of Romanian shepherds using modern information technologies to leverage the mobility of pastoralism in order to reap greater profits (Huband et al., 2004).

The traditional conservation approach, whereby a section of land is essentially alienated from human activities, is drawing increasing criticism, particularly in pastoral areas, as it removes key resources upon which community members depend for their livelihood. This has fueled community resentment and ultimately has undermined support for wildlife conservation. Furthermore, experience is showing that wildlife populations do not appear to thrive under this model, except in situations where the conservancy is very large, which is often not socially and politically possible. In Kenya's Masai land, for example, the greatest diversity and highest concentrations of wildlife have been observed not inside the parks but in the neighborhood of grazing livestock, suggesting that wildlife gain from the presence of pastoralists and their herds. The existence of both land use forms side by side could be most beneficial for the maximization of income and food security in rangeland areas (Norton-Griffiths et al., 2006).

Although the science of opportunism is widely understood and recognized, it has been slow to filter through to field-level outcomes. The wildlife conservation movement has been a driving force in the misunderstanding of rangelands management and has used its allegations of environmental degradation to justify the confiscation of land from pastoralists. Attitudes to conservation are still often dominated by the logic of "fence and fine" instead of opening and enabling, although in recent years conservationists have begun to acknowledge the important role that extensive livestock production can play in maintaining and improving wildlife and plant diversity.

Building Human Capital for Pastoralism

Provision of basic services in pastoral areas is urgently needed to build the capacity of livestock-dependent populations to adapt and diversify and thereby to take advantage of risk-reducing opportunities. For pastoralists to enhance their production system they need to be granted their basic rights; in particular, they require access to health, education, and security services. Clear links have been shown between education and the capacity of rural producers to engage in markets (Strasberg et al., 1999), and it is reasonable to suggest that decentralization, awareness-raising, and education would also improve the adoption or development of new production and processing technologies and could facilitate the adaptation of customary institutions to modern administrative institutions.

The mobility and dispersion of pastoralists, though often essential to manage their natural resource base, creates difficulties in gaining access to services. Pastoralists are often excluded from basic services because of their geographical isolation, and the services that are provided are frequently limited by a lack of staff and the absence of basic infrastructure (Schelling et al., 2005).

Since the late 1980s various development efforts have sought to deliver low-cost, appropriate technologies for bringing services to mobile and remote populations, although such approaches are in their infancy and may not yet achieve the quality of traditional service delivery systems (see Chapter 11). The most notable successes have been in the provision of mobile veterinary services and in particular the training of community animal health workers, or "paravets" (Catley et al., 2002).

Pastoralists' traditional knowledge systems regarding their environment and their production system are very rich, but such skills are not easily transferable outside the pastoral system, and levels of literacy among pastoralists remain low. In Kenya, the literacy rate among pastoralists is below 20%, compared with the national average of 69%, and there are only 2.2 doctors per thousand pastoral people compared with 15 per thousand nationally (Birch and Shuria 2001). In the Afar region of Ethiopia, the overall adult literacy rate was 25% in 1999, but in the rural pastoralist areas it was 8% (UNESCO 2005).

Improving education in pastoral regions is not simply a matter of making schools available and accessible. It also involves making curricula relevant and ensuring that teachers are sensitive to the pastoral culture and that academic calendars are appropriate to the seasonality of pastoral production. Notable models of success in nomadic education are found in Iran (CENESTA 2003), Mongolia (Kratli 2000), Oman (Chatty 1996), and Kenya (UNICEF 2003). Where governments have made efforts to reach nomads and pastoralists in innovative ways, school completion rates have improved (Oxfam 2005).

Health services have been provided, as in East Africa, in areas of settlement within drylands areas, sometimes as part of an explicit program of settlement. Yet evidence strongly suggests that there are many health benefits to the pastoralist way of life, with, for example, lower levels of malnutrition and lower incidence of childhood diseases (Fratkin 1998). It is ironic that pastoralists, in order to benefit from health services, have been encouraged to adopt a way of life that is evidently less healthy.

Success in providing more appropriate human health services have lagged behind education and veterinary services, although efforts have been made in various countries. In Chad, collaborations between public health and veterinary services have included joint vaccination coverage, which has allowed human health service providers to capitalize on the better penetration of veterinarians (Schelling et al., 2005). More remains to be done, however, in terms of providing human health services and persuading health ministries that alternative forms of services delivery are necessary and viable.

There are social consequences associated with improved social service delivery, notably when educated individuals move out of the pastoral system to find paid employment, which can lead in some cases to labor shortfalls but which also provides new sources of income. Economic migration offers opportunities for remittances, reinvestment in the system, and importation of innovative ideas, all of which have fueled agricultural transformations in other societies. A recent U.S. Agency for International Development study estimates that the aggregate value of remittances from migrant laborers across West Africa amounts to $12 billion annually (Orozco and Carana Corporation 2006).

Education services may not have penetrated many pastoral areas, but a number of "ethnic pastoralists" have gained an education and form what is often termed a "pastoral elite." This offers both opportunities and constraints to pastoralism, with, for example, potential for greater political connection and higher rates of services delivery, contrasted with a risk of appropriation of livestock-keepers' "rights" and resources (MacDonald and Azumi 1993, N'Gethe 1992, Mwangi in press). With education comes an increasing recognition and self-awareness of pastoral societies, including a growing cadre of young pastoral professionals and pastoral associations. Simultaneously, and possibly as a result, there is increasing global recognition of the rights of pastoralists and growing attention to human rights–centered development (Ask 2006).

The increase in commoditization of livestock in pastoralist communities is inevitably having an impact on attitudes within the wider pastoralist society. Increased exposure to new ideas and practices in urban centers, increased availability of consumer goods and productive inputs, and educational opportunities are all influencing the aspirations of pastoralists, particularly young people. For example, Ehlers and Kreutzman (2000) found that many young people from traditional herding societies in Pakistan now consider pastoralism to be outmoded and see education as an escape route. If the aspiration of young pastoralists is to gain an education and thereby avoid becoming herders, there is a risk of a widening gap between rural pastoralists and the statutory institutions that govern their lives, increasingly run by those who have left the pastoralist system. Yet, with increasing access to education, it is possible to nurture a generation of formally educated pastoralists who are able to make better-informed decisions about the adoption of new technologies and are better equipped to deal with markets and other institutions.

Low levels of adult literacy and poor health are clearly factors in vulnerability (World Bank 2001), but successful experiences in the delivery of social services to pastoralists are beginning to influence policy and lead to greater provision of such basic human rights (Schmidt 2003). This is long overdue, because the long-term underinvestment in service delivery in pastoral areas has created a situation where pastoralists lag far behind in terms of literacy and life expectancy. Such failure can only serve to fuel the undervaluing of pastoralism, which in turn further discourages investment, creating a potential vicious circle and encouraging investment in alternative systems that may be less economically and environmentally sustainable.

The Future of Sustainable Pastoralism

The trends in pastoral development and the advances made during the past decade make it possible to envisage a future in which pastoralism retains its integral role in maintaining rangeland ecosystem health and biodiversity while also exploiting new economic opportunities and serving the development needs of its practitioners. Indeed, such outcomes are already apparent in many industrial countries, including seven of the leading ones that form the G8.

Despite the social and economic pressures facing pastoral populations in some regions, the sector continues to be vibrant in many countries, with pastoralists contributing strongly to national economies regardless of low investment in the sector. Indeed, there is a growing tendency of governments to protect and invest in pastoralism: a testimony to both the resilience and the logic of pastoral production systems.

As decision makers and government planners accommodate new understandings of rangeland ecosystems and pastoralism, they are able to develop more appropriate policies, yet policy contradictions persist within the same governments. For example, while Tanzania's National Strategy for Growth and Reduction of Poverty (United Republic of Tanzania 2005a) promotes pastoralism as a

sustainable livelihood and is supported by the Wildlife Policy, which promotes "diversified and devolved wildlife utilization," the country's draft Livestock Policy is aligned with the Tanzania Development Vision 2025 in promoting the importation of improved breeds for increasing productivity. And the government maintains that communal use of rangelands promotes overgrazing (United Republic of Tanzania 2005b).

Nevertheless, the trends outlined in this chapter allow a description of good-case scenarios for pastoralism. While these obviously require modifications in different countries, the principles should be applicable in most cases.

Commercialization

Pastoralism is itself changing, adapting to market forces and demographic pressures, as well as being strongly influenced by policies that encourage sedentarization. Many pastoralist communities are indeed settling, sometimes devoting labor to small-scale cultivation, even though the quality and success rate of that cultivation may be low. In Afghanistan, many mobile pastoral households have become semimigratory, with a diversification of household strategies, for example, into off-range labor and agriculture (de Weijer 2005). Yet, as experiences in Pakistan show, with appropriate access to national markets, foodstuffs can be imported to pastoralist regions, and low-quality cultivable drylands can be used to cultivate fodder crops to enhance rather than compete with livestock production.

The process of commercialization will almost certainly continue, driven by both the growing demand for livestock goods and the growing demand among pastoralists for nonpastoral goods. The process may be distorted, however, if the policy emphasis remains on meat and livestock marketing, and policy makers need to explore opportunities for promoting dairying and fiber marketing in pastoral areas, as well as commercialization of nonlivestock rangelands goods and services. Such opportunities need policy support and possibly public-sector investment, for example, in basic infrastructure, in order to encourage private investment. The process of commercialization will be enhanced by continued investment in service provision, most notably the provision of educational, financial, and security services.

Investment

Investment in pastoral areas may be discouraged in some countries by the absence of reliable data on the value and logic of pastoralism. Pastoralism has been shown to contribute significantly to the national economies of a number of developing countries, yet it appears to receive disproportionately low public investment, which makes it that much harder to attract private investment. The nature of pastoralism—its remoteness, mobility, and diversity—presents challenges to gathering the socioeconomic data needed to change attitudes, particularly where governments are not predisposed to pastoralism or lack adequate resources for data collection.

Data are now being assembled that can change perceptions of the value of pastoral systems, particularly when their indirect values are taken into account. This is already having an impact on policy in countries such as Kenya, which depends heavily on tourist revenue and where the lion's share of tourism is to visit game reserves in the rangelands. Similar opportunities exist if the latent value of drylands goods and services is recognized or if the full range of direct values of pastoralism (including milk and fiber) are taken into account.

Human Capital and Risk Management

The capacity to withstand climatic shocks is a defining feature of pastoralism, but pressures on pastoralist systems have undermined the traditional capacities to cope and have not been balanced by advances that compensate for these losses. The loss of pastoralists' capacities to manage risk dynamically has not been adequately compensated by improved human capital (education or health), access to financial capital (savings and credit), or better access to markets and other infrastructure. Pastoralist areas are characterized by disproportionately low infrastructure and service delivery that are poorly adapted to their production system.

In recent years, successes have been achieved in adapting services to the realities of pastoralist areas, and infrastructure has begun to make inroads. Service delivery has been pioneered in many cases by nongovernmental organizations and is not always protected in government policy, but successes in the veterinary sector in East Africa indicate a way to bring government and civil society together to provide durable development solutions. Financial services remain poorly available but will be encouraged as government and the private sector come to realize the investment potential in these areas. This will be further advanced as education comes to be seen as a way to enhance pastoralism rather than an escape route from the rangelands.

Mobility

Sedentarization can, in some cases, bring the benefits of improved access to services without undermining pastoralism, as in Iran, where settlement has been shown to raise access to markets and potentially could reduce transaction costs through improved communication (FAO 2004). But sedentarization of the household does not require sedentarization of the herd, and it can be beneficial to look at household mobility and livestock mobility as two distinct issues. Herd mobility remains, in most rangelands environments, essential for both rangeland ecosystem health and economic viability of the production system.

In some cases, household and herd mobility cannot easily be dissociated, and household sedentarization may not be a realistic option if a sizable labor force is needed to manage the herd, for example, to process fresh milk on site or to provide security. Technologies that could overcome these constraints include labor-saving technologies for milk storage and processing in the field and improved transportation of dairy products from the herd to the household or market. These technologies are routinely used in European mountain pastoral systems, such as those found in the Swiss Alps. With appropriate policy support for the dairy sector, pastoralists in regions such as sub-Saharan Africa will be able to adapt existing technologies or develop new technologies that allow them to capitalize on the mainstay of their dairy economy.

Democratization

Pastoralists remain marginalized in many countries, but the global wave of democratization over the past decade has raised expectations for greater participation, social equity, respect for human rights, and better economic management (Lister 2004). Such democratization also includes devolution (or decentralization) of decision-making authority to lower levels, trends that can help create institutional spaces for pastoralists to negotiate for their rights and organizations.

For pastoralists, a key aspect of this groundswell toward decentralizing governance is the potential to be mainstreamed into national development and economic planning. Moreover, the emphasis on local governance lends momentum to pastoralists' calls for greater support in their efforts to manage natural resources in the midst of socioeconomic change. Pastoralism has an intrinsic global value, and its practitioners have the right of self-determination—to continue to exist and to adapt.

Conclusions

This chapter has deliberately avoided sketching out worst-case scenarios, but they can easily be imagined. The loss of pastoralism on any significant scale will lead to loss of high nature value, or biodiversity, in the rangelands; it will lead to economic loss on a national scale, impoverishment of drylands communities, and social upheaval; it will represent a loss of culture and knowledge as well as a serious violation of human rights. Even if pastoralism survives but continues to be constrained and undermined, land degradation and biodiversity loss will continue in the rangelands. Pastoralists will face growing erosion of their adaptive capacities and will become increasingly susceptible to climatic shocks; their livelihoods will become less resilient and less reliable; and they will face growing levels of conflict and suffering.

Yet this chapter has provided evidence that, despite rapid socioeconomic change and in some cases long-term underinvestment, pastoralism remains an economically viable and adaptive mode of production in extensive rangelands. Pastoralism is modernizing, but the trajectory of its development is influenced by policies and incentives that may distort the logic of the production system. For pastoralism to be viable and sustainable, it needs to be supported by policies that are consistent with the logic of the production system, and it may need significant investment—at least as much as is provided for competing land use options in the rangelands.

In order to enhance extensive livestock production, it is necessary to recognize that pastoralists have been marginalized and that therefore investment must be channeled into education and ongoing processes of decentralization. Policies for rangelands must focus on building the capacities of pastoralists through appropriate training (e.g., mobile education) and the delivery of health and finance services as well as infrastructure development. Governments must take the lead in creating an enabling environment and accountable institutions that seek not to transform pastoralism but instead to enable pastoralists.

Inappropriate policies and disincentives have contributed to poverty and environmental degradation in pastoral areas, but these trends can be reversed. Policy makers must recognize that pastoralism is conducive to environmental stewardship and that a degraded environment often reflects a breakdown of pastoralism, such as the loss of mobility, constraints to resource access, or a weakening of customary management institutions. The final say in the future of pastoralism has to be given to pastoralists themselves. Although they are not the only stakeholders in this debate, they provide the only workable solution to ensuring a sustainable future for the world's rangelands.

References

Abdelali-Martini, M., A. Aw-Hassan, and H. Salahieh. 2006. The role of local institutions in linking small ruminant producers to the market. In *Research workshop on collective action and market access for smallholders*. Cali, Colombia: Capri.

Abdelguerfi, A., and M. Laouar. 2000. Conséquences des changements sur les ressources génétiques du Maghreb. In *Rupture: Nouveaux enjeux, nouvelles fonctions, nouvelle image de l'élevage sur parcours*, ed. A. Bourbouze and M. Qarro. Montpellier: CIHEAM-IAMM. Options Méditerranéennes. Ser. A , No. 39, 77–87.

Abu-Rabia, A. 1994. *The Negev bedouin and livestock rearing*. Oxford: Berg Publishers.

Adams, W. 1992. *Wasting the rain: Rivers, people and planning in Africa*. London: Earthscan Publications Ltd.

Adams, M., J. Berkoff, and E. Daley. 2006. Land-water interactions: Opportunities and threats to water entitlements of the poor in Africa for productive use. In *Human development report 2006* . New York: United Nations Development Programme.

Ait-Baba, A. 2003. *Viande rouge et élevage pastoral au Maroc*. No 106. Centre National de Documentaton du Maroc.

Aklilu, Y. 2002. *An audit of the livestock marketing status in Kenya, Ethiopia and Sudan*. Nairobi: Organization of African Unity/Interafrican Bureau for Animal Resources.

Alary, V., and M. El Mourid. 2005. Les politiques alimentaires au Maghreb et leurs conséquences sur les sociétés agropastorales. *Revue Tiers Monde*. T. XLVI, no. 184.

Anderson, D. M. 1999. Rehabilitation, restocking, and resettlement: Ideology and practice in pastoral development. In *The poor are not us: Poverty and pastoralism*, eds. D. M. Anderson and V. Broch-Due, pp. 240–256. Oxford: James Currey Publishers.

Ask, V. 2006. *UNCCD and food security for pastoralists within a human rights context*. DCG Report No. 43. Oslo: Drylands Coordination Group.

Aw-Hassan, A., F. Shomo, and L. Iñiguez. 2005. Helping small enterprises capture the livestock products market in West Asia and North Africa. *ICARDA Caravan* 22 (June).

Baldock, D., G. Beaufoy, F. Brouwer, and F. Godeschalk. 1996. *Farming at the margins: abandonment or redevelopment of agricultural lands in Europe*. London and the Hague: Institute for European Environmental Policy and Agricultural Economics Research Institute.

Banks, T., C. Richard, L. Ping, and Y. Zhaoli. 2003. *Governing the grasslands of Western China*. https://www.jhf-china.org/cms/index.php?id=69.

Bauer, K. 2004. *High frontiers: Dolpo and the changing world of Himalayan pastoralists*. New York: Columbia University Press.

Bauer, K. 2005. Development and the enclosure movement in pastoral Tibet since the 1980s. *Nomadic Peoples* 9:53–81.

Baxter, P. 1994. *Pastoralists are people: Why development for pastoralists not the development of pastoralism?* The Rural Extension Bulletin No. 4. Brighton: University of Sussex.

BBC (British Broadcasting Corporation). 2001. *Ethiopia: Concerns over animals smuggling*. London: British Broadcasting Corporation.

Behnke, R. H. 1987. Cattle accumulation and the commercialization of the traditional livestock industry in Botswana. *Agricultural Systems* 24:1–29.

Behnke, R. H. 1995. The limits on production and population growth in pastoral economies. *Tropical Agriculture Association Newsletter* 15:2–4.

Behnke, R., and N. Abel. 1996. Revisited: The overstocking controversy in semi-arid Africa. *World Animal Review* 87:4–27.

Berkat, O. 1995. Population structure, dynamics, and regeneration of Artemisia Herba Alba Asso. These Doctorat es Sciences Agronomiques, II.

Birch, I., and H. Shuria. 2001. *Perspectives on pastoral development: A casebook from Kenya*. Oxford: Oxfam Publications.

Blench, R. 1995. *Rangeland degradation and socio-economic changes among the Bedu of Jordan: Results of the 1995 IFAD Survey*. London: Pastoral Development Network, Overseas Development Institute.

Bonfiglioli, A. 1992. *Pastoralists at a crossroads: Survival and development issues in African pastoralism*. Nairobi: UNICEF/UNSO Project for Nomadic Pastoralists in Africa.

Bourbouze, A. 2004. Le métier de berger dans la montagne marocaine. Conference at Agropolis Museum, 10 March 2004, Montpellier, France.

Boutonnet, J. P. 1991. Production de viande ovine en Algérie: Est-elle encore issue des parcours? Proceedings of the Fourth International Rangeland Congress. Montpellier, France.

Breman, H., and C. de Wit. 1983. Rangeland productivity and exploitation in the Sahel. *Science* 221:1341–1347.

Catley, A., S. Blakeway, and T. Leyland. 2002. *Community-based animal healthcare: A practical guide to improving primary veterinary services.* London: ITDG Publishing.

CENESTA (Centre for Sustainable Development). 2003. *Reviving nomadic pastoralism in Iran: Facilitating sustainability of biodiversity and livelihoods.* Centre for Sustainable Development, 15 January, Tehran.

Chatty, D. 1996. *Mobile pastoralists: Development planning and social change in Oman.* New York: Columbia University Press.

Conroy, A. 2001. Maasai agriculture, oxen, and land-use change in Monduli District, Tanzania. PhD Thesis. Department of Natural Resources, University of New Hampshire.

Delgado, C., M. Rosegrant, H. Steinfeld, S. Ehui, and C. Courbois. 1999. *Livestock to 2020: The next food revolution.* Washington, DC, Rome, and Nairobi: International Food Policy Research Institute, Food and Agriculture Organization, and International Livestock Research Institute.

Desta, S., and D. Coppock. 2005. Improving risk management and human welfare among pastoral and agro-pastoral peoples: A pilot outreach and intervention project for the Southern Ethiopian rangelands. In *Semi annual progress report for the Southern Tier Initiative.* 2005 USAID Mission to Ethiopia, Addis Ababa.

de Weijer, F. 2005. *Towards a pastoralist support strategy.* Washington, DC: U.S. Agency for International Development/RAMP.

Diddi, R., and M. Menegay. 1997. *Wool marketing and production: Transition from set-back to new growth.* Washington, DC: Abt Associates, for Asian Development Bank.

Dietz, T., A. A. Nunow, A. W. Roba, and F. Zaal. 2001. Pastoral commercialization: On caloric terms of trade and related issues. In *African Pastoralism, Conflict, Institutions and Government,* ed. M. M Salih, T. Dietz, and A. G. M Ahmed, 194–234. London: Pluto Press.

Dijkman, J. 1998. Carrying capacity: Outdated concept or useful livestock management tool? London: Pastoral Development Network, Overseas Development Institute.

Drabenstott, M. 1995. Agricultural industrialization: Implications for economic development and public policy. *Journal of Agriculture and Applied Economics* 27:13–20.

Dutilly-Diane, C., C. El Koudrim, A. Bouayad, A. Maatougui, A. Bechchari, A. Mahyou, and M. Acherkouk. 2005. Dominance de l'espace pastoral, structuration sociale et évolution des systèmes pastoraux dans l'Oriental Marocain. SDC Fnal Conference on Sustainable Management of the Agropastoral Resource Base in the Maghreb Phase II, November 21–23. Oujda.

Ehlers, E., and H. Kreutzman. 2000. *High mountain pastoralism in northern Pakistan.* Stuttgart, Germany: Franz Steiner Verlag.

English Nature. 2005. The importance of livestock grazing for wildlife conservation. Peterborough, U.K.: English Nature.

European Union. 2005. Support for rural development by the European Agricultural Fund for Rural Development (EAFRD). Council Regulation (EC) No 1698/2005 of 20 September 2005. *Official Journal of the European Union* L277/1.

FAO (Food and Agriculture Organization). 2001a. Pastoralism in the New Millennium. Animal Production and Health Paper No. 150. Rome: Food and Agriculture Organization.

FAO (Food and Agriculture Organization). 2001b. *Crop and food supply assessment mission to Tajikistan,* 7 August. Rome: Food and Agriculture Organization.

FAO (Food and Agriculture Organization). 2003a. *Syrian agriculture at the crossroads.* Rome: Food and Agriculture Organization.

FAO (Food and Agriculture Organization). 2003b. *Pastoralism and mobility in the drylands: The global imperative.* Rome: Food and Agriculture Organization. www.undp.org/drylands/docs/cpapers/PASTORALISM%20PAPER%20FINAL.doc.

FAO (Food and Agriculture Organization). 2004. *The role of local institutions in reducing vulnerability to recurrent natural disasters and in sustainable livelihoods development.* Report by Centre for Sustainable Development, Iran. Rome: Food and Agriculture Organization.

FAO (Food and Agriculture Organization). 2005. *Voluntary guidelines to support the progressive realization of the right to adequate food in the context of national food security.* Cited April 2007. Rome: Food and Agriculture Organization.

FAO (Food and Agriculture Organization). 2006a. FAOSTAT statistical databases. http://faostat.external.fao.org.

FAO (Food and Agriculture Organization). 2006b. *The next thing: Camel milk.* Rome: Food and Agriculture Organization.

Fernandez-Gimenez, M. 2002. Spatial and social boundaries and the paradox of pastoral land tenure: A case study from postsocialist Mongolia. *Human Ecology* 30:49–78.

Fratkin, E. M. 1998. *Ariaal Pastoralists of Kenya: Studying pastoralism, drought and development in Africa's arid lands.* Boston: Pearson.

Fuys, A., E. Mwangi, and S. Dohrn. 2007. *Securing common property regimes in a "modernising" world: Synthesis of 41 case studies on common property regimes from Africa, Asia, Europe and Latin America.* Rome: International Land Coalition.

Gómez Sal, A., and M. Rodríguez-Pascual. 1992. Montaña de León, Cuadernos de la Trashumancia, no. 3. ICONA, Madrid.

Gueydon, A., and N. Roder. 2003. Institutional settings in co-operative pastoral systems in Europe: First results from the LACOPE Research Project. Presented at The commons in transition: property on natural resources in Central and Eastern Europe and the Former Soviet Union. Prague, April 11–13.

Halbach, E., and W. Ahmad. 2005. Prioritizing investments for initiating rural development: The case of rebuilding Afghanistan. International Workshop, Dushanbe, Tajikistan.

Hatfield, R., and J. Davies 2006. *Global review of the economics of pastoralism.* Nairobi: World Conservation Union–IUCN, for World Initiative for Sustainable Pastoralism.

Hawley, S. 2003. *Turning a blind eye: Corruption and the UK export credits guarantee department.* Sturminster Newton, Dorset, U.K.: The Corner House.

Heinz, G., and B. Dugdill. 2000. Highland livestock production systems: Is there a need for specialized livestock product processing and marketing? In *Contribution of livestock to mountain livelihoods: Research and development issues.* ed. P. Tulachan, M. Saleem, J. Maki-Hokkonen, and T. Partap. Kathmandu, Nepal: International Centre for Integrated Mountain Development.

Hesse, C., and J. MacGregor. 2006. *Pastoralism: Drylands' invisible asset?* Issue Paper no. 142. London: International Institute for Environment and Development.

Hesse, C., and B. Thébaud. 2006. Will pastoral legislation disempower pastoralists in the Sahel? *Indigenous Affairs* 1/2006: Africa and the Millennium Development Goals.

Hilali, M. E., L. Iñiguez, and M. Zaklouta. 2005. More yogurt, please! *ICARDA Caravan* 22 (June).

Homewood, K. M. 1993. *Livestock economy and ecology in El Kala, Algeria. Evaluating ecological and economic costs and benefits in pastoralist systems*. London: Pastoral Development Network, Overseas Development Institute.

Huband, S., A. Mertens, and D. McCracken. 2004. An insecure future for transhumance in Romania. *La Cañada* (European Forum for Nature Conservation and Pastoralism) 18:27–30.

Hubbard, M. 1986. *Agricultural exports and economic growth: A study of Botswana's beef industry*. London: Routledge & Kegan Paul.

Hussein, Abdi Abdullahi. 1999. The role of pastoralism in ensuring food security in the Horn of Africa. In *Working together for development in pastoralist communities: A report on the "African Partnership" workshop, 26–29 November*, ed. John Livingstone. Kampala: Pastoral and Environmental Network in the Horn of Africa.

ILRI (International Livestock Research Institute). 2006. Pastoralism: The surest way out of poverty in East African drylands. Fact sheet. Nairobi: International Livestock Research Institute.

Jahnke H. E. 1982. *Livestock production systems and livestock development in tropical Africa*. Kiel: Kieler Wissenschaftsverlag Vauk.

Jefatura del Estado. 1995. Ley 3/95, de 23 de marzo, de Vías Pecuarias. Official State Gazette No. 71, 24 March.

Jodha, N. 2000. Livestock in mountain/highland production systems: Challenges and opportunities. In *Contribution of livestock to mountain livelihoods: Research and development Issues*. ed. P. Tulachan, M. Saleem, J. Maki-Hokkonen, and T. Partap. Kathmandu, Nepal: International Centre for Integrated Mountain Development.

Kay, C., and M. Urioste. 2005. Bolivia's unfinished agrarian reform: Rural poverty and development policies. In *Land rights reform and governance in Africa: How to make it work in the 21st Century?* Land, Poverty and Public Action Policy Paper No. 3. The Hague, Netherlands: United Nations Development Programme and Institute of Social Studies.

Kerven, C. 1992. *Customary commerce: A historical reassessment of pastoral livestock marketing in Africa*. London: Overseas Development Institute.

Kerven, C., ed. 2003. *Prospects for pastoralism in Kazakstan and Turkmenistan: From state farms to private flocks*. London: Routledge Curzon.

Kerven, C. 2006. *Review of the literature on pastoral economics and marketing: Central Asia, China, Mongolia and Siberia*, Report for the World Initiative for Sustainable Pastoralism. Nairobi: World Conservation Union–IUCN.

Kerven, C., J. Channon, and R. Behnke. 1996. *Planning and policies on extensive livestock development in Central Asia*. Working Paper No. 91. London: Overseas Development Institute.

Kerven, C., S. Aryngaziev, N. Malmakov, H. Redden, and A. Smailov. 2005. *Cashmere marketing: A new income source for central Asian livestock farmers*. Davis: Global Livestock Collaborative Research Support Program, University of California, Davis.

Khan, U. N., and M. Iqbal. 2000. *Role and the size of the livestock sector in Afghanistan*. Commissioned by the World Bank, Islamabad.

Kibue, M. 2006. *Challenges in the development of a functioning livestock marketing chain in Kenya—A best practice case study in farming systems and poverty: Making a difference*. Proceedings of the 18th International Farming Systems Association: A Global Learning Opportunity. Rome: International Farming Systems Association.

Kratli, S. 2000. The bias behind nomadic education–Western bias. *UNESCO Courier*, October.

Laaziz, D. 2005. Small ruminant breeds of Algeria. In *Characterization of small ruminant breeds in West Asia and North Africa. Vol 2: North Africa*. ed. L. Iniguez. Aleppo, Syria: International Center for Agricultural Research in Dry Areas.

Lane, C. 1990. *Barabaig natural resource management: Sustainable land use under threat of destruction*. Discussion Paper 12. Geneva: United Nations Research Institute for Social Development.

Lane, C., and R. Moorehead. 1995. New directions in rangeland resource tenure and policy. In *Living with uncertainty: New directions in pastoral development in Africa*, ed. I. Scoones. London: Intermediate Technology Publications.

Leon-Velarde, C., R. Quiroz, P. Zorogastua, and M. Tapia. 2000. Sustainability concerns of livestock-based livelihoods in the Andes. In *Contribution of livestock to mountain livelihoods: Research and development issues*, ed. P. Tulachan, M. Saleem, J. Maki-Hokkonen, and T. Partap. Kathmandu, Nepal: International Centre for Integrated Mountain Development.

Lister, S. 2004. *Pastoralism. Governance, services and productivity: New thinking on pastoralist development*. Brighton, U.K.: Institute of Development Studies.

Little, P. 2001. The global dimensions of cross-border trade in the Somalia borderlands. In *Globalization, democracy, and development in Africa: Future prospects*, ed. T. Assefa, S. Rugumamu, and A. Ahmed. Addis Ababa, Ethiopia: Organization for Social Science Research in Eastern and Southern Africa.

Little, P. D., and H. A. Mahmoud. 2005. *Cross border cattle trade along the Somalia/Kenya and Ethiopia/Kenya borderlands*. Global Livestock Collaborative Research Support Program.

Ly, A., and M. Niamir-Fuller. 2005. La propriété collective et la mobilité pastorale en tant qu'alliées de la conservation—expériences et politiques innovatrices au Ferlo (Sénégal). *Policy Matters* (IUCN Commission on Environmental, Economic and Social Policy) 13:162–173.

MacDonald, M., and E. Azumi. 1993. *Wildebeests and wheat: Crafting a land policy in Kenya's Maasailand*. Nairobi: Ministry of Planning and National Development.

Madani, T., H. Yakhlef, and N. Abbache. 2003. Les races bovines, ovines, caprines et camelines. In *Evaluation des besoins en matière de renforcement des capacités nécessaires à la conservation et l'utilisation durable de la biodiversité importante pour l'agriculture en Algérie, Alger, 22–23/01/2003*. ed. A. Abdelguerfi and S. Ramdane.

MADR (Ministère de l'Agriculture et du Développement Rural du Maroc). 2003. *Atlas sur les réalisations d'aménagement et d'amélioration des terrains de parcours*. Rabat, Morocco.

Malpai Borderlands Group. 2007. Home page. http://www.malpai-borderlandsgroup.org.

Mangas Navas, J. M. 1992. Vías pecuarias. *Cuadernos de la Trashumancia*, n° 0. ICONA. FEPMA. Madrid: Ministry of Environment and Rural Affairs.

Markakis, J. 2004. *Pastoralism on the margin*. London: Minority Rights Group International.

Masri, A. 2001. Country Pasture/Forage Resource Profiles for the Syrian Arab Republic. Rome: Food and Agriculture Organization.

McGahey, D., J. Davies, and E. Barrow. 2008. Pastoralism as conservation in the horn of Africa: Effective policies for conservation outcomes in the drylands of Eastern Africa. *Annals of Arid Zones* 46:353–377.

McPeak, J. 2002. Contrasting income shocks with assets shocks: Livestock sales in northern Kenya. Sixth Annual Conference of the Center for Study of African Economies, Oxford.

McPeak, J. G., and C. B. Barret. 2001. Differential risk exposure and stochastic poverty traps among East African pastoralists. *American Journal of Agricultural Economics* 83:674–679.

McPeak, J., and P. Little. 2006. *Pastoral livestock marketing in eastern Africa: Research and policy challenges.* Essex, U.K.: Intermediate Technology Development Group.

Mearns, R. 1993. Territoriality and land tenure among Mongolian pastoralists: Variation, continuity and change. *Nomadic Peoples* 33:73–103.

Mearns, R., and J. Swift. 1996. Pasture tenure and management in the retreat from a centrally planned economy in Mongolia. In *Rangelands in a Sustainable Biosphere*, ed. N. West, 96–98. Denver, CO: Society for Rangeland Management.

Motzfeldt, G. 2005. *Issue Paper on decentralisation and local governance.* Working Paper. Oslo: The Development Fund.

Muhereza, E., and S. Ossiya. 2004. Pastoralism in Uganda—People, environment, and livestock: Challenges for the PEAP. Kampala. Uganda National NGO Forum and Civil Society Political Task Force.

Mwangi, E. 2007. *Socioeconomic change and land use in Africa: The transformation of property rights in Maasailand.* Basingstoke, U.K.: Palgrave Macmillan Ltd.

Nasr, N., M. Ben Salem, Y. Lalaoui Rachidi, J. Benissad, and Y. Medouni, 2000. Changes in livestock production and collective rangeland management systems in arid region: A case study in El Ouara–Tataouine region of Tunisia. *Science et changements planétaires / Sécheresse* 11:93–100.

Nedjraoui, D. 2001. *FAO country pasture profile for Algeria.* Rome: Food and Agriculture Organization.

N'Gethe, J. C. 1992. Group ranch concept and practice in Kenya with special emphasis on Kajiado District, FAO. In *Proceedings of the Workshop held at Kadoma Ranch Hotel, Zimbabwe 20–23 July 1992*, ed. J. A. Kategile and S. Mubi. Rome: Food and Agriculture Organization.

Niamir-Fuller, M. 1999. *Managing mobility in African rangelands: The legitimization of transhumance.* London: Intermediate Technology Publications Ltd.

Nimbkar, C. 2006. Conservation of livestock biodiversity. Invited lecture in the Biodiversity Awareness Workshop on Animal Genetic Resources and Conservation, National Biodiversity Authority of India. National Bureau of Animal Genetic Resources, Karnal, 22–23 April.

Nori, M., M. Kenyanjui, M. A. Yusuf, and F. H. Mohamed. 2006. Milking drylands: The marketing of camel milk in North-East Somalia. *Nomadic Peoples* 10:5–28.

Norton-Griffiths, M., M. Y. Said, S. Serneels, D. S. Kaelo, M. Coughenour, R. H. Lamprey, D. M. Thompson, and R. S. Reid. 2006. Land use economics in the Mara Area of the Serengeti ecosystem. In *Serengeti 3: Human Wildlife Interactions.* Chicago: Chicago University Press.

Nyariki, D. M. 2004. *The contribution of pastoralism to the local and national economies in Kenya.* Nairobi: RECONCILE/International Institute for Environment and Development.

Orozco, M., and Carana Corporation. 2006. *West African financial flows and opportunities for small businesses.* Washington, DC: United States Agency for International Development.

Oxfam. 2005. *Beyond the mainstream: Education and gender equality series. Programme insights.* Oxford: Oxfam.

Panin, A. 2000. A comparative economic analysis of smallholder cattle and small ruminant production systems in Botswana. *Tropical Animal Health and Production* 32:189–196.

Panin, A., and M. Mahabile. 1996. Profitability and household income contribution of small ruminants to small-scale farmers in Botswana. *Small Ruminant Research* 25:9–15.

Perrier G. 1995. New directions in range management planning in Africa. In *Living with uncertainty: New directions in pastoral development in Africa*, ed. I. Scoones, 47–57. London: Intermediate Technology Publications.

Perry, B., A. N. Pratt, K. Sones, and C. Stevens. 2005. *An appropriate level of risk: Balancing the need for safe livestock products with fair market access for the poor.* PPLPI Working Paper No. 23. Nairobi: International Livestock Research Institute.

Posner, R. A. 1980. Anthropology and economics. *Journal of Political Economy* 88:608–616.

Potkanski, T. 1993: Decollectivization of the Mongolia pastoral economy (1991–1992): Some economic and social consequences. *Nomadic Peoples* 33:123–135.

Pryor, F. L. 1977. *The origins of the economy: A comparative study of distribution in primitive peasant economies.* New York: Academic Press.

Putnam, R. D. 2000. *Bowling alone: The collapse and revival of American community.* New York: Simon Schuster.

Republic of Kenya. 2000. Livestock marketing from pastoral areas: A strategy for pastoralist development. Arid Lands Resources Management Project (ALMP) in conjunction with SNV, OXFAM, and World Concern, Office of the President, Nairobi.

Reusse, E. 1982. Somalia's nomadic livestock economy: Its response to profitable export opportunity. *World Animal Review* 43:2–11.

Sadoulet, E., and A. de Janvry. 1995. *Quantitative development policy analysis.* Baltimore, MD: John Hopkins University Press.

Sandford, S. 1983. *Management of pastoral development in the Third World.* Chichester, U.K.: John Wiley and Sons.

Savory, A. 1999. *Holistic management: A new framework for decision-making.* Washington, DC: Island Press.

Schelling, E., K. Wyss, M. Béchir, D. Moto, and J. Zinsstag. 2005. Synergy between public health and veterinary services to deliver human and animal health interventions in rural low income settings. *British Medical Journal* 331:1264–1267.

Schmidt, S. 2003. The role of herder communities and non-government organizations in shaping the governance of two protected areas in Mongolia. Presentation to the Mobile Peoples Workshop, Governance Stream, World Parks Congress 2003, Durban, South Africa, 8–17 September.

Scoones, I., ed. 1995. *Living with uncertainty: New directions in pastoral development in Africa.* London: Intermediate Technology Publications.

Scoones, I., and W. Woolmer. 2006. Livestock, disease, trade and markets: Policy choices for the livestock sector in Africa. IDS Working Paper 269. Brighton, U.K.: Institute of Development Studies.

Shackleton, S., C. Shackleton, and B. Cousins. 2000. Revaluing the communal lands of southern Africa: New understandings of rural livelihoods. Natural Resource Perspectives No. 62. London: Overseas Development Institute.

Sidahmed A., A. Abdouli, M. Hassani, and M. Nourallah. 2000. *Sheep production systems in the Near East and North Africa (Nena) region: The experience of IFAD in alleviating technical, socio-economic and policy constraints.* Rome: International Fund for Agricultural Development.

Sikana, P., and C. Kerven. 1991. *The impact of commercialization on the role of labor in African pastoral societies.* Pastoral

Development Network Set 31. London: Overseas Development Institute.

Singh, C., and I. Köhler-Rollefson. 2005. *Sheep pastoralism in Rajasthan: Still a viable option?* Workshop report, 31 January–1 February. Rajasthan, India: Lokhit Pashu-Palak Sansthan.

Staal, S., and M. Jabbar. 2000. Markets and livestock in the coming decades: Implications for smallholder highland producers. In *Contribution of livestock to mountain livelihoods: Research and development issues.* ed. P. Tulachan, M. Saleem, J. Maki-Hokkonen, and T. Partap, 57–70. Kathmandu, Nepal: International Centre for Integrated Mountain Development.

Stevens, C., and J. Kennan. 2005. *Botswana beef exports and trade policy.* Brighton, U.K.: Institute of Development Studies.

Stinner, D., B. Stinner, and E. Martsolf. 1997. Biodiversity as an organizing principle in agroecosystem management: Case studies of holistic resource management practitioners in the USA. *Agriculture, Ecosystems and Environment* 62:199–213.

Strasberg, P. J., T. S. Jayne, T. Yamano, J. Nyoro, D. Karanja, and J. Strauss. 1999. *Effects of agricultural commercialization on food crop input use and productivity in Kenya.* International Development Working Paper No. 71. East Lansing: Michigan State University.

Swaid, A. 1997. Syria country paper. In *Global agenda for livestock research: Proceedings of a consultation on setting livestock research priorities in West Asia and North Africa (WANA) region.* ed. E. F. Thomson, R. von Kaufmann, H. Li Pun, T. Treacher, and H. van Houten, 12–16 November 1997. International Center for Agricultural Research in the Dry Areas, Aleppo, Syria.

Swift, J. 1988. *Major issues in pastoral development with special emphasis on selected African countries.* Rome: Food and Agriculture Organization.

Swift, J., T. Heatherington, C. Lussier, T. Farvar, and T. Diao. 2003. *Pastoralism and mobility in the drylands: The global drylands imperative.* New York: United Nations Development Programme.

Thébaud, B. 1990. Politiques d'Hydraulique Pastorale et Getion de l'Espace au Sahel. *Cahier Sciences Humaines* 26:13–31.

Thébaud, B. 2001. Droit du Communage ("Commons") et pastoralism au Sahel: Quel avenir pour les eleveurs saheliens. In *Politics, Property and Production in West African Sahel,* ed. T. A. Benhaminsen and C. Lund, 163–81. Uppsala, Sweden: Nordiska Africainstitutet.

Thomson, E. F. 1997. Small ruminant research in the medium-term plan of ICARDA. In *Global agenda for livestock research: Proceedings of a consultation on setting livestock research priorities in West Asia and North Africa (WANA) Region.* ed. E. F. Thomson, R. von Kaufmann, H. Li Pun, T. Treacher, and H. van Houten, 12–16 November 1997. International Center for Agricultural Research in the Dry Areas, Aleppo, Syria.

Tiviski. 2006. *Tiviski: Welcome to a dairy in the desert.* Nouakchott, Mauritania. http://www.tiviski.com.

Toulmin, C. 2006. *Securing land and property rights in Africa: Improving the investment climate—Global competitiveness report, 2005–06.* Davos, Switzerland: World Economic Forum.

Toulmin, C., P. Lavigne Delville, and S. Traore. 2002. *The dynamics of resource tenure in West Africa.* London: International Institute for Environment and Development/GRET.

Uganda Land Alliance 2000. *Land rights of the Karamojong pastoral minority in Uganda.* Kampala: Uganda Land Alliance.

UNCBD (United Nations Convention on Biological Diversity). 2001. *Lessons learnt from case studies on animal genetic resources.* Montreal: UNCBD Secretariat.

UNDP (United Nations Development Programme). 1998. Cessation of livestock exports severely affects the pastoralist economy of Somali Region, Mission report of 31 March to 7 April. New York: UNDP Emergencies Unit for Ethiopia.

UNDP (United Nations Development Programme). 2003. *Pastoralism and mobility in the drylands.* Global Drylands Imperative Challenge Paper Series. New York: United Nations Development Programme.

UNESCO. 2005. *Education for all: Global monitoring report.* Paris: UNESCO.

UNICEF. 2003. Kenya field trip diary. London: UK Committee for UNICEF.

United Republic of Tanzania. 2005a. National Strategy for Growth and Reduction of Poverty (NSGRP). Vice President's Office, June.

United Republic of Tanzania. 2005b. *The Tanzania Development Vision 2025.* Planning Commission, Government of Tanzania.

Wagle, N., and D. Pathak. 1997. Conservation success behind famine. *Kathmandu Post* September 10.

Werner, W. 2003. Land reform in Namibia: Motor or obstacle of democratic development. Paper presented at a meeting on land reform in southern Africa (Motor or Obstacle of Democratic Development), Friedrich Ebert Foundation, Berlin.

Western, D. 1982. The environment and ecology of pastoralists in arid savannahs. *Development and Change* 13:183–211.

Western, D., and V. Finch. 1986. Cattle and pastoralism: Survival and production in arid lands. *Human Ecology* 14:77–94.

World Bank. 2001. *Toward an understanding of vulnerability in rural Kenya.* http://www1.worldbank.org/sp/safetynets/BBL_VulinKenya_12-01.asp.

World Bank. 2005. *Kazakhstan's livestock sector—Supporting its revival.* Joint Sector Work of the Joint Economic Research Program, Government of Kazakhstan and World Bank.

Yahuza, M. L. 2001. Smallholder dairy production and marketing constraints in Nigeria. In *Smallholder dairy production and marketing: Opportunities and constraints,* eds. Rangneker, D. and w. Thorpe Ed. Proceedings of a South–South workshop held at National Dairy Development Board (NDDB), Anand, India, 13–16 March 2001.

Yirbecho, A., C. Barrett, and G. Gebru. 2004. *Resource conflict in the rangelands: Evidence from Northern Kenya and Southern Ethiopia.* USAID Global Livestock CRSP Research Brief 04-08-PARIMA. Washington, DC: U.S. Agency for International Development.

Zaal, F., and A. Poderman. 2000. *High risk, high benefits: International livestock trade and the commercial system in Eastern and Southern Africa.* Paper presented at the OSSREA Sixth Congress on Globalization, Democracy and Development in Africa: Future Prospects, 24–28 April, Dar es Salaam, Tanzania.

Zhaoli, Y., W. Ning, Y. Dorji, and R. Jia. 2005. A review of rangeland privatization and its implications in the Tibetan plateau, China. *Nomadic Peoples* 9:1–2.

Part III
Responses
Responses to Livestock in a Changing Landscape

Henning Steinfeld

This final section provides an overview of various segments of society and how they deal with and respond to livestock in a changing landscape. Different stakeholders have been responding to each of the six broad issues identified in Part II (Consequences of Livestock Production). The stakeholders can be grouped into the public sector, the corporate private sector, consumers, and civil society. These groups represent different interests, sometimes cooperating and sometimes conflicting, and they use different instruments in pursuit of their objectives.

The public sector response takes the form of rules and regulations, incentive frameworks, institutional development, capacity building, research and development, information, and awareness raising. During the livestock revolution, the public sector has placed most emphasis on facilitating sector economic development and paid considerably less attention to reducing externalities. The overuse of common pool resources, pollution from livestock waste, and emerging diseases are three issues that reflect important market failures but do not receive a vigorous public sector response even in many industrial countries. The issue of smallholder exclusion, in contrast, has received attention from the public sector, and a lot of past and current sector protection is motivated by the desire to keep farmers on their land and reduce competitive pressure between domestic and external producers.

The main mechanism by which the private sector responds to changing demand and resource availability is innovation and adaptation. In fact, most of the production increase has been achieved through productivity advances thanks to the adoption of advanced technology. The corporate sector obviously has an interest in expanding demand and is advertising accordingly. The private sector, however, consists of rather distinctly different subsets of producers—small-scale and large-scale, subsystems and market-oriented—whose objectives and strategies are sometimes opposed.

Civil society and consumers at large form an increasingly powerful group of stakeholders, particularly those in wealthy and middle-income groups. In fact, it can be argued that the balance of power has progressively shifted from producers to consumers over the last

decades. Issues of concern to civil society and consumers include human health and rights protection, animal welfare, environmental protection, the protection of local community identities, and a general unease about large corporations and their dealings. These issues are particularly pronounced in industrial countries but are in no way limited to them. Issues taken up by civil society and consumers often instigate change at the public policy level, feeding back into the production process.

From the discussion in Part II of the many consequences of the livestock revolution, four major issues emerge. While acknowledging the many positive contributions of livestock to economic growth and livelihoods, this chapter pays particular attention to the issues of concern, where more effective responses will be needed. First, livestock production—both the modern, intensive form and the traditional, extensive one—has a large and varied environmental impact. Where the former is not well managed, impacts are characterized by pollution of soil and water from livestock waste as well as the impacts associated with concentrate feed production. In the areas where extensive systems are under greatest pressure, problems emerge from overuse of resources and low efficiency of use, in particular of common property resources such as land and water. Livestock's impact on climate change, caused by both intensive and extensive forms of production, is another important feature. Many of these resource issues reflect market failures, with individuals not bearing the full costs of their actions, which requires correction through public policies. These are being developed only slowly, failing to keep up with the pace of economic development, and it is often through pressure from the neighborhood and environmental groups that such action is instigated. Chapter 16 considers the range of responses to these and other environmental issues.

Second, livestock products play a part in both underconsumption and overconsumption and the associated human health consequences. Livestock products are among the most desired food items in most societies, and the amount and variety of consumption typically increases with income. With exception of specialized and highly livestock-dependent groups like pastoralists, poor people consume insufficiently low amounts of meat, milk, and eggs and are often under- or malnourished. In these situations, some governments have proposed livestock development and increased supply and introduced such measures as school milk programs. In contrast, people with medium or high incomes often consume excessive amounts of food, and some livestock products have been singled out as harmful to human health when consumed in large quantities. The importance of these health issues is reflected in private and public standards setting and in production differentiation. The responses to these two opposite ends of the consumption spectrum are addressed in Chapter 17.

Third, human and animal health issues related to increasing human and animal densities, their close association, growing production intensities, and changing ecologies are another growing concern. In addition, new post-pathogen complexes are developing in association with livestock, and novel diseases in both humans and animals are emerging. Food safety is of growing concern. While this seems to give rise to a certain level of general anxiety, evidenced by consumer reaction in the case of "food scares," specific responses on public human and animal health systems and from producers take time to evolve and adjust. Chapter 18 looks at current and possible responses to these issues.

Fourth, the limited capacity of smallholders, particularly in developed and emerging economies, to participate in livestock sector growth has been well documented. There has been social fallout from the livestock revolution through increasing marginalization of smallholders in the face of scale economies in monogastric systems and heightened market barriers stemming from food safety and quality standards. Indeed, the most common response of smallholder producers to the livestock revolution in fast-growing economies has been to leave the sector. Chapter 19 considers this and other responses to the social issues raised in this volume.

Chapter 20 pulls together the lessons learned to date in dealing with the environmental, health, and social issues of livestock in today's changing landscape, noting that the challenges facing the sector cannot be solved by isolated interventions. The framework for responses of these challenges must be a holistic approach that takes account of the many social, economic, and ecological settings in the livestock sector and that recognizes the social, health, and environmental consequences of change.

16

Responses on Environmental Issues

Henning Steinfeld, Pierre Gerber, and Carolyn Opio

Main Messages

- **Livestock's impact on the environment has been largely negative, and neither policy nor technology has caught up with the problem.** Technical and policy options to mitigate livestock's impact on biodiversity loss are largely untapped, and options to reduce emissions are largely unexploited.

- **Livestock can be part of successful landscape management in extensive forms and can contribute to biodiversity and water management.** However, this role has been largely limited to areas where large herbivores have been part of the evolution of ecosystems. In Africa and Latin America, the pressures on land are such that it will take time and economic incentives to reverse the trends toward land degradation. In industrial countries, while intensification brings its share of environmental challenges, it has also freed up pastureland for reconversion to natural habitats.

- **Increasing water efficiency, water productivity, and prevention of pollution could be greatly improved through improved production practices and regulatory frameworks.**

- **Responses could be improved in several ways, targeted toward specific livestock systems.** In pastoral systems, improved land use rights reflecting greater pressures on land, complemented by technical improvements to improve the productivity of pasture, could help livestock producers adapt to climate change. Landowners in fragile environments of dryland areas and the humid tropics need to be rewarded for conservation and for providing environmental services in the form of forest cover, rather than only being paid for productivity. A shift toward monogastrics, when combined with correct manure management, can reduce greenhouse gas emissions from livestock. This may best be done by reruralizing the periurban market chains, requiring public investment in infrastructure but bringing the possible

benefits of retaining employment in rural areas. To address mixed farming systems where cattle are important, research is needed into suitable technical packages to reduce methane emissions.

- **As awareness and understanding about environmental issues grow, it is possible to imagine a Kyoto follow-up agreement that takes account of the livestock–environment interaction** and enhances the possibility for policies that favor environmentally sound livestock production and processing.

Introduction

The livestock sector depends on a variety of different natural resources. Most important, this includes land of different quality and in different agroecological zones, mostly in the form of grazing land but also as arable land for feed production. Land is further characterized by vegetation and the biodiversity it supports. Water is another natural resource of critical importance, as extensive livestock production is severely constrained by lack of water in many places.

As described in earlier chapters on the consequences of livestock production, livestock affect the global climate, water resources, and biodiversity in major ways. To recap, expansion of pasture is a chief factor in deforestation in Latin America. The production of concentrate feed claims about one-third of total arable land and contributes to water pollution and depletion. Pasture use and the production of feed are associated with land degradation, habitat destruction, and greenhouse gas emissions. Livestock are an important contributor to water pollution, particularly in areas of high animal densities. Both extensive and intensive forms of production contribute to environmental degradation but, overall, extensive production tends to have larger environmental impacts per unit of output.

Extensive production is practiced by many poor producers who use natural grasslands, crop residues, and household and other waste as low-cost feed. A large part of the world's pastureland is degraded, thereby releasing greenhouse gases, negatively affecting water cycles, and affecting vegetation and biodiversity. Pasture expansion into forests, often conducted by smallholders, has important consequences for climate change and biodiversity. Pastoralists in particular are threatened, as they lose access to traditional grazing areas and will be especially affected by climate change. Extensive production is often based on poorly fed ruminants with low productivity, emitting significant amounts of methane.

The environmental problems of intensive production are linked to the production of concentrate feed, the use of fossil fuel, and the disposal of animal waste. Feed grain production usually requires intensively used arable land and the use of fertilizer, pesticides, fossil fuel, and other inputs, affecting the environment in diverse ways. Since more than two-thirds of the nutrients fed to animals are excreted, animal waste is a leading factor in the pollution of land and water resources. Animal waste also emits methane and nitrous oxide, a particularly potent greenhouse gas.

All these environmental issues related to livestock have been largely ignored by policy makers.

Land

Part I documented the rapid expansion of pasture areas in the nineteenth and twentieth centuries and the fact that for most marginal land areas, pasture area has reached its maximum expanse. Pasture area continues to expand into forest area in the humid and subhumid tropics, in particular in Latin America and to a limited extent in Africa.

Policies throughout most of second half of the twentieth century attempted to "sedentarize" mobile livestock production, which was regarded as a backward type of production and lifestyle. More recently, there is increasing recognition of some of the intrinsic advantages of pastoralism and a better understanding of their adaptation to ecosystems in "nonequilibrium." However, as described in Chapter 15, many different factors contribute to the gradual demise of pastoralism, including the expansion of irrigated agriculture, the de facto privatization of land, and increasing conflicts with other groups of land users—conflicts pastoralists tend to lose. As a result, pastoralism is in decline in many areas.

Chapter 10 describes the varied environmental impact on terrestrial ecosystems caused by extensive livestock systems, including pastoral, ranching, and low-input mixed farming systems. In many cases, overgrazing, grazing in unsuitable environments, and deforestation led to erosion and other forms of land degradation.

Chapter 10 also discusses the establishment of access rights to grazing resources as a required key institutional change to reverse the negative environmental impact of livestock on land. Such a rights-based approach needs to be complemented by technical improvements such as reseeding and the introduction of legumes, grazing rotations, and cut-and-carry systems.

In West Asia and North Africa, overgrazing is the most common cause of range degradation. The total area of grazing land is reported to decline in all countries of the region as a result of expansion of crops and desertification. Like elsewhere, land tenure and access rights are key to sustainable management of rangelands.

In Central Asia, fundamental institutional reforms have had a profound impact on livestock production systems since 1990. These policy changes were not uniform; some countries rapidly privatized land, whereas others kept rangelands and even animals under state property. As a result, trends in grazing pressure and pasture degradation are varied; the best results have been obtained in Kazakhstan, where livestock were distributed spatially and where livestock mobility prevents excessive grazing pressure.

In Australia, the introduction of extensive grazing systems and their initial success was followed by high stocking rates. Recurrent droughts led to crashes in vegetation and stocking rates in some places. Here the challenge lies in devising government policies for land stewardship that combine productivity and conservation.

The Common Agricultural Policy of the European Union (EU) has increasingly recognized the role of extensive livestock systems in sustaining environmentally friendly and sustainable rural development. Grazing pressures are not particularly high, but abandonment and the collapse of traditional management practices have led to vegetation changes, in particular shrub encroachment and spontaneous reforestation. Policies in the EU focus less on the productive value of these marginal lands and more on landscape management and cultural heritage.

Livestock and Climate Change

The role of livestock in climate change, through affecting carbon and nitrogen cycles, is increasingly acknowledged, but very little is done to address it. (See Chapters 5 and 6.)

Carbon emissions (carbon dioxide and methane) from livestock production can be reduced in four main ways. First, pasture and feed crop expansion at the expense of forest can be halted or slowed mainly through area protection, selective infrastructure development and appropriate land titling procedures that stem speculation. This is currently facing serious obstacles as favorable prices in combination with insufficient protection and enforcement continue to lead to widespread deforestation, which particularly in Latin America can be related to expansion of pastures and arable area for feed crops (especially soybean). The particular problems with ongoing deforestation are documented for Brazil, and the possible solutions to the deforestation issue are described for Costa Rica in the case study volume (Ibrahim et al., 2009). The latter shows that there is a prospect that with growing population densities and crop cultivation intensities the "forest transition" will set in, whereby the most productive areas continue to be cultivated or grazed, but previously deforested land that is only marginally suitable for agricultural purposes is turned back to forest.

Second, improved management can maintain vegetation cover and enhance carbon sequestration and storage; carefully adjusting grazing pressure to climatic fluctuations, in particular drought events, is especially important. (See Chapter 5.) Such measures face a series of obstacles under current regimes. Lowering grazing pressure may represent a significant loss of revenue in commercial enterprises. Under communal or open access grazing this obstacle is compounded in that restoration benefits do not accrue to the individual, and collective action is required, which is often impeded by institutional weaknesses. Pasture restoration is costly, particularly in drylands, and carbon sequestration rates are low. Payment for environmental services schemes (FAO 2007) may therefore target the conservation of existing carbon pools, which is more cost-effective than recovering them after they are lost. (See Chapter 5.)

Third, productivity-enhancing measures can be used that will reduce the proportion of feed used for maintenance of an animal as opposed to that used for production. This is of particular relevance for animals of low productivity, which are the majority of ruminants kept in traditional livestock production systems in developing countries. High-performing animals, making use of improved genetics, balanced feed and nutrition, and good animal health and husbandry will all help to reduce methane emissions per unit of output. Another way to reduce methane emissions is to switch to monogastric animals (poultry, pigs) as suppliers of meat rather than ruminants (bovins, ovines). This trend is already ongoing practically everywhere—not in response to emission problems but because of the better feed conversion and therefore lower production costs of monogastrics.

Fourth, there is substantial potential in mitigating methane emissions through improved manure management and anaerobic digestion to produce biogas, particularly from intensive livestock operations. Balanced feeding, higher carbon-to-nitrogen ratios in feed, and certain storage methods are among the measures that can be adopted at the production stage to lower methane emissions. However, biogas production through anaerobic digestion is the most effective way of reducing methane emissions from manure, decreasing emissions by 50–75%. The use of biogas is currently limited by the still lower price of alternative sources of energy, and most biogas plants are found in remote areas or where subsidies are being paid.

As for nitrogen, there are numerous ways to contain its load to the environment and reduce the release of nitrous oxide into the atmosphere. Along the food chain, these are the main technical options:

- Increasing the efficiency of nitrogen fertilization of feed crops on arable land, or on pastures, by improved application methods and fertilizer forms.
- Balancing feed rations to minimize the amount of excess nitrogen and the use of enzymes and synthetic amino acids to enhance nitrogen absorption by the animal.
- Using anaerobic digestion (e.g., for biogas production) of animal manure to substantially reduce nitrous oxide emissions, in addition to methane, as already discussed.
- Improving manure application methods on fields, whereby the manure is rapidly incorporated or injected into the soil, and adjusting manure applications in time, amounts, and form to crop physiology and climate.

For grazing animals, nitrous oxide emissions can be limited by avoiding overstocking pastures and preventing late fall and winter grazing.

Measures such as these will not be applied, however, if economic conditions and regulatory frameworks are not supportive of higher nitrogen efficiency. Industrial countries have established detailed regulatory frameworks to limit nitrogen surpluses on agricultural land, such as the EU nitrate directive, and important improvements have been made (see, for instance, the example of Denmark discussed by Mikkelsen et al., 2009). As these are motivated by concerns over water and soil pollution, and not by worries over climate change, they are discussed later in this chapter.

While livestock account for an important share of greenhouse gas emissions in many countries, it is apparent from the country emission reports submitted to the United Nations Framework Convention on Climate Change (UNFCCC) (National Reports, UNFCCC) that mitigation efforts tend to focus on other sectors. This is probably because of the technical difficulties related to assessing and certifying emissions from agricultural and land use, land use change, and forestry (LULUCF) sectors. The Kyoto Protocol included the Clean Development Mechanism (CDM) for creating "certified emissions reductions" (CERs) that can be traded on the carbon market. The CDM is a facility for industrial countries to reduce net carbon emissions by promoting renewable energy, energy efficiency, or carbon sequestration projects in developing countries, receiving CERs in return. Its purpose is to help industrial countries (Annex 1) meet their obligations under the Kyoto Protocol while promoting sustainable development in developing countries. For LULUCF projects, only afforestation or reforestation initiatives are recognized as being permissible during the Kyoto Protocol's first commitment period (2008–12).

A critical factor concerning CDM transactions is an active international market for CERs, which requires partnerships between several agents: project developers, investors, independent auditors, national authorities in host and recipient countries, and the international agencies that are responsible for implementation of the Kyoto Protocol (Mendis and Openshaw 2004).

Since the protocol's ratification in February 2005, a considerable number of projects have been registered. These are mostly based on predefined methodologies. Established methodologies in the livestock sector concern only emissions from the recovery of methane (as a renewable energy source). Scope exists for other types of projects aimed at mitigation of livestock emissions through intensification of production. For example, improving rumen fermentation efficiency through the use of better-quality feed could substantially reduce emissions from the huge Indian dairy sector (Sirohi and Michaelowa 2004). For this, credit (through, e.g., microfinance institutions), effective marketing, incentives, and promotional campaigns are required for broad acceptance of related technologies (Sirohi and Michaelowa 2004).

Afforestation or reforestation initiatives are the only land use change projects that are currently eligible. However, they offer great potential for mitigating livestock's footprint on climate change by returning marginal or degraded pastures back to forest. Other potential methods that could significantly reduce emissions but do not yet qualify for eligibility include forms of pasture restoration and improvement (such as silvopastoral land use), reduced grazing, and technical improvements.

The effects of "leakage" may substantially raise the costs of carbon sequestration (Richards 2004) and undermine the effectiveness of emission reduction schemes. "Leakage" occurs when the effects of a program or project lead to a countervailing response beyond its boundaries. This problem arises from two basic facts. First, land can be shifted back and forth between various forestry and agriculture uses. Second, the overall balance of activities on land will depend on the relative prices in the agriculture and forestry sectors. This is because individual projects and programs do little to change prices or the resulting demand for land. For example, if forestland is preserved from harvest and conversion in one location, the unchanged demand for agricultural land and forest products could lead to increased forest clearing and conversion in another region. Thus the effects of the preservation may be partially or entirely undone by the leakage. Similarly, if agricultural land is converted to forest stands, the underlying demand for farmland may simply cause other forested land to be converted back to agriculture.

The potential for incremental accumulation of organic carbon in soils is huge, and adapting extensive livestock systems is the key to unlocking this potential. Technical options to revert pasture degradation and sequester carbon, particularly in the soil, by building up organic matter in the ground exist, and current pastures are probably the largest terrestrial carbon sink available.

However, the same issues described for afforestation or reforestation activities also apply here: "leakage," the pursuit of multiple goals, sustained government commitment, and so forth. The benefits accrue over a period of decades; in many cases, peak carbon uptake rates occur only after 20–40 years. Landowners who make these investments will no doubt want to know whether the government will still be rewarding carbon sequestration long into the future when their activities come to fruition. Government needs to be able to make credible commitments to provide stable incentives over long periods.

While currently not eligible under the CDM, a most serious effort needs to be made to allow for certified emissions reductions from rehabilitation of degraded land and sustainable management of existing forest, be it under the CDM or in a different framework.

In the context of poorer developing countries, smallholders are a key group in achieving both the necessary scale and developmental as well as environmental goals. In the absence of policy interventions and external financial support, smallholders use improved management practices at individually optimal levels but at socially suboptimal levels. On the basis of case studies, FAO (2004a) concludes that substantial funds from development organizations or carbon investors will be needed if soil carbon sequestration projects in dryland small-scale farming systems are to become a reality.

In addition to these purely economic calculations, there is an ethical concern. Expecting local smallholders to adopt management practices at socially and globally optimal levels implies that they subsidize the rest of society in their respective countries as well as global society. If sustainable agriculture, environmental restoration, and poverty alleviation are to be targeted simultaneously on a large scale and over a longer period, then a more flexible and adaptive management and policy approach is needed. It should generate possibilities to strengthen farmers' own strategies for dealing with uncertainty while providing the necessary incentives.

Just as the livestock sector has large and multiple impacts on both C and N cycles, so are there multiple and effective options for mitigation. Although little is done, much can be done, but it requires a strong involvement of public policy. Most of the options are not cost neutral, and simply enhancing awareness will not lead to widespread adoption. Most mitigation options require the introduction of carbon constraints or equivalents, either in the form of cap-and-trade systems, such as those used in the EU, or taxation of emissions.

For carbon, the environmental focus needs therefore to be on addressing issues of land use change and land degradation. Here the livestock sector offers significant potential for C sequestration, particularly in the form of improved pastures.

In addressing land use change, the challenge lies in slowing and eventually halting and reversing deforestation. Vlek et al. (2004) consider that the only available option to free up the land necessary for C sequestration would be intensification of agricultural production on some of the better lands—for example, by increased fertilizer inputs. They demonstrate that the increased CO_2 emissions related to the extra fertilizer production would be far outweighed by the sequestered or avoided emissions of organic C related to deforestation. Apart from improved fertilizer use, other options for intensification include the use of higher-yielding, better-adapted varieties and improved land and water management.

A huge potential exists for net sequestration of C in cultivated soils. The C sink capacity of the world's agricultural and degraded soils is 50% to 66% of the historic carbon loss from soils of 42 to 78 gigatonnes (Lal 2004). There are proven new practices that can improve soil quality and raise soil organic C levels (e.g., conservation tillage and organic farming), which achieve yields comparable to conventional intensive systems.

Improved grassland management is another major area where soil C losses can be reversed, leading to net sequestration, through the use of trees, improved pasture species, fertilization, and other measures. Because pasture is the largest anthropogenic land use, improved pasture management could potentially sequester more C than any other terrestrial sink (IPCC 2000).

In the humid tropics, silvopastoral systems are one approach to C sequestration and pasture improvement, as shown by the Global Environment Facility and LEAD–sponsored Central American Silvo-pastoral Project (Pagiola et al., 2004). In dryland pastures, some aspects of dryland soils may help in C sequestration. Dry soils are less likely to lose C than wet soils, as lack of water limits soil mineralization and therefore the flux of C to the atmosphere. Consequently, the residence time of carbon in dryland soils is sometimes even longer than in forest soils. Although the rate at which C can be sequestered in these regions is low, it may be cost-effective, particularly taking into account all the side benefits for soil improvement and restoration (FAO 2004a).

The most efficient approach for reducing methane emissions from livestock is by improving the productivity and efficiency of livestock production through better nutrition, genetics, animal health, and general husbandry practices. Greater efficiency means that a larger portion of the energy in the animals' feed is directed toward the creation of useful products, so that CH_4 emissions per unit product are reduced. The trends toward high-performing animals and toward monogastrics and poultry,

in particular, are valuable in this context because they reduce CH_4 per unit of product.

A number of technologies exist to reduce CH_4 release from enteric fermentation. The basic principle is to increase the digestibility of feedstuff, either by modifying feed or by manipulating the digestive process. Examples of improvements in fibrous diets are the use of feed additives or supplements and the increased level of starch or rapidly fermentable carbohydrates in the diet (so as to reduce excess hydrogen and subsequent CH_4 formation).

More advanced technologies that are being studied include reduction of hydrogen production by stimulating acetogenic bacteria, defaunation (eliminating certain protozoa from the rumen), and vaccination (to reduce methanogens).

CH_4 emissions from anaerobic manure management can be readily reduced with existing technologies. Such emissions originate from intensive mixed and industrial systems that usually have the capacity to invest in such technologies. A first obvious option to consider is balanced feeding (increased C to N ratios). Additional measures include anaerobic digestion (producing biogas as an extra benefit), flaring/burning (chemical oxidation, burning), special biofilters (biological oxidation) (Melse and van der Werf 2005, Monteny et al., 2006), composting, and aerobic treatment. It is assumed that biogas can achieve a 50% reduction in emissions in cool climates, and higher in warmer climates, for manures that otherwise would be stored as liquid slurry.

An important nitrogen loss mitigation pathway lies in raising low animal N assimilation efficiency through more balanced feeding by optimizing proteins or amino acids to match the exact requirements. Improved feeding practices also include phased feeding and improving the feed conversion ratio through tailoring feed to physiological requirements. But even with these measures, manure still contains large quantities of nitrogen. The use of an enclosed tank can nearly eliminate N loss during storage and offers an important synergy with respect to mitigating CH_4 emissions and production of biogas; N_2O emissions from the subsequent spread of digested slurry can also be reduced.

The key to reducing N loss resulting from the application/deposition of manure is the fine-tuning of waste application to land with regard to environmental conditions, including timing as well as amounts and form of application in response to crop physiology and climate. Another technological option is the use of nitrification inhibitors that can be added to urea or ammonium compounds (Monteny et al., 2006). Some of these substances can potentially be used on pastures, where they act upon urinary N, an approach being adopted in New Zealand (Di and Cameron 2003).

In conclusion, livestock play an important role in both the C and N cycles. The contribution to the C cycle mainly stems from livestock's land use and role in land use change, in particular deforestation and pasture degradation. Livestock's role in the N cycle is mainly determined by their demand for concentrate feed and by livestock waste storage and disposal. As the scope for pasture expansion and intensification is limited, extensive livestock production is stagnating, but industrial livestock is growing rapidly. Subsequently, there is an ongoing shift toward a growing role of livestock in the N cycle and a stagnating or declining role in the C cycle, albeit from a very high level.

While important steps have been made in the recent past, the impact of livestock production on climate change is still not well understood. More detailed analysis is required to shape policies that can effectively mitigate environmental impact at every relevant step of the various commodity chains and help adapt to climate change. Research and development institutions need to generate substantial additional knowledge, enhance professional and institutional capacity, increase awareness, and develop, test, and replicate novel approaches.

Given the vast area livestock occupy and the strong growth and structural changes the sector is undergoing, animal agriculture offers great potential for high return investments that can simultaneously achieve substantial environmental benefits. In intensive systems, recovery of energy and nutrients from industrial livestock production yields environmental gains while increasing financial returns to producers, once economic distortions and information bottlenecks are removed. On the other hand, extensive grazing systems offer the opportunity to increase the provision of environmental services over vast areas (e.g., carbon sequestration, water, biodiversity conservation) while diversifying income of poor holders through payment schemes.

The areas with high potential returns include the following:

- *Addressing pastoral systems in dryland areas*, including the adaptation to climate change, and income diversification through the provision of environmental services, in particular carbon sequestration in arid rangelands, with specific attention to drought management. Given the large potential of certain grasslands to sequester carbon, such payments for environmental services (PES) could be important supplements for pastoralists and other users/managers of extensive grasslands.

- *Addressing grazing systems in the humid tropics* with a focus on mitigation and especially on the reduction of livestock's contribution to the deforestation and land degradation process. This would need to include the assessment and certification of the carbon sequestration capacity of grazing land under different forms of management and the reduction of methane emissions from enteric fermentation through different approaches, such as improved diets, feed additives, and increased intensification of the production.

- *Addressing mixed systems in sub-Saharan Africa and South Asia* with a focus on reducing methane emissions from dairy cows in low productivity production. This would include an assessment of methane emissions from enteric fermentation in different dairy production systems and the use of suitable technical packages for intensification and productivity growth while significantly reducing methane emissions per liter of milk produced. Through relatively simple measures, methane emissions per liter of milk can be cut by half in traditional dairy systems. An important component would be the development of suitable institutional arrangements—in particular, carbon trading schemes and CDM—to provide a changed incentive framework.

- *Addressing intensive monogastric production worldwide* with a direct focus on recovery and recycling of energy and nutrients. Obliging large-scale commercial producers to adhere to stringent environmental regulations has created a more level playing field where different types of producers compete. Furthermore, pushing industrial producers into rural areas has the advantage of creating economic stimuli and will lead to income and employment in forward and backward linkages. In addition, it will improve the periurban environment, where many poor people currently live.

- *Promotion of knowledge exchange, raising awareness, and transferring technology*, particularly in the areas of rehabilitation of degraded grazing land for poverty reduction and the provision of global and local environmental services (carbon sequestration, biodiversity protection, water); reduction of methane emissions from extensive ruminant production and supporting CDM-like mechanisms; and recovery and recycling of energy and nutrients from industrial livestock production in developing countries.

Biodiversity

Chapter 8 details the impacts that intensive and extensive forms of livestock production have on biodiversity. The impacts are mostly negative and include biodiversity loss from forests as pasture and feed croplands expand; emissions of greenhouse gases contributing to climate change, which then affects biodiversity; diseases spreading to wildlife from livestock; simplified landscape through intensification; water pollution by nutrients, drugs, and sediments affecting aquatic biodiversity; competition between introduced plant and native species; and overfishing to provide fishmeal for livestock. Positive impacts also exist, but they are limited to where moderately grazed pastures result in increased species diversity.

The indirect impacts by way of land use change, climate change, and pollution tend to be larger than direct ones by way of trampling, grazing, and excretion. Chapter 8 calls for a more integrated assessment of total impacts of livestock on biodiversity, making use of tools like life cycle analysis.

The value of the vast majority of animal genetic resources is poorly understood. Livestock output growth in the past century has concentrated on a very small number of breeds worldwide, frequently without due consideration to the local production environment. Exclusive reliance on just a few livestock breeds may carry a risk to global food systems in that the associated erosion of existing diversity may deprive humankind of certain options that may be required in the future, in particular to cope with unforeseen events, such as rapid climate change or changing disease patterns. Extensive smallholder production systems harbor broad genetic diversity, due to century-long exposure to different production environments. Certain genetic traits that these animals possess may be required to sustain intensive production to cope with possible future environmental and disease shocks. Yet the response of stakeholders to livestock's impacts on biodiversity is problematical because the effects often occur far away from consumers and are invisible to them (Deutsch and Folke 2005).

As with other environmental impacts, technical options that would lessen livestock's impact on biodiversity are often available and have significant potential. While sustainable intensification is key to reducing the overall resource requirements of livestock and feed production, related practices could improve biodiversity through a range of measures, including the following:

- Maintaining diversity on farming landscapes by allowing nonagricultural vegetation patches
- Adopting low tillage and conservation agriculture approaches
- Effective pollution control in livestock production
- Appropriate rotational grazing practices, avoiding overgrazing
- Conservation of biodiversity-rich areas like wetlands and riparian corridors.

As noted in Chapter 8, several novel approaches have led to higher compatibility of biodiversity and grazing livestock. These include schemes of payment for environmental services, conservation easements, local conservancies, and public–private partnerships. These provide an incentive and regulatory framework that lessens livestock's impact on biodiversity. Certification and the labeling of livestock products are alternative or complementary strategies.

While biodiversity loss is accelerating, the societal response to the problem has been slow and inadequate. This is caused by a general lack of awareness of the role of biodiversity and the failure of markets to reflect its value and its character as a public good (Loreau and Oteng-Yeboah 2006). It has been suggested that an intergovernmental mechanism akin to the Intergovernmental Panel on Climate Change should be established to link the scientific community to policy making, since the Convention on Biological Diversity is not in a position to mobilize scientific expertise to inform governments (Loreau and Oteng-Yeboah 2006).

The area of biodiversity is intrinsically more complex than other environmental concerns, and it is probably here that the gap between science and policy is largest. However, the scientific understanding of biodiversity and its functions has greatly improved in recent years, which is reflected in increased attention from policy makers. The scope of biodiversity conservation has been broadened to include protected areas and increased protection outside the designated areas based on the fact that whole ecosystems and their services often cannot be conserved by focusing on protected areas alone. New ways of financing biodiversity conservation are being explored to find alternative sources of funding. These include grants or payments from the private sector, conservation trust funds, resource extraction fees, user fees, and debt-for-nature swaps at the governmental level.

A novel mechanism for conservation of biodiversity is the payment for environmental services approach. PES is based on the principle that biodiversity provides a number of economically significant services. Payments need to be made to those who protect biodiversity to ensure the continued provision of these services. Access charges and entrance fees to protected areas are also a form of payment for environmental services—in this case, conservation of biodiversity. They are not new, but recent schemes allow revenues to be used outside the protected areas and also to be returned to local communities to provide incentives for biodiversity conservation (Le Quesne and McNally 2004).

A major challenge for new conservation approaches lies in the fact that in most countries endangered species are considered a public good, while their habitats are often on private land. As a private commodity, land can be transformed and traded. Biodiversity conservation can take place on private land but this relies on the owner's willingness and the land's opportunity cost. The opportunity cost of biodiversity conservation is difficult to estimate, since the value of biodiversity depends on biological resources and ecosystem services.

Biological resources are not fully identified (the total number of species on Earth is still unknown), and information on population numbers and risk status is still missing. However, some progress has been made in the valuation of ecosystem services. According to Boyd et al. (1999) the cost of conserving habitat should be valued as the difference between the value of land in its highest and best private use and its value when used in ways compatible with conservation.

To deal with the issue of ownership, new approaches have been tried with relatively good success (Boyd et al., 1999). Although most of these innovative approaches have been tried in forestry and at the community level, they can also be applied to livestock production.

- The purchase of full property interests involves the transfer of land from an owner who might develop the land to a conservator who will not. In order to purchase the property, the conservator must at least be able to pay the property owner the value of the land in private ownership. This value is the net present value of the land in whatever future use may be made of it, which is its opportunity cost.
- Conservation easements are a contractual agreement between a landowner and a conservator. In exchange for payment (or as a donation that can be tax deductible), landowners agree to extinguish their rights to future land development. This agreement is monitored and enforced by the conservator, which may be a private conservation organization or governmental entity.
- Another way to keep land out of development is for the government to give tax credits or other subsidies equal to the difference in value between developed and undeveloped uses. For instance, if developed land earns $100 more per acre than it does in low-intensity farming, a tax credit of $100 per acre compensates the property owner for not developing the land. The subsidy is a cost borne by taxpayers.
- Tradable development rights imply a restriction on the amount of land that can be developed in a given area. Suppose, for instance, that the government seeks to restrict development by 50% in an area. It can do so by awarding individual landowners the right to develop only 50% of their acreage. These development rights can then be traded. Tradable development rights impose costs on the landowners who have their development rights restricted. The aggregate opportunity cost is, as always, the value of development that is foregone in order to achieve the conservation goal. Though rights will be traded, the initial restriction of development opportunities imposes a cost on landowners.

Livestock production is often structured along the urban to rural gradient, with industrial production systems in the periurban areas, feed crops and mixed farming in rural areas, and extensive systems in the interface with wild habitats. This distribution, common in most countries, often places ruminant production in direct confrontation with wildlife and habitats.

In industrial countries this interface is characterized mostly by wealthy or resource-rich farmers operating under legislation for environmental protection, which is mostly enforced. In developing countries, however, the interface is characterized by a wide range of actors, stretching from resource-rich farmers to subsistence livestock holdings and herders. Even where legislation for environmental protection exists, it is often poorly enforced, if at all. It is not surprising, then, that the major impact of livestock production is on habitat change. Land use changes modify habitats extensively and are an important driver of biodiversity losses.

The current state of thinking prefers landscape-focused conservation over site-focused conservation, particularly as an option to retain biodiversity in human-dominated landscapes (Tabarelli and Gascon 2005). Based on biodiversity conservation in corridors, the fundamental nature of landscape-focused conservation is to engage both conservation needs and economic development by finding mutually beneficial interventions that might not necessarily occur within the buffer zones of protected areas. This might include new protected areas to safeguard watersheds, landscape management that adds value to tourism, and the use of tradable development rights and easements to promote development compatible with the movement of species between protected areas (Sanderson et al., 2003).

The integration of livestock production into landscape management poses many challenges for all policy and decision makers and requires a truly holistic approach. The major challenges from the conservation point of view would be as follows:

- To maintain the resilience of the ecosystem by predicting, monitoring, and managing gradually changing variables affecting resilience such as land use, nutrient stocks, soil properties, and biomass of long-term persistent species (including livestock) rather than merely controlling fluctuations (Sheffer et al., 2001).
- To sustain the functionality of the ecosystem and its capacity to sustain the processes required for maintaining itself, developing, and responding dynamically to constantly occurring environmental changes (Ibisch et al., 2005). This includes the capacity of the ecosystem to provide environmental services.
- To foster conservation efforts for taxa or species outside the protected areas and to include forms of livestock development (land use and management practices) that are compatible with the requirements of such taxa or species.

To fully integrate livestock into landscape management, it is necessary to recognize the multiple functions of livestock at the landscape level. Apart from production objectives, livestock production can have environmental objectives (carbon sequestration, watershed protection) and social and cultural objectives (recreation, aesthetics, and natural heritage) that should also be recognized in order to achieve sustainable production. Livestock production has been proposed as a landscape management tool mostly for natural pasture habitats (Bernués et al., 2005, Gibon 2005, Hadjigeorgiou et al., 2005), as it can constitute a cost-effective instrument to modulate the dynamics of vegetation to maintain landscapes in protected areas and to prevent forest fires (Bernués et al., 2005).

For an effective integration of livestock production into landscape management, radical changes should take place in management practices and land uses at the farm level. Recent research is focusing on new practices in managing grasslands in order to address the relationships between grassland production and nonproduction functions. Research topics include the following:

- How management affects short- and long-term changes in grassland species composition and production—aiming to discover the impact of reduced fertilizer application on animal nutrition and N balance and/or the possibility of sustaining species-rich vegetation.
- The role of pasture vegetation, management practices, and grazing behavior on natural vegetation and faunal diversity, in both marginal and intensive livestock production areas, in relation to biodiversity conservation.
- The spatial organization and dynamics of plant–animal grazing interactions at a variety of scales—with a view to optimizing the management of grazed landscapes so as to balance diversity, heterogeneity, and agricultural performance.
- The production and feeding value of species-rich grasslands—with a view to their integration in livestock production (Gibon 2005).

However, the most important topic in relation to biodiversity conservation will be the issue of intensification because of its effect on habitat change. Intensification and extensification will need to be managed at the landscape level according to socioeconomic and environmental conditions. The optimal approach will probably be a mixture of intensification on land area, extensive grazing, and setting aside land for conservation structured along the gradient: farm–communal area–buffer zone–protected area.

The driving factors that should be addressed at the landscape level are degradation and shrinkage of common land, high livestock densities, lack of common property management, and inequity in the distribution of watershed benefits. Intensification of livestock production can contribute to biodiversity conservation at the watershed level. This would include pasture development; multipurpose trees for fodder, fuel, or timber; and improvement of the genetic capacity of local breeds. It would be accompanied by payments for environmental services (biodiversity protection, carbon sequestration, and water quality) and a rationing system for common property resources (e.g., grazing fees).

Policies are needed to guide the current opportunistic process of livestock development at the landscape level for preservation of biodiversity. One main issue for the formulation of policies is that at the landscape level, property boundaries do not correspond with ecological boundaries. The number of landowners and the mixed set of ownership types (public and private) ensure that individual owners' decisions have an effect on the decisions of neighboring landowners (Perrings and Touza-Montero 2004). Enforcement, auditing, and monitoring mechanisms and decision support tools should be embedded into the policy framework.

Water

Chapters 7 and 9 assess the large and complex interactions of both extensive and intensive livestock production with hydrological cycles, affecting both quality and availability of water. Water required for plant growth for animal feed in the form of roughage or pasture and concentrate feed has been identified as the dominant freshwater resource challenge in the livestock sector. At the same time, water pollution from intensive production units, particularly where their density exceeds the absorptive capacity of the surrounding land, has been identified as a main environmental pressure, with implications for human health.

Chapter 7 points out that feed production affects water flows by way of blue water withdrawal, land cover change, and altered water partitioning and suggests an analytical framework for assessing the effects of hydrological changes due to livestock production. The authors reckon that livestock's total share of global hydrological flows amounts to 10%. In their assessment, they highlight the need to expand the blue water focus toward integrated water resource management that incorporates green water and land use.

Chapter 9, focusing on impacts of rapidly growing intensive livestock production systems, identifies pollution and eutrophication of surface waters, groundwater, and coastal marine ecosystems. Such pollution not only involves major nutrient overloading, such as with nitrogen and phosphorus, but also gaseous emissions and heavy metal accumulation, all affecting human health and biodiversity as well as contributing to climate change.

Responses to these issues have been scarce and are largely limited to addressing the water pollution problem in industrial countries and, to a more limited extent, emerging economies. Improving water use efficiency is a critical objective as water resources become scarcer. The fundamental role of prices is to help allocate resources among competing uses, users, and time periods (Ward and Michelsen 2002) and to encourage efficient use by users. In practice, water for agriculture is in many cases provided free (representing a 100% subsidy). Even in countries where pricing systems have been instituted, water remains greatly underpriced (Norton 2003). In setting water prices, effluent charges, and incentives for pollution control, it is important to estimate the full cost of water used in a particular sector. Prices should signal the true scarcity to users of water and the cost of providing the service; they should provide incentives for more efficient water use and provide service providers and investors with information on the real demand for any needed extension of water services.

Through measures such as pollution charges and water pricing to encourage conservation and improved efficiency, pricing can serve as a means to ensure that actors internalize the environmental externalities that may arise from agricultural activities (Small and Carruthers 1991, Johansson 2000, Bosworth et al., 2002). Adequate pricing can significantly reduce water withdrawals and consumption by agriculture, industry, and households.

The fact that operation and maintenance costs are not recovered, or not fully recovered, amounts to a subsidy for the crop and livestock sectors. Countries' experiences with cost recovery have been mixed. In a comparative study of 22 countries (Dinar and Subramanian 1997), irrigation operation and maintenance cost recovery in developing countries were found to range from a low of 20–30% in India and Pakistan (where the state remains heavily involved in the operation of irrigation systems) to a high of about 75% in Madagascar (where the role of the government is much reduced in favor of water users' associations who have been given responsibility for managing the irrigation systems). In members of the Organization for Economic Cooperation and Development (OECD), the recovery of costs is much higher, with the majority of countries obtaining full cost recovery for operation and maintenance cost. Countries like Australia, France, Japan, Spain, and the Netherlands also recover full supply costs from users (OECD 1999). In the United States, state laws limit the charges that irrigation districts can impose on farmers to no more than their cost. Consequently, water prices are set to cover only the costs of delivery and maintenance (Wahl 1997).

Subsidies can take several forms, including the public provision of water for agriculture at no or low prices and subsidization of irrigation equipment or of energy for pumping groundwater. The removal of these subsidies is of prime importance in order to encourage efficient water use.

With water pollution, the establishment of water quality standards and control measures is central. While the use of uniform standards may simplify enforcement, smaller farms or enterprises may be unable to afford the cost of meeting the regulatory requirements or the waste treatment and relocation costs (FAO 1999). Hence standards can be defined locally or regionally, taking into account environmental and economic viewpoints, as the marginal costs for technical adjustments may vary.

Regulatory mechanisms to control pollution can take a variety of forms:

- Definition of minimum discharge and application standards in order to reduce emissions and effluents to acceptable levels.
- Specification of equipment to be used (effluent treatment) to meet the minimum standards.
- Issuance of permits for the discharge of pollutants, which can also be traded. Tradable permits rely on payment per unit of pollution or the use of credits for reducing pollution. In that case market mechanisms are used to allocate pollution rights once an acceptable overall level of pollution has been established.
- Specification of maximum industrial activity. For example, in livestock production systems limits may be placed on the number of livestock per hectare (FAO 1999).

A set of criteria is used to monitor the impacts of livestock production systems on water quality and to set water quality standards for specific water bodies. Parameters to be monitored to evaluate the impacts of livestock production systems include sediment level; presence of nutrients (nitrogen, phosphorus and organic carbons); water temperature; dissolved oxygen level; pH level; pesticide levels; presence of heavy metals and drug residues; and levels of biological contaminants. The close monitoring of these parameters is a key element in evaluating compliance of production systems with defined standards and codes of practices. The European Commission proposes EU-wide emission controls and environmental quality standards for the substances and measures, its objective being the ultimate cessation, within 20 years, of emission of substances identified as hazardous (Kallis and Butler 2001). Monitoring is costly and may represent a financial burden, especially in countries with limited monitoring capacities.

With regard to pollution from intensive livestock production, Chapter 9 points out that most of the nutrients contained in animal waste are indeed wasted, as only a fraction is recycled for crop production, causing great environmental pressures, particularly where production units are geographically concentrated. The chapter also provides different options that have different pollution abatement potential and different costs. "Industrial production parks" with high-tech manure processing, already practiced in some places like the Netherlands and China, may solve most environmental issues but are knowledge- and capital-intensive. Spatial zoning and related licensing, manure application regulation, or manure quota schemes are well suited to increase the rate of nutrient recycling through cropping but are demanding in institutional capacity. Most European countries and other OECD countries have adopted such policies. Emerging countries, particularly those in East Asia but also in Latin America, have yet to develop related schemes and are currently faced with unparalleled livestock densities and pollution in some locations, particularly close to consumption centers and feed markets.

Practices that pollute water resources are taxed in some places. In Belgium, for example, wastewater from livestock production is either assimilated into domestic wastewater and taxed as such or spread over agricultural land, where it is subject to a special industrial tax (OECD 1999). The EU water policy framework includes a principle of "no direct discharge" to ground water (Kallis and Butler 2001). The EU Nitrates Directive is a good example of regulation for nonpoint source pollution.

Nonpoint source pollution is less easy to regulate. Codes of environmental practices and their enforcement are key elements in ensuring that agricultural activities that generate nonpoint source pollution would need prior authorization or registration based on binding rules (Kallis and Butler 2001).

Extraction levels of groundwater resources are often regulated, especially in industrial countries. Abstraction charges, particularly within OECD countries, aim to control overexploitation of groundwater resources. Countries where such charges are applied include Belgium, Bulgaria, Hungary, the Netherlands (Roth 2001), and Jordan (Chohin-Kuper et al., 2003).

Practices that lead to the provision of environmental services, such as improved water quantity and quality, can be encouraged through payments to the providers. In a watershed context, upstream actors can be considered service providers if their actions result in improved water quality or quantity, for which they are compensated by downstream users. PES schemes require a market where the beneficiaries of these services (downstream water users) buy them from upstream providers. Obviously, this needs to be based on established cause–effect relations between the upstream land use and the downstream water resource conditions (FAO 2004b).

PES schemes related to water services are usually of local importance at the watershed level, with users and providers geographically close to each other. This facilitates their implementation, compared with other types of environmental services with more remote or abstract linkages such as carbon sequestration or biodiversity protection, because of reduced transaction costs and easy information flow among the economic agents (FAO 2004b). Concluding remarks and lessons learned are made in Chapter 20.

References

Bernués, A., J. L. Riedel, M. A. Asensio, M. Blanco, A. Sanz, R. Revilla, and I. Casasús. 2005. An integrated approach to study the role of grazing farming systems in the conservation of rangelands in a protected natural park (Sierra de Guara Spain). *Livestock Production Science* 96: 75–85.

Bosworth, B., G. Cornish, C. Perry, and F. van Steenbergen. 2002. *Water charging in irrigated agriculture: Lessons from the literature.* Report OD 145. Wallingford, U.K.: HR Wallingford.

Boyd, J. W., K. Caballero, and R. D. Simpson. 1999. *The law and economics of habitat conservation: Lessons from an analysis of easement acquisitions.* Discussion Paper 99–32. Washington, DC: Resources for the Future.

Chohin-Kuper, A., T. Rieu, and M. Montginoul. 2003. *Water policy reform: Pricing water, cost recovery, water demand and impact on agriculture. Lessons from the Mediterranean experience.* Water Pricing Seminar Proceedings. Washington, DC: World Bank Institute.

Deutsch, L., and C. Folke. 2005. Ecosystem subsidies to Swedish food consumption from 1962– 1994. *Ecosystems* 8 (5): 512–528.

Di, H. J., and K. C. Cameron. 2003. Mitigation of nitrous oxide emissions in spray-irrigated grazed grassland by treating the soil with dicyandiamide, a nitrification inhibitor. *Soil Use and Management* 19 (4): 284–290.

Dinar, A., and A. Subramanian. 1997. *Water pricing experiences: An international perspective.* World Bank Technical Paper No. 386. Washington, DC: World Bank.

FAO (Food and Agriculture Organization). 1999. *Livestock and environment toolbox.* Livestock, Environment and Development Initiative. Rome.

FAO (Food and Agriculture Organization). 2004a. *Carbon sequestration in dryland soils.* World Soils Resources Reports 102. Rome.

FAO (Food and Agriculture Organization). 2004b. *Payment schemes for environmental services in watersheds.* Regional forum, 9–12 June 2003, Arequipa, Peru. Organized by the FAO Regional Office for Latin America and the Caribbean.

FAO (Food and Agriculture Organization). 2007. *Paying farmers for environmental services.* State of the Food and Agriculture, 2007, Food and Agriculture Organization of the United Nations, Rome.

Gibon, A. 2005. Managing grassland for production, the environment and the landscape. Challenges at the farm and the landscape level. *Livestock Production Science* 96 (1): 11–31.

Hadjigeorgiou, I., K. Osoro, J. P. Fragoso de Almeida, and G. Molle. 2005. Southern European grazing lands: Production, environmental and landscape management aspects. *Livestock Production Science* 96 (1): 51–59.

Ibisch, P., M. D. Jennings, and S. Kreft. 2005. Biodiversity needs the help of global change managers not museum-keepers. *Nature* 438:156.

Ibrahim, M. 2009. Brazil and Costa Rica: Deforestation, pasture expansion and policy implications for landscape restoration and livelihoods of smallholder ranchers. In *Livestock in a Changing Landscape: Experiences and Regional Perspectives.*, ed. P. Gerber, H. Mooney, J. Dijkman, S. Tarawali, and C. de Haan. Washington, DC: Island Press

IPCC (Intergovernmental Panel on Climate Change). 2000. *Land use, land use change and forestry. A special report of the IPCC.* Cambridge, UK: Cambridge University Press.

Johansson, R. C. 2000. *Pricing irrigation water—A literature survey*. Policy Research Working Paper 2249. Washington, DC: World Bank.

Kallis, G., and F. Butler. 2001. The EU water framework directive: measures and implications. *Water Policy* 3 (2): 125–142.

Lal, R. 2004. Soil carbon sequestration impacts on global climate change and food security. *Science* 304 (5677): 1623–1627.

Le Quesne, T., and R. McNally. 2004. *The green buck: Using economic tools to deliver conservation goals. WWF field guide*. The WWF Sustainable Economics Network. Gland, Switzerland: WWF.

Loreau, M., and A. Oteng-Yeboah. 2006. Diversity without representation. *Nature* 442: 245–246.

Melse, R. W., and A. W. van der Werf. 2005. Biofiltration for mitigation of methane emissions from animal husbandry. *Environmental Science and Technology* 39 (14): 5460–5468.

Mendis, M, and K. Openshaw. 2004. The Clean Development Mechanism: Making it operational. *Environment, Development and Sustainability* 6 (12): 183–211.

Mikkelsen, S. A., T. M. Iversen, B. H. Jacobsen, S. S. Kjær. 2009. Denmark–EU: Reducing nutrient losses from intensive livestock operations. In *Livestock in a Changing Landscape: Experiences and Regional Perspectives*, ed. P. Gerber, H. Mooney, J. Dijkman, S. Tarawali, and C. de Haan. Washington DC: Island Press

Monteny, G. J., A. Bannink, and D. Chadwick. 2006. Greenhouse gas abatement strategies for animal husbandry. *Agriculture, Ecosystems and Environment* 112:163–170.

Norton, R. D. 2003. *Agricultural development policy: Concepts and experiences*. Hoboken, NJ: Wiley.

OECD (Organization for Economic Co-operation and Development). 1999. *Agricultural water pricing in OECD countries*. Paris: OECD.

Pagiola, S., P. Agostini, J. Gobbi, C. de Haan, M. Ibrahim, E. Murgueitio, E. Ramirez, M. Rosales, and J. Pablo Ruiz. 2004. *Paying for biodiversity conservation services in agricultural landscapes*. Environment Department Paper No. 96. Washington, DC: World Bank.

Perrings, C., and J. Touza-Montero. 2004. Spatial interactions and forests management: Policy issues. In *Proceedings of the Conference on Policy Instruments for Safeguarding Forest Biodiversity—Legal and economic viewpoints*, 15–16 January 2004, ed. P. Horne, S. Tönnes, and T. Koskela, 15–24. Helsinki: Finnish Forest Research Institute.

Richards, K. 2004. A brief overview of carbon sequestration economics and policy. *Environmental Management*, 33 (4): 545–558.

Roth, E. 2001. *Water pricing in the EU: A review*. Brussels: European Environmental Bureau.

Sanderson, J., K. Alger, G. da Fonseca, C. Galindo-Leal, V. H. Inchausty, and K. Morrison. 2003. *Biodiversity conservation corridors: Planning, implementing, and monitoring sustainable landscapes*. Washington, DC: Conservation International.

Sheffer, M., S. Carpenter, J. A. Foley, C. Folke, and B. Walker. 2001. Catastrophic shifts in ecosystems. *Nature* 413:591–596.

Sirohi, S., and A. Michaelowa. 2004. *CDM potential of dairy sector in India*. HWWA Discussion Paper 273. Hamburg, Germany: Hamburgisches Welt-Wirtschafts-Archiv, Hamburg Institute of International Economics.

Small, L., and I. Carruthers. 1991. *Farmer financed irrigation—Allocating a scarce resource: Water-use efficiency*. Cambridge, U.K.: Cambridge University Press.

Tabarelli, M., and C. Gascon. 2005. Lessons from fragmentation research: Improving management and policy guidelines for biodiversity conservation. *Conservation Biology* 19 (3): 734–739.

Vlek, P. L. G., G. Rodríguez-Kuhl, and R. Sommer. 2004. Energy use and CO_2 production in tropical agriculture and means and strategies for reduction or mitigation. *Environment, Development and Sustainability* 6 (12): 213–233.

Wahl, R. W. 1997. United States. In *Water pricing experiences: An international perspective*, ed. A. Dinar and A. Subramania, 144–148. World Bank Technical Paper No. 386. Washington, DC: World Bank.

Ward, F., and A. Michelsen. 2002. The economic value of water in agriculture: Concepts and policy applications. *Water Policy* 4 (5): 423–446.

17

Responses on Human Nutrition Issues

Carolyn Opio and Henning Steinfeld

Main Messages

- **Food insecurity is a major problem in low-income countries, and targeted interventions need to separately address emergency situations and livestock sector development.** Livestock products provide high-value protein and other nutrients in emergency situations, and livestock remain a support to livelihoods in many food-insecure situations in developing countries.

- **Livestock products or the absence of them play a major role in the double burden of undernutrition and overnutrition in emerging economies.** Targeted nutrition programs can help alleviate the underprovision of protein and other nutrients through, for example, school milk programs and supplementary feeding programs. Much more effort is required to raise awareness and inform the emerging middle class in developing countries about healthy nutrition to counter the emerging food culture that can lead to obesity.

- **Overweight and obesity, and related noncommunicable diseases, are an increasing problem in industrial countries.** These need to be addressed through awareness and educational programs and, if appropriate, be supported through price disincentives.

- **Food safety is generally becoming a shared concern in all countries in response to increasing foodborne illnesses and food scares** (both perceived and real). Livestock and livestock products are of particular concern because of the perishable nature of many products, the potential for disease and parasites to spread from animals to humans, and the impact of animal products on human nutrition. Low-income countries are faced with trade-offs between food security and food safety issues—and food safety is often compromised in an effort to meet basic daily food requirements. In industrial countries, consumer concerns are driving regulatory changes orientated toward addressing food safety across the entire food chain, emphasizing shared responsibility between government, industry, and consumers.

Introduction

Contrary to popular perception, human malnutrition is not only a consequence of inadequate food. Poor diet quality and poor eating habits are increasingly the main causes of malnutrition around the world. Diets can be poor in terms of what they do not contain—minerals and vitamins, for example, or sufficient fruits and vegetables or animal products. They may also be substandard in the sense that they contain too many elements that are harmful when taken in excess, such as saturated fats and sugar (IFPRI 2004a).

Nutrition has generally over the last decades improved in most regions of the developing world. FAO estimates for 2003–05 showed that all four developing regions were making progress in reducing the prevalence of hunger. With the recent upsurge in food prices, however, progress has been reversed in all regions, resulting in increased hunger prevalence for the entire developing world (FAO 2008).

The face of malnutrition varies between countries and regions. In some areas it is characterized by undernutrition and in others by overnutrition. And in an increasing number of countries it is characterized by both—commonly known as the double burden of malnutrition. The coexistence of underweight and overweight in close proximity is common in countries experiencing rapid changes in diet.

Livestock products are important sources of nutrients. They benefit infants and children, promoting steady growth in the early years of life. Animal products also improve the dietary availability of micronutrients in general. Eating meat, for instance, provides many valuable nutrients, such as bioavailable iron and vitamins, not found in plant foods (WHO 2003a; Chapter 12). However, these benefits begin to decline rapidly as intake levels rise above a certain threshold. As noted in Chapter 12, increased consumption of red meats tends to increase the risk of colon cancer, while increased intake of saturated fats and cholesterol from meat, dairy products, and eggs increases the risk of suffering from chronic noncommunicable diseases such as cardiovascular diseases (SCN 2005).

Food safety is another important dimension of health. Consumers are increasingly concerned about issues such as the microbiological and chemical safety of foods, emerging zoonotic diseases and prions, new processing techniques, and the application of biotechnology in relation to the use of genetically modified foods in the human food chain. Food safety is receiving more attention due to several global trends that affect the food system: increasing movement of people, animals, and products across borders, rapid urbanization, changes in production processes, and the emergence of new pathogens. Concurrently, the chain of responsibility has become longer and more complex as have the food supply chains to deliver the products to the consumers. Although food safety issues affect all countries, they vary in nature and scale according to the level of economic development.

Over the past few years, several high-profile food safety issues have been linked with livestock products. The bovine spongiform encephalopathy (BSE) crisis, outbreaks of salmonella and *E. coli* O157: H7, and dioxin contamination of animal feedstuffs in Europe as well as the avian influenza crisis in Asia have heightened public concerns over the safety of foods of animal origin. The recent food scares such as the melamine-contaminated milk crisis in China are an indication of the potential foodborne risks and the systemic weaknesses of the global food system.

This chapter highlights how stakeholders have responded to these various challenges of nutrition. It also addresses questions such as why nutritional problems continue to persist and why little progress has been made in addressing the issues despite the various efforts made at several levels of intervention. The chapter concludes with specific options for dealing with these issues.

Undernutrition and Food Insecurity

Over the last four decades, global agricultural productivity has risen sharply, with food supply improving significantly to meet the food requirements of the world's population. Yet despite the abundance of food, the absolute numbers of hungry people have increased, the vast majority of whom live in developing countries. FAO estimated the number of undernourished people worldwide at 923 million in 2007. And FAO estimates that recent rising prices have driven an additional 41 million people in Asia and 24 million in sub-Saharan Africa below the hunger threshold (FAO 2008). Furthermore, the global toll of people affected by micronutrient deficiencies, often also referred to as "hidden hunger," is estimated to exceed 2 billion.

Progress and gains in reducing food insecurity and undernutrition have not been even across geographic regions and countries. Regions like sub-Saharan Africa and South Asia are no better off today than they were decades ago. In the Near East and North Africa, food insecurity is persistent (although relatively low) and actually rising both in absolute numbers and in prevalence. Latin America, on the other hand, has seen a reduction in the number of undernourished from 59 million in 1990–92 to 52 million in 2001–03. In transition countries, socioeconomic reforms have increased food insecurity because the growth in poverty during the transition increased the number of people who cannot afford an adequate and healthy diet (Ralston 2000).

A wide range of players in both private and public sectors have taken numerous steps to address these problems. Many of the interventions have focused on improving food availability at the household or national level, with a few focusing on longer-term solutions such as social safety net programs, supplementary programs, or nutrition surveillance.

Food aid remains the chief instrument for ensuring the basic nutritional needs of people affected by humanitarian crisis (FAO 2006a). Food aid is commonly of three types: program, project, or emergency (humanitarian). Program aid is generally untargeted distribution sold on recipient-country markets to raise revenue used to support development interventions. Project food aid is targeted at clearly defined beneficiary groups often through supplementary feeding programs or food-for-work schemes. Humanitarian assistance is directed at unanticipated human-made and natural disasters (Barrett 2002b).

A large proportion of food aid is directed at sub-Saharan Africa (IATP 2005a), the only world region where food production increases have lagged behind population growth. The share of all food aid going to these countries since 1990 has gone from one-fifth to one-half.

In 2007, of the total food aid 86% (5.1 million tons) consisted of cereals and 14% (0.85 million tons) of noncereals. Of the latter, 0.4% consisted of animal source foods (dairy products were 0.3% and meat and fish were 0.1%) (WFP INTERFAIS 2007).

Food aid has also been instrumental in enhancing local production of livestock products. Food aid in the form of dairy products (dried skimmed milk powder and butter oil) has been used for the development of the local dairy sector and to complement the low and fluctuating domestic supply levels. Related projects were based on the monetization of dairy food aid, with the use of the proceeds to finance dairy development. Skimmed-milk powder to sub-Saharan Africa provided under the U.S. Department of Agriculture (USDA) program has been used this way. The European Union (EU) also supported dairy aid developmental projects in Bangladesh and Tanzania.

The Operation Flood program in India has shown how food aid (in the form of milk powder and butter oil) can be an effective investment in the development of the dairy sector. Operation Flood, launched in 1970, continued in three phases into the 1990s and is widely cited as a successful model of dairy development initially supported by direct dairy aid. This became the model in which food aid committed on a multiyear basis was sold to a publicly owned dairy industry cooperative structure (OECD 2005).

Food aid still plays an important role in addressing food security even in countries at an intermediate stage of development. And although food aid shipments to these countries have gone down over the past three decades, they have been used to supplement nutritional food gaps. During the 1990s, the United States and EU donated food aid to a number of transition economies, including Albania, Armenia, Azerbaijan, Georgia, Kyrgyzstan, Moldova, Russia, and Tajikistan (ERS-USDA 2004). The Russian Federation received food aid in the form of meat and dairy products (FAO 2003b).

Although food aid can play a crucial role in reducing hunger, it does not offer long-term solutions to the problems of food insecurity and undernutrition and it may even exacerbate nutrition problems (Oxfam 2006). In humanitarian infant and child feeding programs, the use and distribution of milk powder has been heavily criticized because it can act as a disincentive to breastfeeding, perpetuating the likelihood of malnutrition (WHO 2000). It may compromise the immune system of infants, making them susceptible to infectious diseases and, later, to chronic noncommunicable diseases. Provision of food aid with appropriate micronutrient content has also been a long-standing concern surrounding food aid. Far more people suffer from micronutrient deficiencies (e.g., of iron, iodine, and vitamins A or D) that affect health than from calorie undernutrition (Barrett 2002a), and very often food aid has failed to contribute much to needed nutritional variety. The use and distribution of nonfat dried skimmed milk rather than whole milk powder during humanitarian operations has been criticized because of its low fat and iron content, which infants need.

In order to meet the growing nutritional demand for livestock products, countries, particularly in sub-Saharan Africa, have had to rely heavily on imports to bridge the gap between domestic production and demand. Most of the consumption gains in the region have been achieved through imports, with production experiencing slower growth than other developing regions. Despite the opportunity provided by growing demand for livestock products, domestic production in sub-Saharan Africa has not grown enough to meet these demands. (See Chapter 2.) The rapidly growing demand for livestock products in this region is reflected in the rapidly declining net trade balance for livestock products, as imports continue to outpace exports in order to meet demand and nutritional requirements.

In transition economies, the most immediate impact of economic reform was a sharp decline in production, particularly of livestock products, and a rise in imports of meat products. In Russia, livestock production declined more significantly in order to maintain relatively high levels of meat consumption; significant amounts of meat, particularly poultry meat, had to be imported. About half of the chicken and 30% of the meat consumed in the Russian Federation is imported. Similarly, Uzbekistan imports 30% of its meat and 25% of its milk (Wehrheim and Wiesmann 2006).

Imports of livestock products can have mixed impacts on human nutrition and health. On the one hand, importing animal products is a means of bridging the food gap during shortfalls and makes important contributions to food security, but on the other, it can also perpetuate nutrition inequalities and have negative impacts on local production, employment, and livelihoods. Governments in low-income countries have often given priority to supplying urban consumers with cheap subsidized livestock products, yet the vast majority of the undernourished and food-insecure live in the rural areas. Poverty statistics as well as national income trends in low-income countries indicate that food insecurity is mainly related to access to food. An FAO assessment of the impacts of import surges on sub-Saharan Africa found that imported livestock products were primarily sold in the urban areas (Sharma et al., 2005).

Together with other equally important factors such as export subsidies, reductions in tariffs, domestic market liberalization, and market shocks in importing countries, imports of livestock products have in certain instances contributed to the depression of local production, as local producers cannot compete with cheap imports. In sub-Saharan Africa, surges in livestock imports are disrupting local markets and having negative impacts on prices, production, livelihoods, and food security (Sharma et al., 2005). The product characteristics of imported chicken, usually frozen, low priced dark meat, make it difficult for local broiler operations producing fresh and highly perishable whole chickens that have a higher cost of production, to compete. Box 17.1 summarizes findings from a study on the impact of poultry imports in several sub-Saharan African countries.

In contrast to sub-Saharan Africa's reliance on imports to fill the food gap, some low-income countries have embarked on programs to improve domestic production and consumption of animal source foods. Viet Nam, for example, has successfully implemented

Box 17.1. Impacts of Poultry Imports on Several African Countries

An FAO analysis on the impact of poultry imports indicates that rising poultry imports in several African countries have coincided with structural difficulties for local broiler industries, reducing capacity utilization and forcing operations to close.

Country	Poultry Import Trends	Impacts
Ghana	Imports grew from 4,000 tonnes in 1998 to 124,000 tonnes in 2004	Local poultry enterprises operation at low capacities; capacity utilization only 25% for hatcheries, 42% for feed mills, 25% for processing plants; stagnant production
Cameroon	Between 1999 and 2004, imports increased by almost 300%	Loss of rural jobs; lower domestic production; declining retail poultry prices
Côte d'Ivoire	Imports expanded sixfold between 1998 and 2004	Production decreased by 7% in 2002 and by 49% in 2004, in comparison to 2001
Senegal	Imports increased from 1% of consumption in 2000 to over 20% by 2002	Producers leaving industry or converting operations to laying hen production; declining retail prices; 70% of broiler operations reported to have closed
Tanzania	Imports rose 81-fold between 1997 and 2002	Lower capacity utilization and employment losses in the industry

Source: FAO 2006b.

Box 17.2. Programs to Improve the Production and Consumption of Livestock Products in Viet Nam

In 1995 the government of Viet Nam ratified the National Plan of Action for Nutrition with the aim of improving the nutritional status of the Vietnamese people. Prior to this, much of the protein was supplied by rice; consumption of animal source foods was negligible. One component of the program focused on increasing the production and consumption of the main animal source foods—pork and poultry—by investing in breeding, strengthening of veterinary services, and feed technology and processing.

National surveys indicate a vast improvement in the population's dietary intake and a reduction in nutritional deficiencies. Per capita consumption of meats increased from 24.4 to 51.0 g/person/day during the period 1987–2000. Egg and milk consumption increased from 2.9 to 10.3 g/person/day over the same period. Child malnutrition (underweight) has also remarkably dropped from 51.5% in 1985 to 44.5% in 1995 to 33.1% in 2000, while the prevalence of stunting fell from 56.7% to 36.5% in 2000.

Source: Hop et al., 2003.

national programs aimed at improving food intake and the nutritional status of the Vietnamese people by boosting domestic production of animal source foods such as meat and milk. (See Box 17.2.)

Nongovernmental organizations (NGOs) and international organizations (UN agencies) have also taken steps to address food insecurity and nutritional deficiencies in low-income countries through programs aimed at increasing the production and consumption of livestock products at the household level. These programs are often implemented through training on animal husbandry and breeding and are complemented with the provision of veterinary services and improved feed. NGOs like Heifer International, Farm Africa, VEDCO, Helen Keller International (HKI), and Oxfam have taken leading roles in propagating local livestock projects with the aim of improving human nutrition.

In Bangladesh, Cambodia, and Nepal, HKI has promoted production of animal foods at the household level to address micronutrient deficiencies. The program promotes household egg production through the introduction of improved poultry breeds and improved veterinary services, along with adequate housing and feed. It also provides access to better varieties of grass to improve milk production (HKI 2003). Program impacts include increased egg and chicken liver consumption among poor households.

Many of these programs target female-headed households through the donation of animals such as cows, goats, and microlivestock not only to provide milk and meat, but also to serve as a source of income when excess milk is sold. In Kenya, the U.S. Agency for International Development (USAID), in partnership with other organizations, is funding a Dairy Development Project that aims to build a strong private, commercially oriented dairy sector, provide better quality and less expensive products to markets, and improve the health of Kenya's population. The program focuses on promoting increased consumption of milk and dairy products through promotional campaigns (Ferris-Morris 2003).

Social safety net programs to support vulnerable groups can play a fundamental role in reducing food insecurity and prevent people from sliding into chronic poverty in the wake of crisis. The objective of such programs is to reduce the impact of transitional crisis on vulnerable people. The enhancement of resilience to repeated shocks requires that income and assets be protected. In circumstances where there is a shortfall of food, transferring food can prevent recipients from having to sell off their productive assets such as livestock in order to feed themselves. A wide range of safety net instruments has been used, particularly by NGOs and multilateral organizations (such as Oxfam and the World Food Programme [WFP]), including food-for-work or cash-for-work schemes (which may be conditional or unconditional) and subsidies for goods or services (Barrett 2006). Box 17.3 illustrates how safety net programs have been used to intervene early and prevent malnutrition and hunger-related deaths.

During a food crisis, international organizations like UNICEF have adopted a similar approach to address undernutrition in children and supplement their diets with milk. In Niger, in the midst of the 2005 food and nutrition crisis, UNICEF supported a women-led goat project to help vulnerable families in rural areas avoid the effects of the crisis.

Food supplementary programs such as school feeding programs, mother–child feeding programs, and food fortification programs to address malnutrition and micronutrient deficiencies are a common approach used by a wide range of actors—from governments to international agencies, NGOs, and the food industry in both development and emergency situations. Supplementary nutritional programs are most effective when targeted at high-risk individuals and have been used to address food security requirements of children, pregnant women, and people in emergency situations in low- and middle-income countries.

Mother–child supplementary feeding programs are used to tackle problems related to an intergenerational cycle of undernourishment. Several countries have designed programs targeted at maintaining adequate diets for mothers and their infants. Experience in many middle-income countries show that supplementary feeding using conditional transfers has gotten poor people to use nutrition services. Large-scale, long-term mother–child nutrition intervention programs have been implemented in Thailand (Basic Minimum Needs Poverty Alleviation program), India (Integrated Child Development Services), Mexico (Programa de Educación, salud y Alimentación–*Progresa*), and Chile (National Complementary Food Program, or PNAC). (See Box 17.4.) These programs are usually combined with a nutrition and health education component and/or a health component. Many other countries have implemented similar programs, including economies in transition in Central Asia and Russia.

Large-scale school feeding programs are another kind of supplementary feeding intervention oriented toward the reduction of malnutrition. Feeding statistics from the WFP Global School Feeding campaign currently implemented in low-income countries confirm this upward trend. In 1999 WFP school feeding activities for children reached 11.9 million children in 52 countries; by 2005, this figure reached 21.7 million children in 74

Box 17.3. Livelihood Support—Linking Relief and Development

At the peak of the emergency in the Sahel, Oxfam was able to help people avoid falling deeper into debt by paying a normal price for weak animals, which were then slaughtered and used to feed the most vulnerable members of society. This plan was accompanied by a "cash-for-work" program that supported people in northern Mali and southern Niger in exchange for working on community projects. Over 8,000 weak animals were bought and slaughtered, 960 hectares of pasture were replanted, and 30 dams were rehabilitated. Once the peak of the crisis had passed, Oxfam continued to support the same communities through animal and feed fairs, providing hundreds of farmers and herders with options for rebuilding their livelihood.

Source: Oxfam 2006.

Box 17.4. Impacts of the Mother–Child Supplementary Feeding Programs in Thailand and Chile

The national Basic Minimum Needs Poverty Alleviation program in Thailand incorporated a maternal and child nutrient component that consisted of growth monitoring and nutrition education. The program provided coupons to mothers of underweight preschool children to exchange for milk and eggs. It virtually eliminated severe malnutrition over a decade, with significant reductions in underweight in preschool children from a prevalence rate of 50% in 1982 to 14% in 1996.

Chile is often presented as a paradigm of success of supplementary feeding programs. The Chilean National Complementary Food Program (PNAC, from the Spanish) has the main objectives of promoting growth and development in children from conception through 6 years of age by providing food supplements to the mother during pregnancy and lactation and to the child from birth to the age of 6; promoting breastfeeding; preventing low birth weight related to maternal malnutrition; and preventing infant–childhood malnutrition. Chile's PNAC provided fortified food supplements to about 1.2 million children and 200 pregnant and nursing mothers. Of the total amount of supplements, 65% consisted of milk and milk substitutes, 9% milk–cereal blend, and about 26% rice. Hertrampf et al. (1998) reported that the food program in Chile reduced the prevalence of anemia in children aged 12–18 months from 27.3% to 8.8%.

Source: Kain et al., 1994.

beneficiary countries (WFP 2009). Animal source foods such as milk are a main component of school feeding programs. These have been implemented in several developing countries such as China, Indonesia, Philippines, India, Thailand, Peru, Mexico, and Brazil. The Chinese government-sponsored School Milk Program, which started in 2000, involves the supply of milk to selected schools at discounted prices with the aim of encouraging milk consumption. By the end of 2005, the program had expanded to more than 10,000 schools supplying over 2.4 million students—a 26% increase since 2004 (Euromonitor International 2005). In Thailand, the government continues to support free milk to all government supported preschool centers and to underweight preschoolers in communities and public primary schools (Smitasiri and Chotiboriboon 2003). The Thai school milk program reaches about 70% of students between the ages of 3 and 12.

Public–private collaborative efforts have become common. In a collaborative effort, Tetra Pak and a local milk industry and NGO, with support from USAID and the Indonesian government, have supported school milk programs in Indonesia. The milk program reaches about 12% of primary school students between the ages of 6 and 12. The national Iranian School Milk Committee in cooperation with Tetra Pak is running one of the largest school feeding programs in the world. In the school year 2004/05, 8 million children received milk three times a week (Tetra Pak 2006). Similar programs are run in a variety of countries.

Trade-offs in targeting are particularly relevant in the case of food supplementary programs. Various studies on the targeting efficiency of these programs have highlighted the substantial trade-offs involved. Evidence from several middle-income countries, particularly in Latin America, demonstrates this. An FAO survey with data from 19 Latin American countries found that over 20% of the population—approximately 83 million people—received some level of food assistance benefits from nutrition-related programs. In contrast, the number of malnourished in the study countries was estimated at 10 million—that is, 12% of the beneficiaries (Uauy et al., 2001).

Even in the face of the rapid changes in nutrition, several middle-income countries have maintained nutritional programs addressing the problems of diet deficits. As a result, though malnutrition has been on the decline in these countries and obesity on the increase, the original programs have continued to deliver huge amounts of food to beneficiaries who are not malnourished. One consequence is that these programs have served to fuel the overweight-obesity epidemic. A study conducted on the impact of the preschool National Nursery Schools Council Program in Chile found that there was a threefold increase in obesity in children receiving supplementary feeding (Uauy et al., 2001). Furthermore, supplementary

feeding programs such as school feeding usually target children above the age of 3 and therefore often come too late, given that the damage to human growth, brain development, and human capital caused by malnutrition is greatest—and largely irreversible—during gestation and the first two years of life (Cleaver et al., 2006). In low-income countries where school enrollment does not include the entire population of school-aged children, school feeding may exclude the most vulnerable.

Fortification—the addition of nutrients to commonly eaten foods—is used to combat micronutrient malnutrition and is increasingly recognized as an effective medium- and long-term approach to improve the micronutrient status of large populations. Milk and dairy products are common food vehicles in this regard. Today, the business sector is taking the lead in supplying fortified milk and in supporting supplementary feeding programs in low-income countries. For example, Tetra Pak, in partnership with governments, has been involved in providing fortified milk to schoolchildren in Bangladesh (IFPRI 2004b), Nigeria, and Pakistan (World Bank 2007). In Bangladesh, a joint venture between Grameen Bank and Danone corporation has also facilitated the development of low-cost fortified yogurt for sale to rural communities. In support of fortification programs, Argentina, Brazil, Guatemala, Honduras, Malaysia, Mexico, Chile, Venezuela, and the Philippines have established mandatory fortification of milk with vitamins A and D. Public–private partnerships such as the new Global Alliance for Improved Nutrition have supported the provision of fortified food to address the problem of micronutrient deficiency (WHO 2003b).

At the national level, large-scale multisectoral programs have been implemented to address the problems of hunger and undernourishment. Brazil's unique *Projecto Fome Zero* (Zero Hunger project) is one of the most recent and commonly cited programs that integrates structural policies (such as income distribution, production increase, employment generation, agrarian reform, and emergency interventions, among others) aimed at addressing the causes of hunger with policies to increase access to food (Takagi et al., 2006). The project has been instrumental in improving the nutrition status of vulnerable groups in Brazil either through the food card program that targets poor regions with the aim of ensuring access to a daily food intake or through the innovative food purchase program that encourages family farming through the purchase of small farmers' production. Livestock commodities that benefited from the food purchase program include milk, meat, and cheese.

Food-specific subsidies, whether in the form of general subsidies, ration cards, or quotas, have been used to increase consumption of animal source foods, with positive effects on health and nutrition of undernourished populations. China is an example of how subsidies can influence consumption patterns: in the early 1990s the government instituted programs aimed at improving the nutritional status of the population. Foods like meat and eggs were subsidized, and the proportion consumed as part of the diets increased as a result (D'Arcy et al., 2006). In the 1970s and 1980s, the campaign to raise livestock production in the USSR was supported by sizable subsidies to producers for livestock products, particularly beef and pork (FAO 2003b). By 1990, Russian food and agricultural subsidies, primarily for livestock products, reached 11.8% of gross domestic product (GDP) (Dinar and Subramanian 1997). The result of this support was that consumption in central and eastern Europe and the Commonwealth of Independent States was much higher than income levels would suggest.

In many developing countries, food subsidy programs are a major part of nutritional programs. The aim is to improve the accessibility of food to vulnerable sections of the population by ensuring food price stability and lowering price levels. They have been commonly used in several countries and regions: Latin America, South Africa, the Middle East, and North Africa. In Mexico, milk subsidy programs (a subsidy of about 70%) have been used to target poor and low-income neighborhoods in urban areas. In 1997, the LICONSA milk program distributed 11% of total domestic milk consumption, benefiting over 2 million children in rural areas and 3.4 million children in urban areas (ERS-USDA 2000). Although food subsidy policies in the Middle East and North Africa have been scaled backed, they still remain widespread. For the most part, milk has been the most commonly subsidized animal source food in that region. However, other products like eggs and meat have also been included in the subsidy programs; for example, Egypt imports meat from Sudan, which is sold at less than half the price of locally produced meat (FAO 2006c).

In summary, while substantial progress has been made in addressing undernourishment, the continued persistence of malnutrition—particularly in low-income countries—is an indication that the responses used so far are inadequate to achieve food security. Food insecurity is often perceived as a problem of inadequate food supply and production; as a result, emphasis has been placed on increasing national food supplies. This has been clearly demonstrated in South Asia, where, although national food availability and per capita income are higher relative to other low-income developing countries in other regions, undernutrition and micronutrient deficiency remain endemic.

Overnutrition and Diet-Related Diseases

In several countries, human diet, activity patterns, and nutritional status have undergone a sequence of major shifts, with animal source foods playing a major role in the changes. Popkin (1994) characterized these changes as the "nutrition transition," which has been associated with increased consumption of superior grains, higher fat and sugar foods, animal products, and processed foods. People are shifting from consuming traditional diets based on coarse cereals and vegetables to energy-dense diets richer in saturated fat, added sugar, dairy, and animal products (IFPRI 2001).

As a result, overweight and obesity are rapidly growing in all regions, affecting children and adults alike. Globally, there are over a billion people overweight, with at least 300 million of them categorized as obese. These problems are now so common in some developing countries that they are dominating more traditional public health concerns such as undernutrition (SCN 2000).

Differences between socioeconomic groups are blurring, and there is a diminishing or reversal of the differences between income groups and the urban–rural populations. In the more affluent middle-income countries, obesity tends to be associated with lower socioeconomic status. The reverse is true in low-income countries, where obesity is more common in people of higher socioeconomic status and those living in urban areas.

The reasons for the alarming rate at which obesity and diet-related diseases are spreading are manifold. Behavioral factors, including dietary intake, physical activity, and sedentary lifestyles, have been important contributors. Intakes of total fat, animal products, and sugar are increasing just when the consumption of cereals, fruits, and vegetables is declining (World Bank 2006). In addition to genetic predisposition and lifestyle, increasing consumption of energy-dense foods has been a major contributing factor to the obesity epidemic. Another causal factor that has received little attention is the economic environment in which obese and overweight people live and the link between income and diet quality, food choices, and obesity.

Responses to overnutrition and diet-related disease range from the use of regulatory and market instruments to communication and awareness, engaging the international community, governments, and the food industry.

In Latin America, some governments have made significant efforts to restructure existing nutrition programs. In Brazil and Chile, the school feeding programs that were originally conceived to curb undernutrition are currently being restructured into obesity prevention ones. New legislation in Brazil mandates that at least 70% of the program's annual budget (of about $500 million) must be spent on fresh fruits and vegetables, with minimal inclusion of processed foods (Jacoby 2004, Coitinho et al., 2002). Chile is presently adapting its preschool food supplementary program (National Nursery Schools Council Program) by lowering the sugar and saturated fat content of the rations and providing skimmed milk and fresh fruit and vegetables (Uauy et al., 2001).

As the prevalence of overweight, obesity, and nutrition-related chronic diseases and their associated economic costs of treatment increase, governments have taken to preventive measures such as monitoring, nutrition education, and consumer awareness to address nutrition-related chronic diseases and obesity. Most Latin American countries are now aiming to strengthen consumer options through better and simpler labeling of food and through campaigns that provide the best options based upon available resources—stressing the benefits of eating a traditional diet that is high in plant sources and fruits and low in fat and meats. Countries like Chile, China, Cuba, Iran, India, Morocco, South Africa, South Korea, and Thailand have large-scale monitoring systems that assess the prevalence of overweight, obesity, and nutrition-related chronic diseases. Similarly, nutrition education, particularly in schools, coupled with physical activity programs is being implemented by countries such as Brazil, China, Malaysia, Mexico, South Africa, and South Korea.

Countries such as Brazil, Thailand, and Singapore use mass media campaigns to address the problem of overnutrition. As part of the Regional Strategy and Plan of Action on an Integrated Approach to the Prevention and Control of Chronic Diseases, Including Diet, Physical Activity, and Health, the Pan American Health Organization launched a campaign whose objectives are to create awareness and to prompt action to tackle the current epidemics of obesity and chronic disease. In its initial stage (2006–08), the campaign is focusing its efforts on Chile, Colombia, Costa Rica, México, and Venezuela. In Brazil, the use of mass media campaigns directed at obesity has been hailed as one of the important causal factors in the decline in obesity, especially among urban Brazilian women in the upper 75% of the income level, with the incidence of obesity reducing by over 28% from 1989 to 1997 (Monteiro et al., 2002).

Countries in the Pacific islands have taken drastic measures to deal with problems of overweight and obesity. The government of Fiji, concerned about the high fat content of low-quality sheep meat (mutton flaps) and turkey tails and the health consequences of importing such products, imposed an import ban on mutton flaps and instituted a ban on the sale (whether imported or locally produced) of these high-fat foods (Curtis 2004, Nugent and Knaul 2006, WHO–Western Pacific Region 2003a). Following this lead, the Tongan authorities have banned imports of mutton flaps. In 2007, the Samoan government also banned the import of turkey tail meat from the United States in support of health preventive measures to the rapidly expanding problem of obesity and diet-related noncommunicable diseases.

National governments have put in place regulatory frameworks ranging from broad food-based dietary guidelines and national nutritional plans to specifically address the problems of under- and overnutrition. Food-based dietary guidelines aim to promote general nutritional well being as well as to prevent and control both ends of the malnutrition spectrum. The guidelines generally recommend the consumption of livestock products

in moderate amounts, with an emphasis placed on the consumption of fat-free or low-fat dairy products and lean meats (FAO 2009). Industrial countries have developed national food-based dietary guidelines that promote healthy eating. Generally, these advocate replacing saturated fats (animal fats) with unsaturated vegetable oils or soft margarines, replacing fatty meat and meat products with lean meats or white meats, and using milk and dairy products that are low in fat.

Recommended nutrient intake is essential in establishing specific targets or benchmarks against which dietary intake can be assessed and monitored (WHO 2004). WHO and FAO have developed international standards that countries can adopt. Industrial countries have established national recommended nutrient intake for their populations based on these international standards.

Industrial countries have also responded to risk factors for overnutrition through the development of policy framework documents. The European Union has over the past few years developed a series of instruments with the objective of addressing problems of overnutrition. Table 17.1 summarizes the main policy instruments, target groups, and issues addressed.

Many countries recognize the growing problem of obesity as a priority in their national strategies. Some EU countries have developed population-wide national plans of action or public health strategies dealing with obesity and related risk factors while others specifically target children either through action plans or through regulations/guidelines on marketing of food to children. Others have developed separate documents to promote physical activity.

Product promotional campaigns have been used to encourage consumers to switch from full-fat livestock products to low-fat products. The "1% or Less" campaign in the United States is an example of this. The campaign is a health-education program that aims to reduce the total and saturated fat consumption of communities by encouraging adults and children over 2 years of age to switch from drinking whole or 2% milk to 1% or fat-free (skim) milk.

Industrial countries have also used nutrition intervention programs to address noncommunicable diseases, which are a major health burden in these countries. Analyses of major public health programs, such as those in Finland and Norway, have shown that the substantial fall in coronary heart disease rates is predominantly explained by a 15% fall in average serum cholesterol levels as the consumption of milk fat—in milk, butter, and milk products—drops (WHO 2004). Finland provides an excellent example of sustained improvements in cardiovascular disease mortality (Box 17.5).

Other countries have successfully experimented with policies to prevent chronic diseases, such as Poland, which achieved a 6–10% annual reduction in cardiovascular disease deaths in the 1990s by removing price

Table 17.1. EU regional policy instruments addressing overnutrition.

Action framework	Target group	Priority areas
Platform on diet, physical activity, and health (2005)	• Stakeholders (private and public) forum/platform	• Obesity • Noncommunicable diseases
European strategy for the prevention and control of noncommunicable diseases (2006)	• High-risk individuals	• Noncommunicable diseases
Blueprint for action on the promotion of breastfeeding in Europe (2004)	• Mothers • Community • Health and social systems	• Obesity • Infections
European strategy for child and adolescent health and development (2005)	• School-age children	• Obesity
European network for the promotion of health-enhancing physical activity	• Community • School-age children	• Obesity • Cardiovascular diseases • Diabetes

Source: WHO 2004, 2007.

subsidies from butter and increasing the availability of cheaper vegetable oils (CSIS 2006).

Market-driven responses are increasingly emerging from the food industry in response to mounting pressure from consumers to produce more nutritious foods and to market them more responsibly. Industry responses range from the development of healthy products and the reduction in portion sizes to consumer awareness campaigns and support for physical activity.

Box 17.5. Nutrition Intervention in Finland—The North Karelia Project

In the early 1970s, in response to a worryingly high mortality rate from coronary heart disease, the Finnish government, in collaboration with WHO, set up an intervention at the community level in North Karelia that aimed to improve the population's blood pressure and cholesterol levels by reducing saturated fat intake, increasing fruit and vegetable consumption, and reducing salt intake. Following successful results, the measures were implemented nationwide.

The changes in the North Karelian and Finnish diets have been quite remarkable. In the late 1970s, about 60% of Finns reported using mostly butter; by 1998, the proportion was only 5%. This contributed significantly to the decrease in saturated fat intake. In 1978, 40% of the men and 35% of the women used fatty milk; in 1998, the respective proportions were only 9% and 4%. Low-fat milk became the standard milk used in schools as early as the 1970s.

The dietary changes have been associated with a high decline in cholesterol levels, which decreased by 18% between 1972 and 1997 in North Karelia. Mortality rates of ischemic heart disease decreased by 73% in North Karelia and by 65% in the whole country from 1971 to 1995.

Source: North East Public Health Observatory 2005, Puska 2002.

Research conducted by Ethical Investment Research Services (EIRIS 2006) documents that food manufacturers are developing new, calorie-reduced products or are improving existing ones by lowering fat and sugar levels. Food companies like McDonald's have introduced lower-fat dairy options (Lewin et al., 2006). Reducing the size of single-portion products also appears to be a standard way of meeting health concerns, although it is not yet practiced widely.

In order to reduce fat intake, countries have used standards to set limits on the amount of fat in livestock products. USDA has "standards of identity" that limit the fat content of certain processed meats such as ground beef, hot dogs, and pork sausages. Carcass classification in the EU has played an important role in the gradual improvement of meat carcass quality by promoting leaner meat production to meet consumer demand. The EU has encouraged the meat industry to use carcass classification grid systems, whereby carcasses are graded by their conformation and their lean meat content (European Commission 2003).

Nutrition labeling is also a potential measure in the overall strategy to combat overnutrition and the associated diet-related noncommunicable diseases. Clear, concise, and simple labeling is one way of ensuring that consumers have the necessary information and tools to make informed decisions regarding food choices for improved health. Many countries have regulations requiring some form of nutrition labeling. Overall, countries can be characterized as having one of the following types of regulatory environments: mandatory nutrition labeling on prepackaged foods (Brazil, the United States, and Australia); voluntary nutrition labeling that becomes mandatory when a nutrition claim is made (the EU) or for certain foods with special dietary uses (India, Pakistan, and China); or no regulations on nutrition labeling (WHO 2004).

Though this is likely to change in the future, nutrition labeling in the EU is currently not compulsory on food packaging; it only becomes compulsory if a nutrition claim such as "low fat" or "reduces cholesterol" is made on the product label or in an advertising message. In the United States, the requirement for nutrition labeling has stimulated the food industry to reduce the amount of fat in a wide range of products (Schäfer et al., 2006).

In countries like Finland and Poland replacement of products containing saturated fats has been achieved through the creation of markets for healthy products. Generally, there is a move toward the consumption of less fatty livestock products to leaner or low-fat products. In Finland, soft margarines and butter–oil mixtures as well as low-fat spreads have replaced butter—originally the traditional source of fat in the Finnish diet (Puska 2002).

Notwithstanding the efforts made to counter the obesity epidemic, the problem of obesity and diet-related disease is increasing at a faster pace throughout the world. So the question is, Why is the epidemic spreading faster than expected?

While developing countries are only beginning to grapple with the problems of overnutrition, industrial countries have focused on interventions aimed at enhancing consumer awareness, on promoting physical activity and changes in lifestyle, and on developing legislation targeting obesity—all valid entry points if obesity is to be addressed. These actions have also been aimed at reducing the demand for certain types of food. But the broader public health issues raised by agricultural and food policy, the impacts of the globalization of the food system, the role of advertising and information asymmetry, and the link between income and dietary choices have been overlooked by policy makers.

Agricultural policy as well as improvements in agricultural productivity have been key factors in the rising dietary energy supplies in high-income countries. As such, directly or indirectly, these policies have contributed to the obesity epidemic. Agricultural policies in industrial countries have profound and complex effects on food supply as well as on demand through production incentives that stimulate oversupply of certain foods, fueling the obesity epidemic. Food surpluses have induced marketing measures aimed at increasing consumption, and this has led to excessive domestic consumption with negative health effects (WHO 2007).

The EU's Common Agricultural Policy (CAP) is an example of a policy that conflicts with public health policies. Despite reform of the CAP, the livestock sector in the EU still remains heavily supported—support that has stimulated overproduction. (See Box 17.6.)

Despite the continued support to farmers, the EU is increasingly aiming at improving the quality of the product and encouraging more environmentally sustainable farming practices (European Commission 2003).

Box 17.6. EU Subsidies to the Livestock Sector

Within the European Union's CAP, there are a number of different forms of support for the farming sector. A large part of the support is for crops, including fodder crops, but a good deal also goes specifically to animal products. This has been achieved in a variety of ways, many of which are counterproductive to achieving good public health.

Two main types of support in the EU's agriculture budget benefit producers of animal products. In part, what is called direct support is paid out to farmers who have a certain kind of animal or who produce a certain type of animal product. Additionally, there is a kind of support that is referred to as *intervention*, which involves the EU going in and assuring demand for farm products in different ways. Interventions include the EU providing financial aid for the export of a given product to countries outside the EU. The EU also buys up and stores the surplus of a given product at a guaranteed price, called the intervention price, so that the producers are guaranteed income for the product they produce. The EU also provides support for marketing of different animal products so that their sales will increase. In the EU budget for the financial year 2007, interventions and direct support to the livestock sector amounted to 3.5 billion euros.

Source: EU 2007, Holm and Jokkala 2008.

Furthermore, in 2007 the EU Council, in an attempt to reduce childhood obesity, adopted a package of measures that will introduce a flat-rate subsidy (Euro 18.15/100 kg) for the European school milk scheme. This means that for the first time skim milk distribution will have the same financial support as full-fat milk (EPHA 2007, EU 2007).

The pricing of production input commodities is another factor that has circuitously influenced public health in industrial countries. U.S. farm policy is a good case in point: it has been increasingly directed toward driving down the price of a few farm commodities, such as corn and soybeans. By keeping these costs artificially low, U.S. farm policy provides an indirect subsidy to grain-fed livestock and therefore encourages production of cheap livestock products. Furthermore, by enabling the production of low-cost feed grains, agricultural policies create unfair advantage to grain-fed as opposed to grass-fed livestock. The latter have been shown to be higher than grain-fed beef in health-promoting nutrients, omega-3 fatty acids, and cancer-fighting conjugated linoleic acid (IATP 2006). High-grain diets are also said to increase the fat content of livestock. Agricultural policy in industrial countries, by supporting intensive livestock production, also facilitates the increased use of hormones, antibiotics, and so on that may pose numerous health risks.

The globalization of the food system has altered the quantity, type, and cost of food, generating food inequalities across various income groups by making healthy foods accessible to only the richer sections of society. Globalization processes operating throughout the food supply chain therefore have different impacts on the different forms of malnutrition. Globalization may introduce opportunities to address undernutrition by cheapening food, but in so doing it may also introduce risks for overnutrition. (See Box 17.7.)

Promotion campaigns have been used to increase the consumption of livestock products, particularly in the United States. The beef, pork, dairy, and egg industries, with administrative assistance from USDA, have facilitated the promotion of foods that are inconsistent with its own dietary guidelines (e.g., "Pork—the other white meat," "Beef gives strength," and "Got Milk?"). The U.S. National Dairy Council has spent $200 million since 2003 promoting the idea that milk can help people lose weight. The U.S. mandatory Check-off Program, which enforces the collection of more than $600 million annually from producers, is used to fund advertising campaigns aimed at increasing consumption of these commodities. The largest food commodity check-off programs are for meat and dairy products (Wilde 2006).

There is also an imbalance in information between the food industry and consumers, which has resulted in unmatched consumption of energy-dense foods. Most consumers are unaware of the quantities of added fats and sugars in processed foods. And while nutrition labeling addresses this problem, there is the concern that labeling is of benefit to mainly consumers who are in many cases already better educated about nutrition, and it is obviously of limited value in areas where literacy is weak (Hawkes 2006, Haddad 2003).

The voluntary food industry efforts to help tackle the obesity epidemic have been called into question by a recent analysis of the commitments made by 25 of the largest food companies. The report showed that the majority of the 25 had made general statements about diet and health but that less than half had made any policy commitments, with little implementation, and that only 4 had stated support for a voluntary code on advertising to children (Lang et al., 2006). Schäfer et al. (2006) argue that the major problem with involving the food industry in partnerships aimed at reducing obesity is the potential for serious conflicts of interest, especially when their products are recognized as some of the leading causes of overweight and obesity.

Lewin et al. (2006), in a study on the supposed changes in marketing of foods to children in the United States by McDonald's, found that for every positive step taken by the food company, countermeasures were also put in place that could weaken efforts to promote healthy eating. For example, although the company has introduced healthier foods such as salads and low-fat dairy options, it also introduced new food items such as chicken salads high in calories (450–660g) and fat

---

Box 17.7. International Trade in Cheap High-Fat Foods: A Need for Regulation?

Evans et al. (2001) researched the preferences and frequency of consumption of different foods in the small south Pacific island country of Tonga. They concluded that despite people's awareness of the nutritional value of available foods and despite their preference for more nutritious traditional food, economic forces have induced a shift toward consumption of cheap imported meats such as high-fat mutton flaps (sheep bellies). Cost and availability considerations lead to unhealthy food choices that contribute to the high incidence of obesity and diet-related chronic diseases. This problem was considered serious enough in Fiji to lead the government to ban the sale of mutton flaps. The Western Pacific Regional Office for the World Health Organization has produced a manual on how domestic laws can be used to regulate the availability of such foods that contribute to obesity.

Ghana has used a regulatory approach that involves setting a composition standard that prescribes a fat content level for meat and poultry products. Poultry imports must have a fat content of less than 15%; beef, less than 25%; and pork, less than 35%. These regulations have effectively halted U.S. exports to Ghana of turkey tails, which typically contain at least 30% fat.

Source: Evans et al., 2001; WHO-Western Pacific Region 2003a, 2003b; SCN 2004.

(11–29g). McDonald's pricing strategies were also found to promote less healthy options and the use of the "one dollar" menu to sell food that was less healthy.

While the problem of overnutrition has risen to the top of the public agenda in several countries, little attention has been paid to the subtle link between diet, economics of food choices, and people's socioeconomic status. Several recent studies suggest that there is a link between obesity and poverty (Drewnowski and Spector 2004, Drewnowski 2004, Monteiro et al., 2004, Tanumihardjo et al., 2007). Intuitively it would seem that obesity and poverty could not be linked, but obesity has been found to be more prevalent in the lower economic strata in industrial countries. In the United States, the highest rates of obesity and diabetes tend to cluster among the groups with the highest poverty rates (Drewnowski and Darmon 2005). A similar pattern can be found in Europe, with obesity common among socially deprived communities, characterized by lower income and education (WHO 2007). This trend is beginning to emerge in middle-income countries. A recent review by Monteiro and others (2004) of the incidence concluded that at as GNP increases in developing countries, the burden of obesity tends to shift toward groups of lower socioeconomic status. Research on mechanisms that link social economic status to obesity is still lacking in the developing world, and the subject certainly deserves more attention.

Income directly affects obesity through its effects on expenditure and consumption of energy-dense foods. Drewnowski and Spector (2004) argue that limited economic resources may shift dietary choices toward cheap, energy-dense, convenient, and highly palatable diets that provide maximum calories for the least volume and lowest cost. In addition, people in the lower economic strata have to work long hours and often do not have the time to prepare healthful meals or the knowledge about what such a meal consists of.

So far, the policy response to problems of overnutrition has not been bold enough and has focused on single factors or called on individual responsibility through educational campaigns. Coupled with this is the argument of the right of individual "free choice," which is often advanced in industrial countries. In several industrial countries, nutritional habits are seen as a matter of individual choice and hence government intervention is not warranted.

But the obesity pandemic cannot be halted simply by merely encouraging individuals to reduce their risk factors and adopt healthier lifestyles. There is a need to address the whole environment that subsidizes and promotes obesity using a multisectoral approach. High-income countries still tend to focus on consumer behavior—with a reluctance to tackle the more structural drivers of change. Implementing policies that address the root causes of overnutrition implies balancing a multitude of conflicting interests and confronting several forces and institutions—a path that governments are often unwilling to embark on. Policies that are closer to the consumer, such as labeling, marketing regulations, and restrictions, although worthwhile, are subject to being undermined as they affect the incentives of both the upstream and downstream sectors along the agrifood chain: such as the production sector and the processing, marketing, and retail sectors.

Food Safety

Food safety is becoming a dominant theme worldwide as the incidence of foodborne illness increases. This is in part a reflection of the prevalence of foodborne illnesses worldwide and in part consumer concerns about the safety and quality of the food they consume, particularly in industrial countries. Food safety has therefore become a clear objective of governments in many areas, as indicated by the emergence of new standards and methods of regulation, new institutions, and increased investments in the field.

Although foodborne diseases are prevalent in both developing and developed countries in many ways, there is a stark difference between these nations. Consumer attitudes toward risk as well as the food safety risk levels,

priorities, and approaches to safety and quality vary significantly. Countries have responded in different ways to the growing concern of food safety: some have approached the problem from the perspective of domestic consumer welfare while others, those with a strong export orientation, have responded for fear of losing export markets.

Food systems in developing countries are not always as well organized and developed as in the industrialized world. Moreover, problems of growing populations, urbanization, limited resources to deal with this issue, and lower or poorly enforced regulatory standards imply that food systems in developing countries will continue to adversely affect the quality and safety of food supplies. People in developing countries are therefore exposed to a wide range of potential food quality and safety risks (FAO 1999).

Food safety concerns in developing countries typically include the inappropriate use of agricultural chemicals; use of untreated wastewater or manure on crops; the absence of food inspection, including of slaughtering and processing facilities such as those for meat; poor hygiene; the lack of appropriate infrastructure, such as cold chains; and the lack of appropriate information on food safety (FAO 2001). In many low-income developing countries, many livestock products are marketed through informal markets. For example, informal milk markets account for nearly 90% of milk sales in Kenya (Omore et al., 2002).

Developing countries are also being challenged to improve food safety for their growing affluent middle class. As economies develop, food processing and preparation tend to shift outside the household, supermarkets increasingly dominate urban food retailing, and consumer food safety demands begin to drive improvements in food safety. The Chinese government has established "green food" certification for a wide range of products, including beef. This was in response to food safety concerns raised by affluent urban consumers who are willing to pay a premium price. A survey revealed that affluent consumers are prepared to pay premiums of 20–30% for "green foods." At the production level, the certificate prohibits use of growth promoters, imposes withholding periods for some veterinary products, and sets national standards to be met on the use of feed additives and antibiotics (IFPRI 2003).

For developing countries, particularly emerging economies, food safety issues in addition have important implications for export opportunities to countries with low risk tolerance and stringent food safety standards. Food safety concerns in industrial countries have led to the application of strict food standards and regulations, such as the use of microbiological and chemical risk assessment techniques and the integration of food safety and food production policies to develop risk-based, farm-to-table approaches to control foodborne hazards

(WHO 2004). Increased food trade also means that chemical contaminants, toxic materials, pesticides, veterinary drugs, and other agrochemicals are under constant surveillance to ensure food safety along the entire food chain. For example, the EU has established regulations that establish procedures for setting maximum residue limits of veterinary medicinal products in foodstuffs of animal origin. Other examples include the regulation of the use of hormones in meat production in the EU.

In order to benefit from trade, developing countries usually have to adopt importing-country standards, as failure to meet international food safety standards can result in significant financial losses (FAO 2003a). The requirements set in place by industrial countries have in effect transferred regulatory responsibility to other countries—often a prerequisite that has to be met if entry into industrial-country markets is to be assured. A public- and private-sector initiative in Brazil aimed at reducing mycotoxins in animal feed is an example of an intervention that aims at enhancing food safety geared at meeting standards in importing countries (IFPRI 2003).

Similarly, Argentina and Uruguay have implemented traceability systems that indicate whether the animals have been exposed to foot-and-mouth disease. Both countries promote their grass-fed beef production as a "healthier" system, which, combined with bans against feeding meat and bone meal to cattle, reduces bovine spongiform encephalopathy (BSE) concerns among importers. Export markets are important to both countries, and their beef industries are increasingly shaped by the demands of foreign buyers. Argentina and Uruguay implemented mandatory national animal identification systems after having less formal systems in place for several years. These programs now make it possible to track animals from birth to slaughter (McConnell and Mathews 2008).

There have been an increasing number of national and international initiatives to implement the application of food safety management systems in order to ensure food safety and safeguard international trade of animal products. Examples include the official World Organization for Animal Health standards on *Terrestrial Animal Health Code*, which sets out the recommended health standards for international trade in animals and animal products. And regions such as the EU have adopted a new Animal Health Strategy that addresses not only animal health issues but also issues inextricably linked to animal health, such as public health and food safety (European Commission 2007).

In industrial countries there has been a more general shift in how food safety issues are managed, with increasingly shared responsibilities between different stakeholders—governments, food industry, and consumers—characterized by a continuous modification of existing regulatory framework to address the ever-increasing risks. Industrial countries have responded to

the immense public concern regarding food safety by broadening the scope of and tightening standards and regulations across the food chain. Regulation can take various forms—from prohibition, where certain products or processes are banned (e.g., compulsory removal of the spinal cord from beef and sheep carcasses or a ban on the use of meat and bone meal as feed) to prescriptive measures that describe in detail procedures for dealing with recognized problems (e.g., a mandatory hazard analysis and critical control point [HACCP] system for all levels of the food chain) and information regulations such as certification, labeling, and traceability.

Certification, consisting of different standards and regulations relating to food safety, generally aims at providing consumers with better information about the characteristics and quality of food products. In industrial countries, regulatory systems increasingly require that safety assurance actions be taken along the whole food supply chain. Certification schemes can range from traceability to labeling (such as mandatory country-of-origin labeling standards for beef, lamb, and pork in Japan, the United States, New Zealand, and Australia) and quality marks and product branding (such as the U.K. quality standard marks for pork, sausages, bacon, and ham that help consumers distinguish between imported and British products).

Countries may require documentation tracing a food product back through the supply chain to its source or forward through the chain to the consumer. For example, the EU requires mandatory traceability for genetically modified feed in the entire food chain—from farmers through processors up to retailers and restaurants. Similarly, the 2002 Japanese law relating to BSE countermeasures requires mandatory trace-back for cattle from feedlot to the packaging plant.

Surveillance for foodborne diseases is an increasingly high priority issue on the public health agenda in many countries. Such systems help eliminate the burden of foodborne diseases, assess the relative impact on health and economies, evaluate disease prevention and control programs, and allow for rapid detection and response to outbreaks. Examples of foodborne surveillance systems include the EU Antibiotic Resistance Surveillance system, the U.S. National Antimicrobial Resistance Monitoring system, and the Danish Zoonosis center, which has a range of surveillance programs for animal feed, poultry, pigs, and cattle.

During the last several decades, risk assessment, risk management, and risk communication have been formalized and incorporated into a process known as risk analysis. This new approach enables information on hazards in food to be linked directly to data on the risk to human health, a step that was not considered in the past. By providing a science-based approach to improving food safety decision-making processes, risk analysis contributes to a reduction in the incidence of foodborne disease and to continuous improvements in food safety (FAO and WHO 2005). FAO and WHO have been at the forefront of the development of risk-based approaches for the management of public health hazards.

Other tools used by governments in industrial countries to monitor food safety include product bans, recalls, or rejections. In order to address food safety risks, countries have banned the use of products such as meat and bone meal (following the BSE crisis) and imports of livestock and livestock products in the event of disease outbreaks (e.g., the mad cow crisis, foot-and-mouth disease, and avian influenza). In December 1985, for example, the EU—referring to consumer concerns—announced that as of January 1988 all imports of meat raised using hormones would be banned. In March 1988, the EU decided not to use hormones within the EU itself. And in January 1989, the EU began enforcing a total ban on imports of meat raised with growth hormones (Animal Health and Production Compendium).

Food authorities in several industrial countries are already using market forces as a tool by informing consumers of the results of inspections in individual food establishments, with the primary purpose of influencing food safety through "naming and shaming" (WHO 2004). For example, in the United Kingdom the inspection results from all slaughterhouses and meat processors are published according to a special classification system.

The realities of food safety have brought about a new paradigm in stakeholder responsibility characterized by a shift of responsibility away from official public control food systems and toward systems established by producers and food processors. Traditionally, there existed clear-cut distinctions on the roles of the public and private sectors regarding food safety and quality. With increasing food scares and consumer demands for safe food, these differences are being dismantled as co-regulation approaches are used.

Today, the private sector is taking a leading role in incorporating consumer concerns regarding health, quality, and safety by setting voluntary standards that spell out various technical specifications governing quality and safety. Private-sector food labeling practices, product quality standards, and buyer requirements to meet specific food safety standards as a condition for doing business have proliferated.

Traceability is another tool used by the food industry in industrial countries to enforce food safety. Food suppliers in a position to provide traceability information are more likely to draw interest from supermarkets and consumers and to gain competitive advantage. Traceability also permits brand protection and fraud detection. An Irish supermarket that uses DNA testing to trace meat to the animal of origin rather than to the herd or farm is an example of initiative taken by the food industry. Similarly, in the Netherlands a veal processor is responding to consumer preferences for trace-back information

on an Internet site. Traceability schemes are relevant not only for safety and quality but also for essential product differentiation.

Vertical food industry alliances are becoming an important means of assuring food safety and quality. Vertical supply chain alliances between producers, processors, and retailers to enable credible assurances of food safety and quality have emerged as a private-sector response to food safety concerns and changing regulatory environments (Hobbs et al., 2002). *Tracesafe*, a farmer-owned company in southwest England that operates a unique cattle traceability and quality assurance system, is one example of this. The *Tracesafe* Cattle Management System is a network of breeders and finishers that allows the history of individual meat cuts to be traced back to the animal of origin. The beef is targeted at specialist retail outlets and high-quality restaurants, where consumers are willing to pay a premium for the assurance of guaranteed traceability. All grain is supplied from a network of mills contracted to provide specially prepared rations. Independent auditing of breeding and finishing units is carried out under the ISO 9002 accreditation requirements. Complete details of an animal's life, including parentage, medication, feeding, and any movements, are documented (Fearne 1999).

The food industry has shifted away from the traditional focus on end product testing (quality control) and toward quality assurance systems that manage food safety along the entire supply chain, from "farm to fork." Examples of systems operating worldwide include the Codex Alimentarius, Good Agricultural Practice, the ISO series, and HACCP (Animal Health and Production Compendium).

Farming and food processing have been plagued by several food scares, which have given rise to increasing demands from consumers for organic products. Consumers are worried not only about diseases such as BSE and avian influenza but also about other pathogens, hormones, and chemical residues that might be present in food. In addition, with the rapidly increasing rate of obesity, consumers are now focusing on the nutritive value of organic foods. Australia has the largest surface area devoted to organic farming (approximately 10 million hectares), mostly represented by pastures, with milk and meat being important commodities. Europe and North America have, respectively, approximately 5.5 million and 1.5 million hectares of land under organic management (Thilmany 2006).

USDA has developed guidelines for healthier meat and dairy consumption—the *Eat Well Guide* that encourages consumers to buy organic livestock products from animals raised without the use of antibiotics, hormones, genetic engineering, and so forth. (IATP 2005b). In the United States, demand for organic livestock has increased nearly 161% since 1992. Organic poultry sales, for example, have quadrupled since 2003, with annual growth rates projected to be between 23% and 28% by 2010. This increased demand for organic meat can be attributed to the heightened consumer concerns related to food safety issues (ERS-USDA 2006).

Among developing countries, the largest organic sectors are reported in Argentina and Brazil, which mainly supply external markets, particularly in Europe (FAO 2002). Uruguay's grass-fed production system is internationally recognized as an independently certified source of natural beef. The "Certified Natural Meat Program of Uruguay" maintains consumer confidence and differentiates Uruguayan meat through certified compliance with international protocols for animal production and industrial practices (McConnell and Mathews 2008).

Food regulation is usually based on domestic standards and regulations. Superimposed on these, however, is the need for governments to operate within an international framework and rules. The international framework on food safety has developed significantly through the enhanced role of certain international organizations, such as the Codex Alimentarius, under the WHO and FAO, and the World Trade Organization's Agreement on the Application of Sanitary and Phytosanitary Measures. Countries are encouraged to adopt international standards, such as those agreed to by the Commission of the Codex Alimentarius, an international expert body dealing with food safety issue. The main problem for poor countries is the cost of compliance. Several key international organizations offer technical assistance to developing-country governments faced with compliance problems. For example, FAO has proposed a food safety and quality fund to provide grants to the least developed countries to strengthen their systems (FAO 2001).

In developing countries, the lack of technical and institutional capacity to control and ensure compliance with international standards essentially compromises food safety. Inadequate technical infrastructure—in terms of food laboratories, human and financial resources, national legislative and regulatory frameworks, enforcement capacity, management, and coordination—weakens their ability to confront these challenges. Such systemic weaknesses not only threaten public health but also may reduce trade access to global food markets. Umali-Deininger and Sur (2007) have also cited the role of cultural issues, such as religious beliefs, that may further constrain the adoption of appropriate food safety measures. In India, the sacred value attached to cattle puts limits on disease control to address food safety and public health, such as culling to prevent disease spread. In addition, improving food safety in informal supply chains characteristic of developing economies raises questions at both the technical and organizational levels.

The complexity of food safety makes the right policy response difficult to identify, especially where there is limited evidence on the magnitude of the problem. What is certain, however, is that food safety issues will continue to plague food systems in various countries in varying degrees. In developing countries, government interventions to ensure food safety in low-income developing countries should be geared toward providing information and education, as well as infrastructure and training, particularly in low-income countries where informal food markets are predominant. While the risks can be minimized, it is prudent to argue that society cannot expect zero risk when it comes to food safety—implying that policy makers together with scientists and the food industry will have to make efforts to define what an acceptable level of risk may be.

Addressing Nutrition Problems: What Needs to Be Done?

The insufficient gains in reducing food insecurity and undernutrition coupled with the emergence of new diet-related diseases calls for a reform of the current policy regime designed to address overall malnutrition. The complexity and speed of change of nutritional problems today in the context of the rapidly changing global food systems pose important challenges for policy makers. The coexistence of underweight and overweight people in countries throughout the world has widened the scope of nutrition problems. Although these problems often occur in the same society, the common approach has been to deal with them independently. Because of the rapidly changing nature of nutrition-related problems, with countries moving along a continuum from undernutrition to overnutrition, intervention strategies have tended to evolve in a rather haphazard way.

Few countries have launched multisectoral initiatives on the double burden of malnutrition. In several middle-income countries, efforts to eradicate undernutrition have sometimes resulted in an increase in the prevalence of obesity. It is therefore important that public health programs simultaneously address both underweight and overweight. China is one of the few countries that has established a multisectoral approach that addresses both sides of the nutrition problem (ESCAP 2005). (See Box 17.8.)

While nutritional problems faced by both developing and industrial countries are distinct, evidence shows that the progression of nutritional status in all countries follows a similar course. Today there is an overlap of health issues, many of which are interrelated and span both categories of countries.

There are several valid justifications for policy reform specifically designed to address the problem of malnutrition:

- *Scale and pace of malnutrition problems*: The pace of change of nutrition problems is accelerating and the challenges are daunting. There are new dimensions to the problem of malnutrition. The epidemic of obesity and noncommunicable diseases is spreading at a fast rate from industrial to developing countries, with the poorest sections in both groups of countries affected

Box 17.8. China's National Plan of Action for Nutrition

Since the 1990s there has been a coordinated effort between the government of China and experts in multiple sectors to address problems related to nutrition. In 1997 the Chinese authorities elaborated a National Plan of Action for Nutrition aimed at alleviating hunger and nutritional deficiencies by securing adequate food supplies; by reducing the incidence of energy-protein malnutrition; and by preventing, controlling, and eliminating micronutrient deficiencies. At the same time the plan aimed to improve the general nutritional status of people and prevent diet-related noncommunicable diseases through proper guidance on food consumption behavior, improvements of dietary patterns, and promotion of healthy lifestyles.

The plan provides for a comprehensive and integrated set of policies and interventions that go beyond the health sector. It strengthens intersectoral cooperation and sets targets for agriculture and processed food production. It addresses adjustments to the agriculture structure (e.g., to increase poultry and fishery production as well as vegetable, fruit, and soybean production).

Source: ESCAP 2005, Zhai et al., 2002.

most. Many poor countries are beginning to suffer from the double burden of undernutrition and obesity.

- *Meeting the Millennium Developments Goals on poverty and hunger*: Persistent malnutrition is contributing to the failure of the international community to meet the first Goal—halving poverty and hunger. Unless policies and priorities are changed, many countries, particularly those in sub-Saharan Africa and South Asia, where hunger is prevalent, will not be able to achieve the Millennium Development Goals.
- *High economic and social costs*: Malnutrition imposes high economic and social costs. Undernutrition, obesity, and diet-related diseases have reached worrying proportions, intensifying in low-income countries, and the trend indicates that the epidemic has yet to reach its peak. Costs related to malnutrition will be a major burden for future generations.

Policy has a potential role in diminishing problems associated with malnutrition. In general, policy interventions can be categorized into those that either enhance or reduce the production and consumption of certain foods. Interventions may also be targeted at particular stakeholders, such as vulnerable groups, consumers, and the food industry. Table 17.2 gives a general overview of various policy options that can be applied to different nutrition problems.

Given the multifaceted nature of malnutrition there are likely to be trade-offs, implying that policy makers will have to make informed decisions. A single intervention can have different outcomes for people at risk from undernutrition than for those at risk from overnutrition, for urban compared with rural populations, or for the poor relative to the rich.

Diets in middle-income developing and industrial countries are changing or have changed—with livestock products being implicated as central to these changes. On the other hand, diets in low-income countries have not improved much. The importance of livestock products to human diets poses a number of policy challenges, and therefore any policy that aims to increase or decrease the consumption of livestock products requires a careful assessment of the likely impacts on nutritional status of diverse groups within society.

For example, addressing the problem of overweight and obesity in emerging economies should be pursued without losing focus on the urgency of curtailing undernutrition. Food consumption deficits in many of these countries are still widespread, so discouraging the consumption of certain foods such as livestock products will deprive these groups of important micronutrients in livestock products. In addition, certain categories of individuals such as infants and children still need to consume some foods that might be discouraged for others, such as full-fat dairy products. The challenge will therefore be to ensure improved diet quality through consumption of micronutrient-rich livestock products while discouraging excess intake of saturated fats.

Likewise, policies that encourage the consumption of livestock products in low-income countries may fuel additional nutritional problems. The situations in Latin America, where hunger-oriented programs that ignored issues of overnutrition have in fact led to increased obesity, are examples of this.

Conversely, increasing the price of livestock products to reduce consumption of saturated fats may also affect low-income and vulnerable consumers, whose diets largely consist of cheap, high-energy, dense foods rich in saturated fats and sugars. A tax on animal fats may enhance the production and consumption of low-fat animal products, but the excess fat might be returned to the food chain indirectly or exported. A case in point is poultry meat, where industrial countries tend to consume the lean parts while exporting the fatty parts—thereby shifting the burden of disease to other countries in the form of cheap imports. A closer review of how pricing policies are affecting dietary patterns and an identification of how food policies could be reoriented to encourage consumption of healthier foods or discourage excessive consumption should be undertaken. Governments usually use direct and indirect methods to influence food prices, from direct subsidies that lower the purchase price to subsidies on inputs into the production process.

Table 17.2. Matrix of policy options to combat malnutrition classified by nutritional status.

Nutritional status	Policy goal	Policy intervention	Impact
Food insecurity and under-nourishment	• Increase availability	• Improvements in food production and productivity • Investment in agricultural technology	• Increased agricultural income • = ↑ purchasing power = ↑ household food demand • Increase in household food supplies from own production • Increased access to food, higher yields, and reduced cost of food
	• Increase accessibility	• Promotion of income generation activities and employment creation • Fiscal and food price policy to increase the purchasing power of the consumer	• Increased cash income • = ↑ purchasing power = ↑ household food demand and supplies • Increased incomes facilitates diet diversification
	• Improve micronutrient status	• Dietary diversification through production and consumption of micronutrient rich-foods (e.g., livestock products)	• Improved nutrition and health status
	• Increase awareness of nutritional requirements	• Nutrition and health education	• Improved utilization of food • Improved feeding practices and promotion of healthy diets • Promotes diet diversification
Overnutrition	• Alter consumer preferences and choices	• Information campaigns, awareness and nutrition education	• Ability of consumer to make informed and healthy decisions regarding food choices • Compels food industry to initiate change and invest in healthy products
	• Regulation of food industry	• Mandatory food labeling • Advertising and marketing regulations	
	• Affect availability and accessibility	• Policy changes that promote better-quality diets and products • Pricing policies • Food and nutrition standards	• Promotes production and consumption of healthy foods • Standards limit access to unhealthy foods • Decreases inequality in accessing healthier foods that are generally more costly

It is therefore important to understand how such polices affect nutrition, food choices, and health.

Encouraging the production and consumption of healthier but more costly foods such as lean meats, fish, fresh vegetables, and fruits may contribute to food inequalities. It is therefore important to assess the range of nutrients and foods that can be affected by such policy changes. Policy makers must be concerned about the impact of pricing policies on protein intake of poorer consumers.

Specific Responses Options on Food Insecurity and Undernutrition

The sole focus on food insecurity in low-income developing countries is no longer a sufficient response—policy makers need to move beyond hunger to issues such as micronutrient deficiencies, diet quality, and health. Addressing food insecurity and undernutrition also requires much more than just the provision of food (whether from domestic supply, imports, or food aid). It calls for a holistic and multidisciplinary approach.

The multiple causes of malnutrition necessitate the adoption of a life cycle approach to address malnutrition—integrated rather than distinct intervention programs and policies that tackle the problem. This needs to recognize the role of globalization, trade, food production and marketing, pricing, education, urbanization, and so on. Such an approach also acknowledges the important role of different stakeholders such as consumer, civil society, government, and industry in promoting nutrition. It is now a well-established fact that the foundations of malnutrition are laid in early childhood and before birth. There is increasing evidence to support the Baker hypothesis that chronic conditions such as cardiovascular disease, type 2 diabetes, and hypertension later in life may be associated with maternal and fetal undernutrition rather than dietary and lifestyle changes alone (FAO 2006c).

To improve food security, it is essential to promote policies that accelerate agricultural growth through increased investment. Particularly in sub-Saharan Africa—where foreign exchange availability is limited, which in turn limits food imports—increasing local production by focusing on investment in the agricultural sector would translate into a gradual increase in food supplies. Higher agricultural production can improve food security by increasing food supply for consumers, increasing rural incomes, and contributing to economic growth. Based on FAO's projections, total livestock production in sub-Saharan Africa will have to grow at an average annual rate of 4.2% by 2015 in order to meet the needs of the growing population, improve nutrition, and progressively eliminate food imports.

Significant improvements in agricultural performances, however, require innovative technologies to increase productivity. A priority action is therefore to facilitate technology transfer. But technological innovation has limited impact if not accompanied by enabling policies and an institutional framework to facilitate policy implementation. Improving access to inputs and services and developing infrastructure for transporting, processing, and marketing livestock and livestock products not only improves productivity but also encourages private investment in the sector.

It is well established that the diets of those in lower-income groups tend to be less diversified and of a lower overall quality than those of higher-income groups. Income growth can therefore help improve both demand and nutritional outcomes. As the incomes of the poor increase, generally they will purchase more and better food. Brazil's national program *Fome Zero*, which uses structural policies such as income distribution and income generation to eradicate hunger, is a successful model that low-income countries can adapt. However, it should also be noted that improvements in income do not necessarily translate into improved nutrition—especially if the income is spent on less nutritious foods. There is a tendency for low-income consumers to shift to cheap, energy-dense foods as their incomes increase.

Still, poverty is the most binding constraint to improving food security in low-income countries. Policies that therefore improve people's access to food by reducing poverty are likely to have the greatest gains in food security improvements. Low-income countries can draw from several poverty reduction interventions that have had positive impacts on nutritional status. Programs implemented in middle-income countries such as Mexico, Brazil, and Thailand have incorporated nutrition components and have registered successes in improved nutrition.

An obvious and straightforward step to combat malnutrition would seem to be raising awareness regarding nutrition and diets through information and nutrition education. The nutrition situation is exacerbated by a lack of nutritional information and knowledge, not only among consumers but also across a wide range of decision makers. Added to this are undesirable dietary habits and nutrition-related practices, attitudes, perceptions, and sociocultural influences that adversely affect nutritional status. Particularly in low-income countries, increased awareness and nutrition education regarding the health benefits of consuming livestock products, especially meat, would certainly go a long way in addressing several nutrition problems such as underweight and micronutrient deficiencies and consequently result in dietary diversification.

Improving the quality of diets in low-income countries represents a major challenge. Increasing the variety of food, such as foods of animal origin, is important in enhancing dietary quality and reducing micronutrient deficiencies. The conventional approach to hunger has stressed increased supply of cereals as opposed to other foods. This explains why undernutrition and micronutrient deficiency are still widespread in South Asia despite the increase in food availability. Policies that focus on diet quality are therefore important for addressing problems across the whole nutritional spectrum.

Specific Response Options on Overnutrition

The problems of overweight, obesity, and diet-related noncommunicable diseases and the ways to tackle them are much less well understood and more complex than the traditional interventions for undernutrition. Recent research suggests that overweight and obesity often have earlier roots. Malnourished children are likely to become obese later in life, and there is a growing body of evidence that suggests that maternal food deprivation or low birth weight may program a child to be more prone to adulthood obesity and noncommunicable diseases (WHO and World Bank 2002, World Bank 2006, WHO 2007).

The simplistic and classical approach to overnutrition that has tended to characterize the perceptions of policy makers—in which overnutrition is considered to be prevalent among populations with sufficient food—is no longer valid and is indeed a poor guide to policy. The paradox that countries face today is that poverty not only causes food insecurity but may also predispose individuals to become obese and to develop chronic diseases. Altering the food environment has been a major policy response to the obesity epidemic in many countries, with little emphasis being placed on the economic environment of obese persons.

Informing consumers is one of the least contentious policy measures and can be achieved through nutritional labeling and nutrition education. Information regarding nutrition is particularly relevant to countries experiencing the double burden of malnutrition. Nutrition education is important in promoting better nutrition and feeding practices. Many consumers are unaware of the consequences of unhealthy diets; others may be ill informed about the benefits of consuming some foods as opposed to others. And poor dietary habits may be a consequence of improper dietary guidelines and recommendations. Nutritional labeling of foods imposes costs on the food industry and can induce producers to introduce healthier versions such as low-fat foods. It also allows consumers to make food choices based on valid information.

The need to regulate the food industry and foreign domestic investment, particularly in middle-income developing countries, is greatest. These nations, especially emerging economies, offer potentially huge new markets for multinational food companies given the changing incomes, increasing urbanization, and hence increasing demand for food and changes in food preferences with a shift toward more westernized diets. Food processing and marketing policy changes are therefore crucial in determining patterns of food consumption, rather than policies centered on individual responsibility for consumer choice, and it is at this stage that governments should intervene to ensure a supply of healthy products.

Pricing strategies can influence purchasing behavior, an indication that fiscal intervention is a plausible component of policy to respond to overnutrition. Market interventions may consist of either taxation or subsidization to encourage or discourage consumption of certain foods, and this may be necessary to correct the market failures related to obesity. A longitudinal study of food prices and consumption patterns in China found that increases in the price of animal protein foods were associated with decreased consumption of those foods (Guo et al., 1999). The study also found that increasing the price of pork would affect the diet quality of poor consumers. It may not result in the desired effect of reducing fat intake and may in fact reduce the accessibility to important nutrients such as iron or calcium, which are not available in high densities in nonanimal-source foods. Thus governments may have to subsidize low-fat livestock products—lean meats and poultry meats—in order to increase their availability and accessibility to all consumers. Helping low-income consumers obtain high-quality diets at an affordable cost may be the key strategy for curbing the obesity epidemic among disadvantaged populations.

It is often difficult to encourage consumers to change their eating habits toward healthier foods, especially when unhealthy foods are easier to find, convenient, tastier, and cheaper. Both Norway and Finland have successfully reversed high-fat, energy-dense diets in their populations, resulting in significant decreases in serum cholesterol levels and deaths from coronary heart disease (Chopra and Puoane 2003). From 1985 to 2000 the real price of fresh fruits ad vegetables went up almost 40% in the United States, while the real price of livestock commodities such as red meats and poultry declined by about 4% and 5%, respectively (IATP 2006). Economic measures that facilitate healthier food choice and restrict consumption of unhealthy foods—such as increasing the prices of high-density products and reducing the price of healthier products—should be contemplated, taking into consideration their effects on low-income groups.

A wide range of policies can directly affect dietary choice. Particularly absent from policy recommendations targeting obesity in industrial countries is how agricultural policies relate to health and obesity. It will be difficult for these countries to reduce consumption of certain food categories deemed unhealthy, such as livestock products, if agricultural policies encourage excess production of these products and subsidize their use. Current policy incentives for increased production of cheap livestock products should be revised in favor of support for production of healthier foods, such as low-fat milk products and lean meats or pasture-produced meats. Because animal fat promotes heart disease, policies that lower fat content in livestock products would be beneficial. The fat content in milk can be reduced in several ways, such as the use of breeds that produce milk with lower fat, altering feed to lower total fat content of milk, or lowering the saturated fat and increasing the unsaturated fat content.

Consumer pressure is, and will continue to provide, incentive for producers and the food industry to alter production processes in order to provide products that are safe and healthy. In this regard, consumers need to be empowered with information and an effective regulatory framework so they can influence food production and processing. For example, mandatory food labels can give consumers the opportunity to influence how foods are produced, by exerting their consumer choice.

Conclusions

Malnutrition continues to be one of the most serious problems facing both rich and poor countries. Similarly, food safety risks across the entire food chain are increasingly affecting diverse socioeconomic groups in different ways. That the problem of malnutrition is not entirely clear-cut is obvious. And as this chapter has demonstrated, problems related to malnutrition are much more difficult than they appear due to the complex interplay and interactions that shape and influence human diets. It is also true that while nutritional problems are converging across countries, there are important differences in the causal pathways, implying that extrapolation from the experiences in industrial countries to developing countries may not be an appropriate response.

Nutrition trade-offs associated with policies that act as an incentive or disincentive to the production or consumption of livestock products present a challenge to policy makers, particularly in developing countries. Thus nutrition policy will need to carefully consider the likely nutrition impacts of various options. Comprehending these trade-offs as well as evaluating the likely losers and winners is central to addressing the current nutrition challenge.

There is growing evidence that a combination of policy approaches is a prerequisite for addressing malnutrition. Policies that integrate various aspects of health, food production, processing, and marketing as well as nutrition education and awareness will be fundamental to achieving this goal. The identification of multidimensional policies that help address the problems of undernutrition and overnutrition simultaneously are crucial, given the possible nutrition trade-offs associated with efforts to overcome one or the other. To date, no existing programs are expressly designed to address these combined problems. The challenge for policy makers will be how to integrate all the efforts that target these problems individually into a coherent prevention strategy capable of changing the unhealthy aspects of the environment and reversing current trends.

Similarly, strategies to implement these policies are required at all levels: local, national, regional, and global. Different actors and diverse sectors need to be brought together in order to achieve synergies in results, avoid duplication, and save on the already scarce resources as well as raise the profile of nutrition on the policy agenda. More attention needs to be placed on how policies in diverse sectors affect nutrition, food safety, and human health. These efforts must, however, be coupled with appropriate and effective institutional arrangements at various levels. Most important, in order to improve responses to nutritional problems a coherent research agenda is needed to better understand the dynamics of malnutrition.

References

Barrett, C. B. 2002a. Food security and food assistance programs. In *Handbook of agricultural economics*, ed. B. L. Gardner and G. C. Rausser (vol. 2), pp. 2103–2190. Amsterdam: Elsevier.

Barrett, C. B. 2002b. Food aid and commercial international food trade. Background paper prepared for the Trade and Markets Division, Organisation for Economic Co-operation and Development. Paris.

Barrett, C. B. 2006. *Food aid as part of a coherent strategy to advance food security objectives*. Working Paper Series. Ithaca, NY: Department of Applied Economics and Management, Cornell University.

Chopra, M., and T. Puoane. 2003. Prevention of diabetes throughout an obesegenic world. *Diabetes Voice*, vol. 48.

Cleaver, K., N. Okidegbe, and E. De Nys. 2006. Agriculture and rural development: Hunger and malnutrition. World Bank Seminar Series. Washington, DC: World Bank.

Coitinho, D., C. A. Monteiro, and B. M. Popkin. 2002. What Brazil is doing to promote healthy diets and active lifestyles *Public Health Nutrition*, 5 (1A): 263–267.

CSIS (Center for Strategic and International Studies). 2006. *Assessing the global risk of chronic disease*. Washington, DC: CSIS.

Curtis, M. 2004. The obesity epidemic in the Pacific Islands. *Journal of Development and Social Transformation* 1 (1): 37–42.

D'Arcy, M., P. Harduar, M. Orloff, and A. K. Rozas. 2006. The move towards obesity: The nutrition transition in China. Paper prepared for the International Economic Development Program, University of Michigan.

Dinar, A., and A. Subramanian. 1997. *Water pricing experiences: An international perspective*. World Bank Technical Paper No. 386. Washington, DC: World Bank.

Drewnowski, A. 2004. Obesity and the food environment: dietary energy density and diet costs. *American Journal of Preventive Medicine* 27 (3): 154–162.

Drewnowski, A., and N. Darmon. 2005. The economics of obesity: Dietary energy density and energy cost. *American Journal of Clinical Nutrition* 82 (Suppl) : 265S–273S.

Drewnowski, A., and S. E. Spector. 2004. Poverty and obesity: The role of energy density and energy cost. *American Journal of Clinical Nutrition* 79 (1): 6–16.

EIRIS (Ethical Investment Research Services). 2006. *Obesity concerns in the food and beverage industry*. London: EIRIS.

EPHA (European Public Health Alliance). 2007. *A re-balancing of EU subsidies for school milk programmes*. Brussels: EPHA.

ERS-USDA (Economic Research Service–U.S. Department of Agriculture). 2000. *A comparison of food assistance programs in Mexico and the United States*. Food Assistance and Nutrition Research Report, No. 6. Washington, DC: ERS-USDA.

ERS-USDA (Economic Research Service–U.S. Department of Agriculture). 2004. *Food security in Russia: Economic growth and rising incomes are reducing insecurity*. Washington, DC: ERS-USDA.

ERS-USDA (Economic Research Service–U.S. Department of Agriculture). 2006. Organic poultry and eggs capture high price premiums and growing share of specialty markets. *Outlook report from the Economic Research Service*. Washington, DC: ERS-USDA.

ESCAP (Economic and Social Commission for Asia and the Pacific). 2005. *Addressing emerging health risks: Strengthening health promotion*. Bangkok: Committee on Emerging Social Issues, ESCAP.

EU (European Union). 2007. Dairy market: Council backs improvements to school milk scheme and simplification to dairy market. Press Release. Brussels. 29 September.

Euromonitor International. 2005. *Milk experiences a growth spurt in China*. London. 17 January.

European Commission. 2003. *The meat sector in the European Union*. Brussels: Directorate-General for Agriculture.

European Commission. 2007. A new animal health strategy for the European Union (2007–2013) where "prevention is better than cure." Brussels: EU Health and Consumer Protection Directorate-General.

Evans, M., R. Sinclair, C. Fusimalohi, and V. Liav'a. 2001. Globalization, diet, and health: An example from Tonga. *Bulletin of the World Health Organization* 79 (9): 856–862.

FAO (Food and Agriculture Organization). 1999. The importance of food quality and safety for developing countries. FAO Committee on World Food Security, Twenty-fifth Session, Rome.

FAO (Food and Agriculture Organization). 2001. Factsheet on food safety and quality. Rome: FAO Food Quality and Standards service.

FAO (Food and Agriculture Organization). 2002. Market developments for organic meat and dairy products: Implications for developing countries. Rome: Committee on Commodity Problems, Intergovernmental Group on Meat and Dairy Products.

FAO (Food and Agriculture Organization). 2003a. FAO's strategy for a food chain approach to food safety and quality: A framework document for the development of future strategic direction. Rome: Committee on Agriculture.

FAO (Food and Agriculture Organization). 2003b. *Food security in the Russian Federation.* FAO Economic and Social Development Paper No. 153. Rome: FAO.

FAO (Food and Agriculture Organization). 2006a. *State of food insecurity in the world, 2006: Eradicating world hunger–Taking stock ten years after the World Food Summit.* Rome: FAO.

FAO (Food and Agriculture Organization). 2006b. *Surges in developing countries: The case of poultry.* FAO Commodity Brief No. 1. Rome: FAO

FAO (Food and Agriculture Organization). 2006c. *The double burden of malnutrition: Case studies from six developing countries.* FAO Food and Nutrition Paper 84. Rome: FAO.

FAO (Food and Agriculture Organization). 2008. Assessment of the world food security and nutrition situation. Rome: Committee on World Food Security.

FAO (Food and Agriculture Organization). 2009. *Food guidelines by country.* http://www.fao.org/ag/agn/nutrition/education_guidelines_country_en.stm.

FAO (Food and Agriculture Organization) and WHO (World Health Organization). 2005. *Food safety risk analysis: An overview and framework manual.* Rome: FAO.

Fearne, A. 1999. The changing roles of government and industry. Paper presented at Food Safety and International Competitiveness: The Case of Meat. Conference, Banff, Canada, 14–16 April.

Ferris-Morris, M. 2003. Past and present uses of title II non-fat dry milk in humanitarian operations. Background paper. Washington, DC: U.S. Agency for International Development.

Guo, X., B. M. Popkin, T. A. Mroz, and F. Zhai, F. 1999. Food price policy can favorably alter macronutrient intake in China. *Journal of Nutrition* 129:994–1001.

Haddad, L. 2003. Redirecting the diet transition: What can food policy do. *Development Policy Review* 21 (5–6): 599–614.

Hawkes, C. 2006. Uneven dietary development: Linking the policies and processes of globalization with the nutrition transition, obesity and diet-related chronic disease. *Globalization and Health* 2:4.

Hertrampf, E., M. Olivares, F. Pizarro and T. Walter. 1998. High absorption of fortification iron from current infant formulas. *Journal of Pediatric Gastroenterology and Nutrition* 27:425–430.

HKI (Helen Keller International) 2003. Integration of animal husbandry into home gardening to increase vitamin A intake from foods: Bangladesh, Cambodia and Nepal. Special Issue, January 2003. HKI, Asia-Pacific Regional Office.

Hobbs, J. E., A. Fearne, and J. Spriggs. 2002. Incentive structures for food safety and quality assurance: an international comparison. *Food Control* 13 (2): 77–81.

Holm, J., and T. Jokkala. 2008. The livestock industry and climate: EU makes bad worse. *Spectrezine,* 19 November.

Hop, L. T., L. B. Mai, and N. C. Khan. 2003. Trends in food production and food consumption in Vietnam during the period 1980–2000. *Malaysian Journal of Nutrition* 9 (1): 1–5.

IATP (Institute for Agriculture and Trade Policy). 2005a. *U.S food aid: Time to get it right.* Minneapolis, MN: IATP.

IATP (Institute for Agriculture and Trade Policy). 2005b. *Smart meat and dairy guide for parents and children.* Minneapolis, MN: IATP.

IATP (Institute for Agriculture and Trade Policy). 2006. *Food without thought: How U.S. farm policy contributes to obesity.* Minneapolis, MN: IATP.

IFPRI (International Food Policy Research Institute). 2001. *The nutritional and epidemiological transitions. Sustainable food security for all by 2020.* Bonn, Germany, 4–6 September.

IFPRI (International Food Policy Research Institute). 2003. *Food safety in food security and food trade.* IFPRI 2020 Vision Focus Brief No. 10. Washington, DC: IFPRI.

IFPRI (International Food Policy Research Institute). 2004a. *Research and outreach: Food system governance.* Washington, DC: IFPRI.

IFPRI (International Food Policy Research Institute). 2004b. *The impact of feeding children in school: Evidence from Bangladesh.* Washington, DC: IFPRI.

Jacoby, E. 2004. The obesity epidemic in the Americas: Making healthy choices the easiest choices. *Pan American Journal of Public Health* 15 (4): 278–284.

Kain, J., I. Vial, E. Muchnik, and A. Contreras. 1994. An evaluation of the enhanced Chilean supplementary feeding program. *Archivos Latinoamericanos de Nutrición* 44:242–250.

Lang, T., G. Rayner, and E. Kaelin. 2006. *The food industry, diet, physical activity and health: A review of reported commitments and practice of 25 of the world's largest food companies.* London: Center for Food Policy, City University.

Lewin, A., L. Lindstrom, and M. Nestle. 2006. Food industry promises to address childhood obesity: Preliminary evaluation. *Journal of Public Health and Policy* 27:327–348.

McConnell, M., and K. Mathews. 2008. Global market opportunities drive beef production decisions in Argentina and Uruguay. *Amber Waves* (U.S. Department of Agriculture). April.

Monteiro, C. A., W. L. Conde, and B. M. Popkin. 2002. Is obesity replacing or adding to undernutrition? Evidence from different social classes in Brazil. *Public Health Nutrition* 5:105–112.

Monteiro, C. A., E. C. Moura, W. L. Conde, and B. M. Popkin. 2004. Socioeconomic status and obesity in adult populations of developing countries: A review. *Bulletin of the World Health Organization* 82 (12): 940–946.

North East Public Health Observatory. 2005. *Obesity and overweight in Europe and lessons from France and Finland.* Occasional Paper No. 10. Middlesbrough, U.K.: University of Teesside.

Nugent, R., and F. Knaul. 2006. Fiscal policies for health promotion and disease prevention. In *Disease control priorities in developing countries,* 2nd ed., ed. D. T. Jamison, J. G. Breman, and A. R. Measham, 211–224. New York: Oxford University Press.

OECD (Organisation for Economic Co-operation and Development). 2005. *The development effectiveness of food aid: Does tying matter?* Paris: OECD.

Omore, A. O., S. Arimi, S. M., Kang'ethe, E. K., McDermott, J. J. and S. Staal. 2002. Analysis of milk-borne public health risks in milk markets in Kenya. Paper prepared for the Annual Symposium of the Animal Production Society of Kenya, May 9-10, 2002, KARI NAHRS, Naivasha

Oxfam. 2006. Causing hunger: An overview of the food crisis in Africa. Briefing Paper. Oxford: Oxfam.

Popkin, B. M. 1994. The nutrition transition in low-income countries: An emerging crisis. *Nutrition Reviews* 52 (9): 285–298.

Puska, P. 2002. Successful prevention of non-communicable diseases: 25 year experiences with North Karelia Project in Finland. *Public Health Nutrition* 4 (1): 5–7.

Ralston, K. 2000. How government policies and regulations can affect dietary choices. *Government regulation and food choices*. Washington, D.C.: U.S. Department of Agriculture, Economic Research Service.

Schäfer, E. L., K. Lock, and G. Blenkuš. 2006. Public health, food and agricultural policy in the European Union. In *Health in all policies: Prospects and potentials*, ed. T. Stahl, M. Wismar, E. Ollila, E. Lahtinen, and K. Leppo, 93–110. Helsinki: Ministry of Social Affairs and Health European Observatory on Health Systems and Policies.

SCN (Standing Committee on Nutrition). 2000. Ending malnutrition by 2020: An agenda for change in the millennium. A final report to the ACC/SCN by the Commission on the Nutrition challenges of the 21st Century. *Food and Nutrition Bulletin* 21 (3).

SCN (Standing Committee on Nutrition). 2004. *Nutrition for improved development outcomes: Fifth report on the world nutrition situation*. New York: SCN, United Nations.

SCN (Standing Committee on Nutrition). 2005. Overweight and obesity: A new nutrition emergency? *SCN News* No. 29.

Sharma, R., D. Nyange, G. Duteutre, and N. Morgan. 2005. *The impact of import surges: Country case study results for Senegal and Tanzania*. Commodity and Trade Policy Research Working Paper No. 11. Rome: FAO.

Smitasiri, S., and S. Chotiboriboon. 2003. Experience with programs to increase animal source food intake in Thailand. *Journal of Nutrition* 133: 4000S–4005S.

Takagi, M., M. E. Del Grossi, and J. G. da Silva. 2006. The zero hunger programme two years later. Paper delivered at the 2006 Meeting of the Latin American Studies Association, San Juan, Puerto Rico, 15–18 March.

Tanumihardjo, S.A., C. Anderson, M. Kaufer-Horwitz, L. Bode, N. J. Emenaker, A. M. Haqq, J. A. Satia, H. J. Silver, and D. D. Stadler. 2007. Poverty, obesity, and malnutrition: an international perspective recognizing the paradox. *Journal of the American Dietetic Association* 107 (11): 1966–1972.

Tetra Pak. 2006. Food for development: A catalyst to agricultural development and economic growth. Tetra Pak Food for Development Global Office.

Thilmany, D. 2006. *The US organic industry: Important trends and emerging issues for the USDA*. Fort Collins: Colorado State University.

Uauy, R., C. Albala, and J. Kain. 2001. Obesity trends in Latin America: Transition from under-to overweight. *Journal of Nutrition* 131:893S–899S.

Umali-Deininger, D., and M. Mona Sur. 2007. Food safety in a globalizing world: Opportunities and challenges for India. *Agricultural Economics* 37 (S1): 135–147.

Wehrheim, P., and D. Wiesmann. 2006. Food security analysis and policies for transition countries. *Journal of Agricultural and Development Economics* 3 (2): 112–143.

WFP (World Food Programme). 2009. School Feeding Program website. Accessed February (http://www.wfp.org/school-meals).

WFP INTERFAIS. 2007. World Food Programme: International Food Aid Information System. Available at http://www.wfp.org/fais.

WHO (World Health Organization). 2000. *Manual on the management of nutrition in major emergencies*. IFRC/UNHCR/WFP/WHO. Geneva: WHO.

WHO (World Health Organization). 2003a. *Diet, nutrition and prevention of chronic disease. Report of a joint WHO/FAO Expert consultation*. WHO Technical Report Series 916. Geneva: WHO.

WHO (World Health Organization). 2003b. New global alliance brings food fortification to world's poor. Press Release. Geneva: Global Alliance for Improved Nutrition, WHO.

WHO (World Health Organization). 2004. *Food and health in Europe: A new basis for action*. WHO Regional Publications European Series, No. 96. World Health Organization, Europe.

WHO (World Health Organization). 2007. *The challenge of obesity in the WHO European region and the strategies for response*. Copenhagen: WHO Regional Office for Europe.

WHO–Western Pacific Region. 2003a. *Diet, food supply and obesity in the Pacific*. Manila: WHO Regional Office for the Western Pacific.

WHO–Western Pacific Region. 2003b. *Using domestic law in the fight against obesity: An introductory guide for the Pacific*. Manila: WHO Regional Office for the Western Pacific.

WHO (World Health Organization) and World Bank. 2002. *Food policy options: Preventing and controlling nutrition related non-communicable diseases*. Health, Nutrition, and Population Series. Washington, DC: Human Development Network, World Bank.

Wilde, P. 2006. *Nutrition and environmental impacts of federal food advertising*. Medford, MA: Food Policy and Applied Nutrition, Tufts University.

World Bank. 2006. *Repositioning nutrition as central to development: A strategy for large-scale action*. Washington, DC: World Bank.

World Bank. 2007. Private sector tackles malnutrition. Press release, 19 April.

Zhai, F., D. Fu, S. Du, K. Ge, C. Chen, and B. M. Popkin. 2002. What is China doing in policy-making to push back the negative aspects of the nutrition transition? *Public Health Nutrition* 5 (1A): 269–273.

18

Responses on Emerging Livestock Diseases

Anni McLeod, Nick Honhold, and Henning Steinfeld

Main Messages

- **New livestock diseases are emerging and spreading and old diseases are moving to new locations, driven by changes in production and marketing systems, human demographics, and environmental pressures.** The diseases themselves and the measures used to control them have considerable economic and social costs and are in turn leading to changes in the structure and management of the livestock sector. Animal health systems are straining to meet the challenges placed on them.

- **Responding to emerging and changing diseases requires improved capabilities in several areas:** anticipating and understanding problems so that prevention and control measures can be well planned; taking preventive action to minimize risks and impacts of diseases, including biosecurity measures in farms and markets and restructuring of market chains; and responding quickly and effectively to disease threats when they occur so that surprises do not become crises, as well as recovering from animal health emergencies when they have been contained.

- **Existing response systems suffer from weaknesses in several important areas:**
  - *The state has retreated from animal health,* withdrawing funding and failing to update regulations.
  - *There is a large dichotomy between countries in the global export network,* whose systems are relatively well funded and where regulations and approaches to prevention and control are harmonized, and poorer countries, where investment and regulation are inconsistent. Each system has its own problems and each causes problems for the other.
  - *There is also a growing dichotomy between rich and poor livestock keepers,* with the latter being excluded from markets and lacking access to the services that might help them with markets. Investment tends to be skewed toward the richer parts of the livestock sector.

- **An improved response would acknowledge that emerging animal diseases are not susceptible to simple, technology-driven solutions.** It would bring more transparency into surveillance and disease intelligence, engage a wider range of disciplines, increase investments in minimizing risks and impact, and improve the speed of response by a number of measures, especially by developing more robust advance plans and building up surge capacity.

Introduction

The intensification of farm animal production, housing conditions, and the management and handling of animals, together with broader demographic trends such as urbanization and the movement of people and animals, have all contributed to creating an environment in which livestock disease is a constant challenge (see Chapter 11 and Chapter 2). This chapter focuses on the responses to two categories of livestock disease that are causing a great deal of concern and economic damage: emerging diseases and those that are changing their location. Recent experience has shown that even the best-prepared animal health system can be taken by surprise—and that the global animal health system is by no means universally well prepared.

Livestock diseases emerge or change their distribution for reasons related to biology, production and marketing systems, and trade patterns. As described in earlier chapters, globalization, urbanization, and intensification of systems (particularly monogastrics) have been particularly important. Diseases are able to spread widely and to become persistent when animal health systems (public and private) lack the capacity to accommodate changes and respond rapidly and effectively—and sometimes when even good capacity is overwhelmed by a new set of circumstances. In the longer term, there is potential for both globalization and urbanization to contribute to improved animal health, since the wish to stabilize trade necessitates international regulations and private-sector agreements on good practice, but this potential has yet to be realized.

In spite of the efforts that have been made by governments, the international community, and the private sector to contain livestock diseases, they continue to be a problem. As discussed in Chapter 11, three striking examples in the past 10 years of diseases completely new to livestock production or emerging in a new form are bovine spongiform encephalopathy (BSE) in cattle, highly pathogenic avian influenza (HPAI) H5N1 in poultry, and Nipah virus in pigs. A recent example of a known disease being spread to new livestock populations is bluetongue, a disease of sheep and cattle transmitted by *Culicoides* midges. Rift Valley fever is another disease that appears to be changing location. Known diseases that periodically reinfect livestock populations in which they had been eradicated include foot-and-mouth disease (FMD) in ruminants and contagious bovine pleuropneumonia (CBPP) in cattle.

The capacity of animal health systems to respond to these diseases is critical in limiting their spread and persistence and hence improving the productivity of livestock. Systems must be able to deal with the human and institutional dimensions of disease spread as well as the technical challenges. There are three opportunities to respond to a disease that is emerging or changing location. The ideal time is when the disease agent first enters the livestock population. A swift and effective response at this point causes minimum economic and social disruption; however, it is hard to achieve, requiring a very strong response capability and an element of luck. The next possibility is to contain the disease as it is spreading through the livestock population. This tends to be much more expensive and disruptive, as demonstrated, for example, in the eradication of FMD in the United Kingdom in 2001; the disease was eradicated but only with considerable direct damage to livestock owners and collateral damage to tourism. If both opportunities are missed, the disease becomes established and persists in the population, the state known as "endemic." Eradicating an endemic disease is a long, slow process. The way in which the initial response is handled has an important impact on the persistence of the disease. Nipah virus was eradicated in Malaysia, for example, but arguably the response would have been faster if it had been treated as a serious problem from the start.

In some cases, eradication would require actions that could be environmentally damaging or are impractical to implement under prevailing conditions; this is often the case for disease carried by vectors like ticks and midges. Diseases that should be controllable can persist when animal health systems are unprepared, so that the disease is not recognized, reported, and dealt with quickly. This is particularly likely to happen with vectorborne diseases like bluetongue, where eradication of the disease agent can only be assured by eradicating the vector or keeping it away from infected livestock over a prolonged period. The only viable option may be to vaccinate all animals at risk so that although the disease agent remains in the environment for many years, livestock are protected from getting clinical disease.

Current Responses

Responses to livestock disease threats are the actions taken to prevent or control diseases or to reduce the damage from livestock disease. Good response systems are built on underlying capabilities in animal health systems, supported by factors in the wider political economy (Rushton et al., 2006). To reduce the problem and cost of emerging diseases, four types of response capability are needed. They must all be working well and supporting each other if an animal health system is to be effective:

- The capability to anticipate and understand problems so that prevention and control measures can be well planned for an effective response that minimizes economic and social damage
- The capability to take preventive action to minimize risks and impacts of diseases
- The capability to respond quickly and effectively to disease threats when they occur so that surprises do not become crises
- The capability to recover from animal health emergencies.

Anticipating and Understanding Disease Threats

The capability to anticipate and understand livestock disease threats has three elements: predicting problems that may occur, finding problems quickly and reporting them when they do occur, and analyzing the source and spread of disease so that an appropriate response can be made:

- Predicting problems that may occur allows steps to be taken to prevent them and to plan an effective response that minimizes economic and social damage. Prediction may involve qualitative assessments by experts or mathematical modeling, and it requires data together with multidisciplinary expertise and technology to bring together a wide range of variables and understand what they might imply for possible disease trends and shocks.
- No matter how effective the prediction system is, it will never anticipate or prevent every problem, so it is even more important to have the capability to find new disease problems as soon as they do occur and before they spread too far. This means a very effective frontline system for seeing and reporting disease.

- Finally, when a problem is identified it is essential to be able to analyze the nature of the problem accurately and quickly so that the most appropriate control strategies can be put into place. This requires a strong epidemiological capability to make the best technical diagnosis of the problem, together with a capacity for social and economic analysis to assess the likely impact of proposed control measures on different stakeholder groups and suggest how the costs might best be shared.
- The ability to communicate is important throughout these processes, in order to learn from stakeholders what they know and what concerns them and to provide information about risks and threats to anyone who may need to take action.

Predicting and finding problems is accomplished by activities with labels like "horizon scanning," "disease intelligence," and "surveillance." Analyzing problems comes under the heading of "risk analysis," which is well established as a methodology (OIE 2003).

- Horizon scanning is mostly about prediction, and it encompasses the looking-forward exercises carried out by governments, the international community, and large private companies that involve a wide range of parameters, a long time horizon, and a number of scenarios to explore what the future might hold and what this might imply for their policies and strategies. Horizon scanning related to disease control might examine the impact of a major global trend such as climate change (FAO 2008), the possible development scenarios for the global livestock sector (Freeman et al., 2006), or the factors in the economic and social environment that could affect disease risk. The U.K. Department of the Environment, Food and Rural Affairs defines it as "the systematic examination of potential threats, opportunities and likely future developments which are at the margins of current thinking and planning" (DEFRA 2008).
- Surveillance systems look for disease when it occurs and provide information about suspected and actual outbreaks. They are the first line of defense in understanding disease.
- *Disease intelligence* is the term used to cover activities that "boost awareness of disease threats and developments that may otherwise remain unknown" (FAO EMPRES 2002). In this it overlaps with both of the previous activities in that it includes looking for disease, using information from government surveillance systems and other less official sources, as well as analyzing existing disease problems and making predictions of new ones. It tends to be narrower in scope than horizon scanning and focuses mainly on factors with an immediate link to disease threats. The EMPRES watch bulletins put out by the United Nations Food and Agriculture Organization (FAO) are an example of international disease intelligence.

Sources of Information

Horizon scanning, surveillance, disease intelligence, risk analysis—all of these activities need data and information. There are a great many animal health databases of varying quality and accessibility held by national governments, private companies, and international organizations.

All national governments have some kind of animal health information system to store national disease data. Countries that consider international trade a priority invest most and have the best developed systems, while those that are poor and have very dispersed or diverse livestock systems have the least capacity. The way that data are collected and verified and the extent to which decentralized systems are able to transfer data between locations and transform it into useful information affect the value of what is available. Information technology is making systems faster and cheaper, but the critical human element that encourages people to report has received less attention (Dung et al., 2006). The best disease information of all comes from livestock keepers who see their animals regularly, but little of this information finds its way into official systems. Also, many systems do not record negative findings, even though negative information is a good indicator that a system is active and functioning.

Large companies also have their own disease intelligence networks, and well-run commercial farms have herd recording systems that record clinical signs and analyze productivity to indicate when there may be problems. However, these records are private and may not be accessible.

Internationally, several official systems exist, such as the World Animal Health Information Database (WAHID) of the World Organisation for Animal Health (OIE); the Global Early Warning and Response System for Major Animal Diseases, including Zoonoses (GLEWS) collaboration between FAO, OIE, and the World Health Organization; the OFFLU laboratory network (the joint OIE-FAO network of experts on influenza); and the FAO Emergency Prevention System for Transboundary Animal and Plant Pests and Diseases (EMPRES). WAHID stores and makes public information about official outbreak reports; its main function is to collect information from many countries so that disease alerts are widely publicized. The quality of the data is only as good as what countries have available and are prepared to report. The GLEWS information system has a database of official case reports and unofficial information from a variety of sources and is shared between international agencies, with limited access to the most sensitive information. The OFFLU laboratory network allows international reference centers to share information and the workload of testing samples for avian influenza. EMPRES uses disease tracking and modeling to attempt to predict problems. In addition, the informal Promed system is an electronic mailing list where individuals send in reports. A certain amount of peer review goes into the system, and it is widely used.

Communication

Communication is an essential tool for anticipating and understanding emerging disease threats. Surveillance systems rely on individuals and companies communicating with governments when they see something suspicious, yet people are unwilling to report a suspicion of disease if they fear the consequences of doing so, believe that there will be no response, or have no easy way to make a report. In industrial countries and those emerging economies that trade on the world market there may be an initial reluctance to report a suspected new disease until it has been confirmed as a problem, in order not to create panic among consumers or trading partners. There may be social reasons for not wanting to interrupt a market at a critical time such as a major religious festival. The OIE regularly expresses concern to its members about late or insufficient reports of outbreaks.

In disease intelligence and horizon scanning, communication between organizations and disciplines is needed to validate information and relate disease problems to changes in economies, social systems, and ecologies.

Communication is also necessary to inform people who make their living in the livestock sector as well as the wider public about the nature of disease risks. This is important for any emerging disease but particularly critical (and sensitive) for zoonotic diseases, which may pose a direct danger to people. In the first stages of a new zoonotic disease threat, consumers are understandably concerned about the threat to their health; many will severely reduce their consumption of products from the species of animal affected, creating a market shock that may cause more economic damage than the disease, while the poorest consumers may take advantage of low prices and consume potentially unsafe food. The market shock effects have been very marked for HPAI and BSE, manifested as sharp demand drops in some countries and instability in international markets (Morgan 2006). FMD, by contrast, causes market instability because of trade bans but does not lead to market shocks from consumer panic.

Information asymmetries add to the problem when individuals are unable to obtain or interpret information adequately to make informed choices or when one party has access to private information that is not shared. Consumer response and the size of market shock are partly dictated by access to alternative products and food habits and partly by confidence in the food safety system.

The media often exaggerate response, for instance, to zoonotic disease, although in some cases, such as Turkey's response to HPAI, it may be harnessed to provide in-depth information to allow consumers to make a more informed choice.

Existing Capacity to Anticipate and Understand Diseases

There are few examples of consistently good systems for understanding disease threats, and none in developing countries. While a great deal is known, recorded, and stored in different places, overall the systems are not well "joined up," and there are significant gaps both in data and in the analytical expertise applied to it. The most developed systems have dedicated epidemiological or information units staffed with experienced analysts who are consulted by planners and policy makers. Many of the poorer countries have tried to establish these units, but they have failed because of lack of data, lack of interest in their findings, or lack of trained staff to give continuity.

It is also difficult to measure success because a successful outcome often means that something does not happen, while any failures are highly visible. A system may work very well most of the time and occasionally fail very badly. The U.K. government provides just one example of this. It has a generally strong capability for understanding disease problems. It has a comprehensive horizon-scanning program that looks at human demographics, trade patterns, and other factors that can influence the level of disease threat long and short term and has certainly averted many problems or reduced their impact. It also has a good capability for spotting disease problems, as was demonstrated by the rapid discovery of FMD in 2007, and it is capable of very rapid problem diagnosis, as in the case of BSE.

Yet flaws in the system were revealed by the FMD outbreak of 2001. The introduction of FMD into Europe was expected, and countries had been urged in 2000 by the European Commission for the Control of FMD to reappraise their FMD control strategies. But at the time the expert consensus was that the most likely route of introduction was through southeast Europe (FAO EMPRES 2002). In fact, FMD was found in northwest Europe, specifically the United Kingdom, with the most likely source being illegally imported meat. The outbreak was detected approximately three weeks after the initial infection, and by the time a national movement ban was introduced 74 cases were incubating. The subsequent control plan was driven by an inaccurate database of animal locations and movements, which resulted in an unnecessarily extensive culling program (Taylor et al., 2004).

Another less dramatic example comes from Malaysia, a country with a good record of livestock disease control in recent years. The Nipah virus outbreak in 1999 was controlled quite quickly once it was taken seriously, but it took some time to come onto the national radar, for reasons that appear to have been more sociopolitical than technical (Vu 2009).

In contrast to these two examples, from countries where understanding of disease problems is usually quite good, national animal health information systems in most developing countries are patchy and poorly funded. Even when a computer system is in place to store and analyze data, if the front-line surveillance is not well resourced and well run, disease intelligence suffers badly. This lack of capacity has contributed to diseases spreading and becoming endemic. All the countries that currently have endemic HPAI had a combination of complex production systems and an initially slow response. Several of the areas where CBPP is spreading have large populations of pastoralist cattle whose owners have limited confidence in animal health services and are more likely to treat the disease than to report it.

Understanding of global disease problems is especially challenging, particularly where there is considerable scientific disagreement or uncertainty or where high present costs of prevention are balanced against possible problems at some uncertain future date. HPAI was identified as a risk to people in the late 1990s, and the outbreaks in Hong Kong in 1996 and 1997 were a fairly clear warning signal, but the threat was not understood by the global animal health community, and little was done by countries in the region to prepare for the disease. Following the 2003–04 wave of outbreaks in Southeast Asia and heavy losses from poultry deaths and culling, high levels of investment and effort were made by Thailand, Viet Nam, and China, and the international community provided a very large amount of emergency finance. By then, though, the disease was spreading worldwide.

Anticipating emerging diseases or finding a new disease quickly enough to deal with it effectively increases the chances of mounting a rapid response and controlling it with minimum damage. The best response of all takes place in a well-prepared livestock sector that has taken preventive action, as described in the next section.

Minimizing Risks and Impacts

Three measures that can minimize the risk of emerging and changing diseases are preventive vaccination, biosecurity, and restructuring the livestock sector. Each has some potential to reduce the impact of diseases by preventing outbreaks or reducing their spread, but they require different levels of investment and effort and have different impacts for different stakeholders. What is notable about all of them, compared with other activities discussed here, is that they require considerable private-sector commitment.

Preventive Vaccination

Preventive vaccination done by livestock owners to protect their animals is perhaps the simplest measure that can be applied when it is done under the right conditions. Vaccination is widely used by commercial producers against endemic diseases, and in industrial countries preventive vaccination has become more important as the use of antibiotics has been increasingly regulated in pigs and poultry (Glisson 2008). For emerging diseases, however, vaccines are not always available, and if the country is currently disease free or concerned about export, vaccination may be banned. For example, at the time of writing, preventive vaccination against HPAI was used in large commercial flocks in Indonesia and Viet Nam, where flock owners paid for the vaccine, but it was banned in Europe (except with permission) and Thailand. FMD vaccination is forbidden for routine use in countries that belong to the Organization for Economic Cooperation and Development (OECD) because those using vaccination are considered to have a different health status for trade purposes than countries free of disease and not vaccinating.

Biosecurity

Biosecurity is perhaps the most important weapon available to protect livestock populations against emerging and changing diseases, although it is not infallible and must be supported by other actions.

Biosecurity measures can be put into place for a single production unit, a market, a compartment,[1] a geographical zone, or a country. While the practical details vary greatly with the population being protected, the general principles are the same: segregation, which means keeping healthy populations separate from those that might endanger them, together with cleaning and if necessary disinfection of any objects that might carry disease into or out of a population (FAO et al., 2008a).

Regardless of the scale and scope of a biosecurity program, it always requires a plan, investment in infrastructure or equipment, changes in behavior, and incentives and penalties to encourage compliance. If it works well, it will reduce the number of times that disease enters the population and will increase the chance of containing any outbreaks with minimum economic damage and social disruption. It has a preventive effect equivalent to making the disease agent walk through treacle. Biosecurity does, however, require investment and changes of management practice, and they can have the effect of excluding small-scale producers and traders from markets if made mandatory.

Biosecurity is taken very seriously by industrial-country governments with strong, healthy livestock

1. A *compartment* is defined in OIE's Terrestrial Animal Health Code as "an animal *subpopulation* contained in one or more *establishments* under a common biosecurity management system with a distinct health status with respect to a specific *disease* or specific *diseases* for which required *surveillance*, control and biosecurity measures have been applied for the purpose of *international trade*." See www.oie.int/eng/normes/mcode/en_chapitre_1.4.4.htm#rubrique_compartimentation.

sectors and also by large companies, both of which have strong economic incentives to reduce the impact of disease outbreaks.

In OECD countries there is a strong driver toward biosecurity in government regulations on food safety, which relate to both endemic and exotic diseases. For diseases like salmonellosis and bacteria like *E. coli*, which are always present to some extent in animals but which cause illness in humans who eat contaminated products, the approach is to gradually reduce risk to acceptable levels by testing the final product. Farms found to be affected above acceptable levels are cut off from markets or have their prices reduced. These diseases can no longer be controlled by the routine application of antibiotics, so there is a strong market incentive for farmers to apply biosecurity. Slaughterhouses are also subject to rigorous hygiene checks and are required to apply the hazard analysis and critical control point system.

Biosecurity against exotic diseases in OECD countries relies heavily on national border controls, backed up by measures on farms and in transport and slaughter facilities. Diseases that have severe consequences for trade, whether they endanger human health or not, are subject to strict control measures (culling and movement restrictions) when outbreaks occur. This might be expected to create incentives for farmers to apply biosecurity in order to reduce risk. The reality is that good managers do, but others do not. Their behavior may be explained by the fact that regardless of the measures taken by an individual farm, if an outbreak of a notifiable disease occurs in a neighboring farm, their animals may still be culled and their movements will certainly be restricted. Targeted culling restricted to dangerous contacts might increase the incentive for farmers to make their own farms biosecure, although there is no hard evidence to support this supposition. There is tentative evidence to suggest that biosecurity measures can protect farms from infection even during an outbreak, though it has not been widely publicized. The likelihood that farmers will voluntarily apply biosecurity measures depends on their perception of disease risk and their ability to affect it. In the FMD outbreak in the United Kingdom, many farmers saw disease as something affected as much by others (the government and the public) as by their own preventive biosecurity measures (Gunn et al., 2008).

The extent to which biosecurity measures can be applied successfully depends a great deal on the structure of the industry for a particular species. It is easier to implement and monitor biosecurity measures in a sector with small numbers of units organized into integrated market chains, as the whole chain can be monitored. It is also relatively simple in commercial pig and poultry systems, where the animals can be kept in small areas and mostly indoors for their entire life. It is much more difficult in extensive livestock systems.

For the most successful companies, continuous investment in segregation and cleaning is a normal part of business, and they will upgrade their systems as often as needed to keep up with, or even ahead of, market requirements. Those with valuable breeding stock, in particular, apply very strict biosecurity rules. The biosecurity plans of world-class companies can be regarded as the gold standard.

Denmark, for example, has programs to control salmonellosis in broilers, pigs, and layers of table eggs that rely on cooperation between the government and the livestock industry (Bager and Halgaard 2002) and are funded almost entirely by the industry (Wegener et al., 2003). Control in pigs is driven by meat juice testing at abattoirs, with each tested farm given a quality standard that determines how the pigs are slaughtered and the prices that their owners receive. It is possible to do this because pig production is tied into regulated market chains where individual carcasses can be identified from the farm to the end of the slaughter line.

The poultry sectors in industrial countries, consisting mostly of industrialized and highly vertically integrated value chains, have also found it relatively much easier to protect themselves against HPAI outbreaks (although not against indirect trade impacts). By contrast, in countries like Egypt, Indonesia, Nigeria, Viet Nam, and China, where large-scale and small-scale systems intertwine in complex market chains, there is great difficulty in applying adequate biosecurity at enough points along market chains to break disease transmission. In small poultry flocks, biosecurity measures are less likely to be applied and less affordable, and they have received less attention and support than large commercial units (FAO et al., 2008a).

Biosecurity is a very different concept in extensive systems. Here, animals cannot be shut up to keep them away from infectious agents, but they may be segregated by distance or by movement. Almost all extensive grazing systems, whether nomadic on public grazing lands or ranching on large private ranges, use movement as a way of sustaining food supplies and reducing disease threat. Disease is controlled by keeping herds separate, keeping new animals away from existing ones for a quarantine period, and moving animals around. These biosecurity measures break down when systems are put under pressure from reduced grazing land or drought, or when animals are brought together for vaccination campaigns or sale.

Restructuring

The most extreme approach to preventing disease risk is "restructuring" the livestock sector. A simple definition of restructuring is "deliberate action with government involvement to change the shape of market chains." This may imply the following:

- Functional concentration, resulting in fewer but larger production, slaughter, and retail units
- Geographic concentration or relocation of premises
- Integration of market chains, with control in the hands of large companies
- Direct or indirect action for relocating or "zoning" (McLeod 2007).

The important consequences of restructuring include the following:

- Premises are often moved from a place where they are considered to create an animal or human health risk
- Chains of different types, which may create risk for each other, are separated
- People change their function and relation with others in the market chain
- Contractual relationships often become more rigid.

Restructuring can be driven by economic realities, and these have played a large part in the changing shape of the Chinese poultry sector (Bingsheng and Yijun 2007). It may also occur when regulations are changed that make it difficult for certain types of operation to continue. When swill feeding of pigs was regulated and eventually banned in the United Kingdom to reduce the risk of classical swine fever, it made backyard pig keeping uneconomical. The pig sector became functionally concentrated when backyard herds vanished, it become geographically concentrated because larger commercial units tend to be located in arable areas near feed sources, and it has also become more integrated, with many producers operating under contract.

Restructuring may also be driven by zoning regulations; for example in Jakarta, Indonesia's capital, poultry keeping was banned within 25 meters of a residence by a governor's decree issued in early 2007 (ICASEPS 2008). Since most of the flocks kept in city limits were very small, this regulation is contributing to functional and geographic concentration of the sector.

Restructuring combined with good biosecurity and the provision of animal health services changes the nature of emerging disease control, because it means that disease intelligence and initial response activities must deal with fewer premises, but each individual larger premise may pose greater risks of spread and be logistically more difficult to deal with if it becomes infected. If done without adequate preparation in places where livestock are making an important contribution to the livelihoods of the poor, restructuring can cause livelihood distress (ICASEPS 2008). It usually requires investment in basic infrastructure by the government and in upgrading of premises by their owners, and it can be expensive. Egypt, Indonesia, Nigeria, Viet Nam, and China are exploring the possibility of restructuring their poultry market chains to improve biosecurity and make disease prevention and control measures easier, but this carries the risk of excluding small-scale producers from the market—a possible trade-off between the food safety and the food security of poor households.

Responding Quickly to Disease Threats

Regardless of the efforts made to anticipate and understand animal health problems, diseases still emerge and change their location. Even the best prepared countries and industries must be ready for surprises. This means that they must be able to see and verify a problem and admit that they have one, as discussed in the previous section. They must have strong contingency and operational plans in place, supported by legislation. They must be able to mobilize finances, people, and equipment to move very quickly if an outbreak needs to be controlled. And they must have a recovery plan to follow the emergency action.

Contingency Planning

Contingency planning for emerging and changing livestock diseases is a government-led activity, and any emergency response activities are coordinated, managed, and mostly funded by government agencies. Well-organized animal health systems have contingency plans in places for the major known diseases, based on objective risk assessment and backed up by policies and legislation. For example, detailed information on contingency and operational planning can be found on the Animal Health Australia Web site.

Canada has, at the time of writing, a list of 32 reportable livestock diseases under the Health of Animals Act that is published on the Food Inspection Agency Web site. Animal owners, veterinarians, and laboratories are required to immediately report suspect cases, triggering control measures (Canadian Food Inspection Agency 2008). An even longer list of notifiable diseases must be reported but do not trigger emergency action in Canada. The Health of Animals Act covers quarantining of premises; restrictions on import, export, and sale of animals; disposal of carcasses; culling and compensation; setting up of control areas with restricted movement; and the rights and responsibilities of individuals and government agencies for various actions that may be required to control outbreaks. When a new disease is identified as a serious threat, it is added to the list of reportable diseases and is then covered by the Act.

In countries where the livestock sector is industrialized, it has influence over government disease control strategies as well as private business contingency plans. In Australia, for example, the animal health emergency plan details how risks and responsibilities are shared between the government and the private sector.

Contingency plans lay out broad structures and control methods. There is also a need for detailed operational plans (sometimes known as standard operating procedures), which lay out in detail how various key operations are to be carried out, particularly in the field (e.g., Ausvetplan 2008, DEFRA 2007). Some countries do not have comprehensive animal health contingency plans, or if they have plans these are not backed up by legislation or operational plans.

The avian influenza emergency has highlighted this deficiency. Between 2003 and 2007, national contingency plans for HPAI control were prepared in all the most-affected countries and several other developing countries, but by December 2007 legislation and regulations were still lagging behind needs (UNSIC and World Bank 2007). Some of the existing contingency plans are accompanied by full operational plans but many are not, some include compensation and some do not, and most could not be fully funded from national resources. By contrast, industrial countries had plans in place quickly and funding to back them, drawing on previous experience and existing plans for other diseases.

Contingency plans should be developed before they are needed. One reason for this is that plans tend to be better if they are not produced in a hurry. Good contingency plans require good livestock sector data, which may not be immediately available, and they may require models, which need time and field data to be validated. Preparation in advance also leaves time to test plans with simulation exercises, train staff, and ensure that resources will be available when needed. Well-organized veterinary services routinely follow these practices; however, the contingency plans for HPAI were developed in many countries after the first outbreak, and simulation exercises have been initiated only quite recently and not in all countries at risk.

Another reason for advance planning is that it is generally acknowledged that stakeholders should be consulted about policies and plans. In industrial countries, this is often a requirement. Elsewhere, it may not be required, but there is still an increasing awareness that it should be done. In India, for example, the state of Madyha Pradesh is considering an electronic consultation on the next livestock development plan, so that even remote communities can be aware of what is proposed. Some responses in HPAI-affected countries have been hastily implemented with little consultation, and this has resulted in poor implementation and probably unnecessary loss of livelihoods (Geerlings et al., 2007, ICASEPS 2008).

In recently decentralized systems or those that are poorly funded or where the central and decentralized parts of the system do not work well together, there are particular challenges in recognizing and dealing with emerging diseases. The decentralized parts of the system may not have a capacity to deal with an emerging problem, and several layers of reporting and action may be required to mount a sufficient response. There is likely to be a central contingency plan but no local operational plan or funding to back it. For example, Indonesia and Viet Nam, which have decentralized systems, experienced problems with compensation in HPAI control. In Viet Nam, the costs were shared between the center and the provinces. Some provinces elected to pay only a very small amount, resulting in different rates being paid in different places for the same type of culled birds (Tuan 2007). This is being resolved through policy revision. In Indonesia, funds were made available at the central level for compensation but they were hard to get at the local level, and many farmers whose birds were culled did not receive compensation (Vadjnal and Hartono 2006).

The international community supports contingency planning through guidelines and technical support missions. The World Animal Health Organisation's Performance of Veterinary Services (PVS) tool helps countries evaluate the performance of their veterinary services. FAO's Good Emergency Practices toolkit provides guidelines for emergency contingency plans and recommends that animal health emergencies should be recognized in national disaster planning to facilitate rapid mobilization of resources. Guidelines for operations planning are under development. Animal health standards defined through the OIE and laid out in its Terrestrial Animal Health Code are recognized in World Trade Organization trade disputes. They include the standards required for certain laboratory tests and the steps by which a country, zone (geographical area), or compartment can be internationally recognized to be free of particular diseases. Codex Alimentarius performs a similar function for food safety standards, providing reference guidelines toward high standards.

Private standards for food safety imposed by global retail chains can be even stricter than those proposed by Codex, with the aim of creating food chains with close to zero risk of consumer illness or market disruption. The concept of private–public partnerships for sharing risks, responsibilities, and benefits has been popular for a long time, although mostly in national initiatives. Initiatives that encourage collaboration between large industry players and producers in developing countries include the Sustainable Agriculture Initiative platform and Safe Supply of Affordable Food Everywhere.

Assuming that contingency and operational plans are in place, a successful response to a new disease outbreak depends on speed of the initial response; good knowledge of both the disease and the livestock sector, so that the initial response is an appropriate one; and well-defined trigger points to determine when the disease has spread too widely to be eradicated quickly, which will require additional and longer-term measures.

Initial Response

Under ideal conditions it is possible to stamp out a new outbreak rapidly by means of culling and movement control. Ideal conditions would include a well-organized and well-resourced veterinary service, good knowledge of the location and movement of animals, the ability to trace back disease outbreaks to their source, and a disease that is not spread by insect vectors or wind. Since these conditions rarely apply, even stamping-out exercises that are ultimately successful in controlling outbreaks can be protracted, expensive, and highly socially disruptive. Examples include FMD in the United Kingdom (National Audit Office 2002), CBPP in Botswana (Mullins et al., 1999), and classical swine fever (CSF) and HPAI in the Netherlands (Meuwissen et al., 1999, Geerlings 2006). The impact on different stakeholders may also be very different. A prospective analysis of CSF control in Viet Nam, for example, suggested that movement control would affect producers very badly but traders very little (McLeod et al., 2003).

Other examples with happier economic outcomes include the most recent FMD outbreak in the United Kingdom and an FMD outbreak in Uruguay that was controlled partly by use of vaccination (Leslie et al., 1997). It is interesting that both Thailand and Viet Nam, having failed to control HPAI by widespread stamping-out methods, have chosen to change their strategies; Thailand has combined limited culling with detailed surveillance, while Viet Nam introduced wide-scale vaccination. The control of bluetongue in the United Kingdom was based on movement control rather than culling, and vaccination was introduced as soon as a suitable vaccine was available in sufficient quantities.

Governments with poorly resourced animal health systems face many problems in mounting an initial response. Their response tends to be late and limited, allowing diseases to spread and, if conditions permit, to become endemic. As well as facing severe limitations in mobilizing resources, many face problems of large distances and insufficient infrastructure. In many developing countries there is chronic underprovision of information and basic animal health services to the poor, increasing the possibility that a disease may emerge, cause damage, and spread through a large part of the population before it is detected. The fact that many reports are not acted on is also a disincentive to reporting and so finding the disease.

While the first response to an emerging disease should be managed by the government veterinary service, it is often carried out by private sector animal health practitioners. These include former members of the veterinary services filling public-sector gaps, together with paravets who may be fully private, partly supported by local government, or partly supported by nongovernmental organizations (NGOs). Some paravets are very well trained, while others are not; they are not always clearly identified, and their clients have no objective reference on their competence. In many countries paraprofessionals have no clear position in the animal health system. All these factors weaken the emergency response systems.

Buying Time

For a limited number of diseases with a really effective vaccine already available, vaccination can be the main measure to contain an outbreak. For others, vaccination can still be a useful measure to buy time when stamping out is likely to be become ineffective in halting the spread of an outbreak or widespread culling is too economically damaging. Vaccination was used as one of the control measures for the FMD outbreak in the Netherlands in 2001 when it became evident that the veterinary service had insufficient human resources to handle wide-scale culling (Bourma et al., 2003). It works well in the right conditions, which would include an effective delivery system and an easily applied vaccine that confers long-term immunity (McLeod and Rushton 2007). However, even when capital costs are low because the delivery chain is already in place, the recurrent costs can be high, and there may be a delay in producing a vaccine for an emerging disease. For diseases like FMD and HPAI, in which the virus mutates quite frequently, the vaccine strain must be monitored and updated.

The highest costs and lowest coverage occur where the vaccine must be injected rather than given orally and needs to be delivered through a cold chain, as is the case with both FMD and HPAI. For maximum protection, HPAI vaccine must also be injected twice, adding to its cost. Both China and Viet Nam introduced widespread vaccination against HPAI after an initially rapid spread of disease and heavy losses from death and culling and have continued to use it preventively in the face of continued disease threat. Viet Nam is trying to move to a more targeted strategy with a defined exit plan, but it is now experiencing new outbreaks, many of which are reported to have been in unvaccinated ducks. The animal health system is also experiencing "vaccination fatigue," as much of its human and financial resources have been taken up with campaigns. Indonesia also uses vaccination in part of its poultry flock, but it has experienced problems in keeping track of changes in the virus and updating vaccine strains accordingly.

Vaccination was an important and eventually successful part of the program that eradicated rinderpest from Africa. Initially, mass vaccination campaigns were used with very high delivery costs that failed to reach animals in more remote areas. In 1993 the strategy was changed to targeted vaccination using community-based approaches and a thermostable formulation of the vaccine (Roeder 2005), making it easier to deliver vaccination even to remote communities. It took several widespread campaigns and over 20 years to achieve eradication.

Understanding the Livestock Sector

Knowledge of the livestock sector is critical for effective response to disease threats. Outbreaks of FMD in the United Kingdom in 2001 and Algeria in 1999 highlighted the importance of understanding the way that animal movements spread the disease. As previously described, the U.K. outbreak spread very fast through complex chains of interrelated markets and long-distance movements through market chains to final destinations because traders came from far away to buy sheep. The one in Algeria was limited in spread, partly because it occurred during Islamic festivals. While there was a great deal of animal movement at the time, there was also a great deal of slaughter, and animals were not returning from markets to farms (FAO EMPRES 2002).

HPAI spread has also been heavily influenced by factors in the production and marketing system. There appears to have been a very strong correspondence between the presence of dual-crop rice systems and herded ducks and the persistence of HPAI (Gilbert et al., 2007). In Viet Nam, there has also been a temporal pattern to outbreaks, with the heaviest reported incidence corresponding closely (although not entirely) to increased production and movement before the Tet festival. Almost every country that has experienced HPAI found gaps in information about the location of medium- and small-sized poultry premises and the way the market chains worked. Some are now registering farms more comprehensively or mapping poultry market chains in order to be better prepared in the future (Siagian et al., 2006).

Recovery

All but the smallest disease outbreaks have medium- to long-term impacts on people within the livestock sector, and sometimes on those outside. This is why plans for recovery and damage limitation should be part of the response planning for emergency diseases. In Canada, for example, the province of Ontario has a plan to minimize economic damage to the pig and cattle industries should there be a border closure between Canada and any of its main trading partners (Ontario Ministry of Agriculture Food and Rural Affairs 2008).

However, there are numerous examples where this has not been done, perhaps because the length or spread of an outbreak was much greater than expected. In Viet Nam, for example, after the first wave of HPAI outbreaks many small commercial producers did not restock for many months (ACI 2006). In the Netherlands, some companies went out of business after HPAI (Geerlings 2006). A CBPP outbreak in Bostwana took much longer than expected to contain and resulted in the domestic market being flooded with animals bred for export, to the detriment of local producers (Mullins et al., 1999).

It is not easy to organize a successful recovery program. Most of the available options focus on livestock owners and not on others in the market chain who suffer losses, and even options for livestock owners have their challenges. Compensation applies only to direct losses from culling and is not available to all livestock keepers; in Indonesia, at the time of writing, only smallholders were eligible for HPAI compensation, while in Egypt only farms above a certain size could apply. Private livestock insurance is available in only a few places and often tied to good management. Restocking programs are difficult to implement successfully, not least because of the problems of finding sufficiently large numbers of suitable, healthy animals without distorting local markets. Microcredit schemes of the kind offered by some NGOs and the Grameen Bank may be another option, available to smallholders but not the very poorest.

Response and recovery plans have a greater chance of being successful if they are applied to a livestock sector that is already taking some measures to prevent disease and minimize both risk and impact. By defining the conditions under which a disease may become a problem, it may be possible to invest in measures to remove those conditions and protect livestock populations.

Toward an Ideal Response

As this chapter has indicated, the capacity of animal health systems to anticipate new livestock disease problems, prevent them, or respond to them when they occur varies a great deal from place to place. There is certainly room for improvement—room to learn from past experience and build on existing capacity, to create systems that are stronger and more capable of dealing with today's pressures. In order to propose improvements we first highlight strengths and weaknesses in existing systems.

Systemic Weaknesses and Failures

Retreat of the State

Animal health systems work best when they are sustainably funded from a combination of private and public finance and when both public and private sectors pay attention to policy and regulation. If one part breaks down, the system as a whole is weakened.

The countries suffering chronic problems with livestock diseases—those where emerging diseases become persistent—suffer from a combination of limited public-sector funding, limited public-sector attention to regulations, and a private sector that is not commercially or institutionally robust enough to contribute much to the animal health system. It is not one of these factors alone but all of them together that create the conditions for diseases to spread and persist.

A typical pattern can be seen in countries that have been encouraged to introduce a structural adjustment program while their economies are also weak, so that investment in the public sector has been severely and too quickly curtailed but has not been replaced by private-sector growth and investment because of a lack of domestic capacity. Limited operational funds have made it hard for state veterinary services to carry out either preventive or response activities. Animal and public health regulations have been allowed to fall out of date and failed to make provision for private-sector actors, whether veterinarians or paraprofessionals, to step into roles that were formerly occupied by government veterinarians. Regulatory gaps have resulted in a lack of clarity about who is responsible for what, who can supervise or carry out which activities, and the extent of professional liability or protection, particularly for paraprofessionals. At the same time, public-sector veterinarians involved in part-time private practice have been in competition with fully private animal health professionals, offering a short-term solution to fill the gap in service but slowing down the establishment of a sustainable private sector.

Commercial livestock production develops within ranching areas and near large towns, but domestic demand for livestock products is limited, and the sector contributes insufficiently through taxation or public–private funds to make up the deficit in public funding. It also tends not to be sufficiently cohesive or motivated to privately monitor and regulate animal health or food safety. Diseases spread and create a barrier to seeking new markets for livestock products and new sources of revenue to fund animal health, and so the downward spiral continues.

By contrast, the countries that do best in dealing with emerging and spreading diseases are those where the gross domestic product (GDP) and the tax base are sufficient to fund animal health on a sustainable basis, where the commercial livestock sector is strong and well organized and makes an important contribution to the GDP and export earnings, and where there is government commitment to developing and enforcing animal health regulations (Rushton et al., 2006). Not all of these are OECD countries, and there are encouraging examples of emerging economies overcoming the problems caused by the retreat of the state. In Latin American exporting countries, there are examples of large commercial producers subsidizing preventive measures in the small-scale producers surrounding them and of increasing levels of dialogue and cooperation between government and the commercial sector.

Decentralization of the government and civil service in several countries offers the possibility of a locally responsive service that could help promote preventive measures suitable for local circumstances. However, decentralization makes an emergency response to emerging diseases very challenging. It can slow the reporting of suspicious cases if a local decision is made not to report something to the center because of concerns about the disruption that the response may cause. It can create confusion in the way that early responses are funded and

handled—for example, local decisions on compensation rates leading to different rates being used in neighboring provinces, resulting in sick animals being moved and disease being spread.

Dichotomy in Animal Health Systems and in the Livestock Sector

Control of new diseases is easiest where livestock market chains are relatively simple and well regulated and where the necessary financial and human resources exist in the animal health system. These conditions are found in the richest countries and also where the livestock sector in emerging economies has intensified, concentrated, and become export oriented. Disease control is hardest where financial resources are limited and livestock systems are either complex and diverse (as seen in the poultry and pig sectors of many Asian countries) or highly dispersed (as in some of the poorest countries of sub-Saharan Africa). These conditions are found in the poorest countries, among the poorest livestock keepers in many countries, and in transitional situations where livestock systems are responding to growing urban demand. Each of these situations has its own problems, yet some of the difficulties in managing emerging and spreading diseases arise not from the problems in any one system or country but from the large imbalance between the best organized animal health systems and those with the least capacity. This widening dichotomy between the livestock sectors and the animal health systems of the rich and the poor results in each posing risks for the other.

The strongest animal health systems have a demonstrable capacity to find disease outbreaks fast and stop them from spreading, as earlier examples have described. These systems exist in industrial economies and the strongest of the emerging economies, although the fluctuations in the fortunes of the latter render them periodically vulnerable to disease (e.g., the reemergence of FMD in some Latin American countries tends to coincide with economic problems). The countries with the strongest animal health systems are globally networked through livestock trade. When new diseases emerge and spread through legal and illegal trade routes, they all take the same "eradication at all cost" approach, and for the most part this has been successful in halting disease spread, although the economic cost of controlling each outbreak is high. Peer pressure and the possibility of severe economic consequences of noncompliance mean that transparency of disease reporting by governments is high. These countries also have a strong (perhaps disproportionately strong) impact on international regulations for animal health and food safety that are driven strongly by the wish to protect trade. Countries with strong animal health system trading with each other are therefore quite institutionally robust.

The countries characterized by underfinancing of state veterinary services together with incomplete privatization and overdue reform have an ad hoc approach to emerging and spreading disease. Surveillance systems are insensitive, and even when disease is detected, reports are delayed. Effective movement control to limit spread is almost impossible, and preventive vaccination cannot be applied at the necessary level of coverage. In the case of zoonotic diseases, the human health system is equally ineffective at identifying and controlling these because many are diseases of the poor, who have limited access to human health systems and animal health advice. Traditional systems of managing disease, particularly in pastoralist societies, are eroding as these societies are changing. (See Chapter 11.) The countries concerned therefore remain vulnerable to disease invasion, allow emerging diseases to become endemic, and then pose a risk to their neighbors. While there are some attempts to establish cross-border agreements (e.g., in West Africa), these generally suffer from lack of sustained funding or political commitment at high levels.

International regulations have tended to focus on eliminating disease and disease agents from geographical areas, and countries unable to create demonstrably disease-free areas have been excluded from export to countries that can meet these conditions. More recently, the introduction of the concept of compartments has made it possible to create disease-free populations rather than geographical areas, allowing a single integrated market chain the possibility of becoming certified as disease-free, provided that the veterinary services in the country where it is based are considered competent to provide the accreditation.

The imbalance between the stronger and weaker animal health systems causes problems in two ways. The less robust systems pose a constant threat of disease to the stronger ones. At the same time, the stronger ones raise barricades to protect their livestock sectors from disease, with international sanitary and phytosanitary regulations set to high standards that can only be met by consistent effort and investment. There is also a tendency to skew investment in animal health in poorer countries toward a small part of the livestock sector concerned with international trade and thus to create another dichotomy, between richer and poorer livestock keepers. The poorest livestock keepers in poor countries find it hardest to get access to animal health services. Many smallholder producers in developing countries whose flocks were culled for HPAI control found it impossible to register for compensation (World Bank et al., 2006). Changes to biosecurity regulations have sharply restricted the access of small-scale commercial producers to markets (ACI 2006). Rinderpest control has been an exception to the extent that it affected all cattle owners in Africa; however, it is worth asking why a similar level of attention has not been paid to diseases of small ruminants in extensive systems.

Conflicting Paradigms

There are several different constituencies within the international animal health and public health communities—and they do not always speak in one voice or act in harmony. Animal health, like other disciplines and sectors, has been affected by changes in fashions, funding, and governments.

There are those who see animal health or public health problems one disease at a time and those who view them primarily in their social and economic development context. The advantage of the former approach is that it gives strong focus and high visibility to eradication programs; its limitation is that farmers and veterinary services rarely have the luxury of dealing with only one problem, and they invariably benefit from economies of scale and scope when several problems are tackled at once.

There is one viewpoint that livestock diseases can be tackled by straight-line rational approaches based primarily on technology and a contrasting viewpoint that many of today's animal health problems are "messy"; they are caused by institutions and people as much as biology and are not susceptible to simple solutions. The advantage of the straight-line rational approach is that it makes analysis and management of problems rather clear-cut; it puts boundaries around problems and provides measurable objectives against which to monitor progress. Its limitation is that it may exclude many of the factors that are the worst contributors to the problem. Even rinderpest, a relatively straightforward disease against which vaccination was highly effective, had messy elements; the final stages of eradication were only possible through creative approaches to vaccine delivery and epidemiology. Emerging diseases have many messy features—and zoonotic diseases particularly so. They exist in diverse livestock populations where a range of animal health services are needed; they exist at the animal–human interface, so that animal health and human health systems must cooperate to control them; they can be politically embarrassing because they are usually labeled "crises," and simple technological solutions are not usually available.

The people who subscribe to these different viewpoints can always find common aims and areas of interest when they look for them, yet often they seem to be shouting at each other from opposite sides of a large room, making no attempt to listen or truly understand each other's perspective or to admit that two perspectives may both have merit and that the solution may lie in a combination of the two.

Suspicion and Fear

The phrase "lack of transparency" is used often in discussions about animal health problems—and it is agreed that this is undesirable and must be remedied. In practical terms, this usually amounts either to not reporting suspected disease cases or not sharing virus samples. There are perfectly good reasons for either. Reporting may lead to destruction of someone's livestock, to no helpful response at all, or to loss of a market. Virus sharing leads to questions about "benefit sharing" between the countries where the virus was found and those who make money from developing vaccines against it.

There is also suspicion and fear of new ideas, as partially addressed in the previous section. Refusal to examine a problem from different angles because "it wastes time" or "we don't need to reinvent the wheel," or the affirmation that "we don't need to change what we do, just do more of it" can be a perfectly reasonable response to an objective examination of the facts. But such responses can also be a symptom of fear that opening up a problem and reframing it may publicize the fact that a mistake has been made, leading to blame or loss of funding.

Strengths and Opportunities

Partnerships and Networks

There is a recognition of many of the weaknesses just discussed. One response has been a growing tendency for organizations, institutions, and individuals to work together in ways that include formal partnerships and coalitions as well as looser information-sharing networks.

Strong international partnerships have formed for the control of certain diseases. The Global Rinderpest Eradication Programme (GREP) is an example that has had a successful result. Following similar practices and working toward international standards on disease freedom, countries have progressively eliminated rinderpest disease and appear close to eradicating the virus worldwide. HPAI has also been tackled on a global scale. To control the disease in poultry and prevent a human influenza pandemic, a United Nations program coordinated through the UN System Influenza Coordination has worked with the World Bank and a range of national projects, many supported by bilateral donors.

FMD control has been organized largely through regional partnerships and networks. The European Union (EU) has been successful in protecting trade on behalf of member countries and has provided them with economic support to deal with outbreaks of several diseases. The European Commission for the Control of FMD was formed to eradicate FMD from Europe, which it has managed to do with only a few reinfections, and it is making steady progress in harmonizing approaches to FMD control in its 35 member countries. Regionalization of FMD control in other regions has had a modest impact and continues to grow. The Pan American Foot-and-Mouth Disease Center in Latin America and the Southeast Asia Foot and Mouth Disease Campaign

program in the Mekong region encourage progressive zoning to halt FMD spread.

Taking a more general rather than single-disease approach, the West African Economic and Monetary Union is working on harmonization of animal health regulations, while the Southern African Development Community had a flourishing livestock policy network to provide information to members. The Pan African Programme for the Control of Epizootics (PACE), one of a succession of projects that dealt with rinderpest, was designed to look more broadly and also worked on CBPP. The Association of Southeast Asian Nations has an animal health working group that brings member countries together to discuss joint approaches to disease control and harmonization of legislation on various diseases.

There are also bilateral partnerships between countries. Mexico and the United States have worked together successfully to eradicate FMD and screwworm and are together tackling CSF, tuberculosis, brucellosis, and Mediterranean fruit fly.

Also very important are the "public–private partnerships" that exist within countries and internationally. Animal Health Australia is an example of a formal private–public grouping where risks and responsibilities are shared. Many OECD countries have official mechanisms through which government is able to or required to consult industry and sometimes the general public about major policy changes. In most of the major livestock-producing and -exporting countries, a formal or informal mechanism exists whereby industry influences government; one of the signs of a "modernizing" livestock sector with a strong commercial sector is the growth of institutional arrangements through which industry shapes policy.

Networks of various kinds have been formed to facilitate organizations to work together and share information. Some are specific to disease control, like the OFFLU network of laboratories and the GLEWS network that shares disease intelligence of various kinds, while others are focused on particular livestock systems like the Family Poultry Network. Networks of laboratories, epidemiologists, and social scientists are being reviewed or considered through the Partnership for Africa Livestock Development (ALive) program in Africa.

The reasons for partnership and information sharing are many. Response to the demands of donors, financial expediency, advances in technology for sharing information, fashion in management strategy, and an acknowledgment that messy problems need wide involvement to solve them have all played a part. Networks, partnerships, and coalitions offer advantages in economies of scale, having more than one set of professional skills and experience, and the potential to harmonize advice and recommendations and reduce public confusion. When partnerships work well they can partly compensate for the retreat of the state and reduce dichotomies and conflicts. They do, however, require time, effort, resources, and management skill to be successful. And they need to be set up so that each partner gains something that compensates for the time they spend on maintaining the partnership. Imbalance in some of the existing partnerships and networks has contributed to the systemic weaknesses previously mentioned.

Human Resources

Human resources have grown stronger in terms of training and experience, although they are not all used to the fullest extent. Even in poor countries, and even where investments in infrastructure have not been sustainable, the human resource is gaining capacity. Most major initiatives to deal with a particular disease have been underpinned by more general investment in animal health. GREP was underpinned by general investment in strengthening of veterinary services in Africa under the Pan African Rinderpest Campaign (PARC) and PACE. HPAI programs were preceded by EU-financed Strengthening of Veterinary Services programs in Asia, by PARC and PACE in Africa, and by a number of bilateral animal health programs over many years. All of these helped governments build up a core capacity of trained professionals and develop and test new methods, which have remained in the institutional memory and continue to be used and developed. The same people who were trained and gained experience under previous projects are now senior members of HPAI control programs.

Sustainable Funding Mechanisms

Animal health has been far better and more sustainably funded in countries with strong economies and strong livestock sectors than in those that have neither. The strongest systems are financed from a mixture of public and private funds. All public funds are of course financed through taxes. In addition, jointly managed funds like the one managed by Animal Health Australia provide a mechanism for formally sharing risks and costs of animal health control. Large-scale commercial producers and processors directly pay for a lot of preventive and curative animal health services from private veterinarians. Private insurance against the loss of animals or production also exists, although it is uncommon except for very valuable animals.

Where extrabudgetary funding is provided to poor countries in large amounts and over long periods it is also possible to make progress: the JP 15, GREP, PARC, and PACE programs between them appear to have eradicated rinderpest. The large emergency funds poured into HPAI control have also helped to build capacity and reduce the spread of disease. But these funds have mostly been uncertain and uneven, and not backed up by sustainable financing from national budgets. The rise of CBPP has been accelerated by the end of rinderpest

vaccination campaigns, since CBPP vaccination used to be done at the same time with little additional cost.

Reducing Weaknesses and Building on Strengths

An ideal response to emerging diseases would be one that anticipates problems, invests to minimize risks, establishes priorities, and prepares in advance to deal with problems while at the same time retaining a capability to deal with surprises. The precise shape of an effective system would vary according to local needs. The necessary features are all present to some degree within existing systems, but they are neither uniform nor joined up. There are some very weak links in the chain. What, then, ought to be done?

Acknowledge the Problem

HPAI was a wake-up call not only for chief veterinary officers (who already knew they had a problem) but for ministers of agriculture and finance (who have acknowledged the need to make some investment in animal health) and the donor community (which has acknowledged that investment in animal health systems will be needed for some time to come). There is a wide and public acknowledgment that the messy problems posed by emerging and spreading diseases will demand innovative ways of working (FAO et al. 2008b). They will also demand strong political commitment. Fostering these ideas will be important.

Improve Transparency and Multidisciplinarity in Disease Intelligence

A more effective system would need a number of building blocks. It would need more reliable current information, held in accessible databases, generated from networks of contacts and studies, held nationally and shared internationally. The databases largely already exist, but the data they contain vary in quality. A great deal of work is needed to remove the elements of suspicion and fear that prevent individuals and nations from sharing information and replace them with a climate of enlightened self-interest.

Organizations with competing agendas need to find ways to reduce the transaction costs of working together. It must be possible to screen and check data quality and comparability, meaning that countries and institutions must be willing to open up their data collection processes to outside scrutiny. Supporting institutions need to foster a no-blame culture that encourages sharing of information and also to treat all suspicions seriously, regardless of the social group from which they originate, so that emerging disease problems can be dealt with equitably. Farmers and other information stakeholders need to be drawn much more closely into designing knowledge systems and given the incentive to cooperate in providing information through participating in schemes for sharing risks and responsibilities as well as benefits. At the same time, acknowledging that suspicion, fear, and noncooperation will always exist to some extent, disease intelligence systems need to accommodate them. Much more will need to be invested in surveillance systems in the poorest countries, and investment will need to be supported from international funds, since good surveillance on emerging diseases benefits everyone.

An ideal system would also hold diverse information—on livestock populations and diseases, trade movements, human demographics and movements, climate shifts, ecosystem changes—and it would have skilled multidisciplinary analytical teams who are able to examine the information from a range of perspectives. This would mean expanding some databases to accommodate wider ranges of information of different types (e.g., a great deal is now being done with spatial data) and making an expanding range of disciplines available, particularly to assist developing countries with their intelligence—such as legal expertise, expertise in scenario building, and sociological expertise, to name only three.

An effective system would have high-quality technology dispersed around the world to improve data storage, analysis, and communication. More investment would be needed to assist countries whose disease intelligence systems are very poorly supported. Recent developments in information and communication technologies and the spread of communication networks even to the poorest countries make this vision more realistic than it would have been even five years ago.

Any good system would include the ability to communicate straightforwardly about risk with members of the public in order not to cause unnecessary shocks. In the case of emerging zoonotic disease, this requires much closer cooperation between the animal health and human health sectors.

A good disease intelligence system needs the ability to very quickly mount an investigation of a suspicion. In some places the finance and equipment, plans, and standard operating procedures for rapid investigation and the necessary legislation still do not exist and will need to be established. In many places, the investigation of a suspect case is not linked to emergency response capacity because no funds are available.

The resources to fund disease intelligence need not be enormous, provided that they are coordinated and used intelligently. There is a temptation to continually add new systems, but there may be more value in making sure that existing systems can communicate with each other and are maintained and supported. At the same time, there are benefits from some degree of redundancy in the system, with different teams and projects working on the same problem, since leaps in knowledge can come from unexpected sources.

Even the best disease intelligence is never going to predict all new problems. A good system will be moderately successful in predicting problems that might occur and very good at locating and tracking new problems quickly.

Partnerships to Minimize Risks and Impacts

The key to minimizing risk is to provide incentives for private individuals and companies to take actions that serve their own and the public interest. A more robust system will need much stronger collaboration between the public and private sector in a variety of ways. Each actor has different incentives, so that joint action must yield benefits to both and clearly define the responsibilities of each. Structural adjustment was an attempt to do this, but instead of building partnerships it created distances and left large gaps in animal health systems. Everyone now talks about public–private partnerships, but they are not well defined or understood (definitions and agendas vary with the stakeholder), and there are few examples of successful ones in developing countries or even the emerging economies.

Several elements will be needed to improve the situation:

- *A clear definition and understanding of responsibilities.* In countries where the commercial livestock sector is strong and growing, the primary responsibility of the government will lie in working on problems and actions that constitute public goods and in providing and monitoring policies and regulations. In areas where the livestock sector has a large number of small-scale producers or where production is marginal but there is an interest in maintaining rural communities, governments will need to be proactive in providing, funding, and promoting services and in monitoring the impacts of policies on those who do not have the voice to speak for themselves.
- *Stronger processes and institutions that oblige the public and private sectors to come together on a regular basis* to agree on risks and responsibilities.
- *A sensible balance of regulations and incentives to encourage investment in biosecurity,* with the state providing the majority of regulations and the market providing incentives.
- *Formal and well laid out mechanisms to manage joint funding.*
- *Information on appropriate prevention measures that is more easily accessible to everyone working in the livestock industry.* Since a growing number of people have access to the Internet, mobile phones, and television, it is not difficult to make information available. However, it does need to be clearly worded and well thought out, and people need to be able to ask questions about it.
- *Documented examples of successful partnerships of different kinds that all stakeholders can draw on.*

Where emerging diseases have the potential for large externalities, public investment (national and international) as well as investments from private foundations and corporate social investment will continue to be justified.

Recognizing that there are economies of scale and scope in animal health systems and that the priorities of livestock keepers may be different from those of governments, investments in disease prevention and control could usefully take a "many for one" approach, dealing with a range of animal health problems at the same time. This would shift large programs away from a single-disease approach. This idea will cause concern in some camps because managing programs that attempt to address multiple problems holds considerable challenges. The key may be not to shift suddenly from highly focused to very broad approaches but simply to broaden the scope slightly and in well-defined ways. One possibility may be to introduce programs to tackle groups of problems that have the same cause or are susceptible to similar solutions (e.g., multiple animal health problems in extensive small ruminants). Another approach is to focus capacity-building efforts in countries or systems that have similar systems and similar institutional problems. In the recent "One World, One Health" discussions, there were proposals to invest in areas where dense human and animal populations, together with overloaded public services, allow emerging zoonotic disease to take hold and persist—the "hot spot" areas.

Developing and emerging economies would need to find ways to face the challenge of managing the trade-off between reducing the risk and impact of disease and reducing the impacts of excluding certain people from livestock production and trade. When investment is made in biosecurity and restructuring, this can create barriers to participating in the livestock sector. People who are excluded are often those who might previously have invested in livestock as a way out of poverty. An expanding number of approaches and tools are available (zones, compartments, and commodity trade, to name three). Expanding the interpretation of disease freedom to encompass alternative approaches and continuing to look for innovative ways to provide and certify safe livestock products will be important in helping to cushion the effects of market exclusion by allowing stepwise progress toward more biosecure systems.

Improve Response Speeds and Surge Capacity

The key to effective emergency response is to be prepared. Every country needs to have robust and flexible animal health contingency plans that can be adapted for new problems with legislation to support their implementation and clear trigger points that indicate when an emergency has occurred and how to act. These plans and regulations do not exist everywhere even for the diseases that cause the greatest problems. International systems will need to continue to assist by providing well-designed, unambiguous, and accessible guidelines for a range of situations and helping to have them tested. In turn, countries that need help with these preparations will need to be organized and vocal about asking for it. For emerging diseases of global potential, ideally regional and international contingency plans would also be in place, but in practice this will rarely be possible: the priority is to work toward harmonization across and between national contingency plans.

In support of advance planning, up-to-date and accessible information on location and movements of animals is essential. This is rarely the case, and it would be far too ambitious to expect good quantitative data to be available everywhere. One approach that may alleviate this problem is better use of qualitative analyses, expert data, and distant (e.g., satellite) data, all of which may be quite rapidly and cheaply obtained so they can be updated more often. Another useful approach is to share information between countries that have similar livestock sector structures so that one with very poor data can learn from one where the data are much more robust.

Streamlining of finances is a priority. Emergency livestock disease control and the associated investment need to be given greater weight in country budgets to enable response systems to be rapid. Another critical requirement often missing, particularly in poorer countries, is surge capacity when actually responding to a crisis. As well as mounting a first response very rapidly, an effective system has the capability to scale up rapidly in response to recognized triggers, mobilizing finances, people, and equipment on a large scale in the face of an outbreak—the principles that underlie the Incident Command System. This requires measured indicators of performance, a way of predicting what might happen next, and access to resources ahead of when they are needed. A progress-tracking system is essential in order to evaluate and adjust control measures and forecast resource needs. Speed of access to resources is as important as the amount available, because the faster an emerging disease can be stopped from spreading, the cheaper and less disruptive it will be.

The possibility of a global animal health fund is being explored to partly cover this kind of contingency, but national governments also need to build up their own funding sources. In some cases the problem may be resolved by private–public funds, with rules for access agreed in advance and reviewed on a regular basis. There might also be merit in reclassifying major animal health emergencies as natural disasters, allowing fast-track access to earmarked disaster mitigation funds. For the most efficient response in an emergency, it may be necessary to have some resources that are not allocated all the time.

Stakeholder buy-in is critical, since quick response depends on information that is often provided by farmers or members of the public. Communication between the veterinary service and farmers is essential, and regulations that make it possible to publish serious transgressors are useful, although sometimes hard to implement. Two examples are withholding compensation from those who fail to report disease and making professional certification for animal health practitioners contingent on disease reporting.

Fast response needs leadership that has a sufficient degree of central command and authority to cut through red tape in emergency but can also consult with a decentralized system when planning and can delegate authority to local levels as needed. These two requirements come from very different management approaches, and it takes considerable skill to achieve a balance.

Technology and methodologies for outbreak investigation and response are always under review, including models (provided accurate data are available and the output can be validated), geographic information systems, handheld and penside devices, and vaccines in easily applicable forms such as thermostabile or noninjectable to facilitate their use in extensive systems and cut down costs for intensive systems. However, it is unlikely that new scientific methods will come out of the livestock industry, where profit margins are quite low; spillover from medical research would be needed for vaccine development and from human health and military research for electronic technology.

Conclusions

It is easy to describe an ideal response and more difficult to envisage how it might come about. Moving from the current response toward one that is more robust will require strengthening of the best existing systems, boosting weaker ones, and some rethinking and repositioning of the design of animal health programs. Considerable political and financial commitment will be needed for many years to come.

The suggestions made here demonstrate the challenges ahead to build on the best of existing responses and improve the system overall. New pressures on the livestock sector mean that improved disease control is essential. Emerging and changing diseases are not the only important animal health problems, but they are damaging for economies and livestock keepers, and they increase the uncertainty and instability of the sector. The challenge to the global animal health system must somehow be met.

References

ACI (Agrifood Consulting International). 2006. Poultry sector rehabilitation project—Phase I. The impact of avian influenza on poultry sector restructuring and its socio-economic effects. Rome: Food and Agriculture Organization.

Ausvetplan. 2008. *Australian veterinary emergency plan.* Summary document proof for approvals version 3.1 2008. http://www

.animalhealthaustralia.com.au/programs/eadp/ausvetplan/ausvetplan_home.cfm

Bager, F., and C. Halgaard. 2002. Salmonella control programmes in Denmark by Danish Veterinary and Food Administration, Copenhagen. FAO/WHO Global Forum of Food Safety Regulators, Marrakech, Morocco, 28–30 January.

Bingsheng, K, and H. Yijun. 2007. Poultry sector in China: Structural changes during the past decade and future trends. Paper presented at Poultry in the 21st Century: Avian influenza and beyond. Food and Agriculture Organization. Bangkok, 5–7 November.

Bourma, A., A. R. W. Elbers, A. Dekker, A. de Koeijer, C. Bartels, P. Vellema, P. van der Wal, E. M. A. van Rooij, F. H. Pluimers, and M. C. M. de Jong. 2003. The Foot-and-Mouth Disease epidemic in the Netherlands in 2001. *Preventive Veterinary Medicine* 57 (3): 155–166.

Canadian Food Inspection Agency. 2008. http://inspection.gc.ca/english/anima/disemala/guidee.shtml

DEFRA (U.K. Department for Environment Food and Rural Affairs). 2007. DEFRA's framework response plan for exotic animal diseases: For foot and mouth disease, avian influenza, Newcastle disease, classical swine fever, African swine fever, swine vesicular disease, rabies, bluetongue, specified types of equine exotic diseases: Glanders, dourine, infectious anaemia and equine encephalomyelitis (of all types including West Nile virus). Presented to Parliament pursuant to Section 14A of the Animal Health Act 1981 (as amended 2002). 10 December.

DEFRA (U.K. Department for Environment Food and Rural Affairs). 2008. Horizon Scanning & Futures Home. http://horizonscanning.defra.gov.uk.

Dung, D. H, N. M. Taylor, and A. McLeod. 2006. *Improving veterinary surveillance in Viet Nam—A knowledge management approach.* Proceedings of the 11th International Symposium on Veterinary Epidemiology and Economics. Cairns, Australia.

FAO (Food and Agriculture Organization). 2008. Expert meeting on climate-related transboundary pests and diseases including relevant aquatic species. http://www.fao.org/foodclimate/expert/em3/en.

FAO (Food and Agriculture Organization) EMPRES. 2002. Towards a global early warning system for animal diseases. EMPRES Transboundary Animal Disease Bulletin 20/1.

FAO (Food and Agriculture Organization), World Bank, and World Organisation for Animal Health. 2008a. *Biosecurity for HPAI: Issues and options.* Rome: FAO.

FAO (Food and Agriculture Organization), World Organisation for Animal Health), World Health Organization, and World Bank. 2008b. Contributing to one world, one health: A strategic framework for reducing risks of infectious diseases at the animal–human–ecosystems interface. Consultation document. 14 October.

Freeman, A. H., P. K. Thornton, J. ven de Steeg, and A. McLeod. 2006. Future scenarios of livestock systems in developing countries. In *Animal production and animal science worldwide.* World Association for Animal Production. Wageningen, Netherlands: Wageningen Academic Publishers.

Geerlings, E. 2006. Literature review of the Dutch poultry sector and impacts of AI. Internal Report. Rome: Food and Agriculture Organization.

Geerlings, E., L. Albrechtsen, J. Rushton, Z. Ahmed, F. Abd El-Kader Ahmed, S. Ahmed Eldawy, A. Saied, et al. 2007. Highly pathogenic avian influenza: A rapid assessment of the socioeconomic impact on vulnerable households in Egypt. Report produced for FAO. Rome: Food and Agriculture Organization.

Gilbert, M., X. Xiao, P. Chaitaweesub, W. Kalpravidh, S. Premashthira, S. Boles, and J. Slingenbergh. 2007. Avian influenza, domestic ducks and rice agriculture in Thailand. *Agriculture, Ecosystems and Environment* 119:409–415.

Glisson, J. 2008. From Pasteur to genomics: Past, present and future control of poultry diseases. Proceedings of the XXIII World's Poultry Congress, Brisbane, Australia, 30 June–4 July.

Gunn G. J., C. Heffernan, M. Hall, A. McLeod, and M. Hovi. 2008. Measuring and comparing constraints to improved biosecurity amongst GB farmers, veterinarians and the auxiliary industries. *Preventive Veterinary Medicine* 84 (3–4): 310–323.

ICASEPS (Indonesian Center for Agro-socioeconomic and Policy Studies). 2008. *Livelihood and gender impact of rapid changes to bio-security policy in the Jakarta area and lessons learned for future approaches in urban areas.* Rome: ICASEPS in collaboration with Food and Agriculture Organization.

Leslie, J., J. Barozzi, and M. J. Otte. 1997. The economic implications of a change in FMD policy: A case study in Uruguay. In Proceedings of 8th International Symposium on Epidemiology and Economics, Paris, 8–11 July. Published as a special issue of Épidémiologie et Santé Animale 31/32.

McLeod, A. 2007. Social impacts of structural change in the poultry sector. Paper presented at Poultry in the 21st Century: Avian influenza and beyond. Food and Agriculture Organization. Bangkok, 5–7 November.

McLeod, A., and J. Rushton. 2007. Economics of animal vaccination. *Rev. sci. tech. Off. int. Epiz.* 26 (2): 313–326.

McLeod, A., N. Taylor, N. T. Thuy, and L. T. K. Lan. 2003. *Control of Classical Swine Fever in the Red River Delta of Vietnam: A stakeholder analysis and assessment of potential benefits, costs and risks of improved disease control in three provinces.* Phase 3 report to Strengthening of Veterinary Services in Vietnam. Hanoi: Department of Animal Health.

Meuwissen, M. P. M., S. H. Horst, R. B. M. Huirne, and A. A. Dijkhuizen. 1999. A model to estimate the financial consequences of classical swine fever outbreaks: principles and outcomes. *Preventive Veterinary Medicine* 42 (3–4): 249–270.

Morgan, N. 2006. Meating the Market: Outlook and Issues. Presentation made to the International Poultry Council at the seminar on Global Trends in Meat Production and the Impact of Animal Diseases, VIV Europe 2006, 16–18 May 2006, Utrecht.

Mullins, G., B. Fidzani, and M. Koanyane. 1999. At the end of the day: The socio-economic impacts of eradicating contagious bovine pleuropneumonia from Botswana. In *Tropical diseases: Control and prevention in the context of "the new world order."* Proceedings of 5th biennial conference of the Society for Tropical Veterinary Medicine, Key West, 12–16 June. New York Academy of Sciences.

National Audit Office [of the UK Government]. 2002. The 2001 outbreak of Foot and Mouth Disease. Report by the comptroller and auditor general. HC 939 Session 2001–2002: 21 June.

OIE (World Organisation for Animal Health). 2003. *International Animal Health Code: Chapter 1.3.2: Guidelines for import risk analysis.* Paris.

Ontario Ministry of Agriculture Food and Rural Affairs. 2008. *Livestock border closure contingency plan.* Toronto.

Roeder, P. 2005. Eradicating pathogens: The animal story. *BMJ* 331:1262–1264.

Rushton, J., A. McLeod, and J. Lubroth. 2006. Managing transboundary animal disease. In *Livestock Report, 2006,* ed. A. McLeod, pp. 29–44. Rome: Food and Agriculture Organization.

Siagian, A., P. Sembiring, Z. Siregar, M. Tafsin, N. D. Hanafi, Rasmaliah, D. S., and R. Hasibuan. 2006. *Poultry market chain study in North Sumatra.* Jakarta: Food and Agriculture Organization.

Taylor, N. M., N. Honhold, A. D. Paterson, and L. M. Mansley. 2004. Risk of foot-and-mouth disease associated with proximity in space and time to infected premises and the implications for control policy during the 2001 epidemic in Cumbria. *Veterinary Record* 154 (20): 617–626.

Tuan, N. A. 2007. Avian influenza, poultry culling and support policy of the Viet Nam government. In Future of Poultry Farmers in Viet Nam after HPAI, ed. A. McLeod and F. Dolberg, pp. 10–12. Workshop held at Horison Hotel, Hanoi, March 2007. Hanoi: Food and Agriculture Organization

UNSIC (U.N. System Influenza Coordination) and World Bank. 2007. *Responses to avian influenza and state of pandemic readiness: Synopsis of the Third Global Progress Report.* New York and Washington, DC.

Vadjnal, D., and T. Hartono. 2006. *Rapid assessment of the highly pathogenic avian influenza compensation scheme in Indonesia.* Report prepared for Food and Agriculture Organization and cofinanced by World Bank.

Vu, T. 2009. *The political economy of avian influenza response and control in Vietnam.* STEPS Working Paper 19. Brighton, U.K.: STEPS Centre.

Wegener, H. C., T. Hald, D. Lo Fo Wong, M. Madsen, H. Korsgaard, F. Bager, P. Gerner-Smidt, and K. Mølbak. 2003. Salmonella control programs in Denmark. *Emerging Infectious Diseases* 9 (7).

World Bank, Food and Agriculture Organization, International Food Policy Research Institute, and World Organisation for Animal Health. 2006. *Enhancing control of Highly Pathogenic Avian Influenza in developing countries through compensation: Issues and good practice.* Washington, DC: World Bank.

19

Responses on Social Issues

Jeroen Dijkman and Henning Steinfeld

Main Messages

- **Livestock can be instrumental in reducing poverty and can be a growth engine in rural areas.** They can be particularly suitable tool for poverty reduction in poor countries and areas that are not fully exposed to globalized food markets. Smallholder dairy production and certain forms of cooperatives and contract farming provide opportunities for expanding and intensifying smallholder livestock production. In rapidly growing and developed economies and in periurban areas with products primarily aimed at urban markets, economies of scale and market barriers are unlikely to allow smallholders to compete. The speed of change will not be the same everywhere, but the direction of change is clear.

- **Producers that remain within the livestock sector need to be able to innovate constantly.** The sector's capacity to respond and adapt to change needs to be enhanced in ways that allow producers to innovate while safeguarding the livelihoods of poor people linked with the sector. This includes the social and institutional arrangements required to mobilize different sorts of knowledge and support services. It is also important for smallholders to organize by forming, for example, cooperatives, large-scale/smallholder partnership schemes, or other organizational arrangements in order to be able to deliver the volume and type of products that the market demands and to exploit niche markets for specialized products.

- **For those that find livestock production too risky or the barriers to entry too great, alternative livelihoods need to be sought in other sectors.** Policies and programs for targeted social protection need to support the transition process so as to avoid social hardship and prevent rapid loss of livelihoods when alternatives cannot be provided. There may be more opportunity for employment in agrofood industries than in primary production.

- **The objective of pro-poor livestock sector development policies should not be to maintain smallholder production systems at any cost, but to mediate sector transition, in which the role of the poor needs to be considered broadly,** including as consumers, market agents, and employees as well as small-scale producers. From a poverty reduction viewpoint, as long as waged employment can absorb people exiting from the sector, this is not necessarily a bad thing: on the contrary, it should be viewed as a positive development.

Introduction

The increasing connection of rural activities in the developing world to the global environment is diversifying livelihood options and also exposing such activity to rapidly changing patterns of competition, market preferences, and standards. This growing interconnectedness brings not only challenges to those in the changing livestock landscape, but also opportunities arising from new market opportunities, technology, and knowledge, such as biotechnology and information technology, and from convergence among different areas of economic and social activity.

Notwithstanding such radical change in many places, food production and subsistence agriculture remain crucial to a large proportion of the "extreme poor"— those living on less than $1.25 a day (in 2005 dollars). However, the impact of HIV/AIDS and other diseases, conflicts, and an increasing number of climate-related disasters means that people in these production systems also need to tackle an ever-evolving set of production, pest, and disease problems—and often in rapidly declining environmental conditions. Concomitant changes have also occurred in the roles, responsibilities, and working practices of rural actors through decentralization and the privatization of public services, combined with an increasing and strengthening role and influence of the private sector and civil society in development.

These are just some of the factors that shape the fates of the majority of the world's 1.4 billion extreme poor, most of whom depend for their livelihoods on farming or supplying farm labor. Statistics reveal that an estimated 70% of the poor are women, for whom livestock play an important role in the improvement of status and represent one of the most important assets and sources of income (DFID 2000). Many of these people also fully or partially depend on livestock for income or subsistence. Designing appropriate policies to support this segment of the population in all countries is one more challenge facing the livestock sector today.

The Response Context

Although the poor clearly have a major stake in the livestock sector, most of them have not been able to take advantage of the opportunities presented by the demand-led growth of recent decades. This has been due to a combination of global, regional, and national policies, regulations, norms, and values that are defined as the societal "rules of the game." They provide a framework for access to and control of capital assets, influence the political effectiveness of economic interests, and control the political agenda (North 1990).

Chapters 13, 14, and 15 clearly present the social fallout of this situation under different livestock sector scenarios. Although this is clearly an oversimplification, the dynamics of need and potential responses can be summarized under three separate, but not mutually exclusive, livestock sector development scenarios:

- *Reducing vulnerability.* In a number of developing countries, economic growth remains weak and is not driving an expansion in demand for animal products. Here the situation is characterized by large numbers of highly vulnerable rural poor for whom livestock represent one of the few opportunities to support and enhance their livelihoods. Many of the barriers and constraints apply equally to this group, and some of these represent norms and societal rules of the game that derive from international policies. Enhancing livestock-related livelihoods through improved access to and control of capital assets (natural, social, human, physical, and financial) will not only reduce vulnerability and risk, it will also position resource-poor livestock producers to benefit from any upturn in the economy and demand for animal products that occurs.
- *Creating the conditions for growth.* Growing demand for animal products, where it exists, offers substantial opportunities for small-scale producers

to participate and benefit. However, a number of technical, infrastructural, and institutional constraints in these areas impede an appropriate production response to increased demand. To create an enabling environment in which poor producers can take advantage of development opportunities, these productivity, trade, and other barriers will have to be overcome. Barring this, there is a real danger that poor people who depend on the livestock sector will be marginalized further.
- *Coping with growth.* This development scenario applies where economic growth is driving a burgeoning demand for animal products, with a correspondingly dynamic response through a rise in industrial production systems. Here the primary public goods involve issues of equity, environmental pollution, and animal and public health, including the risk of highly pathogenic emerging diseases. Policies, institutions, and technologies are required that enable small producers to benefit from the cost advantages of large-scale production in order to create a more level playing field.

The way these development scenarios are shaped through organizational enforcement of national and international rules of the game supports or prohibits the development of the poor. Such "rules" currently translate into barriers, lack of competitiveness, and risks—all of which prevent the poor from taking advantage of the available development potential. Appropriate targeting of the priority policy changes and institutional reforms and the concomitant technologies required under different sector dynamics is therefore essential if sector dynamics are to contribute to socially desirable outcomes.

Transaction costs for smallholder producers, for example, can be reduced through collective action, such as formation of cooperatives and various forms of contract farming. Such arrangements also have potential to incorporate smallholders in high-value supply chains from which they would otherwise be excluded. Contract arrangements vary but often involve the contractor supplying genetically superior breeds of chicks or piglets, feed, advice and support, and a guaranteed market for the end product.

Thus it appears that smallholder producers can stay in business as long as the opportunity cost of family labor remains low and alternative employment opportunities are limited, and they can benefit from some sort of collective organization and support network. In situations where alternative employment options offer higher wages, such as the more developed parts of China, the competitive advantage of smallholder producers is wiped out and there is likely to be a mass exit from the sector. From a poverty reduction viewpoint, as long as waged employment can absorb those exiting from the sector, this is not necessarily a bad thing: on the contrary it should be viewed as a positive development. In addition, the interests of poor consumers are likely to be different to those of small-scale and poor producers. Medium-sized and semicommercial producers appear to be particularly vulnerable to shocks of all sorts: climatic, disease, changing markets, changing norms and standards, and so on.

Financial services in the form of credit/loans for the purchase of livestock, feed, and health services as well as insurance against the loss of valuable productive assets may play an important role in encouraging investments in new technology and in coping with difficulties such as drought and disease. In developing-country markets, commercial banks rarely offer such services, however. Financial and insurance services are not well developed, and loans, credit, or microcredit and insurance for animal loss are almost always provided by the state. This type of financing often penalizes smallholders because they lack collateral. As a result, most of the loans are taken by wealthier farmers. For these reasons public-sector provision of subsidized loans for agriculture has fallen out of favor. Some nongovernmental group credit schemes exist where animals are loaned or given to poor livestock keepers, and some of these schemes have been very successful. Borrowers are generally required to give back the borrowed funds either in kind (e.g., a calf in the Heifer Project) or in cash (e.g., the Grameen-type credit) (ILRI 2004).

In developing countries, however, the institutional context and weak legal framework, as well as the lack of private providers of livestock insurance, make public intervention often the only alternative. In a number of Asian countries livestock insurance schemes for poor farmers already exist. As part of the national insurance scheme in India, for example, livestock insurance is automatically approved for poor dairy farmers when buying an animal from a specific source (CRISP 2008). Although most of these schemes initially focused on dairy production only, they have since expanded to include other livestock. These schemes have also created strong links with banks (normally a national and a few commercial banks), and when farmers now ask for a loan or credit from the bank, an insurance policy is often obligatory. Public rural financial institutions have been mostly dismantled or privatized. In addition, many have had their role reduced to the provision of insurance for use as collateral for loans. Beyond credit, access to an adequate and affordable supply of animal health and extension services is essential for effective livestock production. Until the roles of the public and private sectors in service provision are clarified, both privatization and decentralization may prove ineffective.

Although growth in the livestock sector in developing countries has been at the center of a number of studies (e.g., Delgado et al., 2008), most of these have largely ignored issues related to the kind of capacities and services that would be required to ensure that these trends are best used to promote equitable wealth creation and sustainable economic development. Most studies have dealt solely with the economic and environmental impact of scale increases and vertical integration of production systems that have accompanied market trends. Studies that have focused on service delivery have generally disregarded the evolving needs in skills and support services and have thus ignored the changes required in the institutional landscape that would enable the delivery of such knowledge and services.

A recent comparative study of the livestock sector at different stages of development and under different political systems in three nations that have witnessed rapid livestock sector growth in the last two decades explored innovation capacity by looking at the changing spectrum of services, actor roles, and interactions required to effectively respond to new opportunities and threats in the sector (Dijkman and Vu 2007). The study maintains that the concept of "response capacity" covers two specific elements of innovation capacity. First, there is a sense of urgency arising from the emergence of new investment opportunities, disease outbreaks, environmental disasters, volatile capital and price shifts, and fast-track schedules of trade liberalization imposed from outside. These all demand quick responses from both producers and governments. Existing response capacity may help, but a common weakness is the ability to respond in a timely fashion. Second, there is the degree of local specificity. For response capacity to be effective, it must be built on an analysis of specific local conditions. These involve not just dynamic markets but also political and institutional changes such as democratizing and decentralizing trends. Again, existing innovation capacity may not sufficiently stress the specific kinds of opportunities and threats facing actors and may generate misguided responses—responses that may be appropriate generally but not in the specific contexts they are meant to address.

Country case studies revealed a number of weaknesses in existing response capacity, including a continued reliance on public livestock services. Often these involve only primarily research and the provision of extension and veterinary services, even though these are well recognized to be woefully inadequate as a response capacity mechanism. In addition, a lack of systemic coherence between the different elements of response capacity—research and other public services, private-sector livestock activity, and the policy and regulatory system—weakens them even further. To respond effectively to new developments and to promote sustainable and equitable growth in the sector, it is suggested that governments revisit existing veterinary services arrangements and identify ways of bringing in new sources of knowledge about emerging challenges and ways of dealing with them. In addition, governments need to find ways of strengthening the patterns of interaction between different actors, policies, and initiatives relevant to the sector and to make sure that stakeholders who can champion a social and environmental agenda are included.

Another critical issue is how to better use agricultural science and technology to enable people-centric, entrepreneurial, and environmentally friendly innovation. Side-stepping this issue will ultimately make poor people poorer.

The Need for Policy and Institutional Change

Although there have been glimpses of responses to the social implications of the livestock sector's changing landscape, as described in the preceding section, much of this has been inadequate. Many of these examples clearly show that technology alone is not going to transform livestock sector development in ways that will necessarily help the poor. The role of policy and institutional change has been flagged in particular as a way of creating the framework conditions in which sector development could be steered toward the needs of the poor.

The history of agricultural development in Europe and North America also shows that it was not primarily hampered by technological constraints but that farmers were only too willing and able to adopt existing technologies once an enabling policy and institutional environment were in place. Such an environment allowed them to obtain new technologies and reap the benefits of their adoption. This enabling environment is influenced by economic and institutional factors that are beyond the households' immediate control (Birner 1999). Thus, to ensure that livestock sector dynamics in developing countries contribute to socially desirable outcomes, institutional and policy reforms are required.

In a world characterized by market, technological, social, and environmental conditions that are evolving rapidly and in often unpredictable ways, the argument is made that it is not the changes in policy and institutional circumstances that need to be assessed. Rather, what should be considered is the underlying processes that bring about such changes and that build the capacity to manage and exploit change. In other words, understanding and promoting policy and institutional change is actually concerned with the underlying capacities for change and this capacity is largely a function of the patterns of linkages and the quality of the associated relationships among actors in the sector.

Obviously, a capacity-strengthening perspective on policy and institutional change has a number of implications on how to deal with the livestock sector's changing landscape. First is the design and implementation of policies and interventions that provide incentives and resources to diverse groups of players to work together in local contexts. This allows relationships to form and experiences about working in new, more collective ways to be gained. This is a process of learning-by-doing and, over time, it changes the way organizations work, as it

is often a lack of experience of working in this manner that prevents these linkages from being established in the first place. Such relationships are particularly useful for connecting research organizations to relevant areas of economic activity.

Capacity can also be built through the use of a range of peer learning mechanisms referred to collectively as a community of practice approach. In the case of livestock sector policy and institutional change, this would involve sharing ideas and lessons among initiatives, organizations, and individuals using different processes and ideas. Such exposure to the knowledge and experience of others can help organizations and policy makers who are struggling to translate these principles into operational strategies relevant to their own contexts. Ideally the sharing of lessons and experiences should be both among similar organizations working in different contexts (for example, research organization working in different sectors or countries) and among different types of organization (such as research organizations and private companies). The inclusion of policy makers in such processes and their exposure to operational lessons is critical, as these can then be reflected in framework conditions.

Building Policy-Relevant Capacity

As noted, the livestock sector needs to focus on the issue of capacity and the nature of capacity under the three identified scenarios. It is not just innovation capacity in a general sense, however, but the aspect of responsiveness to changing contexts and particularly responses to rapidly changing market conditions, as alluded to earlier in this chapter. For natural resource–based industries like livestock in developing countries, this has become a particularly important concern. The higher degree of market integration that accompanied the globalization of the value chains of these products is increasingly exposing farmers and industries to competition in the global marketplace and to changing consumer demands and standards and norms in distant markets. The ever-increasing rate of change in these markets means that responsiveness is likely to be the critical factor in innovation capacity. So what are the specific elements that would help build the necessary response capacity within the three livestock sector development scenarios and lead to desirable outcomes? A useful perspective is the four-point analytical framework developed by the World Bank to investigate agricultural innovation capacity: actors and their roles, patterns of interaction, habits and practices, and the enabling environment (World Bank 2006). The sections that follow illustrate specific characteristics of this capacity and how it may be built.

Actors and Their Roles

The process of livestock sector innovation requires a diversity of actors, including entrepreneurs, research and training organizations, public policy bodies, and civil society organizations. In terms of responsiveness, it is not possible or helpful to be too prescriptive about which actors should be present. However, the following roles are important:

- *Sector coordination* provides coherence to actions of the different actors in the sector. Not only does this facilitate the patterns of interaction needed for innovation, it also helps coordinate and speed up the actions needed to respond. This response might be new groupings of actors, dissemination of new information, or the introduction of new practices. Sometimes coordination might be more concerned with third-party brokerage rather than sector-wide oversight.
- *Entrepreneurship* is important because it is by grasping opportunities and taking risks that sectors innovate in response to changing market opportunities. If this role is not well developed, responses to opportunities will be weak. While this seems obvious, entrepreneurship is not uniformly developed across all sectors and countries. This role could be played by public or private sectors, although the private sector is now usually thought to be better suited to it. Again, this can vary from country to country and sector to sector.
- *Providing knowledge of the future* is important in response capacity, as it is the way in which early warning information about upcoming challenges and opportunities is collected and transmitted to entrepreneurs and policy makers. Buyers and others in market chains might play this role in relation to information about changing market demands and policy and regulatory changes in distant markets. Informal networks linking sectors to policy makers might give early warnings on upcoming policy changes. Farmers' organizations and industry associations might also play this role. Some countries might have formal foresight committees or standing panels on certain export commodities with the specific purpose of collating expert opinions on future market or technology conditions.
- *Research*. Not all innovation requires formal research and, by the same logic, neither does responsiveness. Nevertheless, research plays an important role in the innovation process as a specialist form of knowledge creation for both addressing unforeseen challenges and opportunities and providing opportunities through new technology. Both private and public organizations can play a research role, but in many developing countries this is left to the public sector, and this is often closely connected to tertiary training.
- *Service provision*. A number of services are required for innovation and contribute to responsiveness.

These services include technical advisory services, financial services, and auditing services to help with regulator compliance.

- *Ensuring socially desirable outcomes* in sectors is no longer a nonmarket issue. As global markets become more sensitive to the social and ethical consequences of the goods they purchase, "policing" and assistance with compliance to these standards are important roles in responding to the ever more sophisticated demands of the market. The advocacy role of organizations is also important in this context, alerting society to malpractices and social inequity. Usually civil society organizations play this role, but the public and even the private sector may also do so in certain countries.

Patterns of Interaction

Patterns of interaction will be shaped by the local context and the particular challenges or opportunities that are being addressed. These patterns are dynamic and will change over time as challenges and opportunities evolve. What can be expected is evidence of a loose network of linkages between actors in the livestock sector that provides coherence and acts as a foundation for more concrete forms of linkage and collaboration in times of need. Forms of interactions that allow two-way flows of information are most important. Also, while frequency of interaction is an important consideration, quality of interaction is equally important. Aspects of the patterns of interaction specifically relevant to building response capacity in the livestock sector include the following:

- *Links to consumers*. Consumers provide critical information about preferences and how these are changing. Often, especially in the case of export markets, interaction with buyers (intermediaries) is a critical source of information on consumer preference in distant markets. Trade fairs can facilitate a similar form of interaction.
- *Links to specialist technical knowledge*. Changing market and policy conditions often involve the need for technical upgrading. This might be in terms of production and process technology, production organization, packaging, or marketing. It might require linkage to research organizations in cases where problems and opportunities require new knowledge. More often it will involve linkage to technical services or organizations that can help adapt existing technologies and processes to a new situation.
- *Mechanisms that facilitate interaction*. Coordination bodies exist in some sectors to help foster interaction between different actors. These might be government agencies or industry associations. Since response capacity involves often rapid reconfiguration of patterns of interaction, mechanisms that facilitate

this are vital. One mechanism for achieving this is to have a high degree of organization in the sector—such as farmers' associations, women's groups, self-help groups, industry associations, export associations, and cooperatives.

Habits and Practices

Habits and practices of institutions are probably the most intangible element of response capacity—but undoubtedly the most important. These are the mainly informal rule sets that determine how people behave and the way they do things. Particular attention should be given to the institutions that affect the processes of interaction, information sharing, and learning. In relation to livestock sector response capacity, a number of basic ways of working are likely to be important, and these include the following institutions that underpin them:

- *Attitudes toward change.* Most people tend to be conservative by nature, and fear and avoid change. Organizations in the public and private sectors exhibit varying degrees of reluctance to change. This is most marked in public-sector organizations because of the rigid rules and hierarchies that characterize them. These same characteristics can also be seen in the private sector, but because this attitude seriously undermines companies' ability to respond to markets and competition, such an outfit tends not to remain in business long.
- *Trust* is an important lubricant in social relationships. Since innovation is a social process of interaction and information sharing and learning, trust is a necessary ingredient in the relationships needed to underpin innovation. By the same argument, responsiveness to rapidly changing circumstances requires a sufficient degree of trust across actors in a sector to ease the creation of the new relationships needed for innovation. Trust often manifests itself as willingness to cooperate with other actors.
- *Shared identity.* Another institution that can help coherence and thus aid responsiveness is the degree of shared identity or the sense of belonging that exists among actors in the livestock sector. Increasingly, and particularly for export sectors, competitive pressures come from other countries rather than domestic competitors. Responding to this type of competition requires collaboration and the recognition that survival of the sector is a collective responsibility. One example is the need to respond to quality demands in distant markets. Often this is as much about building a sector's reputation for quality, which requires all actors to recognize this as a shared responsibility. Such a "we are us" mentality is therefore an important aspect of response capacity.
- *"Learningness."* The innovation systems concept gives very strong emphasis to learning and the habits and relationships that enable this process. This is often a difficult set of habits and practices to identify and is best evidenced by identifying how practices and behaviors of actors or sectors have changed as a result of learning from experience, research, interaction, or searching for information. Attitudes that support this include a propensity for experimentation and risk taking, as well as the organizational "space" for trying new things out. Some organizations may have structured learning practices as part of their management strategy, or they may use external evaluation. Since much learning accumulates in the form of tacit knowledge, organizations with low staff turnover may be better placed to learn and build capacity in the longer term. Learning is particularly important in relation to response capacity, for in a dynamic environment the ability to learn how to cope with a continuous succession of largely unpredictable events is an essential attribute for success.
- *National culture.* The origins of the innovations concept are a good indication that there are different national styles of innovation and that they emerge from cultural, historical, and political contexts at a national level. National culture is an emotive subject, and it is important not to dismiss certain national cultural traits as constraints to innovation and responsiveness. National cultures do also evolve, and certain cultural stereotypes can become quickly outdated. Nevertheless, attributes like the degree of social cohesiveness or the degree of social hierarchies found in society as a whole can reflect on innovation responsiveness. Similarly, accepted views on the role of certain actors in society, particularly the public and private sectors, can be reflected in capacity.

Enabling Environment

The enabling environment is the wider set of policies and institutions in which an innovation process is situated. Much of the enabling environment manifests itself through the factors already discussed. For example, agricultural science and technology policy often determine the degree of interaction between researchers and actors in the productive sectors. More specific factors might include monetary policy, infrastructure, the level of corruption, the effectiveness of the legal system, educational practices, regulatory regimes, and sector governance. A final aspect of the enabling environment is the presence of a strong market or, in its absence, of policy triggers that provide the incentives for innovation.

The Road Ahead

To effectively address the constraints that currently prevent the poor from taking advantage of the available livestock sector development potential and from competing, coping, and prospering under evolving market conditions, livestock producers need to be able to constantly innovate. Moreover, to fully exploit the poverty reduction potential of these livestock sector trends, the sector's capacity to respond and adapt to these changes needs to be enhanced in ways that both allow producers to innovate while also safeguarding the livelihoods of poor people linked with the sector.

The transition from subsistence to small-scale commercial production not only opens potential access to more lucrative markets; as noted earlier, the higher degree of market integration that has accompanied globalization also exposes producers and enterprises to rising competition in the global marketplace and to changing consumer demands and standards and norms in distant markets. The ever-increasing rate of change in these markets means that responsiveness to changing contexts and conditions is essential. Without such capacity, such producers and enterprises are unlikely to cope and prosper.

Even if this capacity can be supported and built, managing the exit of a large number of former livestock keepers from the sector—helping to ensure they have a "soft landing," either within the large-scale industrial livestock production sector and associated food processing and distribution industries or, in most cases, in emerging manufacturing and service industries—will be essential. This could be achieved, for example, by ensuring that schools and colleges equip students with the new skills these industries require or by providing suitable training opportunities for those who have already left the formal education system to enable them to adapt to the new order. The objective of pro-poor livestock sector development policies, in these scenarios, should not be to maintain smallholder production systems at any cost but rather to mediate the sector transition, in which the role of the poor needs to be considered broadly, including as consumers, market agents, and employees, as well as small-scale producers.

Evidence from rapidly developing economies such as China suggests that when alternative employment options arise and formal distribution networks for processed livestock products extend beyond major towns and cities, small-scale livestock keepers leave the sector. The need for and viability of keeping a few backyard poultry or pigs diminishes. Better-remunerated, less risky opportunities in the construction, manufacturing, or service sectors, usually in towns and cities, prove irresistible, especially to young people.

In countries where the livestock revolution has not occurred, small-scale "low-input, low-output" producers continue to keep livestock, usually as a supplement to crop farming or other activities. They enjoy multiple benefits from their animals, including enhanced food and nutritional security, and sometimes also draft power and manure to enhance crop production. Livestock also serve important sociocultural roles, such as payment of dowry and traditional fines and as determinants of status. In

these settings an important role of livestock is to provide children with high-quality protein and micronutrients, such as vitamin B_{12}, which are essential to enable normal physical and cognitive development but difficult to supply solely through cereal- and vegetable-based diets. Consumption of even small amounts of livestock products can make a significant difference in this regard. It is thus the nontradable functions of livestock that are of overriding importance in this situation.

It is clearly difficult to forecast what will happen in the future regarding feed and food prices. Most analysts and observers, however, currently believe that in the short to medium term, prices will remain higher than in the recent past and that increased volatility of prices will become the norm. But whether prices remain high or revert to the long-term decreasing trend of the past 50 years or so, the polarization of the livestock sector in the developing world looks set to prevail. In a scenario where staple food and feed prices remain at historic highs, there will be additional pressure to supply meat, milk, and eggs at the lowest possible prices to meet the demand of urban consumers. This is likely to increase the trend toward larger, more-intensive, and more-efficient production units, especially for poultry. At the same time, "something for nothing" production by poor, rural, small-scale livestock producers will become even more important in terms of food and nutritional security as their own and their local customers' purchasing power in nearby food markets decreases. While this could potentially provide market opportunities to smallholders that do not generally rely on bought feed, previously indicated resource constraints and policy and institutional barriers would need to be addressed for this to have a real impact.

If, on the other hand, feed prices start to decrease again, accompanied by drops in the price of livestock products and other foods, this is likely to further boost demand for cheap meat, milk, and eggs and to stimulate expansion of the large-scale industrial livestock sector. It is difficult to see how small-scale poor producers will readily become integrated into increasingly complex, formal supply chains in either scenario.

Policies and institutional arrangements are essential for mediating how the livestock sector develops, how current and former livestock keepers fare, and how well the needs of consumers—rich and poor, rural and urban—are served. In poorer non–livestock revolution zones, policies and institutional arrangements are needed that reduce vulnerability and help to maintain livestock production as a pillar of livelihood and a safety net for poor households, while minimizing risks from zoonotic and foodborne diseases, as well as environmental hazards to the livestock keepers themselves and the wider community.

In areas where economies are beginning to take off and the livestock sector is in early stages of transit, supportive policies and institutional arrangements are needed to manage this transition. Thus far, under these conditions, the potential for the great majority of traditional small-scale producers to supply formal value chains appears to be limited by a number of factors. These include a lack of funds to invest or provide access to credit, small and diminishing landholdings, poor access to input and output services and markets, increasingly stringent food and safety standards, the increasing power of supermarkets, and poor knowledge access and infrastructure (the rural poor tend to live in marginal areas where lack of year-round roads is a major hindrance to development in general, but particularly in relation to perishable commodities).

In rapidly emerging economies, where large-scale integrators increasingly dominate the livestock sector, policies and institutional arrangements are required to ensure that public health and environmental standards are upheld and that the needs of diverse consumers, including the less well off, are met. In Thailand, for example, the introduction of punitive taxation regimes were one factor that caused large-scale poultry enterprises to relocate further from Bangkok, thereby reducing public health and environmental risks caused by having large concentrations of livestock and people in close proximity (Steinfeld et al., 2006). At all levels, however, trade-offs are likely to be necessary between the interests of the diminishing number of poor livestock producers, processors, and market agents and the increasing number of poor, largely urban, consumers.

Encouragingly, there are a number of other examples where pro-poor policies and institutional arrangements and targeted investments have performed well in developing-country livestock sectors. In India, for example, milk production and distribution remain in the hands of huge numbers of very small-scale producers and informal market agents. With supportive government policies, this arrangement has enabled milk consumption in India to increase dramatically over the past few decades while securing the livelihoods of millions of small-scale producers and market agents. In Brazil, however, recent changes to the law have let large integrators stop contracting with small-scale livestock farmers in favor of a focus on larger farms (Delgado et al., 2008)—perhaps reflecting a rebalancing of the needs of consumers versus producers.

At least one potentially positive outcome of the recent food price crisis is that it has put agriculture firmly back on the development agenda. A number of agencies have recognized that rising food prices, in the longer term, may represent an opportunity for farmers. In response, the World Bank is doubling its lending for agriculture in Africa to $800 million from 2010. A big question, of course, remains how the monies will be spent and whether it will make a difference. Traditional investments and interventions to promote innovation in the agricultural sector have recognized the importance of knowledge but have mainly emphasized the knowledge supply side, focusing on supporting research and promoting technology transfer. It is now widely recognized that research arrangements need to pay more attention to technology demand from users, particularly poor people, but also from other key economic actors, such as entrepreneurs and industrialists, who can create new opportunities for growth and welfare.

With today's livestock sector development landscape being determined by rapid and often unpredictable changes in market, technological, social, and environmental circumstances, there appears to be growing consensus that the way to deal with these dynamics is to focus policy and development interventions on supporting a continuous process of creating, obtaining, and using knowledge and information to generate new products, services, production arrangements, and strategies that meet social and economic goals. Only by building such capacity can sectors and countries solve problems and improve practices and, in so doing, cope, prosper, and compete in a world that itself is continuously changing. Thailand is a good example of this adaptive capacity: within months of losing its large export market for frozen poultry due to an outbreak of avian influenza, it had reengineered and reinvented itself as a major exporter of precooked poultry-based ready meals. This ability to mobilize and use knowledge is emerging as a new source of comparative advantage, replacing the traditional importance of natural resource endowments as a source of competitiveness for developing countries. The recent emergence of Chile as a major salmon producer, Viet Nam as a coffee producer, and Indonesia as a furniture producer is an example of this. And it is not just market competition that is innovation based. The ability to mobilize and use knowledge is becoming increasingly important in determining how countries cope with climate change and disease outbreaks and how they seize opportunities arising from new technological, policy, and market opportunities. Innovation in livestock production, processing, utilization, and distribution usually takes place where different players in the sector are well networked, allowing them to make creative use of ideas, technologies, and information from different sources, including from research.

Dealing with these issues requires enhanced mechanisms and institutional capacities to predict, prevent, and rationally respond, without compromising the positive role livestock can play in enhancing rural livelihoods. This will entail identifying and fostering the social and institutional arrangements that can mobilize the requisite knowledge stocks and support services. This can be achieved by explicitly linking action research on livestock sector transition and the incidence and livelihood impacts of environmental and climate change in different livestock production systems to a process that enhances capacity for sound policy formulation and effective policy implementation.

References

Birner, R. 1999. *The role of livestock in agricultural development: Theoretical approaches and their application in the case of Sri Lanka*. Aldershot, U.K.: Ashgate Publishing.

CRISP (Centre for Research on Innovation and Science Policy). 2008. *Tacit knowledge and innovation capacity in the Livestock Sector*. FAO Final Report. Hyderabad, India: CRISP.

Delgado, C. L., C. A. Narrod, M. M. Tiongco, G. Sant'Ana de Camargo Barros, M. A., Catelo, A. Costales, R. Mehta, V. Naranong, and N. Poapongsakorn. 2008. *Determinants and implications of the growing scale of livestock farms in four fast-growing developing countries*. Research Report 157. Washington, DC: International Food Policy Research Institute.

DFID. 2000. Eliminating World Poverty: Making Globalisation Work for the Poor. White Paper on International Development. Presented to Parliament by the Secretary of State for International Development by Command of Her Majesty, December 2000. https://www.dfid.gov.uk/Documents/publications/whitepaper2000.pdf

Dijkman, J. T., and T. Vu. 2007. Institutional strategies to improve producers' and governments' innovation response capacity in Southeast Asia's dynamic livestock markets. Paper presented at the GLOBELICS 2006. Thiruvanathapuram, India, 4–7 October 2006.

ILRI (International Livestock Research Institute). 2004. *Bridging the regulatory gap for small-scale milk traders*. Nairobi: ILRI.

North, D. C. 1990. *Institutions, institutional change and performance*. Cambridge: Cambridge University Press.

Steinfeld, H., P. Gerber, T. Wassenaar, V. Castel, M. Rosales, and C. de Haan. 2006. *Livestock's long shadow: Environmental issues and options*. Rome: Food and Agriculture Organization.

World Bank. 2006. *Enhancing agricultural innovation: How to go beyond the strengthening of research systems*. Washington, DC: World Bank.

20

Livestock in a Changing Landscape

Conclusions and Lessons Learned

Henning Steinfeld, Carolyn Opio, Jeroen Dijkman, Anni McLeod, and Nick Honhold

Main Messages

- **The factors that created the livestock revolution still exist,** although they have been dampened by rising energy and feed prices that have begun to attenuate livestock sector growth and change the pattern of demand.

- **Given the planet's finite land and other resources, there is a continuing and growing need for further efficiency gains in resource use of livestock production through price corrections for inputs and the replacement of current suboptimal production with advanced production methods.** If the demand for animal products continues to grow, it can only be met by intensification. This has generally meant scaling up and concentration of production and marketing units with an increasing dependence on inputs brought in from a distance.

- **Rural activities in the developing world are increasingly connected to the global environment.** This presents an opportunity to diversify livelihood options, but at the same time it exposes all livestock keepers to rapidly changing patterns of competition, market preferences, and standards. The livelihoods of many of the most vulnerable are thus, for better or worse, linked to the changing markets, technologies, and diseases that are transforming the sector.

- **Continued growth and intensification may be inevitable, but they carry environmental consequences.** Livestock play a major role in greenhouse gas emissions. Water is depleted by livestock activities through withdrawal for feed production and through pollution. Water cycles are affected by the presence of livestock in many ecosystems, particularly when grazing pressure is poorly managed. Water pollution in areas of high livestock density is a major factor. At the same time, extensive livestock production will be subject to increasing stress and climatic variability caused by climate change.

Scientists and policy makers have only recently begun to explore the measures needed to deal with these impacts and to understand the positive contribution that livestock can make to the environment by providing environmental services and contributing to the biodiversity of ecozones.

- **The challenges facing the livestock sector cannot be solved by isolated interventions.** Understanding, let alone solving, such complex issues requires a holistic approach that takes account of the variety of social, economic, and ecological settings in which livestock are kept and that recognizes the social, health, and environmental consequences of change. Technologies and policies are both needed to respond to the changes occurring today. The ability to deal with challenges will depend on building capacity for innovation and developing nuanced responses tailored to very diverse livestock systems that are changing over time.

Driven by rising incomes, population growth, and urbanization, the livestock sector has experienced unparalleled growth in the last few decades. Three major factors are responsible for this growth: First, growth in the number of livestock is in part due to newer technologies and the widespread application of improved genetics, animal hygiene, and disease control measures, along with enhanced feeding and nutrition in areas previously characterized by low levels of traditional technology. Second, the rapidly growing scale of operating units has allowed for scale economies and a resultant reduction of production costs and access to growing formal markets. Third, there has been a species shift from ruminants to monogastrics and a concomitant move from fibrous feed material to feed concentrates produced on arable land. These sources of growth strongly resonate at the environmental, social, and health level, affecting human welfare in major ways. Without any major policy changes, the large environmental, social, and health impacts of continuing livestock sector growth will only increase further.

Livestock's impact on the environment has been, but need not be, overwhelmingly negative; research has demonstrated the environmental damage caused when livestock systems are managed without thought of externalities, but more recent research demonstrates the positive impact that livestock can play by providing environmental services.

There are important greenhouse gas (GHG) emissions associated with the sector. Deforestation for pasture and the expansion of feed crop and pasture degradation may not accelerate, but they will continue to release large amounts of carbon dioxide into the atmosphere. Likewise, methane emissions from ruminant livestock and livestock waste will add to anthropogenic GHG emissions. With the exception of some modest attempts to capture some of the methane from waste, there are no attempts to curb these carbon dioxide and methane emissions. Livestock also have a huge impact on biodiversity in a variety of direct and indirect ways. And water is being polluted and aquatic ecosystems compromised through overloading with nutrients, antibiotics, and other substances.

In addition to these impacts, there is the wider concern about resource use and resource intensity in the livestock sector. While it is true that important gains are being made to improve the productivity of livestock in most parts of the world, as evidenced by increasing milk yields and offtake rates, this is achieved to a large extent by shifting to more protein-rich feed concentrate, which requires the cultivation and intensification of arable land. Moreover, these productivity gains only attenuate resource use growth but do not halt or reverse the trend of increasing resource use by livestock. This likewise applies to land, water, and nutrients.

On the positive side of the environmental ledger, however, livestock can be part of successful landscape management in its extensive forms and can contribute to biodiversity and water management. To date this role has been largely limited to areas where large herbivores have been part of the evolution of ecosystems. More work is clearly needed to build on the positive and balance the equation through local management initiatives and policies.

With regard to social aspects, the livestock sector is tremendously important as livelihood support for the majority of the world's poor. For them, livestock provide an important source of nutrition, inputs to crop agriculture in the form of animal draft and manure, and an important asset and form of insurance. These roles are particularly important in areas that are not connected to the global food economy. In successful developing countries, the sector is increasingly integrated into the global food economy and exposed to domestic and international competitive pressure as well as heightened sanitary and food quality standards. These developments usually put smallholder producers at a disadvantage because they lack the knowledge and the economies of scale to compete in the marketplace, face important barriers in getting access to formal markets, and tend to be driven out by more powerful competitors. The number of producers declines, the size of operations increases, and intensive, commercialized producers who are integrated in modern markets dominate. Sector consolidation continues in industrial countries as well, leaving ever fewer producers.

In terms of health, the livestock industry has undoubtedly contributed to more people enjoying a richer and more varied diet than at any point in recent history. For many people in developing countries, food products of animal origin are still a useful and much desired component of otherwise poor diets that are often characterized by protein and nutrient deficiencies. The success of the livestock industry in providing ever larger amounts of food while dramatically improving productivity is arguably its greatest achievement. But that accomplishment is overshadowed by the fact that livestock products play a central role in the diets of overnourished individuals and have been associated with a number of noncommunicable diseases such as diabetes, cardiovascular disease, and certain types of cancer. Equally worrisome are emerging and reemerging diseases of animal origin that may continue to threaten human health and economic stability.

Although the overall picture is dim, there are a number of encouraging signs. Demand will not grow indefinitely, for example. Health, environmental, and animal welfare concerns have already led to somewhat decreased consumption levels of meat in industrial countries, particularly and first among the wealthier segments of the population. There is growing recognition that eating excessive amounts of animal products is unhealthy. In successful developing countries, growth in consumption is also slowing, in particular in China, where it has reached about 60 kg per capita. In Latin America, meat consumption has been traditionally high and markets are partially close to saturation. It is uncertain whether the traditionally vegetarian populations of South Asia will adopt Western-style consumption patterns, however; if so, there would be huge increases, given current population levels there. In less successful developing countries, large increases in consumption can also be expected once these economies grow more strongly and incomes rise. Overall, at the world level it is likely that the largest relative and absolute increases in meat consumption are over, but growth is likely to continue—albeit at lower rates—until at least 2050.

It is likely that the recent price increases for energy and agricultural commodities will be permanent to some extent and will further attenuate demand. Because livestock production depends to a large and growing extent on cereals and other feed concentrates, production costs have recently risen. Beef, which involves large amounts of cereals when produced in feedlots, will have relatively high increases in production costs, while these will be more moderate for efficient feed converters like poultry. Together with health concerns, these price differentials will reinforce the trend toward poultry, which on balance is the meat with the lowest environmental costs.

Recent price increases in agricultural commodities also seem to have encouraged more reduction of subsidies that continue to be the legacy of past low commodity prices. Further trade liberalization and subsidy reduction have the potential to stimulate productivity increases in the livestock sector. Such productivity increases have already reduced resource requirements when measured on a unit of output basis. In fact, because there have been significant increases in both arable crop and livestock productivity, total land requirements have increased only marginally at an aggregate level. Given the low productivity levels that still prevail in most developing countries, there is undoubtedly considerable potential for productivity increases that can be unleashed by applying conventional technology in the areas of animal genetics, feed and nutrition, animal health, and general husbandry. Productivity increases will not only limit resource requirements, they will also reduce GHG emissions, in particular in ruminant livestock. However, such productivity gains will only attenuate total growth in resource use rather than leading to a reduction in the foreseeable future.

The perception that meat carries higher environmental costs than most plant products is spreading. There is growing awareness of the environmental impact of livestock production, especially intensive production, and previous pro–livestock sector growth policies are being reconsidered in this light. Some East Asian and Latin American countries (China, Viet Nam, Thailand, Brazil, and Mexico) have started to address pollution issues related to animal waste much more rigorously.

The widespread efforts to take into account the interests of poor livestock owners and producers and to develop policy frameworks that are pro-poor illustrate that the awareness and political will to improve the position of this large group are growing. The recognition of pastoralists' rights, the development of sanitary regulations that give smallholders access to segments of domestic markets, and the facilitation of collective action in the form of cooperatives or contract farming are all steps in the right direction. Poor livestock owners are increasingly also involved in collective forms of resource management, such as in India and parts of sub-Saharan Africa. And it is increasingly understood that a number of traditional and emerging animal health issues are exacerbated by the dichotomous structure of the livestock sector, with poor subsistence farmers with little biosecurity existing alongside large-scale commercial livestock producers in the same area or connected through market links.

There is growing recognition that human and animal health professions and related services need to communicate and collaborate on many fronts. The recent "One World, One Health" initiative of the Wildlife Conservation Society and other health and environmental groups holds promise for addressing some of the joint human–animal health challenges that are brought about by changing ecologies and host–pathogen complexes.

From this analysis it is clear that public and private responses to the growing challenges in the livestock sector fall short of what is needed in terms of the potential environmental, social, and health benefits. A thorough reform is therefore required that involves all stakeholders in industrial and developing countries alike.

In addition to these challenges, the sector needs to adjust to higher and fluctuating fuel and grain prices. These prices are likely to spur additional investments into research and development and will induce more rapid adoption of productivity-enhancing technologies among producers who are commercially oriented. Producers will also need to comply with ever-growing sanitary and quality standards and adjust their operating and monitoring procedures. They will need to address animal welfare concerns as well and allow for more natural behavior of the animal stock. In addition, producers need to accommodate different consumer demands. These additional pressures and demands will undoubtedly result in higher production costs; they would also slant competition toward producers who are quick to adapt as markets change.

Demand for livestock products is an issue. As noted earlier, a growing share of the global population, still predominantly in industrial countries but rapidly growing in successful developing countries, consumes livestock products in excessive amounts when measured against standards for healthy nutrition. A reduction to more moderate levels is therefore desirable solely from a human health perspective, and public awareness campaigns need to convey this message. Food labeling that reports different ingredients is now commonplace in industrial countries and is rapidly spreading among developing countries. In order to encourage fair pricing of food items, subsidies need to be phased out. Certain countries, particularly industrial ones, have even started considering taxing food items because of their high fat, sugar, or salt content.

Livestock's contribution to climate change is another issue that needs to be addressed with urgency. Although there are a number of uncertainties related to the attribution and measurement of GHGs from this sector—including land use and land use change, feed production, enteric fermentation, and waste—there is no dispute that major reductions are required and can be achieved. While it is too early to predict the shape of a possible Kyoto follow-up agreement, there is some hope that emissions from land use and land use change and from other sources in the livestock sector may be explicitly considered and become part of a clean development mechanism–type arrangement. If so, this could lead to an incentive structure conducive to the application of emission-reducing technologies in the livestock sector, either as part of a cap-and-trade system or in the form of a carbon tax. This will cause the costs of carbon-intensive production to rise and will lead to higher production costs for beef than for poultry.

Commercially oriented livestock products will need to be regulated much more tightly to ensure that environmental damage is accounted for. Regulations and their enforcement will need to be directed at animal waste and nutrient management and at the resultant air and water pollution. De-clustering can be an effective tool to avoid nutrient loading currently affecting areas with high densities of animals and livestock operations. Appropriate zoning and licensing of establishments are required to bring nutrient flows into balance. Where this is impossible, waste treatment and export to nutrient-deficient areas are required, but this likely carries high if not prohibitive costs.

While extensive livestock production also has important environmental consequences, addressing these is complicated by the low financial resources of the many poor producers who keep livestock as part of their livelihood strategy. Public policy needs to ensure that these producers are not marginalized economically and socially. Where the potential exists, smallholders can intensify and expand production, provided market barriers can be minimized. Smallholders, especially pastoralists, can play a major role in the provision of environmental services, such as carbon sequestration on degraded rangelands, landscape maintenance, protection of biodiversity, and water management.

As long as viable livelihood strategies outside the livestock and agricultural sectors do not exist, subsistence livestock keepers need access to land and water resources and to markets that allow them to support themselves. Public policies need to be geared toward protecting their livelihoods by providing them with secure access to key resources, by assisting local communities in protecting livestock against infectious diseases, and by helping them adjust their stock as part of a community-driven drought strategy. Extensive livestock production and the many livelihoods sustained by it will likely be affected by climate change in a major way, and a humanitarian crisis in marginal zones is a daunting prospect that national and international agencies need to prepare for.

The heightened requirements for biosecurity will have important consequences for the structure of commercially oriented livestock production. The establishment of biosecure systems and monitoring has strong economies of scale, and larger operations have an advantage over smaller ones. Ultimately, the key to biosecurity is to reduce physical contact between humans and animals to the utmost degree possible. Automation is likely to play a large role not only because it reduces labor costs but particularly because pathogen incursion can effectively be reduced or prevented by eliminating direct contact. This is bound to increase the dichotomous nature of the sector. While backyard and other extensive forms of production will continue to exist, there will be growing pressure on the "middle ground" (i.e., medium-size systems that operate at intermediate levels of biosecurity with relatively high production costs compared with their large-scale competitors). Growing integration of producers with food chains and large-scale processors and retailers will further accelerate the structural change process.

Structural change driven by economies of scale and sanitary requirements is likely to be inevitable in open economies. Particularly in monogastric production, the number of producers declines but income and employment are created in growing food chains. Public policy efforts, for example, through sector protection, are likely to be ineffective in reversing this process and may at best retard it. This may be justified in places where large segments of the population are engaged in primary livestock production and where alternative sources of income are not readily available.

While structural change, growing intensities, a shift to monogastrics, and the application of productivity-enhancing technologies have negative social implications in some places, they will help reduce overall resource requirements of the livestock sector. Economic forces and sanitary requirements are driving this process, which will only accelerate if food and energy prices stay high.

The bottom line is that extensive production, while environmentally and socially benign in some places, will not be capable of meeting the growing demands of an expanding and more affluent world population. Thus there is no realistic alternative to intensive production. Making intensification more sustainable while taking into account environmental consequences, animal welfare, and human health is the challenge facing the livestock sector in a changing landscape, and it offers exciting opportunities for innovative technology, policy, and institutional development.

Acronyms and Abbreviations

AAP   American Academy of Pediatrics
ACU   adult cattle units
ANPP   aboveground net primary production
ASF   animal source food
BNF   biological nitrogen fixation
BSE   bovine spongiform encephalitis (mad cow disease)
CAP   Common Agricultural Policy (of the EU)
CBD   Convention on Biological Diversity
CBPP   contagious bovine pleuropneumonia
CDM   clean development mechanism
CER   certified emissions reduction
CIRAD   Centre de Coopération Internationale en Recherche Agronomique pour le Développement (French Agricultural Research Centre for International Development)
CIS   Commonwealth of Independent States
CSF   classical swine fever
DALY   disability-adjusted life-year
DIN   dissolved inorganic nitrogen
DON   dissolved organic nitrogen
EMPRES   Emergency Prevention System for Transboundary Animal and Plant Pests and Diseases
ET   evapotranspiration
EU   European Union
FAO   Food and Agriculture Organization, United Nations
FMD   foot-and-mouth disease
GDP   gross domestic product
GHGs   greenhouse gases
GLEWS   Global Early Warning and Response System for Major Animal Diseases, including Zoonoses
GNI   gross national income
GNP   gross national product
GREP   Global Rinderpest Eradication Programme
HACCP   hazard analysis and critical control point
HKI   Helen Keller International
HNV   high nature value
HPAI   highly pathogenic avian influenza
IAPS   industrialized (intensive) animal production systems

ILRI   International Livestock Research Institute
INRA   Institut National de la Recherche Agronomique (French National Institute for Agricultural Research)
IPCC   Intergovernmental Panel on Climate Change
IRD   Institut de Recherche pour le Développement (Research Institute for Development [France])
ISRA   Institut Sénégalais de Recherches Agricoles (Senegalese Institute for Agricultural Research [Senegal])
IUCN   International Union for Conservation of Nature
IVM   integrated vector management
LEAD   Livestock, Environment and Development Initiative (FAO)
LUCC   land use, climate change
LULUCF   land use, land use change, and forestry
MA   Millennium Ecosystem Assessment
MAP   mean annual precipitation
MDG   Millennium Development Goal
ME   multilateral environmental agreement
MFP   meat, fish, and poultry
MND   micronutrient deficiency
MPA   marine protected area
NCRSP   Nutrition Collaborative Research Support Program
NGO   nongovernmental organization
NPP   net primary production
NSF   National Science Foundation, USA
OECD   Organisation for Economic Cooperation and Development
OIE   World Organisation for Animal Health (from original name in French)
PA   protected area
PACE   Pan-African Programme for the Control of Epizootics
PARC   Pan-African Rinderpest Campaign
PEM   protein-energy malnutrition
PES   payment for environmental services
PFP   partial factor productivity
Pg   petagrams
PN   particulate nitrogen

PPP	purchasing power parity
PUFA	polyunsaturated fatty acids
RBO	river basin organization
REML	residual maximum likelihood
SARS	severe acute respiratory syndrome
SCOPE	Scientific Committee on Problems of the Environment
SHL	Swiss College of Agriculture
SOC	soil organic carbon
SPS	sanitary and phytosanitary
TFP	total factor productivity
TRQ	tariff rate quota
UNCCD	United Nations Convention to Combat Desertification
UNDP	United Nations Development Programme

UNEP	United Nations Environment Programme
UNFCCC	United Nations Framework Convention on Climate Change
URAA	Uruguay Round Agreement on Agriculture
USAID	U.S. Agency for International Development
USDA	U.S. Department of Agriculture
USDA–ARS	U.S. Department of Agriculture–Agricultural Research Service
USGS	United States Geological Survey
WAHID	World Animal Health Information Database
WANA	West Asia and North Africa
WFP	World Food Programme
WWF	World Wide Fund for Nature/World Wildlife Fund

Chemical Symbols, Compounds, and Units of Measurement

CH_4 methane

CO   carbon monoxide

CO_2 carbon dioxide

GtC-eq   gigatons of carbon equivalent

N   nitrogen

N_2O nitrous oxide

NOx   nitrogen oxides

ppmv   parts per million by volume

SO_2 sulfur dioxide

Teragram 10^{12} grams

M–Mega   SI system of units denoting a factor of 10 to the sixth power, or 1,000,000 (one million)

Glossary

The note number following each term indicates the source reference at the end of the glossary for the term and definition.

Adaptive management[4] The mode of operation in which an intervention (action) is followed by monitoring (learning), with the information then being used in designing and implementing the next intervention (acting again) to steer the system toward a given objective or to modify the objective itself.

Agroecological classification[1] Based on length of available growing period (LGP), which is defined as the period (in days) during the year when rainfed available soil moisture supply is greater than half potential evapotranspiration (PET). It includes the period required for evapotranspiration of up to 100 mm of available soil moisture stored in the soil profile. It excludes any time interval with daily mean temperatures less than 5°C.
- Arid LGP less than 75 days
- Semiarid LGP in the range 75–180 days
- Subhumid LGP in the range 180–270 days
- Humid LGP greater than 270 days

Agropastoralism[2] A production system where all of the family and livestock are sedentary.

Arid zones[1] The areas where the growing period is less than 75 days, too short for reliable rainfed agriculture. The coefficient of variation of the annual rainfalls is high, up to 30%. Abiotic factors, especially rainfall, determine the state of the vegetation. The *nonequilibrium* theory applies in this environment. The main systems found in these zones are the mobile systems on communal lands. Some cases of ranching are present.

Benefits transfer[4] Economic valuation approach in which estimates obtained (by whatever method) in one context are used to estimate values in a different context. This approach is widely used because of its ease and low cost but is risky because values are context-specific and cannot usually be transferred.

Biodiversity[4] The variability among living organisms from all sources including terrestrial, marine, and other aquatic ecosystems and the ecological complexes of which they are part; this includes diversity within and among species and diversity within and among ecosystems.

Biomass[4] The mass of living tissues in ecosystems.

Biosolids[3] Organic solids resulting from wastewater treatment that can be usefully recycled.

Capacity building[4] A process of strengthening or developing human resources, institutions, or organizations.

Capital value (of an ecosystem)[4] The present value of the stream of future benefits that an ecosystem will generate under a particular management regime. Present values are typically obtained by discounting future benefits and costs; the appropriate rates of discount are often a contested issue, particularly in the context of natural resources.

Change in productivity approach[4] Economic valuation techniques that value the impact of changes in ecosystems by tracing their impact on the productivity of economic production processes. For example, the impact of deforestation could be valued (in part) by tracing the impact of the resulting changes in hydrological flows on downstream water uses such as hydroelectricity production, irrigated agriculture, and potable water supply.

Characteristic scale[4] The typical extent or duration over which a process is most significantly or apparently expressed.

Constituents of well-being[4] The experiential aspects of well-being, such as health, happiness, and freedom to be and do, and, more broadly, basic liberties.

Cultural services[4] The nonmaterial benefits people obtain from ecosystems through spiritual enrichment, cognitive development, reflection, recreation, and aesthetic experience, including, for example, knowledge systems, social relations, and aesthetic values.

Decision maker[4] A person whose decisions and actions can influence a condition, process, or issue under consideration.

Decomposition[4] The ecological process carried out primarily by microbes that leads to a transformation of dead organic matter into inorganic mater; the converse of biological production. For example, the transformation of dead plant material, such as leaf litter and dead wood, into carbon dioxide, nitrogen gas, and ammonium and nitrates.

Driver[4] Any natural or human-induced factor that directly or indirectly causes a change in an ecosystem (system).

Driver, direct[4] A driver that unequivocally influences ecosystem (system) processes and can therefore be identified and measured to differing degrees of accuracy.

Driver, endogenous[4] A driver whose magnitude can be influenced by the decision maker. The endogenous or exogenous characteristic of a driver depends on the organizational scale. Some drivers (e.g., prices) are exogenous to a decision maker at one level (a farmer) but endogenous at other levels (the nation-state).

Driver, exogenous[4] A driver that cannot be altered by the decision maker. See also *endogenous driver.*

Driver, indirect[4] A driver that operates by altering the level or rate of change of one or more direct drivers.

Ecological footprint[4] The area of productive land and aquatic ecosystems required to produce the resources used and to assimilate the wastes produced by a defined population at a specified material standard of living, wherever on Earth that land may be located.

Ecosystem[4] A dynamic complex of plant, animal, and microorganism communities and their nonliving environment interacting as a functional unit.

Ecosystem approach[4] A strategy for the integrated management of land, water, and living resources that promotes conservation and sustainable use in an equitable way. An ecosystem approach is based on the application of appropriate scientific methodologies focused on levels of biological organization, which encompass the essential structure, processes, functions, and interactions among organisms and their environment. It recognizes that humans, with their cultural diversity, are an integral component of many ecosystems.

Ecosystem function[4] An intrinsic ecosystem characteristic related to the set of conditions and processes whereby an ecosystem maintains its integrity (such as primary productivity, food chain, biogeochemical cycles). Ecosystem functions include such processes as decomposition, production, nutrient cycling, and fluxes of nutrients and energy.

Ecosystem health[4] A measure of the stability and sustainability of ecosystem functioning or ecosystem goods and services that depends on an ecosystem being active and maintaining its organization, autonomy, and resilience over time. Ecosystem health contributes to human wellbeing through sustainable ecosystem services and conditions for human health.

Ecosystem interactions[4] Exchanges of materials and energy among ecosystems.

Ecosystem properties[4] The size, biodiversity, stability, degree of organization, internal exchanges of materials and energy among different pools, and other properties that characterize an ecosystem.

Ecosystem services[4] The benefits people obtain from ecosystems. These include provisioning services such as food and water; regulating services such as flood and disease control; cultural services such as spiritual, recreational, and cultural benefits; and supporting services such as nutrient cycling that maintain the conditions for life on Earth. The concept of ecosystem goods and services is synonymous with ecosystem services.

Extensive[5] A livestock production system that uses predominantly noncommercial inputs to the system.

Externality[4] A consequence of an action that affects someone other than the agent undertaking that action and for which the agent is neither compensated nor penalized. Externalities can be positive or negative.

Grazing system[1] The grazing system is predominantly dependent on the natural productivity of grasslands and is therefore defined largely by the agroecological zone. The populations relying on these systems are generally referred to as pastoralist groups, with their main differences defined by their mobility in response to environmental variability. At one extreme the nomadic groups are highly mobile, living in areas with major differences in both seasonal and annual climatic patterns. At the other end agropastoralists and ranchers operate sedentary systems where seasonal and annual climatic variations are minor.

Grazing systems[1] Livestock systems in which more than 90% of dry matter fed to animals comes from rangelands, pastures, annual forages, and purchased feeds and less than 10% of the total value of production comes from nonlivestock farming activities. Annual stocking rates are less than 10 livestock units per hectare of agricultural land. Grazing systems are described for each of the following regions arid, semiarid, subhumid and humid, temperate, and tropical highlands.

Indicator[4] Information based on measured data used to represent a particular attribute, characteristic, or property of a system.

Industrial livestock system[1] Industrial systems are primarily directed at producing high-quality animal protein and other animal products for the urban markets. As a result of this market demand, intensive animal production and processing often take place near urban areas, while primary feed production takes place in distant rural areas. Industrial systems are generally considered modern and efficient, requiring a high level of knowledge and skill. Production techniques are more or less independent of the agroecological zone and of the climate, which explains the worldwide occurrence of the industrial system. These systems have average stocking rates greater than 10 livestock units per hectare of agricultural land, and <10% of the dry matter fed to livestock is produced on the farm. (This is similar to Seré and Steinfeld's classification Landless Livestock Production Systems.) Industrial livestock production systems are associated with a concentration of animals into large units, generally concentrating on a single species. They produce large volumes of waste material, can lead to high animal and human health risks, and pay less attention to animal welfare. Industrial production also occurs in small units operated by specialized smallholders as part of the mixed livestock system.

Intensive[5] A livestock production system that relies on commercial inputs and trade.

Land cover[4] The physical coverage of land, usually expressed in terms of vegetation cover or lack of it. Influenced by but not synonymous with *land use*.

Land use[4] The human utilization of a piece of land for a certain purpose (such as irrigated agriculture or recreation). Influenced by but not synonymous with *land cover*.

Landscape[4] An area of land that contains a mosaic of ecosystems, including human-dominated ecosystems. The term *cultural landscape* is often used when referring to landscapes containing significant human populations.

Livestock mobility[2] Can be divided into macromobility, which refers to large movements (such as transhumance between dry season and wet season pastures) and micromobility, referring to daily or frequent movement between microniches within the same pasture.

Livestock unit (LU)[3] A unit used to compare or aggregate numbers of animals of different species or categories. Equivalences are defined on the feed requirements (or sometimes nutrient excretion). For example for the EU, one 600 kg dairy cow producing 3000 liters of milk per year equals 1 LU, a sow equals 0.45 LU, and a ewe equals 0.18 LU.

Manure supernatant[3] The upper liquid fraction after sedimentation of liquid waste or liquid manure.

Mixed systems[1] Farming systems conducted by households or by enterprises where crop cultivation and livestock rearing are more or less integrated components of one single farming system. The more integrated systems are characterized by interdependency between crop and livestock activities. They are basically resource driven, aiming at an optimal circulation of locally available nutrients (nutrient circulation systems); for example, ecological farming and some, but not all, low external input agriculture (LEIA) systems. The less integrated systems are those where crop and livestock activities can make use of, but not rely on, each other. In general one or both activities are demand driven, supported by external inputs (nutrient throughput systems); for example, high external input agriculture (HEIA) systems. Mixed systems are those systems in which more than 10% of the dry matter fed to livestock comes from crop by-products and/or stubble or more than 10% of the value of production comes from nonlivestock farming activities.

Nomad[2] Production system that is highly mobile but does not necessarily return to a "base" every year and does not include cultivation (e.g., nomads of the Sahara). See also *pastoralism.*

Organic residues[3] Organic material resulting from dead plant material or by-products from processing organic materials of the food industry or other industry.

Organic wastes[3] A general term for any wastes from organic rather than inorganic origin and so containing carbon (e.g., livestock manure, sewage sludge, abattoir wastes).

Pastoralism[2] Predominantly extensive production system that depends on livestock for more than 50% of income; includes nomads, transhumants, and semitranshumants.

Precautionary principle[4] The management concept stating that in cases "where there are threats of serious or irreversible damage, lack of full scientific certainty shall not be used as a reason for postponing cost-effective measures to prevent environmental degradation," as defined in the Rio Declaration.

Prediction (or forecast)[4] The result of an attempt to produce a most likely description or estimate of the actual evolution of a variable or system in the future. See also *projection* and *scenario.*

Primary production[4] Assimilation (gross) or accumulation (net) of energy and nutrients by green plants and by organisms that use inorganic compounds as food.

Private costs and benefits[4] Costs and benefits directly felt by individual economic agents or groups as seen from their perspective. (Externalities imposed on others are ignored.) Costs and benefits are valued at the prices actually paid or received by the group, even if these prices are highly distorted. Sometimes termed "financial" costs and benefits. Compare *social costs and benefits.*

Provisioning services[4] Products obtained from ecosystems, including, for example, genetic resources, food and fiber, and freshwater.

Resilience[4] The capacity of a system to tolerate impacts of drivers without irreversible change in its outputs or structure.

Responses[4] Human actions, including policies, strategies, and interventions, to address specific issues, needs, opportunities, or problems. In the context of ecosystem management, responses may be of a legal, technical, institutional, economic, and behavioral nature and may operate at a local or micro, regional, national, or international level and at various time scales.

Scale[4] The physical dimensions, in either space or time, of phenomena or observations.

Scenario[4] A plausible and often simplified description of how the future may develop, based on a coherent and internally consistent set of assumptions about key driving forces (e.g., rate of technology change, prices) and relationships. Scenarios are neither predictions nor projections and sometimes may be based on a "narrative storyline." Scenarios may be derived from projections but are often based on additional information from other sources.

Semitranshumant[2] Production system where only part of the family and/or livestock is seasonally mobile and the rest is sedentary in one of the seasonal bases, practicing cultivation (e.g., Dinka tribe of Sudan and Karimojong tribe of Uganda).

Solid manure[3] Manure from housed livestock that does not flow under gravity, cannot be pumped but can be stacked in a heap. May include manure from cattle, pigs, poultry, horses, sheep, goats, camelids, and rabbits. There are several types of solid manure arising from different types of livestock housing, manure storage, and treatment.

Stakeholder[4] An actor having a stake or interest in a physical resource, ecosystem service, institution, or social system, or someone who is or may be affected by a public policy.

Stocking rate[3] The number of livestock (or livestock units) per unit area of land.

Sustainability[4] A characteristic or state whereby the needs of the present and local population can be met without compromising the ability of future generations or populations in other locations, in order to meet their needs.

Transhumant[2] Production system that is highly mobile but moves between definite seasonal bases every year (e.g., Samburu of Kenya); it may include a nonsedentary form of cultivation (e.g., Zaghawa of Chad). See also *pastoralism.*

Valuation[4] The process of expressing a value for a particular good or service in a certain context (e.g., of decision making), usually in terms of something that can be counted, often money, but also through methods and measures from other disciplines (sociology, ecology, and so on).

Well-being[4] A context- and situation-dependent state, comprising basic material for a good life, freedom and choice, health, good social relations, and security.

Notes

[1] Excerpts from the LEAD Toolkit (http://www.fao.org/lead/) and Seré and Steinfeld (1996). World livestock production systems: current status, issues and trends. Animal Production and Health Paper No. 127. FAO, Rome.

[2] *Source:* Extract from *Pastoral development in Africa.* Proceedings of the first technical consultation of donor and international development agencies, Paris, December 1993. UNSO/UNDP, 1994.

[3] *Source: RAMIRAN Glossary of terms on manure management 2003.*

[4] *Source:* Millennium Ecosystem Assessment, 2005. *Ecosystems and human well-being: Synthesis.* Island Press, Washington, DC.

[5] Derived from LCL Consultation working group discussion.

List of Editors and Contributors

Editors

Henning Steinfeld
Chief, Livestock Sector Analysis and Policy Branch (AGAL)
Coordinator, Livestock Environment and Development
    Initiative (LEAD)
Food and Agriculture Organization of the United Nations
    (FAO) Headquarters, Rome, Italy

Harold A. Mooney
Professor, Environmental Biology
Department of Biological Sciences
Stanford University, California, USA

Fritz Schneider
Vice Director
Bern University of Applied Sciences
Swiss College of Agriculture, Switzerland

Laurie E. Neville
Coordinator, Livestock in a Changing Landscape
Global Consultation & Integrated Analysis
Department of Biological Sciences
Stanford University, California, USA

Contributors

Justin Ayayi Akakpo
Professor
Ecole Inter-Etats des Sciences et Médecine Vétérinaires
Dakar, Sénégal

Steven R. Archer
Professor
School of Renewable Natural Resources
University of Arizona, Tucson, USA

Andrew Ash
Director
Climate Adaptation Flagship
Commonwealth Scientific and Industrial Research
    Organisation (CSIRO)
St Lucia, Queensland, Australia

Gregory P. Asner
Professor
Department of Global Ecology
Carnegie Institution
Stanford University, California, USA

Kenneth Bauer
President
DROKPA
Norwich, Vermont, USA

Claire Bedelian
International Livestock Research Institute (ILRI)
Nairobi, Kenya

Roy Behnke
The Macaulay Institute
Craigiebuckler, Aberdeen, Scotland

Bassirou Bonfoh
Researcher
Director General
Centre Suisse de Recherches Scientifiques (CSRS)
Abidjan, Côte d'Ivoire

A. F. Bouwman
Netherlands Environmental Assessment Agency
Bilthoven, The Netherlands

Marshall Burke
Program Manager
Program on Food Security and Environment
The Freeman Spogli Institute for International Studies
Stanford University, California, USA

Colin Burton
Agriculture and Environmental Engineering Research
    (Cemagref)
Rennes, France

Vincent Castel
Economist
Development Research Department
African Development Bank Group
Tunis, Tunisia

Achilles C. Costales
Livestock Economist
Pro-Poor Livestock Policy Initiative (PPLPI)
Livestock Information, Sector Analysis and Policy Branch
    (AGAL)
Food and Agriculture Organization of the United Nations
    (FAO)
Rome, Italy

Jonathan Davies
Regional Drylands Coordinator, Eastern and Southern Africa
International Union for Conservation of Nature (IUCN)
Nairobi, Kenya

Christopher Delgado
Strategy and Policy Adviser for Agriculture and Rural
  Development
World Bank, Washington, D.C., USA

Montague W. Demment
Department of Plant Sciences
University of California at Davis, California, USA

Frank Dentener
European Commission, Joint Research Centre
Institute for Environment and Sustainability, Climate Change
  Unit
Ispra, Italy

Lisa Deutsch
Stockholm Resilience Centre
Stockholm University, Sweden

Jeroen Dijkman
Livestock Development Officer (Pro-Poor Livestock Sector
  Strategies)
Pro-Poor Livestock Policy Initiative (PPLPI)
Livestock Information, Sector Analysis and Policy Branch
  (AGAL)
Food and Agriculture Organization of the United Nations
  (FAO), Headquarters, Rome, Italy

Amadou Tamsir Diop
Institut Sénégalais de Recherches Agricoles (ISRA)
Dakar, Senegal

Natalie Drorbaugh
School of Public Health
University of California at Los Angeles, USA

Egon Dumont
Hydrological Modeler
Center for Ecology and Hydrology
Wallingford, Oxfordshire, United Kingdom

Céline Dutilly-Diane
International Center for Agricultural Research in the Dry
  Areas
Centre de Coopération Internationale en Recherche
  Agronomique pour le Développement (CIRAD)
Campus de Baillarguet
Montpellier, France

Malin Falkenmark
Stockholm International Water Institute
Stockholm Resilience Centre
Stockholm University, Sweden

Carl Folke
Beijer Institute for Ecological Economics
Stockholm Resilience Centre and Department of Systems
  Ecology
Stockholm University, Sweden

Gianluca Franceschini
Consultant
Global Livestock Information System
Food and Agriculture Organization of the United Nations
  (FAO), Rome, Italy

Maryam Niamir-Fuller
Director
Division of Global Environment Facility (GEF) Coordination
United Nations Environmental Programme (UNEP)
Nairobi, Kenya

James Galloway
Chair, International Nitrogen Initiative (INI)
Professor, Department of Environmental Sciences
University of Virginia, Charlottesville, USA

Kathleen A. Galvin
Department of Anthropology and Natural Resource Ecology
  Laboratory
Colorado State University, Fort Collins, USA

Didier Genin
Institut de Recherche pour le Développement (IRD/LPED)
France

Pierre Gerber
Livestock Policy Officer
Livestock, Environment and Development (LEAD)
Livestock Information, Sector Analysis and Policy Branch
  (AGAL)
Food and Agriculture Organization of the United Nations
  (FAO), Rome, Italy

Annick Gibon
Institut National de la Recherche Agronomique (INRA)
France

Line Gordon
Stockholm Resilience Centre and Department of Systems
  Ecology
Stockholm Environment Institute
Stockholm University, Sweden

Cees de Haan
Retired Senior Livestock Development Advisor
Consultant, World Bank
Washington, D.C., USA

Jörg Hartung
Director
Institute of Animal Hygiene, Welfare and Behaviour of Farm
  Animals Foundation
University of Veterinary Medicine at Hanover, Germany

Nick Honhold
Animal Production and Health Division
Food and Agriculture Organization of the United Nations
(FAO)

Muhammad Ibrahim
Centro Agronómico Tropical de Investigación y Enseñanza/
Tropical Agricultural Research and Higher Education
Center (CATIE)
Turrialba, Costa Rica

Alexandre Ickowicz
Vétérinaire, Zootechnicien Pastoraliste
Unité de Recherche en Partenariat, French Agricultural
Research Centre for International Development/Centre de
Coopération Internationale en Recherche Agronomique
pour le Développement (CIRAD)
Campus de Baillarguet, Montpellier, France

Carol Kerven
Co-Scientific Project Coordinator (Range Enclosure on
the Tibetan Plateau, Impacts on Pastoral Livelihoods,
Marketing, Livestock Productivity and Rangeland
Biodiversity)
The Macaulay Institute
Craigiebuckler, Aberdeen, Scotland

Richard A. Kohn
Professor
University of Maryland, College Park, USA

Carolien Kroeze
Professor
Environmental Systems Analysis Group
Wageningen University, The Netherlands

Russell L. Kruska
GIS Scientist
International Livestock Research Institute (ILRI)
Nairobi, Kenya

Audrey Maretzki
Professor Emeritus of Food Science and Nutrition
Department of Food Science
Pennsylvania State University, University Park, USA

Rogerio Martins Mauricio
Fundação Ezequiel Dias, Fazenda Experimental
Belo Horizonte, Minas Gerais, Brazil

Ellen McCullough
Associate Program Officer
Agricultural Development
Bill & Melinda Gates Foundation
Seattle, Washington, USA

Anni McLeod
Senior Officer
Livestock Policy, Sector Analysis and Policy Branch (AGAL)
Food and Agriculture Organization of the United Nations
(FAO), Headquarters, Rome, Italy

Harald Menzi
Professor
Swiss College of Agriculture
Zollikofen, Switzerland

François-Xavier Meslin
World Health Organization
Geneva, Switzerland

Clare Narrod
Senior Research Fellow
International Food Policy Research Institute (IFPRI)
Markets, Trade, and Institutions Division
Washington, D.C., USA

Rudy Nayga
Professor of Food Policy Economics
University of Arkansas, USA

Charlotte G. Neumann
School of Public Health and The David Geffen School of
Medicine
University of California at Los Angeles, USA

Oene Oenema
Professor
Wageningen University & Research Centre
Alterra
Wageningen, The Netherlands

Jennifer Olson
Associate Professor, Department of Geography, Michigan
State University
Scientist, International Livestock Research Institute (ILRI),
Nairobi, Kenya

Carolyn Opio
Consultant
Livestock Information, Sector Analysis and Policy Branch
(AGAL)
Food and Agriculture Organization of the United Nations
(FAO), Headquarters, Rome, Italy

Joachim Otte
Coordinator
Livestock Policy Analyst, Pro-Poor Livestock Policy Initiative
(PPLPI)
Livestock Information, Sector Analysis and Policy Branch
(AGAL)
Food and Agriculture Organization of the United Nations
(FAO), Rome, Italy

Ugo Pica-Ciamarra
Livestock Policy Analyst, Pro-Poor Livestock Policy Initiative
(PPLPI)
Livestock Information, Sector Analysis and Policy Branch
(AGAL)
Food and Agriculture Organization of the United Nations
(FAO), Rome, Italy

Prabhu Pingali
Deputy Director
Agricultural Development
Bill & Melinda Gates Foundation
Seattle, Washington, USA

Allan Rae
Professor of Agricultural Economics
Director, Centre for Agribusiness Policy and Strategy
College of Business
Massey University, Manawatu Campus
New Zealand

Robin S. Reid
Director, Center for Collaborative Conservation
Director, Sustainable Rangelands Roundtable
Senior Research Scientist, Natural Resource Ecology Lab
Faculty Affiliate, Dept. of Human Dimensions of Natural
    Resources, Dept. of Forest, Rangeland and Watershed
    Stewardship, Colorado State University, Fort Collins, USA

Tim Robinson
Livestock Information Officer
Pro-Poor Livestock Policy Initiative (PPLPI)
Livestock Sector Analysis and Policy Branch (AGAL)
Food and Agriculture Organization of the United Nations
    (FAO), Rome, Italy

Johan Rockström
Stockholm Resilience Centre at Stockholm University
Stockholm Environment Institute
Sweden

Mohammed Y. Said
System Ecologist
International Livestock Research Institute (ILRI)
Nairobi, Kenya

Esther Schelling
International Livestock Research Institute (ILRI)
Nairobi, Kenya

Karin Schwabenbauer
Project Coordinator, Senior Veterinary Policy Adviser
Animal Production and Health Division
Food and Agriculture Organization of the United Nations
    (FAO), Rome, Italy

Sybil Seitzinger
Director
Rutgers/National Oceanic and Atmospheric Administration,
    Cooperative Education and Marine Research Program
    (NOAA/CMER) Program
Institute of Marine and Coastal Sciences
The State University of New Jersey, New Brunswick, USA

Oleg Shipin
Associate Professor and Coordinator
Environmental Engineering & Management Field of Study
Asian Institute of Technology (AIT)
Pathumthani,Thailand

Vijay Kumar Taneja
Vice-Chancellor, Guru Angad Veterinary and Animal Sciences
    University (GADVASU)
Deputy Director General of Animal Science at ICAR
New Delhi, India

Marcel Tanner
Professor of Epidemiology and Public Health
University of Basel
Swiss Tropical Institute
Basel, Switzerland

Philip K. Thornton
Senior Scientist and Systems Analyst
International Livestock Research Institute (ILRI)
Nairobi, Kenya

Marites Tiongco
Research Fellow
International Food Policy Research Institute (IFPRI)
Markets, Trade, and Institutions Division
Washington, D.C., USA

Bernard Toutain
Chercheur en Pastoralisme, Centre de Coopération
    Internationale en Recherche Agronomique pour le
    Développement (CIRAD)
Département d'Elevage et de Médecine Vétérinaire, Unité
    de Recherche en Partenariat, Pôle Pastoral Zones Sèches,
    (Cirad-Cse-Enea-Isra-Ucad)
Dakar, Sénégal

Rea Tschopp
Swiss Tropical Institute
Basel, Switzerland

David Wallinga
Director, Food and Health
Institute for Agriculture and Trade Policy (IATP)
Minneapolis, Minnesota, USA

Tom Wassenaar
Agro-Environmental Scientist (CIRAD)
Environmental Risks of Recycling Research Unit
Reunion Island, France

Jakob Zinsstag
Department of Public Health and Epidemiology
Swiss Tropical Institute
Basel, Switzerland

Index

Page numbers in italics indicate figures, tables, and boxes.